STUDENT'S SOLUTIONS MANUAL
to accompany

COLLEGE ALGEBRA
Second Edition

Arnold Steffensen
Northern Arizona University

L. Murphy Johnson
Northern Arizona University

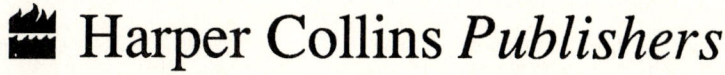

FOR THE STUDENT

This *Student Solutions Manual*, written by the authors, provides complete, step-by-step solutions to all of the Practice Exercises, to the exercises in Exercises A, to the Chapter Review Exercises and the practice Chapter Tests, and to the Final Review Exercises in the text.

ACKNOWLEDGEMENTS

Special thanks go to Barbara Johnson, Pam Johnson, Barbara Steffensen, Becky Steffensen, Cindy Steffensen, and Joseph Mutter for their participation in the preparation of the supplements for this text.

Student Solutions Manual to accompany COLLEGE ALGEBRA, Second Edition
By Steffensen/Johnson

Copyright © 1992 by HarperCollins Publishers Inc.

All rights reserved. Printed in the United States of America. No part of this book may be used or reproduced in any manner whatsoever without written permission with the following exception: testing materials may be copied for classroom testing. For information, address HarperCollins Publishers, Inc. 10 E. 53rd St., New York, NY 10022

ISBN: 0-673-46456-3
     95 96 9 8 7 6

CONTENTS

1 Review of Fundamental Concepts 1

1.1 The Real Number System 1
1.2 Integer Exponents and Scientific Notation 4
1.3 Algebraic Expressions and Polynomials 7
1.4 Factoring Polynomials 13
1.5 Rational Expressions 17
1.6 Radical Expressions 25
1.7 Rational Exponents 30
    Chapter 1 Review Exercises 34
    Chapter Test 41

2 Equations, Inequalities, and Problem Solving 43

2.1 Linear and Absolute Value Equations 43
2.2 Problem Solving and Applications of Linear Equations 49
2.3 Complex Numbers 54
2.4 Quadratic Equations 58
2.5 Equations that Result in Quadratic Equations 64
2.6 Problem Solving and Applications of Quadratic Equations 68
2.7 Linear and Absolute Value Inequalities 74
    Chapter 2 Review Exercises 78
    Chapter 2 Test 85

3 Relations, Functions, and Graphs 87

3.1 The Rectangular Coordinate System 87
3.2 Linear Equations 92
3.3 Relations and Functions 100
3.4 Properties of Functions and Transformations 105
3.5 Composite and Inverse Functions 112
3.6 Quadratic Functions 119
3.7 Mathematical Modeling 128
3.8 Variation 133
    Chapter 3 Review Exercises 139
    Chapter 3 Test 152

4 Polynomial and Rational Functions 155

4.1 Polynomials and Synthetic Division 155
4.2 The Remainder and Factor Theorems 160
4.3 More Theorems Involving Polynomials 163
4.4 Bounds and The Rational Root Theorem 168
4.5 Graphing Polynomial Functions 176
4.6 Rational Functions 182
4.7 Polynomial and Rational Inequalities 191
   Chapter 4 Review Exercises 200
   Chapter 4 Test 208

5 Exponential and Logarithmic Functions 211

5.1 Introduction to Logarithms 211
5.2 Exponential and Logarithmic Functions 215
5.3 Properties of Logarithms 221
5.4 Common and Natural Logarithms 226
5.5 Exponential and Logarithmic Equations 230
5.6 More Applications of Exponentials and Logarithms 238
   Chapter 5 Review Exercises 245
   Chapter 5 Test 253

6 Systems of Equations and Inequalities 255

6.1 Linear Systems in Two Variables 255
6.2 Linear Systems in More than Two Variables 262
6.3 Problem Solving Using Systems of Equations 270
6.4 Linear Systems of Inequalities 281
6.5 Linear Programming 289
6.6 Partial Fractions 296
   Chapter 6 Review Exercises 304
   Chapter 6 Test 315

7 Matrices and Determinants   320

7.1 Matrices   320
7.2 Matrix Multiplication   325
7.3 Solving Systems of Equations Using Matrices   331
7.4 The Inverse of a Square Matrix   339
7.5 Determinants and Cramer's Rule   350
7.6 More on Determinants   361
    Chapter 7 Review Exercises   368
    Chapter 7 Test   380

8 Topics in Analytic Geometry   384

8.1 Conic Sections and the Circle   384
8.2 The Ellipse   389
8.3 The Hyperbola   396
8.4 The Parabola   402
8.5 Nonlinear Systems   409
    Chapter 8 Review Exercises   417
    Chapter 8 Test   427

9 Sequences, Series, and Probability   430

9.1 Sequences and Series   430
9.2 Arithmetic Sequences and Series   436
9.3 Geometric Sequences and Series   443
9.4 Infinite Geometric Sequences and Series   451
9.5 Mathematical Induction   457
9.6 Permutations and Combinations   463
9.7 The Binomial Theorem   469
9.8 Probability   473
    Chapter 9 Review Exercises   479
    Chapter 9 Test   489

Final Review Exercises   492

PRACTICE EXERCISES   SECTION 1.1   1

CHAPTER 1 REVIEW OF FUNDAMENTAL CONCEPTS

SECTION 1.1 The Real Number System

1. (a) Only 500 is a natural number.

   (b) Only 500 is a whole number.

   (c) Only −23 and 500 are integers.

   (d) The rational numbers in the list are 0.01, −23, −1.09, and 500.

   (e) The irrational numbers are $\sqrt{13}$ and $\frac{\pi}{2}$.

2. (a) The reflexive property states that every number is equal to itself. Thus, $x + 1 = \underline{x + 1}$.

   (b) The transitive property states that if $a < b$ and $b < c$, then $a < c$. Thus, if $y < 5$ and $5 < w$, then $\underline{y < w}$.

   (c) The substitution property states that if two quantities are equal, then either can be substituted for the other in an expression without changing the expression. If $a = 3$, then 3 can be substituted for a in any expression. Thus, if $a = 3$ then $a − 5 = \underline{3 − 5}$, or $a − 5 = \underline{−2}$.

   (d) The trichotomy property states that for any two real numbers a and b, then $a < b$, $a > b$, or $a = b$. Thus if z is a real number, 0 is also a real number, so $z < 0$, $z > 0$, or $\underline{z = 0}$.

3. (a) $|\sqrt{3}| = \sqrt{3}$, the distance $\sqrt{3}$ is from 0 on a number line.

   (b) $\left|-\frac{1}{4}\right| = \frac{1}{4}$, the distance $-\frac{1}{4}$ is from 0 on a number line.

4. (a) $-1.2 + (-3.5) = -(1.2 + 3.5) = -4.7$

   (b) $13 + (-25) = -(25 - 13) = -12$

   (c) $(-2\pi) + \pi = -(2\pi - \pi) = -\pi$

5. (a) $(-1.3)(2.5) = -[(1.3)(2.5)]$
         $= -[3.25] = -3.25$

   (b) $(-8)(-3) = +(8)(3) = +24 = 24$

   (c) $(0)(-\sqrt{5}) = 0$ since any number times 0 is 0.

6. (a) By the commutative property of multiplication, $ab = ba$. Thus, $2(a + b) = \underline{(a + b)2}$.

   (b) By the additive identity property, if a is any real number, then $a + 0 = a$. Thus, $(2x) + 0 = \underline{2x}$.

   (c) By the property of reciprocals (or the multiplicative inverse property), any nonzero number times its reciprocal is 1. Thus, $6\left(\frac{1}{6}\right) = 1$.

7. (a) $6 - (-5) = 6 + (-(-5)) = 6 + 5 = 11$

   (b) $-\sqrt{2} - 0 = -\sqrt{2} + (-0) = -\sqrt{2} + 0 = -\sqrt{2}$

   (c) $-\frac{1}{2} \div \left(-\frac{1}{4}\right) = \left(-\frac{1}{2}\right) \cdot \left(-\frac{4}{1}\right) = \frac{4}{2} = 2$

   (d) $\left|6 \div -\frac{1}{5}\right| = |6 \cdot (-5)| = |-30| = 30$

8. (a) $1 - (-a) = 1 + a$

   (b) $0 \div (1 + z) = 0$

   (c) $(x + y) - (-x - y) = x + y + x + y$
       $= 2x + 2y$

   (d) $\frac{-a - (-b)}{b - a} = \frac{-a + b}{b - a} = \frac{b - a}{b - a} = 1$

CHAPTER 1 REVIEW OF FUNDAMENTAL CONCEPTS

SECTION 1.1 The Real Number System

1. true

2. false (the natural numbers start with 1)

3. false ($\sqrt{5}$ is an irrational number)

4. true

5. true

6. false (an irrational number cannot be an integer which is rational)

7. true

8. false (for example, $1 \div 2 = \frac{1}{2}$ is not a whole number but 1 and 2 are)

9. true

10. true

11. true

12. false (when $x = 0$, then $\frac{1}{x}$ is not a real number)

13. commutative property of addition

14. symmetric property of equality

15. 0 is the additive identity

16. closure property of multiplication

17. $(x + 3) + 7$

18. $a < 9$

19. -15

20. π

21. $-(-(-3)) = -(+3) = -3$

22. $6 - (x - 4) = 6 - x + 4 = 10 - x$

23. $-[-(1 - x)] = -[-1 + x] = 1 - x$

24. $-[x - (x + y)] = -[x - x - y] = -[-y] = y$

25. $(-14) - (-3) = (-14) + 3 = -(14 - 3) = -11$

26. $17 + (-5) = 17 - 5 = 12$

27. $-\frac{1}{9} + \left(-\frac{1}{3}\right) = -\frac{1}{9} + \left(-\frac{3}{9}\right) = \frac{-1 - 3}{9} = -\frac{4}{9}$

28. $\frac{2}{3} - \left(-\frac{3}{4}\right) = \frac{2}{3} + \frac{3}{4} = \frac{8}{12} + \frac{9}{12} = \frac{8 + 9}{12} = \frac{17}{12}$

29. $\left(-\frac{2}{3}\right)\left(\frac{3}{4}\right) = -\left(\frac{2}{3}\right)\left(\frac{3}{4}\right) = -\frac{2}{4} = -\frac{1}{2}$

30. $\left(-\frac{2}{3}\right) \div \left(-\frac{3}{4}\right) = \left(\frac{2}{3}\right)\left(\frac{4}{3}\right) = \frac{8}{9}$

31. Since zero divided by any number (other than zero) is always zero, $0 \div (-5\pi) = 0$.

32. Since any number divided by zero is undefined, $\frac{7}{8} \div 0$ is undefined.

33. $-2.5 - 0.3 = -(2.5 + 0.3) = -2.8$

34. $\left|-\frac{3}{4}\right| = -\left(-\frac{3}{4}\right) = \frac{3}{4}$

35. If $x < y$, then $x - y < 0$ so that $|x - y| = -(x - y) = -x + y = y - x$.

36. $|-7 + 4| = |-3| = 3$

37. $|-5| - |-8| = 5 - 8 = -3$

38. $\frac{|-12|}{|3|} = \frac{12}{3} = 4$

EXERCISES A   SECTION 1.1   3

39. $|(-3)(-2) - (-4)| = |6 + 4| = |10| = 10$

If a and b are real numbers, the distance between a and b is defined to be $|a - b|$. Use this definition in Exercises 40-42.

40. $|-3 - 2| = |-5| = 5$

41. $|7 - y| = -(7 - y) = -7 + y = y - 7$, since if $y > 7$, then $7 - y < 0$.

42. $|x - y| = -(x - y) = -x + y = y - x$, since if $x < 0$ and $y > 0$, then $x - y < 0$.

43. One set of numbers is $a = 1$, $b = 2$, and $c = 3$. Then $(a - b) - c = (1 - 2) - 3 = -1 - 3 = -4$, but $a - (b - c) = 1 - (2 - 3) = 1 - (-1) = 1 + 1 = 2$.

44. One example is $4 \div 2 = 2$, but $2 \div 4 = \dfrac{2}{4} = \dfrac{1}{2}$.

CHAPTER 1 REVIEW OF FUNDAMENTAL CONCEPTS

SECTION 1.2 Integer Exponents and Scientific Notation

1. (a) $3y^2 y^4 y^5 = 3y^{2+4+5} = 3y^{11}$

   (b) $\dfrac{2x^8 y^4}{x^3 y} = 2x^{8-3} y^{4-1} = 2x^5 y^3$

   (c) $(z^4)^5 = z^{4 \cdot 5} = z^{20}$

2. (a) $(-2a^2 b^5)^3 = (-2)^3 (a^2)^3 (b^5)^3 = -8a^6 b^{15}$

   (b) $\left(\dfrac{-3z^2}{u^3}\right)^4 = \dfrac{(-3)^4 (z^2)^4}{(u^3)^4} = \dfrac{81 z^8}{u^{12}}$

3. (a) $(-5)^0 = 1$ Any number except 0 raised to the 0 power is 1.

   (b) $(-3)^{-2} = \dfrac{1}{(-3)^2} = \dfrac{1}{9}$

   (c) $(2x)^{-3} = \dfrac{1}{(2x)^3} = \dfrac{1}{2^3 x^3} = \dfrac{1}{8x^3}$

   (d) $\dfrac{1}{x^{-5}} = \dfrac{1}{\frac{1}{x^5}} = 1 \cdot \dfrac{x^5}{1} = x^5$

4. (a) $(-2w)^{-2} = \dfrac{1}{(-2w)^2} = \dfrac{1}{(-2)^2 w^2} = \dfrac{1}{4w^2}$

   (b) $-2w^{-2} = -2 \cdot \dfrac{1}{w^2} = -\dfrac{2}{w^2}$

   (c) $\left(\dfrac{a^5}{3x^{-3}}\right)^{-2} = \dfrac{(a^5)^{-2}}{3^{-2}(x^{-3})^{-2}} = \dfrac{a^{-10}}{3^{-2} x^6}$
   $= \dfrac{3^2}{a^{10} x^6} = \dfrac{9}{a^{10} x^6}$

   (d) $\dfrac{1}{a^{-1} + b^{-1}} = \dfrac{1}{\frac{1}{a} + \frac{1}{b}} = \dfrac{1}{\frac{a+b}{ab}} = \dfrac{ab}{a+b}$

   (e) $\left[\left(\dfrac{5u^{-3} v^3}{u^{-1} v^{-2}}\right)^{-2}\right]^{-1} = \left(\dfrac{5u^{-3} v^3}{u^{-1} v^{-2}}\right)^2 = (5u^{-3-(-1)} v^{3-(-2)})^2$
   $= (5u^{-2} v^5)^2 = 5^2 (u^{-2})^2 (v^5)^2$
   $= 25 u^{-4} v^{10} = \dfrac{25 v^{10}}{u^4}$

5. (a) $142{,}000{,}000 = 1.42 \times 10^8$

   (b) $0.0000000142 = 1.42 \times 10^{-8}$

   (c) $52 = 5.2 \times 10^1 = 5.2 \times 10$

   (d) $0.003 = 3 \times 10^{-3}$

6. (a)
   $\dfrac{(0.0000000042)^3}{(24{,}000{,}000)(0.00000035)}$
   $= \dfrac{(4.2 \times 10^{-9})^3}{(2.4 \times 10^7)(3.5 \times 10^{-7})}$
   $= \dfrac{(4.2)^3}{(2.4)(3.5)} \times \dfrac{10^{-27}}{10^7 \times 10^{-7}}$
   $= 8.82 \times \dfrac{10^{-27}}{10^0}$
   $= 8.82 \times 10^{-27}$

   (b)
   $\dfrac{(1.24 \times 10^{10})(3.45 \times 10^{-9})}{(7.18 \times 10^{-7})^2}$
   $= \dfrac{(1.24)(3.45) \times 10^{10} \times 10^{-9}}{(7.18)^2 \times 10^{-14}}$
   $= \dfrac{(1.24)(3.45)}{(7.18)^2} \times \dfrac{10^1}{10^{-14}}$
   $= 0.082983527 \times 10^{15}$
   $= 8.2983527 \times 10^{-2} \times 10^{15}$
   $\approx 8.30 \times 10^{13}$

EXERCISES A SECTION 1.2 5

CHAPTER 1 REVIEW OF FUNDAMENTAL CONCEPTS

SECTION 1.2 Integer Exponents and Scientific Notation

1. $4 \cdot 4yyyyy = 4^2 y^5$

2. $(7a)(7a)(7a) = (7a)^3$

3. $(x+y)(x+y)(x+y)(x+y) = (x+y)^4$

4. $a^2 b^3 c^4 = aabbbcccc$

5. $x^2 + y^2 = xx + yy$

6. $5b^5 = 5bbbbb$

7. $a^2 a^{-3} a^4 = a^{2+(-3)+4} = a^3$

8. $3x^3 x^{-2} = 3x^{3+(-2)} = 3x^1 = 3x$

9. $\dfrac{x^2 x^{-5}}{x^3} = x^{2+(-5)-3} = x^{-6} = \dfrac{1}{x^6}$

10. $(b^{-3})^{-2} = b^{(-3)(-2)} = b^6$

11. $\dfrac{a^3}{b^2}$ cannot be simplified further

12. $5^0 = 1$

13. 0^0 is undefined

14. $(3y)^{-1} = \dfrac{1}{3y}$

15. $3y^{-1} = 3\dfrac{1}{y} = \dfrac{3}{y}$

16. $\dfrac{2x^2}{x^{-2}} = 2x^{2-(-2)} = 2x^{2+2} = 2x^4$

17. $\left(\dfrac{2x^2}{y^{-3}}\right)^{-2} = \dfrac{2^{-2}(x^2)^{-2}}{(y^{-3})^{-2}} = \dfrac{2^{-2}x^{-4}}{y^6} = \dfrac{1}{2^2 x^4 y^6} = \dfrac{1}{4x^4 y^6}$

18. $\left(\dfrac{2x^2}{y^{-3}}\right)^{2} = \dfrac{2^2(x^2)^2}{(y^{-3})^2} = \dfrac{2^2 x^4}{y^{-6}} = 4x^4 y^6$

19. $\dfrac{1}{x^{-1}+y^{-1}} = \dfrac{1}{\dfrac{1}{x}+\dfrac{1}{y}} = \dfrac{1}{\dfrac{y+x}{xy}} = \dfrac{xy}{x+y}$

20. $\dfrac{1}{(x+y)^{-2}} = \dfrac{1}{\dfrac{1}{(x+y)^2}} = (x+y)^2$

21. $\dfrac{3x^3 y^{-5}}{12x^{-4} y^{-1}} = \dfrac{x^3 y^{-5}}{4x^{-4} y^{-1}} = \dfrac{x^{3-(-4)} y^{-5-(-1)}}{4} = \dfrac{x^7 y^{-4}}{4} = \dfrac{x^7}{4y^4}$

22. $\left(\dfrac{6^0 x^6 y}{3x^{-2} y^{-1}}\right)^3 = \left(\dfrac{(1)x^{6-(-2)} y^{1-(-1)}}{3}\right)^3 = \left(\dfrac{x^8 y^2}{3}\right)^3$
$= \dfrac{(x^8)^3 (y^2)^3}{3^3} = \dfrac{x^{24} y^6}{27}$

23. $\left(\dfrac{2x^{-2} y^{-3}}{x^4 y^{-1}}\right)^{-2} = (2x^{-2-4} y^{-3-(-1)})^{-2} = (2x^{-6} y^{-2})^{-2}$
$= 2^{-2}(x^{-6})^{-2}(y^{-2})^{-2} = \dfrac{1}{2^2} x^{12} y^4$
$= \dfrac{x^{12} y^4}{4}$

24. $\left[\left(\dfrac{2a^{-2} b^{-3}}{a^{-4} b^2}\right)^{-2}\right]^{-1} = \left(\dfrac{2a^{-2} b^{-3}}{a^{-4} b^2}\right)^2$
$= (2a^{-2-(-4)} b^{-3-2})^2$
$= (2a^2 b^{-5})^2$
$= 2^2 (a^2)^2 (b^{-5})^2$
$= 4a^4 b^{-10}$
$= \dfrac{4a^4}{b^{10}}$

25. $456{,}000{,}000 = 4.56 \times 10^8$

26. $0.00000000000321 = 3.21 \times 10^{-12}$

27. $0.01 = 1 \times 10^{-2}$

28. $100 = 1 \times 10^2$

29. $2.35 \times 10^8 = 235,000,000$

30. $8.62 \times 10^2 = 862$

31. $4.17 \times 10^{-8} = 0.0000000417$

32. $3.2 \times 10^{-1} = 0.32$

33. $(0.000273)(428,000)$
$= (2.73 \times 10^{-4})(4.28 \times 10^5)$
$= (2.73)(4.28) \times 10^1$
$= 11.6844 \times 10^1$
$= 1.16844 \times 10^1 \times 10^1$
$= 1.17 \times 10^2$

34. $(0.0000000182)(0.00726)$
$= (1.82 \times 10^{-8})(7.26 \times 10^{-3})$
$= (1.82)(7.26) \times 10^{-11}$
$= 13.2132 \times 10^{-11}$
$= 1.32132 \times 10^1 \times 10^{-11}$
$= 1.32 \times 10^{-10}$

35. $\dfrac{(86,000)^2(0.000000392)}{(0.00157)^2}$

$= \dfrac{(8.6 \times 10^4)^2(3.92 \times 10^{-7})}{(1.57 \times 10^{-3})^2}$

$= \dfrac{(8.6)^2(3.92) \times 10^8 \times 10^{-7}}{(1.57)^2 \times 10^{-6}}$

$= 117.6206743 \times 10^7$
$= 1.176206743 \times 10^2 \times 10^7$
$= 1.18 \times 10^9$

36. $\dfrac{(0.00000562)^2(369,000)^2}{183,000,000,000}$

$= \dfrac{(5.62 \times 10^{-6})^2(3.69 \times 10^5)^2}{1.83 \times 10^{11}}$

$= \dfrac{(5.62)^2(3.69)^2 \times 10^{-12} \times 10^{10}}{1.83 \times 10^{11}}$

$= 235.0034693 \times 10^{-13}$
$= 2.35 \times 10^2 \times 10^{-13}$
$= 2.35 \times 10^{-11}$

37. The amount of debt per person is found by dividing $6,250,000,000 by 2,120,000. Changing to scientific notation we have

$$\dfrac{6.25 \times 10^9}{2.12 \times 10^6}$$

which becomes

2.948113208×10^3 or $\$2.95 \times 10^3$ or $2950.

38. The profit on each mixer is determined by dividing $8,620,000 by 925,000. Changing to scientific notation we have

$$\dfrac{8.62 \times 10^6}{9.25 \times 10^5}$$

which becomes

0.931891891×10^1 or $\$9.32 \times 10^0$ or $9.32.

39. The distance to the sun in meters is determined by multiplying 92,900,000 by 1.61×10^3. Changing to scientific notation we have
$(9.29 \times 10^7)(1.61 \times 10^3)$
which becomes

14.9569×10^{10} or 1.50×10^{11} meters.

40. To find the time required for light to reach the earth from the sun, divide the distance from the sun to the earth in meters, 1.50×10^{11} (from Exercise 39) by 3.00×10^8, which becomes

$0.5 \times 10^3 = 5.00 \times 10^2$ sec (8.33 min).

41. true

42. false (for example, $2 \div 5$ is not an integer but 2 and 5 are)

43. true

44. true

PRACTICE EXERCISES SECTION 1.3

CHAPTER 1 REVIEW OF FUNDAMENTAL CONCEPTS

SECTION 1.3 Algebraic Expressions and Polynomials

1. (a) Substitute 4 for x and -3 for y in:
$$-2y^3 + 3x = -2(-3)^3 + 3(4)$$
$$= -2(-27) + 12$$
$$= 54 + 12$$
$$= 66$$

   (b) Substitute 4 for x and -3 for y in:
$$5[(x + y)^3 - 2] = 5[((4) + (-3))^3 - 2]$$
$$= 5[(1)^3 - 2]$$
$$= 5[1 - 2]$$
$$= 5[-1]$$
$$= -5$$

   (c) Substitute 4 for x and -3 for y in:
$$\frac{x^2 - 2y^2 + 2}{2(1 - y)} = \frac{(4)^2 - 2(-3)^2 + 2}{2(1 - (-3))}$$
$$= \frac{16 - 2(9) + 2}{2(1 + 3)}$$
$$= \frac{16 - 18 + 2}{2(4)}$$
$$= \frac{0}{8} = 0$$

2. Substitute 11.25 for t in the following formula.
$$h = -16t^2 + 180t$$
$$= -16(11.25)^2 + 180(11.25)$$
$$= -16(126.5625) + 2025$$
$$= -2025 + 2025 = 0$$

   Thus, 11.25 seconds into the flight, the rocket has returned to the ground, when its height is 0 ft. Values for t larger than 11.25 would be meaningless in this problem.

3. (a)
$$(y^4 - 5y^3 + 3y - 9) + (7y^4 + y^3 - 4y^2 - y + 12)$$
$$= y^4 - 5y^3 + 3y - 9 + 7y^4 + y^3 - 4y^2 - y + 12$$
$$= y^4 + 7y^4 - 5y^3 + y^3 - 4y^2 + 3y - y - 9 + 12$$
$$= 8y^4 - 4y^3 - 4y^2 + 2y + 3$$

   (b)
$$(a^2b^2 + 3a^3b - 2ab^3 - ab) +$$
$$(4a^2b^2 - a^3b + 4ab^3 + ab)$$
$$= a^2b^2 + 3a^3b - 2ab^3 - ab +$$
$$4a^2b^2 - a^3b + 4ab^3 + ab$$
$$= a^2b^2 + 4a^2b^2 + 3a^3b - a^3b -$$
$$2ab^3 + 4ab^3 - ab + ab$$
$$= 5a^2b^2 + 2a^3b + 2ab^3$$

4. (a)
$$(4a^2b^3 + 7ab^2 - 3b^2 + 8) - (-3a^2b^3 + 9ab^2 - 5)$$
$$= 4a^2b^3 + 7ab^2 - 3b^2 + 8 + 3a^2b^3 - 9ab^2 + 5$$
$$= 4a^2b^3 + 3a^2b^3 + 7ab^2 - 9ab^2 - 3b^2 + 8 + 5$$
$$= 7a^2b^3 - 2ab^2 - 3b^2 + 13$$

   (b)
$$(3u^2v^2 - 2uv^2) - (uv + 2u^2v^2) -$$
$$(4uv^2 - uv - 7u^2v^2)$$
$$= 3u^2v^2 - 2uv^2 - uv - 2u^2v^2 - 4uv^2 + uv + 7u^2v^2$$
$$= 3u^2v^2 - 2u^2v^2 + 7u^2v^2 - 2uv^2 - 4uv^2 - uv + uv$$
$$= 8u^2v^2 - 6uv^2$$

5. (a) $-2x^2y(3xy^3 - 5xy)$
$$= (-2x^2y)(3xy^3) + (-2x^2y)(-5xy)$$
$$= (-2)(3)x^3yy^3 + (-2)(-5)x^3xyy$$
$$= -6x^3y^4 + 10x^3y^2$$

   (b) $(2x - 3y)(x + 5y)$
$$= (2x)(x) + (2x)(5y) + (-3y)(x) + (-3y)(5y)$$
$$= 2x^2 + 10xy - 3xy - 15y^2$$
$$= 2x^2 + 7xy - 15y^2$$

6.
$$\begin{array}{r} 2a^3 - 3a^2 + a - 5 \\ a - 1 \\ \hline 2a^4 - 3a^3 + a^2 - 5a \\ -2a^3 + 3a^2 - a + 5 \\ \hline 2a^4 - 5a^3 + 4a^2 - 6a + 5 \end{array}$$

7. (a) $(a - 3b)(a + 3b) = (a)^2 - (3b)^2 = a^2 - 9b^2$

   (b) $(u + 2v)(u + 2v) = u^2 + 2u(2v) + (2v)^2$
$$= u^2 + 4uv + 4v^2$$

   (c) $(2x^3 + y)(2x^3 - y) = (2x^3)^2 - (y)^2$
$$= 4x^6 - y^2$$

   (d) $(2x^3 - y)(2x^3 - y) = (2x^3)^2 - 2(2x^3)(y) + y^2$
$$= 4x^6 - 4x^3y + y^2$$

8. (a)
$$\frac{-7y^5 - 14y^3 + 21y}{-7y^2}$$
$$= \frac{-7y^5}{-7y^2} + \frac{-14y^3}{-7y^2} + \frac{21y}{-7y^2}$$
$$= y^3 + 2y - \frac{3}{y}$$

(b)

$$\frac{3a^3b^4 + 6a^2b^3 - 9ab^5 + 3ab}{3a^2b^2}$$

$$= \frac{3a^3b^4}{3a^2b^2} + \frac{6a^2b^3}{3a^2b^2} + \frac{-9ab^5}{3a^2b^2} + \frac{3ab}{3a^2b^2}$$

$$= ab^2 + 2b - \frac{3b^3}{a} + \frac{1}{ab}$$

9. (a)

$$\begin{array}{r} y + 5 \\ y - 3 \overline{\smash{)}\, y^2 + 2y - 15} \\ \underline{y^2 - 3y} \\ 5y - 15 \\ \underline{5y - 15} \\ 0 \end{array}$$

Thus, the quotient is $y + 5$.

(b)

$$\begin{array}{r} a^2 - 2a + 4 \\ a + 2 \overline{\smash{)}\, a^3 + 8} \\ \underline{a^3 + 2a^2} \\ -2a^2 \\ \underline{-2a^2 - 4a} \\ 4a + 8 \\ \underline{4a + 8} \\ 0 \end{array}$$

Thus, the qoutient is $a^2 - 2a + 4$.

EXERCISES A

CHAPTER 1 REVIEW OF FUNDAMENTAL CONCEPTS

SECTION 1.3 Algebraic Expressions and Polynomials

1. $2a^2 = 2(-2)^2 = 2(4) = 8$

2. $(2a)^2 = (2(-2))^2 = (-4)^2 = 16$

3. $-2a^2 = -2(-2)^2 = -2(4) = -8$

4. $(-2a)^2 = (-2(-2))^2 = (4)^2 = 16$

5. $(a - c)^{-3} = ((-2) - (-1))^{-3}$
   $= (-2 + 1)^{-3}$
   $= (-1)^{-3}$
   $= \dfrac{1}{(-1)^3} = \dfrac{1}{-1} = -1$

6. $(c - a)^3 = ((-1) - (-2))^3$
   $= (-1 + 2)^3$
   $= (1)^3 = 1$

7. $-a^2 + b = -(-2)^2 + (3)$
   $= -(4) + (3) = -4 + 3 = -1$

8. $3[a - (b + c)] = 3[(-2) - (3 + (-1))]$
   $= 3[-2 - 2]$
   $= 3[-4] = -12$

9. $\dfrac{a^2 - b^2}{c} = \dfrac{(-2)^2 - (3)^2}{-1}$
   $= \dfrac{4-9}{-1} = \dfrac{-5}{-1} = 5$

10. $\sqrt{6a^2 + 1} - b = \sqrt{6(-2)^2 + 1} - (3)$
    $= \sqrt{6(4) + 1} - 3$
    $= \sqrt{25} - 3 = 5 - 3 = 2$

11. $\sqrt{(a + b)^2} = \sqrt{((-2) + (3))^2}$
    $= \sqrt{(1)^2}$
    $= \sqrt{1} = 1$

12. $b + 3[c - (4 - a)] = (3) + 3[(-1) - (4 - (-2))]$
    $= 3 + 3[-1 - (4 + 2)]$
    $= 3 + 3[-1 - 6]$
    $= 3 + 3[-7]$
    $= 3 + (-21)$
    $= 3 - 21 = -18$

13. $2y - 3 + 5y + 7$
    $= 2y + 5y - 3 + 7$
    $= 7y + 4$

14. $2x - [x - 2(x - 1)]$
    $= 2x - [x - 2x + 2]$
    $= 2x - [-x + 2]$
    $= 2x + x - 2$
    $= 3x - 2$

15. $-3[4w - (1 + w)]$
    $= -3[4w - 1 - w)]$
    $= -3[3w - 1]$
    $= -9w + 3$

16. $x^3 - 6x$ has two terms, x^3 and $-6x$, so it is a binomial. Since the degree of the term x^3 is 3, and the degree of the term $-6x$ is 1, the binomial has degree 3, the same as the degree of the term of highest degree.

17. $a^3b^3 - 6a^2bc^4 + 8ab$ has three terms, a^3b^3 of degree $3 + 3 = 6$, $-6a^2bc^4$ of degree $2 + 1 + 4 = 7$, and $+8ab$ of degree $1 + 1 = 2$, so it is a trinomial. Since the term of highest degree 7 is $-6a^2bc^4$, the trinomial also has degree 7.

18. -2 has one term so it is a monomial. Since the degree of this term is 0, the degree of the monomial is also 0.

19. $(3x^2 + 2x - 5) + (-2x^2 + 5)$
    $= 3x^2 + 2x - 5 - 2x^2 + 5$
    $= 3x^2 - 2x^2 + 2x - 5 + 5$
    $= x^2 + 2x$

20. $(6a^2b^2 - 3ab + 2) + (4a^2b^2 + 8ab - 5)$
    $= 6a^2b^2 - 3ab + 2 + 4a^2b^2 + 8ab - 5$
    $= 6a^2b^2 + 4a^2b^2 - 3ab + 8ab + 2 - 5$
    $= 10a^2b^2 + 5ab - 3$

21. $(6x^2 - 4x + 1) - (-2x^2 + 3x - 4)$
    $= 6x^2 - 4x + 1 + 2x^2 - 3x + 4$
    $= 6x^2 + 2x^2 - 4x - 3x + 1 + 4$
    $= 8x^2 - 7x + 5$

22. $(2a^2b^2 - 6ab - 3) - (-7a^2b^2 + 6ab - 3)$
    $= 2a^2b^2 - 6ab - 3 + 7a^2b^2 - 6ab + 3$
    $= 2a^2b^2 + 7a^2b^2 - 6ab - 6ab - 3 + 3$
    $= 9a^2b^2 - 12ab$

23. $(a^2 + 3a - 4) + (a^3 - a^2 - a + 4)$
    $= a^2 + 3a - 4 + a^3 - a^2 - a + 4$
    $= a^3 + a^2 - a^2 + 3a - a - 4 + 4$
    $= a^3 + 2a$

24. $(y^5 + 4y^2 - 2) - (2y^2 - y^5 + 6)$
    $= y^5 + 4y^2 - 2 - 2y^2 + y^5 - 6$
    $= y^5 + y^5 + 4y^2 - 2y^2 - 2 - 6$
    $= 2y^5 + 2y^2 - 8$

25. $(5a^4b^2 - 7a^2b^4 + 3) - (6a^4b^2 + 6a^2b^4 - 7)$
    $= 5a^4b^2 - 7a^2b^4 + 3 - 6a^4b^2 - 6a^2b^4 + 7$
    $= 5a^4b^2 - 6a^4b^2 - 7a^2b^4 - 6a^2b^4 + 3 + 7$
    $= -a^4b^2 - 13a^2b^4 + 10$

26. $(4a^2b^2 - 2ab + 3) + (2a^2b^2 + ab - 7) -$
    $(-a^2b^2 - ab - 6)$
    $= 4a^2b^2 - 2ab + 3 + 2a^2b^2 + ab - 7 +$
    $a^2b^2 + ab + 6$
    $= 4a^2b^2 + 2a^2b^2 + a^2b^2 - 2ab + ab + ab +$
    $3 - 7 + 6$
    $= 7a^2b^2 + 2$

27. $(-6x^2y^3)(-3x^4y) = (-6)(-3)x^2x^4y^3y = 18x^6y^4$

28. $4ab^3(-3a^2b^2 + 5ab)$
    $= (4ab^3)(-3a^2b^2) + (4ab^3)(5ab)$
    $= (4)(-3)aa^2b^3b^2 + (4)(5)aab^3b$
    $= -12a^3b^5 + 20a^2b^4$

29. $(a - b)(a + 2b)$
    $= (a)(a) + (a)(2b) + (-b)(a) + (-b)(2b)$
    $= a^2 + 2ab - ab - 2b^2$
    $= a^2 + ab - 2b^2$

30. $(5xy + 3)(2xy - 7)$
    $= (5xy)(2xy) + (5xy)(-7) + (3)(2xy) + (3)(-7)$
    $= 10x^2y^2 - 35xy + 6xy - 21$
    $= 10x^2y^2 - 29xy - 21$

31. $(x - 2y)(x + 2y)$
    $= (x)^2 - (2y)^2$
    $= x^2 - 4y^2$

32. $(x + 2y)^2$
    $= (x)^2 + 2(x)(2y) + (2y)^2$
    $= x^2 + 4xy + 4y^2$

33. $(x - 2y)^2$
    $= (x)^2 - 2(x)(2y) + (2y)^2$
    $= x^2 - 4xy + 4y^2$

34. $(3a^2 - b^2)(3a^2 + b^2)$
    $= (3a^2)^2 - (b^2)^2$
    $= 9a^4 - b^4$

35. $(6x - 7y + 2)(2x - 3y)$
    $= (6x - 7y + 2)(2x) + (6x - 7y + 2)(-3y)$
    $= (6x)(2x) + (-7y)(2x) + (2)(2x) + (6x)(-3y)$
    $+ (-7y)(-3y) + (2)(-3y)$

$= 12x^2 - 14xy + 4x - 18xy + 21y^2 - 6y$
$= 12x^2 - 14xy - 18xy + 21y^2 + 4x - 6y$
$= 12x^2 - 32xy + 21y^2 + 4x - 6y$

36. $(3a - 7b)(4a^2 - 2ab + 7b^2)$
    $= (3a - 7b)(4a^2) + (3a - 7b)(-2ab)$
    $+ (3a - 7b)(7b^2)$
    $= (3a)(4a^2) + (-7b)(4a^2) + (3a)(-2ab)$
    $+ (-7b)(-2ab) + (3a)(7b^2) + (-7b)(7b^2)$
    $= 12a^3 - 28a^2b - 6a^2b + 14ab^2 + 21ab^2 - 49b^3$
    $= 12a^3 - 34a^2b + 35ab^2 - 49b^3$

37. $(x + 2y - 5)^2 = [(x + 2y) - 5]^2$
    $= (x + 2y)^2 - 2(5)(x + 2y) + 5^2$
    $= x^2 + 4xy + 4y^2 - 10x - 20y + 25$

38. $[(x - 2y) + 5][(x - 2y) - 5]$
    $= (x - 2y)^2 - (5)^2$
    $= x^2 - 2(x)(2y) + (2y)^2 - 25$
    $= x^2 - 4xy + 4y^2 - 25$

39. $x^{2n}y^n(x^n - y^n)$
    $= x^{2n}x^ny^n - x^{2n}y^ny^n$
    $= x^{2n+n}y^n + x^{2n}y^{n+n}$
    $= x^{3n}y^n + x^{2n}y^{2n}$

40. $(x^n - y^n)(x^n + y^n)$
    $= (x^n)^2 - (y^n)^2$
    $= x^{2n} - y^{2n}$

41. $(x^{n+1} + y^{n+1})^2$
    $= (x^{n+1})^2 + 2(x^{n+1})(y^{n+1}) + (y^{n+1})^2$
    $= x^{2n+2} + 2x^{n+1}y^{n+1} + y^{2n+2}$

42. $(x^2 - 3xy + 2y^2)(2x^2 + 4xy - 3y^2)$
    $= (x^2 - 3xy + 2y^2)(2x^2)$
    $+ (x^2 - 3xy + 2y^2)(4xy)$
    $+ (x^2 - 3xy + 2y^2)(-3y^2)$
    $= 2x^4 - 6x^3y + 4x^2y^2$
    $+ 4x^3y - 12x^2y^2 + 8xy^3$
    $- 3x^2y^2 + 9xy^3 - 6y^4$
    $= 2x^4 - 2x^3y - 11x^2y^2 + 17xy^3 - 6y^4$

43. $\dfrac{5y^5 - 30y^4 + 25y^3 - 5}{5y^2}$
    $= \dfrac{5y^5}{5y^2} + \dfrac{-30y^4}{5y^2} + \dfrac{25y^3}{5y^2} + \dfrac{-5}{5y^2}$
    $= y^3 - 6y^2 + 5y - \dfrac{1}{y^2}$

EXERCISES A SECTION 1.3 11

44. $(15x^4y^3 - 5x^3y^4 + 10x^5y^5) \div (-5x^2y^3)$

$$= \frac{15x^4y^3}{-5x^2y^3} + \frac{-5x^3y^4}{-5x^2y^3} + \frac{10x^5y^5}{-5x^2y^3}$$

$$= -3x^2 + xy - 2x^3y^2$$

45.
$$\begin{array}{r} a^2 + a + 2 \\ a-1\overline{\smash{\big)}\,a^3 + a - 3} \\ \underline{a^3 - a^2 } \\ a^2 + a \\ \underline{a^2 - a } \\ 2a - 3 \\ \underline{2a - 2} \\ -1 \end{array}$$

Thus, we obtain $a^2 + a + 2 - \dfrac{1}{a-1}$.

46.
$$\begin{array}{r} 4y^2 + 2y + 1 \\ 2y-1\overline{\smash{\big)}\,8y^3 - 1} \\ \underline{8y^3 - 4y^2 } \\ 4y^2 \\ \underline{4y^2 - 2y } \\ 2y - 1 \\ \underline{2y - 1} \\ 0 \end{array}$$

Thus we obtain $4y^2 + 2y + 1$.

47.
$$\begin{array}{r} x + 3y \\ 2x-y\overline{\smash{\big)}\,2x^2 + 5xy - 3y^2} \\ \underline{2x^2 - xy } \\ 6xy - 3y^2 \\ \underline{6xy - 3y^2} \\ 0 \end{array}$$

Thus we obtain $x + 3y$.

48.
$$\begin{array}{r} x^3 - x^2 + 3x - 1 \\ x^2-5\overline{\smash{\big)}\,x^5 - x^4 - 2x^3 + 4x^2 - 15x + 5} \\ \underline{x^5 - 5x^3 } \\ -x^4 + 3x^3 + 4x^2 \\ \underline{-x^4 + 5x^2 } \\ +3x^3 - x^2 - 15x \\ \underline{+3x^3 - 15x } \\ -x^2 + 5 \\ \underline{-x^2 + 5} \\ 0 \end{array}$$

Thus we obtain $x^3 - x^2 + 3x - 1$.

49.
$$\begin{array}{r} x^2 - 2x + 3 \\ x+1\overline{\smash{\big)}\,x^3 - x^2 + x + m} \\ \underline{x^3 + x^2 } \\ -2x^2 + x \\ \underline{-2x^2 - 2x } \\ +3x + m \\ \underline{+3x + 3} \\ m - 3 \end{array}$$

Thus, for the remainder to be zero, $m - 3$ must be zero. If $m - 3 = 0$, then $m = 3$.

50.
$$\begin{array}{r} x^2 - 2 \\ 2x+1\overline{\smash{\big)}\,2x^3 + x^2 - 4x + m} \\ \underline{2x^3 + x^2 } \\ -4x + m \\ \underline{-4x - 2} \\ m + 2 \end{array}$$

Thus, for the remainder to be 8, $m + 2$ must be 8. If $m + 2 = 8$, then $m = 6$.

51. Use the formula $C = \dfrac{5}{9}(F - 32)$ and substitute 5 for F.

$$C = \tfrac{5}{9}(5 - 32)$$
$$= \tfrac{5}{9}(-27)$$
$$= 5(-3) = -15$$

Thus, the the temperature is $-15°C$.

52. Use the formula $V = \pi r^2 h$ to find the volume of the bottle.
$$V = (3.14)(2.1)^2(5.6)$$
$$= 77.54544$$

To find the cost, multiply the volume by the cost per cubic centimeter, $52.50, to obtain

$$(77.54544)(52.50) = 4071.1356.$$

Thus, the cost of the bottle is about $4071, to the nearest dollar.

53. (a) $C = M + W$
$= (4u^2 - u + 5) + (2u^2 + u + 6)$
$= 4u^2 - u + 5 + 2u^2 + u + 6$
$= 4u^2 + 2u^2 - u + u + 5 + 6$
$= 6u^2 + 11$

(b) Substitute 15 for u and evaluate C.
$C = 6(15)^2 + 11$
$= 6(225) + 11 = 1350 + 11 = 1361$
Thus the total cost is $1361 when 15 units are produced and sold.

54. $\left(\dfrac{3x^3}{y^{-2}}\right)^{-2} = \dfrac{3^{-2}(x^3)^{-2}}{(y^{-2})^{-2}} = \dfrac{3^{-2}x^{-6}}{y^4} = \dfrac{1}{3^2 x^6 y^4} = \dfrac{1}{9x^6 y^4}$

55. $\dfrac{4x^5 y^{-2}}{10x^{-2} y^{-4}} = \dfrac{2x^{5-(-2)} y^{-2-(-4)}}{5} = \dfrac{2x^7 y^2}{5}$

56. $\left(\dfrac{2^0 x^4 y^{-9}}{3(xy)^{-6}}\right)^{-2} = \left(\dfrac{x^4 y^{-9}}{3x^{-6} y^{-6}}\right)^{-2}$
$= \left(\dfrac{x^{4-(-6)} y^{-9-(-6)}}{3}\right)^{-2}$
$= \left(\dfrac{x^{10} y^{-3}}{3}\right)^{-2}$
$= \dfrac{(x^{10})^{-2}(y^{-3})^{-2}}{3^{-2}}$
$= \dfrac{x^{-20} y^6}{3^{-2}}$
$= \dfrac{3^2 y^6}{x^{20}} = \dfrac{9y^6}{x^{20}}$

PRACTICE EXERCISES

CHAPTER 1 REVIEW OF FUNDAMENTAL CONCEPTS

SECTION 1.4 Factoring Polynomials

1. (a) The terms have a common factor of $5y$.
$$5y^3 - 15y = (5y)(y^2) - (5y)(3)$$
$$= 5y(y^2 - 3)$$

   (b) The terms have a common factor of $4xy^3$.
$$12x^2y^4 + 4x^4y^3 - 8xy^3$$
$$= (4xy^3)(3xy) + (4xy^3)(x^3) - (4xy^3)(2)$$
$$= 4xy^3(3xy + x^3 - 2)$$

2. (a) $3x^2 + 6xy + 5x + 10y$
$$= 3x(x + 2y) + 5(x + 2y)$$
$$= (3x + 5)(x + 2y)$$

   (b) $6ab^2 - 3ab - 14b + 7$
$$= 3ab(2b - 1) - 7(2b - 1)$$
$$= (3ab - 7)(2b - 1)$$

3. (a) $u^2 + 7u + 12$
The factors of 12 that add to give 7 are 4 and 3. Since all the signs are positive, the signs in the factors must all be positive. Thus,
$$u^2 + 7u + 12 = (u + 4)(u + 3).$$

   (b) $u^2 - 7u + 12$
The factors of 12 that add to give -7 are -4 and -3. Both factors must use negative signs since the middle term is negative and the last term is positive. Thus,
$$u^2 - 7u + 12 = (u - 4)(u - 3).$$

   (c) $u^2 + u - 12$
The factors of -12 that add to give 1 are 4 and -3. The factors must have opposite signs for the product to be negative 12. Since the sign of the middle term is plus, the larger number must have the plus sign. Thus,
$$u^2 + u - 12 = (u + 4)(u - 3).$$

   (d) $u^2 - u - 12$
The factors of -12 that add to give -1 are -4 and 3. The factors must have opposite signs for the product to be negative 12. Since the sign of the middle term is minus, the larger number must have the minus sign. Thus,
$$u^2 - u - 12 = (u - 4)(u + 3).$$

4. (a) $6x^2 - 19x + 10$
The factors of 6 and 10 that combine to give -19 are 3,2 and $-2,-5$. Thus,
$$6x^2 - 19x + 10 = (3x - 2)(2x - 5).$$

   (b) $-24u^2 + 6u + 9$
First remove the common factor -3.
$$-24u^2 + 6u + 9 = -3(8u^2 - 2u - 3)$$
The factors of 8 and -3 that combine to give -2 are 2,4 and 1,-3. Thus,
$$-24u^2 + 6u + 9 = -3(2u + 1)(4u - 3).$$

5. (a) $12x^2 + 17x + 6$
Since $a = 12$ and $c = 6$, $ac = 72$. The factors of 72 that add to give 17 are 9 and 8. Rewrite the trinomial.
$$12x^2 + 9x + 8x + 6$$
Now factor by grouping.
$$3x(4x + 3) + 2(4x + 3)$$
$$= (3x + 2)(4x + 3)$$

   (b) $-12y^2 + 22y - 6$
First remove the common factor -2.
$$-12y^2 + 22y - 6 = -2(6y^2 - 11y + 3)$$
Since $a = 6$ and $b = 3$, $ab = 18$. The factors of 18 that add to give -11 are -2 and -9. Rewrite the trinomial.
$$6y^2 - 2y - 9y + 3$$
Now factor by grouping.
$$2y(3y - 1) - 3(3y - 1)$$
$$= (2y - 3)(3y - 1)$$
Don't forget the common factor -2 in the final factorization.
$$-12y^2 + 22y - 6 = -2(2y - 3)(3y - 1)$$

6. (a) $18x^6 - 8y^4$
First remove the common factor 2.
$$18x^6 - 8y^4 = 2(9x^6 - 4y^4)$$
$$= 2[(3x^3)^2 - (2y^2)^2]$$
$$= 2(3x^3 + 2y^2)(3x^3 - 2y^2)$$

   (b) $9x^2 + 30xy + 25y^2$
$$= (3x)^2 + 2(3x)(5y) + (5y)^2$$
$$= (3x + 5y)^2$$

   (c) $x^2 - 12x + 36 = x^2 - 2(x)(6) + 6^2$
$$= (x - 6)^2$$

   (d) $2x^3 + 16y^3 = 2[x^3 + 8y^3] = 2[x^3 + (2y)^3]$
$$= 2(x + 2y)(x^2 - 2xy + 4y^2)$$

   (e) $2u^3 - 250v^3 = 2[u^3 - 125v^3] = 2[u^3 - (5v)^3]$
$$= 2(u - 5v)(u^2 + 5uv + 25v^2)$$

CHAPTER 1 REVIEW OF FUNDAMENTAL CONCEPTS

SECTION 1.4 Factoring Polynomials

1. $6x^2y^2 - 3xy = (3xy)(2xy) - (3xy)(1)$
   $= 3xy(2xy - 1)$

2. $8x^3y^2 + 4x^2y^3 - 12x^2y^2$
   $= (4x^2y^2)(2x) + (4x^2y^2)(y) + (4x^2y^2)(-3)$
   $= 4x^2y^2(2x + y - 3)$

3. $5a^2b - 10ab + 3a - 6$
   $= 5ab(a - 2) + 3(a - 2)$
   $= (5ab + 3)(a - 2)$

4. $8a^3b^3 - 2a^2b^2 - 12abc^2 + 3c^2$
   $= 2a^2b^2(4ab - 1) - 3c^2(4ab - 1)$
   $= (2a^2b^2 - 3c^2)(4ab - 1)$

5. $x^2 + 8x + 7$
   The factors of 7 that add to give 8 are 1 and 7. Thus
   $x^2 + 8x + 7 = (x + 1)(x + 7)$.

6. $x^2 - 8x + 7$
   The factors of 7 that add to give -8 are -1 and -7. Thus
   $x^2 - 8x + 7 = (x - 1)(x - 7)$.

7. $x^2 + 6x - 7$
   The factors of -7 that add to give 6 are -1 and 7. Thus
   $x^2 + 6x - 7 = (x - 1)(x + 7)$.

8. $x^2 - 6x - 7$
   The factors of -7 that add to give -6 are 1 and -7. Thus
   $x^2 - 6x - 7 = (x + 1)(x - 7)$.

9. $a^2 + 9a + 20$
   The factors of 20 that add to give 9 are 4 and 5. Thus
   $a^2 + 9a + 20 = (a + 4)(a + 5)$.

10. $x^2 - 2x - 35$
    The factors of -35 that add to give -2 are -7 and 5. Thus
    $x^2 - 2x - 35 = (x - 7)(x + 5)$.

11. $3x^2 - 5x - 2$
    The factors of 3 and -2 that combine to give -5 are 3, 1 and 1, -2. Thus
    $3x^2 - 5x - 2 = (3x + 1)(x - 2)$.

12. $a^2 - 4ab - 5b^2$
    The factors of -5 that add to give -4 are 1 and -5. Thus
    $a^2 - 4ab - 5b^2 = (a + b)(a - 5b)$.

13. $16x^2 + 10xy - 21y^2$
    The factors of 16 and -21 that combine to give 10 are 2, 8 and 3, -7. Thus
    $16x^2 + 10xy - 21y^2 = (2x + 3y)(8x - 7y)$.

14. $4x^2 - 9y^2$
    This is a special form, the difference of two squares, which factors into a sum and difference.
    $4x^2 - 9y^2 = (2x)^2 - (3y)^2$
    $= (2x + 3y)(2x - 3y)$

15. $4x^2 + 12xy + 9y^2$
    This is a special form, a perfect square trinomial.
    $4x^2 + 12xy + 9y^2 = (2x)^2 + 2(2x)(3y) + (3y)^2$
    $= (2x + 3y)(2x + 3y)$
    $= (2x + 3y)^2$

16. $4x^2 - 12xy + 9y^2$
    This is a special form, a perfect square trinomial.
    $4x^2 - 12xy + 9y^2 = (2x)^2 - 2(2x)(3y) + (3y)^2$
    $= (2x - 3y)(2x - 3y)$
    $= (2x - 3y)^2$

17. $27u^3 - v^3$
    This is a special form, the difference of two cubes, and uses $a^3 - b^3 = (a - b)(a^2 + ab + b^2)$.
    $27u^3 - v^3 = (3u)^3 - (v)^3$
    $= (3u - v)(9u^2 + 3uv + v^2)$

18. $27u^3 + v^3$
    This is a special form, the sum of two cubes, and uses $a^3 + b^3 = (a + b)(a^2 - ab + b^2)$.
    $27u^3 + v^3 = (3u)^3 + (v)^3$
    $= (3u + v)(9u^2 - 3uv + v^2)$

19. $28x^2 - 58xy - 30y^2$
    First remove the common factor 2.
    $28x^2 - 58xy - 30y^2 = 2(14x^2 - 29xy - 15y^2)$
    The factors of 14 and -15 that combine to give -29 are 2, 7 and -5, 3. Thus
    $28x^2 - 58xy - 30y^2 = 2(2x - 5y)(7x + 3y)$.

20. $6x^2 - 6y^2$
    First remove the common factor 6. The resulting factor is the difference of two squares.
    $6x^2 - 6y^2 = 6(x^2 - y^2)$
    $= 6(x + y)(x - y)$

EXERCISES A

21. $25x^2 + 10xy + y^2$
This is a special form, a perfect square trinomial.
$$25x^2 + 10xy + y^2 = (5x)^2 + 2(5x)(y) + (y)^2$$
$$= (5x + y)(5x + y)$$
$$= (5x + y)^2$$

22. $-4u^2 - 34u - 70$
First factor out the common factor -2.
$$-4u^2 - 34u - 70 = -2(2u^2 + 17u + 35)$$
The factors of 2 and 35 that combine to give 17 are 2, 1 and 7, 5. Thus
$$-4u^2 - 34u - 70 = -2(2u + 7)(u + 5).$$

23. $5u^2v^2 - 11uv + 2$
The factors of 5 and 2 that combine to give -11 are 1, 5 and $-2, -1$. Thus
$$5u^2v^2 - 11uv + 2 = (uv - 2)(5uv - 1).$$

24. $3a^2 - 60ab + 300b^2$
First factor out the common factor 3.
$$3a^2 - 60ab + 300b^2 = 3(a^2 - 20ab + 100b^2)$$
The remaining factor is a perfect square trinomial.
$$3a^2 - 60ab + 300b^2 = 3(a - 10b)(a - 10b).$$
$$= 3(a - 10b)^2$$

25. $32x^3 + 4y^9$
First factor out the common factor 4.
$$32x^3 + 4y^9 = 4(8x^3 + y^9)$$
The resulting factor is a sum of cubes.
$$32x^3 + 4y^9 = 4[(2x)^3 + (y^3)^3]$$
$$= 4(2x + y^3)(4x^2 - 2xy^3 + y^6)$$

26. $x^2 + 2xy + 2$
This trinomial cannot be factored.

27. $-5u^2 + 70uv - 240v^2$
First factor out the common factor -5.
$$-5u^2 + 70uv - 240v^2 = -5(u^2 - 14uv + 48v^2)$$
The factors of 48 that add to give -14 are -6 and -8. Thus
$$-5u^2 + 70uv - 240v^2 = -5(u - 6v)(u - 8v).$$

28. $2x^2 + 2y^2$
First factor out the common factor 2. The resulting factor is the sum of squares which cannot be factored further. Thus
$$2x^2 + 2y^2 = 2(x^2 + y^2).$$

29. $(u - v)^2 - 16$
This is a special form $a^2 - b^2 = (a + b)(a - b)$ with $a = u - v$ and $b = 4$. Substituting we obtain
$$(u - v)^2 - (4)^2 =$$
$$= [(u - v) + 4][(u - v) - 4]$$
$$= (u - v + 4)(u - v - 4).$$

30. $(u - v)^2 - 4(u - v) + 4$
This is a special form $a^2 - 2ab + b^2 = (a - b)(a - b) = (a - b)^2$ with $a = u - v$ and $b = 2$. Substituting we have
$$(u - v)^2 - 4(u - v) + 4$$
$$= [(u - v) - 2][(u - v) - 2]$$
$$= (u - v - 2)^2$$

31. $9u^6 - 30u^3v^2 + 25v^4$
This is a perfect square trinomial.
$$9u^6 - 30u^3v^2 + 25v^4$$
$$= (3u^3)^2 - 2(3u^3)(5v^2) + (5v^2)^2$$
$$= (3u^3 - 5v^2)^2$$

32. $u^6 + 2u^3v^3 + v^6$
This is a perfect square trinomial.
$$u^6 + 2u^3v^3 + v^6 = (u^3 + v^3)(u^3 + v^3)$$
$$= (u^3 + v^3)^2$$
But each of these factors can be factored again since each is the sum of cubes.
$$u^3 + v^3 = (u + v)(u^2 - uv + v^2)$$
Thus
$$u^6 + 2u^3v^3 + v^6 = [(u + v)(u^2 - uv + v^2)]^2$$
$$= (u + v)^2(u^2 - uv + v^2)^2.$$

33. $x^6 + 7x^3y^3 - 8y^6$
The factors of -8 that add to give 7 are -1 and 8. Thus
$$x^6 + 7x^3y^3 - 8y^6 = (x^3 - y^3)(x^3 + 8y^3).$$
Then each of these factors can be factored again since the first is a difference of cubes and the second is a sum of cubes.
$$x^6 + 7x^3y^3 - 8y^6 =$$
$$(x - y)(x^2 + xy + y^2)(x + 2y)(x^2 - 2xy + 4y^2)$$

34. $4x^2 - 9y^2 - 2x - 3y$
Group the first two terms and the last two terms and factor.
$$(4x^2 - 9y^2) - (2x + 3y)$$
$$= (2x + 3y)(2x - 3y) - 1(2x + 3y)$$
This gives two terms with the common factor $(2x + 3y)$ which can be removed.
$$= [(2x - 3y) - 1](2x + 3y)$$
$$= (2x - 3y - 1)(2x + 3y)$$

35. $x^2 - y^2 - 4y - 4x$
Group the first two terms and the last two terms and factor each.
$$(x^2 - y^2) - 4(y + x)$$
$$= (x - y)(x + y) - 4(x + y)$$
The result has $(x + y)$ as a common factor which can be removed.
$$= [(x - y) - 4](x + y)$$
$$= (x - y - 4)(x + y)$$

36. $x^3 - y^3 - x^2y + xy^2$
    Group the first two terms and the last two terms.
    The first group is the difference of cubes.
    $(x^3 - y^3) + (-x^2y + xy^2)$
    $= (x - y)(x^2 + xy + y^2) - xy(x - y)$
    The two terms now have a common factor $(x - y)$ which can be removed.
    $= (x - y)[(x^2 + xy + y^2) - xy]$
    $= (x - y)(x^2 + y^2)$

37. (a) Find the area of the circle with radius R, which is πR^2, and subtract the area of the inner circle with radius r, which is πr^2. Thus, the polynomial that represents tha area of the garden to be planted is $\pi R^2 - \pi r^2$.

    (b) $\pi R^2 - \pi r^2 = \pi(R^2 - r^2) = \pi(R + r)(R - r)$

    (c) Substitute 9.25 for R and 2.05 for r and use 3.14 for π
    $3.14(9.25 + 2.05)(9.25 - 2.05)$
    $= 3.14(11.3)(7.2)$
    $= 255.4704$
    Thus, to two decimal places, the area of the garden is 255.47 m^2.

    (d) Since each plant requires approximately 0.75 m^2 to grow properly, divide the area of the garden, 255.47 m^2, by 0.75 to obtain the number of plants to purchase.
    $255.47 \div 0.75 = 340.6266667$
    Thus, approximately 341 plants should be purchased.

38. (a) To find the volume of the remaining region, subtract the volume of the inner cylinder, $\pi r^2 h$, from the volume of the outer cylinder, $\pi R^2 h$. Thus the volume of the region is $\pi R^2 h - \pi r^2 h$.

    (b) $\pi R^2 h - \pi r^2 h = \pi h(R^2 - r^2)$
    $= \pi h(R + r)(R - r)$

    (c) Substitute 3.28 for R, 1.18 for r, 8.15 for h, and 3.14 for π and evaluate.
    $\pi h(R + r)(R - r)$
    $= (3.14)(8.15)(3.28 + 1.18)(3.28 - 1.18)$
    $= (3.14)(8.15)(4.46)(2.1)$
    $= 239.685306$
    Thus the volume of the ramaining metal is about 239.69 cm^3.

39. $(2y^3 - y^2 + 3) + (4y^3 - 6y - 3)$
    $= 2y^3 - y^2 + 3 + 4y^3 - 6y - 3$
    $= 2y^3 + 4y^3 - y^2 - 6y + 3 - 3$
    $= 6y^3 - y^2 - 6y$

40. $(6a^2 - a + 3) - (3a^2 - a + 1)$
    $= 6a^2 - a + 3 - 3a^2 + a - 1$
    $= 6a^2 - 3a^2 - a + a + 3 - 1$
    $= 3a^2 + 2$

41. $(3u - v)^2 = (3u)^2 - 2(3u)(v) + (v)^2$
    $= 9u^2 - 6uv + v^2$

42. $\dfrac{4x^3 - 2x^2 + 6x}{-2x} = \dfrac{4x^3}{-2x} + \dfrac{-2x^2}{-2x} + \dfrac{6x}{-2x}$
    $= -2x^2 + x - 3$

PRACTICE EXERCISES   SECTION 1.5   17

CHAPTER 1 REVIEW OF FUNDAMENTAL CONCEPTS

SECTION 1.5 Rational Expressions

1. (a) $\dfrac{y+3}{y+1}$

    Set the denominator equal to 0 to find the values for y to exclude.
    $$y + 1 = 0$$
    $$y = -1$$
    Thus the value to exclude is $y = -1$.

    (b) $\dfrac{x}{x^2 + 3x - 10}$

    Set the denominator equal to 0.
    $$x^2 + 3x - 10 = 0$$
    $$(x - 2)(x + 5) = 0$$
    $$x - 2 = 0 \qquad x + 5 = 0$$
    $$x = 2 \qquad x = -5$$
    Thus the values to exclude are $x = 2$ and $x = -5$.

2. Equivalent fractions are obtained when the numerator and denominator of a fraction are multiplied or divided by the same nonzero number, **not** when both have the same nonzero number added. Thus, since $\dfrac{2a+5}{3y+5}$ has been obtained from $\dfrac{2a}{3y}$ by adding 5 to both the numerator and denominator, these two fractions are not equivalent.

3. $\dfrac{8x^4 + 2x}{2x} = \dfrac{2x(4x^3 + 1)}{2x} = 4x^3 + 1$

4. $\dfrac{x^2 - 5x + 6}{x^2 + x - 6} \cdot \dfrac{x+3}{x-2} = \dfrac{(x-2)(x-3)}{(x+3)(x-2)} \cdot \dfrac{x+3}{x-2}$
    $$= \dfrac{(x-2)(x-3)(x+3)}{(x+3)(x-2)(x-2)}$$
    $$= \dfrac{x-3}{x-2}$$

5. (a) $\dfrac{x^2}{x+2} \div \dfrac{x}{x^2 + 4x + 4} = \dfrac{x^2}{x+2} \cdot \dfrac{x^2 + 4x + 4}{x}$
    $$= \dfrac{x^2}{x+2} \cdot \dfrac{(x+2)(x+2)}{x}$$
    $$= \dfrac{x^2(x+2)(x+2)}{(x+2)x}$$
    $$= x(x+2)$$

    (b)
    $$\dfrac{a^2 - b^2}{a^2 + a - 2} \cdot \dfrac{a^2 + 2a}{a^2 + 2ab + b^2} \div \dfrac{a^3 - b^3}{a^2 - a + ba - b}$$
    $$= \dfrac{a^2 - b^2}{a^2 + a - 2} \cdot \dfrac{a^2 + 2a}{a^2 + 2ab + b^2} \cdot \dfrac{a^2 - a + ba - b}{a^3 - b^3}$$
    $$= \dfrac{(a-b)(a+b)}{(a+2)(a-1)} \cdot \dfrac{a(a+2)}{(a+b)(a+b)} \cdot \dfrac{a(a-1) + b(a-1)}{(a-b)(a^2 + ab + b^2)}$$
    $$= \dfrac{(a-b)(a+b)a(a+2)(a+b)(a-1)}{(a+2)(a-1)(a+b)(a+b)(a-b)(a^2 + ab + b^2)}$$
    $$= \dfrac{a}{a^2 + ab + b^2}$$

6. (a) Factor the denominators of each fraction.
    $$\dfrac{a+1}{a^3 + 2a^2 + a} = \dfrac{a+1}{a(a+1)(a+1)}$$
    $$= \dfrac{1}{a(a+1)}$$

    $$\dfrac{5-a}{a^2 - 4a - 5} = \dfrac{(-1)(a-5)}{(a-5)(a+1)}$$
    $$= -\dfrac{1}{a+1}$$

    Thus, the LCD must consist of one factor of a and one factor of $(a+1)$.
    $$\text{The LCD} = a(a+1).$$

    (b) Factor the denominators of each fraction.
    $$\dfrac{3x+2}{6x^4 - 12x^3 + 6x^2} = \dfrac{3x+2}{6x^2(x^2 - 2x + 1)}$$
    $$= \dfrac{3x+2}{6x^2(x-1)^2}$$

    $$\dfrac{4x}{x^2 - 7x + 6} = \dfrac{4x}{(x-1)(x-6)}$$

    $$\dfrac{5x^2 + 1}{8(x^5 - 12x^4 + 36x^3)} = \dfrac{5x^2 + 1}{8x^3(x-6)^2}$$

    Thus, the LCD must consist of one factor of 24, one factor of x^3, two factors of $(x-1)$, and two factors of $(x-6)$.
    $$\text{The LCD} = 24x^3(x-1)^2(x-6)^2.$$

7. (a)
$$\frac{3x+y}{x^2+4xy-21y^2} + \frac{x+y}{9y^2-x^2}$$
$$= \frac{3x+y}{(x+7y)(x-3y)} + \frac{x+y}{(3y+x)(3y-x)}$$
$$= \frac{3x+y}{(x+7y)(x-3y)} + \frac{(-1)(x+y)}{(-1)(3y+x)(3y-x)}$$
$$= \frac{3x+y}{(x+7y)(x-3y)} + \frac{-x-y}{(3y+x)(x-3y)}$$
$$= \frac{(3x+y)(3y+x)}{(x+7y)(x-3y)(3y+x)} + \frac{(-x-y)(x+7y)}{(3y+x)(x-3y)(x+7y)}$$
$$= \frac{(3x^2+10xy+3y^2)+(-x^2-8xy-7y^2)}{(x-3y)(x+3y)(x+7y)}$$
$$= \frac{3x^2+10xy+3y^2-x^2-8xy-7y^2}{(x-3y)(x+3y)(x+7y)}$$
$$= \frac{2x^2+2xy-4y^2}{(x-3y)(x+3y)(x+7y)}$$
$$= \frac{2(x+2y)(x-y)}{(x-3y)(x+3y)(x+7y)}$$

(b)
$$\frac{y^2-1}{y^3-1} - \frac{y}{2y^2+2y+2}$$
$$= \frac{(y-1)(y+1)}{(y-1)(y^2+y+1)} - \frac{y}{2(y^2+y+1)}$$
$$= \frac{(y+1)}{(y^2+y+1)} - \frac{y}{2(y^2+y+1)}$$
$$= \frac{2(y+1)}{2(y^2+y+1)} - \frac{y}{2(y^2+y+1)}$$
$$= \frac{2y+2-y}{2(y^2+y+1)}$$
$$= \frac{y+2}{2(y^2+y+1)}$$

8. $$\frac{\frac{b}{a+b}-1}{\frac{b}{a-b}+1} = \frac{\frac{b}{a+b}-\frac{a+b}{a+b}}{\frac{b}{a-b}+\frac{a-b}{a-b}}$$
$$= \frac{\frac{b-(a+b)}{a+b}}{\frac{b+(a-b)}{a-b}}$$
$$= \frac{\frac{b-a-b}{a+b}}{\frac{b+a-b}{a-b}}$$
$$= \frac{\frac{-a}{a+b}}{\frac{a}{a-b}}$$
$$= \frac{-a}{a+b} \cdot \frac{a-b}{a}$$
$$= \frac{(-1)(a-b)}{a+b} = \frac{b-a}{a+b}$$

9. $$\frac{a-\frac{4}{a}}{\frac{16}{a^3}-a} = \frac{a^3\left[a-\frac{4}{a}\right]}{a^3\left[\frac{16}{a^3}-a\right]}$$
$$= \frac{a^4-4a^2}{16-a^4}$$
$$= \frac{a^2(a+2)(a-2)}{(4-a^2)(4+a^2)}$$
$$= \frac{a^2(a+2)(a-2)}{(2-a)(2+a)(4+a^2)}$$
$$= \frac{(-1)a^2}{4+a^2}$$
$$= \frac{-a^2}{4+a^2}$$

10. D = distance hiked down (and out) of the canyon,

$\dfrac{D}{5}$ = time required to hike down the canyon,

$\dfrac{D}{1.5}$ = time required to hike out of the canyon.

Then the average rate is:

$$\frac{\text{total distance hiked}}{\text{total time of the hike}} = \frac{2D}{\dfrac{D}{5}+\dfrac{D}{1.5}}$$

By simplifying this complex fraction we can obtain the desired rate.

$$\frac{2D}{\dfrac{D}{5}+\dfrac{D}{1.5}} = \frac{2D}{\dfrac{1.5D+5D}{(5)(1.5)}}$$
$$= \frac{2D}{1} \cdot \frac{(5)(1.5)}{1.5D+5D}$$
$$= \frac{(2D)(5)(1.5)}{6.5D}$$
$$= \frac{(2)(5)(1.5)}{6.5}$$
$$= \frac{15}{6.5} \approx 2.307692308$$

Thus, the average rate on Kathy's hike was about 2.3 mph.

EXERCISES A   SECTION 1.5   19

CHAPTER 1 REVIEW OF FUNDAMENTAL CONCEPTS

SECTION 1.5 Rational Expressions

1. Set the denominator, $x - 5$, equal to zero and solve for x.
$$x - 5 = 0$$
$$x = 5$$

   Thus the value to exclude is $x = 5$.

2. Set the denominator, $(y^2 - 5y + 6)(x + 7)$, equal to zero which means that each factor must be zero. Solve for y and x.
$$y^2 - 5y + 6 = 0$$
$$(y - 2)(y - 3) = 0$$
$$y - 2 = 0 \text{ or } y - 3 = 0$$
$$y = 2 \qquad\qquad y = 3$$

$$x + 7 = 0$$
$$x = -7$$

   Thus the values to exclude are $x = -7$, $y = 2$, and $y = 3$.

3. Set the denominator, $a^2 + ab$, equal to zero and solve.
$$a^2 + ab = a(a + b) = 0$$

   Then $a = 0$ or $a + b = 0$. If $a + b = 0$, then $a = -b$. Thus the values to exclude are $a = 0$ and $a = -b$.

4. Yes. The second fraction can be obtained from the first by dividing the numerator and denominator by x. (Provided $x \neq 0$.)

5. No. Neither fraction can be obtained from the other by multiplying or dividing the numerator and denominator by the same expression. Note that equivalent fractions do not result when the same expression is *added* to both the numerator and denominator.

6. Yes. The second fraction can be obtained from the first by dividing numerator and denominator by $a - b$. (Provided $a \neq b$ and $a \neq -b$.)

7. $\dfrac{77x^2y^4}{33x^3y^3} = \dfrac{7}{3}x^{2-3}y^{4-3} = \dfrac{7}{3}x^{-1}y^1 = \dfrac{7y}{3x}$

8. $\dfrac{a + b}{a^2 + b^2}$ cannot be reduced further since the denominator cannot be factored.

9. $\dfrac{a^3 - b^3}{a - b} = \dfrac{(a - b)(a^2 + ab + b^2)}{a - b}$
$$= a^2 + ab + b^2$$

10. $\dfrac{8x^3}{9x} \cdot \dfrac{45}{16} = \dfrac{8x^2 x(9)(5)}{9x(8)(2)} = \dfrac{5x^2}{2}$

11. $\dfrac{3(x + y)}{6x^2y^2} \cdot x^3y^3 = \dfrac{(x + y)x^3y^3}{2x^2y^2} = \dfrac{(x + y)xy}{2}$

12. $\dfrac{2a + 2}{a - 3} \cdot \dfrac{a^2 - 9}{4a + 4} = \dfrac{2(a + 1)}{a - 3} \cdot \dfrac{(a + 3)(a - 3)}{4(a + 1)}$
$$= \dfrac{2(a + 1)(a + 3)(a - 3)}{4(a - 3)(a + 1)}$$
$$= \dfrac{a + 3}{2}$$

13.
$$\dfrac{(a - 3)^2}{35(a + 3)} \cdot \dfrac{5(a^2 + 6a + 9)}{a^2 - 9}$$
$$= \dfrac{(a - 3)^2 (5)(a + 3)^2}{(5)(7)(a + 3)(a + 3)(a - 3)}$$
$$= \dfrac{a - 3}{7}$$

14.
$$\dfrac{a^2 - 4}{a^2 - 4a + 4} \cdot \dfrac{a^2 - 9a + 14}{a^3 + 2a^2}$$
$$= \dfrac{(a + 2)(a - 2)}{(a - 2)(a - 2)} \cdot \dfrac{(a - 2)(a - 7)}{a^2(a + 2)}$$
$$= \dfrac{(a + 2)(a - 2)(a - 2)(a - 7)}{a^2(a - 2)(a - 2)(a + 2)}$$
$$= \dfrac{a - 7}{a^2}$$

15. $\dfrac{x^2 - x - 12}{x^2 - 9} \div \dfrac{x^2 - 16}{5x - 5}$

$= \dfrac{x^2 - x - 12}{x^2 - 9} \cdot \dfrac{5x - 5}{x^2 - 16}$

$= \dfrac{(x - 4)(x + 3)}{(x - 3)(x + 3)} \cdot \dfrac{5(x - 1)}{(x - 4)(x + 4)}$

$= \dfrac{5(x - 4)(x + 3)(x - 1)}{(x - 3)(x + 3)(x - 4)(x + 4)}$

$= \dfrac{5(x - 1)}{(x - 3)(x + 4)}$

16. $\dfrac{a^3 + 64}{2a^2 + 18a + 40} \div \dfrac{a^2 - 4a + 16}{a^2 + 4a}$

$= \dfrac{a^3 + 64}{2a^2 + 18a + 40} \cdot \dfrac{a^2 + 4a}{a^2 - 4a + 16}$

$= \dfrac{(a + 4)(a^2 - 4a + 16)}{2(a + 4)(a + 5)} \cdot \dfrac{a(a + 4)}{a^2 - 4a + 16}$

$= \dfrac{a(a + 4)}{2(a + 5)}$

17. $\dfrac{uv - uw + xv - xw}{v - w} \div \dfrac{v^2 - 2vw + w^2}{xv - xw}$

$= \dfrac{u(v - w) + x(v - w)}{v - w} \cdot \dfrac{xv - xw}{v^2 - 2vw + w^2}$

$= \dfrac{(u + x)(v - w)}{v - w} \cdot \dfrac{x(v - w)}{(v - w)(v - w)}$

$= \dfrac{(u + x)(v - w)x(v - w)}{(v - w)(v - w)(v - w)}$

$= \dfrac{x(u + x)}{v - w}$

18. $\dfrac{y^2 - 1}{2y + 2} \cdot \dfrac{y^2 - 8y + 12}{y^2 - 4y + 4} \cdot \dfrac{y + 2}{y - 6}$

$= \dfrac{(y + 1)(y - 1)}{2(y + 1)} \cdot \dfrac{(y - 2)(y - 6)}{(y - 2)(y - 2)} \cdot \dfrac{y + 2}{y - 6}$

$= \dfrac{(y + 1)(y - 1)(y - 2)(y - 6)(y + 2)}{2(y + 1)(y - 2)(y - 2)(y - 6)}$

$= \dfrac{(y - 1)(y + 2)}{2(y - 2)}$

19. $\dfrac{x^2 - y^2}{x^2 - xy + y^2} \cdot \dfrac{2x^2 - 3xy - 2y^2}{x^2 + 2xy + y^2} \div \dfrac{2x^2 - xy - y^2}{x^3 + y^3}$

$= \dfrac{(x + y)(x - y)}{x^2 - xy + y^2} \cdot \dfrac{(2x + y)(x - 2y)}{(x + y)(x + y)} \cdot \dfrac{(x + y)(x^2 - xy + y^2)}{(2x + y)(x - y)}$

$= x - 2y$

20. $\dfrac{a}{35} = \dfrac{a}{(5)(7)}$ and $\dfrac{a^2}{50} = \dfrac{a^2}{(5)(5)(2)}$

With the denominators now factored, we recognize that the LCD must consist of two factors of 5, one factor of 2, and one factor of 7. Thus the LCD = $(5)(5)(2)(7) = 350$.

21. $\dfrac{a + 2}{a^2 - 5a + 6}$ and $\dfrac{a}{a^2 + a - 12}$

Factor the denominators.
$a^2 - 5a + 6 = (a - 2)(a - 3)$
$a^2 + a - 12 = (a + 4)(a - 3)$

The LCD must consist of one factor of $(a - 2)$, one factor of $(a - 3)$, and one factor of $(a + 4)$. Thus,

$\quad$ LCD $= (a - 2)(a - 3)(a + 4)$.

22. $\dfrac{x + 7}{x^5 - 27x^2} = \dfrac{x + 7}{x^2(x - 3)(x^2 + 3x + 9)}$

and $\dfrac{x - 3}{5x^3 + 15x^2 + 45x} = \dfrac{x - 3}{5x(x^2 + 3x + 9)}$

With both denominators in factored form, we can see that the LCD must consist of one factor of 5, two factors of x, one factor of $(x - 3)$, and one factor of $(x^2 + 3x + 9)$. Thus, the LCD = $5x^2(x - 3)(x^2 + 3x + 9)$.

23. $\dfrac{6}{5x - 10} + \dfrac{3}{x + 2} = \dfrac{6}{5(x - 2)} + \dfrac{3}{x + 2}$

$= \dfrac{(6)(x + 2)}{5(x - 2)(x + 2)} + \dfrac{(3)(5)(x - 2)}{5(x - 2)(x + 2)}$

$= \dfrac{6x + 12 + 15x - 30}{5(x - 2)(x + 2)}$

$= \dfrac{21x - 18}{5(x - 2)(x + 2)}$

24. $\dfrac{3a}{a - 4} + \dfrac{5a}{4 - a} = \dfrac{3a}{a - 4} + \dfrac{(-1)(5a)}{(-1)(4 - a)}$

$= \dfrac{3a}{a - 4} + \dfrac{-5a}{a - 4}$

$= \dfrac{3a - 5a}{a - 4}$

$= \dfrac{-2a}{a - 4}$

EXERCISES A

25. $\dfrac{y}{y^2-4} - \dfrac{2}{y+2} = \dfrac{y}{(y+2)(y-2)} - \dfrac{2}{y+2}$

$= \dfrac{y}{(y-2)(y+2)} - \dfrac{2(y-2)}{(y+2)(y-2)}$

$= \dfrac{y - 2(y-2)}{(y-2)(y+2)}$

$= \dfrac{y - 2y + 4}{(y-2)(y+2)}$

$= \dfrac{4 - y}{(y-2)(y+2)}$

26. $\dfrac{a-3}{a^2-9} + \dfrac{1}{(a+3)^2}$

$= \dfrac{a-3}{(a+3)(a-3)} + \dfrac{1}{(a+3)^2}$

$= \dfrac{1}{a+3} + \dfrac{1}{(a+3)^2}$

$= \dfrac{a+3}{(a+3)^2} + \dfrac{1}{(a+3)^2}$

$= \dfrac{a+3+1}{(a+3)^2}$

$= \dfrac{a+4}{(a+3)^2}$

27. $\dfrac{5x}{x^3-16x} - \dfrac{4}{x-4}$

$= \dfrac{5x}{x(x+4)(x-4)} - \dfrac{4}{(x-4)}$

$= \dfrac{5}{(x+4)(x-4)} - \dfrac{4}{(x-4)}$

$= \dfrac{5}{(x+4)(x-4)} - \dfrac{4(x+4)}{(x+4)(x-4)}$

$= \dfrac{5 - 4(x+4)}{(x+4)(x-4)}$

$= \dfrac{5 - 4x - 16}{(x+4)(x-4)} = \dfrac{-4x - 11}{(x+4)(x-4)}$

28. $\dfrac{6a}{a^3-27} - \dfrac{4}{2(a^2+3a+9)}$

$= \dfrac{6a}{(a-3)(a^2+3a+9)} - \dfrac{2}{(a^2+3a+9)}$

$= \dfrac{6a}{(a-3)(a^2+3a+9)} - \dfrac{2(a-3)}{(a-3)(a^2+3a+9)}$

$= \dfrac{6a - 2(a-3)}{(a-3)(a^2+3a+9)}$

$= \dfrac{6a - 2a + 6}{(a-3)(a^2+3a+9)}$

$= \dfrac{4a + 6}{(a-3)(a^2+3a+9)}$

$= \dfrac{2(2a+3)}{a^3 - 27}$

29. $\dfrac{2ab}{a^2-b^2} - \dfrac{b}{a-b} + 5$

$= \dfrac{2ab}{(a+b)(a-b)} - \dfrac{b(a+b)}{(a+b)(a-b)} + \dfrac{5(a+b)(a-b)}{(a+b)(a-b)}$

$= \dfrac{2ab - b(a+b) + 5(a^2 - b^2)}{(a+b)(a-b)}$

$= \dfrac{2ab - ab - b^2 + 5a^2 - 5b^2}{(a+b)(a-b)}$

$= \dfrac{5a^2 + ab - 6b^2}{(a+b)(a-b)}$

$= \dfrac{(5a + 6b)(a - b)}{(a+b)(a-b)}$

$= \dfrac{5a + 6b}{a + b}$

30. $\dfrac{x^2}{x^2-2xy+y^2} + \dfrac{1}{x^2-xy} - \dfrac{x}{x-y}$

$= \dfrac{x^2}{(x-y)(x-y)} + \dfrac{1}{x(x-y)} - \dfrac{x}{x-y}$

$= \dfrac{x^3}{x(x-y)(x-y)} + \dfrac{(x-y)}{x(x-y)(x-y)} - \dfrac{x^2(x-y)}{x(x-y)(x-y)}$

$= \dfrac{x^3 + (x-y) - x^2(x-y)}{x(x-y)(x-y)}$

$= \dfrac{x^3 + x - y - x^3 + x^2 y}{x(x-y)(x-y)}$

$= \dfrac{x^2 y + x - y}{x(x-y)^2}$

31. $\dfrac{\dfrac{1}{2y} - \dfrac{1}{3y}}{1 + \dfrac{1}{4y}} = \dfrac{12y\left[\dfrac{1}{2y} - \dfrac{1}{3y}\right]}{12y\left[1 + \dfrac{1}{4y}\right]}$

$= \dfrac{12y\left(\dfrac{1}{2y}\right) - (12y)\left(\dfrac{1}{3y}\right)}{12y(1) + (12y)\left(\dfrac{1}{4y}\right)}$

$= \dfrac{6 - 4}{12y + 3}$

$= \dfrac{2}{3(4y + 1)}$

32. $\dfrac{\dfrac{1}{a}+\dfrac{2}{a^2}}{\dfrac{1}{4}-\dfrac{1}{a^2}} = \dfrac{4a^2\left[\dfrac{1}{a}+\dfrac{2}{a^2}\right]}{4a^2\left[\dfrac{1}{4}-\dfrac{1}{a^2}\right]}$

$= \dfrac{4a^2\left(\dfrac{1}{a}\right)+4a^2\left(\dfrac{2}{a^2}\right)}{4a^2\left(\dfrac{1}{4}\right)-4a^2\left(\dfrac{1}{a^2}\right)}$

$= \dfrac{4a+8}{a^2-4}$

$= \dfrac{4(a+2)}{(a-2)(a+2)}$

$= \dfrac{4}{a-2}$

33. $\dfrac{x+6+\dfrac{8}{x}}{x+4+\dfrac{4}{x}} = \dfrac{x\left[x+6+\dfrac{8}{x}\right]}{x\left[x+4+\dfrac{4}{x}\right]}$

$= \dfrac{x^2+6x+x\left(\dfrac{8}{x}\right)}{x^2+4x+x\left(\dfrac{4}{x}\right)}$

$= \dfrac{x^2+6x+8}{x^2+4x+4}$

$= \dfrac{(x+2)(x+4)}{(x+2)(x+2)}$

$= \dfrac{x+4}{x+2}$

34. $\dfrac{1+\dfrac{5}{a}+\dfrac{4}{a^2}}{1+\dfrac{1}{a+3}} = \dfrac{\dfrac{a^2+5a+4}{a^2}}{\dfrac{a+3+1}{a+3}}$

$= \dfrac{\dfrac{(a+4)(a+1)}{a^2}}{\dfrac{a+4}{a+3}}$

$= \dfrac{(a+4)(a+1)}{a^2}\cdot\dfrac{a+3}{a+4}$

$= \dfrac{(a+1)(a+3)}{a^2}$

35. $\dfrac{1}{1+\dfrac{1}{1+\dfrac{1}{x}}} = \dfrac{1}{1+\dfrac{1}{\dfrac{x+1}{x}}}$

$= \dfrac{1}{1+\dfrac{x}{x+1}}$

$= \dfrac{1}{\dfrac{x+1+x}{x+1}}$

$= \dfrac{1}{\dfrac{2x+1}{x+1}}$

$= \dfrac{x+1}{2x+1}$

36. $a-\dfrac{a}{1-\dfrac{a}{1-a}} = a-\dfrac{a}{\dfrac{1-a-a}{1-a}}$

$= a-\dfrac{a}{\dfrac{1-2a}{1-a}}$

$= a-\dfrac{a(1-a)}{1-2a}$

$= \dfrac{a(1-2a)-a(1-a)}{1-2a}$

$= \dfrac{a-2a^2-a+a^2}{1-2a}$

$= \dfrac{-a^2}{1-2a}$

$= \dfrac{(-1)(-a^2)}{(-1)(1-2a)}$

$= \dfrac{a^2}{2a-1}$

37. $\dfrac{\dfrac{1}{(x+h)}-\dfrac{1}{x}}{h} = \dfrac{\dfrac{x}{x(x+h)}-\dfrac{(x+h)}{x(x+h)}}{h}$

$= \dfrac{\dfrac{x-x-h}{x(x+h)}}{h}$

$= \dfrac{\dfrac{-h}{x(x+h)}}{h}$

$= \dfrac{-h}{hx(x+h)}$

$= \dfrac{-1}{x(x+h)}$

EXERCISES A SECTION 1.5 23

38. $\dfrac{\dfrac{1}{xy} + \dfrac{1}{yz} + \dfrac{1}{xz}}{\dfrac{x+y+z}{xyz}} = \dfrac{xyz\left[\dfrac{1}{xy} + \dfrac{1}{yz} + \dfrac{1}{xz}\right]}{xyz\left[\dfrac{x+y+z}{xyz}\right]}$

$= \dfrac{z + x + y}{x + y + z}$

$= \dfrac{x + y + z}{x + y + z} = 1$

39. $\dfrac{x^{-1} + y^{-1}}{x^{-1}y^{-1}} = \dfrac{\dfrac{1}{x} + \dfrac{1}{y}}{\dfrac{1}{xy}}$

$= \dfrac{\dfrac{y + x}{xy}}{\dfrac{1}{xy}}$

$= \dfrac{x + y}{xy} \cdot \dfrac{xy}{1}$

$= \dfrac{x + y}{1} = x + y$

40. $\dfrac{a^{-1} - b^{-1}}{a^{-1} + b^{-1}} = \dfrac{\dfrac{1}{a} - \dfrac{1}{b}}{\dfrac{1}{a} + \dfrac{1}{b}}$

$= \dfrac{ab\left[\dfrac{1}{a} - \dfrac{1}{b}\right]}{ab\left[\dfrac{1}{a} + \dfrac{1}{b}\right]}$

$= \dfrac{b - a}{b + a}$

41. Remember to multiply and divide before adding and subtracting.

$\dfrac{x}{x^2 - y^2} - \dfrac{y}{x^2 - y^2} \cdot \dfrac{x + y}{x - y}$
$+ \dfrac{y(x + y)}{x^2 - 3xy - 4y^2} \div \dfrac{(x + y)^2}{x - 4y}$

$= \dfrac{x}{x^2 - y^2} - \left[\dfrac{y}{(x - y)(x + y)} \cdot \dfrac{(x + y)}{(x - y)}\right]$

$+ \left[\dfrac{y(x + y)}{(x + y)(x - 4y)} \div \dfrac{(x + y)^2}{x - 4y}\right]$

$= \dfrac{x}{(x - y)(x + y)} - \left[\dfrac{y}{(x - y)(x - y)}\right]$

$+ \left[\dfrac{y(x + y)}{(x + y)(x - 4y)} \cdot \dfrac{(x - 4y)}{(x + y)(x + y)}\right]$

$= \dfrac{x}{(x - y)(x + y)} - \dfrac{y}{(x - y)(x - y)} + \dfrac{y}{(x + y)(x + y)}$

$= \dfrac{x(x - y)(x + y)}{(x - y)^2(x + y)^2} - \dfrac{y(x + y)^2}{(x - y)^2(x + y)^2} + \dfrac{y(x - y)^2}{(x - y)^2(x + y)^2}$

$= \dfrac{x^3 - xy^2 - yx^2 - 2xy^2 - y^3 + yx^2 - 2xy^2 + y^3}{(x - y)^2(x + y)^2}$

$= \dfrac{x^3 - 5xy^2}{(x - y)^2(x + y)^2}$

$= \dfrac{x(x^2 - 5y^2)}{(x - y)^2(x + y)^2}$

42. $\dfrac{2D}{\dfrac{D}{50} + \dfrac{D}{65}} = \dfrac{2D}{\dfrac{65D + 50D}{(50)(65)}}$

$= \dfrac{2D(50)(65)}{65D + 50D}$

$= \dfrac{6500D}{115D}$

$= \dfrac{6500}{115} = 56.52173913$

Thus, to the nearest tenth, the average rate for the trip was 56.5 mph.

43. $\dfrac{\dfrac{1}{p}\left(1 + \dfrac{p}{q}\right)}{1 + (1 + p)\dfrac{q}{p}} = \dfrac{p\left[\dfrac{1}{p}\left(1 + \dfrac{p}{q}\right)\right]}{p\left[1 + (1 + p)\dfrac{q}{p}\right]}$

$= \dfrac{1 + \dfrac{p}{q}}{p + (1 + p)q}$

$= \dfrac{\dfrac{q + p}{q}}{p + q + pq}$

$= \dfrac{q + p}{q(p + q + pq)}$

$= \dfrac{q + p}{pq + q^2 + pq^2}$

44. First remove the common factor 25. The remaining factor is the difference of two squares.

$100a^2 - 25b^4 = 25(4a^2 - b^4)$
$ = 25(2a + b^2)(2a - b^2)$

45. The factors of 15 and −28 that combine to give 1 are 5, 3 and 7, −4. Thus,

$15x^2 + xy - 28y^2 = (5x + 7y)(3x - 4y)$

46. This is the sum of two cubes.

$$27u^3 + 64v^3 = (3u)^3 + (4v)^3$$
$$= (3u + 4v)(9u^2 - 12uv + 16v^2)$$

47. false ($\sqrt{7}$ is an irrational number)

48. true

49. false (integers are always rational numbers, never irrational)

50. false (for example, $2 - 5 = -3$, but 2 and 5 are natural numbers and -3 is not)

PRACTICE EXERCISES

CHAPTER 1 REVIEW OF FUNDAMENTAL CONCEPTS

SECTION 1.6 Radical Expressions

1. (a) $\sqrt{(-5)^2} = |-5| = 5$

   (b) $\sqrt[5]{(-5)^5} = -5$

   (c) $\sqrt{-25}$ is not a real number.

   (d) $-\sqrt{(6)^2} = -|6| = -6$

   (e) $\sqrt[3]{27y^3} = \sqrt[3]{3^3 y^3} = \sqrt[3]{(3y)^3} = 3y$

2. (a) $\sqrt[3]{24y^5} = \sqrt[3]{3 \cdot 2^3 \cdot y^3 \cdot y^2} = \sqrt[3]{2^3 y^3}\sqrt[3]{3y^2} = 2y\sqrt[3]{3y^2}$

   (b) $\sqrt{50x^3 y}\sqrt{6x^5 y^3} = \sqrt{5^2 \cdot 2^2 \cdot 3 \cdot x^8 y^4}$
   $= \sqrt{5^2 \cdot 2^2 \cdot x^8 y^4}\sqrt{3}$
   $= 5 \cdot 2 \cdot x^4 y^2 \sqrt{3}$
   $= 10 x^4 y^2 \sqrt{3}$

3. (a) $\dfrac{\sqrt[3]{40x^5}}{\sqrt[3]{5x}} = \sqrt[3]{\dfrac{40x^5}{5x}} = \sqrt[3]{8x^4}$
   $= \sqrt[3]{2^3 x^3 x} = \sqrt[3]{2^3 x^3}\sqrt[3]{x}$
   $= 2x\sqrt[3]{x}$

   (b) $\sqrt[4]{\dfrac{48u^2 v^4}{3u^{-2}v^{-1}}} = \sqrt[4]{16u^4 v^5} = \sqrt[4]{2^4 \cdot u^4 v^4 v}$
   $= \sqrt[4]{2^4 u^4 v^4}\sqrt[4]{v} = 2uv\sqrt[4]{v}$

4. $\sqrt[4]{\sqrt{256a^8 b^{11}}} = \sqrt[8]{256a^8 b^{11}} = \sqrt[8]{2^8 a^8 b^8 b^3}$
   $= \sqrt[8]{2^8 a^8 b^8}\sqrt[8]{b^3} = 2ab\sqrt[8]{b^3}$

5. (a) $5\sqrt{147} - 9\sqrt{75} = 5\sqrt{49 \cdot 3} - 9\sqrt{25 \cdot 3}$
   $= 5\sqrt{49}\sqrt{3} - 9\sqrt{25}\sqrt{3}$
   $= 5 \cdot 7\sqrt{3} - 9 \cdot 5\sqrt{3}$
   $= 35\sqrt{3} - 45\sqrt{3}$
   $= (35 - 45)\sqrt{3}$
   $= -10\sqrt{3}$

   (b)
   $-3\sqrt{125a^5 b^2} + 5a\sqrt{80a^3 b^2}$
   $= -3\sqrt{25 \cdot 5 \cdot a^4 ab^2} + 5a\sqrt{16 \cdot 5a^2 ab^2}$
   $= -3\sqrt{25a^4 b^2}\sqrt{5a} + 5a\sqrt{16a^2 b^2}\sqrt{5a}$
   $= -3(5)a^2 b\sqrt{5a} + 5a(4)ab\sqrt{5a}$
   $= -15a^2 b\sqrt{5a} + 20a^2 b\sqrt{5a}$
   $= (-15a^2 b + 20a^2 b)\sqrt{5a}$
   $= 5a^2 b\sqrt{5a}$

6. (a) $\sqrt{\dfrac{36x^2}{y}} = \dfrac{\sqrt{36x^2}}{\sqrt{y}} = \dfrac{6x}{\sqrt{y}} = \dfrac{6x\sqrt{y}}{\sqrt{y}\sqrt{y}} = \dfrac{6x\sqrt{y}}{y}$

   (b) $\dfrac{\sqrt[5]{2y}}{\sqrt[5]{x^2}} = \dfrac{\sqrt[5]{2y}}{\sqrt[5]{x^2}} \cdot \dfrac{\sqrt[5]{x^3}}{\sqrt[5]{x^3}} = \dfrac{\sqrt[5]{2x^3 y}}{\sqrt[5]{x^5}} = \dfrac{\sqrt[5]{2x^3 y}}{x}$

   (c) $\dfrac{x - y}{\sqrt{x} - \sqrt{y}} = \dfrac{(x-y)(\sqrt{x} + \sqrt{y})}{(\sqrt{x} - \sqrt{y})(\sqrt{x} + \sqrt{y})}$
   $= \dfrac{(x-y)(\sqrt{x} + \sqrt{y})}{(\sqrt{x})^2 - (\sqrt{y})^2}$
   $= \dfrac{(x-y)(\sqrt{x} + \sqrt{y})}{(x - y)}$
   $= \sqrt{x} + \sqrt{y}$

7. Substitute 39 for L and evaluate.

   $T = 2\pi\sqrt{\dfrac{L}{32}}$
   $= 2\pi\sqrt{\dfrac{39}{32}}$
   $= 2\pi\sqrt{1.21875}$
   ≈ 6.936448764

   Thus, rounded to the nearest tenth, the time for one complete swing of the pendulum is approximately 6.9 sec.

CHAPTER 1 REVIEW OF FUNDAMENTAL CONCEPTS

SECTION 1.6 Radical Expressions

1. $\sqrt{36} = \sqrt{6^2} = 6$

2. $\sqrt[3]{-27} = \sqrt[3]{(-3)^3} = -3$

3. $-\sqrt{121} = -\sqrt{(11)^2} = -11$

4. $-\sqrt[5]{-32} = -\sqrt[5]{(-2)^5} = -(-2) = 2$

5. $\sqrt{4a^8} = \sqrt{2^2(a^4)^2} = \sqrt{2^2}\sqrt{(a^4)^2} = 2a^4$

6. $\sqrt[3]{8x^6} = \sqrt[3]{2^3(x^2)^3} = \sqrt[3]{2^3}\sqrt[3]{(x^2)^3} = 2x^2$

7. $\sqrt[5]{-32x^5y^{10}} = \sqrt[5]{(-2)^5 x^5 (y^2)^5}$
  $= \sqrt[5]{-2)^5}\sqrt[5]{x^5}\sqrt[5]{(y^2)^5}$
  $= -2xy^2$

8. $\sqrt[4]{81x^8y^4} = \sqrt[4]{3^4(x^2)^4 y^4}$
  $= \sqrt[4]{3^4}\sqrt[4]{(x^2)^4}\sqrt[4]{y^4}$
  $= 3x^2y$

9. $\sqrt{x^2 - 10x + 25} = \sqrt{(x-5)^2} = x - 5$

10. $\sqrt{4x^2 + 12x + 9} = \sqrt{(2x+3)^2}$
  $= 2x + 3$

11. $2\sqrt{2}\sqrt{6} = 2\sqrt{2 \cdot 6}$
  $= 2\sqrt{2^2 \cdot 3}$
  $= 2\sqrt{2^2}\sqrt{3}$
  $= 2 \cdot 2\sqrt{3}$
  $= 4\sqrt{3}$

12. $4\sqrt{5}\sqrt{15}\sqrt{3} = 4\sqrt{5 \cdot 15 \cdot 3}$
  $= 4\sqrt{(15)^2} = 4 \cdot 15 = 60$

13. $4\sqrt[3]{24} = 4\sqrt[3]{2^3 \cdot 3} = 4 \cdot 2\sqrt[3]{3} = 8\sqrt[3]{3}$

14. $3\sqrt{72} = 3\sqrt{6^2 \cdot 2} = 3 \cdot 6\sqrt{2} = 18\sqrt{2}$

15. $\sqrt[3]{3xy^2}\sqrt[3]{9x^2y^4} = \sqrt[3]{(3xy^2)(9x^2y^4)}$
  $= \sqrt[3]{3^3 \cdot x^3 \cdot (y^2)^3}$
  $= 3xy^2$

16. $\sqrt[4]{4x^3y}\sqrt[4]{4xy^5} = \sqrt[4]{(4x^3y)(4xy^5)}$
  $= \sqrt[4]{2^4 \cdot x^4 \cdot y^4 \cdot y^2}$
  $= 2xy\sqrt[4]{y^2} = 2xy\sqrt{y}$

17. $\dfrac{\sqrt{8x^3z}}{\sqrt{2xz^3}} = \sqrt{\dfrac{8x^3z}{2xz^3}} = \sqrt{\dfrac{4x^2}{z^2}} = \dfrac{2x}{z}$

18. $\dfrac{\sqrt[3]{16w^4z^2}}{\sqrt[3]{2wz^5}} = \sqrt[3]{\dfrac{16w^4z^2}{2wz^5}} = \sqrt[3]{\dfrac{8w^3}{z^3}} = \dfrac{2w}{z}$

19. $\dfrac{\sqrt[3]{2^{-1}x^{-3}y^{-2}}}{\sqrt[3]{16^{-1}x^{-4}y^{-8}}} = \sqrt[3]{\dfrac{2^{-1}x^{-3}y^{-2}}{16^{-1}x^{-4}y^{-8}}} = \sqrt[3]{8xy^6} = 2y^2\sqrt[3]{x}$

20. $\sqrt[3]{\sqrt[4]{a^{24}b^{48}}} = \sqrt[3]{\sqrt[4]{(a^6)^4(b^{12})^4}} = \sqrt[3]{a^6b^{12}} = a^2b^4$

21. $\sqrt[3]{\sqrt[3]{11u^{25}v^{19}}} = \sqrt[9]{11u^{25}v^{19}} = \sqrt[9]{(u^2)^9 \cdot (v^2)^9 \cdot 11u^7v}$
  $= u^2v^2\sqrt[9]{11u^7v}$

22. $3\sqrt{27} - 4\sqrt{12} = 3\sqrt{3^2 \cdot 3} - 4\sqrt{2^2 \cdot 3}$
  $= 3 \cdot 3\sqrt{3} - 4 \cdot 2\sqrt{3}$
  $= 9\sqrt{3} - 8\sqrt{3}$
  $= (9-8)\sqrt{3} = \sqrt{3}$

23. $2\sqrt[3]{16} + 4\sqrt[3]{54} = 2\sqrt[3]{2^3 \cdot 2} + 4\sqrt[3]{3^3 \cdot 2}$
  $= 2 \cdot 2\sqrt[3]{2} + 4 \cdot 3\sqrt[3]{2}$
  $= 4\sqrt[3]{2} + 12\sqrt[3]{2}$
  $= (4 + 12)\sqrt[3]{2} = 16\sqrt[3]{2}$

EXERCISES A   **SECTION 1.6**

24. $5\sqrt[4]{32} - 3\sqrt[4]{2} = 5\sqrt[4]{2^4 \cdot 2} - 3\sqrt[4]{2}$
$= 5 \cdot 2\sqrt[4]{2} - 3\sqrt[4]{2}$
$= 10\sqrt[4]{2} - 3\sqrt[4]{2}$
$= (10 - 3)\sqrt[4]{2} = 7\sqrt[4]{2}$

25. $2\sqrt{\dfrac{25}{4}} - 5\dfrac{\sqrt{8}}{\sqrt{12}} = 2\left(\dfrac{5}{2}\right) - 5\sqrt{\dfrac{8}{12}}$
$= 5 - 5\sqrt{\dfrac{4}{6}}$
$= 5 - 5\left(\dfrac{2}{\sqrt{6}}\right)$
$= 5 - \dfrac{10}{\sqrt{6}}$
$= 5 - \dfrac{10\sqrt{6}}{6}$
$= 5 - \dfrac{5\sqrt{6}}{3}$
$= \dfrac{15 - 5\sqrt{6}}{3}$

26. $7\sqrt{27a^2b} - 3a\sqrt{3b} = 7 \cdot 3a\sqrt{3b} - 3a\sqrt{3b}$
$= 21a\sqrt{3b} - 3a\sqrt{3b}$
$= (21a - 3a)\sqrt{3b}$
$= 18a\sqrt{3b}$

27. $\sqrt{5a^2 + 10a + 5} + \sqrt{20a^2 + 40a + 20}$
$= \sqrt{5(a^2 + 2a + 1)} + \sqrt{20(a^2 + 2a + 1)}$
$= \sqrt{5(a+1)^2} + \sqrt{5 \cdot 4(a+1)^2}$
$= (a+1)\sqrt{5} + 2(a+1)\sqrt{5}$
$= 3(a+1)\sqrt{5}$

28. $\dfrac{2 + \sqrt{8}}{2} = \dfrac{2 + \sqrt{4 \cdot 2}}{2}$
$= \dfrac{2 + 2\sqrt{2}}{2}$
$= \dfrac{2(1 + \sqrt{2})}{2} = 1 + \sqrt{2}$

29. $\dfrac{5 - \sqrt{50}}{10} = \dfrac{5 - \sqrt{25 \cdot 2}}{10}$
$= \dfrac{5 - 5\sqrt{2}}{10}$
$= \dfrac{5(1 - \sqrt{2})}{5 \cdot 2} = \dfrac{1 - \sqrt{2}}{2}$

30. $\sqrt{\dfrac{125}{45}} = \sqrt{\dfrac{25 \cdot 5}{9 \cdot 5}} = \dfrac{\sqrt{25}}{\sqrt{9}} = \dfrac{5}{3}$

31. $\dfrac{\sqrt[4]{5}}{\sqrt[4]{2}} = \dfrac{\sqrt[4]{5}\sqrt[4]{2^3}}{\sqrt[4]{2}\sqrt[4]{2^3}} = \dfrac{\sqrt[4]{5 \cdot 8}}{\sqrt[4]{2^4}} = \dfrac{\sqrt[4]{40}}{2}$

32. $\dfrac{\sqrt{3x}}{\sqrt{6z}} = \sqrt{\dfrac{3x}{6z}} = \sqrt{\dfrac{x}{2z}} = \dfrac{\sqrt{x}}{\sqrt{2z}}$
$= \dfrac{\sqrt{x}\sqrt{2z}}{\sqrt{2z}\sqrt{2z}} = \dfrac{\sqrt{2xz}}{2z}$

33. $\sqrt{\dfrac{4a^3}{ab}} = \sqrt{\dfrac{4a^2}{b}} = \dfrac{\sqrt{4a^2}}{\sqrt{b}}$
$= \dfrac{2a}{\sqrt{b}} = \dfrac{2a\sqrt{b}}{\sqrt{b}\sqrt{b}}$
$= \dfrac{2a\sqrt{b}}{b}$

34. $\dfrac{\sqrt[4]{32a^5}}{\sqrt[4]{2b^3}} = \dfrac{\sqrt[4]{2^4 a^4 \cdot 2a}\sqrt[4]{2^3 b}}{\sqrt[4]{2b^3}\sqrt[4]{2^3 b}}$
$= \dfrac{2a\sqrt[4]{2a \cdot 2^3 b}}{\sqrt[4]{2^4 b^4}}$
$= \dfrac{4a\sqrt[4]{ab}}{2b} = \dfrac{2a\sqrt[4]{ab}}{b}$

35. $\sqrt[3]{\dfrac{72x^2}{3xy}} = \sqrt[3]{\dfrac{24x}{y}} = \dfrac{\sqrt[3]{24x}}{\sqrt[3]{y}}$
$= \dfrac{2\sqrt[3]{3x}\sqrt[3]{y^2}}{\sqrt[3]{y}\sqrt[3]{y^2}} = \dfrac{2\sqrt[3]{3xy^2}}{y}$

36. $\dfrac{\sqrt{5}}{\sqrt{3} - \sqrt{5}} = \dfrac{\sqrt{5}(\sqrt{3} + \sqrt{5})}{(\sqrt{3} - \sqrt{5})(\sqrt{3} + \sqrt{5})}$
$= \dfrac{\sqrt{5}(\sqrt{3} + \sqrt{5})}{(\sqrt{3})^2 - (\sqrt{5})^2}$
$= \dfrac{\sqrt{5}(\sqrt{3} + \sqrt{5})}{3 - 5}$
$= \dfrac{\sqrt{5}(\sqrt{3} + \sqrt{5})}{-2}$
$= \dfrac{\sqrt{15} + 5}{-2}$

37. $\dfrac{\sqrt{7}+\sqrt{2}}{\sqrt{7}-\sqrt{2}} = \dfrac{(\sqrt{7}+\sqrt{2})(\sqrt{7}+\sqrt{2})}{(\sqrt{7}-\sqrt{2})(\sqrt{7}+\sqrt{2})}$
$= \dfrac{(\sqrt{7})^2 + 2\sqrt{7}\sqrt{2} + (\sqrt{2})^2}{(\sqrt{7})^2 - (\sqrt{2})^2}$
$= \dfrac{7 + 2\sqrt{14} + 2}{7 - 2}$
$= \dfrac{9 + 2\sqrt{14}}{5}$

38. $\dfrac{\sqrt{x}+1}{\sqrt{x}-1} = \dfrac{(\sqrt{x}+1)(\sqrt{x}+1)}{(\sqrt{x}-1)(\sqrt{x}+1)}$
$= \dfrac{(\sqrt{x})^2 + 2\sqrt{x} + 1}{(\sqrt{x})^2 - 1}$
$= \dfrac{x + 2\sqrt{x} + 1}{x - 1}$

39. $2\sqrt{18x} + \dfrac{3\sqrt{x}}{\sqrt{2}} = 2\cdot 3\sqrt{2x} + \dfrac{3\sqrt{x}\sqrt{2}}{\sqrt{2}\sqrt{2}}$
$= 6\sqrt{2x} + \dfrac{3\sqrt{2x}}{2}$
$= \left(6 + \dfrac{3}{2}\right)\sqrt{2x}$
$= \left(\dfrac{12}{2} + \dfrac{3}{2}\right)\sqrt{2x}$
$= \dfrac{15}{2}\sqrt{2x} = \dfrac{15\sqrt{2x}}{2}$

40. $\dfrac{1+\sqrt{2}}{1-\sqrt{2}} = \dfrac{(1+\sqrt{2})(1-\sqrt{2})}{(1-\sqrt{2})(1-\sqrt{2})}$
$= \dfrac{1^2 - (\sqrt{2})^2}{1 - 2\sqrt{2} + (\sqrt{2})^2}$
$= \dfrac{1 - 2}{1 - 2\sqrt{2} + 2}$
$= \dfrac{-1}{3 - 2\sqrt{2}}$
$= \dfrac{(-1)(-1)}{(-1)(3 - 2\sqrt{2})}$
$= \dfrac{1}{-3 + 2\sqrt{2}} = \dfrac{1}{2\sqrt{2} - 3}$

41. $\dfrac{\sqrt{a}+\sqrt{b}}{\sqrt{ab}} = \dfrac{(\sqrt{a}+\sqrt{b})(\sqrt{a}-\sqrt{b})}{\sqrt{ab}(\sqrt{a}-\sqrt{b})}$
$= \dfrac{(\sqrt{a})^2 - (\sqrt{b})^2}{\sqrt{ab}\sqrt{a} - \sqrt{ab}\sqrt{b}}$
$= \dfrac{a - b}{a\sqrt{b} - b\sqrt{a}}$

42. $\dfrac{\sqrt{a+b}+\sqrt{a}}{b} = \dfrac{(\sqrt{a+b}+\sqrt{a})(\sqrt{a+b}-\sqrt{a})}{b(\sqrt{a+b}-\sqrt{a})}$
$= \dfrac{(\sqrt{a+b})^2 - (\sqrt{a})^2}{b(\sqrt{a+b}-\sqrt{a})}$
$= \dfrac{a + b - a}{b(\sqrt{a+b}-\sqrt{a})}$
$= \dfrac{b}{b(\sqrt{a+b}-\sqrt{a})} = \dfrac{1}{\sqrt{a+b}-\sqrt{a}}$

43. Substitute 3000 for h and evaluate using a calculator to obtain the viewing distance d.

$$d = 1.4\sqrt{h} = 1.4\sqrt{3000} \approx 76.68115805$$

Since the viewing distance is about 77 miles, the island, at a distance of 90 miles, cannot be seen.

44. Substitute 250 for h and evaluate using a calculator to obtain the time t.

$$t = \dfrac{\sqrt{h}}{4} = \dfrac{\sqrt{250}}{4} \approx 3.952847075$$

Thus, to the nearest tenth, the time required for the object to hit the ground is about 4.0 sec.

45. The distance from home plate to second base is the hypotenuse of a right triangle with equal legs measuring 90 ft. Let x be the distance from home plate to second base. Then we must find x in the following.

$$x^2 = (90)^2 + (90)^2$$
$$x^2 = 2(90)^2$$
$$x = \sqrt{2(90)^2}$$
$$x = 90\sqrt{2}$$
$$x \approx 127.2792206$$

Thus, to the nearest tenth, the distance from home plate to second base is about 127.3 ft.

EXERCISES A

46.

$$\frac{6y^2 - xy - x^2}{2y^2 - 3xy + x^2} \div \frac{3y^2 - 5xy - 2x^2}{x^2 - y^2}$$

$$= \frac{6y^2 - xy - x^2}{2y^2 - 3xy + x^2} \cdot \frac{x^2 - y^2}{3y^2 - 5xy - 2x^2}$$

$$= \frac{(3y + x)(2y - x)}{(2y - x)(y - x)} \cdot \frac{(x - y)(x + y)}{(3y + x)(y - 2x)}$$

$$= \frac{(3y + x)(2y - x)(x - y)(x + y)}{(2y - x)(y - x)(3y + x)(y - 2x)}$$

$$= \frac{(x - y)(x + y)}{(y - x)(y - 2x)}$$

$$= (-1)\frac{(x + y)}{(y - 2x)} = \frac{x + y}{2x - y}$$

47.

$$\frac{3}{x + 1} + \frac{4 - 2x}{x^2 - 1}$$

$$= \frac{3}{x + 1} + \frac{4 - 2x}{(x - 1)(x + 1)}$$

$$= \frac{3(x - 1)}{(x + 1)(x - 1)} + \frac{4 - 2x}{(x - 1)(x + 1)}$$

$$= \frac{3(x - 1) + (4 - 2x)}{(x - 1)(x + 1)}$$

$$= \frac{3x - 3 + 4 - 2x}{(x - 1)(x + 1)}$$

$$= \frac{x + 1}{(x - 1)(x + 1)} = \frac{1}{x - 1}$$

48. $(2x^2y^2 - 5xy^2 + 8xy) - (3x^2y^2 - 2xy^2 + 4xy)$
$= 2x^2y^2 - 5xy^2 + 8xy - 3x^2y^2 + 2xy^2 - 4xy$
$= 2x^2y^2 - 3x^2y^2 - 5xy^2 + 2xy^2 + 8xy - 4xy$
$= -x^2y^2 - 3xy^2 + 4xy$

49.

$$\frac{u^{-1} + v^{-1}}{1 - (uv)^{-1}} = \frac{\frac{1}{u} + \frac{1}{v}}{1 - \frac{1}{uv}}$$

$$= \frac{uv\left[\frac{1}{u} + \frac{1}{v}\right]}{uv\left[1 - \frac{1}{uv}\right]}$$

$$= \frac{v + u}{uv - 1}$$

50. The error was made when $x + 1$ was subtracted from 2 without using parentheses. The numerator should have been $2 - (x + 1) = 2 - x - 1$. The correct answer is obtained below.

$$\frac{2}{x^2 - 1} - \frac{1}{x - 1}$$

$$= \frac{2}{(x - 1)(x + 1)} - \frac{1}{x - 1}$$

$$= \frac{2}{(x - 1)(x + 1)} - \frac{(x + 1)}{(x - 1)(x + 1)}$$

$$= \frac{2 - (x + 1)}{(x - 1)(x + 1)}$$

$$= \frac{2 - x - 1}{(x - 1)(x + 1)}$$

$$= \frac{1 - x}{(x - 1)(x + 1)}$$

$$= \frac{1 - x}{x - 1} \cdot \frac{1}{x + 1}$$

$$= (-1) \cdot \frac{1}{x + 1} = -\frac{1}{x + 1}$$

CHAPTER 1 REVIEW OF FUNDAMENTAL CONCEPTS

SECTION 1.7 Rational Exponents

1. (a) $a^{1/4} = \sqrt[4]{a}$

   (b) $8^{-1/3} = \dfrac{1}{8^{1/3}} = \dfrac{1}{\sqrt[3]{8}} = \dfrac{1}{2}$

   (c) $36^{-3/2} = \dfrac{1}{36^{3/2}} = \dfrac{1}{(\sqrt{36})^3} = \dfrac{1}{6^3} = \dfrac{1}{216}$

2. (a) $\sqrt[5]{a^3} = a^{3/5}$

   (b) $\dfrac{1}{\sqrt[3]{2}} = \dfrac{1}{2^{1/3}} = 2^{-1/3}$

   (c) $\sqrt[8]{y^6} = y^{6/8} = y^{3/4}$

3. (a) $x^{1/2} \cdot x^{1/6} = x^{1/2 + 1/6}$
   $= x^{3/6 + 1/6}$
   $= x^{4/6} = x^{2/3} = \sqrt[3]{x^2}$

   (b) $\dfrac{w^{1/2}}{w^{2/3}} = w^{1/2 - 2/3}$
   $= w^{3/6 - 4/6}$
   $= w^{-1/6}$
   $= \dfrac{1}{w^{1/6}} = \dfrac{1}{\sqrt[6]{w}}$

   (c) $(m^{3/2})^{-2/3} = m^{(3/2)(-2/3)} = m^{-1} = \dfrac{1}{m}$

   (d) $(y^{1/2} - y^{3/2})^2 = (y^{1/2})^2 - 2y^{1/2}y^{3/2} + (y^{3/2})^2$
   $= y^{(1/2)(2)} - 2y^{1/2 + 3/2} + y^{(3/2)(2)}$
   $= y^1 - 2y^{4/2} + y^3$
   $= y - 2y^2 + y^3$

4. (a) $\sqrt[4]{16a^8b^{24}} = (16a^8b^{24})^{1/4}$
   $= 16^{1/4}a^{8/4}b^{24/4} = 2a^2b^6$

   (b) $\left(\dfrac{243a^4b^{-10}}{64a^{-6}b^5}\right)^{-1/5} = \left(\dfrac{3^5 a^{4-(-6)}b^{-10-5}}{2^5 \cdot 2}\right)^{-1/5}$
   $= \left(\dfrac{3^5 a^{10} b^{-15}}{2^5 \cdot 2}\right)^{-1/5}$
   $= \dfrac{(3^5)^{-1/5}(a^{10})^{-1/5}(b^{-15})^{-1/5}}{(2^5)^{-1/5} 2^{-1/5}}$
   $= \dfrac{3^{-1}a^{-2}b^3}{2^{-1}2^{-1/5}}$
   $= \dfrac{2b^3 \cdot 2^{-1/5}}{3a^2} = \dfrac{2b^3 \sqrt[5]{2}}{3a^2}$

   (c) $\sqrt[4]{\sqrt[3]{a^2}} = \left(\sqrt[3]{a^2}\right)^{1/4} = (a^{2/3})^{1/4}$
   $= a^{(2/3)(1/4)} = a^{1/6} = \sqrt[6]{a}$

   (d) $\sqrt[3]{\sqrt[4]{x+y}} = \left(\sqrt[4]{x+y}\right)^{1/3} = ((x+y)^{1/4})^{1/3}$
   $= (x+y)^{(1/4)(1/3)} = (x+y)^{1/12}$
   $= \sqrt[12]{x+y}$

5. The formula for the length of the edge of a cube with volume V is

   $$e = \sqrt[3]{V} = V^{1/3}.$$

   Substitute 36.8 for V and evaluate using the y^x button on a calculator.

   $$e = V^{1/3} = (36.8)^{1/3} \approx 3.326206994$$

   Thus, the length of the edge of the cube is approximately 3.3 in, correct to the nearest tenth.

EXERCISES A SECTION 1.7   31

CHAPTER 1 REVIEW OF FUNDAMENTAL CONCEPTS

SECTION 1.7   Rational Exponents

1. $7^{2/5} = \sqrt[5]{(7)^2} = \sqrt[5]{49}$

2. $y^{3/4} = \sqrt[4]{y^3}$

3. $16^{1/4} = \sqrt[4]{16} = \sqrt[4]{2^4} = 2$

4. $4^{-3/2} = \dfrac{1}{4^{3/2}} = \dfrac{1}{\sqrt{4^3}} = \dfrac{1}{4\sqrt{4}} = \dfrac{1}{8}$

5. $32^{-2/5} = \dfrac{1}{32^{2/5}} = \dfrac{1}{\left(\sqrt[5]{32}\right)^2} = \dfrac{1}{\left(\sqrt[5]{(2)^5}\right)^2}$
$= \dfrac{1}{2^2} = \dfrac{1}{4}$

6. $(ab^6)^{1/2} = \sqrt{ab^6} = \sqrt{b^6}\sqrt{a} = b^3\sqrt{a}$

7. $\sqrt[5]{3^2} = 3^{2/5}$

8. $\dfrac{1}{\sqrt{x}} = \dfrac{1}{x^{1/2}} = x^{-1/2}$

9. $\sqrt[6]{a^2} = a^{2/6} = a^{1/3}$

10. $\sqrt[3]{\sqrt{5}} = \left(\sqrt{5}\right)^{1/3} = \left(5^{1/2}\right)^{1/3} = 5^{1/6}$

11. $\sqrt{a}\,\sqrt[4]{a} = a^{1/2}a^{1/4} = a^{1/2+1/4} = a^{3/4}$

12. $\sqrt[3]{27^{-1}} = 27^{-1/3} = \dfrac{1}{27^{1/3}} = \dfrac{1}{(3^3)^{1/3}} = \dfrac{1}{3}$

13. $w^{1/3} \cdot w^{1/6} = w^{1/3 + 1/6} = w^{3/6}$
$= w^{1/2} = \sqrt{w}$

14. $\dfrac{a^{3/4}}{a^{1/4}} = a^{3/4 - 1/4} = a^{1/2} = \sqrt{a}$

15. $2x^{1/3}x^{2/3}x^{1/5} = 2x^{1/3 + 2/3 + 1/5} = 2x^{18/15}$
$= 2x^{6/5} = 2\sqrt[5]{x^6} = 2x\sqrt[5]{x}$

16. $(m^{3/4})^{2/3} = m^{2/4} = m^{1/2} = \sqrt{m}$

17. $(y^{1/4} + y^{3/4})y^{7/4} = y^{1/4}y^{7/4} + y^{3/4}y^{7/4}$
$= y^{1/4 + 7/4} + y^{3/4 + 7/4}$
$= y^2 + y^{10/4}$
$= y^2 + y^{4/4}y^{4/4}y^{2/4}$
$= y^2 + y \cdot y \cdot y^{1/2}$
$= y^2 + y^2\sqrt{y}$

18. $(a^{1/3} + a^{2/3})^2 = (a^{1/3})^2 + 2a^{1/3}a^{2/3} + (a^{2/3})^2$
$= a^{2/3} + 2a^{1/3 + 2/3} + a^{4/3}$
$= a^{2/3} + 2a + a^{4/3}$
$= \sqrt[3]{a^2} + 2a + \sqrt[3]{a^4}$
$= \sqrt[3]{a^2} + 2a + a\sqrt[3]{a}$

19. $\dfrac{x^{1/4}y^{2/3}}{x^{1/8}y^{4/3}} = x^{1/4 - 1/8}y^{2/3 - 4/3}$
$= x^{1/8}y^{-2/3}$
$= \dfrac{x^{1/8}}{y^{2/3}} = \dfrac{\sqrt[8]{x}}{\sqrt[3]{y^2}} = \dfrac{\sqrt[8]{x}\sqrt[3]{y}}{y}$

20. $\left(\dfrac{x^6}{8y^3}\right)^{-1/3} = \dfrac{(x^6)^{-1/3}}{8^{-1/3}(y^3)^{-1/3}}$
$= \dfrac{x^{-2}}{(2^3)^{-1/3}y^{-1}}$
$= \dfrac{x^{-2}}{2^{-1}y^{-1}} = \dfrac{2y}{x^2}$

21. $(16m^4n^{-2})^{-1/4} = 16^{-1/4}(m^4)^{-1/4}(n^{-2})^{-1/4}$
$= (2^4)^{-1/4}m^{-1}n^{1/2}$
$= 2^{-1}m^{-1}\sqrt{n} = \dfrac{\sqrt{n}}{2m}$

22. $\sqrt{125x^2} = (125x^2)^{1/2} = (5^2 \cdot 5 \cdot x^2)^{1/2}$
$= (5^2)^{1/2} \cdot 5^{1/2} \cdot (x^2)^{1/2} = 5x\sqrt{5}$

23. $\sqrt[6]{64a^{18}b^{24}} = (64a^{18}b^{24})^{1/6} = (64)^{1/6}(a^{18})^{1/6}(b^{24})^{1/6}$
$= (2^6)^{1/6}a^{18/6}b^{24/6} = 2a^3b^4$

24. $\sqrt{75x^5y^7} = (75x^5y^7)^{1/2} = (75)^{1/2}(x^5)^{1/2}(y^7)^{1/2}$
$= (5^2 \cdot 3)^{1/2}(x^4x)^{1/2}(y^6y)^{1/2}$
$= (5^2)^{1/2}(x^4)^{1/2}(y^6)^{1/2}3^{1/2}x^{1/2}y^{1/2}$
$= 5x^2y^3(3xy)^{1/2}$
$= 5x^2y^3\sqrt{3xy}$

25. $\sqrt{\dfrac{25a^3b^4}{4a^{-3}b^2}} = \sqrt{\dfrac{25a^6b^2}{4}}$
$= \left(\dfrac{25a^6b^2}{4}\right)^{1/2}$
$= \dfrac{25^{1/2}(a^6)^{1/2}(b^2)^{1/2}}{4^{1/2}}$
$= \dfrac{(5^2)^{1/2}a^3b}{(2^2)^{1/2}}$
$= \dfrac{5a^3b}{2}$

26. $\sqrt[4]{32(x+y)^5} = \sqrt[4]{2^5(x+y)^5}$
$= (2^4 \cdot 2(x+y)^4(x+y))^{1/4}$
$= (2^4)^{1/4}[(x+y)^4]^{1/4}2^{1/4}(x+y)^{1/4}$
$= 2(x+y)\sqrt[4]{2(x+y)}$

27. $\sqrt[4]{(16x^{12}y^{-8})^{-3}} = (16x^{12}y^{-8})^{-3/4}$
$= (2^4)^{-3/4}(x^{12})^{-3/4}(y^{-8})^{-3/4}$
$= 2^{-3}x^{-9}y^6$
$= \dfrac{y^6}{2^3x^9} = \dfrac{y^6}{8x^9}$

28. $\sqrt[3]{\sqrt{10}} = (\sqrt{10})^{1/3} = (10^{1/2})^{1/3} = 10^{1/6} = \sqrt[6]{10}$

29. $\sqrt[4]{\sqrt[3]{x^5}} = (\sqrt[3]{x^5})^{1/4} = (x^{5/3})^{1/4} = x^{5/12} = \sqrt[12]{x^5}$

30. $\sqrt{m}\sqrt[3]{m^2} = m^{1/2}m^{2/3} = m^{7/6} = m^{6/6}m^{1/6} = m\sqrt[6]{m}$

31. $\sqrt[7]{y^2\sqrt{y^3}} = (y^2y^{3/2})^{1/7} = y^{2/7}y^{3/14}$
$= y^{4/14 + 3/14} = y^{7/14} = y^{1/2} = \sqrt{y}$

32. $\sqrt[5]{\sqrt[3]{a+b}} = (\sqrt[3]{a+b})^{1/5} = [(a+b)^{1/3}]^{1/5}$
$= (a+b)^{1/15} = \sqrt[15]{a+b}$

33. $\dfrac{\sqrt[4]{a}\sqrt[3]{a^2}}{\sqrt{a}} = \dfrac{a^{1/4}a^{2/3}}{a^{1/2}} = a^{1/4 + 2/3 - 1/2}$
$= a^{3/12 + 8/12 - 6/12} = a^{5/12} = \sqrt[12]{a^5}$

34. Use the y^x key with $y = 41$ and $x = 1/4 = 0.25$.
$$\sqrt[4]{41} = 41^{1/4} = 2.53$$

35. Use the y^x key with $y = 7$ and $x = 2/3$.
$$\sqrt[3]{7^2} = 7^{2/3} = 3.66$$

36. Use the y^x key with $y = 18.5$ and $x = 3/5 = 0.6$.
$$18.5^{3/5} = 5.76$$

37. Use the y^x key with $y = 37$ and $x = 2/7$.
$$\left(\sqrt[7]{37}\right)^2 = [(37)^{1/7}]^2 = 37^{2/7} = 2.81$$

38. Substitute 0.521 for u and evaluate with a calculator using the y^x key.
$$I = 0.712(1 + u)^{1/3}$$
$$= 0.712(1 + 0.521)^{1/3}$$
$$= 0.712(1.521)^{1/3}$$
$$= 0.818822431$$

Thus, to the nearest hundredth, the manufacturing index is 0.82.

39. Substitute 2500 for P, 7200 for A, and 5 for t and evaluate with a calculator using the y^x key.
$$r = \left(\dfrac{A}{P}\right)^{1/t} - 1$$
$$= \left(\dfrac{7200}{2500}\right)^{1/5} - 1$$
$$= (2.88)^{0.2} - 1$$
$$= 0.235601701$$

Thus, the annual rate of return on the investment is about 23.6%.

40. Substitute 40 for S, 185 for l, 62 for L, and evaluate with a calculator using the square root key.
$$s = S\left(\dfrac{l}{L}\right)^{1/2}$$
$$= 40\left(\dfrac{185}{62}\right)^{1/2}$$
$$= 40\sqrt{\dfrac{185}{62}}$$
$$= 69.09553928$$

Thus, the stolen car was traveling about 69 mph at the time of the crash.

EXERCISES A

41. $\sqrt{\dfrac{a}{b}} = \left(\dfrac{a}{b}\right)^{1/2} = \dfrac{a^{1/2}}{b^{1/2}} = \dfrac{\sqrt{a}}{\sqrt{b}}$

42. $[(-5)^2]^{1/2}$ is the same as $\sqrt{(-5)^2} = \sqrt{25}$ which is the principal root, 5, **NOT** -5.

43. Any negative value of x will work. For example, if $x = -2$, then
$$(x^2)^{1/2} = [(-2)^2]^{1/2} = 4^{1/2} = \sqrt{4} = 2 \ne x.$$

44. $\sqrt{x^6} = \sqrt{(x^3)^2} = |x^3|$

45. $\sqrt[3]{x^3 y^3} = \sqrt[3]{(xy)^3} = xy$

46. $\sqrt[4]{x^4 y^8} = \sqrt[4]{(xy^2)^4} = |xy^2|$

47. $\sqrt[5]{32x^{20}} = \sqrt[5]{2^5 (x^4)^5} = \sqrt[5]{2^5}\sqrt[5]{(x^4)^5} = 2x^4$

48. Substitute 30.5 for L and evaluate with a calculator using the square root key.

$$T = 2\pi\sqrt{\dfrac{L}{32}}$$
$$= 2\pi\sqrt{\dfrac{30.5}{32}}$$
$$= 6.134155751$$

Thus, to the nearest tenth, the time required for the pendulum to make one complete swing is 6.1 sec.

CHAPTER 1 REVIEW OF FUNDAMENTAL CONCEPTS

CHAPTER 1 Review Exercises

1. false ($\sqrt{5}$ is an irrational number)

2. true

3. false (for example, 2 is a rational number, but $\sqrt{2}$ is not rational)

4. true

5. associative property of addition

6. transitive property of <

7. $-[a - (b + a)] = -[a - b - a]$
   $= -[-b]$
   $= b$

8. $|-6| - |-13| = 6 - 13 = -(13 - 6) = -7$

9. If $x < 0$, then $-x > 0$ so $|-x| = -x$.

10. The distance between a and b is $|a - b|$. Thus the distance between $-\frac{1}{5}$ and $\frac{3}{10}$ is:

    $\left|-\frac{1}{5} - \frac{3}{10}\right| = \left|-\frac{2}{10} - \frac{3}{10}\right| = \left|-\frac{5}{10}\right| = \left|-\frac{1}{2}\right| = \frac{1}{2}$

11. The distance between $\sqrt{7}$ and $-3\sqrt{7}$ is:

    $|\sqrt{7} - (-3\sqrt{7})| = |\sqrt{7} + 3\sqrt{7}| = |4\sqrt{7}| = 4\sqrt{7}$

12. $(-8) + (-4) = -(8 + 4) = -12$

13. $(-5) - (-4) = (-5) + 4 = -(5 - 4) = -1$

14. $(-8) \div 0$ is undefined

15. $0 \div (-8) = 0$

16. $\left(-\frac{3}{4}\right) \cdot \left(\frac{1}{9}\right) = -\frac{3 \cdot 1}{4 \cdot 9} = -\frac{1}{4 \cdot 3} = -\frac{1}{12}$

17. $\left(-\frac{2}{3}\right) \div \left(-\frac{4}{3}\right) = \left(-\frac{2}{3}\right) \cdot \left(-\frac{3}{4}\right) = \frac{2 \cdot 3}{3 \cdot 4} = \frac{1}{2}$

18. $5x^3 \cdot x^{-5} = 5x^{3+(-5)} = 5x^{-2} = \dfrac{5}{x^2}$

19. $3y^{-1} = 3 \cdot \dfrac{1}{y} = \dfrac{3}{y}$

20. $\left(\dfrac{8^0 x^{-1}}{y^3}\right)^{-2} = \left(\dfrac{x^{-1}}{y^3}\right)^{-2} = \dfrac{(x^{-1})^{-2}}{(y^3)^{-2}} = \dfrac{x^2}{y^{-6}} = x^2 y^6$

21. $\left(\dfrac{5x^2 y}{x^{-1} y^2}\right)^{-3} = (5x^{2-(-1)} y^{1-2})^{-3}$
    $= (5x^3 y^{-1})^{-3}$
    $= 5^{-3}(x^3)^{-3}(y^{-1})^{-3}$
    $= 5^{-3} x^{-9} y^3$
    $= \dfrac{y^3}{5^3 x^9} = \dfrac{y^3}{125 x^9}$

22. $\dfrac{(0.000571)^2}{992,000} = \dfrac{(5.71 \times 10^{-4})^2}{9.92 \times 10^5}$
    $= \dfrac{(5.71)^2}{9.92} \times \dfrac{10^{-8}}{10^5}$
    $= 3.286703629 \times 10^{-13}$
    $\approx 3.29 \times 10^{-13}$

23. $\dfrac{(0.0000216)(8,360,000)}{(0.0000000115)}$
    $= \dfrac{(2.16 \times 10^{-5})(8.36 \times 10^6)}{(1.15 \times 10^{-8})}$
    $= \dfrac{(2.16)(8.36)}{1.15} \times \dfrac{10^{-5} \times 10^6}{10^{-8}}$
    $= 15.70226087 \times 10^9$
    $\approx 1.57 \times 10^1 \times 10^9$
    $= 1.57 \times 10^{10}$

24. We must first find the number of seconds in a year.
    1 year = 365 days
    = (365)(24) hours
    = (365)(24)(60) minutes
    = (365)(24)(60)(60) seconds
    Multiply this number by the speed of light, 3.00×10^5.

$$(365)(24)(60)(60)(3.00 \times 10^5)$$
$$= 94{,}608{,}000 \times 10^5$$
$$= 9.46 \times 10^7 \times 10^5$$
$$= 9.46 \times 10^{12}$$

Thus, one light year is about 9.46×10^{12} km.

Substitute -2 for a, 3 for b, and -1 for c in each expression in Exercises 25–30.

25. $(-2ab)^{-1} = [-2(-2)(3)]^{-1}$
$\phantom{(-2ab)^{-1}} = [12]^{-1}$
$\phantom{(-2ab)^{-1}} = \dfrac{1}{12}$

26. $(3a)^2 = [3(-2)]^2 = [-6]^2 = 36$

27. $3a^2 = 3(-2)^2 = 3(4) = 12$

28. $(a + b + c)^2 = [(-2) + (3) + (-1)]^2 = [0]^2 = 0$

29. $a^2 + bc = (-2)^2 + (3)(-1) = 4 + (-3) = 1$

30. $(c - b)^3 = [(-1) - (3)]^3 = [-4]^3 = -64$

31. $(13x^2y^2 - 7xy + 8) + (-2x^2y^2 + 3xy + 1)$
$= 13x^2y^2 - 7xy + 8 - 2x^2y^2 + 3xy + 1$
$= 13x^2y^2 - 2x^2y^2 - 7xy + 3xy + 8 + 1$
$= 11x^2y^2 - 4xy + 9$

32. $(2x^2 - 5y^2 - 6x + 2y) - (3x^2 + 2y^2 - 5x + 4y)$
$= 2x^2 - 5y^2 - 6x + 2y - 3x^2 - 2y^2 + 5x - 4y$
$= 2x^2 - 3x^2 - 5y^2 - 2y^2 - 6x + 5x + 2y - 4y$
$= -x^2 - 7y^2 - x - 2y$

33. $(-4a^2b^2 - 3ab + 5) - (4a^2b^2 - 4ab - 6)$
$= -4a^2b^2 - 3ab + 5 - 4a^2b^2 + 4ab + 6$
$= -4a^2b^2 - 4a^2b^2 - 3ab + 4ab + 5 + 6$
$= -8a^2b^2 + ab + 11$

34. $(8u^3v^2 - 2u^2v^2) + (-4u^2v^3 + 2u^2v^2) - (u^2v^2 - 2)$
$= 8u^3v^2 - 2u^2v^2 - 4u^2v^3 + 2u^2v^2 - u^2v^2 + 2$
$= 8u^3v^2 - 4u^2v^3 - u^2v^2 + 2$

35. $5x^3y^2(-2xy + 7x^2y)$
$= (5x^3y^2)(-2xy) + (5x^3y^2)(7x^2y)$
$= -10x^4y^3 + 35x^5y^3$

36. $(6u - v)(4u + v)$
$= (6u)(4u) + (6u)(v) + (-v)(4u) + (-v)(v)$
$= 24u^2 + 6uv - 4uv - v^2$
$= 24u^2 + 2uv - v^2$

37. $(7x - 2y)^2 = (7x)^2 - 2(7x)(2y) + (2y)^2$
$ = 49x^2 - 28xy + 4y^2$

38. $(5x + 7y)^2 = (5x)^2 + 2(5x)(7y) + (7y)^2$
$ = 25x^2 + 70xy + 49y^2$

39. $(3a + 8b)(3a - 8b) = (3a)^2 - (8b)^2 = 9a^2 - 64b^2$

40. $(2a^2 + b)(5a^2 - 2b)$
$= (2a^2)(5a^2) + (2a^2)(-2b) + (b)(5a^2) + (b)(-2b)$
$= 10a^4 - 4a^2b + 5a^2b - 2b^2$
$= 10a^4 + a^2b - 2b^2$

41. $\dfrac{14a^3b^4 - 28a^2b^3 + 49ab^5}{-7a^2b^3}$
$= \dfrac{14a^3b^4}{-7a^2b^3} + \dfrac{-28a^2b^3}{-7a^2b^3} + \dfrac{49ab^5}{-7a^2b^3}$
$= -2ab + 4 - \dfrac{7b^2}{a}$

42.
$$\begin{array}{r}
x^2 - xy - y^2 \\
x - 3y \overline{\smash{)}\,x^3 - 4x^2y + 2xy^2 - 3y^3} \\
\underline{x^3 - 3x^2y} \\
-x^2y + 2xy^2 \\
\underline{-x^2y + 3xy^2} \\
-xy^2 - 3y^3 \\
\underline{-xy^2 + 3y^3} \\
-6y^3
\end{array}$$

Thus, we obtain $x^2 - xy - y^2 - \dfrac{6y^3}{x - 3y}$.

43. Substitute 3.25 for t and evaluate to find the height of the rocket.
$-16t^2 + 240t = -16(3.25)^2 + 240(3.25)$
$ = -16(10.5625) + 780$
$ = 611$
Thus, the rocket is 611 ft high after 3.25 seconds.

44. $10x^2y^3 - 15xy^2 = (5xy^2)(2xy) - (5xy^2)(3)$
$ = 5xy^2(2xy - 3)$

45. $2x^2 - 4xy - 3x + 6y$
$= 2x(x - 2y) - 3(x - 2y)$
$= (2x - 3)(x - 2y)$

46. $u^2 - 7uv + 6v^2$
The factors of 6 that add to give -7 are -1 and -6. Thus,
$$u^2 - 7uv + 6v^2 = (u - v)(u - 6v).$$

47. $u^2 + 13u + 42$
The factors of 42 that add to give 13 are 6 and 7. Thus,
$$u^2 + 13u + 42 = (u + 6)(u + 7).$$

48. $9a^2 - 16b^2$
   This is a special form, the difference of two squares. Thus,
   $9a^2 - 16b^2 = (3a)^2 - (4b)^2 = (3a - 4b)(3a + 4b)$.

49. $3a^2 + ab - 14b^2$
   The factors of 3 and -14 that combine to give 1 are 3,1 and 7,-2. Thus,
   $3a^2 + ab - 14b^2 = (3a + 7b)(a - 2b)$.

50. $54x^3 + 2y^6$
   First factor out the common factor 2.
   $54x^3 + 2y^6 = 2(27x^3 + y^6)$
   The remaining factor is the sum of cubes.
   $$27x^3 + y^6 = (3x)^3 + (y^2)^3$$
   $$= (3x + y^2)(9x^2 - 3xy^2 + y^4)$$
   Thus,
   $54x^3 + 2y^6 = 2(3x + y^2)(9x^2 - 3xy^2 + y^4)$.

51. $25x^2 + 30xy + 9y^2$
   This is a special form, a perfect square trinomial.
   $$25x^2 + 30xy + 9y^2$$
   $$= (5x)^2 + 2(5x)(3y) + (3y)^2$$
   $$= (5x + 3y)^2$$

52. $8u^2 - 2uv - 15v^2$
   The factors of 8 and -15 that combine to give -2 are 2,4 and -3,5. Thus,
   $8u^2 - 2uv - 15v^2 = (2u - 3v)(4u + 5v)$.

53. $125x^3 - 64y^3$
   This is a special form, the difference of cubes.
   $$125x^3 - 64y^3 = (5x)^3 - (4y)^3$$
   $$= (5x - 4y)(25x^2 + 20xy + 16y^2)$$

54. $x^2 + 4y^2$
   This cannot be factored since a sum of squares, unlike the difference of squares, is not factorable.

55. $16x^2 - 56xy + 49y^2$
   This is a special form, a perfect square trinomial.
   $$16x^2 - 56xy + 49y^2 = (4x)^2 - 2(4x)(7y) + (7y)^2$$
   $$= (4x - 7y)^2$$

56. $\dfrac{36a^2b^2c^2}{75ab^4c^2} = \dfrac{3 \cdot 12aab^2c^2}{3 \cdot 25ab^2b^2c^2} = \dfrac{12a}{25b^2}$

57. $\dfrac{x-5}{5-x} = \dfrac{(-1)(5-x)}{(5-x)} = -1$

58. $\dfrac{x^2 + 5x - 24}{x^2 - 10x + 21} = \dfrac{(x+8)(x-3)}{(x-7)(x-3)}$
   $= \dfrac{x+8}{x-7}$

59. $\dfrac{a^2 - 5a + 6}{a^2 - 6a + 9} \cdot \dfrac{a^2 + 4a - 21}{a^2 + 2a - 35}$
   $= \dfrac{(a-2)(a-3)}{(a-3)(a-3)} \cdot \dfrac{(a+7)(a-3)}{(a+7)(a-5)}$
   $= \dfrac{(a-2)(a-3)(a+7)(a-3)}{(a-3)(a-3)(a+7)(a-5)}$
   $= \dfrac{a-2}{a-5}$

60. $\dfrac{a^3 - b^3}{a^2 + 4a - 5} \div \dfrac{a^2 + ab + b^2}{a^2 + 10a + 25}$
   $= \dfrac{a^3 - b^3}{a^2 + 4a - 5} \cdot \dfrac{a^2 + 10a + 25}{a^2 + ab + b^2}$
   $= \dfrac{(a-b)(a^2 + ab + b^2)}{(a+5)(a-1)} \cdot \dfrac{(a+5)(a+5)}{a^2 + ab + b^2}$
   $= \dfrac{(a-b)(a^2 + ab + b^2)(a+5)(a+5)}{(a+5)(a-1)(a^2 + ab + b^2)}$
   $= \dfrac{(a-b)(a+5)}{a-1}$

61. $\dfrac{5-x}{x^2 - 12x + 35} + \dfrac{x}{x^2 - 14x + 49}$
   $= \dfrac{(-1)(x-5)}{(x-5)(x-7)} + \dfrac{x}{(x-7)(x-7)}$
   $= \dfrac{-1}{x-7} + \dfrac{x}{(x-7)^2}$
   $= \dfrac{(-1)(x-7)}{(x-7)^2} + \dfrac{x}{(x-7)^2}$
   $= \dfrac{-x + 7 + x}{(x-7)^2}$
   $= \dfrac{7}{(x-7)^2}$

62. $\dfrac{3}{y^2 + y} - \dfrac{2}{y+1} = \dfrac{3}{y(y+1)} - \dfrac{2}{y+1}$
   $= \dfrac{3}{y(y+1)} - \dfrac{2y}{y(y+1)}$
   $= \dfrac{3 - 2y}{y(y+1)}$

63. $\dfrac{8ab}{a^3 - b^3} = \dfrac{8ab}{(a-b)(a^2 + ab + b^2)}$ and $\dfrac{7}{(a-b)^2}$
   The LCD consists of one factor of $(a^2 + ab + b^2)$ and two factors of $(a-b)$. Thus, the
   LCD $= (a-b)^2(a^2 + ab + b^2)$.

64. $\dfrac{a+1}{a^2-7a+6} = \dfrac{a+1}{(a-1)(a-6)}$

$\dfrac{a+5}{a^2-12a+36} = \dfrac{a+5}{(a-6)^2}$

The LCD consists of one factor of $(a-1)$ and two factors of $(a-6)$. Thus, the
LCD $= (a-1)(a-6)^2$.

65. $\dfrac{\frac{a}{b}-1}{\frac{b}{a}-1} = \dfrac{ab\left[\frac{a}{b}-1\right]}{ab\left[\frac{b}{a}-1\right]}$

$= \dfrac{a^2-ab}{b^2-ab}$

$= \dfrac{a(a-b)}{b(b-a)}$

$= \dfrac{a}{b}(-1) = -\dfrac{a}{b}$

66. $\dfrac{a-7+\frac{10}{a}}{a-10+\frac{25}{a}} = \dfrac{a\left[a-7+\frac{10}{a}\right]}{a\left[a-10+\frac{25}{a}\right]}$

$= \dfrac{a^2-7a+10}{a^2-10a+25}$

$= \dfrac{(a-2)(a-5)}{(a-5)(a-5)}$

$= \dfrac{a-2}{a-5}$

67. $\dfrac{4x^3(4x) - (2x^2-1)12x^2}{16x^6}$

$= \dfrac{16x^4 - 24x^4 + 12x^2}{16x^6}$

$= \dfrac{-8x^4 + 12x^2}{16x^6}$

$= \dfrac{4x^2(-2x^2+3)}{4x^2 \cdot 4x^4}$

$= \dfrac{-2x^2+3}{4x^4}$

68. $\sqrt[3]{-27} = \sqrt[3]{(-3)^3} = -3$

69. $\sqrt[5]{-1} = \sqrt[5]{(-1)^5} = -1$

70. $\sqrt[6]{-64}$ is not a real number.

71. $\sqrt{9a^4} = \sqrt{(3a^2)^2} = 3a^2$

72. $\sqrt[3]{8x^6y^9} = \sqrt[3]{2^3(x^2)^3(y^3)^3} = 2x^2y^3$

73. $\sqrt[4]{81a^8b^{20}} = \sqrt[4]{3^4(a^2)^4(b^5)^4} = 3a^2b^5$

74. $\sqrt{a^2-14a+49} = \sqrt{(a-7)^2} = a-7$

75. $3\sqrt{4x}\sqrt{12xy^2} = 3\sqrt{(4x)(12xy^2)}$

$= 3\sqrt{4^2x^2y^2 \cdot 3}$

$= 3 \cdot 4 \cdot x \cdot y\sqrt{3}$

$= 12xy\sqrt{3}$

76. $\dfrac{\sqrt{20a^3b^7}}{\sqrt{5ab}} = \sqrt{\dfrac{20a^3b^7}{5ab}}$

$= \sqrt{4a^2b^6}$

$= \sqrt{2^2a^2(b^3)^2}$

$= 2ab^3$

77. $\dfrac{\sqrt[4]{48a^5b^{10}}}{\sqrt[4]{243ab^5}} = \sqrt[4]{\dfrac{48a^5b^{10}}{243ab^5}}$

$= \sqrt[4]{\dfrac{3 \cdot 2^4 a^4 b^5}{3 \cdot 3^4}}$

$= \dfrac{\sqrt[4]{2^4 a^4 b^4 b}}{\sqrt[4]{3^4}}$

$= \dfrac{2ab\sqrt[4]{b}}{3}$

78. $4\sqrt{125} + 5\sqrt{80} = 4\sqrt{5^2 \cdot 5} + 5\sqrt{4^2 \cdot 5}$

$= 4 \cdot 5\sqrt{5} + 5 \cdot 4\sqrt{5}$

$= 20\sqrt{5} + 20\sqrt{5}$

$= 40\sqrt{5}$

79. $-7\sqrt[3]{54} + 10\sqrt[3]{128} = -7\sqrt[3]{3^3 \cdot 2} + 10\sqrt[3]{4^3 \cdot 2}$

$= (-7)(3)\sqrt[3]{2} + (10)(4)\sqrt[3]{2}$

$= -21\sqrt[3]{2} + 40\sqrt[3]{2}$

$= 19\sqrt[3]{2}$

80. $4\sqrt[3]{3xy^3} - 6y\sqrt[3]{81x} = 4y\sqrt[3]{3x} - 6y\sqrt[3]{3^3 \cdot 3x}$
    $= 4y\sqrt[3]{3x} - 6y \cdot 3\sqrt[3]{3x}$
    $= 4y\sqrt[3]{3x} - 18y\sqrt[3]{3x}$
    $= -14y\sqrt[3]{3x}$

81. $\dfrac{5 - \sqrt{175}}{10} = \dfrac{5 - \sqrt{5^2 \cdot 7}}{10}$
    $= \dfrac{5 - 5\sqrt{7}}{5 \cdot 2}$
    $= \dfrac{5(1 - \sqrt{7})}{5 \cdot 2}$
    $= \dfrac{1 - \sqrt{7}}{2}$

82. $\dfrac{\sqrt{8} - \sqrt{3}}{\sqrt{2} + \sqrt{3}}$
    $= \dfrac{(\sqrt{8} - \sqrt{3})(\sqrt{2} - \sqrt{3})}{(\sqrt{2} + \sqrt{3})(\sqrt{2} - \sqrt{3})}$
    $= \dfrac{(\sqrt{8})(\sqrt{2}) - (\sqrt{8})(\sqrt{3}) - (\sqrt{3})(\sqrt{2}) + (\sqrt{3})^2}{(\sqrt{2})^2 - (\sqrt{3})^2}$
    $= \dfrac{\sqrt{16} - \sqrt{24} - \sqrt{6} + 3}{2 - 3}$
    $= \dfrac{4 - 2\sqrt{6} - \sqrt{6} + 3}{-1}$
    $= \dfrac{7 - 3\sqrt{6}}{-1}$
    $= 3\sqrt{6} - 7$

83. $a^{7/5} = \sqrt[5]{a^7} = \sqrt[5]{a^5 a^2} = a\sqrt[5]{a^2}$

84. $(-8)^{2/3} = \left(\sqrt[3]{-8}\right)^2 = (-2)^2 = 4$

85. $(a^2 b^4)^{1/2} = (a^2)^{1/2}(b^4)^{1/2} = a^{2(1/2)}b^{4(1/2)} = ab^2$

86. $(y^{3/4})^{1/3} = y^{(3/4)(1/3)} = y^{1/4} = \sqrt[4]{y}$

87. $\left(\dfrac{8a^5 b^7}{a^{-4} b^{-3}}\right)^{1/3} = (8a^{5-(-4)}b^{7-(-3)})^{1/3}$
    $= (2^3 a^9 b^{10})^{1/3}$
    $= 2^{3/3} a^{9/3} b^{10/3}$
    $= 2a^3 b^{9/3} b^{1/3}$
    $= 2a^3 b^3 \sqrt[3]{b}$

88. $\left(\dfrac{27^{2/3} x^{-5/3} y^2}{x^{-10/3} y^{-2}}\right)^{-3} = (27^{2/3} x^{-5/3-(-10/3)} y^{2-(-2)})^{-3}$
    $= (27^{2/3} x^{5/3} y^4)^{-3}$
    $= (27^{2/3})^{-3}(x^{5/3})^{-3}(y^4)^{-3}$
    $= 27^{-2} x^{-5} y^{-12}$
    $= \dfrac{1}{27^2 x^5 y^{12}}$
    $= \dfrac{1}{729 x^5 y^{12}}$

89. To find the values to exclude in
$$\dfrac{x + 7}{x^2 - 3x + 2}$$
set the denominator equal to zero and solve for x.
$x^2 - 3x + 2 = 0$
$(x - 1)(x - 2) = 0$
$x - 1 = 0 \qquad x - 2 = 0$
$x = 1 \qquad x = 2$
Thus, the values to exclude are $x = 1$ and $x = 2$.

90. There are no values to exclude in
$$\dfrac{x^2 + 5x - 7}{5}$$
since the denominator, 5, can never be zero.

91. To find the values to exclude in
$$\dfrac{4a^2 + 5b}{(a^2 - 4)(b + 7)}$$
set the denominator equal to 0 and solve.
$(a^2 - 4)(b + 7) = 0$
$a^2 - 4 = 0 \qquad b + 7 = 0$
$(a + 2)(a - 2) = 0 \qquad b = -7$
$a + 2 = 0 \qquad a - 2 = 0$
$a = -2 \qquad a = 2$
Thus, the values to exclude are $a = 2$, $a = -2$, and $b = -7$.

92. $\dfrac{x + y}{x^2 - y^2}$ and $\dfrac{1}{x - y}$ are equivalent since the first can be obtained from the second by multiplying numerator and denominator by $(x + y)$, assuming that $x + y \neq 0$, that is, that $x \neq -y$.

CHAPTER 1 REVIEW REVIEW EXERCISES 39

93. $\dfrac{a^2+1}{a^2}$ and $\dfrac{a^2+2}{a^2+1}$ are not equivalent since one is obtained from the other by adding or subtracting the same number to both numerator and denominator, a process that does not result in equivalent fractions.

94. $\dfrac{x-2}{x^2+x-6}$ and $\dfrac{1}{x+3}$ are equivalent since the first can be obtained from the second by multiplying numerator and denominator by $x-2$, provided $x-2 \neq 0$, that is, provided that $x \neq 2$.

95. $5{,}860{,}000 = 5.86 \times 10^6$

96. $0.00000000586 = 5.86 \times 10^{-9}$

97. $47.2 = 4.72 \times 10^1$

98. $5.86 \times 10^{-3} = 0.00586$

99. $5.86 \times 10^4 = 58{,}600$

100. $3.81 \times 10^{-5} = 0.0000381$

101. Let s = the employee's former salary.
If each receives a 12% raise in pay, then the raise is given by $0.12s$. The new salary is the former salary plus the raise. Thus, the new salary for an employee is given by $s + 0.12s = (1 + 0.12)s = 1.12s$. To find Pat's new salary, substitute 24,500 for s and evaluate.

$$1.12(24{,}500) = \$27{,}440$$

Substitute -3 for a, -1 for b, and 5 for c in each espression in Exercises 102-107.

102. $-(a - c) = -[(-3) - (5)] = -[-8] = 8$

103. $\begin{aligned} 3a + b - c &= 3(-3) + (-1) - (5) \\ &= -9 - 1 - 5 = -15 \end{aligned}$

104. $\begin{aligned} |2a - (-b) + c| &= |2(-3) - (-(-1)) + 5| \\ &= |-6 - 1 + 5| = |-2| = 2 \end{aligned}$

105. $a^2 = (-3)^2 = 9$

106. $-3a^2 = -3(-3)^2 = -3(9) = -27$

107. $-a^2 = -(-3)^2 = -(9) = -9$

108. $\dfrac{4x^2+4x+1}{16x+8} \cdot \dfrac{x^2-8x+12}{2x^2-11x-6}$
$= \dfrac{(2x+1)(2x+1)}{8(2x+1)} \cdot \dfrac{(x-2)(x-6)}{(2x+1)(x-6)}$
$= \dfrac{(2x+1)(2x+1)(x-2)(x-6)}{8(2x+1)(2x+1)(x-6)}$
$= \dfrac{x-2}{8}$

109. $\dfrac{x+2}{x-3} + \dfrac{2x-1}{x-3} = \dfrac{(x+2)+(2x-1)}{x-3}$
$= \dfrac{x+2+2x-1}{x-3}$
$= \dfrac{3x+1}{x-3}$

110. $\dfrac{2xy}{x^2-y^2} + \dfrac{y}{y-x} + 3$
$= \dfrac{2xy}{(x-y)(x+y)} + \dfrac{-y}{x-y} + 3$
$= \dfrac{2xy}{(x-y)(x+y)} + \dfrac{(-y)(x+y)}{(x-y)(x+y)} + \dfrac{3(x-y)(x+y)}{(x-y)(x+y)}$
$= \dfrac{2xy + (-y)(x+y) + 3(x-y)(x+y)}{(x-y)(x+y)}$
$= \dfrac{2xy - xy - y^2 + 3x^2 - 3y^2}{(x-y)(x+y)}$
$= \dfrac{3x^2 + xy - 4y^2}{(x-y)(x+y)}$
$= \dfrac{(3x+4y)(x-y)}{(x-y)(x+y)} = \dfrac{3x+4y}{x+y}$

111. $\dfrac{2x^2+x-15}{x^2-x-12} \div \dfrac{2x^2+3x-20}{16x-x^3}$
$= \dfrac{2x^2+x-15}{x^2-x-12} \cdot \dfrac{16x-x^3}{2x^2+3x-20}$
$= \dfrac{(2x-5)(x+3)}{(x-4)(x+3)} \cdot \dfrac{x(4-x)(4+x)}{(2x-5)(x+4)}$
$= \dfrac{(2x-5)(x+3)x(-1)(x-4)(x+4)}{(x-4)(x+3)(2x-5)(x+4)}$
$= (-1)x = -x$

112. $(2u^2 - 3uv + v^2)(5u + 8v)$
$= (2u^2)(5u + 8v) + (-3uv)(5u + 8v)$
$\quad + (v^2)(5u + 8v)$
$= 10u^3 + 16u^2v - 15u^2v - 24uv^2$
$\quad + 5uv^2 + 8v^3$
$= 10u^3 + u^2v - 19uv^2 + 8v^3$

113. $(3x - 2y - 5)^2 = [(3x - 2y) - 5]^2$
$= (3x - 2y)^2 - (2)(5)(3x - 2y) + (5)^2$
$= 9x^2 - 12xy + 4y^2 - 30x + 20y + 25$

114. $\dfrac{\sqrt{x}+\sqrt{y}}{2\sqrt{x}-\sqrt{y}} = \dfrac{(\sqrt{x}+\sqrt{y})(2\sqrt{x}+\sqrt{y})}{(2\sqrt{x}-\sqrt{y})(2\sqrt{x}+\sqrt{y})}$

$= \dfrac{2(\sqrt{x})^2 + \sqrt{x}\sqrt{y} + 2\sqrt{x}\sqrt{y} + (\sqrt{y})^2}{(2\sqrt{x})^2 - (\sqrt{y})^2}$

$= \dfrac{2x + 3\sqrt{xy} + y}{4x - y}$

115. $5\sqrt{40a} + \dfrac{8\sqrt{2a}}{\sqrt{5}} = 5 \cdot 2\sqrt{10a} + \dfrac{8\sqrt{2a}\sqrt{5}}{\sqrt{5}\sqrt{5}}$

$= 10\sqrt{10a} + \dfrac{8\sqrt{10a}}{5}$

$= \left(10 + \dfrac{8}{5}\right)\sqrt{10a}$

$= \left(\dfrac{58}{5}\right)\sqrt{10a} = \dfrac{58\sqrt{10a}}{5}$

116. $\sqrt{\dfrac{125a^5b^{-2}}{4a^{-2}b^{-6}}} = \sqrt{\dfrac{5^2 \cdot 5 \cdot a^{5-(-2)}b^{-2-(-6)}}{2^2}}$

$= \dfrac{\sqrt{5^2 \cdot 5 \cdot a^7 b^4}}{\sqrt{2^2}}$

$= \dfrac{5a^3 b^2 \sqrt{5a}}{2}$

117. $\dfrac{\sqrt[3]{-81x^3 y^{-3}}}{\sqrt[3]{3x^{-3}y^3}} = \sqrt[3]{\dfrac{-81x^3 y^{-3}}{3x^{-3}y^3}}$

$= \sqrt[3]{\dfrac{-27x^6}{y^6}}$

$= \dfrac{\sqrt[3]{-27}\sqrt[3]{x^6}}{\sqrt[3]{y^6}}$

$= \dfrac{-3x^2}{y^2}$

118. $\sqrt[4]{m}\sqrt[3]{m^2} = m^{1/4}m^{2/3} = m^{1/4+2/3} = m^{11/12} = \sqrt[12]{m^{11}}$

119. $\sqrt[5]{\sqrt{a^3}} = \left(\sqrt{a^3}\right)^{1/5} = \left(a^{3/2}\right)^{1/5} = a^{3/10} = \sqrt[10]{a^3}$

120. $10a^2 + 3ab - 18b^2$

The factors of 10 and 18 that combine to give 3 are 5,2 and -6,3. Thus,
$10a^2 + 3ab - 18b^2 = (5a - 6b)(2a + 3b)$.

121. $5x^3y^3 - 40z^3 = 5(x^3y^3 - 8z^3) = 5[(xy)^3 - (2z)^3]$
The remaining factor is a difference of cubes. Thus,
$5x^3y^3 - 40z^3$
$= 5(xy - 2z)(x^2y^2 + 2xyz + 4z^2)$.

122. $ab + xb - ay - xy = (a + x)b - (a + x)y$
$= (a + x)(b - y)$

123. $2u^4 - u^2v^2 - v^4$
The factors of 2 and -1 that combine to give -1 are 2,1 and 1,-1. Thus,
$2u^4 - u^2v^2 - v^4 = (2u^2 + v^2)(u^2 - v^2)$.
But the second factor is a difference of squares.
$2u^4 - u^2v^2 - v^4 = (2u^2 + v^2)(u + v)(u - v)$

124. $x^2 - y^2 + 10y - 25 = x^2 - (y^2 - 10y + 25)$
$= x^2 - (y - 5)^2$
$= [x + (y - 5)][x - (y - 5)]$
$= (x + y - 5)(x - y + 5)$

125. $9x^{2n} - y^{2n} = (3x^n)^2 - (y^n)^2$
$= (3x^n + y^n)(3x^n - y^n)$

Use the y^x button on a calculator to evaluate each expression in Exercises 126-129, correct to the nearest hundredth.

126. $13^{1/3} = 2.351334686 \approx 2.35$ ($y = 13$, $x = 1/3$)

127. $\sqrt[4]{30} = 30^{1/4} = 2.340347319 \approx 2.34$
($y = 30$, $x = 1/4$)

128. $\sqrt[3]{(15)^2} = 15^{2/3} = 6.082201985 \approx 6.08$
($y = 15$, $x = 2/3$)

129. $\sqrt[5]{\dfrac{4}{3}} = \left(\dfrac{4}{3}\right)^{1/5} = 1.059223841 \approx 1.06$
($y = 4/3$, $x = 1/5$)

130. Substitute 150 for g and evaluate using a calculator to find r.

$r = 7.75\sqrt{g} = 7.75\sqrt{150} = 94.91772753$

Thus, the radius of the spill, correct to the nearest tenth, is 94.9 yd.

CHAPTER 1 REVIEW OF FUNDAMENTAL CONCEPTS

CHAPTER 1 Test

1. false (for example, $\frac{1}{2}$ is a rational number but not an integer)

2. false (this is the commutative property of addition)

3. $-(-(-8)) = -(+8) = -(8) = -8$

4. $|3 \cdot 0 + 5(-4)| = |0 - 20| = |-20| = 20$

5. $(2a^{-2}b^3)^{-4} = 2^{-4}(a^{-2})^{-4}(b^3)^{-4}$
   $= 2^{-4}a^8b^{-12} = \dfrac{a^8}{2^4 b^{12}} = \dfrac{a^8}{16b^{12}}$

6. $\left(\dfrac{2x^2y}{x^{-1}y^2}\right)^{-3} = (2x^{2-(-1)}y^{1-2})^{-3} = (2x^3y^{-1})^{-3}$
   $= 2^{-3}(x^3)^{-3}(y^{-1})^{-3} = 2^{-3}x^{-9}y^3$
   $= \dfrac{y^3}{2^3 x^9} = \dfrac{y^3}{8x^9}$

7. Divide $\$86{,}500{,}000{,}000 = \$8.65 \times 10^{10}$ by $3{,}920{,}000 = 3.92 \times 10^6$ to find the amount of debt per person.

   $\dfrac{8.65 \times 10^{10}}{3.92 \times 10^6} = \dfrac{8.65}{3.92} \times 10^4$
   $= 2.206632653 \times 10^4$
   $\approx 2.21 \times 10^4$

   Thus, the amount of debt per person is about $\$2.21 \times 10^4$.

8. $(2a^2b^2 + ab - 3) - (a^2b^2 - 3ab + 1)$
   $\qquad - (3a^2b^2 - ab + 6)$
   $= 2a^2b^2 + ab - 3 - a^2b^2 + 3ab - 1$
   $\qquad - 3a^2b^2 + ab - 6$
   $= 2a^2b^2 - a^2b^2 - 3a^2b^2 + ab + 3ab + ab$
   $\qquad - 3 - 1 - 6$
   $= -2a^2b^2 + 5ab - 10$

9. $(2x^2 - 3y)^2 = (2x^2)^2 - (2)(2x^2)(3y) + (3y)^2$
   $= 4x^4 - 12x^2y + 9y^2$

10. $(2x + y)(x - 3y)$
    $= (2x)(x) + (2x)(-3y) + (y)(x) + (y)(-3y)$
    $= 2x^2 - 6xy + xy - 3y^2$
    $= 2x^2 - 5xy - 3y^2$

11.
$$\begin{array}{r}
a^2 + 4ab + 2b^2 \\
a - 2b \overline{\smash{)}a^3 + 2a^2b - 6ab^2 - 4b^3} \\
\underline{a^3 - 2a^2b} \\
4a^2b - 6ab^2 \\
\underline{4a^2b - 8ab^2} \\
2ab^2 - 4b^3 \\
\underline{2ab^2 - 4b^3} \\
0
\end{array}$$

Thus, the quotient is $a^2 + 4ab + 2b^2$.

12. $4a^2 - 4ab - 3b^2$
    The factors of 4 and -3 that combine to give -4 are 2,2 and $-3,1$. Thus,
    $4a^2 - 4ab - 3b^2 = (2a - 3b)(2a + b)$.

13. $u^2 - 4v^2$
    This is a special form, the difference of two squares. Thus,
    $u^2 - 4v^2 = (u)^2 - (2v)^2$
    $= (u - 2v)(u + 2v)$.

14. $2x^3 + 250 = 2(x^3 + 125)$
    The remaining factor is a special form, the difference of cubes. Thus,
    $2x^3 + 250 = 2(x^3 + 5^3)$
    $= 2(x + 5)(x^2 - 5x + 25)$.

15. $a^2 - 4b^2 + 2a + 4b$
    $= (a^2 - 4b^2) + (2a + 4b)$
    $= (a - 2b)(a + 2b) + 2(a + 2b)$
    $= [(a - 2b) + 2](a + 2b)$
    $= (a - 2b + 2)(a + 2b)$

16. $C = M + W$
    $= (2u^2 + u - 6) + (3u^2 + u + 1)$
    $= 2u^2 + 3u^2 + u + u - 6 + 1$
    $= 5u^2 + 2u - 5$
    Substitute 5 for u.
    $C = 5(5)^2 + 2(5) - 5$
    $= 5(25) + 10 - 5$
    $= 125 + 10 - 5$
    $= 130$
    Thus, the total cost is $130 when $u = 5$.

17. $\dfrac{x^2 - 4}{x^2 - x} \cdot \dfrac{x^2 + 5x - 6}{x^2 + 5x + 6} \div \dfrac{x^2 + 3x - 10}{x^2 + 3x}$

$= \dfrac{x^2 - 4}{x^2 - x} \cdot \dfrac{x^2 + 5x - 6}{x^2 + 5x + 6} \cdot \dfrac{x^2 + 3x}{x^2 + 3x - 10}$

$= \dfrac{(x+2)(x-2)}{x(x-1)} \cdot \dfrac{(x+6)(x-1)}{(x+2)(x+3)} \cdot \dfrac{x(x+3)}{(x+5)(x-2)}$

$= \dfrac{(x+2)(x-2)(x+6)(x-1)x(x+3)}{x(x-1)(x+2)(x+3)(x+5)(x-2)}$

$= \dfrac{x+6}{x+5}$

18. $\dfrac{x}{x^2 - y^2} + \dfrac{y}{x^2 - 2xy + y^2}$

$= \dfrac{x}{(x-y)(x+y)} + \dfrac{y}{(x-y)(x-y)}$

$= \dfrac{x(x-y)}{(x-y)^2(x+y)} + \dfrac{y(x+y)}{(x-y)^2(x+y)}$

$= \dfrac{x(x-y) + y(x+y)}{(x-y)^2(x+y)}$

$= \dfrac{x^2 - xy + xy + y^2}{(x-y)^2(x+y)}$

$= \dfrac{x^2 + y^2}{(x-y)^2(x+y)}$

19. $\dfrac{2a+b}{a^2 - ab} - \dfrac{3b}{a^2 - b^2}$

$= \dfrac{2a+b}{a(a-b)} - \dfrac{3b}{(a-b)(a+b)}$

$= \dfrac{(2a+b)(a+b)}{a(a-b)(a+b)} - \dfrac{3ab}{a(a-b)(a+b)}$

$= \dfrac{2a^2 + 3ab + b^2 - 3ab}{a(a-b)(a+b)}$

$= \dfrac{2a^2 + b^2}{a(a-b)(a+b)}$

20. $\dfrac{a - \dfrac{a}{b}}{b - \dfrac{b}{a}} = \dfrac{ab\left[a - \dfrac{a}{b}\right]}{ab\left[b - \dfrac{b}{a}\right]}$

$= \dfrac{a^2 b - a^2}{ab^2 - b^2}$

$= \dfrac{a^2(b-1)}{b^2(a-1)}$

21. $\sqrt[3]{2ab^2}\,\sqrt[3]{16a^2b^5} = \sqrt[3]{(2ab^2)(16a^2b^5)}$

$= \sqrt[3]{2^3 a^3 (b^2)^3 \cdot 4b}$

$= 2ab^2 \sqrt[3]{4b}$

22. $\left(\dfrac{9x^{-1}y}{x^2 y^3}\right)^{-1/2} = (9x^{-1-2}y^{1-3})^{-1/2} = (3^2 x^{-3} y^{-2})^{-1/2}$

$= (3^2)^{-1/2}(x^{-3})^{-1/2}(y^{-2})^{-1/2} = 3^{-1} x^{3/2} y^1$

$= \dfrac{x^{3/2} y}{3} = \dfrac{x^{2/2} y x^{1/2}}{3} = \dfrac{xy\sqrt{x}}{3}$

23. $\dfrac{\sqrt[3]{2xy}}{\sqrt[3]{y^2}} = \dfrac{\sqrt[3]{2xy}\,\sqrt[3]{y}}{\sqrt[3]{y^2}\,\sqrt[3]{y}} = \dfrac{\sqrt[3]{2xy^2}}{y}$

24. $\dfrac{\sqrt{a} + 1}{\sqrt{a} - 1} = \dfrac{(\sqrt{a}+1)(\sqrt{a}+1)}{(\sqrt{a}-1)(\sqrt{a}+1)}$

$= \dfrac{(\sqrt{a})^2 + 2\sqrt{a} + 1}{(\sqrt{a})^2 - 1}$

$= \dfrac{a + 2\sqrt{a} + 1}{a - 1}$

25. $\sqrt[4]{m^3}\,\sqrt[3]{m} = m^{3/4} m^{1/3} = m^{3/4 + 1/3} = m^{13/12}$

$= m^{12/12} m^{1/12} = m\,\sqrt[12]{m}$

26. $\sqrt{\sqrt[4]{(a+b)^7}} = \left(\sqrt[4]{(a+b)^7}\right)^{1/2} = \left((a+b)^{7/4}\right)^{1/2}$

$= (a+b)^{7/8} = \sqrt[8]{(a+b)^7}$

27. Substitute 0.448 for u and use a calculator with a y^x button to evaluate I.

$I = 0.435(1+u)^{1/4}$
$= 0.435(1 + 0.448)^{1/4}$
$= 0.435(1.448)^{1/4}$
≈ 0.477179082

Thus, correct to three decimal places, the manufacturing index when $u = 0.448$ is about 0.477.

28. Substitute 1.75 for L and use a calculator to evaluate T.

$T = 2\pi \sqrt{\dfrac{L}{32}}$

$= 2\pi \sqrt{\dfrac{1.75}{32}}$

≈ 1.46934542

Thus, correct to the nearest tenth, the time required for the pendulum to make one swing is about 1.5 sec.

PRACTICE EXERCISES SECTION 2.1

CHAPTER 2 EQUATIONS, INEQUALITIES, AND PROBLEM SOLVING

SECTION 2.1 Linear and Absolute Value Equations

1.(a)
$$6x + 1 - 4x = 3(4 - x) - 11 + x$$
$$2x + 1 = 12 - 3x - 11 + x$$
$$2x + 1 = 1 - 2x$$
$$4x = 0$$
$$x = \frac{0}{4} = 0$$

(b)
$$1.5w + 2.5 = 0.5w - 7.5$$
$$10(1.5w + 2.5) = 10(0.5w - 7.5)$$
$$15w + 25 = 5w - 75$$
$$10w + 25 = -75$$
$$10w = -100$$
$$w = \frac{-100}{10} = -10$$

(c)
$$\frac{5}{3}z + 1 = \frac{2}{3}z + 1 + z$$
$$3\left(\frac{5}{3}z + 1\right) = 3\left(\frac{2}{3}z + 1 + z\right)$$
$$5z + 3 = 2z + 3 + 3z$$
$$5z + 3 = 5z + 3$$
$$3 = 3$$

This identity implies that every real number is a solution.

(d)
$$(x - 1)^2 + 3x = x^2 - 7$$
$$x^2 - 2x + 1 + 3x = x^2 - 7$$
$$-2x + 1 + 3x = -7$$
$$x + 1 = -7$$
$$x = -8$$

2.
$$3\sqrt{2y - 5} - \sqrt{y + 23} = 0$$
$$3\sqrt{2y - 5} = \sqrt{y + 23}$$
$$(3\sqrt{2y - 5})^2 = (\sqrt{y + 23})^2$$
$$9(2y - 5) = y + 23$$
$$18y - 45 = y + 23$$
$$17y - 45 = 23$$
$$17y = 68$$
$$y = \frac{68}{17} = 4$$

3.
$$\sqrt{w + 2} + \sqrt{w + 6} = 4$$
$$\sqrt{w + 2} = 4 - \sqrt{w + 6}$$
$$(\sqrt{w + 2})^2 = (4 - \sqrt{w + 6})^2$$
$$w + 2 = 16 - 8\sqrt{w + 6} + w + 6$$
$$w + 2 = 22 + w - 8\sqrt{w + 6}$$
$$-20 = -8\sqrt{w + 6}$$
$$5 = 2\sqrt{w + 6}$$
$$5^2 = (2\sqrt{w + 6})^2$$
$$25 = 4(w + 6)$$
$$25 = 4w + 24$$
$$1 = 4w$$
$$\frac{1}{4} = w$$

4.
$$\frac{y - 2}{y + 1} = \frac{y}{y - 3}$$
$$(y - 2)(y - 3) = y(y + 1)$$
$$y^2 - 5y + 6 = y^2 + y$$
$$-5y + 6 = y$$
$$6 = 6y$$
$$1 = y$$

5.
$$\frac{x}{x + 4} = \frac{4}{x - 4} + \frac{x^2 + 16}{x^2 - 16}$$

$$LCD = (x + 4)(x - 4)$$

$$x(x - 4) = 4(x + 4) + x^2 + 16$$
$$x^2 - 4x = 4x + 16 + x^2 + 16$$
$$x^2 - 4x = 4x + 32 + x^2$$
$$-4x = 4x + 32$$
$$-8x = 32$$
$$x = \frac{32}{-8} = -4$$

If -4 is substituted for x in the original problem, two denominators are 0. Thus there is no solution.

6.(a)
$$|1 - 2y| = 3$$
$$1 - 2y = 3 \quad \text{or} \quad 1 - 2y = -3$$
$$-2y = 2 \qquad\qquad -2y = -4$$
$$y = -1 \qquad\qquad\; y = 2$$

(b)
$$|a| = |3a + 4|$$
$$a = 3a + 4 \quad \text{or} \quad -a = 3a + 4$$
$$-2a = 4 \qquad\qquad -4a = 4$$
$$a = -2 \qquad\qquad\; a = -1$$

(c) $|5 - 4y| = -10$

Since $|5 - 4y| \geq 0$ for every y, there are no values of y that will make $|5 - 4y| = -10$. There is no solution.

(d) $|7a + 21| = 0$

$\quad 7a + 21 = 0 \quad$ or $\quad 7a + 21 = -0$
$\quad\quad 7a = -21 \quad\quad\quad\quad 7a = -21$
$\quad\quad\quad a = -3 \quad\quad\quad\quad\quad a = -3$

7. $\quad a^2 = \dfrac{3}{b + c}$

$a^2(b + c) = 3$
$a^2 b + a^2 c = 3$
$\quad\quad a^2 c = 3 - a^2 b$
$\quad\quad\quad c = \dfrac{3 - a^2 b}{a^2}$

8. $\quad A = \dfrac{1}{2}(b_1 + b_2)h$

$\quad 2A = (b_1 + b_2)h$
$\quad 2A = b_1 h + b_2 h$
$2A - b_1 h = b_2 h$
$\dfrac{2A - b_1 h}{h} = b_2$

EXERCISES A

CHAPTER 2 EQUATIONS, INEQUALITIES, AND PROBLEM SOLVING

SECTION 2.1 Linear and Absolute Value Equations

1. $3y + 5 = 7 - y$
   $4y + 5 = 7$
   $4y = 2$
   $y = \dfrac{2}{4} = \dfrac{1}{2}$

2. $1.8x - 4.1 = 6.5 + 0.2x$
   $10(1.8x - 4.1) = 10(6.5 + 0.2x)$
   $18x - 41 = 65 + 2x$
   $16x - 41 = 65$
   $16x = 106$
   $x = \dfrac{106}{16} = 6.625$

3. $\dfrac{z}{2} + \dfrac{3}{8} - \dfrac{z}{4} = \dfrac{5}{4}$
   $8\left(\dfrac{z}{2} + \dfrac{3}{8} - \dfrac{z}{4}\right) = 8\left(\dfrac{5}{4}\right)$
   $4z + 3 - 2z = 10$
   $2z + 3 = 10$
   $2z = 7$
   $z = \dfrac{7}{2}$

4. $4(y + 2) = 7(3 - y)$
   $4y + 8 = 21 - 7y$
   $11y + 8 = 21$
   $11y = 13$
   $y = \dfrac{13}{11}$

5. $5 + 7x = 7x + 4$
   $5 + 7x - 7x = 7x - 7x + 4$
   $5 = 4$

   This contradiction implies that there is no solution.

6. $5 + 7z = 7z + 5$
   $5 + 7z - 7z = 7z - 7z + 5$
   $5 = 5$

   This identity implies that every real number is a solution.

7. $3y - (y - 2) = 8$
   $3y - y + 2 = 8$
   $2y + 2 = 8$
   $2y = 6$
   $y = \dfrac{6}{2} = 3$

8. $2(3x - 2) - 2(4x + 3) = 4$
   $6x - 4 - 8x - 6 = 4$
   $-2x - 10 = 4$
   $-2x = 14$
   $x = \dfrac{14}{-2} = -7$

9. $z^2 + 2 = (z - 1)(z + 1) + z$
   $z^2 + 2 = z^2 - 1 + z$
   $2 = -1 + z$
   $3 = z$

10. $2[z - (1 - 3z)] = 2(3 + 2z)$
    $2[z - 1 + 3z] = 6 + 4z$
    $2[4z - 1] = 6 + 4z$
    $8z - 2 = 6 + 4z$
    $4z - 2 = 6$
    $4z = 8$
    $z = 2$

11. $\sqrt{2z + 5} - 3\sqrt{z - 1} = 0$
    $\sqrt{2z + 5} = 3\sqrt{z - 1}$
    $\left(\sqrt{2z + 5}\right)^2 = \left(3\sqrt{z - 1}\right)^2$
    $2z + 5 = 9(z - 1)$
    $2z + 5 = 9z - 9$
    $-7z + 5 = -9$
    $-7z = -14$
    $z = 2$

    2 does check in the original equation.

12. $\sqrt{x^2 + 8} - x - 4 = 0$
    $\sqrt{x^2 + 8} = x + 4$
    $\left(\sqrt{x^2 + 8}\right)^2 = (x + 4)^2$
    $x^2 + 8 = x^2 + 8x + 16$
    $8 = 8x + 16$
    $-8 = 8x$
    $-1 = x$

    -1 does check in the original equation.

13. $\sqrt{z - 4} + \sqrt{z - 8} = -2$

    Since $\sqrt{z - 4} \geq 0$ and $\sqrt{z - 8} \geq 0$, their sum cannot be -2 for any z. There is no solution.

14. $\sqrt{2x+10} - \sqrt{2x-5} = 3$
$\sqrt{2x+10} = \sqrt{2x-5} + 3$
$(\sqrt{2x+10})^2 = (\sqrt{2x-5}+3)^2$
$2x+10 = (\sqrt{2x-5})^2 + 6\sqrt{2x-5} + 3^2$
$2x+10 = 2x-5 + 6\sqrt{2x-5} + 9$
$2x+10 = 2x+4 + 6\sqrt{2x-5}$
$10 = 4 + 6\sqrt{2x-5}$
$6 = 6\sqrt{2x-5}$
$1 = \sqrt{2x-5}$
$(1)^2 = (\sqrt{2x-5})^2$
$1 = 2x-5$
$6 = 2x$
$3 = x$

Since 3 does check in the original equation, it is a solution.

15. $\dfrac{z-2}{z+1} = \dfrac{z}{z-3}$
$(z-2)(z-3) = z(z+1)$
$z^2 - 5z + 6 = z^2 + z$
$-5z + 6 = z$
$6 = 6z$
$1 = z$

1 does check in the original equation.

16. $\dfrac{2x}{x-3} - \dfrac{3}{x+4} = 2$
$(x-3)(x+4)\left[\dfrac{2x}{x-3} - \dfrac{3}{x+4}\right] = 2(x-3)(x+4)$
$2x(x+4) - 3(x-3) = 2(x^2 + x - 12)$
$2x^2 + 8x - 3x + 9 = 2x^2 + 2x - 24$
$2x^2 + 5x + 9 = 2x^2 + 2x - 24$
$5x + 9 = 2x - 24$
$3x + 9 = -24$
$3x = -33$
$x = -11$

-11 does check.

17. $\dfrac{3}{z+5} - \dfrac{2}{z-2} = \dfrac{4}{z^2+3z-10}$
$\dfrac{3}{z+5} - \dfrac{2}{z-2} = \dfrac{4}{(z+5)(z-2)}$
$3(z-2) - 2(z+5) = 4$
$3z - 6 - 2z - 10 = 4$
$z - 16 = 4$
$z = 20$

20 does check.

18. $\dfrac{1}{z^2-z-2} - \dfrac{3}{z^2-2z-3} = \dfrac{1}{z^2-5z+6}$
$\dfrac{1}{(z-2)(z+1)} - \dfrac{3}{(z-3)(z+1)} = \dfrac{1}{(z-2)(z-3)}$

Multiply by the LCD $= (z+1)(z-2)(z-3)$ and simplify.

$(z-3) - 3(z-2) = z+1$
$z - 3 - 3z + 6 = z + 1$
$-2z + 3 = z + 1$
$-3z + 3 = 1$
$-3z = -2$
$z = \dfrac{2}{3}$

This does check in the original equation.

19. $\sqrt[3]{a+1} = \sqrt[3]{2a+7}$
$(\sqrt[3]{a+1})^3 = (\sqrt[3]{2a+7})^3$
$a + 1 = 2a + 7$
$-a + 1 = 7$
$-a = 6$
$a = -6$

-6 does check and is a solution.

20. $\sqrt[3]{3a+2} + 4 = 6$
$\sqrt[3]{3a+2} = 2$
$(\sqrt[3]{3a+2})^3 = 2^3$
$3a + 2 = 8$
$3a = 6$
$a = 2$

2 does check.

21. $\sqrt[4]{3x} - \sqrt[4]{4-x} = 0$
$\sqrt[4]{3x} = \sqrt[4]{4-x}$
$(\sqrt[4]{3x})^4 = (\sqrt[4]{4-x})^4$
$3x = 4 - x$
$4x = 4$
$x = 1$

This does check in the original equation.

22. $|x+1| = 3$
$x + 1 = 3$ or $x + 1 = -3$
$x = 2$ $x = -4$

23. $|x-1| = -2$

Since $|x-1| \geq 0$ for all x, $|x-1|$ cannot equal -2, and there is no solution.

24. $|y+2| = 0$
$y + 2 = 0$ or $y + 2 = -0$
$y = -2$ $y = -2$

The only solution is -2.

EXERCISES A

25. $|z| = z + 1$

    If $z \geq 0$, then $z = z + 1$ or $0 = 1$. This contradiction implies that there is no solution for $z \geq 0$. If $z < 0$, then

    $-z = z + 1$
    $-2z = 1$
    $z = -\dfrac{1}{2}$.

26. $|y - 1| = y - 1$ implies that $y - 1 \geq 0$ or $y \geq 1$. Thus any $y \geq 1$ is a solution.

27. $|7 - 10x| = 7$

    $7 - 10x = 7$ or $7 - 10x = -7$
    $-10x = 0$ $\qquad -10x = -14$
    $x = 0$ $\qquad\qquad x = \dfrac{7}{5}$

28. $3a + 2b = 5c$
    $3a = 5c - 2b$
    $a = \dfrac{5c - 2b}{3}$

29. $3uvw = p$
    $\dfrac{3uvw}{3uw} = \dfrac{p}{3uw}$
    $v = \dfrac{p}{3uw}$

30. $\sqrt{\dfrac{x}{y}} = z$
    $\left(\sqrt{\dfrac{x}{y}}\right)^2 = z^2$
    $\dfrac{x}{y} = z^2$
    $x = yz^2$
    $\dfrac{x}{z^2} = y$

31. $xy = w - y$
    $xy + y = w$
    $y(x + 1) = w$
    $y = \dfrac{w}{x + 1}$

32. $\sqrt{a + b} + \sqrt{b} = \sqrt{c}$
    $\sqrt{a + b} = \sqrt{c} - \sqrt{b}$
    $(\sqrt{a + b})^2 = (\sqrt{c} - \sqrt{b})^2$
    $a + b = c - 2\sqrt{cb} + b$
    $a = c - 2\sqrt{cb}$

33. $1 = \dfrac{1}{x} + \dfrac{1}{y}$
    $xy(1) = xy\left(\dfrac{1}{x} + \dfrac{1}{y}\right)$
    $xy = y + x$
    $xy - x = y$
    $x(y - 1) = y$
    $x = \dfrac{y}{y - 1}$

34. $A = \dfrac{1}{2}(b_1 + b_2)h$
    $2A = (b_1 + b_2)h$
    $\dfrac{2A}{b_1 + b_2} = h$

35. $A = P(1 + rt)$
    $A = P + Prt$
    $A - P = Prt$
    $\dfrac{A - P}{Pt} = r$

36. $\dfrac{1}{f} = \dfrac{1}{d_1} + \dfrac{1}{d_2}$
    $\dfrac{fd_1d_2}{f} = \dfrac{fd_1d_2}{d_1} + \dfrac{fd_1d_2}{d_2}$
    $d_1d_2 = fd_2 + fd_1$
    $d_1d_2 = f(d_2 + d_1)$
    $\dfrac{d_1d_2}{d_2 + d_1} = f$

37. $A = 2\pi rh + 2\pi r^2$
    $A - 2\pi r^2 = 2\pi rh$
    $\dfrac{A - 2\pi r^2}{2\pi r} = h$

38. $\dfrac{3a + 5}{6} = \dfrac{4 + 3a}{5}$
    $5(3a + 5) = 6(4 + 3a)$
    $15a + 25 = 24 + 18a$
    $-3a + 25 = 24$
    $-3a = -1$
    $a = \dfrac{1}{3}$

39. $\dfrac{3a+5}{6} + \dfrac{4+3a}{5}$
$= \dfrac{5(3a+5)}{6 \cdot 5} + \dfrac{6(4+3a)}{6 \cdot 5}$
$= \dfrac{5(3a+5) + 6(4+3a)}{30}$
$= \dfrac{15a + 25 + 24 + 18a}{30}$
$= \dfrac{33a + 49}{30}$

40. $\dfrac{2}{x+1} - \dfrac{3}{x} = 0$
$x(x+1)\left[\dfrac{2}{x+1} - \dfrac{3}{x}\right] = 0 \cdot x(x+1)$
$2x - 3(x+1) = 0$
$2x - 3x - 3 = 0$
$-x - 3 = 0$
$-x = 3$
$x = -3$

41. $\dfrac{2}{x+1} - \dfrac{3}{x}$
$= \dfrac{2x}{x(x+1)} - \dfrac{3(x+1)}{x(x+1)}$
$= \dfrac{2x - 3(x+1)}{x(x+1)}$
$= \dfrac{2x - 3x - 3}{x(x+1)}$
$= \dfrac{-x - 3}{x(x+1)}$

42. Substitute 9 for a in $\sqrt{a} + 3 = 0$.

$\sqrt{a} + 3 = \sqrt{9} + 3 = 3 + 3 = 6 \neq 0$

This contradiction implies that $a = 9$ is not equivalent to $\sqrt{a} + 3 = 0$.

43. The equation $x = 5$ is not equivalent to

$\dfrac{1}{x-5} + \dfrac{1}{x} = \dfrac{5}{x^2 - 5x}$

since if 5 is substituted for x, two of the denominators are 0.

44. $(b^{-4})^{-3} = b^{(-4)(-3)} = b^{12}$

45. $(3y)^{-1} = \dfrac{1}{3y}$

46. $3y^{-1} = \dfrac{3}{y}$

47. $-3y^{-1} = \dfrac{-3}{y}$

48. $(6u - v)(4u + v)$
$= (6u)(4u) + (6u)(v) - (4u)(v) - (v)(v)$
$= 24u^2 + 6uv - 4uv - v^2$
$= 24u^2 + 2uv - v^2$

49. $\dfrac{2x^2 + x - 15}{x^2 - x - 12} \div \dfrac{2x^2 + 3x - 20}{16x - x^3}$
$= \dfrac{2x^2 + x - 15}{x^2 - x - 12} \cdot \dfrac{16x - x^3}{2x^2 + 3x - 20}$
$= \dfrac{(2x - 5)(x + 3)(-x)(x - 4)(x + 4)}{(x - 4)(x + 3)(2x - 5)(x + 4)}$
$= -x$

50. $\left(\dfrac{125a^5 b^{-2}}{4a^{-2}b^{-6}}\right)^{\frac{1}{2}}$
$= \left(\dfrac{5^3 a^{5-(-2)} b^{-2-(-6)}}{2^2}\right)^{\frac{1}{2}}$
$= \left(\dfrac{5^3 a^7 b^4}{2^2}\right)^{\frac{1}{2}}$
$= \dfrac{(5^3)^{\frac{1}{2}} (a^7)^{\frac{1}{2}} (b^4)^{\frac{1}{2}}}{(2^2)^{\frac{1}{2}}}$
$= \dfrac{5^{\frac{3}{2}} a^{\frac{7}{2}} b^2}{2}$
$= \dfrac{5 \cdot 5^{\frac{1}{2}} a^3 a^{\frac{1}{2}} b^2}{2}$
$= \dfrac{5a^3 b^2 \sqrt{5a}}{2}$

51. $\sqrt[4]{w} \, \sqrt[5]{w^8} = w^{\frac{1}{4}} w^{\frac{8}{5}} = w^{\frac{1}{4} + \frac{8}{5}}$
$= w^{\frac{5}{20} + \frac{32}{20}} = w^{\frac{37}{20}} = w \sqrt[20]{w^{17}}$

52. $\sqrt{\sqrt[3]{x^2}} = \left[\sqrt[3]{x^2}\right]^{\frac{1}{2}} = \left[(x^2)^{\frac{1}{3}}\right]^{\frac{1}{2}}$
$= (x^2)^{\frac{1}{6}} = x^{\frac{1}{3}} = \sqrt[3]{x}$

PRACTICE EXERCISES — SECTION 2.2

CHAPTER 2 EQUATIONS, INEQUALITIES, AND PROBLEM SOLVING

SECTION 2.2 Problem Solving and Appilcations of Linear Equations

1. *Analysis:* A ratio is the quotient of numbers. Thus, we must divide 3 more than the number by 5 less than the number to obtain $-13/3$.

   Tabulation: Let x = the number
   $x + 3$ = 3 more than the number
   $x - 5$ = 5 less than the number

   Translation:
   $$\frac{x+3}{x-5} = -\frac{13}{3}$$
   $$3(x+3) = -13(x-5)$$
   $$3x + 9 = -13x + 65$$
   $$16x + 9 = 65$$
   $$16x = 56$$
   $$x = \frac{56}{16} = \frac{7}{2}$$

   Approximation: If 3 is added to $7/2$, we obtain a positive fraction and if 5 is subtracted from $7/2$, we obtain a negative fraction. The ratio of these fractions should be about $-13/3$.

   Check:
   $$\frac{\frac{7}{2}+3}{\frac{7}{2}-5} = \frac{\frac{7}{2}+\frac{6}{2}}{\frac{7}{2}-\frac{10}{2}} = \frac{\frac{13}{2}}{-\frac{3}{2}} = -\frac{13}{3}$$

2. *Analysis:* We will use the same equation

   commission = (comission rate)(total sales)

   as in Example 2 but with the commission rate unknown.

   Tabulation: Let x = her commission rate

   Translation: $143{,}500.00x = 9327.50$
   $$x = \frac{9327.50}{143{,}500.00}$$
   $$= 0.065$$

   The commission rate is 6.5%.

   Approximation: Since 6.5% is a reasonable commission rate, we can proceed to the check.

   Check: 6.5% of $143,500.00 is $9327.50. Thus, Nancy's commission rate is 6.5%.

3. *Analysis:* total price = retail price + sales tax
   sales tax = (0.055)(retail price)

   Tabulation: Let x = retail price (price before tax)
   $0.055x$ = sales tax
   $x + 0.055x$ = total price

   Translation: $x + 0.055x = 400.90$
   $$1.055x = 400.90$$
   $$x = \frac{400.90}{1.055} = 380.00$$

   The price before tax was $380.00.

   Approximation: Since this is about $20 less than the total price, it seems like a reasonable answer.

   Check: $380.00 + (0.055)(380.00) = 380.00 + 20.90$
   $= 400.90$

4. *Analysis:* For a rectangle

   perimeter = 2(width) + 2(length).

   Tabulation: Let w = width of the garden
   $5w - 2$ = length of the garden

   Translation:
   $$2w + 2l = 104$$
   $$2w + 2(5w - 2) = 104$$
   $$2w + 10w - 4 = 104$$
   $$12w - 4 = 104$$
   $$12w = 108$$
   $$w = \frac{108}{12} = 9$$
   $$l = 5w - 2$$
   $$= 5(9) - 2$$
   $$= 45 - 2 = 43$$

   The garden is 43 yd by 9 yd.

   Approximation: Since 43 is about 5 times 9 and the sum of 43 and 9 is half the perimeter, the answers seem reasonable.

   Check: $5(9) - 2 = 43$ and $2(9) + 2(43) = 104$. Thus the length is 43 yd and the width is 9 yd.

5. *Analysis:* This is a work problem and we must use the work equation.

   Tabulation: We have the following:
   Let t = minutes for Jane working alone,

   $\dfrac{1}{15}$ = amount sorted together in 1 minute,

   $\dfrac{1}{21}$ = amount sorted by Mary in 1 minute,

   $\dfrac{1}{t}$ = amount sorted by Jane is 1 minute.

   Translation: We write the work equation and note that it has LCD = 105t.

   $$\dfrac{1}{21} + \dfrac{1}{t} = \dfrac{1}{15}$$
   $$\dfrac{105t}{21} + \dfrac{105t}{t} = \dfrac{105t}{15}$$
   $$5t + 105 = 7t$$
   $$105 = 2t$$
   $$\dfrac{105}{2} = t$$

   It would take Jane $52\dfrac{1}{2}$ minutes working alone.

   Approximation: Jane's time is longer than the time together and does seem reasonable since Mary's time is close to the time working together.

   Check: $\dfrac{1}{21} + \dfrac{1}{\frac{105}{2}} = \dfrac{1}{21} + \dfrac{2}{105}$
   $$= \dfrac{5}{105} + \dfrac{2}{105}$$
   $$= \dfrac{7}{105} = \dfrac{1}{15}$$

6. *Analysis:* We use the formula $d = rt$.

   Tabulation: Let d = distance between boats at 3:00 P.M.,
   r = rate of slower boat,
   $r + 5$ = rate of faster boat,
   7 = hours traveled.

   Translation: $d = 7(r + 5) - 7r$
   $= 7r + 35 - 7r = 35$

   The boats are 35 miles apart at 3:00 P.M.

   Approximation: This number seems reasonable enough for us to check.

   Check: Note that the boats travel for 7 hours at speeds that differ by 5 mph and that (7)(5) = 35.

7. Let r = speed of the wind
   $350 + r$ = speed of the plane with the wind
   $350 - r$ = speed of the plane against the wind

   $$\dfrac{2400}{350+r} = \dfrac{1800}{350-r}$$
   $$(350+r)(350-r)\dfrac{2400}{350+r} = \dfrac{1800}{350-r}(350+r)(350-r)$$
   $$2400(350-r) = 1800(350+r)$$
   $$(2400)(350) - 2400r = (1800)(350) + 1800r$$
   $$(2400)(350) - (1800)(350) = 2400r + 1800r$$
   $$\dfrac{210{,}000}{4200} = r$$
   $$50 = r$$

   The speed of the wind is 50 mph.

8. Let x = number of defective drives in 900.

   $$\dfrac{x}{900} = \dfrac{2}{100}$$
   $$(100)(900)\dfrac{x}{900} = \dfrac{2}{100}(100)(900)$$
   $$100x = (2)(900)$$
   $$x = \dfrac{1800}{100} = 18$$

   Expect 18 defective drives in a shipment of 900.

9. Let x = number of trout in the lake.

   $$\dfrac{200}{x} = \dfrac{8}{100}$$
   $$(200)(100) = 8x$$
   $$\dfrac{(200)(100)}{8} = x$$
   $$2500 = x$$

   There are approximately 2500 trout in the lake.

EXERCISES A   SECTION 2.2   51

CHAPTER 2 EQUATIONS, INEQUALITIES, AND PROBLEM SOLVING

SECTION 2.2 Problem Solving and Application of Linear Equations

1. *Analysis:* Five times a number minus 7 must be set equal to two times the number plus 10.

   Tabulation: Let $\quad x =$ the number,
   $\quad 5x - 7 = 7$ less than 5 times the number
   $\quad 2x + 10 = 10$ more than 2 times the number

   Translation: $5x - 7 = 2x + 10$
   $\quad\quad\quad\quad 5x = 2x + 17$
   $\quad\quad\quad\quad 3x = 17$
   $\quad\quad\quad\quad x = \dfrac{17}{3}$

   Approximation and Check: This does seem reasonable and does check.

2. *Analysis:* We add three consecutive odd integers and set their sum equal to 4 times the least minus 29.

   Tabulation: Let $\quad n =$ the first integer,
   $\quad n + 2 =$ the second integer,
   $\quad n + 4 =$ the third integer.

   Translation: $n + (n + 2) + (n + 4) = 4n - 29$
   $\quad\quad\quad\quad n + n + 2 + n + 4 = 4n - 29$
   $\quad\quad\quad\quad 3n + 6 = 4n - 29$
   $\quad\quad\quad\quad 3n = 4n - 35$
   $\quad\quad\quad\quad -n = -35$
   $\quad\quad\quad\quad n = 35$
   $\quad\quad\quad\quad n + 2 = 37$
   $\quad\quad\quad\quad n + 4 = 39$

   Approximation: 35, 37, and 39 are consecutive odd integers that seem to satisfy the other conditions.

   Check: $\quad 35 + 37 + 39 = 111$
   $\quad\quad\quad 4(35) - 29 = 140 - 29 = 111$

3. *Analysis:* To average four scores, add them and divide by 4. This should equal 80.

   Tabulation: Let $x =$ the fourth score.

   Translation: $\dfrac{61 + 89 + 86 + x}{4} = 80$
   $\quad\quad\quad (4)\left[\dfrac{236 + x}{4}\right] = (80)(4)$
   $\quad\quad\quad 236 + x = 320$
   $\quad\quad\quad x = 84$

   Approximation and Check: 84 seems reasonable and
   $\dfrac{61 + 89 + 86 + 84}{4} = \dfrac{320}{4} = 80.$

4. Let $\quad x =$ the first piece,
   $\quad 2x =$ the second piece,
   $\quad 2x + 3 =$ the third piece.

   $x + 2x + (2x + 3) = 23$
   $x + 2x + 2x + 3 = 23$
   $5x + 3 = 23$
   $5x = 20$
   $x = 4$
   $2x = 8$
   $2x + 3 = 11$

   The pieces have lengths of 4 ft, 8 ft, and 11 ft.

5. Let $\quad x =$ the tax.
   $\quad x = (0.04)(52.50) = 2.10$
   The tax is $2.10.

6. Let $\quad x =$ Maria's salary before the raise,
   $x + 0.12x =$ her salary after the raise.

   $x + (0.12)x = 25{,}760$
   $1.12x = 25{,}760$
   $x = \dfrac{25{,}760}{1.12} = 23{,}000$

   Her salary before the raise was $23,000.

7. Let $\quad x =$ the second angle,
   $3x + 5 =$ the first angle,
   $x - 10 =$ the third angle.
   $(3x + 5) + x + (x - 10) = 180$
   $3x + 5 + x + x - 10 = 180$
   $5x - 5 = 180$
   $5x = 185$
   $x = 37$
   $3x + 5 = 116$
   $x - 10 = 27$
   The angles are 116°, 37°, and 27°.

8. Let $\quad x =$ the width of the field,
   $x + 40 =$ the length of the field.
   $2x + 2(x + 40) = 1760$
   $2x + 2x + 80 = 1760$
   $4x + 80 = 1760$
   $4x = 1680$
   $x = 420$
   $x + 40 = 460$
   The field is 420 m wide and 460 m long.

9. Let t = number of days required working together.

$$\frac{1}{8} + \frac{1}{9} = \frac{1}{t}$$
$$72t\left[\frac{1}{8} + \frac{1}{9}\right] = 72t\left[\frac{1}{t}\right]$$
$$9t + 8t = 72$$
$$17t = 72$$
$$t = \frac{72}{17} \approx 4.2$$

It will take about 4.2 days if they work together.

10. Let t = number of hours required working together.

$$\frac{1}{25} + \frac{1}{20} = \frac{1}{t}$$
$$100t\left[\frac{1}{25} + \frac{1}{20}\right] = 100t\left[\frac{1}{t}\right]$$
$$4t + 5t = 100$$
$$9t = 100$$
$$t = \frac{100}{9} \approx 11.1$$

It will take about 11.1 hours if they work together.

11. Let t = minutes for student trainer to do the job working alone.

$$\frac{1}{30} + \frac{1}{t} = \frac{1}{20}$$
$$60t\left[\frac{1}{30} + \frac{1}{t}\right] = 60t\left[\frac{1}{20}\right]$$
$$2t + 60 = 3t$$
$$60 = t$$

It will take the student trainer 60 minutes to do the job if he works alone.

12. Let t = hours to fill pool with both pipes open.

$$\frac{1}{2} - \frac{1}{10} = \frac{1}{t}$$
$$10t\left[\frac{1}{2} - \frac{1}{10}\right] = 10t\left[\frac{1}{t}\right]$$
$$5t - t = 10$$
$$4t = 10$$
$$t = \frac{10}{4} = 2.5$$

It will take 2.5 hours to fill the empty pool with both pipes open.

13. Let r = the rate of the slower plane,
$r + 50$ = the rate of the faster plane.

$$3r + 3(r + 50) = 3030$$
$$3r + 3r + 150 = 3030$$
$$6r + 150 = 3030$$
$$6r = 2880$$
$$r = 480$$
$$r + 50 = 530$$

The slower plane is traveling 480 mph and the faster 530 mph.

14. Let r = the speed of the wind,
$135 + r$ = the speed of the plane with the wind,
$135 - r$ = the speed of the plane against the wind.

$$\frac{480}{135 + r} = \frac{330}{135 - r}$$
$$480(135 - r) = 330(135 + r)$$
$$64{,}800 - 480r = 44{,}550 + 330r$$
$$64{,}800 = 44{,}550 + 810r$$
$$20{,}250 = 810r$$
$$25 = r$$

The speed of the wind is 25 mph.

15. Let r = rate of the slower press,
$r + 30$ = rate of the faster press.

$$210r + 210(r + 30) = 31{,}500$$
$$210r + 210r + 6300 = 31{,}500$$
$$420r + 6300 = 31{,}500$$
$$420r = 25{,}200$$
$$r = 60$$
$$r + 30 = 90$$

The slower press produces 60 fliers per minute and the faster produces 90 fliers per minute.

16. Let x = liters of water that must be used.

$$\frac{21}{x} = \frac{7}{3}$$
$$63 = 7x$$
$$9 = x$$

9 liters of water must be used.

17. Let x = the height of the tower.

$$\frac{5}{12} = \frac{x}{252}$$
$$\frac{(5)(252)}{12} = x$$
$$105 = x$$

The tower is 105 feet high.

EXERCISES A   SECTION 2.2   53

18. Let $x = $ the length of the tangent.

$$\frac{49}{x} = \frac{x}{4}$$
$$(49)(4) = x^2$$
$$\pm\sqrt{(49)(4)} = x$$
$$(7)(2) = x$$
$$14 = x$$

We used only the positive square root since a length is positive. The tangent is 14 cm long.

19. $\sqrt{4a^2 + 8} - 2a - 4 = 0$
$$\sqrt{4a^2 + 8} = 2a + 4$$
$$\left(\sqrt{4a^2 + 8}\right)^2 = (2a + 4)^2$$
$$4a^2 + 8 = 4a^2 + 8a + 8a + 16$$
$$4a^2 + 8 = 4a^2 + 16a + 16$$
$$8 = 16a + 16$$
$$-8 = 16a$$
$$-\frac{1}{2} = a$$

This does check in the original equation.

20. $\frac{7}{x-6} + \frac{5}{x-8} = \frac{2}{x^2 - 14x + 48}$

$\frac{7}{x-6} + \frac{5}{x-8} = \frac{2}{(x-6)(x-8)}$

$7(x - 8) + 5(x - 6) = 2$
$7x - 56 + 5x - 30 = 2$
$12x - 86 = 2$
$12x = 88$
$x = \frac{88}{12} = \frac{22}{3}$

This does check in the original equation.

21. $|6 - 5z| = 1$ translates to the following:

$6 - 5z = 1$ or $6 - 5z = -1$
$-5z = -5 \qquad\quad -5z = -7$
$z = 1 \qquad\qquad z = \frac{-7}{-5} = \frac{7}{5}$

22. $S = \frac{a_1}{1 - r}$
$S(1 - r) = a_1$
$S - Sr = a_1$
$-Sr = a_1 - S$
$r = \frac{a_1 - S}{-S} = \frac{S - a_1}{S}$

23. $(4 + 2\sqrt{2})(-3 - 5\sqrt{2})$
$= (4)(-3) + (4)(-5\sqrt{2}) + (2\sqrt{2})(-3) + (2\sqrt{2})(-5\sqrt{2})$
$= -12 - 20\sqrt{2} - 6\sqrt{2} - 20$
$= -32 - 26\sqrt{2}$

24. $(2 - 3\sqrt{2})(2 - 3\sqrt{2})$
$= 4 - 6\sqrt{2} - 6\sqrt{2} + 18$
$= 22 - 12\sqrt{2}$

25. $\frac{1}{5 + 4\sqrt{2}} = \frac{1}{5 + 4\sqrt{2}} \cdot \frac{5 - 4\sqrt{2}}{5 - 4\sqrt{2}}$
$= \frac{5 - 4\sqrt{2}}{(5)^2 - (4\sqrt{2})^2}$
$= \frac{5 - 4\sqrt{2}}{25 - 32}$
$= \frac{5 - 4\sqrt{2}}{-7} = \frac{4\sqrt{2} - 5}{7}$

CHAPTER 2 EQUATIONS, INEQUALITIES, AND PROBLEM SOLVING

SECTION 2.3 Complex Numbers

1.(a) $\sqrt{-121} = \sqrt{121(-1)} = \sqrt{121}\sqrt{-1} = 11i$

(b) $\sqrt{-2}\sqrt{-6} = \sqrt{2(-1)}\sqrt{6(-1)}$
$= \sqrt{2}\sqrt{-1}\sqrt{6}\sqrt{-1}$
$= \sqrt{2}\cdot i\cdot\sqrt{6}\cdot i$
$= \sqrt{12}\,i^2$
$= \sqrt{4\cdot 3}(-1) = -2\sqrt{3}$

(c) $\dfrac{\sqrt{-50}}{\sqrt{-10}} = \dfrac{\sqrt{50(-1)}}{\sqrt{10(-1)}} = \dfrac{\sqrt{50}\sqrt{-1}}{\sqrt{10}\sqrt{-1}}$
$= \dfrac{\sqrt{50}\,i}{\sqrt{10}\,i} = \sqrt{\dfrac{50}{10}} = \sqrt{5}$

(d) $\sqrt{-25} - \sqrt{-49} = \sqrt{25}\,i - \sqrt{49}\,i = 5i - 7i = -2i$

2.(a) $i^{63} = (i^4)^{15}\,i^3 = (1)^{15}i^3 = (1)(-i) = -i$

(b) $i^{128} = (i^4)^{32} = (1)^{32} = 1$

3.(a) If $x + yi = 3 - 7i$, then $x = 3$ and $y = -7$.

(b) If $p + qi = -3$, then $p = -3$ and $q = 0$.

(c) If $m + ni = \sqrt{5}\,i$, then $m = 0$ and $n = \sqrt{5}$.

(d) If $3x - yi = x + 1 + 2yi$, then

$3x = x + 1$
$2x = 1$
$x = \dfrac{1}{2}$.

Also,

$-y = 2y$
$-3y = 0$
$y = 0$.

4.(a) $(3 - 8i) + (-2 + i) = (3 - 2) + (-8 + 1)i = 1 - 7i$

(b) $(3 + i\sqrt{2}) - (1 - i\sqrt{2}) = (3 - 1) + (\sqrt{2} + \sqrt{2})$
$= 2 + 2i\sqrt{2}$

5.(a) $(2 - 5i)(1 + 3i) = 2 + 6i - 5i - 15i^2$
$= 2 - 15(-1) + 6i - 5i$
$= (2 + 15) + (6 - 5)i$
$= 17 + i$

(b) $(4 - 3i)(4 + 3i) = 16 + 12i - 12i - 9i^2$
$= 16 - 9(-1)$
$= 16 + 9 = 25$

6. $\dfrac{3 + 7i}{2 - 5i} = \dfrac{(3 + 7i)(2 + 5i)}{(2 - 5i)(2 + 5i)}$
$= \dfrac{6 + 15i + 14i + 35i^2}{4 + 10i - 10i - 25i^2}$
$= \dfrac{6 + 29i - 35}{4 + 25}$
$= \dfrac{-29 + 29i}{29} = -1 + i$

7. $\dfrac{1}{3 - 2i} = \dfrac{(1)(3 + 2i)}{(3 - 2i)(3 + 2i)}$
$= \dfrac{3 + 2i}{9 + 4} = \dfrac{3 + 2i}{13}$
$= \dfrac{3}{13} + \dfrac{2}{13}i$

8. $(2 - i)x + 3i = 4ix + 5$
$(2 - i)x = 4ix + 5 - 3i$
$(2 - i)x - 4ix = 5 - 3i$
$(2 - 5i)x = 5 - 3i$
$x = \dfrac{5 - 3i}{2 - 5i}$
$= \dfrac{(5 - 3i)(2 + 5i)}{(2 - 5i)(2 + 5i)}$
$= \dfrac{10 + 25i - 6i - 15i^2}{4 - 25i^2}$
$= \dfrac{25 + 19i}{29} = \dfrac{25}{29} + \dfrac{19}{29}i$

EXERCISES A   SECTION 2.3

CHAPTER 2 EQUATIONS, INEQUALITIES, AND PROBLEM SOLVING

SECTION 2.3 Complex Numbers

1. $\sqrt{-16} = \sqrt{16}\sqrt{-1} = 4i$

2. $\sqrt{-24} = \sqrt{4}\sqrt{-1}\sqrt{6} = 2i\sqrt{6}$

3. $\sqrt{-3}\sqrt{-11} = \sqrt{-1}\sqrt{3}\sqrt{-1}\sqrt{11} = i\sqrt{3}\,i\sqrt{11}$
   $= i^2\sqrt{3}\sqrt{11} = -\sqrt{33}$

4. $-\sqrt{-8}\sqrt{-2} = -\sqrt{-1}\sqrt{8}\sqrt{-1}\sqrt{2}$
   $= -i^2\sqrt{16}$
   $= -(-1)(4) = 4$

5. $\dfrac{\sqrt{-45}}{\sqrt{-9}} = \dfrac{\sqrt{(-9)(5)}}{\sqrt{-9}} = \dfrac{\sqrt{-9}\sqrt{5}}{\sqrt{-9}} = \sqrt{5}$

6. $\dfrac{\sqrt{-64}}{\sqrt{-8}} = \dfrac{\sqrt{(-8)(8)}}{\sqrt{-8}} = \dfrac{\sqrt{-8}\sqrt{8}}{\sqrt{-8}} = \sqrt{8} = 2\sqrt{2}$

7. $\dfrac{-\sqrt{-121}}{11} = \dfrac{-\sqrt{(11)^2(-1)}}{11} = \dfrac{-11i}{11} = -i$

8. $\sqrt{-9 \cdot 36} = \sqrt{-45} = \sqrt{(9)(-1)(5)} = 3i\sqrt{5}$

9. $\sqrt{-9} - \sqrt{-36} = 3i - 6i = -3i$

10. If $5x + (3y + 2)i = -i$, then $5x = 0$, which means $x = 0$, and
    $$3y + 2 = -1$$
    $$3y = -3$$
    $$y = -1.$$

11. $x + yi = \sqrt{9} - \sqrt{-9}$
    $x + yi = 3 - 3i$
    Thus, $x = 3$ and $y = -3$.

12. $(4 - 3i) + (-2 - 5i) = 4 - 3i - 2 - 5i$
    $= 4 - 2 - 3i - 5i$
    $= 2 - 8i$

13. $(-1 - 2i) - (5 - 4i) = -1 - 2i - 5 + 4i$
    $= -1 - 5 - 2i + 4i$
    $= -6 + 2i$

14. $(3 - 2i) + 8i = 3 - 2i + 8i$
    $= 3 + 6i$

15. $(1 - 3i)(3 + 5i) = 3 + 5i - 9i - 15i^2$
    $= 3 - 4i - 15(-1)$
    $= 18 - 4i$

16. $(4 - i)(-2 + 5i) = -8 + 20i + 2i - 5i^2$
    $= -8 + 22i - 5(-1)$
    $= -3 + 22i$

17. $2i(3 - 7i) = 6i - 14i^2$
    $= 6i - 14(-1)$
    $= 14 + 6i$

18. $3(5 - 12i) = 15 - 36i$

19. $(4 - 3i)^2 = 16 - 24i + (3i)^2$
    $= 16 - 24i + 9(-1)$
    $= 7 - 24i$

20. $(-1 + i)^3 = (-1 + i)(-1 + i)^2$
    $= (-1 + i)(1 - 2i + i^2)$
    $= (-1 + i)(-2i)$
    $= 2i - 2i^2$
    $= 2i - 2(-1) = 2 + 2i$

21. $i^{15} = i^{12}i^3 = (i^4)^3 i^3 = (1)^3(-i) = -i$

22. $i^{37} = i^{36}i = (i^4)^9 i = (1)^9 i = i$

23. $i^{74} = i^{72}i^2 = (i^4)^{18}i^2 = (1)^{18}(-1) = -1$

24. The conjugate of $a - bi$ is $a + bi$. Thus the conjugate of $3 - 10i$ is $3 + 10i$.

25. The conjugate of $15i = 0 + 15i$ is $0 - 15i = -15i$.

26. The conjugate of $-3 = -3 + 0i$ is $-3 - 0i = -3$.

27. The conjugate of $4 - i\sqrt{7}$ is $4 + i\sqrt{7}$.

28. $\dfrac{4-5i}{1+i} = \dfrac{4-5i}{1+i} \cdot \dfrac{1-i}{1-i}$
$= \dfrac{4 - 4i - 5i + 5i^2}{1 - i^2}$
$= \dfrac{4 - 9i + 5(-1)}{1 - (-1)}$
$= \dfrac{-1 - 9i}{2} = -\dfrac{1}{2} - \dfrac{9}{2}i$

29. $\dfrac{-2+3i}{4+3i} = \dfrac{-2+3i}{4+3i} \cdot \dfrac{4-3i}{4-3i}$
$= \dfrac{-8 + 6i + 12i - 9i^2}{16 - 9i^2}$
$= \dfrac{-8 + 18i - 9(-1)}{16 - 9(-1)}$
$= \dfrac{1 + 18i}{25} = \dfrac{1}{25} + \dfrac{18}{25}i$

30. $\dfrac{1+i}{(1-i)^2} = \dfrac{1+i}{1 - 2i + i^2}$
$= \dfrac{1+i}{1 - 2i + (-1)}$
$= \dfrac{1+i}{-2i} \cdot \dfrac{2i}{2i}$
$= \dfrac{2i + 2i^2}{-4i^2}$
$= \dfrac{2i + 2(-1)}{-4(-1)}$
$= \dfrac{-2 + 2i}{4} = -\dfrac{1}{2} + \dfrac{1}{2}i$

31. $\dfrac{1}{2-i} = \dfrac{1}{2-i} \cdot \dfrac{2+i}{2+i}$
$= \dfrac{2+i}{4 - i^2} = \dfrac{2+i}{4 - (-1)}$
$= \dfrac{2+i}{5} = \dfrac{2}{5} + \dfrac{1}{5}i$

32. $\dfrac{1}{4i} = \dfrac{1}{4i} \cdot \dfrac{-4i}{-4i} = \dfrac{-4i}{16} = -\dfrac{1}{4}i$

33. The reciprocal of $\dfrac{3-i}{1+i}$ is $\dfrac{1+i}{3-i}$.

$\dfrac{1+i}{3-i} = \dfrac{1+i}{3-i} \cdot \dfrac{3+i}{3+i}$
$= \dfrac{3 + i + 3i + i^2}{9 - i^2}$
$= \dfrac{3 + 4i + (-1)}{9 - (-1)} = \dfrac{2 + 4i}{10}$
$= \dfrac{2}{10} + \dfrac{4}{10}i = \dfrac{1}{5} + \dfrac{2}{5}i$

34. $3ix + 2 = 5 - 2ix$
$3ix + 2ix + 2 = 5 - 2ix + 2ix$
$5ix + 2 = 5$
$5ix = 3$
$x = \dfrac{3}{5i} = \dfrac{3}{5} \cdot \dfrac{1}{i} \cdot \dfrac{-i}{-i} = -\dfrac{3}{5}i$

35. $(1 + i)x - 2i = 3ix + 5$
$(1 + i)x - 3ix = 5 + 2i$
$(1 - 2i)x = 5 + 2i$
$x = \dfrac{5 + 2i}{1 - 2i}$
$= \dfrac{5 + 2i}{1 - 2i} \cdot \dfrac{1 + 2i}{1 + 2i}$
$= \dfrac{5 + 10i + 2i + 4i^2}{1 - 4i^2}$
$= \dfrac{1 + 12i}{5} = \dfrac{1}{5} + \dfrac{12}{5}i$

36. (a) $|3 + 4i| = \sqrt{3^2 + 4^2} = \sqrt{25} = 5$

(b) $|1 - i| = \sqrt{(1)^2 + (-1)^2} = \sqrt{1 + 1} = \sqrt{2}$

(c) $|-9i| = \sqrt{0^2 + (-9)^2} = \sqrt{81} = 9$

37. Let x = Henry's score on the other quiz.

$\dfrac{82 + 63 + 92 + x}{4} = 76$
$82 + 63 + 92 + x = (76)(4)$
$237 + x = 304$
$x = 67$

The other score was 67.

38. Let s = the speed of Mary's mother,
$s + 12$ = Mary's speed.

$\dfrac{250}{s} = \dfrac{310}{s + 12}$
$250(s + 12) = 310s$
$250s + 3000 = 310s$
$3000 = 60s$
$50 = s$
$62 = s + 12$

Mary's speed is 62 mph and her mother's speed if 50 mph.

EXERCISES A SECTION 2.3 57

39. Let r = speed of the boat in still water,
$r + 8$ = speed of the boat going downstream,
$r - 8$ = speed of the boat going upstream.

$$\frac{105}{r+8} = \frac{49}{r-8}$$
$$105(r-8) = 49(r+8)$$
$$105r - 840 = 49r + 392$$
$$105r - 49r = 840 + 392$$
$$56r = 1232$$
$$r = 22$$

The speed of the boat in still water is 22 km/hr.

40.
$$6(x + 5) - 2(4x - 1) = 0$$
$$6x + 30 - 8x + 2 = 0$$
$$-2x + 32 = 0$$
$$-2x = -32$$
$$x = 16$$

41. $x^2 - 12x + 36 = x^2 - 2(6)x + 6^2 = (x - 6)^2$

42. $x^2 + \frac{2}{3}x + \frac{1}{9} = x^2 + 2\left(\frac{1}{3}\right)x + \left(\frac{1}{3}\right)^2 = \left(x + \frac{1}{3}\right)^2$

43. $\sqrt{b^2 - 4ac} = \sqrt{(5)^2 - 4(-1)(-4)} = \sqrt{25 - 16} = \sqrt{9} = 3$

CHAPTER 2 EQUATIONS, INEQUALITIES, AND PROBLEM SOLVING

SECTION 2.4 Quadratic Equations

1. (a) $\quad 2y^2 - 8y = 10$
$\quad\quad 2y^2 - 8y - 10 = 0$
$\quad\quad y^2 - 4y - 5 = 0$
$\quad\quad (y-5)(y+1) = 0$
$\quad\quad y - 5 = 0$ or $y + 1 = 0$
$\quad\quad\quad y = 5 \quad\quad\quad y = -1$

(b) $z^2 - 7z = 0$
$\quad z(z-7) = 0$
$\quad z = 0$ or $z - 7 = 0$
$\quad\quad\quad\quad\quad z = 7$

(c) $\quad x + \dfrac{1}{4} = -x^2$
$\quad\quad x^2 + x + \dfrac{1}{4} = 0$
$\quad\quad 4x^2 + 4x + 1 = 0$
$\quad\quad (2x+1)(2x+1) = 0$
$\quad\quad 2x + 1 = 0$ or $2x + 1 = 0$
$\quad\quad 2x = -1 \quad\quad 2x = -1$
$\quad\quad x = -\dfrac{1}{2} \quad\quad x = -\dfrac{1}{2}$

The only solution is -1/2.

2. (a) $3x^2 - 24 = 0$
$\quad 3x^2 = 24$
$\quad x^2 = 8$
$\quad x = \pm\sqrt{8}$
$\quad x = \pm 2\sqrt{2}$

(b) $2y^2 + 18 = 0$
$\quad 2y^2 = -18$
$\quad y^2 = -9$
$\quad y = \pm\sqrt{-9}$
$\quad y = \pm\sqrt{(9)(-1)}$
$\quad y = \pm 3i$

3. $x^2 - 5x - 24 = 0$
$\quad x^2 - 5x = 24$
$\quad x^2 - 5x + \dfrac{25}{4} = 24 + \dfrac{25}{4}$
$\quad \left(x - \dfrac{5}{2}\right)^2 = \dfrac{96}{4} + \dfrac{25}{4}$
$\quad x - \dfrac{5}{2} = \pm\sqrt{\dfrac{121}{4}}$
$\quad x = \dfrac{5}{2} \pm \dfrac{11}{2}$

Thus, the solutions are 8 and −3 which do check by factoring since $x^2 - 5x - 24 = (x-8)(x+3)$.

4. $\quad 3y^2 + y - 1 = 0$
$\quad y^2 + \dfrac{1}{3}y = \dfrac{1}{3}$
$\quad y^2 + \dfrac{1}{3}y + \dfrac{1}{36} = \dfrac{1}{3} + \dfrac{1}{36}$
$\quad \left(y + \dfrac{1}{6}\right)^2 = \dfrac{12}{36} + \dfrac{1}{36}$
$\quad y + \dfrac{1}{6} = \pm\sqrt{\dfrac{13}{36}}$
$\quad y = -\dfrac{1}{6} \pm \dfrac{\sqrt{13}}{6}$
$\quad\quad = \dfrac{-1 \pm \sqrt{13}}{6}$

5. (a) Since $x^2 - 9x + 18 = 0$, $a = 1$, $b = -9$, and $c = 18$.
$\quad x = \dfrac{-b \pm \sqrt{b^2 - 4ac}}{2a}$
$\quad\quad = \dfrac{-(-9) \pm \sqrt{(-9)^2 - 4(1)(18)}}{2(1)}$
$\quad\quad = \dfrac{9 \pm \sqrt{81 - 72}}{2}$
$\quad\quad = \dfrac{9 \pm 3}{2}$

The solutions are 3 and 6.

(b) $\quad x^2 - 1 = \dfrac{5}{3}x$
$\quad 3x^2 - 3 = 5x$
$\quad 3x^2 - 5x - 3 = 0$

$a = 3$, $b = -5$, and $c = -3$.

$\quad x = \dfrac{-b \pm \sqrt{b^2 - 4ac}}{2a}$
$\quad\quad = \dfrac{-(-5) \pm \sqrt{(-5)^2 - 4(3)(-3)}}{2(3)}$
$\quad\quad = \dfrac{5 \pm \sqrt{25 + 36}}{6}$
$\quad\quad = \dfrac{5 \pm \sqrt{61}}{6}$

(c) $a = 2$, $b = 1$, and $c = 1$.
$\quad x = \dfrac{-b \pm \sqrt{b^2 - 4ac}}{2a}$
$\quad\quad = \dfrac{-1 \pm \sqrt{(1)^2 - 4(2)(1)}}{2(2)} = \dfrac{-1 \pm \sqrt{1 - 8}}{4}$
$\quad\quad = \dfrac{-1 \pm \sqrt{-7}}{4} = \dfrac{-1 \pm i\sqrt{7}}{4}$

PRACTICE EXERCISES

6.(a) $a = 5$, $b = -2$, and $c = 3$.

$$b^2 - 4ac = (-2)^2 - 4(5)(3)$$
$$= 4 - 60 = -56 < 0$$

There are two complex solutions.

(b) $a = 3$, $b = -8$, and $c = 1$.

$$b^2 - 4ac = (-8)^2 - 4(3)(1)$$
$$= 64 - 12 = 52 > 0$$

There are two real solutions.

(c) $a = 9$, $b = 12$, and $c = 4$.

$$b^2 - 4ac = (12)^2 - 4(9)(4)$$
$$= 144 - 144 = 0$$

There is one real solution.

7.
$$x^3 + 27 = 0$$
$$(x + 3)(x^2 - 3x + 9) = 0$$
$$x + 3 = 0 \quad \text{or} \quad x^2 - 3x + 9 = 0$$
$$x = -3$$

$$x = \frac{-b \pm \sqrt{b^2 - 4ac}}{2a}$$
$$= \frac{-(-3) \pm \sqrt{(-3)^2 - 4(1)(9)}}{2(1)}$$
$$= \frac{3 \pm \sqrt{9 - 36}}{2}$$
$$= \frac{3 \pm \sqrt{-27}}{2} = \frac{3 \pm 3i\sqrt{3}}{2}$$

8. $a = 2$, $b = -3$, and $c = 6$.

Sum: $-\dfrac{b}{a} = -\dfrac{-3}{2} = \dfrac{3}{2}$

Product: $\dfrac{c}{a} = \dfrac{6}{2} = 3$

9.(a)
$$(x - x_1)(x - x_2) = 0$$
$$(x - (-4))(x - 1) = 0$$
$$(x + 4)(x - 1) = 0$$
$$x^2 + 3x - 4 = 0$$

(b)
$$(x - x_1)(x - x_2) = 0$$
$$(x - (2 - i))(x - (2 + i)) = 0$$
$$x^2 - (2 + i)x - (2 - i)x + (2 - i)(2 + i) = 0$$
$$x^2 - 4x + 5 = 0$$

10. Use the quadratic formula to indicate the solutions.

$$x = \frac{-(-25.2) \pm \sqrt{(-25.2)^2 - 4(0.3)(1.4)}}{2(0.3)}$$

Calculate $\sqrt{(-25.2)^2 - 4(0.3)(1.4)}$ as in Example 10 and store it in your calculator. Add this to $-(-25.2) = 25.2$ and divide by $2(0.3)$. Then subtract from 25.2 and divide by $2(0.3)$. The two solutions are 83.944408 and 0.0555924.

CHAPTER 2 EQUATIONS, INEQUALITIES, AND PROBLEM SOLVING

SECTION 2.4 Quadratic Equations

1. $$x^2 - 3x - 10 = 0$$
$$(x-5)(x+2) = 0$$
$$x - 5 = 0 \text{ or } x + 2 = 0$$
$$x = 5 \qquad x = -2$$

2. $$2y^2 - 10y = 12$$
$$2y^2 - 10y - 12 = 0$$
$$y^2 - 5y - 6 = 0$$
$$(y-6)(y+1) = 0$$
$$y - 6 = 0 \text{ or } y + 1 = 0$$
$$y = 6 \qquad y = -1$$

3. $$y^2 + y = -\frac{1}{4}$$
$$y^2 + y + \frac{1}{4} = 0$$
$$4y^2 + 4y + 1 = 0$$
$$(2y+1)^2 = 0$$
$$2y + 1 = 0$$
$$2y = -1$$
$$y = -\frac{1}{2}$$

4. $$2x^2 + 3x = 0$$
$$x(2x+3) = 0$$
$$x = 0 \text{ or } 2x + 3 = 0$$
$$2x = -3$$
$$x = -\frac{3}{2}$$

5. $$5y(y+2) = 7y$$
$$5y^2 + 10y = 7y$$
$$5y^2 + 3y = 0$$
$$y(5y+3) = 0$$
$$x = 0 \text{ or } 5y + 3 = 0$$
$$5y = -3$$
$$y = -\frac{3}{5}$$

6. $$4(x-1)(x+1) = -3$$
$$4(x^2 - 1) = -3$$
$$4x^2 - 4 = -3$$
$$4x^2 - 1 = 0$$
$$(2x-1)(2x+1) = 0$$
$$2x - 1 = 0 \text{ or } 2x + 1 = 0$$
$$2x = 1 \qquad 2x = -1$$
$$x = \frac{1}{2} \qquad x = -\frac{1}{2}$$

7. $$x^2 - 36 = 0$$
$$x^2 = 36$$
$$x = \pm\sqrt{36}$$
$$x = \pm 6$$

8. $$3y^2 + 75 = 0$$
$$3y^2 = -75$$
$$y^2 = -25$$
$$y = \pm\sqrt{-25}$$
$$y = \pm 5i$$

9. $$(y+1)^2 = 9$$
$$y + 1 = \pm\sqrt{9}$$
$$y + 1 = \pm 3$$
$$y = -1 \pm 3$$
$$y = -4, 2$$

10. We must take one-half the coefficient of $b = -8$ and square it to complete the square.
$$\left(\frac{b}{2}\right)^2 = \left(\frac{-8}{2}\right)^2 = (-4)^2 = 16$$
Thus, $x^2 - 8x + 16 = (x-4)^2$.

11. $$\left(\frac{b}{2}\right)^2 = \left(\frac{-1}{2}\right)^2 = \frac{1}{4}$$
Thus, $x^2 - x + \frac{1}{4} = \left(x - \frac{1}{2}\right)^2$.

12. $$\left(\frac{1}{2} \cdot b\right)^2 = \left(\frac{1}{2} \cdot \frac{1}{3}\right)^2 = \left(\frac{1}{6}\right)^2 = \frac{1}{36}$$
Thus, $y^2 + \frac{1}{3}y + \frac{1}{36} = \left(y + \frac{1}{6}\right)^2$.

13. $$x^2 - 5x - 24 = 0$$
$$x^2 - 5x + \left(\frac{-5}{2}\right)^2 = \left(\frac{-5}{2}\right)^2 + 24$$
$$x^2 - 5x + \frac{25}{4} = \frac{25}{4} + 24$$
$$\left(x - \frac{5}{2}\right)^2 = \frac{25}{4} + \frac{96}{4} = \frac{121}{4}$$
$$x - \frac{5}{2} = \pm\sqrt{\frac{121}{4}} = \pm\frac{11}{2}$$
$$x = \frac{5}{2} \pm \frac{11}{2} = 8, -3$$

EXERCISES A SECTION 2.4 61

14.
$$3y^2 - 5y + 1 = 0$$
$$y^2 - \frac{5}{3}y + \frac{1}{3} = 0$$
$$y^2 - \frac{5}{3}y + \left(-\frac{5}{6}\right)^2 = \left(-\frac{5}{6}\right)^2 - \frac{1}{3}$$
$$y^2 - \frac{5}{3}y + \frac{25}{36} = \frac{25}{36} - \frac{12}{36}$$
$$\left(y - \frac{5}{6}\right)^2 = \frac{13}{36}$$
$$y - \frac{5}{6} = \pm\sqrt{\frac{13}{36}} = \pm\frac{\sqrt{13}}{6}$$
$$y = \frac{5}{6} \pm \frac{\sqrt{13}}{6} = \frac{5 \pm \sqrt{13}}{6}$$

15. For $x^2 - 5x - 24 = 0$, $a = 1$, $b = -5$, and $c = -24$. Now use the quadratic formula.
$$x = \frac{-b \pm \sqrt{b^2 - 4ac}}{2a}$$
$$= \frac{-(-5) \pm \sqrt{(-5)^2 - 4(1)(-24)}}{2(1)}$$
$$= \frac{5 \pm \sqrt{25 + 96}}{2} = \frac{5 \pm \sqrt{121}}{2}$$
$$= \frac{5 \pm 11}{2} = 8, -3$$

16. For $3y^2 - 5y + 1 = 0$, $a = 3$, $b = -5$, and $c = 1$.
$$y = \frac{-(-5) \pm \sqrt{(-5)^2 - 4(3)(1)}}{2(3)}$$
$$= \frac{5 \pm \sqrt{25 - 12}}{6} = \frac{5 \pm \sqrt{13}}{6}$$

17. Multiply by 4 to clear fractions and write in general form.
$$\frac{1}{2}x^2 - \frac{3}{4}x = 1$$
$$2x^2 - 3x = 4$$
$$2x^2 - 3x - 4 = 0$$
Thus, $a = 2$, $b = -3$, and $c = -4$.
$$x = \frac{-(-3) \pm \sqrt{(-3)^2 - 4(2)(-4)}}{2(2)}$$
$$= \frac{3 \pm \sqrt{9 + 32}}{4} = \frac{3 \pm \sqrt{41}}{4}$$

18.
$$(y + 1)(y - 1) + 2y = 0$$
$$y^2 - 1 + 2y = 0$$
$$y^2 + 2y - 1 = 0$$
Thus, $a = 1$, $b = 2$, and $c = -1$.
$$y = \frac{-2 \pm \sqrt{(2)^2 - 4(1)(-1)}}{2(1)}$$
$$= \frac{-2 \pm \sqrt{4 + 4}}{2} = \frac{-2 \pm \sqrt{8}}{2}$$
$$= \frac{-2 \pm 2\sqrt{2}}{2} = \frac{2(-1 \pm \sqrt{2})}{2}$$
$$= -1 \pm \sqrt{2}$$

19. $x^2 - 25 = 0$
$$x^2 = 25$$
$$x = \pm\sqrt{25} = \pm 5$$

20.
$$3(y^2 + y) = -10y - 4$$
$$3y^2 + 3y = -10y - 4$$
$$3y^2 + 13y + 4 = 0$$
$$(3y + 1)(y + 4) = 0$$
$$3y + 1 = 0 \quad \text{or} \quad y + 4 = 0$$
$$3y = -1 \qquad\qquad y = -4$$
$$y = -\frac{1}{3}$$

21.
$$z^2 = 3(z + 1)$$
$$z^2 = 3z + 3$$
$$z^2 - 3z - 3 = 0$$
Use the quadratic formula with $a = 1$, $b = -3$, and $c = -3$.
$$z = \frac{-(-3) \pm \sqrt{(-3)^2 - 4(1)(-3)}}{2(1)}$$
$$= \frac{3 \pm \sqrt{9 + 12}}{2} = \frac{3 \pm \sqrt{21}}{2}$$

22. Write the equation as $y^2 - 4y + 1 = 0$ and use the quadratic formula.
$$y = \frac{-(-4) \pm \sqrt{(-4)^2 - 4(1)(1)}}{2(1)}$$
$$= \frac{4 \pm \sqrt{16 - 4}}{2} = \frac{4 \pm \sqrt{12}}{2}$$
$$= \frac{4 \pm 2\sqrt{3}}{2} = \frac{2(2 \pm \sqrt{3})}{2}$$
$$= 2 \pm \sqrt{3}$$

23. Write the equation in the form $3x^2 - x - 1 = 0$ and use the quadratic formula.

$$x = \frac{-(-1) \pm \sqrt{(-1)^2 - 4(3)(-1)}}{2(3)}$$
$$= \frac{1 \pm \sqrt{1 + 12}}{6} = \frac{1 \pm \sqrt{13}}{6}$$

24. Write the equation as $y^2 - 3y = y(y - 3) = 0$ and use the zero-product rule to obtain $y = 0$ or $y = 3$.

25. $3z^2 + 12 = 0$
$3z^2 = -12$
$z^2 = -4$
$z = \pm\sqrt{-4} = \pm 2i$

26. Write the equation in the form $x^2 - x + 1 = 0$ and use the quadratic formula.

$$x = \frac{-(-1) \pm \sqrt{(-1)^2 - 4(1)(1)}}{2(1)}$$
$$= \frac{1 \pm \sqrt{-3}}{2} = \frac{1 \pm i\sqrt{3}}{2}$$

27. $z(2z - 1) = -7$
$2z^2 - z = -7$
$2z^2 - z + 7 = 0$

Use the quadratic formula.

$$z = \frac{-(-1) \pm \sqrt{(-1)^2 - 4(2)(7)}}{2(2)}$$
$$= \frac{1 \pm \sqrt{-55}}{4} = \frac{1 \pm i\sqrt{55}}{4}$$

28. $x^2 - 36a^2 = 0$
$x^2 = 36a^2$
$x = \pm\sqrt{36a^2} = \pm 6a$

29. $x^2 - (a + b)^2 = 0$
$x^2 = (a + b)^2$
$x = \pm\sqrt{(a + b)^2}$
$= \pm(a + b)$

30. $x^2 - ax - 2a^2 = (x - 2a)(x + a) = 0$. Now use the zero-product rule.
$x - 2a = 0$ or $x + a = 0$
$x = 2a$ $\qquad$ $x = -a$

31. $y^3 - 1 = (y - 1)(y^2 + y + 1) = 0$. Using the zero-product, $y - 1 = 0$ or $y^2 + y + 1 = 0$. The first equation gives $y = 1$, and we use the quadratic formula to solve the other equation.

$$y = \frac{-(1) \pm \sqrt{(1)^2 - 4(1)(1)}}{2(1)}$$
$$= \frac{-1 \pm \sqrt{-3}}{2} = \frac{-1 \pm i\sqrt{3}}{2}$$

These two answers together with $y = 1$ give three solutions.

32. $x^3 + 4x = x(x^2 + 4) = 0$. The zero-product rule gives $x = 0$ and $x^2 + 4 = 0$.

$x^2 + 4 =$
$x^2 = -4$
$x = \pm\sqrt{-4} = \pm 2i$

The three solutions are 0, $2i$, and $-2i$.

33. $z^4 - 1 = (z^2 - 1)(z^2 + 1) = 0$. Use the zero-product rule to obtain $z^2 - 1 = 0$ or $z^2 + 1 = 0$. Using the zero-product rule again on the first of these equations, we have $z = 1$ or $z = -1$. Taking roots on the other equation gives $z = i$ and $z = -i$. The four solutions are 1, -1, i, and $-i$.

34. Write the equation in the form $3x^2 - 5x + 8 = 0$ to show that $a = 3$, $b = -5$, and $c = 8$.

$b^2 - 4ac = (-5)^2 - 4(3)(8) = 25 - 96 = -71$

Since the discriminant is negative, the equation has two complex solutions.

35. Since $x^2 - 10x + 25 = 0$, $a = 1$, $b = -10$, and $c = 25$.

$b^2 - 4ac = (-10)^2 - 4(1)(25) = 100 - 100 = 0$

Since the discriminant is zero, the equation has one real solution.

36. Since $-5x^2 + 2x + 1 = 0$, $a = -5$, $b = 2$, and $c = 1$.

$b^2 - 4ac = (2)^2 - 4(-5)(1) = 4 + 20 = 24$

Since the discriminant is positive, the equation has two real solutions.

EXERCISES A SECTION 2.4

37. Write $x^2 - 10x = -25$ as $x^2 - 10x + 25 = 0$. $a = 1$, $b = -10$, and $c = 25$.

$$x_1 + x_2 = -\frac{b}{a} = -\frac{-10}{1} = 10$$

$$x_1 x_2 = \frac{c}{a} = \frac{25}{1} = 25$$

38. Multiplying the equation by 6 to clear fractions we obtain $3x^2 - 2 = 6x$ or $3x^2 - 6x - 2 = 0$. $a = 3$, $b = -6$, and $c = -2$.

$$x_1 + x_2 = -\frac{b}{a} = -\frac{-6}{3} = 2$$

$$x_1 x_2 = \frac{c}{a} = \frac{-2}{3} = -\frac{2}{3}$$

39. For the equation $x^2 - 3x + 2 = 0$, $a = 1$, $b = -3$, and $c = 2$.

$$-\frac{b}{a} = -\frac{-3}{1} = 3 = 2 + 1 = x_1 + x_2$$

$$\frac{c}{a} = \frac{2}{1} = 2 = (2)(1) = x_1 x_2$$

Thus, 2 and 1 are solutions to the quadratic equation.

40. For $4x^2 + 11x - 3 = 0$, $a = 4$, $b = 11$, and $c = -3$.

$$-\frac{b}{a} = -\frac{11}{4} \neq \frac{15}{4} = x_1 + x_2$$

Since the sum does not check, we know the given numbers are not solutions. We do not have to check the product in this case.

41.
$$(x - (-2))(x - 3) = 0$$
$$(x + 2)(x - 3) = 0$$
$$x^2 - 3x + 2x - 6 = 0$$
$$x^2 - x - 6 = 0$$

42.
$$(x - 5i)(x - (-5i)) = 0$$
$$(x - 5i)(x + 5i) = 0$$
$$x^2 - (5i)^2 = 0$$
$$x^2 - 25i^2 = 0$$
$$x^2 - 25(-1) = 0$$
$$x^2 + 25 = 0$$

43.
$$(x - (5 - i))(x - (5 + i)) = 0$$
$$(x - 5 + i)(x - 5 - i) = 0$$
$$((x - 5) + i)((x - 5) - i) = 0$$
$$(x - 5)^2 - i^2 = 0$$
$$x^2 - 10x + 25 - (-1) = 0$$
$$x^2 - 10x + 26 = 0$$

44. $(2 + 5i) - (3 - 8i) = 2 + 5i - 3 + 8i$
$$= -1 + 13i$$

45. $\dfrac{1}{2 - 3i} = \dfrac{1}{2 - 3i} \cdot \dfrac{2 + 3i}{2 + 3i}$

$$= \frac{2 + 3i}{(2)^2 - (3i)^2}$$

$$= \frac{2 + 3i}{4 - 9i^2} = \frac{2 + 3i}{4 + 9}$$

$$= \frac{2 + 3i}{13} = \frac{2}{13} + \frac{3}{13}i$$

46. $(4 + 3i)(2 - i) = 8 - 4i + 6i - 3i^2$
$$= 8 + 2i - 3(-1)$$
$$= 11 + 2i$$

47. $(5 - 3i)^2 = 25 - 30i + (3i)^2$
$$= 25 - 30i + 9i^2$$
$$= 25 - 30i + 9(-1)$$
$$= 16 - 30i$$

CHAPTER 2 EQUATIONS, INEQUALITIES, AND PROBLEM SOLVING

SECTION 2.5 Equations That Result in Quadratic Equations

1. Let $u = z - 4$.

$$u^2 - 5u + 6 = 0$$
$$(u - 2)(u - 3) = 0$$

Use the zero-product rule and then substitute $z - 4$ for u.

$u - 2 = 0$ or $u - 3 = 0$
$u = 2 \qquad\qquad u = 3$
$z - 4 = 2 \qquad z - 4 = 3$
$z = 6 \qquad\qquad z = 7$

2. $\quad y^{2/3} - 5y^{1/3} - 6 = 0$
$(y^{1/3})^2 - 5y^{1/3} - 6 = 0$
$u^2 - 5u - 6 = 0$
$(u + 1)(u - 6) = 0$

Use the zero-product rule and then substitute $y^{1/3}$ for u.

$u + 1 = 0$ or $u - 6 = 0$
$u = -1 \qquad\qquad u = 6$
$y^{1/3} = -1 \qquad y^{1/3} = 6$
$(y^{1/3})^3 = (-1)^3 \qquad (y^{1/3})^3 = (6)^3$
$y = -1 \qquad\qquad y = 216$

3. $\qquad \dfrac{x}{3} = 1 + \dfrac{6}{x}$

$3x\left[\dfrac{x}{3}\right] = 3x\left[1 + \dfrac{6}{x}\right]$
$\qquad x^2 = 3x + 18$
$x^2 - 3x - 18 = 0$
$(x + 3)(x - 6) = 0$
$x + 3 = 0$ or $x - 6 = 0$
$x = -3 \qquad\qquad x = 6$

Both of the solutions check.

4. $\qquad \dfrac{3}{x + 1} = 1 - \dfrac{5}{x - 1}$

$(x + 1)(x - 1)\left[\dfrac{3}{x + 1}\right] = \left[1 - \dfrac{5}{x - 1}\right](x + 1)(x - 1)$
$\qquad 3(x - 1) = (x + 1)(x - 1) - 5(x + 1)$
$\qquad 3x - 3 = x^2 - 1 - 5x - 5$
$\qquad 0 = x^2 - 8x - 3$

Use the quadratic formula.

$x = \dfrac{-(-8) \pm \sqrt{(-8)^2 - 4(1)(-3)}}{2(1)}$

$= \dfrac{8 \pm \sqrt{64 + 12}}{2}$

$= \dfrac{8 \pm \sqrt{76}}{2} = \dfrac{8 \pm 2\sqrt{19}}{2}$

$= 4 \pm \sqrt{19}$

5. $\sqrt{a^2 - 7a + 15} - \sqrt{4a - 13} = 0$
$\qquad \sqrt{a^2 - 7a + 15} = \sqrt{4a - 13}$
$\qquad \left(\sqrt{a^2 - 7a + 15}\right)^2 = \left(\sqrt{4a - 13}\right)^2$
$\qquad a^2 - 7a + 15 = 4a - 13$
$\qquad a^2 - 11a + 28 = 0$
$\qquad (a - 7)(a - 4) = 0$
$x - 7 = 0$ or $x - 4 = 0$
$x = 7 \qquad\qquad x = 4$

6. $\sqrt{w - 1} + \sqrt{3w + 3} = 4$
$\qquad \sqrt{w - 1} = 4 - \sqrt{3w + 3}$
$\qquad \left(\sqrt{w - 1}\right)^2 = \left(4 - \sqrt{3w + 3}\right)^2$
$\qquad w - 1 = 16 - 8\sqrt{3w + 3} + 3w + 3$
$\qquad -2w - 20 = -8\sqrt{3w + 3}$
$\qquad w + 10 = 4\sqrt{3w + 3}$
$\qquad (w + 10)^2 = \left(4\sqrt{3w + 3}\right)^2$
$w^2 + 20w + 100 = 16(3w + 3)$
$w^2 + 20w + 100 = 48w + 48$
$w^2 - 28w + 52 = 0$
$(w - 2)(w - 26) = 0$

Using the zero-product rule we obtain solutions of 2 and 26, but the 26 does not check. The only solution is 2.

EXERCISES A

CHAPTER 2 EQUATIONS, INEQUALITIES, AND PROBLEM SOLVING

SECTION 2.5 Equations That Result in Quadratic Equations

1. $$x^4 - 10x^2 + 9 = 0$$
$$(x^2 - 1)(x^2 - 9) = 0$$

    Use the zero-product rule to give $x^2 - 1 = 0$ or $x^2 - 9 = 0$. These two equations give the four solutions 1, -1, 3, and -3.

2. To solve $(y + 2)^2 - 13(y + 2) + 42 = 0$ let $u = y + 2$.

$$u^2 - 13u + 42 = 0$$
$$(u - 6)(u - 7) = 0$$

$u - 6 = 0$ or	$u - 7 = 0$
$u = 6$	$u = 7$
$y + 2 = 6$	$y + 2 = 7$
$y = 4$	$y = 5$

3. Let $u = z^{1/3}$. Then $u^2 = z^{2/3}$.

$$u^2 - 5u + 6 = 0$$
$$(u - 2)(u - 3) = 0$$

$u - 2 = 0$ or	$u - 3 = 0$
$u = 2$	$u = 3$
$z^{1/3} = 2$	$z^{1/3} = 3$
$z = (2)^3$	$z = (3)^3$
$= 8$	$= 27$

4. Let $u = y^2 + 4y$.

$$u^2 - u - 20 = 0$$
$$(u - 5)(u + 4) = 0$$

$u - 5 = 0$ or	$u + 4 = 0$
$u = 5$	$u = -4$

    If $u = 5$, then $y^2 + 4y = 5$.

$$y^2 + 4y - 5 = 0$$
$$(y + 5)(y - 1) = 0$$

$y + 5 = 0$ or	$y - 1 = 0$
$y = -5$	$y = 1$

    If $u = -4$, then $y^2 + 4y = -4$.

$$y^2 + 4y + 4 = 0$$
$$(y + 2)^2 = 0$$
$$y + 2 = 0$$
$$y = -2$$

    There are three solutions, -5, 1, and -2.

5. Let $u = z^{-1}$. Then $u^2 = z^{-2}$.

$$u^2 + 2u - 15 = 0$$
$$(u - 3)(u + 5) = 0$$

$u - 3 = 0$ or	$u + 5 = 0$
$u = 3$	$u = -5$

    If $u = 3$, then $z^{-1} = 3$ and $z = 1/3$.
    If $u = -5$, then $z^{-1} = -5$ and $z = -1/5$.

6. Let $u = \dfrac{2x - 1}{x}$.

$$2u^2 + 7u - 4 = 0$$
$$(2u - 1)(u + 4) = 0$$

$2u - 1 = 0$ or	$u + 4 = 0$
$2u = 1$	$u = -4$
$u = \dfrac{1}{2}$	$\dfrac{2x - 1}{x} = -4$
$\dfrac{2x - 1}{x} = \dfrac{1}{2}$	$2x - 1 = -4x$
$2(2x - 1) = x$	$6x = 1$
$4x - 2 = x$	$x = \dfrac{1}{6}$
$3x = 2$	
$x = \dfrac{2}{3}$	

7. Multiplying by $y^{1/2}$ gives $y + 6 = 5y^{1/2}$.
Let $u = y^{1/2}$. Then $u^2 = y$.

$$u^2 + 6 = 5u$$
$$u^2 - 5u + 6 = 0$$
$$(u - 2)(u - 3) = 0$$

$u - 2 = 0$ or	$u - 3 = 0$
$u = 2$	$u = 3$
$y^{1/2} = 2$	$y^{1/2} = 3$
$y = 4$	$y = 9$

8. If $u = \sqrt[6]{a} = a^{1/6}$, then $u^2 = (a^{1/6})^2 = a^{1/3} = \sqrt[3]{a}$.

$$u + u^2 - 6 = 0$$
$$u^2 + u - 6 = 0$$
$$(u - 2)(u + 3) = 0$$

$u - 2 = 0$ or	$u + 3 = 0$
$u = 2$	$u = -3$
$\sqrt[6]{a} = 2$	$\sqrt[6]{a} = -3$
$a = 64$	

    Since a root with an ever index cannot be negative, the only solution is 64.

9. If we multiply the equation by x^2, we obtain the quadratic equation

$$x^2 - 2x = 3.$$
$$x^2 - 2x - 3 = 0$$
$$(x - 3)(x + 1) = 0$$
$$x - 3 = 0 \text{ or } x + 1 = 0$$
$$x = 3 \qquad x = -1$$

10.
$$\frac{1}{y} = \frac{-6}{y^2 + 5}$$
$$\frac{y(y^2 + 5)}{y} = \frac{(-6)(y)(y^2 + 5)}{y^2 + 5}$$
$$y^2 + 5 = -6y$$
$$y^2 + 6y + 5 = 0$$
$$(y + 5)(y + 1) = 0$$

Using the zero-product rule, $y = -5$ and $y = -1$. Both of these solutions check in the original equation.

11.
$$3y - \frac{2}{y+1} = \frac{4}{y+1}$$
$$3y(y + 1) - 2 = 4$$
$$3y^2 + 3y - 6 = 0$$
$$y^2 + y - 2 = 0$$
$$(y - 1)(y + 2) = 0$$

The zero-product rule gives $y = 1$ and $y = -2$. Both of these solutions check in the original equation.

12. Multiply both sides of the equation by the LCD $= (2z + 1)(2z - 1)$.

$$\frac{3z}{2z+1} = \frac{2}{4z^2-1} + \frac{z}{2z-1}$$
$$3z(2z - 1) = 2 + z(2z + 1)$$
$$6z^2 - 3z = 2 + 2z^2 + z$$
$$4z^2 - 4z - 2 = 0$$
$$2z^2 - 2z - 1 = 0$$

Since this does not factor, use the quadratic formula.

$$z = \frac{-(-2) \pm \sqrt{(-2)^2 - 4(2)(-1)}}{2(2)}$$
$$= \frac{2 \pm \sqrt{4+8}}{4} = \frac{2 \pm \sqrt{12}}{4}$$
$$= \frac{2 \pm 2\sqrt{3}}{4} = \frac{1 \pm \sqrt{3}}{2}$$

These solutions will check.

13. $x^2 - x - 2 = (x - 2)(x + 1)$
$x^2 - 5x + 6 = (x - 2)(x - 3)$
$x^2 - 2x - 3 = (x - 3)(x + 1)$
Multiply by the LCD $= (x - 2)(x - 3)(x + 1)$.

$$\frac{x}{(x-2)(x+1)} - \frac{2}{(x-2)(x-3)} = \frac{-3}{(x-3)(x+1)}$$
$$x(x - 3) - 2(x + 1) = -3(x - 2)$$
$$x^2 - 3x - 2x - 2 = -3x + 6$$
$$x^2 - 2x - 8 = 0$$
$$(x + 2)(x - 4) = 0$$

The zero-product rule gives solutions of -2 and 4. Both check in the original equation.

14. $\sqrt{y + 4} + 8 = y$
$\sqrt{y + 4} = y - 8$
$(\sqrt{y + 4})^2 = (y - 8)^2$
$y + 4 = y^2 - 16y + 64$
$0 = y^2 - 17y + 60$
$0 = (y - 5)(y - 12)$

The zero-product rule gives $y = 5$ and $y = 12$, but the only solution that checks is 12.

15. $2\sqrt{3y - 2} + \sqrt{3y^2 + 2y} = 0$
$2\sqrt{3y - 2} = -\sqrt{3y^2 + 2y}$
$(2\sqrt{3y - 2})^2 = (-\sqrt{3y^2 + 2y})^2$
$4(3y - 2) = 3y^2 + 2y$
$12y - 8 = 3y^2 + 2y$
$0 = 3y^2 - 10y + 8$
$0 = (3y - 4)(y - 2)$

The zero-product rule gives $y = 4/3$ and $y = 2$, but neither of these numbers make both radicals zero to give a sum of zero. Thus, there is no solution.

16. In order for the sum of two nonnegative radicals to be 0 there must be an x that makes both 0. The first is 0 for $x = 0$ and for $x = -6$, but neither of these numbers makes the second radical 0. Thus, there is no solution.

17. $\sqrt{3a + 1} - 3 = \sqrt{a - 4}$
$(\sqrt{3a + 1} - 3)^2 = (\sqrt{a - 4})^2$
$3a + 1 - 6\sqrt{3a + 1} + 9 = a - 4$
$2a + 14 = 6\sqrt{3a + 1}$
$a + 7 = 3\sqrt{3a + 1}$
$(a + 7)^2 = (3\sqrt{3a + 1})^2$
$a^2 + 14a + 49 = 9(3a + 1)$
$a^2 + 14a + 49 = 27a + 9$
$a^2 - 13a + 40 = 0$
$(a - 5)(a - 8) = 0$

The zero-product rule gives $a = 5$ and $a = 8$. Both check in the original equation.

EXERCISES A

18. $\sqrt[3]{x-1} = 2$
$\left(\sqrt[3]{x-1}\right)^3 = (2)^3$
$x - 1 = 8$
$x = 9$

The solution 9 does check.

19. $(2x + 11)^{1/2} - x - 4 = 0$
$(2x + 11)^{1/2} = x + 4$
$[(2x + 11)^{1/2}]^2 = (x + 4)^2$
$2x + 11 = x^2 + 8x + 16$
$0 = x^2 + 6x + 5$
$0 = (x + 1)(x + 5)$

The zero-product rule gives solutions of -1 and -5, but the only number that checks in the original equation is -1.

20. $(3x + 3)^{1/2} = 4 - (x - 1)^{1/2}$
$[(3x + 3)^{1/2}]^2 = [4 - (x - 1)^{1/2}]^2$
$3x + 3 = 16 - 8(x - 1)^{1/2} + x - 1$
$2x - 12 = -8(x - 1)1/2$
$x - 6 = -4(x - 1)^{1/2}$
$(x - 6)^2 = [-4(x - 1)^{1/2}]^2$
$x^2 - 12x + 36 = 16(x - 1)$
$x^2 - 12x + 36 = 16x - 16$
$x^2 - 28x + 52 = 0$
$(x - 2)(x - 26) = 0$

Using the zero-product rule $x = 2$ and $x = 26$ are possible solutions, but the only one that checks in the original equation is 2.

21. $x^2 + 5x = 21\left(\dfrac{1}{2} + \dfrac{x}{2}\right)$
$2(x^2 + 5x) = 21(2)\left(\dfrac{1}{2} + \dfrac{x}{2}\right)$
$2x^2 + 10x = 21 + 21x$
$2x^2 - 11x - 21 = 0$
$(x - 7)(2x + 3) = 0$

By the zero-product rule the two solutions are 7 and $-3/2$.

22. $(y + 2)(y - 2) + 3y = -4$
$y^2 - 4 + 3y = -4$
$y^2 + 3y = 0$
$y(y + 3) = 0$

The solutions are 0 and -3.

CHAPTER 2 EQUATIONS, INEQUALITIES, AND PROBLEM SOLVING

SECTION 2.6 Problem Solving and Applications of Quadratic Equations

1. *Analysis:* Follow the instructions given. Square Mary's age and subtract 250. Set this equal to seven times her age plus 10.

   Tabulation: Let x = Mary's age,
   $x^2 - 250$ = square of Mary's age decreased by 250,
   $7x + 10$ = 10 more than seven times her age

   Translation:
   $$x^2 - 250 = 7x + 10$$
   $$x^2 - 7x - 260 = 0$$
   $$(x - 20)(x + 13) = 0$$
   $$x - 20 = 0 \quad \text{or} \quad x + 13 = 0$$
   $$x = 20 \qquad\qquad x = -13$$

   Approximation: The answer 20 is reasonable for an age, but the -13 must be discarded.

   Check: $(20)^2 - 250 = 400 - 250 = 150$
   $7(20) + 10 = 140 + 10 = 150$

   Thus Mary is 20 years old.

2. *Analysis:* Use the formula $A = \frac{1}{2}bh$ and the fact that the base is 4 more than the height.

   Tabulation: Let h = the height,
   $h + 4$ = the base.

   Translation:
   $$\frac{1}{2}bh = 48$$
   $$bh = 96$$
   $$(h + 4)h = 96$$
   $$h^2 + 4h = 96$$
   $$h^2 + 4h - 96 = 0$$
   $$(h - 8)(h + 12) = 0$$
   $$h - 8 = 0 \quad \text{or} \quad h + 12 = 0$$
   $$h = 8 \qquad\qquad h = -12$$

   Approximation: The height is 8m since -12 cannot be a height. The base is $b = h + 4 = 8 + 4 = 12$.

   Check: $\frac{1}{2}(12)(8) = (6)(8) = 48$. Thus the base is 12m and the height is 8m.

3. Let r = the interest rate.
   $$A = P(1 + r)^2$$
   $$1798.54 = 1500(1 + r)^2$$
   $$1.1990267 \approx (1 + r)^2$$
   $$\pm\sqrt{1.1990267} \approx 1 + r$$
   $$\pm 1.0950008 \approx 1 + r$$
   $$-1 \pm 1.095 \approx r$$
   $$0.095, -2.095 \approx r$$

   Since r cannot be negative, discard -2.095 and thus the interest rate is $0.095 = 9.5\%$.

4. *Analysis:* The basic equation is:

   total cost = (no. of lots)(cost/lot)

   Tabulation: Let n = original no. of lots,
   c = original cost/lot.
   Thus $45{,}000 = nc$. Also,
   $n - 4$ = no. of lots sold,
   $c + 4000$ = selling price per lot.
   Thus $45{,}000 = (n - 4)(c + 4000)$.

   Translation: Since $45{,}000 = nc$ then $c = \dfrac{45{,}000}{n}$.

   Also, since both are equal to 45,000 equate nc and $(n - 4)(c + 4000)$ and substitute $\dfrac{45{,}000}{n}$ for c.

   $$nc = (n - 4)(c + 4000)$$
   $$nc = nc + 4000n - 4c - 16{,}000$$
   $$0 = 4000n - 4\left[\frac{45{,}000}{n}\right] - 16{,}000$$
   $$0 = 4000n^2 - 180{,}000 - 16{,}000n$$
   $$0 = n^2 - 4n - 45$$
   $$0 = (n - 9)(n + 5)$$
   $$n - 9 = 0 \quad \text{or} \quad n + 5 = 0$$
   $$n = 9 \qquad\qquad n = -5$$

   Approximation and Check: Discard the -5 as a number of lots and the answer of 9 lots does check.

PRACTICE EXERCISES

5. Let t = number of hours required by Zach,
   $t - 2$ = number of hours required by Andy.

$$\frac{1}{t} + \frac{1}{t-2} = \frac{1}{4}$$

$$4t(t-2)\left[\frac{1}{t} + \frac{1}{t-2}\right] = 4t(t-2)\left[\frac{1}{4}\right]$$

$$4(t-2) + 4t = t(t-2)$$
$$4t - 8 + 4t = t^2 - 2t$$
$$0 = t^2 - 10t + 8$$

Now use the quadratic formula to find the approximate value of t.

$$t = \frac{-(-10) \pm \sqrt{(-10)^2 - 4(1)(8)}}{2(1)}$$

$$= \frac{10 \pm \sqrt{100 - 32}}{2}$$

$$= \frac{10 \pm \sqrt{68}}{2}$$

$$\approx \frac{10 \pm 8.2}{2}$$

$$= 9.1, 0.9$$

Discard the 0.9 since it would make Andy's time negative. It takes 9.1 hours for Zach to do the job.

6. Let r = rate with a full load,
   $r + 11$ = rate empty.

$$\frac{825}{r} + \frac{825}{r+11} = 27.5$$

$$r(r+11)\left[\frac{825}{r} + \frac{825}{r+11}\right] = 27.5r(r+11)$$

$$825(r+11) + 825r = 27.5r(r+11)$$
$$30(r+11) + 30r = r(r+11)$$
$$30r + 330 + 30r = r^2 + 11r$$
$$r^2 - 49r - 330 = 0$$
$$(r-55)(r+6) = 0$$
$$r - 55 = 0 \quad \text{or} \quad r + 6 = 0$$
$$r = 55 \qquad\qquad r = -6$$

Since -6 cannot be a rate for the truck, the rate of speed with a full load is 55 mph.

7. Use the equation in the text with $h = 0$. Solve for t.

$$h = -16t^2 + v_0 t + h_0$$
$$0 = -16t^2 + 440t + 10,000$$
$$0 = 2t^2 - 55t - 1250$$

Now use the quadratic formula.

$$x = \frac{55 \pm \sqrt{(-55)^2 - 4(2)(-1250)}}{2(2)}$$

$$x = \frac{55 \pm \sqrt{3025 + 10,000}}{4}$$

$$x = \frac{55 \pm \sqrt{13,025}}{4}$$

$$x \approx 42.3$$

We used the positive radical since the negative gives a negative time. The hatch hit the water in approximately 42.3 seconds.

CHAPTER 2 EQUATIONS, INEQUALITIES, AND PROBLEM SOLVING

SECTION 2.6 Problem Solving and Applications of Quadratic Equations

1. *Analysis:* Use a variable to represent the first of two consecutive positive odd intergers. Determine the second in terms of the first. Take one-third of their product and set it equal to 65.

   Tabulation: Let n = the first integer,
   $n + 2$ = the next consecutive odd integer

   Tabulation:
   $$\frac{1}{3}n(n+2) = 65$$
   $$n(n+2) = (3)(65)$$
   $$n(n+2) = 195$$
   $$n^2 + 2n - 195 = 0$$
   $$(n-13)(n+15) = 0$$

   Using the zero-product rule we have $n = 13$ or $n = -15$. Since we are looking for positive integers, the -15 is discarded. The consecutive odd integers are 13 and $n + 2 = 15$.

   Approximation: The integers 13 and 15 are consecutive and odd.

   Check: $\frac{1}{3}(13)(15) = (13)(5) = 65$

   Thus, 13 and 15 are the integers.

2. *Anslysis:* We must set the positive square root of a number plus 4 equal to the number minus 8.

   Tabulation: Let x = the number.

   Translation:
   $$\sqrt{x+4} = x - 8$$
   $$\left(\sqrt{x+4}\right)^2 = (x-8)^2$$
   $$x + 4 = x^2 - 16x + 64$$
   $$0 = x^2 - 17x + 60$$
   $$0 = (x-5)(x-12)$$

   The zero-product rule gives $x = 5$ or $x = 12$.

   Approximation: The 5 will give a negative value on the right side of the equation and must be discarded since the radical is positive. We need only check the 12.

   Check: $\sqrt{12 + 4} = \sqrt{16} = 4 = 12 - 8$
   The 12 does check.

3. *Analysis:* The area of a rectangle is given by the formula $A = lw$.

   Tabulation: Let w = the width,
   $w + 11$ = the length.

   Translation: $w(w + 11) = 80$
   $$w^2 + 11w - 80 = 0$$
   $$(w - 5)(w + 16) = 0$$

   The zero-product rule gives $w = 5$ since the other solution is negative. The length is $w + 11 = 16$.

   Approximation and Check: The numbers 5 and 16 differ by 11 and $(5)(16) = 80$.

4. *Analysis:* The base of the finished box will be $(2)(4) = 8$ inches less than the length of the side of the cardboard, and the height of the box will be 4 inches. The volume is the square of the base times the height.

   Tabulation: Let x = the length of the side of the cardboard,
   $x - 8$ = the length of the side of the box.

   Translation: $4(x - 8)^2 = 400$
   $$(x - 8)^2 = 100$$
   $$x^2 - 16x + 64 = 100$$
   $$x^2 - 16x - 36 = 0$$
   $$(x - 18)(x + 2) = 0$$

   The zero-product rule gives $x = 18$ since -2 is not a length.

   Approximation and Check: $x - 8 = 18 - 8 = 10$ and $4(10)^2 = 400$. Thus the cardboard must be 18 inches on a side.

5. *Analysis:* Each of the dimensions of the original garden will decreased by twice the width of the sidewalk and the product of these numbers will be 216 yd^2.

   Tabulation: Let x = width of the sidewalk,
   $18 - 2x$ = width of new garden,
   $24 - 2x$ = length of new garden.

EXERCISES A   SECTION 2.6   71

Translation: $(18 - 2x)(24 - 2x) = 216$
$432 - 84x + 4x^2 = 216$
$4x^2 - 84x + 216 = 0$
$x^2 - 21x + 54 = 0$
$(x - 3)(x - 18) = 0$

The zero-product rule gives $x = 3$ or $x = 18$, but 18 yd would be too wide for a sidewalk around a garden with one dimension 18 yd. We only need to check the 3 yd.

Approximation and Check: The 3 yards is a reasonable result and

$(18 - 2(3))(24 - 2(3)) = (18 - 6)(24 - 6)$
$= (12)(18) = 216.$

The sidewalk is 3 yd wide.

6. $A = P(1 + r)^2$
$12{,}321 = 10{,}000(1 + r)^2$
$1.2321 = (1 + r)^2$
$\pm\sqrt{1.2321} = 1 + r$
$\pm 1.11 = 1 + r$
$1.11 - 1 = r$
$0.11 = r$

We used only the 1.11 since −1.11 would have given a negative r. The interest rate is 11%.

7.   Let c = original price per toy,
$c - 2$ = new price per toy,
n = original number sold per week,
$n + 50$ = new number sold per week.

Since we are given that $cn = 2000$, we know that $n = 2000/c$.

$(c - 2)(n + 50) = 2000$
$(c - 2)\left(\dfrac{2000}{c} + 50\right) = 2000$
$2000 + 50c - \dfrac{4000}{c} - 100 = 2000$
$50c - \dfrac{4000}{c} - 100 = 0$
$50c^2 - 4000 - 100c = 0$
$c^2 - 2c - 80 = 0$
$(c - 10)(c + 8) = 0$

The only positive answer given by the zero-product rule is 10. This does check and the original price per toy was $10.

8. $c = 10x^2 - 100x - 2000$
$10{,}000 = 10x^2 - 100x - 2000$
$0 = 10x^2 - 100x - 12{,}000$
$0 = x^2 - 10x - 1200$
$0 = (x - 40)(x + 30)$

By the zero-product rule $x = 40$ or $x = -30$. Since x must be positive, there were 40 items made during the month when the cost was $10,000.

9. Since 3080 is not divisible by 25, we assume that more than 250 sweaters were purchased.
Let n = total number of sweaters purchased.

$n(25 - 0.05n) = 3080$
$25n - 0.05n^2 = 3080$
$-0.05n^2 + 25n - 3080 = 0$
$n^2 - 500n + 61{,}600 = 0$
$(n - 220)(n - 280) = 0$

The zero-product rule gives $n = 220$ or $n = 280$. Since the number purchased must be more than 250, the bill was for 280 sweaters. This does check.

10. Let t = number of hours for Jim's father,
$t + 9$ = number of hours required by Jim.

$\dfrac{1}{t} + \dfrac{1}{t + 9} = \dfrac{1}{20}$

$20t(t + 9)\left[\dfrac{1}{t} + \dfrac{1}{t+9}\right] = \dfrac{1}{20}(20)(t)(t+9)$
$20(t + 9) + 20t = t(t + 9)$
$20t + 180 + 20t = t^2 + 9t$
$0 = t^2 - 31t - 180$
$0 = (t + 5)(t - 36)$

The only positive answer given by the zero-product rule is 36. Thus, Jim's father can build the cabinet in 36 hours, and Jim can do the job in $t + 9 = 36 + 9 = 45$ hours.

11. Let t = hours for older printer to do the job,
$t - 5$ = hours for newer printer to do the job.

$\dfrac{1}{t} + \dfrac{1}{t - 5} = \dfrac{1}{10}$
$10(t - 5) + 10t = t(t - 5)$
$10t - 50 + 10t = t^2 - 5t$
$0 = t^2 - 25t + 50$

Use the quadratic formula to solve the equation.

$t = \dfrac{-(-25) \pm \sqrt{(-25)^2 - 4(1)(50)}}{2(1)} = \dfrac{25 \pm \sqrt{425}}{2}$

The plus sign gives the answer that checks, 22.8 hr.

12. Let r = rate of the slower truck,
$r + 10$ = rate of the faster truck,
$2r$ = miles traveled by slower truck in two hours,
$2(r + 10)$ = miles traveled by faster truck in two hours.

Since the trucks are traveling at right angles, a right triangle is formed and we can use the Pythagorean theorem.

$$[2r]^2 + [2(r + 10)]^2 = (100)^2$$
$$4r^2 + 4(r^2 + 20r + 100) = 10{,}000$$
$$4r^2 + 4r^2 + 80r + 400 = 10{,}000$$
$$8r^2 + 80r - 9600 = 0$$
$$r^2 + 10r - 1200 = 0$$
$$(r - 30)(r + 40) = 0$$

The zero-product rule gives $r = 30$ since -40 cannot be a speed. Thus, the slower truck is is traveling 30 mph and the faster $r + 10 = 40$ mph.

13. Let r = rate of speed on the trip,
$r + 10$ = rate 10 mph faster,

$$\frac{400}{r} = \text{time required for trip,}$$

$$\frac{400}{r + 10} = \text{time which is two hours less.}$$

$$\frac{400}{r + 10} + 2 = \frac{400}{r}$$
$$400r + 2r(r + 10) = 400(r + 10)$$
$$400r + 2r^2 + 20r = 400r + 4000$$
$$2r^2 + 20r - 4000 = 0$$
$$r^2 + 10r - 2000 = 0$$
$$(r - 40)(r + 50) = 0$$

Since r must be positive, the coach's speed was 40 mph.

14. We use the equation $h = -16t^2 + v_0 t + h_0$. The initial height, $h_0 = 1472$, the height of the building. The initial velocity, $v_0 = 0$, since the coin was dropped and not thrown. $h = 0$ when the coin hits the ground. We must solve the following equation.

$$0 = -16t^2 + 1472$$
$$16t^2 = 1472$$
$$t^2 = 92$$
$$t = \sqrt{92} \approx 9.6$$

Since we are calculating a time, we use the positive square root and the time is approximately 9.6 sec.

15. We use the equation $h = -16t^2 + v_0 t + h_0$ with $h = 3000$ ft, $v_0 = 2400$ ft/sec, and $h_0 = 0$ ft.

$$3000 = -16t^2 + 2400t + 0$$
$$16t^2 - 2400t + 3000 = 0$$
$$2t^2 - 300t + 375 = 0$$

We use the quadratic formula.

$$t = \frac{-(-300) \pm \sqrt{(-300)^2 - 4(2)(375)}}{2(2)}$$
$$= \frac{300 \pm \sqrt{90{,}000 - 3000}}{4}$$
$$= \frac{300 \pm \sqrt{87{,}000}}{4} \approx \frac{300 \pm 294.96}{4}$$

Using the plus sign we get approximately 148.78 and the minus sign gives 1.26. The bullet will hit the helicopter in approximately 1.26 seconds. The larger number is the time that would be required for the bullet to be at 3000 feet on the way down if it did not hit the helicopter.

16. $h = -16t^2 + v_0 t + h_0$ with $h = 0$ ft, $h_0 = 256$ ft, and $v_0 = 16$ ft/sec.

$$0 = -16t^2 + 16t + 256$$
$$0 = t^2 - t - 16$$
$$t = \frac{-(-1) \pm \sqrt{(-1)^2 - 4(1)(-16)}}{2(1)}$$
$$= \frac{1 \pm \sqrt{65}}{2} \approx 4.53$$

Since the minus sign gives a negative answer, we use the plus sign to give 4.53 seconds.

17. Let $u = \dfrac{x - 1}{x + 2}$.

$$u^2 + u = 6$$
$$u^2 + u - 6 = 0$$
$$(u - 2)(u + 3) = 0$$

The zero-product rule gives $u = 2$ or $u = -3$.

$u = 2$ or $u = -3$

$\dfrac{x-1}{x+2} = 2$ $\dfrac{x-1}{x+2} = -3$

$x - 1 = 2(x + 2)$ $x - 1 = -3(x + 2)$

$x - 1 = 2x + 4$ $x - 1 = -3x - 6$

$-5 = x$ $4x = -5$

$x = -\dfrac{5}{4}$

EXERCISES A SECTION 2.6

18. In order for there to be a solution to
$$\sqrt{x^2+1} + \sqrt{x+1} = 0$$
both radicals must be zero. But $x^2 + 1$ cannot be zero. The equation has no solution.

19. We will use the quadratic formula to determine the solutions of $x^2 + bx + c = 0$ and then add them to obtain $-b$.
$$x = \frac{-b \pm \sqrt{b^2 - 4(1)c}}{2(1)} = \frac{-b \pm \sqrt{b^2 - 4c}}{2}$$

Now we add the solutions.
$$\frac{-b + \sqrt{b^2 - 4c}}{2} + \frac{-b - \sqrt{b^2 - 4c}}{2} = \frac{-2b}{2} = -b$$

20. In order for $3x^2 + 3x + m = 0$ to have only one real solution the discriminant must be zero.
$$b^2 - 4ac = 0$$
$$(3)^2 - 4(3)(m) = 0$$
$$9 - 12m = 0$$
$$9 = 12m$$
$$m = \frac{9}{12} = \frac{3}{4}$$

21. Although the sum of the two solution does equal $-b/a$, the product of solutions is not c/a.
$$(5 + i)(5 - i) = 25 - i^2$$
$$= 25 - (-1)$$
$$= 26 \neq -52/2 = -26$$

22. Subtract each of the solution from x, and set their product equal to zero. Multiply by 2 to clear fractions.
$$(x - (-3))\left(x - \frac{1}{2}\right) = 0$$
$$(x + 3)(2x - 1) = 0$$
$$2x^2 - x + 6x - 3 = 0$$
$$2x^2 + 5x - 3 = 0$$

CHAPTER 2 EQUATIONS, INEQUALITIES, AND PROBLEM SOLVING

SECTION 2.7 Linear and Absolute Value Inequalities

1. $2y + 3 \leq 4y - 9$
    $2y \leq 4y - 12$
    $-2y \leq -12$
    $y \geq 6$

2. $x - (1 - 3x) < 3(2x + 5) - 8$
    $x - 1 + 3x < 6x + 15 - 8$
    $4x - 1 < 6x + 7$
    $4x < 6x + 8$
    $-2x < 8$
    $x > -4$

3. (a) $2(y - 3) \geq y - (1 - y)$
    $2y - 6 \geq y - 1 + y$
    $2y - 6 \geq 2y - 1$
    $-6 \geq -1$

    Since the inequality is false, we have a contradiction and there is no solution.

   (b) $5y - (4 - 2y) < 3(2y - 1) + y$
    $5y - 4 + 2y < 6y - 3 + y$
    $7y - 4 < 7y - 3$
    $-4 < -3$

    Since this is a true inequality, every real number is a solution.

4. *Analysis:* The sum of all the scores divided by the total possible points, 450, must be greater than or equal to $90\% = 0.90$.

    Tabulation: Let $s =$ Jay's score on the final.

    Translation: $\dfrac{85 + 96 + 92 + s}{450} \geq 0.90$
    $\dfrac{273 + s}{450} \geq 0.90$
    $450\left[\dfrac{273 + s}{450}\right] \geq (0.90)(450)$
    $273 + s \geq 405$
    $s \geq 132$

    Approximation and Check: It does seem reasonable that Jay's score should be greater than or equal to 132, and this does check.

5. $1 \geq 1 - 3x > -5$
    $1 - 1 \geq 1 - 1 - 3x > -5 - 1$
    $0 \geq -3x > -6$
    $\dfrac{0}{-3} \leq \dfrac{-3x}{-3} < \dfrac{-6}{-3}$
    $0 \leq x < 2$

    Notice that when we divided by a negative number we reversed the inequalities. The solution in interval notation is $[0,2)$. The graph is given below.

    ◄—┼—┼—┼—┼—┼—[—┼—)—┼—┼—┼—►
    -5-4-3-2-1 0 1 2 3 4 5

6. $2x + 3 \leq -3$ or $2x + 3 > 1$
    $2x \leq -6$ $2x > -2$
    $x \leq -3$ $x > -1$

    In interval notation the solution is $(-\infty, -3]$ or $(-1, \infty)$. The graph is given below.

    ◄—┼—┼—]—┼—┼—┼—(—┼—┼—┼—┼—►
    -5-4-3-2-1 0 1 2 3 4 5

7. (a) $|y| \leq 4$ translates to $-4 \leq y \leq 4$. In interval notation this is $[-4, 4]$.

   (b) $|y| > 4$ translates to $y < -4$ or $y > 4$. In interval notation this is $(-\infty, -4)$ or $(4, \infty)$.

   (c) $|y| > 0$ translates to $y < -0$ or $y > 0$. Since we know that $-0 = 0$, the solution is all real numbers except 0. In interval notation this is $(-\infty, 0)$ or $(0, \infty)$.

   (d) $|y| \geq 0$ is true for all real numbers. In interval notation this is $(-\infty, \infty)$.

   (e) $|y| > -4$ is true for all real numbers. In interval notation this is $(-\infty, \infty)$.

8. (a) $|2x + 1| \leq 5$ translates to the following:

    $-5 \leq 2x + 1 \leq 5$
    $-6 \leq 2x \leq 4$
    $-3 \leq x \leq 2$

    In interval notation this is $[-3, 2]$.

   (b) $|3 - x| > 1$ translates to the following:
    $3 - x > 1$ or $3 - x < -1$
    $-x > -2$ $-x < -4$
    $x < 2$ $x > 4$

    In interval notation this is $(-\infty, 2)$ or $(4, \infty)$.

   (c) Since $|5x + 10| \geq 0$ for all x, $|5x + 10| < -2$ has no solution.

CHAPTER 2 EQUATIONS, INEQUALITIES, AND PROBLEM SOLVING

SECTION 2.7 Linear and Absolute Value Inequalities

1. $4x - 2 \leq 6$
   $4x \leq 8$
   $x \leq 2$

   In interval notation, $(-\infty, 2]$.

2. $2y + 5 < 4y - 9$
   $-2y < -14$
   $y > 7$

   In interval notation, $(7, \infty)$.

3. $(z - 1)^2 < z(z + 2)$
   $z^2 - 2z + 1 < z^2 + 2z$
   $1 < 4z$
   $\frac{1}{4} < z$

   In interval notation, $\left(\frac{1}{4}, \infty\right)$.

   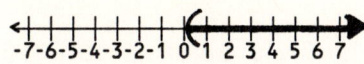

4. $(4y + 1)(y - 3) \geq (2y + 1)(2y - 1)$
   $4y^2 - 12y + y - 3 \geq 4y^2 - 1$
   $-11y - 3 \geq -1$
   $-11y \geq 2$
   $y \leq -\frac{2}{11}$

   In interval notation, $\left(-\infty, -\frac{2}{11}\right]$.

5. For the fraction to be positive, the denominator must be positive.

   $2y + 4 > 0$
   $2y > -4$
   $y > -2$

   In interval notation, $(-2, \infty)$.

6. For $(z + 4)^{-1}$ to be positive, then $z + 4 > 0$ or $z > -4$. In interval notation, $(-4, \infty)$.

7. $\frac{y + 1}{3} \leq \frac{y + 1}{2}$
   $2(y + 1) \leq 3(y + 1)$
   $2y + 2 \leq 3y + 3$
   $-1 \leq y$

   In interval notation, $[-1, \infty)$.

   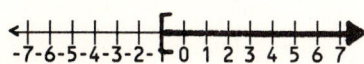

8. $4 \leq 2x - 6 < 10$
   $10 \leq 2x < 16$
   $5 \leq x < 8$

   In interval notation, $[5, 8)$.

   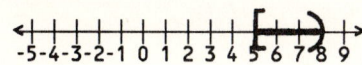

9. $0 \leq \frac{z + 1}{2} < 1$
   $0 \leq z + 1 < 2$
   $-1 \leq z < 1$

   In interval notation, $[-1, 1)$.

10. $2y - 6 < 4$ or $2y - 6 \geq 10$
    $2y < 10$ $2y \geq 16$
    $y < 5$ $y \geq 8$

    In interval notation, $(-\infty, 5)$ or $[8, \infty)$.

11. $2z + 1 > z$ or $z + 3 < 0$
    $z > -1$ $z < -3$

    In interval notation, $(-\infty, -3)$ or $(-1, \infty)$.

    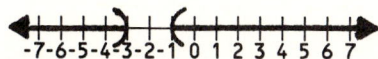

12. $y < 2y + 1$ and $2y + 1 < 1 - y$
    $-1 < y$ $3y < 0$
                          $y < 0$

    In interval notation, $(-1, 0)$.

    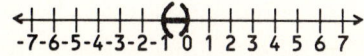

13. For $3(x + 5)$ to be positive $x + 5$ must be positive or $x + 5 > 0$. Thus $x > -5$.

14. For $-3(x + 5)$ to be positive $x + 5$ must be negative or $x + 5 < 0$. Thus $x < -5$.

15. $(x + 5)^2$ is positive for all values of x except where $x + 5 = 0$ or $x = -5$. This can be expressed as $x < -5$ or $x > -5$.

16. For $3(x + 5)$ to be negative $x + 5$ must be negative or $x + 5 < 0$. Thus $x < -5$.

17. For $-3(x + 5)$ to be negative $x + 5$ must be positive or $x + 5 > 0$. Thus $x > -5$.

18. $-(x + 5)^2$ is negative for all values of x except where $x + 5 = 0$ or $x = -5$. This can be expressed a $x < -5$ or $x > -5$.

19. $$-15 \leq C \leq 30$$
$$-15 \leq \frac{5}{9}(F - 32) \leq 30$$
$$-135 \leq 5(F - 32) \leq 270$$
$$-27 \leq F - 32 \leq 54$$
$$5 \leq F \leq 86$$

Thus $5° \leq F \leq 86°$.

20. $$S > C$$
$$200x > 150x + 300$$
$$50x > 300$$
$$x > 6$$

Thus at least 7 recorders must be sold to make a profit.

21. $$2.75 \leq F \leq 6.60$$
$$2.75 \leq 5.5x \leq 6.60$$
$$\frac{2.75}{5.5} \leq \frac{5.5x}{5.5} \leq \frac{6.60}{5.5}$$
$$0.5 \leq x \leq 1.2$$

Thus 0.5 in $\leq x \leq 1.2$ in.

22. $|x| < 7$ translates to $-7 < x < 7$ or $(-7,7)$.

23. $|z| \leq 0$ is true only if $z = 0$.

24. $|y| > 0$ is true for all values of y except 0. This can be written $y < 0$ or $y > 0$. In interval notation, $(-\infty,0)$ or $(0,\infty)$.

25. $|x + 2| > 3$ translates to $x + 2 > 3$ or $x + 2 < -3$. Thus $x > 1$ or $x < -5$. In interval notation $(-\infty,-5)$ or $(1,\infty)$.

26. $|1 - 5z| < 6$ translates to the following inequalities.
$$-6 < 1 - 5z < 6$$
$$-7 < -5z < 5$$
$$\frac{7}{5} > z > -1$$

This can be written as $(-1, 7/5)$.

27. $\left|\frac{y + 7}{3}\right| > 3$ translates to the following inequalities.

$\frac{y+7}{3} > 3$ or $\frac{y+7}{3} < -3$
$y + 7 > 9$ $y + 7 < -9$
$y > 2$ $y < -16$

This can be written as $(-\infty,-16)$ or $(2,\infty)$.

28. Since $|2x - 7| \geq 0$ for all x, $|2x - 7| < -1$ has no solution.

29. Since $|2z - 7| \geq 0$ for all z, $|2z - 7| \leq 0$ has a solution only for $2z - 7 = 0$ or $z = 7/2$.

30. Let n = number of shares purchased,
$n + 5$ = 5 more than number of shares,
c = cost per share,
$c - 3$ = 3 dollars less than cost per share.

Since $cn = 300$, the $c = 300/n$.

$$(c - 3)(n + 5) = 300$$
$$cn + 5c - 3n - 15 = 300$$
$$\left(\frac{300}{n}\right)n + 5\left(\frac{300}{n}\right) - 3n - 15 = 300$$
$$300 + \frac{1500}{n} - 3n - 15 = 300$$
$$\frac{1500}{n} - 3n - 15 = 0$$
$$1500 - 3n^2 - 15n = 0$$
$$n^2 + 5n - 500 = 0$$
$$(n - 20)(n + 25) = 0$$

n must be positive and thus the original purchase was 20 shares.

EXERCISES A   SECTION 2.7   77

31. The equation to use is $h = -16t^2 + 128t$.
  (a) When the rocket returns to the ground, $h = 0$.

$$0 = -16t^2 + 128t$$
$$0 = -16t(t - 8)$$

Since $t = 0$ is when the rocket was fired from ground level, $t = 8$ sec is the time when it hits the ground.

  (b) Set $h = 240$ ft and solve.

$$240 = -16t^2 + 128t$$
$$0 = -16t^2 + 128t - 240$$
$$0 = t^2 - 8t + 15$$
$$0 = (t - 3)(t - 5)$$

Use the zero-product rule to show that $t = 3$ or $t = 5$. Thus, the rocket will be at 240 ft in 3 seconds on the way up and in 5 seconds on the way down.

  (c)
$$256 = -16t^2 + 128t$$
$$0 = -16t^2 + 128t - 256$$
$$0 = t^2 - 8t + 16$$
$$0 = (t - 4)^2$$

The only solution is 4 sec. This is the time of the maximun height of the rocket and 256 ft is the maximun height.

  (d) The maximum height of the rocket is 256 ft and thus the rocket will not reach 300 ft. If we solve quadratic equation with $h = 300$ ft, we would get complex number solutions.

32. Let $u = a^2 - 2$.

$$u^2 - 6u - 7 = 0$$
$$(u - 7)(u + 1) = 0$$

Thus $u = 7$ or $u = -1$.
$a^2 - 2 = 7$ or $a^2 - 2 = -1$
   $a^2 = 9$ $a^2 = 1$
   $a = \pm 3$ $a = \pm 1$

33. Let $u = y^2$ and then $u^2 = y^4$.

$$u^2 - 2u - 63 = 0$$
$$(u - 9)(u + 7) = 0$$

If $u = 9$, then $y^2 = 9$ and $y = \pm 3$.

If $u = -7$, then $y^2 = -7$ and $y = \pm i\sqrt{7}$.

CHAPTER 2 EQUATIONS, INEQUALITIES, AND PROBLEM SOLVING

CHAPTER 2 Review Exercises

1. $6z - 5 = 2z + 8$
   $4z = 13$
   $z = \dfrac{13}{4}$

2. $\dfrac{2y}{3} - \dfrac{5}{6} = -\dfrac{3y}{2} - \dfrac{1}{3}$
   $4y - 5 = -9y - 2$
   $13y = 3$
   $y = \dfrac{3}{13}$

3. $6.5 + 2.5z = 3z + \dfrac{3}{2}$
   $65 + 25z = 30z + 15$
   $50 = 5z$
   $10 = z$

4. $3x - 2 = x + (2x + 1)$
   $3x - 2 = x + 2x + 1$
   $3x - 2 = 3x + 1$
   $-2 = 1$

   This contradiction implies that there is no solution.

5. $\left(\dfrac{z}{2} - 1\right) - \left(z - \dfrac{1}{2}\right) = 1$
   $\dfrac{z}{2} - 1 - z + \dfrac{1}{2} = 1$
   $z - 2 - 2z + 1 = 2$
   $-z - 1 = 2$
   $-z = 3$
   $z = -3$

6. $4(x - 2) + 3(3 - x) = 3x - 5$
   $4x - 8 + 9 - 3x = 3x - 5$
   $x + 1 = 3x - 5$
   $6 = 2x$
   $3 = x$

7. If $|2x + 3| = 3$, then $2x + 3 = 3$ or $2x + 3 = -3$.
   If $2x + 3 = 3$, then $x = 0$ and if $2x + 3 = -3$, then $x = -3$. The solutions are 0 and -3.

8. If $|3 - 4x| = 0$, then $3 - 4x = 0$ and $x = 3/4$.

9. Since $|3 - 4x| \geq 0$ for all x, $|3 - 4x| = -1$ has no solution.

10. $\dfrac{z - 2}{z - 5} = \dfrac{z + 1}{z + 3}$
    $(z - 2)(z + 3) = (z + 1)(z - 5)$
    $z^2 + z - 6 = z^2 - 4z - 5$
    $z - 6 = -4z - 5$
    $5z = 1$
    $z = \dfrac{1}{5}$

    The 1/5 does check in the original equation.

11. $\dfrac{3}{x + 6} - \dfrac{4}{x - 2} = \dfrac{2}{x^2 + 4x - 12}$
    $\dfrac{3}{x + 6} - \dfrac{4}{x - 2} = \dfrac{2}{(x + 6)(x - 2)}$
    $3(x - 2) - 4(x + 6) = 2$
    $3x - 6 - 4x - 24 = 2$
    $-x - 30 = 2$
    $-32 = x$

    The -32 does check in the original equation.

12. $2\sqrt[3]{a - 6} - 3 = -2$
    $2\sqrt[3]{a - 6} = 1$
    $\left(2\sqrt[3]{a - 6}\right)^3 = (1)^3$
    $8(a - 6) = 1$
    $8a - 48 = 1$
    $8a = 49$
    $a = \dfrac{49}{8}$

    The answer does check.

13. $\sqrt{z - 5} - \sqrt{z + 3} = -2$
    $\sqrt{z - 5} = \sqrt{z + 3} - 2$
    $\left(\sqrt{z - 5}\right)^2 = \left(\sqrt{z + 3} - 2\right)^2$
    $z - 5 = z + 3 - 4\sqrt{z + 3} + 4$
    $-12 = -4\sqrt{z + 3}$
    $3 = \sqrt{z + 3}$
    $(3)^2 = \left(\sqrt{z + 3}\right)^2$
    $9 = z + 3$
    $6 = z$

    The 6 does check in the original equation.

14. Let $n = $ the first integer,
    $n + 2 = $ the next consecutive even integer,
    $n + 4 = $ the third integer.

    $n + (n + 2) + (n + 4) = 2(n + 4) + 40$
    $3n + 6 = 2n + 48$
    $n = 42$

    The integers are 42, 44, and 46.

CHAPTER 2 REVIEW — REVIEW EXERCISES

15. Let x = Jane's weight.

$$\frac{x + 65 + 84 + 88}{4} = 78$$
$$x + 65 + 84 + 88 = 4(78)$$
$$x + 237 = 312$$
$$x = 75$$

Jane weighs 75 kg.

16. Let x = Mary's age,
$x + 11$ = Joe's age.

$$2(x + 11) + 3x = 172$$
$$2x + 22 + 3x = 172$$
$$5x = 150$$
$$x = 30$$
$$x + 11 = 41$$

Mary is 30 and Joe is 41.

17. Let x = Gloria's former salary,
$0.08x$ = her raise,
$x + 0.08x$ = her new salary.

$$x + 0.08x = 30{,}240$$
$$1.08x = 30{,}240$$
$$x = 28{,}000$$

Gloria now makes $28,000.

18. Let w = width of rectangle,
$2w - 1$ = length of rectangle.

$$P = 2w + 2l$$
$$70 = 2w + 2(2w - 1)$$
$$70 = 2w + 4w - 2$$
$$72 = 6w$$
$$12 = w$$

The width is 12 m and the length is
$2w - 1 = 2(12) - 1 = 23$ m.

19. Let x = the first angle,
$6x - 5$ = the second angle,
$x - 15$ = the third angle.

$$x + 6x - 5 + x - 15 = 180$$
$$8x - 20 = 180$$
$$8x = 200$$
$$x = 25$$
$$6x - 5 = 145$$
$$x - 15 = 10$$

The angles are 25°, 145°, and 10°.

20. Let n = number of days for Beth working alone.

$$\frac{1}{10} + \frac{1}{n} = \frac{1}{6}$$
$$3n + 30 = 5n$$
$$30 = 2n$$
$$15 = n$$

It will take Beth 15 days to do the job working alone.

21. Let x = the length of one piece,
$64 - x$ = the length of the other piece.

$$\frac{x}{64 - x} = \frac{11}{5}$$
$$5x = 11(64 - x)$$
$$5x = 704 - 11x$$
$$16x = 704$$
$$x = 44$$
$$64 - x = 20$$

The lengths are 44 ft and 20 ft.

22.
$$\frac{3 - 5i}{-2 + i} = \frac{3 - 5i}{-2 + i} \cdot \frac{-2 - i}{-2 - i}$$
$$= \frac{-6 - 3i + 10i + 5i^2}{(-2)^2 - i^2}$$
$$= \frac{-6 - 3i + 10i + 5(-1)}{4 - (-1)}$$
$$= \frac{-11 + 7i}{5} = -\frac{11}{5} + \frac{7}{5}i$$

23.
$$\frac{1}{4 - 3i} = \frac{1}{4 - 3i} \cdot \frac{4 + 3i}{4 + 3i}$$
$$= \frac{4 + 3i}{(4)^2 + (3)^2}$$
$$= \frac{4 + 3i}{25} = \frac{4}{25} + \frac{3}{25}i$$

24. $(2 - 3i) + (5 - i) = 2 - 3i + 5 - i$
$= 7 - 4i$

25. $[-(\sqrt{-4} - \sqrt{-9})]^{31} = [-(2i - 3i)]^{31}$
$= [-(-i)]^{31} = i^{31}$
$= i^{28} i^3 = (1)(-i) = -i$

26. $|-4 + 3i| = \sqrt{(-4)^2 + (3)^2}$
$= \sqrt{16 + 9} = \sqrt{25} = 5$

27. $(2 - i)(3 + 2i) = 6 + 4i - 3i - 2i^2$
$= 6 + 4i - 3i - 2(-1)$
$= 8 + i$

28. $(1 - 4i)^2 = 1 - 8i + 16i^2$
$= 1 - 8i - 16 = -15 - 8i$

29.
$$x^2 - 2x - 48 = 0$$
$$(x - 8)(x + 6) = 0$$

The zero-product rule gives solutions of 8 and −6.

30.
$$y(y + 1) = 6(3 - y)$$
$$y^2 + y = 18 - 6y$$
$$y^2 + 7y - 18 = 0$$
$$(y - 2)(y + 9) = 0$$

The zero-product rule gives solutions of 2 and −9.

31. To solve $3x^2 - x - 1 = 0$ we use the quadratic formula.

$$x = \frac{-(-1) \pm \sqrt{(-1)^2 - 4(3)(-1)}}{2(3)}$$
$$= \frac{1 \pm \sqrt{1 + 12}}{6} = \frac{1 \pm \sqrt{13}}{6}$$

32.
$$2y(y - 2) = -1$$
$$2y^2 - 4y = -1$$
$$2y^2 - 4y + 1 = 0$$

Now use the quadratic formula.

$$y = \frac{-(-4) \pm \sqrt{(-4)^2 - 4(2)(1)}}{2(2)}$$
$$= \frac{4 \pm \sqrt{16 - 8}}{4} = \frac{4 \pm \sqrt{8}}{4}$$
$$= \frac{4 \pm 2\sqrt{2}}{4} = \frac{2 \pm \sqrt{2}}{2}$$

33.
$$3x^2 + 15 = 0$$
$$3x^2 = -15$$
$$x^2 = -5$$
$$x = \pm\sqrt{-5} = \pm i\sqrt{5}$$

34. To solve $2x^2 - x + 1 = 0$ use the quadratic formula.

$$x = \frac{-(-1) \pm \sqrt{(-1)^2 - 4(2)(1)}}{2(2)}$$
$$= \frac{1 \pm \sqrt{1 - 8}}{4} = \frac{1 \pm \sqrt{-7}}{4}$$
$$= \frac{1 \pm i\sqrt{7}}{4}$$

35. $b^2 - 4ac = (0)^2 - 4(2)(-3) = 0 + 24 = 24$

Since $24 > 0$, the equation has two real solutions.

36. Write the equation as $4x^2 - 4x + 1 = 0$.

$b^2 - 4ac = (-4)^2 - 4(4)(1) = 16 - 16 = 0$

The equation has exactly one real solution.

37. $b^2 - 4ac = (3)^2 - 4(3)(1) = 9 - 12 = -3$

Since $-3 < 0$, there are two complex (nonreal) solutions.

38. Let $u = y^2$ and then $u^2 = y^4$.

$$u^2 - 2u - 63 = 0$$
$$(u - 9)(u + 7) = 0$$

$u - 9 = 0$ or $u + 7 = 0$
$u = 9$ $u = -7$
$y^2 = 9$ $y^2 = -7$
$y = \pm 3$ $y = \pm\sqrt{-7} = \pm i\sqrt{7}$

39. Let $u = \dfrac{y - 3}{y + 1}$.

$$u^2 - 6u + 8 = 0$$
$$(u - 2)(u - 4) = 0$$

$u - 2 = 0$ or $u - 4 = 0$
$u = 2$ $u = 4$
$\dfrac{y-3}{y+1} = 2$ $\dfrac{y-3}{y+1} = 4$
$y - 3 = 2(y + 1)$ $y - 3 = 4(y + 1)$
$y - 3 = 2y + 2$ $y - 3 = 4y + 4$
$-5 = y$ $-7 = 3y$
  $-\dfrac{7}{3} = y$

40.
$$2x + \frac{x}{x + 7} = \frac{8}{x + 7}$$
$$2x(x + 7) + x = 8$$
$$2x^2 + 14x + x = 8$$
$$2x^2 + 15x - 8 = 0$$
$$(2x - 1)(x + 8) = 0$$

Using the zero-product rule $x = 1/2$ or $x = -8$.

41.
$$\sqrt{3y + 1} - \sqrt{y + 4} = 1$$
$$\sqrt{3y + 1} = \sqrt{y + 4} + 1$$
$$(\sqrt{3y + 1})^2 = (\sqrt{y + 4} + 1)^2$$
$$3y + 1 = y + 4 + 2\sqrt{y + 4} + 1$$
$$2y - 4 = 2\sqrt{y + 4}$$
$$y - 2 = \sqrt{y + 4}$$
$$(y - 2)^2 = (\sqrt{y + 4})^2$$
$$y^2 - 4y + 4 = y + 4$$
$$y^2 - 5y = 0$$
$$y(y - 5) = 0$$

0 does not check. The only solution is 5.

42. Let r = rate of the first car,
    $r - 15$ = rate of the second car,
    $2r$ = distance traveled by first car,
    $2(r - 15)$ = distance traveled by second car.

Since a right triangle is involved, use the Pythagorean theorem.

$$[2r]^2 + [2(r - 15)]^2 = (150)^2$$
$$4r^2 + 4(r^2 - 30r + 225) = 22{,}500$$
$$4r^2 + 4r^2 - 120r + 900 = 22{,}500$$
$$8r^2 - 120r - 21{,}600 = 0$$
$$r^2 - 15r - 2700 = 0$$
$$(r - 60)(r + 45) = 0$$

Since -45 is not a possible answer, the rate of the first is 60 mph and the second is $r - 15 = 60 - 15 = 45$ mph.

43. Let n = initial number of men,
    $n - 2$ = number after 2 dropped out,
    c = initial cost per person,
    $c + 11$ = final cost per person.

We have initially that $cn = 132$ and thus $c = 132/n$.

$$(c + 11)(n - 2) = 132$$
$$cn - 2c + 11n - 22 = 132$$
$$\left(\frac{132}{n}\right)n - 2\left(\frac{132}{n}\right) + 11n - 22 = 132$$
$$132 - \frac{264}{n} + 11n - 22 = 132$$
$$-\frac{264}{n} + 11n - 22 = 0$$
$$-264 + 11n^2 - 22n = 0$$
$$n^2 - 2n - 24 = 0$$
$$(n - 6)(n + 4) = 0$$

The zero-product rule shows that there were 6 men initially planning the trip.

44. Let t = minutes for Jan to do job,
    $t + 12$ = minutes for her mom to do job.

$$\frac{1}{t} + \frac{1}{t + 12} = \frac{1}{8}$$
$$8(t + 12) + 8t = t(t + 12)$$
$$8t + 96 + 8t = t^2 + 12t$$
$$0 = t^2 - 4t - 96$$
$$0 = (t - 12)(t + 8)$$

The only positive answer is 12. Jan can do the job in 12 minutes and her mother takes 24 minutes.

45. 984 ft is approximately 0.186 mi. In the right triangle in the figure one leg is 4000 mi and the hypotenuse is approximately 4000.186 mi.

Use the Pythagorean theorem to find the other leg. Let x = the distance she can see.

$$x^2 + (4000)^2 = (4000.186)^2$$
$$x^2 = (4000.186)^2 - (4000)^2$$
$$x = \sqrt{(4000.186)^2 - (4000)^2}$$
$$x \approx 38.6$$

She can see approximately 38.6 miles.

46. $5(x + 2) - 3x \leq x + 7$
    $5x + 10 - 3x \leq x + 7$
    $2x + 10 \leq x + 7$
    $x \leq -3$

In interval notation this is $(-\infty, -3]$. The graph is given below.

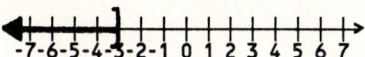

47. $3(x - 1) + x < 4(x + 1)$
    $3x - 3 + x < 4x + 4$
    $4x - 3 < 4x + 4$
    $-3 < 4$

This true inequality implies that every real number is a solution, the interval $(-\infty, \infty)$. The graph is all the number line given below.

48. $-3 < 2x - 3 < 3$
    $0 < 2x < 6$
    $0 < x < 3$

In interval notation, $(0, 3)$.

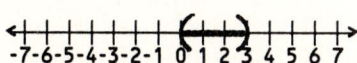

49. $2x - 3 \leq -3$ or $2x - 3 \geq 3$
    $2x \leq 0$ $\qquad$ $2x \geq 6$
    $x \leq 0$ $\qquad$ $x \geq 3$

In interval notation, $(-\infty, 0]$ or $[3, \infty)$.

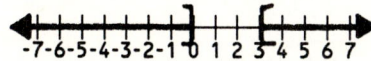

50. $|2 - x| < 1$ translates to the following inequalities.

$-1 < 2 - x < 1$
$-3 < -x < -1$
$3 > x > 1$

In interval notation, $(1, 3)$.

51. $|4x - 3| \geq 1$ translates to the following inequalities.

$$\begin{array}{ll} 4x - 3 \geq 1 & \text{or} \quad 4x - 3 \leq -1 \\ 4x \geq 4 & \quad\quad 4x \leq 2 \\ x \geq 1 & \quad\quad x \leq \dfrac{1}{2} \end{array}$$

In interval notation, $(-\infty, 1/2]$ or $[1, \infty)$.

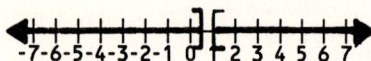

52. Since $|2 - x| \geq 0$ for all x, $|2 - x| \leq 0$ is true only when $2 - x = 0$ or $x = 2$. The graph is a single point on the graph below.

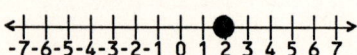

53. Since $|2 - x| \geq 0$ for all x, $|2 - x| < -1$ has no solution. There is no graph.

54. $(3y - 2) - (y + 1) = 0$
$3y - 2 - y - 1 = 0$
$2y - 3 = 0$
$2y = 3$
$y = \dfrac{3}{2}$

55. $5\sqrt{x - 1} - 3\sqrt{2x + 5} = 0$
$5\sqrt{x - 1} = 3\sqrt{2x + 5}$
$(5\sqrt{x - 1})^2 = (3\sqrt{2x + 5})^2$
$25(x - 1) = 9(2x + 5)$
$25x - 25 = 18x + 45$
$7x = 70$
$x = 10$

This does check in the original equation.

56. $\sqrt{z^2 + 7} - z - 1 = 0$
$\sqrt{z^2 + 7} = z + 1$
$(\sqrt{z^2 + 7})^2 = (z + 1)^2$
$z^2 + 7 = z^2 + 2z + 1$
$7 = 2z + 1$
$6 = 2z$
$3 = z$

This does check in the original equation.

57. $\dfrac{x - 3}{x - 1} - \dfrac{5}{x} = 1$
$x(x - 3) - 5(x - 1) = x(x - 1)$
$x^2 - 3x - 5x + 5 = x^2 - x$
$-8x + 5 = -x$
$5 = 7x$
$\dfrac{5}{7} = x$

This solution does check.

58. $\dfrac{z}{z + 2} - \dfrac{2}{z - 2} = \dfrac{z^2 + 4}{z^2 - 4}$
$z(z - 2) - 2(z + 2) = z^2 + 4$
$z^2 - 2z - 2z - 4 = z^2 + 4$
$-4z = 8$
$z = -2$

Since -2 makes two of the denominators in the original equation zero, there is no solution.

59. Use the quadratic formula to solve the equation $5x^2 + 2x + 1 = 0$.

$$x = \dfrac{-2 \pm \sqrt{2^2 - 4(5)(1)}}{2(5)}$$
$$= \dfrac{-2 \pm \sqrt{4 - 20}}{10} = \dfrac{-2 \pm \sqrt{-16}}{10}$$
$$= \dfrac{-2 \pm 4i}{10} = \dfrac{-1 \pm 2i}{5}$$

60. Let $u = x^{1/3}$, then $u^2 = x^{2/3}$.

$u^2 + 2u + 1 = 0$
$(u + 1)^2 = 0$
$u + 1 = 0$
$u = -1$
$x^{1/3} = -1$
$x = -1$

The only solution is -1.

61. $\dfrac{3y}{y + 1} = \dfrac{2}{y^2 - 1} + \dfrac{y}{y - 1}$
$3y(y - 1) = 2 + y(y + 1)$
$3y^2 - 3y = 2 + y^2 + y$
$2y^2 - 4y - 2 = 0$
$y^2 - 2y - 1 = 0$

Use the quadratic formula.

$$y = \dfrac{-(-2) \pm \sqrt{(-2)^2 - 4(1)(-1)}}{2(1)}$$
$$= \dfrac{2 \pm \sqrt{4 + 4}}{2} = \dfrac{2 \pm \sqrt{8}}{2}$$
$$= \dfrac{2 \pm 2\sqrt{2}}{2} = 1 \pm \sqrt{2}$$

62. $\sqrt{x^2+2} - \sqrt{3x+6} = 0$
$$\sqrt{x^2+2} = \sqrt{3x+6}$$
$$\left(\sqrt{x^2+2}\right)^2 = \left(\sqrt{3x+6}\right)^2$$
$$x^2 + 2 = 3x + 6$$
$$x^2 - 3x - 4 = 0$$
$$(x-4)(x+1) = 0$$

The zero-product rule gives solutions of 4 and −1 which both check.

63. To solve $|x+5| = |5-x|$ we must solve $x + 5 = 5 - x$ and $x + 5 = -(5 - x)$. The first of these equations gives $x = 0$ and the second has no solution.

64. $\dfrac{1}{x} - \dfrac{1}{y} = \dfrac{1}{z}$
$$yz - xz = xy$$
$$yz - xy = xz$$
$$y(z - x) = xz$$
$$y = \dfrac{xz}{z-x}$$

65. $ab = ac + d$
$$ab - ac = d$$
$$a(b - c) = d$$
$$a = \dfrac{d}{b-c}$$

66. $2x^2 - 8 = 0$
$$2x^2 = 8$$
$$x^2 = 4$$
$$x = \pm 2$$

67. $2x^2 + 8 = 0$
$$2x^2 = -8$$
$$x^2 = -4$$
$$x = \pm 2i$$

68. $3(x+1) - 2x \geq 2(x+1)$
$$3x + 3 - 2x \geq 2x + 2$$
$$x + 3 \geq 2x + 2$$
$$1 \geq x$$

In interval notation, $(-\infty, 1]$.

69. $2 + 3x < -1$ or $2 + 3x \geq 5$
$\quad 3x < -3 \quad\quad\quad 3x \geq 3$
$\quad x < -1 \quad\quad\quad\quad x \geq 1$

In interval notation, $(-\infty, -1)$ or $[1, \infty)$.

70. $|4 - x| < 1$ translates to $-1 < 4 - x < 1$. Subtracting 4 gives $-5 < -x < -3$. Multiply by -1 to obtain $5 > x > 3$. In interval notation this is $(3, 5)$.

71. $|4 - x| \geq 1$ translates to $4 - x \leq -1$ or $4 - x \geq 1$. Thus, $-x \leq -5$ or $-x \geq -3$ which becomes $x \geq 5$ or $x \leq 3$. In interval notation this is $(-\infty, 3]$ or $[5, \infty)$.

72. $(8 - 6i) - (-3 + i) = 8 - 6i + 3 - i$
$\quad\quad\quad\quad\quad\quad\quad\quad\quad = 11 - 7i$

73. $2i - \sqrt{-25} = 2i - 5i = -3i$

74. $\dfrac{1}{3-i} = \dfrac{1}{3-i} \cdot \dfrac{3+i}{3+i}$
$$= \dfrac{3+i}{(3)^2 - i^2} = \dfrac{3+i}{9 - (-1)}$$
$$= \dfrac{3+i}{10} = \dfrac{3}{10} + \dfrac{1}{10}i$$

75. $x_1 + x_2 = -\dfrac{b}{a} = -\dfrac{(-10)}{5} = 2$
$$x_1 x_2 = \dfrac{c}{a} = \dfrac{25}{5} = 5$$

76. Let t = time for two together.
$$\dfrac{1}{24} + \dfrac{1}{18} = \dfrac{1}{t}$$
$$3t + 4t = 72$$
$$7t = 72$$
$$t = \dfrac{72}{7} \approx 10.3$$

They can do the job together in approximately 10.3 minutes.

77. Let t = time of travel.
$$50t + 56t = 159$$
$$106t = 159$$
$$t = \dfrac{159}{106} = 1.5$$

Since they travel 1.5 hours, they will meet at 9:30 A.M.

78. Let s = speed of the plane in still air,
$s + 20$ = speed of the plane with the wind,
$s - 20$ = speed of the plane against the wind.

$$\frac{700}{s + 20} = \frac{500}{s - 20}$$
$$700(s - 20) = 500(s + 20)$$
$$700s - 14{,}000 = 500s + 10{,}000$$
$$200s = 24{,}000$$
$$s = \frac{24{,}000}{200} = 120$$

The speed of the plane is still air is 120 mph.

79. Let x = the height of the tree.

$$\frac{x}{32} = \frac{100}{64}$$
$$x = \frac{(32)(100)}{64} = 50$$

The tree is 50 feet high.

80.
$$h = -16t^2 + 64t + 6$$
$$54 = -16t^2 + 64t + 6$$
$$16t^2 - 64t + 48 = 0$$
$$t^2 - 4t + 3 = 0$$
$$(t - 1)(t - 3) = 0$$

The zero-product rule gives $t = 1$ or $t = 3$. The ball will be at 54 feet at 1 second and at 3 seconds.

81. Let n = number of antelope in the preserve.

$$\frac{x}{100} = \frac{50}{5}$$
$$x = \frac{(100)(50)}{5} = 1000$$

There are approximately 1000 antelope in the preserve.

82.
$$x = \frac{-(-3.2) \pm \sqrt{(-3.2)^2 - 4(7.3)(-8.4)}}{2(7.3)}$$
$$= \frac{3.2 \pm \sqrt{10.24 + 245.28}}{14.6}$$
$$= \frac{3.2 \pm \sqrt{255.52}}{14.6}$$

A calculator gives solutions of 1.31 and -0.88.

83.
$$\sqrt{\frac{x+y}{x}} = z$$
$$\left[\sqrt{\frac{x+y}{x}}\right]^2 = z^2$$
$$\frac{x+y}{x} = z^2$$
$$x + y = xz^2$$
$$y = xz^2 - x$$
$$y = x(z^2 - 1)$$
$$\frac{y}{z^2 - 1} = x$$

84.
$$2x^2 - 3y + 4z^2 = 9$$
$$2x^2 + 4z^2 - 9 = 3y$$
$$\frac{2x^2 + 4z^2 - 9}{3} = y$$

85.
$$(x - (1 - 2i))(x - (1 + 2i)) = 0$$
$$((x - 1) + 2i)((x - 1) - 2i) = 0$$
$$(x - 1)^2 - 4i^2 = 0$$
$$x^2 - 2x + 1 - 4(-1) = 0$$
$$x^2 - 2x + 5 = 0$$

86.
$$x^2 - 4x + 1 = 0$$
$$x^2 - 4x + 4 = -1 + 4$$
$$(x - 2)^2 = 3$$
$$x - 2 = \pm\sqrt{3}$$
$$x = 2 \pm \sqrt{3}$$

CHAPTER 2 EQUATIONS, INEQUALITIES, AND PROBLEM SOLVING

CHAPTER 2 Test

1. $\sqrt{3x+2} - \sqrt{3x-1} = 1$
$$\sqrt{3x+2} = \sqrt{3x-1} + 1$$
$$(\sqrt{3x+2})^2 = (\sqrt{3x-1} + 1)^2$$
$$3x + 2 = 3x - 1 + 2\sqrt{3x-1} + 1$$
$$2 = 2\sqrt{3x-1}$$
$$1 = \sqrt{3x-1}$$
$$(1)^2 = (\sqrt{3x-1})^2$$
$$1 = 3x - 1$$
$$2 = 3x$$
$$\frac{2}{3} = x$$

2. $|3x + 1| = 4$ translates to $3x + 1 = 4$ or $3x + 1 = -4$. The two solutions are 1 and $-5/3$.

3. $az = ay + w$
$az - ay = w$
$a(z - y) = w$
$a = \dfrac{w}{z - y}$

4. Let $t =$ time working together.
$$\frac{1}{30} + \frac{1}{20} = \frac{1}{t}$$
$$2t + 3t = 60$$
$$5t = 60$$
$$t = \frac{60}{5} = 12$$

It will take 12 minutes if they work together.

5. Let $s =$ speed of the plane in still air,
$s + 40 =$ speed with the wind,
$s - 40 =$ speed against the wind.
$$\frac{680}{s+40} = \frac{520}{s-40}$$
$$680(s - 40) = 520(s + 40)$$
$$680s - 27{,}200 = 520s + 20{,}800$$
$$160s = 48{,}000$$
$$s = \frac{48{,}000}{160} = 300$$

The speed of the plane in still air is 300 km/hr.

6. Let $x =$ length of one piece,
$104 - x =$ length of the other piece.
$$\frac{x}{104 - x} = \frac{9}{4}$$
$$4x = 9(104 - x)$$
$$4x = 936 - 9x$$
$$13x = 936$$
$$x = \frac{936}{13} = 72$$
$$104 - x = 104 - 72 = 32$$

7. $x = \dfrac{-6 \pm \sqrt{6^2 - 4(1)(-1)}}{2(1)}$
$= \dfrac{-6 \pm \sqrt{36 + 4}}{2} = \dfrac{-6 \pm \sqrt{40}}{2}$
$= \dfrac{-6 \pm 2\sqrt{10}}{2} = -3 \pm \sqrt{10}$

8. Let $u = x^{-1}$, then $u^2 = x^{-2}$.
$$15u^2 + 2u - 1 = 0$$
$$(5u - 1)(3u + 1) = 0$$

The zero-product rule gives $u = 1/5$ or $u = -1/3$. Thus, $x = 5$ or $x = -3$.

9. $\dfrac{1}{x+1} + \dfrac{1}{12} = \dfrac{1}{x}$
$$12x + x(x+1) = 12(x+1)$$
$$12x + x^2 + x = 12x + 12$$
$$x^2 + x - 12 = 0$$
$$(x - 3)(x + 4) = 0$$

The zero-product rule gives solutions of 3 and -4.

10. Let $n =$ number of shares bought,
$n + 5 = 5$ more than number purchased,
$c =$ cost per share,
$c - 2 = \$2$ less than cost per share.

Since $cn = 200$, then $c = 200/n$.

$$(c-2)(n+5) = 200$$
$$cn + 5c - 2n - 10 = 200$$
$$\left(\frac{200}{n}\right)n + 5\left(\frac{200}{n}\right) - 2n - 10 = 200$$
$$200 + \frac{1000}{n} - 2n - 10 = 200$$
$$\frac{1000}{n} - 2n - 10 = 0$$
$$1000 - 2n^2 - 10n = 0$$
$$n^2 + 5n - 500 = 0$$
$$(n - 20)(n + 25) = 0$$

The zero-product rule gives 20 and −25, but n must be positive. Thus, 20 shares were purchased.

11. Let r = rate of the slower train,
$r + 10$ = rate of the faster train,
$r(1)$ = distance traveled by slower train in one hour,
$(r + 10)(1)$ = distance traveled by faster train in one hour.

$$r^2 + (r + 10)^2 = (50)^2$$
$$r^2 + r^2 + 20r + 100 = 2500$$
$$2r^2 + 20r - 2400 = 0$$
$$r^2 + 10r - 1200 = 0$$
$$(r - 30)(r + 40) = 0$$

Since r is not negstive, the rate of the slower train is 30 mph and the rate of the faster is 40 mph.

12. $$x = \frac{-(-3) \pm \sqrt{(-3)^2 - 4(2)(2)}}{2(2)}$$
$$= \frac{3 \pm \sqrt{9-16}}{4} = \frac{3 \pm \sqrt{-7}}{4}$$
$$= \frac{3 \pm i\sqrt{7}}{4}$$

13. $$\frac{4-i}{2+3i} = \frac{4-i}{2+3i} \cdot \frac{2-3i}{2-3i}$$
$$= \frac{8 - 12i - 2i + 3i^2}{(2)^2 - 9i^2}$$
$$= \frac{8 - 14i + 3(-1)}{4 - 9(-1)}$$
$$= \frac{5 - 14i}{13} = \frac{5}{13} - \frac{14}{13}i$$

14. $b^2 - 4ac = (-2)^2 - 4(1)(5) = 4 - 20 = -16$

Since $-16 < 0$, there are two complex (nonreal) solutions.

15. $$3x - 3(x-1) \geq 5 - x$$
$$3x - 3x + 3 \geq 5 - x$$
$$3 \geq 5 - x$$
$$x \geq 2$$

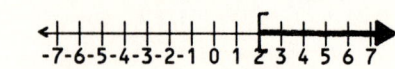

16. $2x - 4 < 6$ or $2x - 4 \geq 12$
$2x < 10$ $2x \geq 16$
$x < 5$ $x \geq 8$

In interval notation, $(-\infty, 5)$ or $[8, \infty)$.

17. $|x + 4| < 11$ translates to $-11 < x + 4 < 11$ or $-15 < x < 7$. In interval notation $(-15, 7)$.

18. $|3x - 1| > 10$ translstes to $3x - 1 < -10$ or $3x - 1 > 10$. Thus, $x < -3$ or $x > 11/3$. In interval notation, $(-\infty, -3)$ or $(11/3, \infty)$.

19. $|x + 1| = |x - 1|$ translates to $x + 1 = x - 1$ or $x + 1 = -(x - 1)$. The first equation has no solution and the second gives $x = 0$.

20. $$h = -16t^2 + v_0 t + h_0$$
$$0 = -16t^2 + (0)t + 500$$
$$0 = -16t^2 + 500$$
$$16t^2 = 500$$
$$t^2 = \frac{500}{16} = 31.25$$
$$t = \sqrt{31.25} \approx 5.6$$

The object will hit the ground in about 5.6 sec.

CHAPTER 3 RELATIONS, FUNCTIONS, AND GRAPHS

SECTION 3.1 The Rectangular Coordinate System

1. Make a table of values by selecting values for x and finding the corresponding values for y. For example, if $x = 0$, then $y = (0)^2 + 3 = 3$, so one pair is $(0,3)$. For $x = 1$, then $y = (1)^2 + 3 = 4$, so another pair is $(1,4)$. Additional ordered pairs can be found in the same manner. The graph is given below.

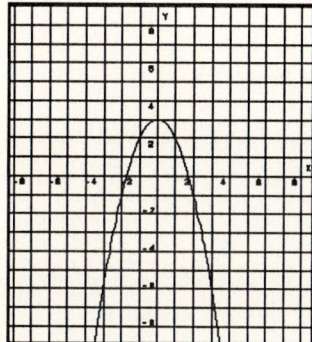

2. Given the triangle with vertices $D(2,1)$, $E(-3,2)$, and $F(-1,-4)$. Use the distance formula to find the length of each side, and then consider the Pythagorean theorem.

$$d_{DE} = \sqrt{(2-(-3))^2 + (1-2)^2}$$
$$= \sqrt{5^2 + (-1)^2} = \sqrt{25+1} = \sqrt{26}$$
$$d_{DF} = \sqrt{(2-(-1))^2 + (1-(-4))^2}$$
$$= \sqrt{3^2 + 5^2} = \sqrt{9+25} = \sqrt{34}$$
$$d_{EF} = \sqrt{(-3-(-1))^2 + (2-(-4))^2}$$
$$= \sqrt{(-2)^2 + 6^2} = \sqrt{4+36} = \sqrt{40}$$

For the triangle to be a right triangle, the square of the longest side (the possible hypotenuse) would have to be equal to the sum of the squares of the other two sides. But since

$$40 = \left(\sqrt{40}\right)^2 \neq \left(\sqrt{26}\right)^2 + \left(\sqrt{34}\right)^2 = 26 + 34 = 60$$

we can see that the triangle does not satisfy the conditions of the Pythagorean theorem, hence is not a right triangle.

3. Use the midpoint formula and substitute $(-3,5)$ for $(\bar{x},\bar{y})$, $(a,6)$ for (x_1,y_1), and $(-1,b)$ for (x_2,y_2). Set the x-coordinates equal and solve for a, and set the y-coordinates equal and solve for b.

$$(\bar{x},\bar{y}) = \left(\frac{x_1+x_2}{2}, \frac{y_1+y_2}{2}\right)$$
$$(-3,5) = \left(\frac{a+(-1)}{2}, \frac{6+b}{2}\right)$$
$$(-3,5) = \left(\frac{a-1}{2}, \frac{b+6}{2}\right)$$

$$\frac{a-1}{2} = -3 \qquad \frac{b+6}{2} = 5$$
$$a-1 = -6 \qquad b+6 = 10$$
$$a = -5 \qquad b = 4$$

CHAPTER 3 RELATIONS, FUNCTIONS, AND GRAPHS

SECTION 3.1 The Rectangular Coordinate System

1. The graph of $\{(2,5),(-1,3)\}$ is given below.

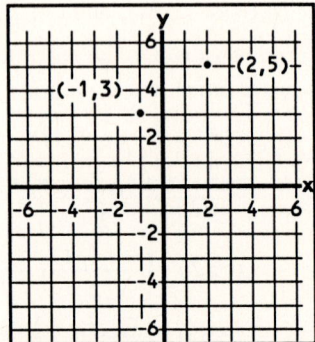

2. The graph of $\{(-3,4),(-3,2),(-3,0),(-3,-2),(-3,-3)\}$ is given below.

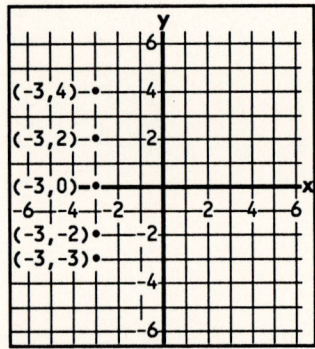

3. The point $(-2,1)$ is in quadrant II.

4. The point $(\sqrt{2},-\sqrt{2})$ is in quadrant IV.

5. The point $\left(-\frac{1}{2},-\frac{1}{4}\right)$ is in quadrant III.

6. The point $(88,90)$ is in quadrant I.

7. The graph of $y - x = 0$ has intercepts $(0,0)$, passes through the point $(1,1)$, and is is given below.

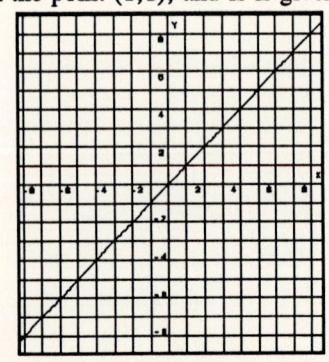

8. The graph of $y = -\frac{1}{2}x + 4$ with intercepts $(0,4)$ and $(8,0)$ is given below.

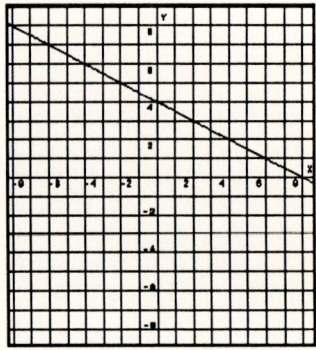

9. The graph of $2x = -5$, or $x = -5/2$, a vertical line passing through x-intercept $(-5/2,0)$, is given below.

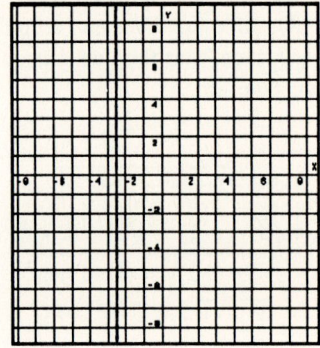

10. The graph of $2y = 7$, or $y = 7/2$, a horizontal line passing through y-intercept $(0,7/2)$, is given below.

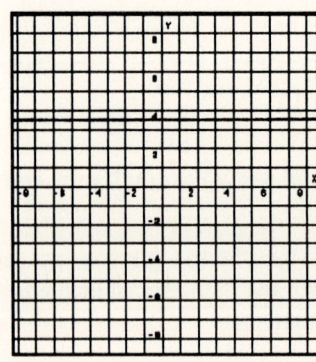

EXERCISES A

SECTION 3.1   89

11. The graph of $y + x^2 = 0$ is a *parabola*, a U-shaped curve, passing through $(0,0)$, $(2,-4)$, and $(-2,-4)$. The graph is given below.

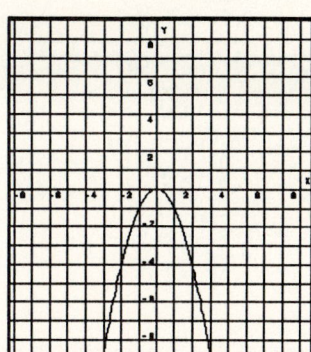

12. The graph of $y^2 = x + 2$ is a parabola opening to the right and passing through $(-2,0)$, $(2,2)$, and $(2,-2)$. The graph is given below.

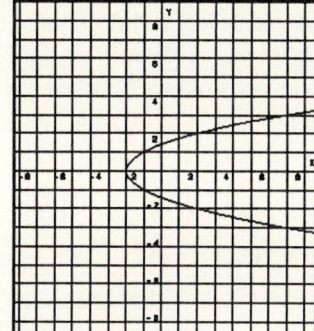

13. $d = \sqrt{(x_1 - x_2)^2 + (y_1 - y_2)^2}$
$= \sqrt{(4 - 1)^2 + (3 - 7)^2}$
$= \sqrt{(3)^2 + (-4)^2}$
$= \sqrt{9 + 16}$
$= \sqrt{25} = 5$

$(\bar{x}, \bar{y}) = \left(\dfrac{x_1 + x_2}{2}, \dfrac{y_1 + y_2}{2}\right)$
$= \left(\dfrac{4 + 1}{2}, \dfrac{3 + 7}{2}\right)$
$= \left(\dfrac{5}{2}, \dfrac{10}{2}\right) = \left(\dfrac{5}{2}, 5\right)$

14. $d = \sqrt{(x_1 - x_2)^2 + (y_1 - y_2)^2}$
$= \sqrt{(7 - 5)^2 + ((-1) - 8)^2}$
$= \sqrt{(2)^2 + (-9)^2}$
$= \sqrt{4 + 81} = \sqrt{85}$

$(\bar{x}, \bar{y}) = \left(\dfrac{x_1 + x_2}{2}, \dfrac{y_1 + y_2}{2}\right)$
$= \left(\dfrac{7 + 5}{2}, \dfrac{(-1) + 8}{2}\right)$
$= \left(\dfrac{12}{2}, \dfrac{7}{2}\right) = \left(6, \dfrac{7}{2}\right)$

15. $d = \sqrt{(x_1 - x_2)^2 + (y_1 - y_2)^2}$
$= \sqrt{((-6) - (-6))^2 + (2 - (-5))^2}$
$= \sqrt{(0)^2 + (7)^2} = \sqrt{49} = 7$

$(\bar{x}, \bar{y}) = \left(\dfrac{x_1 + x_2}{2}, \dfrac{y_1 + y_2}{2}\right)$
$= \left(\dfrac{(-6) + (-6)}{2}, \dfrac{2 + (-5)}{2}\right)$
$= \left(\dfrac{-12}{2}, \dfrac{-3}{2}\right) = \left(-6, -\dfrac{3}{2}\right)$

16. $d = \sqrt{(x_1 - x_2)^2 + (y_1 - y_2)^2}$
$= \sqrt{(a - 2a))^2 + (3 - 4)^2}$
$= \sqrt{(-a)^2 + (-1)^2}$
$= \sqrt{a^2 + 1}$

$(\bar{x}, \bar{y}) = \left(\dfrac{x_1 + x_2}{2}, \dfrac{y_1 + y_2}{2}\right)$
$= \left(\dfrac{a + 2a}{2}, \dfrac{3 + 4}{2}\right)$
$= \left(\dfrac{3a}{2}, \dfrac{7}{2}\right)$

17. $d = \sqrt{(x_1 - x_2)^2 + (y_1 - y_2)^2}$
$= \sqrt{(2\sqrt{x} - (-\sqrt{x}))^2 + (3\sqrt{y} - 2\sqrt{y})^2}$
$= \sqrt{(3\sqrt{x})^2 + (\sqrt{y})^2}$
$= \sqrt{9x + y}$

$(\bar{x}, \bar{y}) = \left(\dfrac{x_1 + x_2}{2}, \dfrac{y_1 + y_2}{2}\right)$
$= \left(\dfrac{2\sqrt{x} + (-\sqrt{x})}{2}, \dfrac{3\sqrt{y} + 2\sqrt{y}}{2}\right)$
$= \left(\dfrac{\sqrt{x}}{2}, \dfrac{5\sqrt{y}}{2}\right)$

18. $d = \sqrt{(x_1 - x_2)^2 + (y_1 - y_2)^2}$
$= \sqrt{(1.634 - 7.812)^2 + (-2.147 - 4.115)^2}$
$= \sqrt{(-6.178)^2 + (-6.262)^2}$
$= \sqrt{77.380328} \approx 8.797$

$$(\bar{x},\bar{y}) = \left(\frac{x_1 + x_2}{2}, \frac{y_1 + y_2}{2}\right)$$
$$= \left(\frac{1.634 + 7.812}{2}, \frac{-2.147 + 4.115}{2}\right)$$
$$= \left(\frac{9.446}{2}, \frac{1.968}{2}\right) \approx (4.723, 0.984)$$

19. First find the representation for the midpoint of the line segment joining $(a,8)$ and $(7,b)$.

$$(\bar{x},\bar{y}) = \left(\frac{x_1 + x_2}{2}, \frac{y_1 + y_2}{2}\right)$$
$$= \left(\frac{a+7}{2}, \frac{8+b}{2}\right)$$

Since the midpoint is equal to $(3,2)$ solve the following equations.

$$\frac{a+7}{2} = 3 \qquad \frac{8+b}{2} = 2$$
$$a + 7 = 6 \qquad 8 + b = 4$$
$$a = -1 \qquad b = -4$$

Thus, $a = -1$ and $b = -4$.

20. Substitute 5 for d, $(a,-2)$ for (x_1,y_1), and $(8,1)$ for (x_2,y_2) in the distance formula and solve for b.

$$d = \sqrt{(x_1 - x_2)^2 + (y_1 - y_2)^2}$$
$$5 = \sqrt{(a - 8)^2 + (-2 - 1)^2}$$
$$5 = \sqrt{(a - 8)^2 + (-3)^2}$$
$$25 = (a - 8)^2 + 9$$
$$16 = (a - 8)^2$$
$$\pm\sqrt{16} = \pm\sqrt{(a - 8)^2}$$
$$\pm 4 = a - 8$$
$$8 \pm 4 = a$$

Thus $a = 8 + 4 = 12$ or $a = 8 - 4 = 4$. The desired points are $(12,-2)$ and $(4,-2)$.

21. Use the distance formula to find the lengths of the sides of the triangle. Let d_{AB} be the length of the side joining A and B, d_{AC} the length of the side joining A and C, and d_{BC} the length of the side joining B and C.

$$d_{AB} = \sqrt{(5 - (-2))^2 + (6 - (-1))^2}$$
$$= \sqrt{(7)^2 + (7)^2}$$
$$= \sqrt{2(7)^2} = 7\sqrt{2}$$
$$d_{AC} = \sqrt{(5 - 2)^2 + (6 - (-3))^2}$$
$$= \sqrt{(3)^2 + (9)^2}$$
$$= \sqrt{10(3)^2} = 3\sqrt{10}$$
$$d_{BC} = \sqrt{(-2 - 2)^2 + (-1 - (-3))^2}$$
$$= \sqrt{(-4)^2 + (2)^2}$$
$$= \sqrt{5(2)^2} = 2\sqrt{5}$$

Since the lengths of the sides do not satisfy the Pythagorean theorem, the triangle is not a right triangle.

22. By plotting the points it is easy to see that if the four points are the vertices of a rectangle, then the diagonal joining $(-7,2)$ and $(6,-1)$ and the diagonal joining $(-2,7)$ and $(1,-6)$ will have to be equal in length. Since the distance between $(-7,2)$ and $(6,-1)$ is

$$d = \sqrt{(-7 - 6)^2 + (2 - (-1))^2}$$
$$= \sqrt{(-13)^2 + (3)^2}$$
$$= \sqrt{169 + 9} = \sqrt{178}$$

and the distance between $(-2,7)$ and $(1,-6)$ is

$$d = \sqrt{(-2 - 1)^2 + (7 - (-6))^2}$$
$$= \sqrt{(-3)^2 + (13)^2}$$
$$= \sqrt{9 + 169} = \sqrt{178}$$

these two diagonals are equal. Thus we have a rectangle since a four-sided figure with diagonals equal must be a rectangle.

23. We must find the distance between the two points with coordinates $(4,7)$ and $(9,3)$.

$$d = \sqrt{(x_1 - x_2)^2 + (y_1 - y_2)^2}$$
$$= \sqrt{(4 - 9)^2 + (7 - 3)^2}$$
$$= \sqrt{(-5)^2 + (4)^2}$$
$$= \sqrt{25 + 16} = \sqrt{41} \approx 6.403124237$$

Thus, the distance between the two herds is about 6.4 mi.

EXERCISES A

24. Let the month represent the x-coordinate and the number of inches of rain represent the y-coordinate. Then the information given corresponds to the three ordered pairs $(J,2.5)$, $(F,3.7)$, and $(M,6.3)$.

25. $x^2 + 5x + 6 > 0$

First solve the related equation.

$$x^2 + 5x + 6 = 0$$
$$(x + 2)(x + 3) = 0$$
$$x + 2 = 0 \text{ or } x + 3 = 0$$
$$x = -2 \qquad x = -3$$

Since there are two critical points, the number line is divided into three intervals,
$$(-\infty,-3), (-3,-2), \text{ and } (-2,\infty)$$
Test points -4, -2.5, and 0 in these intervals show that the solution to the inequality is the intervals $(-\infty,-3)$ or $(-2,\infty)$, or using inequalities, $x < -3$ or $x > -2$.

26. $\dfrac{x-5}{2x+3} \leq 0$

First find the critical points by setting the numerator and denominator equal to zero and solving for x.

$$x - 5 = 0 \qquad 2x + 3 = 0$$
$$x = 5 \qquad\quad 2x = -3$$
$$x = -\tfrac{3}{2}$$

The critical points divide the number line into three intervals,

$$\left(-\infty, -\tfrac{3}{2}\right), \left(-\tfrac{3}{2}, 5\right), (5, \infty)$$

Test points -2, 0, and 6 show that the solution consists of the numbers in the interval

$$\left(-\tfrac{3}{2}, 5\right).$$

Since the inequality is $\leq$, we also include the number that makes the fraction 0, that is that makes the numerator 0, which is 5. Note that the value that makes the denominator 0 is never included. Thus, the solution is

$$\left(-\tfrac{3}{2}, 5\right] \quad \text{or} \quad -\tfrac{3}{2} < x \leq 5.$$

27. $|4x - 3| \geq 1$

Translate to the compound inequality

$$4x - 3 \leq -1 \quad \text{or} \quad 4x - 3 \geq 1$$

and solve.

$$4x \leq 2 \quad \text{or} \quad 4x \geq 4$$
$$x \leq \tfrac{1}{2} \quad \text{or} \quad x \geq 1$$

In interval notation, the solution is

$$\left(-\infty, \tfrac{1}{2}\right] \quad \text{or} \quad [1, \infty).$$

28. (a) If $x = 0$, substitute and solve for y.

$$(0) + 2y - 6 = 0$$
$$2y - 6 = 0$$
$$2y = 6$$
$$y = 3$$

(b) if $y = 0$, substitute and solve for x.

$$x + 2(0) - 6 = 0$$
$$x - 6 = 0$$
$$x = 6$$

(c) $x + 2y - 6 = 0$
$$2y = -x + 6$$
$$y = -\tfrac{1}{2}x + \tfrac{6}{2}$$
$$y = -\tfrac{1}{2}x + 3$$

CHAPTER 3 RELATIONS, FUNCTIONS, AND GRAPHS

SECTION 3.2 Linear Equations

1. **(a)** Given the points $(4,-7)$ and $(6,3)$. The slope of the line through these points is

    $$m = \frac{y_2 - y_1}{x_2 - x_1}$$
    $$= \frac{-7 - 3}{4 - 6} = \frac{-10}{-2} = 5.$$

    (b) Given the points $(-2,1)$ and $(3,-9)$. The slope of the line through these points is

    $$m = \frac{y_2 - y_1}{x_2 - x_1}$$
    $$= \frac{1 - (-9)}{-2 - 3} = \frac{10}{-5} = -2.$$

    (c) Given the points $(1,7)$ and $(-3,7)$. The slope of the line through these points is

    $$m = \frac{y_2 - y_1}{x_2 - x_1}$$
    $$= \frac{7 - 7}{1 - (-3)} = \frac{0}{4} = 0.$$

    (d) Given the points $(-4,9)$ and $(-4,6)$. The slope of the line through these points is

    $$m = \frac{y_2 - y_1}{x_2 - x_1}$$
    $$= \frac{9 - 6}{(-4) - (-4)} = \frac{3}{0} \text{ which is undefined.}$$

2. To find the general form of the equation of the line through $(-5,6)$ and $(-1,-4)$, first use the slope formula to find the slope.

    $$m = \frac{y_2 - y_1}{x_2 - x_1}$$
    $$= \frac{6 - (-4)}{(-5) - (-1)} = \frac{10}{-4} = -\frac{5}{2}$$

    Now use the point-slope form with $(x_1, y_1) = (-5,6)$ and $m = -\frac{5}{2}$.

    $$y - y_1 = m(x - x_1)$$
    $$y - 6 = -\frac{5}{2}(x - (-5))$$
    $$2(y - 6) = -5(x + 5)$$
    $$2y - 12 = -5x - 25$$
    $$5x + 2y + 13 = 0$$

3. To find the slope and y-intercept for $6x - 2y + 22 = 0$, write the equation in slope-intercept form by solving for y. The coefficient of x is the slope, and the constant term is the y-coordinate of the y-intercept.

    $$6x - 2y + 22 = 0$$
    $$-2y = -6x - 22$$
    $$y = \frac{-6}{-2}x - \frac{22}{-2}$$
    $$y = 3x + 11$$

    Thus, $m = 3$ and $(0,b) = (0,11)$.

4. To determine if two lines are parallel or perpendicular, solve each equation for y (write each in slope-intercept form) and compare the slopes. If the slopes are equal (and the y-intercepts are unequal) the lines are parallel. If the slopes are negative reciprocals, the lines are perpendicular.

    (a) $5x - 2y + 8 = 0$ and $-10x + 4y - 12 = 0$

    $$5x - 2y + 8 = 0$$
    $$-2y = -5x - 8$$
    $$y = \frac{5}{2}x + 4$$
    $$m = \frac{5}{2} \quad (0,b) = (0,4)$$

    $$-10x + 4y - 12 = 0$$
    $$4y = 10x + 12$$
    $$y = \frac{10}{4}x + 3$$
    $$y = \frac{5}{2}x + 3$$
    $$m = \frac{5}{2} \quad (0,b) = (0,3)$$

    Since the slopes are equal, the lines are parallel.

PRACTICE EXERCISES

(b) $-3x - 6y + 7 = 0$ and $10x - 5y - 2 = 0$

$$-3x - 6y + 7 = 0$$
$$-6y = 3x - 7$$
$$y = -\frac{1}{2}x + \frac{7}{6}$$
$$m = -\frac{1}{2} \qquad (0,b) = \left(0, \frac{7}{6}\right)$$
$$10x - 5y - 2 = 0$$
$$-5y = -10x + 2$$
$$y = 2x - \frac{2}{5}$$
$$m = 2 \qquad (0,b) = \left(0, -\frac{2}{5}\right)$$

Since the slopes are negative reciprocals, the lines are perpendicular.

5. To find the general form of the equation of the line through $(-1,6)$ and perpendicular to $x - 5y + 4 = 0$, begin by finding the slope of the given line by writing the equation in slope-intercept form.

$$x - 5y + 4 = 0$$
$$-5y = -x - 4$$
$$y = \frac{1}{5}x + \frac{4}{5}$$

Then this line has slope $\frac{1}{5}$, so the desired perpendicular line has slope the negative reciprocal, -5. Use this slope and the point $(-1,6)$ in the point-slope form.

$$y - y_1 = m(x - x_1)$$
$$y - 6 = -5(x - (-1))$$
$$y - 6 = -5(x + 1)$$
$$y - 6 = -5x - 5$$
$$5x + y - 1 = 0$$

Thus, the desired line has general form

$$5x + y - 1 = 0.$$

6. To find the perpendicular bisector of the line segment joining $(4,-7)$ and $(-2,-1)$, first use the midpoint formula to find the midpoint of the segment.

$$(\bar{x}, \bar{y}) = \left(\frac{x_1 + x_2}{2}, \frac{y_1 + y_2}{2}\right)$$
$$= \left(\frac{4 + (-2)}{2}, \frac{-7 + (-1)}{2}\right)$$
$$= \left(\frac{2}{2}, \frac{-8}{2}\right) = (1, -4)$$

Next find the slope of the line segment joining the two points. The negative reciprocal of this slope is the slope of the desired perpendicular bisector.

$$m = \frac{y_2 - y_1}{x_2 - x_1}$$
$$= \frac{-7 - (-1)}{4 - (-2)} = \frac{-6}{6} = -1$$

The negative reciprocal of -1 is 1. Use the point-slope form with $m = 1$ and $(x_1, y_1) = (1, -4)$.

$$y - y_1 = m(x - x_1)$$
$$y - (-4) = 1(x - 1)$$
$$y + 4 = x - 1$$
$$0 = x - y - 5$$

Thus, the general form of the equation of the perpendicular bisector of the segment is

$$x - y - 5 = 0.$$

CHAPTER 3 RELATIONS, FUNCTIONS, AND GRAPHS

SECTION 3.2 Linear Equations

1. $3x - 2y - 6 = 0$

   When $x = 0$, $-2y = 6$ making $y = -3$. The y-intercept is $(0,-3)$. When $y = 0$, $3x = 6$ making $x = 2$. The x-intercept is $(2,0)$. The graph is given below.

   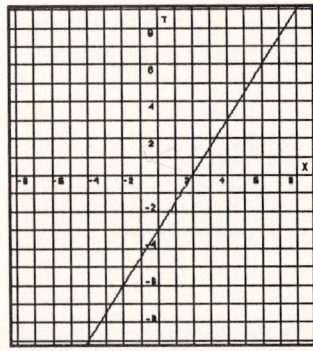

2. $4x + 3y = 12$

   When $x = 0$, $3y = 12$ making $y = 4$. The y-intercept is $(0,4)$. When $y = 0$, $4x = 12$ making $x = 3$. The x-intercept is $(3,0)$. The graph is given below.

   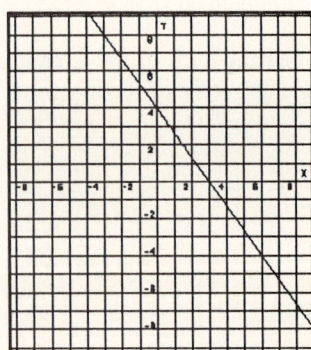

3. $2x - 3 = 0$

   This is equivalent to $x = 3/2$, and has a vertical line through x-intercept $(3/2,0)$ as its graph.

   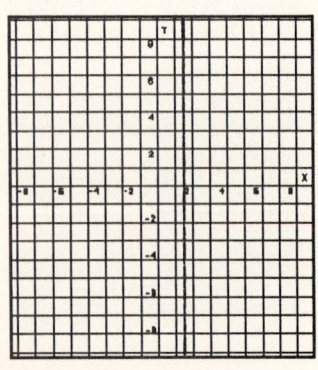

4. $y = 3$

   The graph is a horizontal line through y-intercept $(0,3)$, as shown below.

   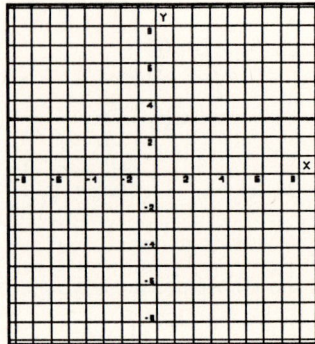

5. $3x - 2y = 0$

   When $x = 0$, $y = 0$ so the intercepts are both $(0,0)$. Another point is $(2,3)$. The graph is given below.

   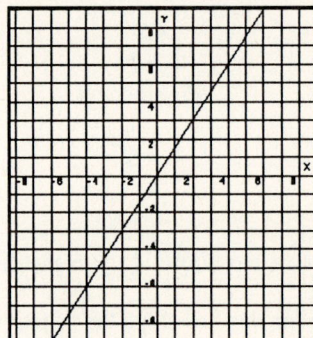

6. $-5x + 2y = 10$

   When $x = 0$, $2y = 10$ making $y = 5$. The y-intercept is $(0,5)$. When $y = 0$, $-5x = 10$ making $x = -2$. The x-intercept is $(-2,0)$. The graph is given below.

   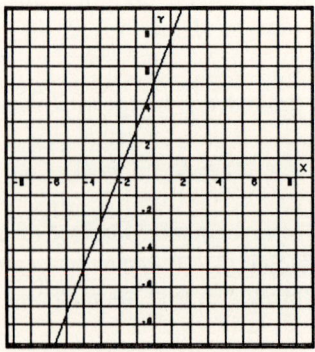

EXERCISES A SECTION 3.2 95

7. Substitute the coordinates of the points, (6,−2) and (4,1) into the slope formula.

$$m = \frac{y_2 - y_1}{x_2 - x_1}$$
$$= \frac{1 - (-2)}{4 - 6}$$
$$= \frac{3}{-2} = -\frac{3}{2}$$

8. Substitute the coordinates of the points, (−3,2) and (−3,−4) into the slope formula.

$$m = \frac{y_2 - y_1}{x_2 - x_1}$$
$$= \frac{-4 - 2}{-3 - (-3)}$$
$$= \frac{-6}{0} \quad \text{which is undefined}$$

9. Substitute the coordinates of the points, (−4,5) and (1,5) into the slope formula.

$$m = \frac{y_2 - y_1}{x_2 - x_1}$$
$$= \frac{5 - 5}{1 - (-4)}$$
$$= \frac{0}{5} = 0$$

10. Substitute −2 for m and (−3,5) for (x_1,y_1) into the point-slope form and simplify the result writing it in general form.

$$y - y_1 = m(x - x_1)$$
$$y - 5 = -2(x - (-3))$$
$$y - 5 = -2(x + 3)$$
$$y - 5 = -2x - 6$$
$$2x + y + 1 = 0$$

Thus, the general form is $2x + y + 1 = 0$.

11. Substitute $\frac{4}{3}$ for m and (3,0) for (x_1,y_1) into the point-slope formula and simplify to obtain the general form.

$$y - y_1 = m(x - x_1)$$
$$y - 0 = \frac{4}{3}(x - 3)$$
$$3y = 4(x - 3)$$
$$3y = 4x - 12$$
$$0 = 4x - 3y - 12$$

Thus, the general form is $4x - 3y - 12 = 0$.

12. A horizontal line is described by an equation of the form $y = c$, that is, all the y-coordinates are constant, and the x-coordinates can be any real number. For the line to pass through (6,−5), all the y-coordinates must be −5. Thus, the equation of the line is $y = -5$, or in general form, $y + 5 = 0$.

13. First find the slope of the line passing through the points (−3,4) and (2,6).

$$m = \frac{y_2 - y_1}{x_2 - x_1}$$
$$= \frac{6 - 4}{2 - (-3)} = \frac{2}{5}$$

Use $\frac{2}{5}$ for m and either point, say (2,6), for (x_1,y_1) in the point-slope form and simplify to obtain the general form.

$$y - y_1 = m(x - x_1)$$
$$y - 6 = \frac{2}{5}(x - 2)$$
$$5(y - 6) = 2(x - 2)$$
$$5y - 30 = 2x - 4$$
$$0 = 2x - 5y + 26$$

Thus, the general form is $2x - 5y + 26 = 0$.

14. First solve $5x - 10y + 3 = 0$ for y to find the slope of this line, the coefficient of x.

$$5x - 10y + 3 = 0$$
$$-10y = -5x - 3$$
$$y = \frac{-5}{-10}x + \frac{-3}{-10}$$
$$y = \frac{1}{2}x + \frac{3}{10}$$

Since perpendicular lines have equal slopes that are negative reciprocals, the slope of the desired line must be -2. Use -2 for m and $(-3,7)$ for (x_1,y_1) in the point-slope form and simplify to obtain the general form.

$$y - y_1 = m(x - x_1)$$
$$y - 7 = -2(x - (-3))$$
$$y - 7 = -2(x + 3)$$
$$y - 7 = -2x - 6$$
$$2x + y - 1 = 0$$

Thus, the general form is $2x + y - 1 = 0$.

15. Since $2y - 3 = 0$, or equivalently, $y = \frac{3}{2}$, is a horizontal line passing through y-intercept $\left(0,\frac{3}{2}\right)$, the desired line is a horizontal line passing through the point $(2,3)$. Since a horizontal line has the form $y = c$, the desired line has equation $y = 3$, or in general form, $y - 3 = 0$.

16. Since $7x + 4 = 0$, or equivalently, $x = -\frac{4}{7}$, is a vertical line passing through x-intercept $\left(-\frac{4}{7},0\right)$, the desired line is a vertical line passing through the point $(-5,0)$. Since a vertical line has the form $x = c$, the desired line has equation $x = -5$, or in general form, $x + 5 = 0$.

17. To find the slope and y-intercept of the line, write the equation in slope-intercept form, that is, solve for y.

$$2x - 7y + 5 = 0$$
$$-7y = -2x - 5$$
$$y = \frac{-2}{-7}x - \frac{5}{-7}$$
$$y = \frac{2}{7}x + \frac{5}{7}$$

Then the slope, the coefficient of x, is $\frac{2}{7}$, and the y-intercept is $\left(0,\frac{5}{7}\right)$.

18. Write the equation in slope-intercept form, that is, solve the equation for y. The desired slope is the coefficient of x, and the y-intercept has y-coordinate equal to the constant term.

$$8x = -3y + 7$$
$$8x - 7 = -3y$$
$$\frac{8}{-3}x - \frac{7}{-3} = y$$
$$-\frac{8}{3}x + \frac{7}{3} = y$$

Thus, the slope is $-\frac{8}{3}$, and the y-intercept is $\left(0,\frac{7}{3}\right)$.

19. Solving $-3y + 6 = 0$ for y, we obtain $-3y = -6$, or $y = 2$. Since this equation can be written as $y = 0x + 2$, we see that the slope is 0 and the y-intercept is $(0,2)$. Note that this is the equation of a horizontal line.

20. Write each equation in slope-intercept form by solving each for y.

$$6x - 2y + 7 = 0$$
$$-2y = -6x - 7$$
$$y = 3x + \frac{7}{2}$$

This line has slope 3.

EXERCISES A

$$x + 3y - 8 = 0$$
$$3y = -x + 8$$
$$y = -\frac{1}{3}x + \frac{8}{3}$$

This line has slope $-\frac{1}{3}$. Since the slopes are negative reciprocals, the lines are perpendicular.

21. Write each equation in slope-intercept form by solving each for y.

$$4x - 2y = -6$$
$$-2y = -4x - 6$$
$$y = 2x + 3$$

The slope of this line is 2.

$$10x = 5y + 8$$
$$10x - 8 = 5y$$
$$2x - \frac{8}{5} = y$$

The slope of this line is also 2. Since the two slopes are equal, the two lines are parallel.

22. Since $3y - 8 = 0$ is a horizontal line with y-intercept $\left(0, \frac{8}{3}\right)$, and $4x + 2 = 0$ is a vertical line with x-intercept $\left(-\frac{1}{2}, 0\right)$, the lines are perpendicular.

23. Find the slope of the line through P and Q, and then find the slope of the line through Q and R. Since the two lines share a common point, Q, if the slopes are equal, the lines must coincide making the three points collinear.

$$m_{PQ} = \frac{3-0}{4-2} = \frac{3}{2}$$
$$m_{QR} = \frac{0-(-6)}{2-(-2)} = \frac{6}{4} = \frac{3}{2}$$

Thus, since the two slopes are equal, the three points are collinear.

24. The two-point form is easy to derive by substituting the expression for the slope into the point-slope form.

That is, if $\dfrac{y_2 - y_1}{x_2 - x_1}$ is substituted for m in the equation $y - y_1 = m(x - x_1)$, we obtain the two-point form

$$y - y_1 = \frac{y_2 - y_1}{x_2 - x_1}(x - x_1).$$

Use this form with $(x_1, y_1) = (-3, 6)$ and $(x_2, y_2) = (2, -4)$ to find the equation of the line through these two points.

$$y - y_1 = \frac{y_2 - y_1}{x_2 - x_1}(x - x_1)$$
$$y - 6 = \frac{-4 - 6}{2 - (-3)}(x - (-3))$$
$$y - 6 = \frac{-10}{5}(x + 3)$$
$$y - 6 = -2(x + 3)$$
$$y - 6 = -2x - 6$$
$$2x + y = 0$$

25. First find the midpoint of the segment joining $(1, 7)$ and $(-3, -5)$. The perpendicular bisector of the segment must pass through this point.

$$(\bar{x}, \bar{y}) = \left(\frac{x_1 + x_2}{2}, \frac{y_1 + y_2}{2}\right)$$
$$= \left(\frac{1 + (-3)}{2}, \frac{7 + (-5)}{2}\right)$$
$$= \left(\frac{-2}{2}, \frac{2}{2}\right)$$
$$= (-1, 1)$$

Next find the slope of the segment joining the given points. The perpendicular bisector must have slope the negative reciprocal.

$$m = \frac{7 - (-5)}{1 - (-3)} = \frac{12}{4} = 3$$

Thus, the perpendicular bisector has slope $-\frac{1}{3}$.

Use the point slope form with $m = -\frac{1}{3}$ and $(x_1, y_1) = (-1, 1)$.

$$y - y_1 = m(x - x_1)$$
$$y - 1 = -\frac{1}{3}(x - (-1))$$
$$3(y - 1) = -1(x + 1)$$
$$3y - 3 = -x - 1$$
$$x + 3y - 2 = 0$$

Thus the perpendicular bisector of the segment has equation $x + 3y - 2 = 0$.

26. Letting x represent the year and y represent the value of the house in that year, we have two data points with coordinates (0, 85,000) and (5, 105,000). We must find the equation of the line that passes through these two points. First find the slope of the line using the slope formula.

$$m = \frac{y_2 - y_1}{x_2 - x_1}$$
$$= \frac{105{,}000 - 85{,}000}{5 - 0}$$
$$= \frac{20{,}000}{5} = 4000$$

Now use the point-slope form with $m = 4000$ and either point, say (0, 85,000) as (x_1, y_1).

$$y - y_1 = m(x - x_1)$$
$$y - 85{,}000 = 4000(x - 0)$$
$$y = 4000x + 85{,}000$$

To find the value of the house 7 years after it was purchased, substitute 7 for x.

$$y = 4000(7) + 85{,}000$$
$$= 28{,}000 + 85{,}000 = 113{,}000$$

Thus, the value of the house after 7 years is $113,000.

27. To find the number of years for the house in Exercise 26 to have a value of $125,000, substitute 125,000 for y and solve for x.

$$125{,}000 = 4000x + 85{,}000$$
$$40{,}000 = 4000x$$
$$\frac{40{,}000}{4000} = x$$
$$10 = x$$

Thus, in 10 years the value will be $125,000.

28. (a) Substitute 3 for n and simplify.

$$S = 1200(1 + 0.09(3))$$
$$= 1200(1 + 0.27)$$
$$= 1200(1.27) = 1524$$

Thus, $S = \$1524$ when $n = 3$.

(b) Substitute 1740 for S and solve for n.

$$1740 = 1200(1 + 0.09n)$$
$$1740 = 1200 + 108n$$
$$540 = 108n$$
$$\frac{540}{108} = n$$
$$5 = n$$

Thus, when $S = \$1740$, n is 5 yr.

29. We are given two data points, (16,3) and (36,8), where (t,g) represents t minutes and g grams.

(a) Find the equation of the line passing through these two points. First find the slope of the line.

$$m = \frac{8 - 3}{36 - 16}$$
$$= \frac{5}{20} = \frac{1}{4}$$

Now use the point-slope form with slope $\frac{1}{4}$ and the point (16,3).

$$g - 3 = \frac{1}{4}(t - 16)$$
$$g - 3 = \frac{1}{4}t - 4$$
$$g = \frac{1}{4}t - 1$$

(b) Substitute 20 for t and evaluate g.

$$g = \frac{1}{4}(20) - 1$$
$$= 5 - 1 = 4$$

Thus, 4 grams would be produced in 20 min.

(c) When $t = 0$, $g = -1$, which is impossible.

EXERCISES A SECTION 3.2

30. $d = \sqrt{(x_1 - x_2)^2 + (y_1 - y_2)^2}$
    $= \sqrt{(-1 - 4)^2 + (2 - 6)^2}$
    $= \sqrt{(-5)^2 + (-4)^2}$
    $= \sqrt{25 + 16} = \sqrt{41}$

    $(\bar{x}, \bar{y}) = \left(\dfrac{x_1 + x_2}{2}, \dfrac{y_1 + y_2}{2}\right)$
    $= \left(\dfrac{-1 + 4}{2}, \dfrac{2 + 6}{2}\right)$
    $= \left(\dfrac{3}{2}, \dfrac{8}{2}\right) = \left(\dfrac{3}{2}, 4\right)$

31. $d = \sqrt{(x_1 - x_2)^2 + (y_1 - y_2)^2}$
    $= \sqrt{(4 - (-4))^2 + (-5 - 2)^2}$
    $= \sqrt{(8)^2 + (-7)^2}$
    $= \sqrt{64 + 49} = \sqrt{113}$

    $(\bar{x}, \bar{y}) = \left(\dfrac{x_1 + x_2}{2}, \dfrac{y_1 + y_2}{2}\right)$
    $= \left(\dfrac{4 + (-4)}{2}, \dfrac{-5 + 2}{2}\right)$
    $= \left(\dfrac{0}{2}, \dfrac{-3}{2}\right) = \left(0, -\dfrac{3}{2}\right)$

32. The given expression will not be a real number when the denominator is 0, that is when

    $$(x + 1)(x - 5) = 0.$$

    Setting each factor equal to 0 and solving, we obtain $x = -1$ and $x = 5$. Thus, the expression will not be a real number when x is either of these two values.

33. The expression will not be a real number when the radicand is less than or equal to 0. Solve the inequality.

    $$x + 3 \leq 0$$
    $$x \leq -3$$

    Thus, the expression will not be real whenever $x \leq -3$.

CHAPTER 3 RELATIONS, FUNCTIONS, AND GRAPHS

SECTION 3.3 Relations and Functions

1. (a) Given $f(x) = 2x - 4$.

   $f(2) = 2(2) - 4 = 4 - 4 = 0$

   $f(-1) = 2(-1) - 4 = -2 - 4 = -6$

   $f(a - 1) = 2(a - 1) - 4$
   $= 2a - 2 - 4 = 2a - 6$

   $f(a + h) = 2(a + h) - 4$
   $= 2a + 2h - 4$

   (b) Given $f(x) = -5$.

   $f(2) = -5$

   $f(-1) = -5$

   $f(a - 1) = -5$

   $f(a + h) = -5$

   (c) Given $f(x) = -x$.

   $f(2) = -2$

   $f(-1) = -(-1) = 1$

   $f(a - 1) = -(a - 1) = -a + 1$

   $f(a + h) = -(a + h) = -a - h$

   (d) Given $f(x) = 3x^2 - 2x + 8$.

   $f(2) = 3(2)^2 - 2(2) + 8$
   $= 3(4) - 4 + 8$
   $= 12 - 4 + 8 = 16$

   $f(-1) = 3(-1)^2 - 2(-1) + 8$
   $= 3(1) + 2 + 8$
   $= 3 + 2 + 8 = 13$

   $f(a - 1) = 3(a - 1)^2 - 2(a - 1) + 8$
   $= 3(a^2 - 2a + 1) - 2a + 2 + 8$
   $= 3a^2 - 6a + 3 - 2a + 2 + 8$
   $= 3a^2 - 8a + 13$

   $f(a + h) = 3(a + h)^2 - 2(a + h) + 8$
   $= 3(a^2 + 2ah + h^2) - 2a - 2h + 8$
   $= 3a^2 + 6ah + 3h^2 - 2a - 2h + 8$

2. (a) Given $g(x) = \dfrac{x + 2}{x - 5}$.

   Then g is defined for all real numbers except those that make the denominator 0. Set $x - 5 = 0$ and solve for x.

   $x - 5 = 0$
   $x = 5$

   Thus, g is defined for all real numbers $x \neq 5$.

   (b) Given $f(x) = \dfrac{x + 1}{\sqrt{1 - x}}$.

   Then f is defined for all real numbers except those that make the denominator 0 and those that make the radicand negative. Set $1 - x > 0$ and solve for x.

   $1 - x > 0$
   $1 > x$

   Thus, f is defined for all $x < 1$.

   (c) Given $h(x) = \dfrac{\sqrt{2 + x}}{(x + 6)(x - 7)}$.

   There are three conditions that must be satisfied in order for h to be defined.

   $2 + x \geq 0$ or $x \geq -2$
   and
   $x + 6 \neq 0$ or $x \neq -6$
   and
   $x - 7 \neq 0$ or $x \neq 7$

   Thus, the domain of h is all real numbers x such that $x \geq -2$, $x \neq -6$, and $x \neq 7$.

3. A retailer buys dresses for $65.50 each.

   (a) If the markup is x percent, then the selling price is $65.50 + 65.50x = 65.50(1 + x)$. The tax rate is 6%, so the sales tax is $0.06(65.50)(1 + x) = 3.93(1 + x)$. Then the total price paid, the selling price plus the sales tax is given by the function

PRACTICE EXERCISES

$$T(x) = 65.50(1 + x) + 3.93(1 + x)$$
$$= [65.50 + 3.93](1 + x)$$
$$= 69.43(1 + x)$$

(b) Find $T(0.32)$, that is find the total price paid when the markup is 32%.

$$T(0.32) = 69.43(1 + 0.32)$$
$$= 69.43(1.32)$$
$$= 91.6476$$

Thus, the total price in this case is $91.65.

4. (a) The volume of a sphere is given by $V = \frac{4}{3}\pi r^3$.

If each cubic foot of liquid costs $0.75, the cost of filling the sphere with liquid is given by

$$(0.75)\frac{4}{3}\pi r^3.$$

The surface area of a sphere is given by $A = 4\pi r^2$. If each square foot of insulation costs $1.50, the cost to insulate the sphere is given by

$$(1.50)4\pi r^2.$$

Then the total cost for the project as a function of the radius r of the sphere is given by:

$$C(r) = (0.75)\frac{4}{3}\pi r^3 + (1.50)4\pi r^2$$
$$= \pi r^2[(0.75)\frac{4}{3}r + (1.50)4]$$
$$= \pi r^2(r + 6)$$

(b) Find $C(2.5)$, that is find the total cost of the project for a sphere of radius 2.5 ft.

$$C(2.5) = \pi(2.5)^2(2.5+6)$$
$$= \pi(6.25)(8.5)$$
$$= \pi(53.125)$$
$$\approx 166.8971097$$

Thus the total cost is about $166.90.

5. (a) Given the relation

$$\{(1,1),(2,3),(1,0)\}.$$

Since $(1,1)$ and $(1,0)$ have the same first coordinate but different second coordinates, 1 is related to both 1 and 0. Thus, the relation is not a function.

(b) Given the relation

$$\{(1,4),(5,4)\}.$$

Since no two ordered pairs have the same first coordinate and different second coordinates, this relation is a function. Note that 1 and 5 are both related to 4, but this is allowed for a relation to be a function.

CHAPTER 3 RELATIONS, FUNCTIONS, AND GRAPHS

SECTION 3.3 Relations and Functions

1. The relation $x = 2y + 5$ describes a function since for any given value of x, there is exactly one value of y associated with it.

2. The relation $y = 5x^2$ describes a function since for any given value of x, there is exactly one value of y associated with it.

3. The relation $y = \dfrac{3x^2 - 2}{x + 1}$ describes a function since for any given value of x, there is exactly one value of y associated with it.

4. To find the range of $f(x) = 5x - 3$ with domain $\{1, 2, 3\}$, substitute each of the values in the domain into the equation to find the corresponding value in the range.

$$f(1) = 5(1) - 3 = 5 - 3 = 2$$
$$f(2) = 5(2) - 3 = 10 - 3 = 7$$
$$f(3) = 5(3) - 3 = 15 - 3 = 12$$

Thus, the range of the function is $\{2, 7, 12\}$.

5. To find the range of $g(x) = \sqrt{x + 2}$, $x \geq -2$, note that the radical represents a nonnegative real number. Thus, the range of the function is all nonnegative real numbers.

6. To find the range of $h(x) = \dfrac{1}{x - 3}$, with domain all real numbers except 3, note that when x is any real number except 3, $\dfrac{1}{x - 3}$ can represent any real number except 0. Thus, the range of the function is all real numbers except 0.

7. Since $7x - 8$ is always a real number for any real number x, $f(x) = 7x - 8$ is defined for all real numbers.

8. The function

$$g(x) = \sqrt{2 - x}$$

is defined for all real numbers for which the radicand is nonnegative, that is, for all real numbers x such that $2 - x \geq 0$, or for which $x \leq 2$.

9. For the function

$$k(x) = \dfrac{\sqrt{x + 8}}{(x + 2)(x - 3)}$$

to be defined, the radicand, $x + 8$ must be nonnegative and the denominator cannot be zero. If $x + 8 \geq 0$, then $x \geq -8$, and if $(x + 2)(x - 3) = 0$, then $x = -2$ and $x = 3$. Thus the function will be defined when $x \geq -8$ and $x \neq -2$ and $x \neq 3$.

10. Since x is to the first power, $f(x) = x$ is a linear function. But in fact, in this case we can say even more, since the value of $f(x)$ for any given x is identically that x, f is the identity function.

11. Since x is to the first power, $h(x) = 1 - x$ is a linear function.

12. Since $k(x) = 14$, the value of the function for any x is the constant 14. Thus, k is a constant function.

13. Given $f(x) = 5x - 3$.

$$f(0) = 5(0) - 3 = 0 - 3 = -3$$
$$f(1) = 5(1) - 3 = 5 - 3 = 2$$
$$f(-3) = 5(-3) - 3 = -15 - 3 = -18$$
$$f(a) = 5(a) - 3 = 5a - 3$$
$$f(a - 1) = 5(a - 1) - 3$$
$$= 5a - 5 - 3$$
$$= 5a - 8$$

14. Given $f(x) = 7$. Since $f(x)$ is always 7 no matter what x is, $f(0) = 7$, $f(1) = 7$, $f(-3) = 7$, $f(a) = 7$, and $f(a - 1) = 7$. That is, f is a constant function.

15. Given $f(x) = |x - 5|$.

$$f(0) = |0 - 5| = |-5| = 5$$
$$f(1) = |1 - 5| = |-4| = 4$$
$$f(-3) = |-3 - 5| = |-8| = 8$$
$$f(a) = |a - 5|$$
$$f(a - 1) = |(a - 1) - 5| = |a - 6|$$

EXERCISES A

16. Given $h(x) = 3x - 7$.

$$h(a^2) = 3(a^2) - 7 = 3a^2 - 7$$
$$h\left(\frac{1}{a}\right) = 3\left(\frac{1}{a}\right) - 7 = \frac{3}{a} - 7$$
$$\frac{1}{h(a)} = \frac{1}{3a - 7}$$

17. Given $h(x) = \frac{1}{x} + 6$.

$$h(a^2) = \frac{1}{a^2} + 6$$
$$h\left(\frac{1}{a}\right) = \frac{1}{\frac{1}{a}} + 6 = a + 6$$
$$\frac{1}{h(a)} = \frac{1}{\frac{1}{a} + 6} = \frac{1}{\frac{1 + 6a}{a}} = \frac{a}{1 + 6a}$$

18. Given $f(x) = x + 5$.

$$f(1 + h) = (1 + h) + 5 = 1 + h + 5 = h + 6$$
$$\frac{f(1+h) - f(1)}{h} = \frac{[h+6] - [(1)+5]}{h}$$
$$= \frac{[h+6] - [6]}{h}$$
$$= \frac{h + 6 - 6}{h}$$
$$= \frac{h}{h} = 1$$
$$f(x + h) = (x + h) + 5 = x + h + 5$$
$$\frac{f(x+h) - f(x)}{h} = \frac{[x+h+5] - [x+5]}{h}$$
$$= \frac{x + h + 5 - x - 5}{h}$$
$$= \frac{h}{h} = 1$$

19. Given $f(x) = x^2 + 2$.

$$f(1 + h) = (1 + h)^2 + 2 = (1 + 2h + h^2) + 2 = h^2 + 2h + 3$$
$$\frac{f(1+h) - f(1)}{h} = \frac{[h^2 + 2h + 3] - [(1)^2 + 2]}{h}$$
$$= \frac{[h^2 + 2h + 3] - [1 + 2]}{h}$$
$$= \frac{h^2 + 2h + 3 - 3}{h} = h + 2$$

$$f(x + h) = (x + h)^2 + 2 = (x^2 + 2xh + h^2) + 2 = x^2 + 2xh + h^2 + 2$$
$$\frac{f(x+h) - f(x)}{h} = \frac{[x^2 + 2xh + h^2 + 2] - [x^2 + 2]}{h}$$
$$= \frac{x^2 + 2xh + h^2 + 2 - x^2 - 2}{h}$$
$$= \frac{2xh + h^2}{h}$$
$$= \frac{h(2x + h)}{h} = 2x + h$$

20. Given $\{(1,2),(-1,3)\}$.

Since no two ordered pairs have the same first coordinate and different second coordinates, this relation is a function.

21. Given $\{(2,5),(4,6),(2,7)\}$.

Since 2 is paired with two different values, 5 and 7, this relation is not a function.

22. Given $\{(3,-1),(4,-1),(5,-1)\}$.
Since no two ordered pairs have the same first coordinate and different second coordinates, this relation is a function.

23. Given $\{(4,3),(4,2),(4,8),(4,9)\}$.

Since 4 is paired with four different values, 3, 2, 8, and 9, this relation is not a function.

24. (a) Let c be the cost of one dress. Then 28.5% of c, or $(0.285)c$ is the mark up. The function S that represents the selling price is
$$S(c) = c + (0.285)c = (1 + 0.285)c = 1.285c.$$

(b) Substitute 56 for c and evaluate.
$$S(56) = 1.285(56) = 71.96$$
Thus the selling price would be $71.96.

25. (a) The markup on pants is 25%. Let x be the discount percent applied to those that do not sell. Suppose a pair of pants cost $40. Then the markup would be 25% of $40, which is $0.25(40) = \$10$. Then the selling price is $40 + $10 = $50. If the pants do not sell, they are discounted by x percent, that is by $50x$, making the sale price $50 - 50x$. Thus,
$$S(x) = 50 - 50x = 50(1 - x)$$
is the function used to describe the discounted price of the pants.

(b) If the sales-tax rate is 5%, the tax on the discounted pants would be would be 5% of $50(1 - x)$ which is

$$0.05(50)(1 - x) = 2.5(1 - x).$$

The total paid, including the tax is

$$\begin{aligned}T(x) &= 50(1 - x) + 2.5(1 - x) \\ &= [50 + 2.5](1 - x) \\ &= 52.5(1 - x)\end{aligned}$$

(c) If the sale price of the pants (excluding sales tax) were to be the same as the original cost, then

$$50(1 - x) = 40.$$

Solve this equation for x.

$$\begin{aligned}50(1 - x) &= 40 \\ 50 - 50x &= 40 \\ -50x &= -10\end{aligned}$$

$$x = \frac{-10}{-50} = \frac{10}{50} = \frac{1}{5} = 0.20$$

Thus, the discount rate would have to be 20%.

26. (a) The surface area of a sphere with radius r is given by

$$A = 4\pi r^2.$$

Since 1 ft^2 of insulation costs $1.25, the function

$$T(r) = (1.25)(4\pi r^2) = 5\pi r^2$$

represents the cost for the insulation job as a function of the radius r.

(b) Substitute 9.6 for r.

$$T(9.6) = 5\pi(9.6)^2 \approx 1447.645895$$

Thus, the cost to insulate a tank with radius 9.6 ft is about $1447.65.

27. (a) Since $265.9x^2 - 36.8$ increases as x varies from 0.75 to 0.95, we can find the range by evaluating the function at 0.75 and 0.95.

$$265.9(0.75)^2 - 36.8 \approx 112.76875$$

$$265.9(0.95)^2 - 36.8 \approx 203.17475$$

Thus, the range is given by

$$\$112.77 \le C(x) \le \$203.17.$$

(b) If x is 0.30, that is if $x = 30\%$, then

$$C(0.30) = 265.9(0.30)^2 - 36.8 \approx -12.869$$

Since $C(0.30)$ is negative, this is meaningless as the cost for extracting metal from a ton of ore. Note that 0.30 is not in the domain of the function.

28. First find the slope of the line passing through the points $(-1,3)$ and $(2,4)$.

$$m = \frac{y_2 - y_1}{x_2 - x_1} = \frac{4 - 3}{2 - (-1)} = \frac{1}{3}$$

Use the point-slope form with $m = \frac{1}{3}$ and $(x_1, y_1) = (2, 4)$.

$$\begin{aligned}y - y_1 &= m(x - x_1) \\ y - 4 &= \tfrac{1}{3}(x - 2) \\ 3(y - 4) &= x - 2 \\ 3y - 12 &= x - 2 \\ 0 &= x - 3y + 10\end{aligned}$$

Thus, the line has equation $x - 3y + 10 = 0$.

29. The line with equation $2y - 7 = 0$ is a horizontal line. Thus, the line through $(8,5)$ that is perpendicular to this line must be a vertical line with all x-coordinates equal to 8, that is the line with equation $x = 8$, or in general form, $x - 8 = 0$.

30. The line $2x - 5 = 0$, or $x = \frac{5}{2}$, is a vertical line with x-intercept $\left(\frac{5}{2}, 0\right)$ and no y-intercept.

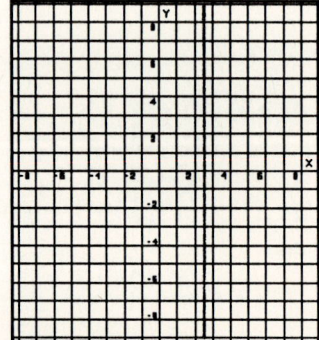

PRACTICE EXERCISES SECTION 3.4

CHAPTER 3 RELATIONS, FUNCTIONS, AND GRAPHS

SECTION 3.4 Properties of Functions and Transformations

1. First graph $h(x) = x^2 - 3$ by making a table of values. Note that the graph is the reflection in the x-axis of the graph given in Figure 3.23. The graph is given below.

   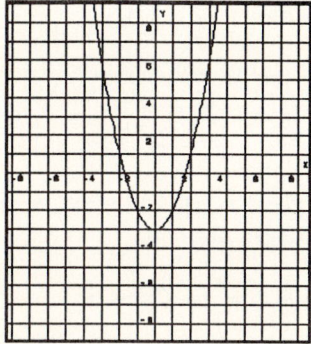

   From the graph we can see that h is increasing for $x \geq 0$ and decreasing for $x \leq 0$.

2. First graph $f(x) = -|x|$ by making a table of values. Note that the graph is the reflection in the x-axis of the graph given in Figure 3.22. The graph is given below.

   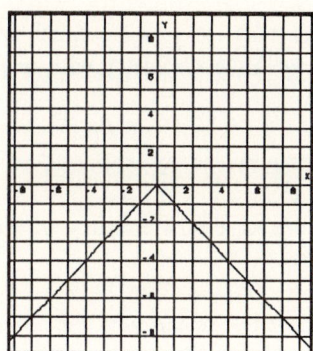

   (a) To obtain the graph of $g(x) = -|x - 1| + 2$ from the graph of $f(x) = -|x|$, shift the graph of f up two units (from the + 2) then right 1 unit (from the - 1). The graph of g is given below.

   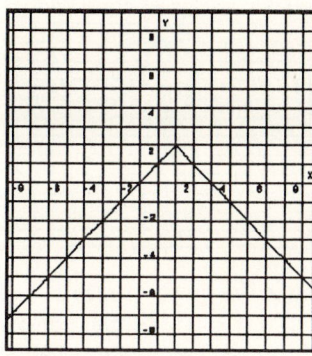

   (b) To obtain the graph of $h(x) = -|x + 2| - 1$ from the graph of $f(x) = -|x|$, shift the graph of f down one unit (from the - 1) then left 2 units (from the + 2). The graph of h is given below.

   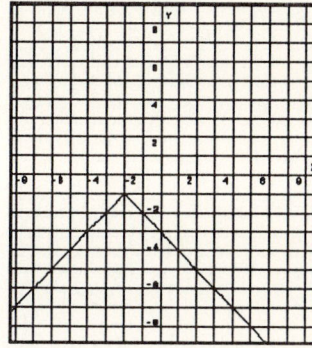

3. First graph $f(x) = x^2$ by making a table of values. The graph is given below.

   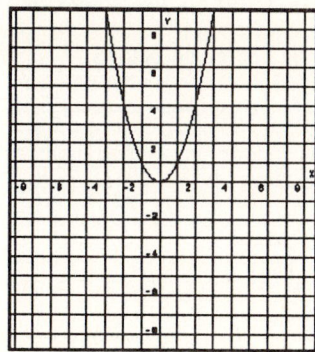

   (a) To obtain the graph of $g(x) = 2x^2$ from the graph of $f(x) = x^2$, all the ordinates (y-coordinates) are multiplied by 2. The graph of g is given below.

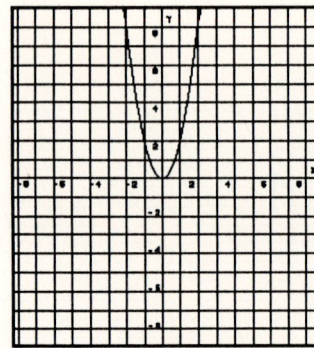

(b) To obtain the graph of $h(x) = -2x^2$ from the graph of $f(x) = x^2$, all the ordinates (y-coordinates) are multiplied by -2. The graph can also be obtained from your graph in part **(a)** by reflecting the graph in the x-axis. The graph of h is given below.

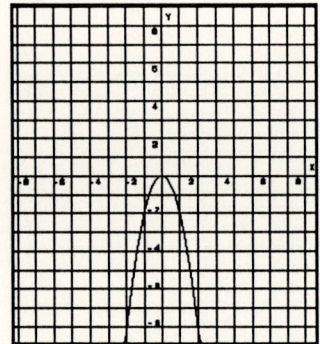

(c) To obtain the graph of $k(x) = \frac{1}{4}x^2$ from the graph of $f(x) = x^2$, all the ordinates (y-coordinates) are multiplied by $\frac{1}{4}$. The graph of k is given below.

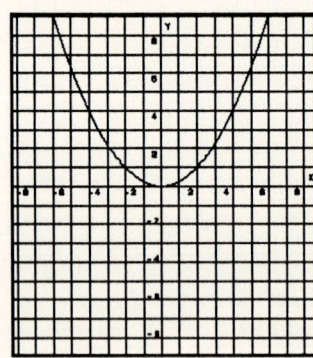

4. Recall that a function f is even if $f(-x) = f(x)$ and odd if $f(-x) = -f(x)$.

(a) Given $f(x) = x^4 + 2x^2$.

$$f(-x) = (-x)^4 + 2(-x)^2$$
$$= x^4 + 2x^2$$
$$= f(x)$$

Thus f is an even function.

(b) Given $g(x) = 2x^3 + 1$.

$$g(-x) = 2(-x)^3 + 1$$
$$= 2(-x^3) + 1$$
$$= -2x^3 + 1$$

Since $g(-x) \neq g(x)$ and $g(-x) \neq -g(x)$, g is neither even nor odd.

(c) Given $h(x) = x - 2x^3 + x^5$.

$$h(-x) = (-x) - 2(-x)^3 + (-x)^5$$
$$= -x + 2x^3 - x^5$$
$$= -(x - 2x^3 + x^5)$$
$$= -h(x)$$

Thus h is an odd function.

5. (a) Given $y = 1 - x^2$.

Replace x by $-x$: $y = 1 - (-x)^2 = 1 - x^2$
Thus, we have symmetry with respect to the y-axis.

Replace y by $-y$: $-y = 1 - x^2$, or $y = -1 + x^2$
Thus, we have no symmetry with respect to the x-axis.

Replace x by $-x$ and y by $-y$: $-y = 1 - (-x)^2$, or $y = -1 + x^2$
Thus, we have no symmetry with respect to the origin.

(b) Given $x = 1 - y^2$.

Replace x by $-x$: $-x = 1 - y^2$, or $x = -1 + y^2$
Thus, we have no symmetry with respect to the y-axis.

Replace y by $-y$: $x = 1 - (-y)^2$, or $x = 1 - y^2$
Thus, we have symmetry with respect to the x-axis.

PRACTICE EXERCISES

Replace x by $-x$ and y by $-y$: $-x = 1 - (-y)^2$, or $x = -1 + y^2$
Thus, we have no symmetry with respect to the origin.

(c) Given $|x| + |y| = 1$.

Replace x by $-x$: $|-x| + |y| = 1$, or $|x| + |y| = 1$
Thus, we have symmetry with respect to the y-axis.

Replace y by $-y$: $|x| + |-y| = 1$, or $|x| + |y| = 1$
Thus, we have symmetry with respect to the x-axis.

Replace x by $-x$ and y by $-y$: $|-x| + |-y| = 1$, or $|x| + |y| = 1$
Thus, we have symmetry with respect to the origin.

(c) Given $x^2y + xy^2 = 5$.

Replace x by $-x$: $(-x)^2y + (-x)y^2 = 5$ or $x^2y - xy^2 = 5$
Thus, we have no symmetry with respect to the y-axis.

Replace y by $-y$: $x^2(-y) + x(-y)^2 = 5$ or $-x^2y + xy^2 = 5$
Thus, we have no symmetry with respect to the x-axis.

Replace x by $-x$ and y by $-y$:
$(-x)^2(-y) + (-x)(-y)^2 = 5$ or $-x^2y - xy^2 = 5$
Thus, we have no symmetry with respect to the origin.

CHAPTER 3 RELATIONS, FUNCTIONS, AND GRAPHS

SECTION 3.4 Properties of Functions and Transformations

1. Since every vertical line passes through exactly one point on the graph, the vertical line test shows that the graph is the graph of a function.

2. Since it is possible to pass a vertical line through the graph and cross the curve in more than one point, in fact in two points, the vertical line test shows that the graph is not the graph of a function.

3. Since every vertical line passes through exactly one point on the graph, the vertical line test shows that the graph is the graph of a function.

4. From the graph of $f(x) = 5x - 3$ given below, we can see that whenever $a < b$, $f(a) < f(b)$, so the function increases for all x.

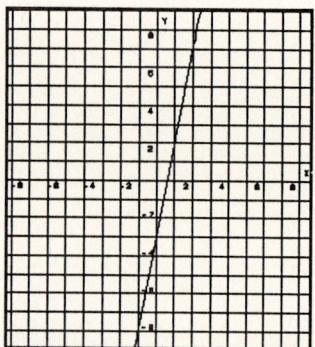

5. From the graph of $h(x) = -4$ given below, we can see that whenever $a < b$, $h(a) = h(b) = -4$, so the function is constant for all x. That is, it neither increases nor decreases for all x.

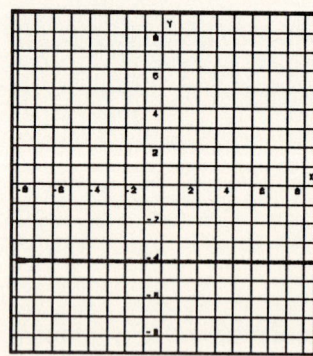

6. From the graph of $k(x) = |x| - 2$ given below, we can see that whenever $a < b \le 0$, $k(a) > k(b)$ so the function is decreasing for $x \le 0$. Also, whenever $0 \le a < b$, $k(a) < k(b)$ so the function is increasing for $x \ge 0$.

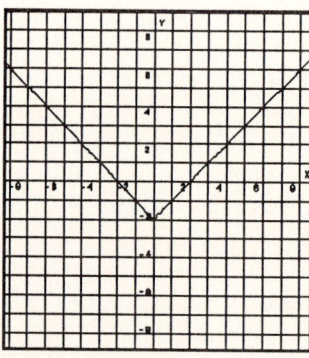

7. From the graph of $g(x) = -(x - 3)^2$ given below, we can see that whenever $a < b \le 3$, $g(a) < g(b)$ so the function is increasing for $x \le 3$. Also, whenever $3 \le a < b$, $g(a) > g(b)$ so the function is decreasing for $x \ge 3$.

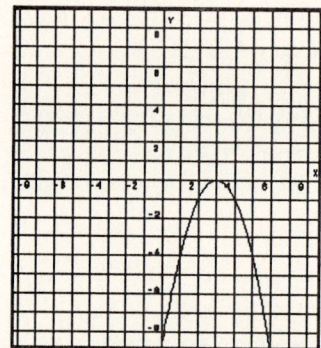

8. To graph $y = f(x) + 3$, slide the graph of $y = f(x)$ up 3 units as shown below.

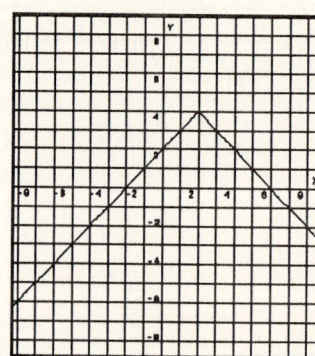

9. To graph $y = f(x + 3)$, slide the graph of $y = f(x)$ left 3 units as shown below.

EXERCISES A

SECTION 3.4

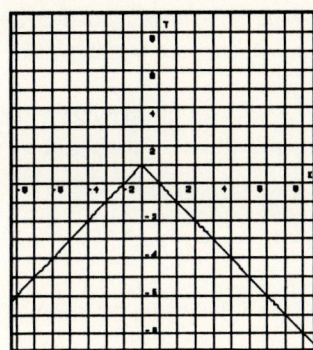

10. To graph $y = -f(x)$, all of the y-values of the graph of $y = f(x)$ must be multiplied by -1. This "reflects" the graph in the x-axis, as shown below.

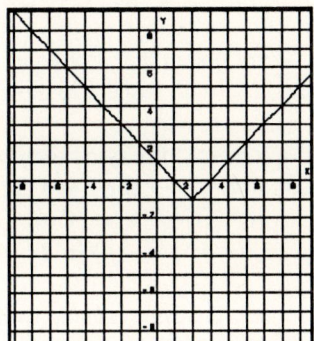

11. To graph $y = f(x + 3) - 2$, slide the graph of $y = f(x)$ left 3 units and then down 2 units as shown below.

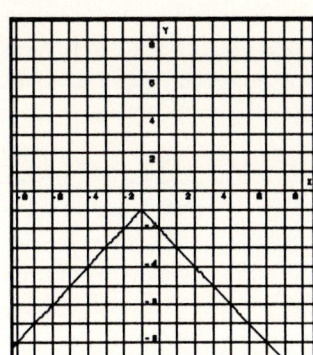

12. To graph $y = g(x) - 1$, slide the graph of $y = g(x)$ down 1 unit as shown below.

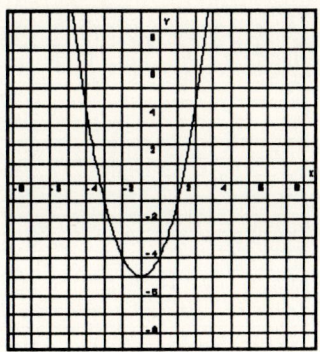

13. To graph $y = g(x - 3)$, slide the graph of $y = g(x)$ right 3 units as shown below.

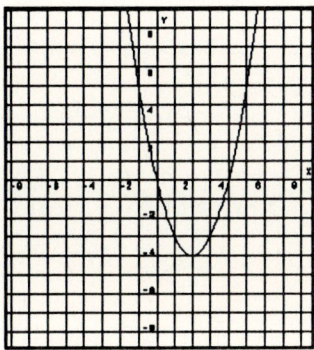

14. To graph $y = \frac{1}{2} g(x)$, all of the y-values of $y = g(x)$ are multiplied by $\frac{1}{2}$. This has the effect of "shrinking" the graph vertically as shown below.

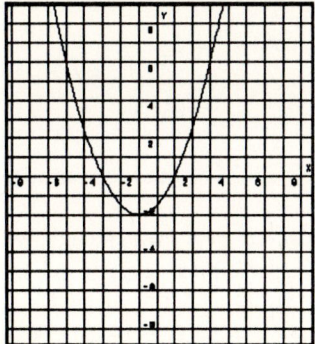

15. To graph $y = g(x - 2) + 3$, slide the graph of $y = g(x)$ right 2 units then up 3 units as shown below.

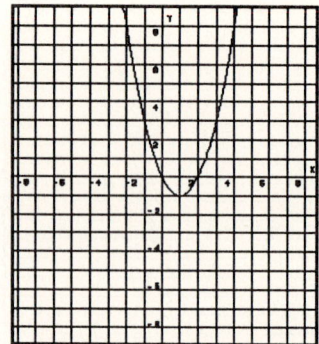

16. Since $y = g(x)$ is 3 units directly below $y = f(x)$, $g(x) = f(x) - 3$.

17. Since $y = h(x)$ is 4 units right of $y = g(x)$ then up from this position 3 units, $h(x) = g(x - 4) + 3$.

18. Given $f(x) = 2x^3 - 3x$. Substitute $-x$ for x and simplify.

$$f(-x) = 2(-x)^3 - 3(-x)$$
$$= -2x^3 + 3x$$
$$= -f(x)$$

Since $f(-x) = -f(x)$, the function is odd.

19. Given $h(x) = 0$. Substitute $-x$ for x to obtain $h(-x) = 0$. Since $h(-x) = h(x)$, the function is even. But also, $-h(x) = -0 = 0$, so $h(-x) = -h(x)$ making h odd. Thus, h is both even and odd.

20. Given $k(x) = 3|x| - 5$. Substitute $-x$ for x and simplify.

$$k(-x) = 3|-x| - 5$$
$$= 3|x| - 5$$
$$= k(x)$$

Since $k(-x) = k(x)$, the function is even.

21. To test $y^2 + x^2 = 8$ for symmetry with respect to the x-axis, substitute $-y$ for y and simplify.

$$(-y)^2 + x^2 = y^2 + x^2 = 8$$

Since we obtain the same equation the graph is symmetric with respect to the x-axis.

To test $y^2 + x^2 = 8$ for symmetry with respect to the y-axis, substitute $-x$ for x and simplify.

$$y^2 + (-x)^2 = y^2 + x^2 = 8$$

Since we obtain the same equation the graph is symmetric with respect to the y-axis.

To test $y^2 + x^2 = 8$ for symmetry with respect to the origin, substitute $-x$ for x and $-y$ for y and simplify.

$$(-y)^2 + (-x)^2 = y^2 + x^2 = 8$$

Since we obtain the same equation, the graph is symmetric with respect to the origin.

22. To test $y = 3x^2 + x^4$ for symmetry with respect to the x-axis, substitute $-y$ for y and simplify.

$$-y = 3x^2 + x^4 \text{ or } y = -3x^2 - x^4$$

Since we do not obtain the same equation, the graph is not symmetric with respect to the x-axis.

To test $y = 3x^2 + x^4$ for symmetry with respect to the y-axis, substitute $-x$ for x and simplify.

$$y = 3(-x)^2 + (-x)^4 = 3x^2 + x^4$$

Since we obtain the same equation, the graph is symmetric with respect to the y-axis.

To test for symmetry with respect to the origin, substitute $-x$ for x and $-y$ for y and simplify.

$$-y = 3(-x)^2 + (-x)^4 \text{ or } y = -3x^2 - x^4$$

Since we do not obtain the same equation, the graph is not symmetric with respect to the origin.

23. When $-x$ is substituted for x, or when $-y$ is substituted for y, or when both $-x$ is substituted for x and $-y$ is substituted for y at the same time, in all cases the resulting equation is not the same as $y = -3x^2 + 2x$. Thus the graph is not symmetric with respect to either axis or the origin.

24. When $-x$ is substituted for x, or when $-y$ is substituted for y, the resulting equation is not the same as the original, $x^2y + xy^2 = 0$. However, when $-x$ is substituted for x and $-y$ is substituted for y at the same time, the resulting equation is $-x^2y - xy^2 = 0$, which can be transformed to the original by multiplying both sides by -1. Thus the graph is symmetric with respect to the origin.

EXERCISES A

SECTION 3.4   111

25. To graph $f(x)$, plot $f(x) = 3$ for all values of x satisfying $x \leq 1$ using a solid circle at the point $(1,3)$. Then plot $f(x) = -2$ for all values of x satisfying $x > 1$ using an open circle at the point $(1,-2)$. The graph of the function is given below.

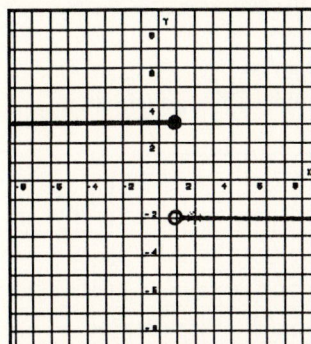

26. As long as $x \neq -2$, we can reduce the fraction.

$$\frac{x^2 - 4}{x + 2} = \frac{(x + 2)(x - 2)}{x + 2} = x - 2$$

Thus for all values of $x \neq -2$, $h(x) = x - 2$ which is simply linear function with graph a straight line with one point removed, the point with coordinates $(-2,-4)$. When $x = -2$, $h(-2) = -1$, so the point with x-coordinate -2 that is on the graph of the function is $(-2,-1)$. The graph is given below.

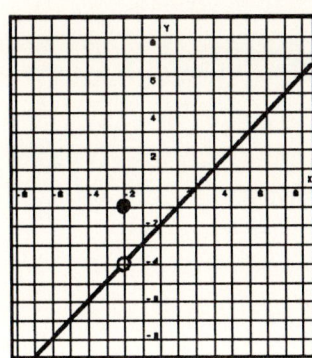

27. The graph of $k(x) = [x] + 2$ can be obtained from the graph of the greatest integer function $y = [x]$ by sliding the graph 2 units up as shown below.

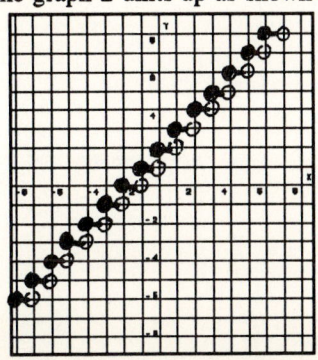

28. Given $f(x) = x^2 - 3$.

$$\begin{aligned} f(-2) &= (-2)^2 - 3 \\ &= 4 - 3 = 1 \end{aligned}$$

$$\begin{aligned} f(x - 1) &= (x - 1)^2 - 3 \\ &= (x^2 - 2x + 1) - 3 \\ &= x^2 - 2x - 2 \end{aligned}$$

$$\begin{aligned} f(x^2) &= (x^2)^2 - 3 \\ &= x^4 - 3 \end{aligned}$$

29. (a) Let x represent the previous salary of each employee. Then an 8% increase would amount to 8% of x which is $0.08x$. If in addition each received a $1200 increase, the function that represents the new salary would be

$$\begin{aligned} S(x) &= x + 0.08x + 1200. \\ &= 1.08(x) + 1200. \end{aligned}$$

(b) If a laboratory technician was making $18,000, her new salary would be given by $S(18,000)$.

$$\begin{aligned} S(18,000) &= 1.08(18,000) + 1200 \\ &= 19440 + 1200 \\ &= 20,640 \end{aligned}$$

Thus, her new salary would be $20,640.

30. $\begin{aligned} d &= \sqrt{(x_2 - x_1)^2 + (y_2 - y_1)^2} \\ &= \sqrt{(-2 - 5)^2 + (6 - 3)^2} \\ &= \sqrt{(-7)^2 + (3)^2} \\ &= \sqrt{49 + 9} = \sqrt{58} \end{aligned}$

CHAPTER 3 RELATIONS, FUNCTIONS, AND GRAPHS

SECTION 3.5 Composite and Inverse Functions

1. Given $f(x) = 1 - x$ and $g(x) = -2x^2 + 3$.

   $$\begin{aligned} g(f(x)) &= g(1 - x) \\ &= -2(1 - x)^2 + 3 \\ &= -2(1 - 2x + x^2) + 3 \\ &= -2 + 4x - 2x^2 + 3 \\ &= -2x^2 + 4x + 1 \end{aligned}$$

   $$\begin{aligned} f(g(x)) &= f(-2x^2 + 3) \\ &= 1 - (-2x^2 + 3) \\ &= 1 + 2x^2 - 3 \\ &= 2x^2 - 2 \end{aligned}$$

2. Given $r(t) = 2t$ and $C(r) = 2630r^2 + 5820r + 9320$.

   (a) To find C as a function of time t, we must evaluate $C(r(t))$.

   $$\begin{aligned} C(r(t)) &= C(2t) \\ &= 2630(2t)^2 + 5820(2t) + 9320 \\ &= 2630(4t^2) + 5820(2t) + 9320 \\ &= 10{,}520t^2 + 11{,}640t + 9320 \end{aligned}$$

   (b) The cost of cleaning up the spill when $t = 40$ is $C(40)$.

   $$\begin{aligned} C(40) &= 10{,}520(40)^2 + 11{,}640(40) + 9320 \\ &= 10{,}520(1600) + 465{,}600 + 9320 \\ &= 16{,}832{,}000 + 465{,}600 + 9320 \\ &= 17{,}306{,}920 \end{aligned}$$

   Thus, the cost of cleaning up the spill after 40 hours is \$17,306,920.

3. To determine whether a function is one-to-one, it is easier if we can look at the graph of the function.

   (a) Given $f(x) = 1 - x^2$.
   The graph of f is given below.

   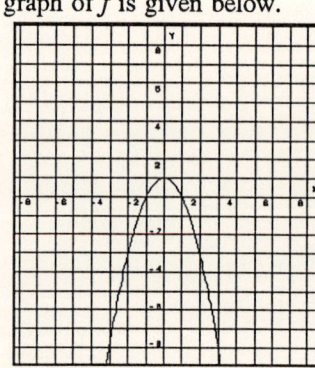

   From the graph we can see that f is increasing for $x \leq 0$ and decreasing for $x \geq 0$ and that the graph does not pass the horizontal line test. Thus, f is not a one-to-one function.

   (b) Given $f(x) = |x| + 1, x > 0$.
   The graph of f is given below.

   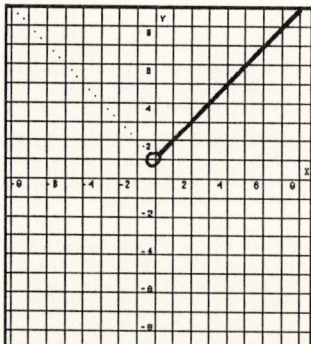

   Since f is an increasing function, f is one-to-one. Notice that the graph also passes the horizontal line test.

4. To find the inverse of a function, interchange x and y and solve the result for y. Keep in mind that the function must be one-to-one to have an inverse.

   (a) Given $f(x) = -2x + 6$.

   Remember that $y = f(x)$, so interchange x and y.

   $$x = -2y + 6$$

   Now solve for y.

   $$x - 6 = -2y$$

   $$-\tfrac{1}{2}x + 3 = y$$

   Thus, $f^{-1}(x) = -\tfrac{1}{2}x + 3$.

   (b) Given $g(x) = \sqrt{2 - x}, \ x \leq 2$.

   Since $y = g(x)$, notice that $y \geq 0$. Interchange x and y.

   $$x = \sqrt{2 - y}, \quad y \leq 2, \ x \geq 0$$

PRACTICE EXERCISES

Now solve for y.

$$x^2 = 2 - y$$
$$x^2 - 2 = -y$$
$$-x^2 + 2 = y$$

Thus, remember that $x \geq 0$, so the inverse is

$$g^{-1}(x) = -x^2 + 2, \quad x \geq 0.$$

(c) Given $h(x) = (x + 1)^2$. Since h is decreasing for $x \leq -1$ and increasing for $x \geq -1$, h is not one-to-one. Thus, h does not have an inverse.

(d) Given $k(x) = x^3 - 3$.

Remember that $y = k(x)$, so interchange x and y.

$$x = y^3 - 3$$

Now solve for y.

$$x + 3 = y^3$$
$$\sqrt[3]{x + 3} = y$$

Thus, $k^{-1}(x) = \sqrt[3]{x + 3}$.

CHAPTER 3 RELATIONS, FUNCTIONS, AND GRAPHS

SECTION 3.5 Composite and Inverse Functions

1. Given $f(x) = 5x$ and $g(x) = 3x - 2$.

    $(f + g)(x) = (5x) + (3x - 2)$
    $= 5x + 3x - 2$
    $= 8x - 2$

    $(f - g)(x) = (5x) - (3x - 2)$
    $= 5x - 3x + 2$
    $= 2x + 2$

    $(fg)(x) = (5x)(3x - 2)$
    $= 15x^2 - 10x$

    $\left(\dfrac{f}{g}\right)(x) = \dfrac{5x}{3x - 2}$

    The quotient will not be defined when $3x - 2 = 0$, that is, when $3x = 2$, or when $x = \dfrac{2}{3}$.

2. Given $f(x) = 2x^2$ and $g(x) = -x$.

    $(f + g)(x) = (2x^2) + (-x)$
    $= 2x^2 - x$

    $(f - g)(x) = (2x^2) - (-x)$
    $= 2x^2 + x$

    $(fg)(x) = (2x^2)(-x)$
    $= -2x^3$

    $\left(\dfrac{f}{g}\right) = \dfrac{2x^2}{-x}$
    $= -2x \quad \text{if } x \neq 0$

3. Given $f(x) = -2x + 1$ and $g(x) = x^2 + 2$.

    $(f + g)(x) = (-2x + 1) + (x^2 + 2)$
    $= -2x + 1 + x^2 + 2$
    $= x^2 - 2x + 3$

    $(f - g)(x) = (-2x + 1) - (x^2 + 2)$
    $= -2x + 1 - x^2 - 2$
    $= -x^2 - 2x - 1$

    $(fg)(x) = (-2x + 1)(x^2 + 2)$
    $= -2x^3 - 4x + x^2 + 2$
    $= -2x^3 + x^2 - 4x + 2$

    $\left(\dfrac{f}{g}\right)(x) = \dfrac{-2x + 1}{x^2 + 2}$

    Since the denominator, $x^2 + 2$, is never 0 for any real value of x, there are no values to exclude.

4. Given $f(x) = x^3 + 1$ and $g(x) = \sqrt[3]{x - 1}$.

    $(f + g)(x) = x^3 + 1 + \sqrt[3]{x - 1}$
    $(f - g)(x) = x^3 + 1 - \sqrt[3]{x - 1}$
    $(fg)(x) = (x^3 + 1)\sqrt[3]{x - 1}$
    $\left(\dfrac{f}{g}\right) = \dfrac{x^3 + 1}{\sqrt[3]{x - 1}} \quad \text{if } \sqrt[3]{x - 1} \neq 0, \ (x \neq 1)$

5. Given $f(x) = 5x$ and $g(x) = 3x - 2$.

    Since $g(2) = 3(2) - 2 = 6 - 2 = 4$,

    $f(g(2)) = f(4) = 5(4) = 20$.

    Since $f(-1) = 5(-1) = -5$,

    $g(f(-1)) = g(-5) = 3(-5) - 2 = -15 - 2 = -17$.

    $f(g(x)) = f(3x - 2)$
    $= 5(3x - 2)$
    $= 15x - 10$

    $g(f(x)) = g(5x)$
    $= 3(5x) - 2$
    $= 15x - 2$

6. Given $f(x) = 2x^2$ and $g(x) = -x$.

    Since $g(2) = -(2) = -2$,

    $f(g(2)) = f(-2) = 2(-2)^2 = 2(4) = 8$.

    Since $f(-1) = 2(-1)^2 = 2(1) = 2$,

    $g(f(-1)) = g(2) = -(2) = -2$.

    $f(g(x)) = f(-x)$
    $= 2(-x)^2$
    $= 2x^2$

EXERCISES A

$$g(f(x)) = g(2x^2)$$
$$= -(2x^2)$$
$$= -2x^2$$

7. Given $f(x) = -2x + 1$ and $g(x) = x^2 + 2$.

   Since $g(2) = (2)^2 + 2 = 4 + 2 = 6$,

   $$f(g(2)) = f(6)$$
   $$= -2(6) + 1$$
   $$= -12 + 1 = -11.$$

   Since $f(-1) = -2(-1) + 1 = 2 + 1 = 3$,

   $$g(f(-1)) = g(3)$$
   $$= (3)^2 + 2$$
   $$= 9 + 2 = 11.$$

   $$f(g(x)) = f(x^2 + 2)$$
   $$= -2(x^2 + 2) + 1$$
   $$= -2x^2 - 4 + 1$$
   $$= -2x^2 - 3$$

   $$g(f(x)) = g(-2x + 1)$$
   $$= (-2x + 1)^2 + 2$$
   $$= 4x^2 - 4x + 1 + 2$$
   $$= 4x^2 - 4x + 3$$

8. Given $f(x) = x^3 + 1$ and $g(x) = \sqrt[3]{x-1}$.

   Since $g(2) = \sqrt[3]{2-1} = \sqrt[3]{1} = 1$,

   $$f(g(2)) = f(1)$$
   $$= (1)^3 + 1$$
   $$= 1 + 1 = 2.$$

   Since $f(-1) = (-1)^3 + 1 = -1 + 1 = 0$,

   $$g(f(-1)) = g(0) = \sqrt[3]{0-1} = \sqrt[3]{-1} = -1$$

   $$f(g(x)) = f(\sqrt[3]{x-1})$$
   $$= (\sqrt[3]{x-1})^3 + 1$$
   $$= x - 1 + 1 = x$$

   $$g(f(x)) = g(x^3 + 1)$$
   $$= \sqrt[3]{(x^3 + 1) - 1}$$
   $$= \sqrt[3]{x^3} = x$$

9. Since for any two different values of x, $f(x)$ will always have different values, f is a one-to-one function.

10. Since $h(0) = |0 + 1| = |1| = 1$, and $h(-2) = |-2 + 1| = |-1| = 1$, two different values of x can have the same functional value. Thus, h is not one-to-one.

11. Since for any two different values of x, $k(x)$ will always have different values, k is a one-to-one function.

12. The function $\{(4,5),(6,7)\}$ is one-to-one since no two pairs have the same second coordinate and different first coordinates.

13. The function $\{(8,9),(9,9)\}$ is not one-to-one since the two pairs have the same second coordinate and different first coordinates.

14. The function $\{(2,1),(3,4),(4,1)\}$ is not one-to-one since the pairs $(2,1)$ and $(4,1)$ have the same second coordinates but different first coordinates.

15. Given $f(x) = \sqrt{x-2}$ and $g(x) = x^2 + 2$, $x \geq 0$.

    Since

    $$f(g(x)) = f(x^2 + 2)$$
    $$= \sqrt{(x^2 + 2) - 2}$$
    $$= \sqrt{x^2}$$
    $$= |x| = x \quad (x \geq 0),$$

    and

    $$g(f(x)) = g(\sqrt{x-2})$$
    $$= (\sqrt{x-2})^2 + 2$$
    $$= x - 2 + 2 = x,$$

    f and g are inverses. The graphs, given below, can be seen to be symmetric with respect to the line $y = x$.

    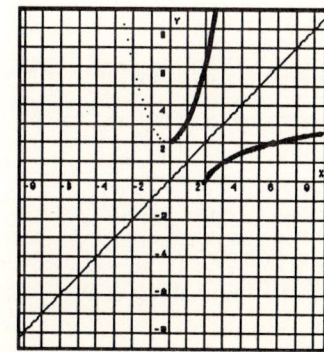

16. Given $f(x) = \sqrt{x+4}$ and $g(x) = x^2 - 4$, $x \geq 0$. Since

$$\begin{aligned} f(g(x)) &= f(x^2 - 4) \\ &= \sqrt{(x^2 - 4) + 4} \\ &= \sqrt{x^2} = |x| = x \quad (x \geq 0), \end{aligned}$$

and

$$\begin{aligned} g(f(x)) &= g(\sqrt{x+4}) \\ &= (\sqrt{x+4})^2 - 4 \\ &= x + 4 - 4 = x, \end{aligned}$$

f and g are inverses. The graphs, given below, can be seen to be symmetric with respect to the line $y = x$.

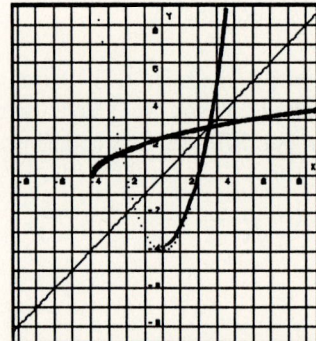

17. Given $f(x) = 2x + 3$. Then $y = 2x + 3$. Interchange x and y, $x = 2y + 3$, and solve for y.

$$\begin{aligned} 2y + 3 &= x \\ 2y &= x - 3 \\ y &= \frac{x-3}{2} \end{aligned}$$

Thus, $f^{-1}(x) = \dfrac{x-3}{2}$.

18. Given $h(x) = \sqrt{x-3}$, then $y = \sqrt{x-3}$ and $y \geq 0$. Interchange x and y and solve for y.

$$\begin{aligned} x &= \sqrt{y-3} \\ x^2 &= y - 3 \\ x^2 + 3 &= y \end{aligned}$$

Then $h^{-1}(x) = x^2 + 3$, for $x \geq 0$.

19. Given $k(x) = 2x^3$, then $y = 2x^3$. Interchange x and y and solve for y.

$$\begin{aligned} x &= 2y^3 \\ \frac{x}{2} &= y^3 \\ \sqrt[3]{\frac{x}{2}} &= y \end{aligned}$$

Then $k^{-1}(x) = \sqrt[3]{\dfrac{x}{2}}$.

20. Given $g(x) = \sqrt{1-x^2}$, with $0 \leq x \leq 1$. Then $y = \sqrt{1-x^2}$. Interchange x and y and solve for y.

$$\begin{aligned} x &= \sqrt{1-y^2} \\ x^2 &= 1 - y^2 \\ 1 - x^2 &= y^2 \\ \pm\sqrt{1-x^2} &= y \end{aligned}$$

Since $0 \leq y \leq 1$, we use the positive root only.

Thus, $g^{-1}(x) = \sqrt{1-x^2}$, for $x \geq 0$.

21. Given $\{(8,8),(9,9)\}$. The inverse of this function is formed by interchanging the x- and y-coordinates to obtain $\{(8,8),(9,9)\}$, the same as the original function.

22. Given $\{(6,4),(7,8),(8,3)\}$. The inverse of this function is formed by interchanging the x- and y-coordinates to obtain $\{(4,6),(8,7),(3,8)\}$.

23. Given $f(x) = mx + b$, or $y = mx + b$. Interchange x and y and solve for y.

$$\begin{aligned} x &= my + b \\ x - b &= my \\ \frac{x-b}{m} &= y \end{aligned}$$

Then $f^{-1}(x) = \dfrac{x-b}{m}$ and the only restriction is that $m \neq 0$.

EXERCISES A SECTION 3.5

24. Given $r(t) = 2t^2 - 1$, which gives the radius of the circular grass fire at any time t.

(a) Since the circumference of a circle is $C = 2\pi r$,
$$C(t) = 2\pi(2t^2 - 1)$$
gives the circumference as a function of time t.

(b) Since the area of a circle is $A = \pi r^2$,
$$A(t) = \pi(2t^2 - 1)^2$$
gives the area as a function of time t.

25. Given $d(t) = 3t^2 + 2t + 1$, which is the distance that the spill travels down a river in terms of time t.

(a) The area of a rectangle is given by $A = lw$, and in this case, we can use the distance $d(t)$ traveled down river for the length and 22 m for the width. Since $d(t)$ is given in kilometers, first change to meters by multiplying by 1000, then
$$A(t) = 1000(3t^2 + 2t + 1)(22)$$
$$= 22{,}000(3t^2 + 2t + 1).$$

Thus, $A(t)$ is the area of the spill in square meters as a function of time t.

(b) To find the area after 2 hours, substitute 2 for t and evaluate.
$$A(2) = 22{,}000[3(2)^2 + 2(2) + 1]$$
$$= 22{,}000[12 + 4 + 1]$$
$$= 22{,}000[17]$$
$$= 374{,}000 \text{ m}^2$$

26. (a) Since $C(d) = 2220d + 9530$ and $d(t) = 3t^2 + 2t + 1$, substitute and simplify to find C as a function of time t.
$$C(d(t)) = C(3t^2 + 2t + 1)$$
$$= 2220(3t^2 + 2t + 1) + 9530$$
$$= 6660t^2 + 4440t + 2220 + 9530$$
$$= 6660t^2 + 4440t + 11{,}750$$

(b) To find the cost after 2 hours, substitute 2 for t and evaluate.
$$C(d(2)) = 6660(2)^2 + 4440(2) + 11{,}750$$
$$= 6660(4) + 8880 + 11{,}750$$
$$= 26{,}640 + 8880 + 11{,}750$$
$$= 47{,}270$$
Thus the cost is $47,270.

27. To obtain the graph of $y = f(x - 4) + 2$ from $y = f(x)$, slide the graph of $y = f(x)$ right 4 units and then up 2 units.

28. To obtain the graph of $y = f(x + 4) - 2$ from $y = f(x)$, slide the graph of $y = f(x)$ left 4 units and then down 2 units.

29. Given $f(x) = 4x^3 + 2x$. Then
$$f(-x) = 4(-x)^3 + 2(-x)$$
$$= -4x^3 - 2x$$
$$= -f(x),$$
so f is an odd function.

30. Given $g(x) = |x^3|$. Then
$$g(-x) = |(-x)^3|$$
$$= |-x^3|$$
$$= |(-1)x^3|$$
$$= |-1||x^3|$$
$$= |x^3|$$
$$= g(x),$$
so g is an even function.

31. First find the slope of the line $3x - 2y + 5 = 0$ by writing the equation in slope intercept form, that is, by solving for y.

$$3x - 2y + 5 = 0$$
$$-2y = -3x - 5$$
$$y = \frac{3}{2}x + \frac{5}{2}$$

The slope of this line is $\frac{3}{2}$, so the slope of the desired parallel line is also $\frac{3}{2}$. Use $\frac{3}{2}$ for m and $(-6, 2)$ as (x_1, y_1) in the point-slope form.

$$y - y_1 = m(x - x_1)$$
$$y - 2 = \frac{3}{2}(x - (-6))$$
$$2(y - 2) = 3(x + 6)$$
$$2y - 4 = 3x + 18$$
$$0 = 3x - 2y + 22$$

Thus, the equation of the desired line, in general form, is $3x - 2y + 22 = 0$.

32. First find the slope of the segment joining (4,6) and (−8,−2).

$$m = \frac{y_2 - y_1}{x_2 - x_1}$$
$$= \frac{-2 - 6}{-8 - 4} = \frac{-8}{-12} = \frac{2}{3}$$

Next find the midpoint of this segment.

$$(\bar{x}, \bar{y}) = \left(\frac{x_1 + x_2}{2}, \frac{y_1 + y_2}{2}\right)$$
$$= \left(\frac{4 + (-8)}{2}, \frac{6 + (-2)}{2}\right)$$
$$= \left(\frac{-4}{2}, \frac{4}{2}\right) = (-2, 2)$$

Then use $-\frac{3}{2}$ for m and $(-2, 2)$ for (x_1, y_1) in the point-slope form.

$$y - y_1 = m(x - x_1)$$
$$y - 2 = -\frac{3}{2}(x - (-2))$$
$$2(y - 2) = -3(x + 2)$$
$$2y - 4 = -3x - 6$$
$$3x + 2y + 2 = 0$$

Thus, the equation of the perpendicular bisector of the segment, in general form, is

$$3x + 2y + 2 = 0.$$

33. (a) To complete the square on $x^2 - 8x$, add half the coefficient of x, squared. That is, add

$$\left(\frac{1}{2}(-8)\right)^2 = (-4)^2 = 16$$

since $x^2 - 8x + 16 = (x - 4)^2$.

(b) To complete the square on $x^2 + 3x$, add half the coefficient of x, squared. That is, add

$$\left(\frac{1}{2}(3)\right)^2 = \left(\frac{3}{2}\right)^2 = \frac{9}{4}$$

since $x^2 + 3x + \frac{9}{4} = \left(x + \frac{3}{2}\right)^2$.

34. To solve $2x^2 + 9x - 5 = 0$ factor the left side and use the zero-product rule.

$$2x^2 + 9x - 5 = 0$$
$$(2x - 1)(x + 5) = 0$$
$$2x - 1 = 0 \quad \text{or} \quad x + 5 = 0$$
$$2x = 1 \qquad\qquad x = -5$$
$$x = \tfrac{1}{2}$$

Thus, the solutions are −5 and $\frac{1}{2}$.

PRACTICE EXERCISES

CHAPTER 3 RELATIONS, FUNCTIONS, AND GRAPHS

SECTION 3.6 Quadratic Functions

1. The vertex (h,k) of $g(x) = -2x^2 + 4x - 7$ has the property that $h = -\dfrac{b}{2a}$. Since $a = -2$ and $b = 4$, substitute to find h.

$$h = -\frac{b}{2a} = -\frac{4}{2(-2)} = -\frac{4}{-4} = 1$$

To find the y-coordinate of the vertex, k, find $g(h) = g(1)$.

$$\begin{aligned}g(1) &= -2(1)^2 + 4(1) - 7 \\ &= -2(1) + 4 - 7 \\ &= -2 + 4 - 7 \\ &= -5\end{aligned}$$

Thus, the vertex is $(1,-5)$.

2. (a) Given $f(x) = x^2 - 4x + 3$.

    Since $a = 1 > 0$, the graph opens up.
    The vertex (h,k) of $f(x) = x^2 - 4x + 3$ has the property that $h = -\dfrac{b}{2a}$. Substitute 1 for a and -4 for b to find h.

    $$h = -\frac{b}{2a} = -\frac{-4}{2(1)} = -\frac{-4}{2} = 2$$

    To find the y-coordinate of the vertex, k, find $f(h) = f(2)$.

    $$\begin{aligned}f(2) &= (2)^2 - 4(2) + 3 \\ &= 4 - 8 + 3 \\ &= -1\end{aligned}$$

    Thus, the vertex is $(2,-1)$.

    To find find the x-intercepts, solve:

    $$\begin{aligned}x^2 - 4x + 3 &= 0 \\ (x - 1)(x - 3) &= 0 \\ x - 1 = 0 \quad &\text{or} \quad x - 3 = 0 \\ x = 1 \quad\quad\quad &\quad\quad\quad x = 3\end{aligned}$$

    Thus, the x-intercepts are $(1,0)$ and $(3,0)$. The graph is given below.

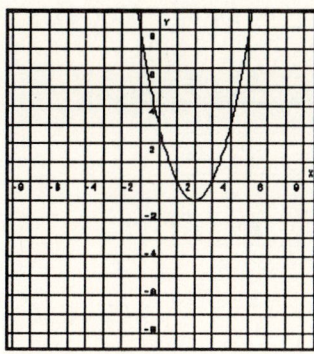

(b) Given $g(x) = -2x^2 - 6x + 1$.

Since $a = -2 < 0$, the graph opens down.
The vertex (h,k) of $g(x) = -2x^2 - 6x + 1$ has the property that $h = -\dfrac{b}{2a}$. Substitute -2 for a and -6 for b to find h.

$$h = -\frac{b}{2a} = -\frac{-6}{2(-2)} = -\frac{-6}{-4} = -\frac{3}{2}$$

To find the y-coordinate of the vertex, k, find $g(h) = g\left(-\dfrac{3}{2}\right)$.

$$\begin{aligned}g\left(-\frac{3}{2}\right) &= -2\left(-\frac{3}{2}\right)^2 - 6\left(-\frac{3}{2}\right) + 1 \\ &= -2\left(\frac{9}{4}\right) + 9 + 1 \\ &= -\frac{9}{2} + 10 \\ &= -\frac{9}{2} + \frac{20}{2} = \frac{11}{2}\end{aligned}$$

Thus, the vertex is $\left(-\dfrac{3}{2}, \dfrac{11}{2}\right)$.

To find find the x-intercepts, solve:

$$-2x^2 - 6x + 1 = 0$$

or

$$2x^2 + 6x - 1 = 0$$

Since the left side will not factor, use the quadratic formula.

$$x = \frac{-b \pm \sqrt{b^2 - 4ac}}{2a}$$

$$= \frac{-6 \pm \sqrt{(6)^2 - 4(2)(-1)}}{2(2)}$$

$$= \frac{-6 \pm \sqrt{36 + 8}}{4}$$

$$= \frac{-6 \pm \sqrt{44}}{4}$$

$$= \frac{-6 \pm 2\sqrt{11}}{4}$$

$$= \frac{-3 \pm \sqrt{11}}{2} \approx 0.16, -3.16$$

Thus, the x-intercepts are approximately $(0.16, 0)$ and $(-3.16, 0)$. The graph is given below.

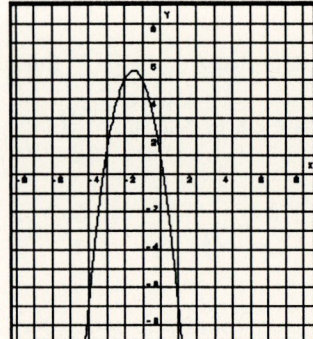

3. The height of the ball after t seconds is given by the quadratic function $h(t) = -16t^2 + 64t$. The maximum height will be the the second coordinate of the vertex. Begin by finding the first coordinate of the vertex which is

$$-\frac{b}{2a} = -\frac{64}{2(-16)} = -\frac{64}{-32} = 2.$$

Then the maximum height is given by:

$$h(2) = -16(2)^2 + 64(2)$$
$$= -16(4) + 128$$
$$= -64 + 128$$
$$= 64$$

Thus, the maximum height is 64 ft which occurs after 2 sec.

4. The problem is exactly like the one set up in the example except the revenue function is given by:

$$R(n) = (20 + n)(32 - n)$$
$$= 640 + 12n - n^2$$

Since $a = -1 < 0$, the function has a maximum. Find h in (h, k).

$$h = -\frac{b}{2a} = -\frac{12}{2(-1)} = -\frac{12}{-2} = 6$$

Thus, the maximum revenue will occur when the daily rental fee is reduced by $6. This means that the new fee should be $32 - $6 = $26.

EXERCISES A SECTION 3.6 121

CHAPTER 3 RELATIONS, FUNCTIONS, AND GRAPHS

SECTION 3.6 Quadratic Functions

1. To find the vertex of $f(x) = x^2 - 5x + 6$, use the vertex formula. The x-coordinate of the vertex is

$$-\frac{b}{2a} = -\frac{-5}{2(1)} = -\frac{-5}{2} = \frac{5}{2}.$$

To find the y-coordinate of the vertex, evaluate

$$f\left(-\frac{b}{2a}\right) = f\left(\frac{5}{2}\right) = \left(\frac{5}{2}\right)^2 - 5\left(\frac{5}{2}\right) + 6$$
$$= \frac{25}{4} - \frac{25}{2} + 6$$
$$= \frac{25}{4} - \frac{50}{4} + \frac{24}{4}$$
$$= -\frac{1}{4}.$$

Thus, the vertex is $\left(\frac{5}{2}, -\frac{1}{4}\right)$. To find the x-intercepts, solve the following equation.

$$x^2 - 5x + 6 = 0$$
$$(x - 2)(x - 3) = 0$$
$$x - 2 = 0 \quad \text{or} \quad x - 3 = 0$$
$$x = 2 \quad\quad\quad x = 3$$

Thus, the x-intercepts are (2,0) and (3,0).

2. To find the vertex of $h(x) = x^2 + 2x + 2$, use the vertex formula. The x-coordinate of the vertex is

$$-\frac{b}{2a} = -\frac{2}{2(1)} = -\frac{2}{2} = -1.$$

To find the y-coordinate of the vertex, evaluate

$$h\left(-\frac{b}{2a}\right) = h(-1) = (-1)^2 + 2(-1) + 2$$
$$= 1 - 2 + 2 = 1.$$

Thus, the vertex is (-1,1). To find the x-intercepts, solve the following equation.

$$x^2 + 2x + 2 = 0$$

But since the discriminant,
$b^2 - 4ac = 2^2 - 4(1)(2) = 4 - 8 = -4 < 0$,
the solutions are complex (nonreal), so there are no x-intercepts.

3. To find the vertex of $k(x) = x^2 + 8x$, use the vertex formula. The x-coordinate of the vertex is

$$-\frac{b}{2a} = -\frac{8}{2(1)} = -\frac{8}{2} = -4.$$

To find the y-coordinate of the vertex, evaluate

$$k\left(-\frac{b}{2a}\right) = k(-4) = (-4)^2 + 8(-4)$$
$$= 16 - 32 = -16.$$

Thus, the vertex is (-4,-16). To find the x-intercepts, solve the following equation.

$$x^2 + 8x = 0$$
$$x(x + 8) = 0$$
$$x = 0 \quad \text{or} \quad x + 8 = 0$$
$$x = -8$$

Thus, the x-intercepts are (0,0) and (-8,0).

4. To find the vertex of $m(x) = 2x^2 - 4$, use the vertex formula. The x-coordinate of the vertex is

$$-\frac{b}{2a} = -\frac{0}{2(2)} = -\frac{0}{4} = 0.$$

To find the y-coordinate of the vertex, evaluate

$$m\left(-\frac{b}{2a}\right) = m(0) = 2(0)^2 - 4$$
$$= 0 - 4 = -4.$$

Thus, the vertex is (0,-4). To find the x-intercepts, solve the following equation.

$$2x^2 - 4 = 0$$
$$2x^2 = 4$$
$$x^2 = 2$$
$$x = \pm\sqrt{2}$$

Thus, the x-intercepts are $\left(\sqrt{2}, 0\right)$ and $\left(-\sqrt{2}, 0\right)$.

5. To find the vertex of $f(x) = 2x^2 + 8x - 5$, use the vertex formula. The x-coordinate of the vertex is

$$-\frac{b}{2a} = -\frac{8}{2(2)} = -\frac{8}{4} = -2$$

To find the y-coordinate of the vertex, evaluate

$$f\left(-\frac{b}{2a}\right) = f(-2) = 2(-2)^2 + 8(-2) - 5$$
$$= 2(4) - 16 - 5$$
$$= 8 - 16 - 5 = -13$$

Thus, the vertex is $(-2, -13)$. To find the x-intercepts, solve $2x^2 + 8x - 5 = 0$.

$$x = \frac{-b \pm \sqrt{b^2 - 4ac}}{2a}$$
$$= \frac{-(8) \pm \sqrt{(8)^2 - 4(2)(-5)}}{2(2)}$$
$$= \frac{-8 \pm \sqrt{64 + 40}}{4}$$
$$= \frac{-8 \pm \sqrt{104}}{4}$$
$$= \frac{-8 \pm 2\sqrt{26}}{4}$$
$$= \frac{-4 \pm \sqrt{26}}{2}$$

Thus, the x-intercepts are $\left(\frac{-4 + \sqrt{26}}{2}, 0\right)$ and $\left(\frac{-4 - \sqrt{26}}{2}, 0\right)$.

6. To find the vertex of $h(x) = \frac{1}{3}x^2 - 2x + 5$, use the vertex formula. The x-coordinate of the vertex is

$$-\frac{b}{2a} = -\frac{-2}{2\left(\frac{1}{3}\right)} = \frac{2}{\frac{2}{3}} = \frac{6}{2} = 3.$$

To find the y-coordinate of the vertex, evaluate

$$h\left(-\frac{b}{2a}\right) = h(3) = \frac{1}{3}(3)^2 - 2(3) + 5$$
$$= 3 - 6 + 5$$
$$= 2.$$

Thus, the vertex is $(3, 2)$. To find the x-intercepts, solve $\frac{1}{3}x^2 - 2x + 5 = 0$. But since

$$b^2 - 4ac = (-2)^2 - 4\left(\frac{1}{3}\right)(5) = 4 - \frac{20}{3} = -\frac{8}{3} < 0$$

the solutions are complex (nonreal). Thus, there are no x-intercepts.

7. To find the vertex of $g(x) = -x^2 - 2x + 3$, use the vertex formula. The x-coordinate of the vertex is

$$-\frac{b}{2a} = -\frac{-2}{2(-1)} = -\frac{-2}{-2} = -1.$$

To find the y-coordinate of the vertex, evaluate

$$g\left(-\frac{b}{2a}\right) = g(-1) = -(-1)^2 - 2(-1) + 3$$
$$= -1 + 2 + 3 = 4.$$

Thus, the vertex is $(-1, 4)$. To find the x-intercepts, solve the following equation.

$$-x^2 - 2x + 3 = 0$$
$$x^2 + 2x - 3 = 0$$
$$(x - 1)(x + 3) = 0$$
$$x - 1 = 0 \quad \text{or} \quad x + 3 = 0$$
$$x = 1 \qquad\qquad x = -3$$

Thus, the x-intercepts are $(1, 0)$ and $(-3, 0)$, and the graph is given below.

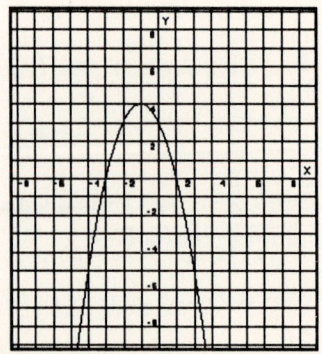

EXERCISES A

SECTION 3.6

8. To find the vertex of $k(x) = -x^2 + 5x - 2$, use the vertex formula. The x-coordinate of the vertex is

$$-\frac{b}{2a} = -\frac{5}{2(-1)} = -\frac{5}{-2} = \frac{5}{2}.$$

To find the y-coordinate of the vertex, evaluate

$$k\left(-\frac{b}{2a}\right) = k\left(\frac{5}{2}\right) = -\left(\frac{5}{2}\right)^2 + 5\left(\frac{5}{2}\right) - 2$$
$$= -\frac{25}{4} + \frac{25}{2} - 2$$
$$= -\frac{25}{4} + \frac{50}{4} - \frac{8}{4}$$
$$= \frac{17}{4}.$$

Thus, the vertex is $\left(\frac{5}{2}, \frac{17}{4}\right)$. To find the x-intercepts, solve $-x^2 + 5x - 2 = 0$.

$$x = \frac{-b \pm \sqrt{b^2 - 4ac}}{2a}$$
$$= \frac{-(5) \pm \sqrt{(5)^2 - 4(-1)(-2)}}{2(-1)}$$
$$= \frac{-5 \pm \sqrt{25 - 8}}{-2}$$
$$= \frac{-5 \pm \sqrt{17}}{-2}$$
$$= \frac{5 \pm \sqrt{17}}{2} \approx 4.56, \ 0.44$$

Thus, the x-intercepts are approximately (4.56, 0) and (0.44, 0), and the graph is given below.

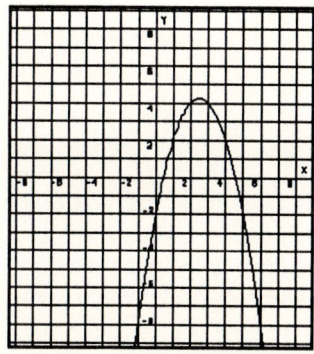

9. To find the vertex of $f(x) = x^2 + 4x + 5$, use the vertex formula. The x-coordinate of the vertex is

$$-\frac{b}{2a} = -\frac{4}{2(1)} = -\frac{4}{2} = -2.$$

To find the y-coordinate of the vertex, evaluate

$$f\left(-\frac{b}{2a}\right) = f(-2) = (-2)^2 + 4(-2) + 5$$
$$= 4 - 8 + 5 = 1$$

Thus, the vertex is $(-2, 1)$. To find the x-intercepts, solve $x^2 + 4x + 5 = 0$. But the solutions are complex (nonreal) so there are no x-intercepts. Two points on the graph are $(0, 5)$ and $(-4, 5)$. The graph is given below.

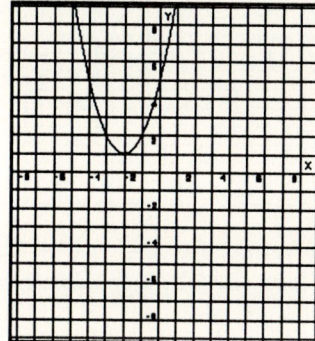

10. To find the vertex of $h(x) = \frac{1}{2}x^2 - 7x + 20$, use the vertex formula. The x-coordinate of the vertex is

$$-\frac{b}{2a} = -\frac{-7}{2\left(\frac{1}{2}\right)} = -\frac{-7}{1} = 7.$$

To find the y-coordinate of the vertex, evaluate

$$g\left(-\frac{b}{2a}\right) = g(7) = \frac{1}{2}(7)^2 - 7(7) + 20$$
$$= \frac{49}{2} - 49 + 20$$
$$= \frac{49}{2} - \frac{98}{2} + \frac{40}{2}$$
$$= -\frac{9}{2}.$$

Thus, the vertex is $\left(7, -\frac{9}{2}\right)$. To find the x-intercepts, solve $\frac{1}{2}x^2 - 7x + 20 = 0$, which simplifies to $x^2 - 14x + 40 = 0$.

$$x^2 - 14x + 40 = 0$$
$$(x - 10)(x - 4) = 0$$
$$x - 10 = 0 \qquad x - 4 = 0$$
$$x = 10 \qquad x = 4$$

Thus, the x-intercepts are (10,0) and (4,0), and the graph is given below.

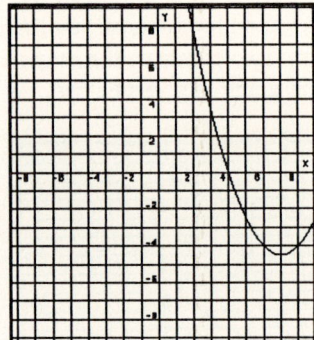

11. To find the vertex of $k(x) = -4x^2 + 8x - 4$, use the vertex formula. The x-coordinate of the vertex is

$$-\frac{b}{2a} = -\frac{8}{2(-4)} = -\frac{8}{-8} = 1.$$

To find the y-coordinate of the vertex, evaluate

$$k(1) = -4(1)^2 + 8(1) - 4$$
$$= -4 + 8 - 4 = 0$$

Thus, the vertex is (1,0). To find the x-intercepts, solve the equation $-4x^2 + 8x - 4 = 0$, which simplifies to

$$x^2 - 2x + 1 = 0$$
$$(x - 1)(x - 1) = 0$$
$$x - 1 = 0 \quad \text{or} \quad x - 1 = 0$$
$$x = 1 \qquad x = 1$$

Thus, there is only one x-intercept, (1,0), which is the same as the vertex. Two additional points on the graph are (0,-4) and (2,-4). The graph is given below.

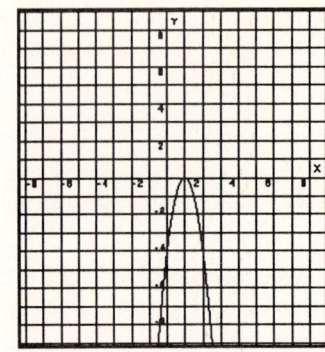

12. Since $m(x) = 2(x + 1)^2 + 2$ is given in standard form, we can read off the vertex as (-1,2). Since the graph opens up, and the vertex is (-1,2), there are no x-intercepts. That is, if we were to try and find the intercepts we would discover they are complex, not real. Two additional points on the graph are (-2,4) and (0,4). The graph is given below.

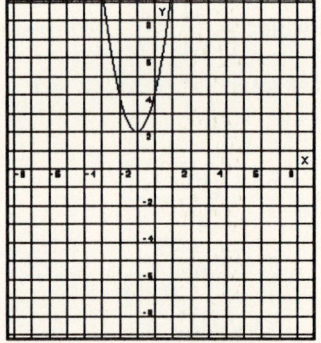

13. Let x = the length of the pen,
y = the width of the pen.

Since the amount of fencing available is 1000 yd, the perimeter of the fence is 1000 yd. Thus,

$$2x + 2y = 1000$$
$$x + y = 500$$
$$y = 500 - x.$$

Since the area of the pen is $A = xy$, we have

$$A = x(500 - x)$$
$$= -x^2 + 500x.$$

Since this is a quadratic function of the area of the pen in terms of the length x, and since the graph of this function open down, the vertex will provide the information necessary to determine the maximum area. Find the first coordinate of the vertex using the formula.

EXERCISES A

$$-\frac{b}{2a} = -\frac{500}{2(-1)} = -\frac{500}{-2} = 250$$

When $x = 250$, $y = 500 - 250 = 250$, so the dimensions of the largest pen are 250 yd by 250 yd. Then the area of the largest pen is

$$A = (250)(250) = 62{,}500 \text{ yd}^2.$$

14. Since the cable follows the curve with equation

$$f(x) = 0.01x^2 + 20,$$

and the towers are 100 feet apart, the height on each tower where the cable is attached can be found by evaluating

$$\begin{aligned}f(50) &= 0.01(50)^2 + 20 \\&= 0.01(2500) + 20 \\&= 25 + 20 \\&= 45\end{aligned}$$

Then the height on each tower where the cable is attached is 45 ft.

15. The function $h(t) = -16t^2 + 256t$ gives the height of a rocket t seconds after it is fired.

    (a) The maximum height of the rocket will occur at the vertex of the graph since the function is a quadratic function with graph a parabola opening down. The t-coordinate of the vertex is

    $$-\frac{b}{2a} = -\frac{256}{2(-16)} = -\frac{256}{-32} = 8$$

    Thus the rocket will reach its maximum height at 8 sec.

    (b) The maximum height will be the h-coordinate of the vertex which is

    $$\begin{aligned}h(8) &= -16(8)^2 + 256(8) \\&= -16(64) + 2048 \\&= -1024 + 2048 \\&= 1024.\end{aligned}$$

    Thus, the maximum height attained is 1024 ft.

    (c) When the rocket is on the ground, $h(t) = 0$. Thus, to find the time when it is on the ground solve:

$$\begin{aligned}-16t^2 + 256t &= 0 \\-16t(t - 16) &= 0 \\t(t - 16) &= 0 \\t = 0 \text{ or } t - 16 &= 0 \\t &= 16\end{aligned}$$

Thus, the rocket is on the ground at 0 seconds (right before it is fired) and again at 16 seconds, when it returns to the ground.

16. The vertex of the quadratic function
    $$P(x) = -x^2 + 120x - 1000$$
    will provide the maximum profit (the second coordinate) which occurs when the number of parts (the first coordinate) is produced. Use the formula to find the first coordinate.

    $$-\frac{b}{2a} = -\frac{120}{2(-1)} = -\frac{120}{-2} = 60$$

    To find the profit when 60 parts are produced, evaluate the following (the second coordinate of the vertex).

    $$\begin{aligned}P(60) &= -(60)^2 + 120(60) - 1000 \\&= -3600 + 7200 - 1000 \\&= 2600\end{aligned}$$

    Thus, the maximum profit of $2600 occurs when 60 parts are produced.

17. Start with the standard form of a quadratic function,
    $$C(x) = a(x - h)^2 + k.$$
    Since the minimum cost occurs at the vertex (30,200), $h = 30$ and $k = 200$. Substituting we have
    $$C(x) = a(x - 30)^2 + 200.$$
    Since the curve also goes through the point (0,2000), substitute 0 for x and 2000 for y (which is $f(x)$) and find the value of a.
    $$\begin{aligned}2000 &= a(0 - 30)^2 + 200 \\2000 &= a(900) + 200 \\1800 &= 900a \\2 &= a\end{aligned}$$
    Thus, the desired quadratic function is
    $$\begin{aligned}C(x) &= 2(x - 30)^2 + 200 \\&= 2(x^2 - 60x + 900) + 200 \\&= 2x^2 - 120x + 1800 + 200 \\&= 2x^2 - 120x + 2000.\end{aligned}$$

18. Let x = the number of components ordered over 200,

$40 - 0.20x$ = the price of each component over 200, The revenue produced on the sale of 200 components is $(200)(40) = \$8000$. Then the total revenue produced on the sale of components is given by

$$R(x) = 8000 + x(40 - 0.20x)$$
$$= 8000 + 40x - 0.20x^2$$

To find the largest revenue, determine the vertex of the graph of this quadratic function. The x-coordinate is given by

$$-\frac{b}{2a} = -\frac{40}{2(-0.20)} = \frac{40}{0.4} = 100.$$

Thus the total order that would give the largest revenue is $100 + 200 = 300$ components (100 at $40 plus the 100 over this amount found above).

19. Given $f(x) = -3x + 4$ and $g(x) = 2x - 5$.

First find $g(0) = 2(0) - 5 = -5$. Then
$f(g(0)) = f(-5) = -3(-5) + 4 = 15 + 4 = 19$.

First find $f(-2) = -3(-2) + 4 = 6 + 4 = 10$. Then
$g(f(-2)) = g(10) = 2(10) - 5 = 20 - 5 = 15$.

$$\begin{aligned}f(g(x)) &= f(2x - 5)\\ &= -3(2x - 5) + 4\\ &= -6x + 15 + 4\\ &= -6x + 19\end{aligned}$$

$$\begin{aligned}g(f(x)) &= g(-3x + 4)\\ &= 2(-3x + 4) - 5\\ &= -6x + 8 - 5\\ &= -6x + 3\end{aligned}$$

20. Given $f(x) = |x + 5|$ and $g(x) = x^2$.

First find $g(0) = (0)^2 = 0$. Then
$f(g(0)) = f(0) = |0 + 5| = |5| = 5$.

First find $f(-2) = |-2 + 5| = |3| = 3$. Then
$g(f(-2)) = g(3) = (3)^2 = 9$.

$$\begin{aligned}f(g(x)) &= f(x^2)\\ &= |x^2 + 5|\\ &= x^2 + 5 \quad \text{(Note that } x^2 + 5 > 0)\end{aligned}$$

$$\begin{aligned}g(f(x)) &= g(|x + 5|)\\ &= (|x + 5|)^2\\ &= (x + 5)^2\\ &= x^2 + 10x + 5\end{aligned}$$

21. Given $f(x) = 7x + 8$, or $y = 7x + 8$. Interchange x and y and solve for y to find the inverse.

$$x = 7y + 8$$
$$x - 8 = 7y$$
$$\frac{x - 8}{7} = y$$

Thus, $f^{-1}(x) = \dfrac{x - 8}{7}$.

22. Given $g(x) = \sqrt{x + 4}$, or $y = \sqrt{x + 4}$. Then $y \geq 0$. Interchange x and y and solve for y.

$$x = \sqrt{y + 4}$$
$$x^2 = y + 4$$
$$x^2 - 4 = y$$

Thus the inverse is $g^{-1}(x) = x^2 - 4$, for $x \geq 0$.

23. Given $h(x) = \sqrt[3]{x + 1}$, or $y = \sqrt[3]{x + 1}$. Interchange x and y and solve for y.

$$x = \sqrt[3]{y + 1}$$
$$x^3 = y + 1$$
$$x^3 - 1 = y$$

Thus the inverse is $h^{-1}(x) = x^3 - 1$.

24. Consider the graph of $g(x) = -5x - 3$ given below.

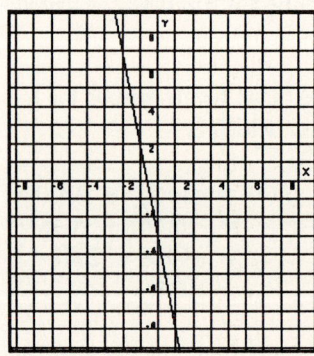

Then g is decreasing for all values of x.

EXERCISES A

25. Consider the graph of $f(x) = |x| + 3$ given below.

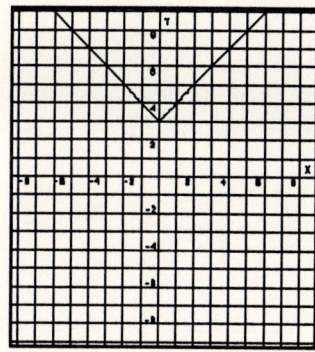

Then f is decreasing whenever $x \leq 0$ since if $a < b < 0$, $f(a) > f(b)$. Also, f is increasing whenever $x \geq 0$ since if $0 < a < b$, $f(a) < f(b)$.

128 SECTION 3.7 PRACTICE EXERCISES

CHAPTER 3 RELATIONS, FUNCTIONS, AND GRAPHS

SECTION 3.7 Mathematical Modeling

1. (a) We must find the equation of the line passing through the two points (0,1500) and (25,3600). First find the slope of the line.

$$m = \frac{y_2 - y_1}{x_2 - x_1}$$
$$= \frac{3600 - 1500}{25 - 0}$$
$$= \frac{2100}{25}$$
$$= 84$$

Use the point (0,1500) with the slope $m = 84$ in the point-slope form.

$$s - 1500 = 84(t - 0)$$
$$s - 1500 = 84t$$
$$s = 84t + 1500$$

Thus the model in this case is

$$s(t) = 84t + 1500.$$

(b) Find $s(9)$ using the model above.

$$s(9) = 84(9) + 1500$$
$$= 756 + 1500$$
$$= 2256$$

Thus, when $t = 9$, that is the year was 1979, the average monthly salary was about $2256.

(c) Use the model above to extimate $s(t)$ for the year 1965, which would correspond to $t = -5$.

$$s(-5) = 84(-5) + 1500$$
$$= -420 + 1500$$
$$= 1080$$

Thus, according to this model, the average monthly salary was about $1080 in the year 1965.

2. Suppose $N(t) = -12t^2 + 120t$ represents the number of customers in the store as a function of time t, where $t = 0$ corresponds to 10:00 A.M., the time the store opens.

(a) To find the time the store has the maximum number of customers, find the first coordinate of the vertex of the quadratic function $N(t) = -12t^2 + 120t$.

$$-\frac{b}{2a} = -\frac{120}{2(-12)} = -\frac{120}{-24} = 5$$

Then $t = 5$ means that 5 hours after the store opened, at 3:00 P.M., the maximum number of customers were present.

(b) The maximum number of customers is the second coordinate of the vertex given by:

$$N(5) = -12(5)^2 + 120(5)$$
$$= -300 + 600$$
$$= 300$$

Thus, the maximum number of customers is 300.

(c) Since N is 0 at closing time (also at opening time when $t = 0$), we must solve

$$0 = -12t^2 + 120t$$
$$0 = -12t(t - 10)$$
$$0 = t(t - 10)$$
$$t = 0 \text{ or } t - 10 = 0$$
$$t = 10$$

Thus, the number of customers is 0 at $t = 0$, or at opening time 10:00 A.M., and at $t = 10$, which corresponds to 8:00P.M., which must be the closing time.

EXERCISES A

CHAPTER 3 RELATIONS, FUNCTIONS, AND GRAPHS

SECTION 3.7 Mathematical Modeling

1. (a) The information in the table has been plotted below, and a straight line approximation of this data is given.

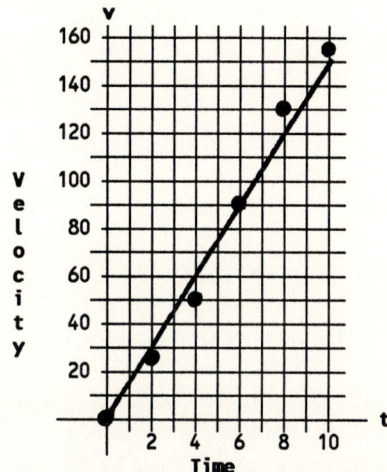

A possible linear model might pass through the points (0,0) and (6,90). Using these we can find the slope of the line.

$$m = \frac{v_2 - v_1}{t_2 - t_1}$$
$$= \frac{90 - 0}{6 - 0} = 15$$

Use the point-slope form with $m = 15$ and $(t_1, v_1) = (0,0)$.
$$v - v_1 = m(t - t_1)$$
$$v - 0 = 15(t - 10)$$
$$v = 15t$$

Thus, a linear model for the velocity in terms of time is $v(t) = 15t$.

(b) Substitute 5 for t to find the velocity at time 5 sec.
$$v(5) = 15(5)$$
$$= 75 \text{ ft/sec}$$

(c) From the table of collected values, we conclude that the domain of this function is $0 \leq t \leq 10$. For values of t outside of this domain, the model is of no value. For example, if $t = 20$, then $v = 300$ ft/sec, but this is probably beyond the car's capability.

2. (a) The information in the table has been plotted below, and a straight line approximation of this data is given.

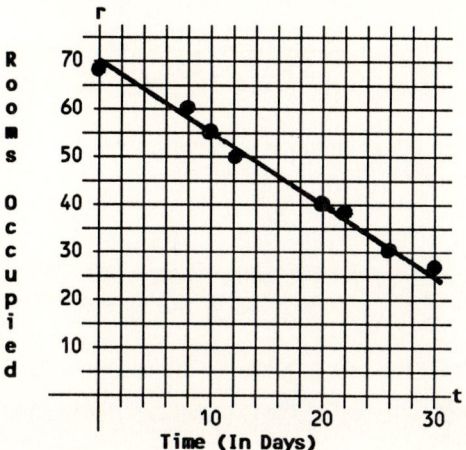

A possible linear model might pass through the points (10,55) and (20,40). Using these we can find the slope of the line.

$$m = \frac{r_2 - r_1}{t_2 - t_1}$$
$$= \frac{55 - 40}{10 - 20} = \frac{15}{-10} = -1.5$$

Use the point-slope form with $m = -1.5$ and $(t_1, r_1) = (10, 55)$.
$$r - r_1 = m(t - t_1)$$
$$r - 55 = -1.5(t - 10)$$
$$r - 55 = -1.5t + 15$$
$$r = -1.5t + 70$$

Thus, a linear model for the number of rooms occupied in terms of time is $r(t) = -1.5t + 70$.

(b) Set $r = 10$ and solve for t.
$$10 = -1.5t + 70$$
$$-60 = -1.5t$$
$$40 = t$$

Since $t = 0$ corresponds of August 15, $t = 40$ corresponds to September 26. Thus, the closing date will be September 26.

(c) No, since the ski season will probably bring more new tourists.

3. It costs $220 to produce 8 small appliances and $360 to produce 18 appliances. The two data points (8,220) and (18,360) can be used to determine a linear model.

   (a) First find the slope of the line through the points.

   $$m = \frac{360 - 220}{18 - 8} = \frac{140}{10} = 14$$

   Use the point-slope form with $m = 14$ and the point (8,220).

   $$C - 220 = 14(x - 8)$$
   $$C - 220 = 14x - 112$$
   $$C = 14n + 108$$

   Thus, the cost is given by the linear model $C(x) = 14x + 108$.

   (b) To find the cost of producing 12 appliances, substitute 12 for x.

   $$C(12) = 14(12) + 108$$
   $$= 168 + 108$$
   $$= 276$$

   Thus it costs $276 to produce 12 appliances.

   (c) The fixed costs, that is the costs for producing no units can be found by substituting 0 for x.

   $$C(0) = 14(0) + 108$$
   $$= 108$$

   Thus, the fixed costs amount to $108.

4. With 310 trees on 20 acres and 640 trees on 50 acres, this can be thought of as two data points of the form (a,t), (20,310) and (50,640).

   (a) First find the slope of the line through these points.

   $$m = \frac{640 - 310}{50 - 20} = \frac{330}{30} = 11$$

   Use the point-slope form with $m = 11$ and the point (20,310).

   $$t - 310 = 11(a - 20)$$
   $$t - 310 = 11a - 220$$
   $$t = 11a + 90$$

   Then a linear model for the number of trees t in terms of the number of acres a is
   $$t(a) = 11a + 90.$$

   (b) To estimate the number of trees on 70 acres, substitute 70 for a.

   $$t(70) = 11(70) + 90$$
   $$= 770 + 90$$
   $$= 860$$

   Thus, there are about 860 trees on 70 acres.

   (c) No. Since the model would give 90 trees on 0 acres, it would probably not be accurate for a small number of acres such as 1 or 2.

5. Let x represent the number of units produced.

   (a) Since the fixed cost in the operation is $1500, and the cost of producing one unit is $12, the cost of producing x units is given by the function

   $$C(x) = 12x + 1500.$$

   The cost of producing 27 units is

   $$C(27) = 12(27) + 1500$$
   $$= 324 + 1500$$
   $$= \$1824.$$

   (b) Since the revenue produced by selling one unit is $22, the revenue produced by selling x units is given by the function

   $$R(x) = 22x.$$

   Then the revenue produced by selling 27 units is

   $$R(27) = 22(27) = \$594.$$

   (c) Since profit is revenue minus cost, the profit function is given by

   $$P(x) = 22x - (12x + 1500)$$
   $$= 22x - 12x - 1500$$
   $$= 10x - 1500.$$

   (d) A break-even point is a solution to the equation $P(x) = 0$.

   $$10x - 1500 = 0$$
   $$10x = 1500$$
   $$x = 150$$

   Thus, the break-even point is 150 units.

EXERCISES A
SECTION 3.7

When $P(x) > 0$, a profit is made. To find the number of units necessary for a profit, solve

$$10x - 1500 > 0 \text{ which is } x > 150.$$

When there is a loss, $P(x) < 0$. To find the number of units that produce a loss, solve

$$10x - 1500 < 0 \text{ which is } x < 150.$$

But since x must be nonnegative, that is $x \geq 0$, the actual number of units that yield a loss is

$$0 \leq x < 150.$$

6. If n is the number of items sold in a wholesale operation, then empirical data has shown that the cost function is

$$C(n) = n^2 - 70n + 900$$

and the revenue function is

$$R(n) = 2n^2 - 80n - 300.$$

(a) Find the profit function.

$$\begin{aligned} P(n) &= R(n) - C(n) \\ &= (2n^2 - 80n - 300) - (n^2 - 70n + 900) \\ &= 2n^2 - 80n - 300 - n^2 + 70n - 900 \\ &= n^2 - 10n - 1200 \end{aligned}$$

(b) Find the break-even point, when $P(n) = 0$.

$$\begin{aligned} n^2 - 10n - 1200 &= 0 \\ (n - 40)(n + 30) &= 0 \\ n - 40 = 0 \quad &\text{or} \quad n + 30 = 0 \\ n = 40 \quad & \quad n = -30 \end{aligned}$$

Since n must be positive, the break-even point is 40 items.

7. Given that the height of a projectile at time t is given by

$$h(t) = \tfrac{1}{2}gt^2 + v_0 t + h_0$$

Where $g = -32$ ft/sec^2, v_0 is the initial velocity, and h_0 is the initial height.

(a) Substitute 0 for h_0, 240 for v_0, and -32 for g.

$$\begin{aligned} h(t) &= \tfrac{1}{2}(-32)t^2 + (240)t + 0 \\ &= -16t^2 + 240t \end{aligned}$$

(b) This is a quadratic function with graph a parabola opening down. To find the domain begin by finding those values for t for which $h(t) = 0$.

$$\begin{aligned} -16t^2 + 240t &= 0 \\ -16t(t - 15) &= 0 \\ t(t - 15) &= 0 \\ t = 0 \quad \text{or} \quad t - 15 &= 0 \\ t &= 15 \end{aligned}$$

The interval over which h is nonnegative, the domain of the function, is $0 \leq t \leq 15$. That is, at time 0, the object is at a height of 0 ft, it is thrown upward, and reaches the ground at time 15 seconds.

(b) To find the range of this function, we must find the maximum height that the object reaches. Clearly the smallest height is 0, when it is on the ground at time 0 seconds and 15 seconds. The maximum height will occur at the second coordinate of the vertex. Begin by finding the first coordinate of the vertex using the formula.

$$-\frac{b}{2a} = -\frac{240}{2(-16)} = -\frac{240}{-32} = 7.5$$

The second coordinate of the vertex gives the maximum height which occurs when $t = 7.5$ seconds.

$$\begin{aligned} h(7.5) &= -16(7.5)^2 + 240(7.5) \\ &= -900 + 1800 \\ &= 900 \end{aligned}$$

Thus, the range of the function is

$$0 \leq h(t) \leq 900.$$

8. From the laws of motion we know that $d = rt$, that is, distance equals rate times time.

(a) Find $d(x)$ if $r(x) = 3x - 6$ and $t(x) = 4x - 20$, where $0 \leq x \leq 10$.

$$\begin{aligned} d(x) &= r(x) \cdot t(x) \\ &= (3x - 6)(4x - 20) \\ &= 12t^2 - 84x + 120 \end{aligned}$$

(b) Find the values of x for which $d(x)$ is 0.

$$12x^2 - 84x + 120 = 0$$
$$12(x^2 - 7x + 10) = 0$$
$$x^2 - 7x + 10 = 0$$
$$(x - 2)(x - 5) = 0$$
$$x - 2 = 0 \qquad x - 5 = 0$$
$$x = 2 \qquad x = 5$$

Thus, $d(x)$ is 0 when x is 2 or 5.

(c) Since the graph of $d(x)$ is a parabola that opens up, to determine the range it will help to identify the vertex. It can be shown that the vertex is $(3.5, -27)$. Recall that the domain of the function is $0 \leq x \leq 10$, and the vertex is in the domain. Thus, the smallest value in the range is -27. The largest value in the range will occur at the value 10, the right endpoint of the domain, since it is farther from the x-coordinate of the vertex than the left endpoint 0. Since $d(10) = 480$, the range of the function is

$$-27 \leq d(x) \leq 480.$$

9. Given $f(x) = 2x^2 + 4x - 48$. Use the formula to find the x-coordinate of the vertex.

$$-\frac{b}{2a} = -\frac{4}{2(2)} = -\frac{4}{4} = -1$$

Substitute to find the y-coordinate of the vertex.

$$f(-1) = 2(-1)^2 + 4(-1) - 48$$
$$= 2 - 4 - 48 = -50$$

Thus the vertex is $(-1, -50)$. To find the x-intercepts, solve the equation

$$2x^2 + 4x - 48 = 0.$$
$$x^2 + 2x - 24 = 0$$
$$(x - 4)(x + 6) = 0$$
$$x - 4 = 0 \quad \text{or} \quad x + 6 = 0$$
$$x = 4 \qquad x = -6$$

Thus, the x-intercepts are $(4,0)$ and $(-6,0)$.

10. Given $g(x) = -x^2 + 16x - 63$. Use the formula to find the x-coordinate of the vertex.

$$-\frac{b}{2a} = -\frac{16}{2(-1)} = \frac{16}{2} = 8$$

Substitute to find the y-coordinate of the vertex.

$$g(8) = -(8)^2 + 16(8) - 63$$
$$= -64 + 128 - 63$$
$$= 1$$

Thus the vertex is $(8,1)$. To find the x-intercepts, solve the equation

$$-x^2 + 16x - 63 = 0.$$
$$x^2 - 16x + 63 = 0$$
$$(x - 7)(x - 9) = 0$$
$$x - 7 = 0 \quad \text{or} \quad x - 9 = 0$$
$$x = 7 \qquad x = 9$$

Thus, the x-intercepts are $(7,0)$ and $(9,0)$.

PRACTICE EXERCISES

CHAPTER 3 RELATIONS, FUNCTIONS, AND GRAPHS

SECTION 3.8 Variation

1. **(a)** First translate the phrase *u varies directly as v*.

    $$u = cv$$

    Substitute 1.4 for v and 28 for u to find the value of c.

    $$28 = c(1.4)$$
    $$\frac{28}{1.4} = c$$
    $$20 = c$$

    Thus, the equation of variation is $u = 20v$.

    (b) From the distance formula we know that $d = rt$, and in this case r is the constant of variation. Substitute 55 for d and 0.11 fot t to find r.

    $$55 = r(0.11)$$
    $$\frac{55}{0.11} = r$$
    $$500 = r$$

    Thus, the variation in this case is $d = 500t$.

2. First translate the *phrase S varies directly as the square of the radius r*.

    $$S = cr^2$$

    Substitute 32.5 for S and 2.6 for r to find c.

    $$32.5 = c(2.6)^2$$
    $$\frac{32.5}{(2.6)^2} = c$$
    $$4.81 \approx c$$

    Now substitute 17.6 for r in the variation equation.

    $$S = (4.81)r^2$$
    $$= (4.81)(17.6)^2$$
    $$\approx 1489.9456$$

    Thus, the surface area is about 1490 m².

3. First translate *V varies inversely as P* into the variation equation.

    $$V = \frac{c}{P}$$

    Substitute 0.44 for V and 5.26 for P to find c.

    $$0.44 = \frac{c}{5.26}$$
    $$(0.44)(5.26) = c$$
    $$2.3144 = c$$

    Substitute 2.81 for V in the variation equation and solve for P.

    $$2.81 = \frac{2.3144}{P}$$
    $$P = \frac{2.3144}{2.81}$$
    $$P \approx 0.823629893$$

    Thus the pressure is about 0.82 lb/ft³.

4. First translate *V varies directly as T and inversely as P* into the variation equation.

    $$V = \frac{cT}{P}$$

    Substitute 480 for T, 16 for P and 22 for V to find c.

    $$22 = \frac{c(480)}{16}$$
    $$\frac{(22)(16)}{480} = c$$
    $$\frac{11}{15} = c$$

    Substitute 24 for P and 11 for V and find T.

    $$11 = \frac{\frac{11}{15}T}{24}$$
    $$\frac{(24)(11)(15)}{11} = T$$
    $$360 = T$$

    Thus, the temperature is 360°.

CHAPTER 3 RELATIONS, FUNCTIONS, AND GRAPHS

SECTION 3.8 Variation

1. First give the equation described by *y varies directly as x*.

$$y = cx$$

Determine the value of c by substituting 16 for x and 8 for y.

$$8 = c(16)$$
$$\tfrac{1}{2} = c$$

The equation of variation is $y = \tfrac{1}{2}x$.

2. First give the equation described by *z varies inversely as w*.

$$z = \frac{c}{w}$$

Determine the value of c by substituting 12 for z and 5 for w.

$$12 = \frac{c}{5}$$
$$(12)(5) = c$$
$$60 = c$$

The equation of variation is $z = \dfrac{60}{w}$.

3. First give the equation described by *z varies jointly as x and y*.

$$z = cxy$$

Determine the value of c by substituting 2 for x, 7 for y, and 42 for z.

$$42 = c(2)(7)$$
$$42 = c(14)$$
$$3 = c$$

The equation of variation is $z = 3xy$.

4. First give the equation described by *a is directly proportional to the square of b and inversely proportional to the square root of d*.

$$a = \frac{cb^2}{\sqrt{d}}$$

Determine the value of c by substituting 3 for b, 25 for d, and 405 for a.

$$405 = \frac{c(3)^2}{\sqrt{25}}$$
$$405 = \frac{9c}{5}$$
$$\frac{(405)(5)}{9} = c$$
$$225 = c$$

The equation of variation is $a = \dfrac{225b^2}{\sqrt{d}}$.

5. First give the equation described by *w varies jointly as the square of x and the cube root of y and inversely as the square root of z*.

$$w = \frac{cx^2 \sqrt[3]{y}}{\sqrt{z}}$$

Determine the value of c by substituting 40 for x, 125 for y, 400 for z, and 330 for w.

$$330 = \frac{c(40)^2 \sqrt[3]{125}}{\sqrt{400}}$$
$$330 = \frac{c(1600)(5)}{20}$$
$$\frac{(20)(330)}{(1600)(5)} = c$$
$$0.825 = c$$

The equation of variation is $w = \dfrac{0.825 x^2 \sqrt[3]{y}}{\sqrt{z}}$.

EXERCISES A SECTION 3.8 135

6. First give the equation described by *I varies directly as A*.

$$I = cA$$

Determine the value of c by substituting 250 for I and 2000 for A.

$$250 = c(2000)$$
$$\frac{250}{2000} = c$$
$$0.125 = c$$

Use this value of c in the equation of variation along with 9250 for A to find the required amount of interest I.

$$I = 0.125(9250) = 1156.25$$

Thus, the interest to be paid would be $1156.25.

7. First give the equation described by *C varies directly as n*.

$$C = cn$$

Determine the value of c by substituting 25,200 for C and 4100 for n.

$$25,200 = c(4100)$$
$$\frac{25,200}{4100} = c$$
$$\frac{252}{41} = c$$

Use this value of c together with 6400 for n and find the value of C.

$$C = \left(\frac{252}{41}\right)(6400)$$
$$\approx 39,336.58537$$

Thus, to the nearest dollar, the cost is $39,337 when 6400 units are produced.

8. First give the equation described by *P varies directly as the square of e and inversely as the square root of d*.

$$P = \frac{ce^2}{\sqrt{d}}$$

Determine the value of c by substituting 7.6 for P, 7 for e, and 16 for d.

$$7.6 = \frac{c(7)^2}{\sqrt{16}}$$
$$7.6 = \frac{c(49)}{4}$$
$$\frac{(7.6)(4)}{49} = c$$

Using this value of c, substitute 8 for e and 9 for d and evaluate to obtain P.

$$P = \frac{\frac{(7.6)(4)}{49}(8)^2}{\sqrt{9}}$$
$$= \frac{7.6)(4)(64)}{(49)(3)} \approx 13.23537415$$

Thus, the productivity is approximately 13.2.

9. First find the equation described by *V varies directly as T and inversely as P*.

$$V = \frac{cT}{P}$$

Determine the value of c by substituting 16 for V, 480 for T, and 45 for P.

$$16 = \frac{c(480)}{45}$$
$$\frac{(16)(45)}{480} = c$$
$$1.5 = c$$

Use this value of c together with 1200 for T and 15 for P to find the value of V.

$$V = \frac{1.5(1200)}{15}$$
$$= 120$$

Thus, the volume is 120 in³.

10. First find the equation described by *w is inversely proportional to the square of its distance d from the center of the earth*.

$$w = \frac{c}{d^2}$$

Substitute $2d$ for d and simplify.

$$w = \frac{c}{(2d)^2} = \frac{c}{4d^2} = \frac{1}{4}\left(\frac{c}{d^2}\right)$$

Thus, the weight is multiplied by $\frac{1}{4}$, or divided by 4.

11. First find the equation described by *P varies directly as the square root of l.*

$$P = c\sqrt{l}$$

Determine the value of c by substituting 1.56 for P and 9.25 for l.

$$1.56 = c\sqrt{9.25}$$
$$\frac{1.56}{\sqrt{9.25}} = c$$

Use this value for c and substitute 16.8 for l to find the value of P.

$$P = \frac{1.56}{\sqrt{9.25}}\sqrt{16.8} \approx 2.102366234$$

Thus, the period is about 2.10 sec.

12. First find the equation described by *I varies inversely as the square of d.*

$$I = \frac{c}{d^2}$$

Substitute $2I$ for I, and determine the effect on d,

$$2I = \frac{c}{d^2}$$
$$I = \frac{c}{2d^2}$$
$$I = \frac{c}{(\sqrt{2}d)^2}$$

For this to equal $\frac{c}{d^2}$, d would have to be replaced with $\frac{1}{\sqrt{2}}d$. Thus, to double the intensity, the distance must be multiplied by $\frac{1}{\sqrt{2}}$.

13. First find the equation described by *T varies directly as A.*

$$T = cA$$

Substitute 25 for T and 35 for A (a 3 by 5 enlargment has area $(3)(5) = 35$ square units).

$$25 = c(35)$$
$$\frac{25}{35} = c$$
$$\frac{5}{7} = c$$

Use this value for c in the variation equation with 154 for A to find the new value for T.

$$T = \frac{5}{7}(154) = 110$$

Thus, the time to make an 11-by-14 enlargement is 110 sec.

14. First find the equation described by *s is directly proportional to the square root of l.*

$$s = c\sqrt{l}$$

Find the value of c by substituting 45 for s and 65 for l.

EXERCISES A   **SECTION 3.8**   **137**

$$45 = c\sqrt{65}$$
$$\frac{45}{\sqrt{65}} = c$$

Use this value for c, substitute 120 for l, and find the value of s.

$$s = \frac{45}{\sqrt{65}}\sqrt{l}$$
$$= \frac{45}{\sqrt{65}}\sqrt{120}$$
$$\approx 61.14295984$$

Thus, the speed of the car is about 61 mph.

15. First find the equation described by m varies jointly with w and the square of t and inversely as l.

$$m = \frac{cwt^2}{l}$$

Substitute $5w$ for w, $\frac{1}{2}t$ for t, and $3l$ for l to find the relationship between the former safe load m and the new safe load M.

$$M = \frac{c(5w)\left(\frac{1}{2}t\right)^2}{3l}$$
$$= \frac{\frac{5}{4}cwt^2}{3l}$$
$$= \frac{5}{12}\left(\frac{cwt^2}{l}\right)$$
$$= \frac{5}{15}m$$

Thus, the new safe load is $\frac{5}{12}$ the former safe load.

16. First find the equation described by $f(x)$ varies directly with x.

$$f(x) = cx$$

Find the value of c by substituting 6.5 for x and 13.0 for $f(x)$.

$$13.0 = c(6.5)$$
$$\frac{13.0}{6.5} = c$$
$$2 = c$$

Thus, the model is $f(x) = 2x$.

17. First find the equation described by $f(x)$ is inversely proportional to x.

$$f(x) = \frac{c}{x}$$

Find the value of c by substituting 0.25 for $f(x)$ and 8 for x.

$$0.25 = \frac{c}{8}$$
$$(8)(0.25) = c$$
$$2 = c$$

Thus, the model is $f(x) = \frac{2}{x}$.

18. (a) First find the equation described by C varies directly as n.

$$C = C(n) = cn$$

Substitute 25 for n and 100 for $C(n)$ to find the value of c.

$$100 = c(25)$$
$$\frac{100}{25} = c$$
$$4 = c$$

Thus, the model for C is

$$C(n) = 4n.$$

(b) First find the equation described by R varies directly as the square of n.

$$R = R(n) = cn^2$$

Substitute 25 for n and 125 for $R(n)$ to find the value of c.

$$125 = c(25)^2$$
$$\frac{125}{(25)^2} = c$$
$$\frac{1}{5} = c$$
$$0.2 = c$$

Thus, the quadratic model for R is

$$R(n) = 0.2n^2.$$

(c) The break-even points are the values of n for which the profit function

$$P(n) = R(n) - C(n) = 0.$$

Solve this equation.

$$0.2n^2 = 4n$$
$$2n^2 = 40n$$
$$n^2 = 20n$$
$$n^2 - 20n = 0$$
$$n(n - 20) = 0$$
$$n = 0 \quad n = 20$$

Thus, the two break-even points are $n = 0$ and $n = 20$.

CHAPTER 3 RELATIONS, FUNCTIONS, AND GRAPHS

CHAPTER 3 Review Exercises

1. The intercepts of the graph of $3x - 2y = 6$ are $(0,-3)$ and $(2,0)$, and the graph is given below.

   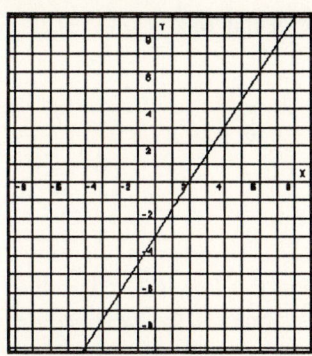

2. $d = \sqrt{(x_1 - x_2)^2 + (y_1 - y_2)^2}$
   $= \sqrt{(-2 - 7)^2 + (6 - 3)^2}$
   $= \sqrt{(-9)^2 + (3)^2}$
   $= \sqrt{81 + 9}$
   $= \sqrt{90} = \sqrt{9 \cdot 10} = 3\sqrt{10}$

3. Substitute $(7,2)$ for the midpoint of (a,b) and $(8,5)$ in the midpoint formula.

   $(\bar{x},\bar{y}) = \left(\dfrac{x_1 + x_2}{2}, \dfrac{y_1 + y_2}{2}\right)$

   $(7,2) = \left(\dfrac{a + 8}{2}, \dfrac{b + 5}{2}\right)$

   $7 = \dfrac{a + 8}{2} \qquad 2 = \dfrac{b + 5}{2}$

   $14 = a + 8 \qquad 4 = b + 5$

   $6 = a \qquad\qquad -1 = b$

   Thus, $(a,b) = (6,-1)$.

4. Substitute 5 for d, $(a,2)$ for (x_1,y_1), and $(2,-1)$ for (x_2,y_2) in the distance formula and solve for a.

   $d = \sqrt{(x_1 - x_2)^2 + (y_1 - y_2)^2}$
   $5 = \sqrt{(a - 2)^2 + (2 - (-1))^2}$
   $5 = \sqrt{(a - 2)^2 + 9}$
   $25 = (a - 2)^2 + 9$
   $16 = (a - 2)^2$
   $\pm\sqrt{16} = a - 2$
   $\pm 4 = a - 2$
   $2 \pm 4 = a$

   Thus, $a = 6$ and $a = -2$, so the two points are $(6,2)$ and $(-2,2)$.

5. $m = \dfrac{y_2 - y_1}{x_2 - x_1}$
   $= \dfrac{-1 - 8}{-3 - (-2)}$
   $= \dfrac{-1 - 8}{-3 + 2} = \dfrac{-9}{-1} = 9$

6. First find the slope of the line through the two points $(-3,-7)$ and $(-1,2)$.

   $m = \dfrac{y_2 - y_1}{x_2 - x_1}$
   $= \dfrac{2 - (-7)}{-1 - (-3)}$
   $= \dfrac{2 + 7}{-1 + 3} = \dfrac{9}{2}$

   Use $\dfrac{9}{2}$ for m and $(-1,2)$ for (x_1,y_1) in the point-slope form.

   $y - y_1 = m(x - x_1)$
   $y - 2 = \dfrac{9}{2}(x - (-1))$
   $y - 2 = \dfrac{9}{2}(x + 1)$
   $2(y - 2) = 9(x + 1)$
   $2y - 4 = 9x + 9$
   $0 = 9x - 2y + 13$

   Thus, the general form of the desired line is
   $9x - 2y + 13 = 0$.

7. First find the slope of the line with equation $4x - 3y + 6 = 0$ by writing the equation in slope-intercept form (by solving for y).

   $4x - 3y + 6 = 0$
   $-3y = -4x - 6$
   $y = \dfrac{4}{3}x + 2$

Then the slope of this line is $\frac{4}{3}$, the same as the slope of every parallel line. Use this slope and the point $(-2,0)$ in the point-slope form.

$$y - y_1 = m(x - x_1)$$
$$y - 0 = \frac{4}{3}(x - (-2))$$
$$y = \frac{4}{3}(x + 2)$$
$$3y = 4(x + 2)$$
$$3y = 4x + 8$$
$$0 = 4x - 3y + 8$$

Thus, the desired line has equation

$$4x - 3y + 8 = 0.$$

8. Solve $7x - 2y = 4$ for y, that is, write the equation in slope-intercept form. The coefficient of x is the slope, and the constant term is the y-coordinate of the y-intercept.

$$7x - 2y = 4$$
$$-2y = -7x + 4$$
$$y = \frac{7}{2}x - 2$$

Thus, the slope is $\frac{7}{2}$, and the y-intercept is $(0,-2)$.

9. Write each equation in slope-intercept form to identify the slopes.

$$8x + 4y - 7 = 0$$
$$4y = -8x + 7$$
$$y = -2x + \frac{7}{4}$$

The slope of this line is -2.

$$-12x - 6y + 5 = 0$$
$$-6y = 12x - 5$$
$$y = -2x + \frac{5}{6}$$

This line also has slope -2. Since the slopes are equal, the lines are parallel.

10. Write each equation in slope-intercept form to identify the slopes.

$$x + 3y - 2 = 0$$
$$3y = -x + 2$$
$$y = -\frac{1}{3}x + \frac{2}{3}$$

The slope of this line is $-\frac{1}{3}$.

$$6x - 2y + 7 = 0$$
$$-2y = -6x - 7$$
$$y = 3x + \frac{7}{2}$$

Ths slope of this line is 3. Since the two slopes are negative reciprocals, the two lines are perpendicular.

11. There are several ways to show that the triangle is a right triangle. Perhaps the simplest is to graph the points to determine which of the two sides are most likely to be perpendicular. Doing so it would appear that side AB might be perpendicular to side BC. Find the slope of the lines containing these points.

$$m_{AB} = \frac{6-1}{5-(-2)} = \frac{5}{7}$$
$$m_{BC} = \frac{1-(-6)}{-2-3} = \frac{7}{-5} = -\frac{7}{5}$$

Since the two slopes are negative reciprocals, the two sides are perpendicular and the triangle is a right triangle.

12. Begin by finding the midpoint of the segment joining $(-2,-4)$ and $(4,8)$.

$$(\bar{x},\bar{y}) = \left(\frac{x_1 + x_2}{2}, \frac{y_1 + y_2}{2}\right)$$
$$= \left(\frac{-2+4}{2}, \frac{-4+8}{2}\right)$$
$$= \left(\frac{2}{2}, \frac{4}{2}\right) = (1,2)$$

Next find the slope of the segment. The slope of the desired line will be the negative reciprocal.

$$m = \frac{y_2 - y_1}{x_2 - x_1} = \frac{8-(-4)}{4-(-2)} = \frac{12}{6} = 2$$

Use $-\frac{1}{2}$ for m and the midpoint, $(1,2)$, for (x_1, y_1) in the point-slope form.

$$y - y_1 = m(x - x_1)$$
$$y - 2 = -\frac{1}{2}(x - 1)$$
$$2(y - 2) = -1(x - 1)$$
$$2y - 4 = -x + 1$$
$$x + 2y - 5 = 0$$

Thus, the desired perpendicular bisector has equation
$$x + 2y - 5 = 0.$$

13. Given $y = 2x^2 + 3$. Since a given value for x determines exactly one value for y, this equation represents a function.

14. Given $x = 2y^2 + 3$. Notice that when x is 5, y is both 1 and -1. That is, both pairs $(5,1)$ and $(5,-1)$ are solutions to the equation. Thus, this equation does not represent a function.

15. Given the function $f(x) = \dfrac{1}{\sqrt{x+1}}$. For this function to be defined, the denominator cannot be 0 and the radicand must be positive. Thus, $x + 1 > 0$, or $x > -1$ gives the values for x for which the function is defined.

16. Given the function $g(x) = \dfrac{x+5}{x^2-9} = \dfrac{x+5}{(x+3)(x-3)}$.

For this function to be defined, the denominator cannot be 0. This means that x cannot be 3 nor -3. Thus, the values for which the function is defined are all numbers $x \neq \pm 3$.

17. Given $g(x) = 1 - x$. Since x is raised to the first power, g is a linear function.

18. Given $h(x) = 16$. Since all the values of the function are the same constant, 16, this is a constant function.

19. Given the function $f(x) = 2x^2 + 3x$.

$$f(0) = 2(0)^2 + 3(0)$$
$$= 2(0) + 0$$
$$= 0$$

$$f(-2) = 2(-2)^2 + 3(-2)$$
$$= 2(4) - 6 = 8 - 6 = 2$$
$$f\left(\frac{1}{2}\right) = 2\left(\frac{1}{2}\right)^2 + 3\left(\frac{1}{2}\right)$$
$$= \frac{1}{2} + \frac{3}{2} = \frac{4}{2} = 2$$
$$f(a) = 2(a)^2 + 3(a) = 2a^2 + 3a$$
$$f(a+1) = 2(a+1)^2 + 3(a+1)$$
$$= 2(a^2 + 2a + 1) + 3a + 3$$
$$= 2a^2 + 4a + 2 + 3a + 3$$
$$= 2a^2 + 7a + 5$$

20. Given $g(x) = 3x - 5$.

$$g(a^3) = 3(a^3) - 5 = 3a^3 - 5$$
$$g\left(\frac{1}{a}\right) = 3\left(\frac{1}{a}\right) - 5$$
$$= \frac{3}{a} - 5 = \frac{3 - 5a}{a}$$
$$g(\sqrt{a}) = 3(\sqrt{a}) - 5 = 3\sqrt{a} - 5$$

21. Since a vertical line can be drawn through the graph that passes through more than one point (in fact two points), this graph is not the graph of a function.

22. Since every vertical line drawn through the graph passes through exactly one point, this graph is the graph of a function.

23. Given the function $g(x) = -\frac{1}{3}x - 2$. From the graph, given below, we can see that the function is decreasing for all values of x.

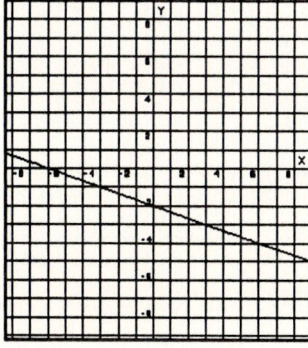

24. Given the function $k(x) = \dfrac{4}{x}$. From the graph, given below, we can see that the function decreases for all values of x where it is defined ($x \neq 0$).

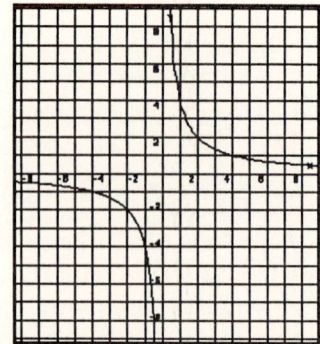

25. To obtain the graph of $y = -f(x)$ from the graph of $y = f(x)$ that is given, all of the y-coordinates must be multiplied by -1. This is the same as "reflecting" the graph of $y = f(x)$ in the x-axis. The graph is given below.

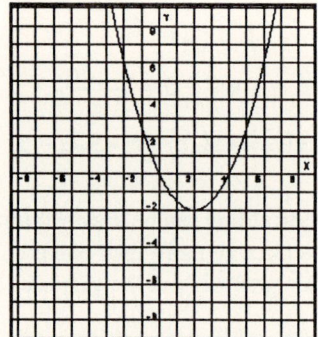

26. To obtain the graph of $y = 2f(x)$ from the graph of $y = f(x)$ that is given, all of the y-coordinates must be multiplied by 2. This is the same as "stretching" the graph of $y = f(x)$ vertically. The graph is given below.

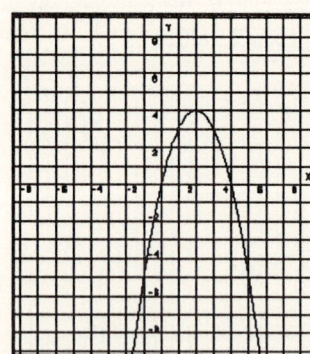

27. To obtain the graph of $y = f(x + 1) - 2$ from the graph of $y = f(x)$ that is given, shift the graph 1 unit left (from the $+ 1$) and then 2 units down (from the $- 2$). The graph is given below.

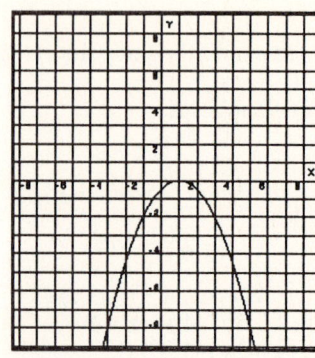

28. To obtain the graph of $y = f(x - 1) + 2$ from the graph of $y = f(x)$ that is given, shift the graph 1 unit right (from the $- 1$) and then 2 units up (from the $+ 2$). The graph is given below.

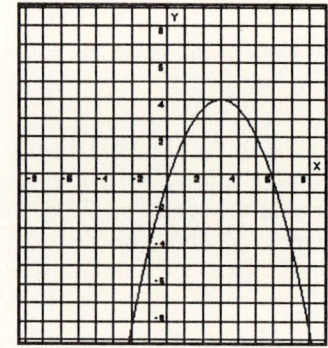

29. Given the function $f(x) = 2x^2 - 3$.

$$f(-x) = 2(-x)^2 - 3$$
$$= 2x^2 - 3$$
$$= f(x)$$

Since $f(-x) = f(x)$, f is an even function.

30. Given the function $g(x) = 4x^3 + 2x$.

$$g(-x) = 4(-x)^3 + 2(-x)$$
$$= -4x^3 - 2x$$
$$= -g(x)$$

Since $g(-x) = -g(x)$, g is an odd function.

31. Given the function $h(x) = |x| + x$.

$$h(-x) = |-x| + (-x)$$
$$= |x| - x$$

Since $h(-x) \neq h(x)$ and $h(-x) \neq -h(x)$, h is neither even nor odd.

32. Given $3x^2 + 2y^2 = 5$. When x is replaced with $-x$, since x is squared, the equation is the same. Thus, the graph is symmetric with respect to the y-axis. When y is replaced with −y, since y is squared, the equation remains the same. Thus, the graph is symmetric with respect to the x-axis. Similarly, when both x and y are replaced with −x and −y, the equation remains the same so the graph is symmetric with respect to the origin.

33. Given $8xy = 1$. When x is replaced with −x, or when y is replaced with −y, the equation becomes $-8xy = 1$, which is different from the given equation. Thus, the graph is not symmetric with respect to either axis. However, when both x and y are replaced with −x and −y, respectively, the equation becomes $8(-x)(-y) = 8xy = 1$, which is the same. Thus, the graph is symmetric with respect to the origin.

34. Given $y^2 + 2x = 5$. Since y is squared, when y is replaced with −y, the equation remains unchanged. Thus, the graph is symmetric with respect to the x-axis. However, when x is replaced with −x (alone or in conjunction with replacing y with −y), the equation will be changed. Thus, the graph is not symmetric with respect to the y-axis nor with respect to the origin.

35. Think of graphing f in three parts. When $x \leq -2$, graph the function $f(x) = -x$. When $-2 < x \leq 1$, graph the function $f(x) = -2$. And when $x > 1$, graph the function $f(x) = 2x - 1$. The graph of f is given below.

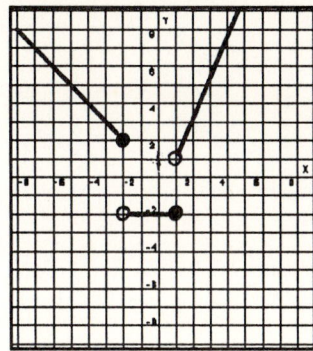

36. Given the function $g(x) = \frac{1}{2}[x]$. The graph of g can be obtained from the graph of the greatest integer function $y = [x]$ by multiplying all the y-values by $\frac{1}{2}$. The graph is given below.

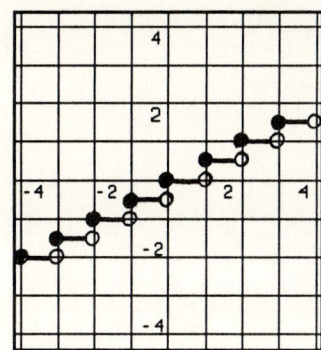

37. Given $f(x) = 3x^2 + 2$ and $g(x) = x - 5$.

First find $g(3)$.

$$g(3) = (3) - 5 = 3 - 5 = -2$$

Then find $f(g(3))$.

$$\begin{aligned} f(g(3)) &= f(-2) \\ &= 3(-2)^2 + 2 \\ &= 3(4) + 2 \\ &= 12 + 2 = 14 \end{aligned}$$

First find $f(3)$.

$$\begin{aligned} f(3) &= 3(3)^2 + 2 \\ &= 3(9) + 2 \\ &= 27 + 2 = 29 \end{aligned}$$

Then find $g(f(3))$.

$$\begin{aligned} g(f(3)) &= g(29) \\ &= (29) - 5 = 29 - 5 = 24 \end{aligned}$$

$$\begin{aligned} f(g(x)) &= f(x - 5) \\ &= 3(x - 5)^2 + 2 \\ &= 3(x^2 - 10x + 25) + 2 \\ &= 3x^2 - 30x + 75 + 2 \\ &= 3x^2 - 30x + 77 \end{aligned}$$

$$\begin{aligned} g(f(x)) &= g(3x^2 + 2) \\ &= (3x^2 + 2) - 5 \\ &= 3x^2 + 2 - 5 \\ &= 3x^2 - 3 \end{aligned}$$

38. Given $f(x) = |x - 2|$ and $g(x) = 1 - x^2$.

First find $g(3)$.

$$\begin{aligned} g(3) &= 1 - (3)^2 \\ &= 1 - 9 = -8 \end{aligned}$$

Next find $f(g(3))$.

$$f(g(3)) = f(-8)$$
$$= |-8 - 2| = |-10| = 10$$

First find $f(3)$.

$$f(3) = |3 - 2| = |1| = 1$$

Next find $g(f(3))$.

$$g(f(3)) = g(1)$$
$$= 1 - (1)^2 = 1 - 1 = 0$$

$$f(g(x)) = f(1 - x^2)$$
$$= |(1 - x^2) - 2|$$
$$= |-1 - x^2|$$
$$= |(-1)(1 + x^2)|$$
$$= |-1||1 + x^2|$$
$$= |1 + x^2|$$
$$= 1 + x^2 \quad \text{Note that } 1 + x^2 > 0$$

$$g(f(x)) = g(|x - 2|)$$
$$= 1 - (|x - 2|)^2$$
$$= 1 - (x^2 - 4x + 4)$$
$$= 1 - x^2 + 4x - 4$$
$$= -x^2 + 4x - 3$$

39. Given $f(x) = \sqrt{x - 4}$. As x increases, y also increases so f is an increasing function. As a result, f is one-to-one.

40. Given $f(x) = |x - 4|$. If $x = 1$, then $f(1) = 3$. Also, if $x = 7$, then $f(7) = 3$. Thus, two values of x, 1 and 7, correspond to the same value of y, 3. As a result, f is not a one-to-one function.

41. Given the function $f(x) = \frac{1}{2}x + 3$, or $y = \frac{1}{2}x + 3$.

To find the inverse, first interchange x and y, then solve for y.

$$x = \frac{1}{2}y + 3$$
$$2x = y + 6$$
$$2x - 6 = y$$

Thus, $f^{-1}(x) = 2x - 6$.

42. Given the function $g(x) = \sqrt{x + 5}$, or $y = \sqrt{x + 5}$.

Then $y \geq 0$ and $x \geq -5$. To find the inverse, first interchange x and y, then solve for y.

$$x = \sqrt{y + 5}$$
$$x^2 = y + 5$$
$$x^2 - 5 = y$$

Thus, $g^{-1}(x) = x^2 - 5$, for $x \geq 0$.

43. Given the function $h(x) = \frac{1}{2}x^2 - 2$, $x \geq 0$, or

$y = \frac{1}{2}x^2 - 2$, $x \geq 0$. First interchange x and y, then solve for y.

$$x = \frac{1}{2}y^2 - 2$$
$$2x = y^2 - 4$$
$$2x + 4 = y^2$$
$$\pm\sqrt{2x + 4} = y$$

But since $y \geq 0$, we use the positive root only.

Thus, $h^{-1}(x) = \sqrt{2x + 4}$.

44. Given the function $k(x) = \frac{1}{2}x^3$, or $y = \frac{1}{2}x^3$. First interchange x and y, then solve for y.

$$x = \frac{1}{2}y^3$$
$$2x = y^3$$
$$\sqrt[3]{2x} = y$$

Thus, $k^{-1}(x) = \sqrt[3]{2x}$.

45. Given the function $f(x) = 2x^2 + 9x - 5$. To find the x-coordinate of the vertex, use the formula:

$$-\frac{b}{2a} = -\frac{9}{2(2)} = -\frac{9}{4}$$

Substitute this value to find the y-coordinate of the vertex.

$$f\left(-\frac{9}{4}\right) = 2\left(-\frac{9}{4}\right)^2 + 9\left(-\frac{9}{4}\right) - 5$$
$$= \frac{162}{16} - \frac{81}{4} - 5$$
$$= \frac{162}{16} - \frac{324}{16} - \frac{80}{16}$$
$$= -\frac{242}{16} = -\frac{121}{8}$$

Thus, the vertex is $\left(-\frac{9}{4}, -\frac{121}{8}\right)$.

To find the x-intercepts, solve the following equation.

$$2x^2 + 9x - 5 = 0$$
$$(2x - 1)(x + 5) = 0$$
$$\begin{aligned} 2x - 1 &= 0 & x + 5 &= 0 \\ 2x &= 1 & x &= -5 \end{aligned}$$
$$x = \frac{1}{2}$$

Thus, the intercepts are $(-5, 0)$ and $\left(\frac{1}{2}, 0\right)$.

46. Given the function $g(x) = -3x^2 - 2x + 8$. To find the x-coordinate of the vertex, use the formula:

$$-\frac{b}{2a} = -\frac{-2}{2(-3)} = -\frac{2}{6} = -\frac{1}{3}$$

Substitute to find the y-coordinate of the vertex.

$$f\left(-\frac{1}{3}\right) = -3\left(-\frac{1}{3}\right)^2 - 2\left(-\frac{1}{3}\right) + 8$$
$$= -\frac{1}{3} + \frac{2}{3} + 8$$
$$= -\frac{1}{3} + \frac{2}{3} + \frac{24}{3}$$
$$= \frac{25}{3}$$

Thus, the vertex is $\left(-\frac{1}{3}, \frac{25}{3}\right)$.

To find the x-intercepts, solve the following equation.

$$-3x^2 - 2x + 8 = 0$$
$$3x^2 + 2x - 8 = 0$$
$$(3x - 4)(x + 2) = 0$$
$$\begin{aligned} 3x - 4 &= 0 & x + 2 &= 0 \\ 3x &= 4 & x &= -2 \end{aligned}$$
$$x = \frac{4}{3}$$

Thus, the intercepts are $(-2, 0)$ and $\left(\frac{4}{3}, 0\right)$.

47. Given the function $f(x) = -2x^2 + 8x - 5$. To find the x-coordinate of the vertex, use the formula:

$$-\frac{b}{2a} = -\frac{8}{2(-2)} = \frac{8}{4} = 2$$

Substitute to find the y-coordinate of the vertex.

$$\begin{aligned} f(2) &= -2(2)^2 + 8(2) - 5 \\ &= -8 + 16 - 5 \\ &= 3 \end{aligned}$$

Thus, the vertex is $(2, 3)$.

To find the x-intercepts, solve the equation $-2x^2 + 8x - 5 = 0$, or $2x^2 - 8x + 5 = 0$.

$$\begin{aligned} x &= \frac{-b \pm \sqrt{b^2 - 4ac}}{2a} \\ &= \frac{-(-8) \pm \sqrt{(-8)^2 - 4(2)(5)}}{2(2)} \\ &= \frac{8 \pm \sqrt{64 - 40}}{4} \\ &= \frac{8 \pm \sqrt{24}}{4} \\ &= \frac{8 \pm 2\sqrt{6}}{4} \\ &= \frac{4 \pm \sqrt{6}}{2} \\ &\approx 3.23, \ 0.78 \end{aligned}$$

Note that $a = -2 < 0$, so the graph, a parabola opens down. The graph is given below.

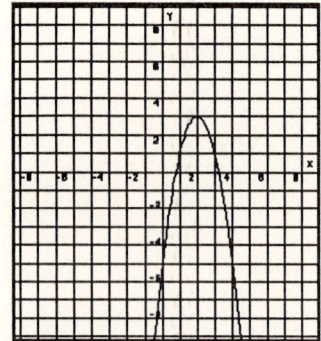

48. Given the function $g(x) = \frac{1}{2}x^2 + 3x - 2$. To find the x-coordinate of the vertex, use the formula:

$$-\frac{b}{2a} = -\frac{3}{2\left(\frac{1}{2}\right)} = -\frac{3}{1} = -3$$

Substitute to find the y-coordinate of the vertex.

$$\begin{aligned}g(-3) &= \frac{1}{2}(-3)^2 + 3(-3) - 2\\ &= \frac{9}{2} - 9 - 2\\ &= \frac{9}{2} - 11\\ &= \frac{9}{2} - \frac{22}{2}\\ &= -\frac{13}{2} = -6.5\end{aligned}$$

Thus, the vertex is $(-3, -6.5)$.

To find the x-intercepts, solve the equation
$\frac{1}{2}x^2 + 3x - 2 = 0$, or $x^2 + 6x - 4 = 0$.

$$\begin{aligned}x &= \frac{-b \pm \sqrt{b^2 - 4ac}}{2a}\\ &= \frac{-(6) \pm \sqrt{(6)^2 - 4(1)(-4)}}{2(1)}\\ &= \frac{-6 \pm \sqrt{36 + 16}}{2}\\ &= \frac{-6 \pm \sqrt{52}}{2}\\ &= \frac{-6 \pm 2\sqrt{13}}{2}\\ &= -3 \pm \sqrt{13}\\ &\approx 0.61,\ -6.61\end{aligned}$$

Note that $a = \frac{1}{2} > 0$, so the graph, a parabola opens up. The graph is given below.

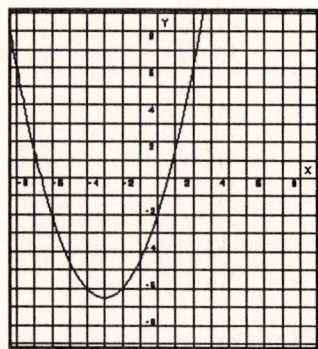

49. From the figure, note that the 120 ft of fence must equal $x + x + x + y = 3x + y$. Thus, $3x + y = 120$, or $y = 120 - 3x$. The total area to be enclosed is given by $A = xy$. Substitute $120 - 3x$ for y to obtain the quadratic function

$$A = xy = x(120 - 3x) = -3x^2 + 120x.$$

Since the graph of this function is a parabola that opens down ($a = -3 < 0$), the maximum value for A, the area, will be the second coordinate of the vertex of the parabola. Use the formula to find the first coordinate.

$$-\frac{b}{2a} = -\frac{120}{2(-3)} = \frac{120}{6} = 20$$

Substitute to find the value of A when $x = 20$.

$$\begin{aligned}A &= -3(20)^2 + 120(20)\\ &= -1200 + 2400\\ &= 1200\end{aligned}$$

The maximum area is 1200 ft². Now find the value of y when $x = 20$.

$$\begin{aligned}y &= 120 - 3(20)\\ &= 120 - 60\\ &= 60\end{aligned}$$

Thus, the dimensions for x and y that will yield the maximum area of the storage area are $x = 20$ ft and $y = 60$ ft.

50. The cost function is $C(x) = 2x^2 - 80x + 2200$, where x is the number of trucks loaded. Since the graph of this function is a parabola that opens up

($a = 2 > 0$), the minimum cost will be the second coordinate of the vertex. First find the x-coordinate of the vertex using the formula.

$$-\frac{b}{2a} = -\frac{-80}{2(2)} = \frac{80}{4} = 20$$

Substitute to find the second coordinate.

$$\begin{aligned} C(20) &= 2(20)^2 - 80(20) + 2200 \\ &= 800 - 1600 + 2200 \\ &= 1400 \end{aligned}$$

Thus, the minimum cost is $1400 when 20 trucks are used.

51. A markup of 30% on a coat costing $120 amounts to $0.30(120) = 36$. Thus, the marked price would be $120 + $36 = $156. If the coat is then marked down x percent, the markdown would be $156x$. Then the marked down price of the coat would be

$$156 - 156x = 156(1 - x).$$

If a sales tax is added on and the rate is 5%, the tax on the coat would be

$$0.05(156)(1 - x) = 7.8(1 - x).$$

The actual amount paid, including the tax would be

$$156(1 - x) + 7.8(1 - x) = 163.8(1 - x).$$

Thus, the function that serves as a model for this total price, including tax, would be

$$T(x) = 163.8(1 - x).$$

Substitute 0.4 for x to find the total price of the coat, including tax, if the markdown rate is 40%.

$$\begin{aligned} T(0.4) &= 163.8(1 - 0.4) \\ &= 163.8(0.6) \\ &= 98.28 \end{aligned}$$

Thus the total price would be $98.28.

52. Use the standard form of the equation of a quadratic function, $y = a(x - h)^2 + k$, where (h,k) is the vertex, and substitute 150 for h and 850 for k.

$$y = a(x - 150)^2 + 850$$

Since the function also passes through the point (130,50), substitute 130 for x and 50 for y to find the value of a.

$$\begin{aligned} 50 &= a(130 - 150)^2 + 850 \\ 50 &= a(-20)^2 + 850 \\ -800 &= a(400) \\ \frac{-800}{400} &= a \\ -2 &= a \end{aligned}$$

Thus, replacing y with $f(x)$, the quadratic model is given by:

$$\begin{aligned} f(x) &= -2(x - 150)^2 + 850 \\ &= -2(x^2 - 300x + 22{,}500) + 850 \\ &= -2x^2 + 600x - 44{,}150 \end{aligned}$$

53. We must find the vertex of the quadratic function

$$C(x) = 923.6x^2 - 1756x + 46.8.$$

The x-coordinate is found using the formula.

$$-\frac{b}{2a} = -\frac{-1756}{2(923.6)} = \frac{1756}{1847.2} \approx 0.950627977$$

Thus, the percent which minimizes the cost is about 95.1%.

54. The formula for the surface area of a sphere is $A = 4\pi r^2$. Since the cost of painting one square meter of surface area is $3.20, multiply the cost per square meter by the number of square meters to obtain the cost function:

$$C(r) = (3.20)(4\pi r^2) = 12.8\pi r^2$$

The cost of painting a sphere with radius 3.2 m is given by

$$C(3.2) = 12.8\pi(3.2)^2 \approx 411.7748323.$$

Thus, the cost would be about $411.77.

55. First translate *u varies directly as the square of v and inversely as w* into the variation equation

$$u = \frac{cv^2}{w}.$$

Substitute 2 for v, 5 for w, and 20 for u to find the value of c.

$$20 = \frac{c(2)^2}{5}$$
$$100 = c(4)$$
$$25 = c$$

Thus, the variation equation is $u = \frac{25v^2}{w}$.

56. First translate *z varies jointly as x and the square root of y and inversely as the cube of r* into the variation equation

$$z = \frac{cx\sqrt{y}}{r^3}.$$

Substitute 6 for x, 9 for y, 3 for r, and 8 for z to find the value of c.

$$8 = \frac{c(6)\sqrt{9}}{3^3}$$
$$8 = \frac{c(6)(3)}{27}$$
$$8 = \frac{2c}{3}$$
$$24 = 2c$$
$$12 = c$$

Thus, the equation of variation is $z = \frac{12x\sqrt{y}}{r^3}$.

57. First translate *R varies directly as n* into the variation equation

$$R = cn.$$

Substitute 2500 for R and 320 for n to find the value of c.

$$2500 = c(320)$$
$$7.8125 = c$$

Thus the variation equation is $R = 7.8125n$. Substitute 920 for n to find the revenue produced when 920 units are sold.

$$R = (7.8125)(920) = 7187.5$$

Thus, the revenue produced on 920 items is $7187.50.

58. First translate *T varies jointly as w and d and inversely as p* into the variation equation

$$T = \frac{cwd}{p}.$$

Substitute 24 for T, 3 for p, 500 for w, and 60 for d to find the value of c.

$$24 = \frac{c(500)(60)}{3}$$
$$24 = c(500)(20)$$
$$24 = 10{,}000c$$
$$\frac{24}{10{,}000} = c$$
$$0.0024 = c$$

Thus, the variation equation is

$$T = \frac{0.0024wd}{p}.$$

Substitute 700 for w, 100 for d, and 35 for T, and solve for p.

$$35 = \frac{0.0024(700)(100)}{p}$$
$$35 = \frac{168}{p}$$
$$p = \frac{168}{35} = 4.8$$

Thus, 4.8 horsepower would be required.

59. Given the function $f(x) = x^2 - 7x + 6$. Use the formula to find the x-coordinate of the vertex.

$$-\frac{b}{2a} = -\frac{-7}{2(1)} = \frac{7}{2}$$

Then the y-coordinate of the vertex is

$$f\left(\frac{7}{2}\right) = -\frac{25}{4}.$$ Thus, the vertex is $\left(\frac{7}{2}, -\frac{25}{4}\right)$.

To find the x-intercepts, solve the following equation:

$$x^2 - 7x + 6 = 0$$
$$(x - 1)(x - 6) = 0$$
$$x - 1 = 0 \qquad x - 6 = 0$$
$$x = 1 \qquad x = 6$$

Thus, the x-intercepts are $(1,0)$ and $(6,0)$.

60. Given the function $f(x) = -x^2 - 12x - 11$. The x-coordinate of the vertex is

$$-\frac{b}{2a} = -\frac{-12}{2(-1)} = -\frac{-12}{-2} = -6.$$

The y-coordinate is $f(-6) = 25$. Thus, the vertex is $(-6, 25)$.

To find the x-intercepts, solve the following equation:

$$-x^2 - 12x - 11 = 0$$
$$x^2 + 12x + 11 = 0$$
$$(x + 1)(x + 11) = 0$$
$$x + 1 = 0 \qquad x + 11 = 0$$
$$x = -1 \qquad x = -11$$

Thus, the x-intercepts are $(-1, 0)$ and $(-11, 0)$.

61. Given the function $g(x) = -2x^2 + 3x + 20$. The x-coordinate of the vertex is

$$-\frac{b}{2a} = -\frac{3}{2(-2)} = -\frac{3}{-4} = \frac{3}{4}.$$

The y-coordinate is $g\left(\frac{3}{4}\right) = \frac{169}{8}$. Thus, the vertex is $\left(\frac{3}{4}, \frac{169}{8}\right)$.

To find the x-intercepts, solve the following equation:
$$-2x^2 + 3x + 20 = 0$$
$$2x^2 - 3x - 20 = 0$$
$$(2x + 5)(x - 4) = 0$$
$$2x + 5 = 0 \qquad x - 4 = 0$$
$$2x = -5 \qquad x = 4$$
$$x = -\frac{5}{2}$$

Thus, the x-intercepts are $(4, 0)$ and $\left(-\frac{5}{2}, 0\right)$.

62. Given the function $g(x) = 3x^2 + 4x - 4$. The x-coordinate of the vertex is

$$-\frac{b}{2a} = -\frac{4}{2(3)} = -\frac{4}{6} = -\frac{2}{3}.$$

The y-coordinate is $g\left(-\frac{2}{3}\right) = -\frac{16}{3}$. Thus, the vertex is $\left(-\frac{2}{3}, -\frac{16}{3}\right)$.

To find the x-intercepts, solve the following equation:

$$3x^2 + 4x - 4 = 0$$
$$(3x - 2)(x + 2) = 0$$
$$3x - 2 = 0 \qquad x + 2 = 0$$
$$3x = 2 \qquad x = -2$$
$$x = \frac{2}{3}$$

Thus, the x-intercepts are $(-2, 0)$ and $\left(\frac{2}{3}, 0\right)$.

63. Given the function $h(x) = 4x^2 - 4x + 1$. The x-coordinate of the vertex is

$$-\frac{b}{2a} = -\frac{-4}{2(4)} = -\frac{-4}{8} = \frac{1}{2}.$$

The y-coordinate is $h\left(\frac{1}{2}\right) = 0$. Thus, the vertex is $\left(\frac{1}{2}, 0\right)$.

To find the x-intercepts, solve the following equation:

$$4x^2 - 4x + 1 = 0$$
$$(2x - 1)(2x - 1) = 0$$
$$2x - 1 = 0 \qquad 2x - 1 = 0$$
$$2x = 1 \qquad 2x = 1$$
$$x = \frac{1}{2} \qquad x = \frac{1}{2}$$

Thus, there is only one x-intercept, $\left(\frac{1}{2},0\right)$, which is the same as the vertex.

64. Given the function $h(x) = -4x^2 + 2x - 2$. The x-coordinate of the vertex is

$$-\frac{b}{2a} = -\frac{2}{2(-4)} = -\frac{2}{-8} = \frac{1}{4}.$$

The y-coordinate is $h\left(\frac{1}{4}\right) = -\frac{7}{4}$. Thus, the vertex is $\left(\frac{1}{4},-\frac{7}{4}\right)$.

Notice that the graph opens down ($a = -4 < 0$), and since the vertex is below the x-axis, there can be no x-intercepts. If we try to solve the equation

$$-4x^2 + 2x - 2 = 0,$$

the solutions will be complex (nonreal).

65. Use the point-slope form with $m = -8$ and $(x_1,y_1) = (-4,0)$.

$$y - y_1 = m(x - x_1)$$
$$y - 0 = -8(x - (-4))$$
$$y = -8(x + 4)$$
$$y = -8x - 32$$
$$8x + y + 32 = 0$$

Thus, the general form of the desired equation is

$$8x + y + 32 = 0.$$

66. Since parallel lines have equal slopes, use the point-slope form with $m = \frac{2}{3}$ and $(x_1,y_1) = (4,-2)$.

$$y - y_1 = m(x - x_1)$$
$$y - (-2) = \frac{2}{3}(x - 4)$$
$$3(y + 2) = 2(x - 4)$$
$$3y + 6 = 2x - 8$$
$$0 = 2x - 3y - 14$$

Thus, the desired line has general form

$$2x - 3y - 14 = 0.$$

67. First find the slope of the line $2x - 5y = 4$ by writing it in slope-intercept form, that is, by solving for y.

$$2x - 5y = 4$$
$$-5y = -2x + 4$$
$$y = \frac{2}{5}x - \frac{4}{5}$$

The slope of this line is $\frac{2}{5}$, so the slope of the desired line is the negative reciprocal, $-\frac{5}{2}$. Use the point-slope form with $m = -\frac{5}{2}$ and $(x_1,y_1) = (3,0)$.

$$y - y_1 = m(x - x_1)$$
$$y - 0 = -\frac{5}{2}(x - 3)$$
$$2y = -5(x - 3)$$
$$2y = -5x + 15$$
$$5x + 2y - 15 = 0$$

Thus, the desired equation, in general form, is

$$5x + 2y - 15 = 0.$$

68. First find the slope of the line with equation $3x - 9y = 5$ by writing it in slope-intercept form.

$$3x - 9y = 5$$
$$-9y = -3x + 5$$
$$y = \frac{1}{3}x - \frac{5}{9}$$

The slope of this line is then $\frac{1}{3}$, the same as the slope of the desired line. Use the point-slope form.

$$y - y_1 = m(x - x_1)$$
$$y - (-8) = \frac{1}{3}(x - 0)$$
$$3(y + 8) = 1 \cdot x$$
$$3y + 24 = x$$
$$0 = x - 3y - 24$$

Thus, the general form of the desired equation is

$$x - 3y - 24 = 0.$$

CHAPTER 3 REVIEW

69. Note that 11% simple interest for 3 years on x dollars amounts to $3(0.11)x = 0.33x$. Then the amount to be paid back is the original amount borrowed, x plus the interest. This gives the model

$$S = x + 0.33x = 1.33x.$$

When $2200 is borrowed, substitute 2200 for x to find the amount to be paid back.

$$S = 1.33(2200) = 2926$$

Thus, $2926 must be paid back.

70. Given that the radius of the spill is $r(t) = 2t + 1$, where t is time in hours, and that the cost of cleaning the spill is $C(r) = 30r^2 + 8000$, we are asked to find the composition of these two functions.

$$\begin{aligned} C(r(t)) &= C(2t + 1) \\ &= 30(2t + 1)^2 + 8000 \\ &= 30(4t^2 + 4t + 1) + 8000 \\ &= 120t^2 + 120t + 30 + 8000 \\ &= 120t^2 + 120t + 8030 \end{aligned}$$

Thus, $C(t) = 120t^2 + 120t + 8030$. To find the cost of cleaning the spill after 3 hr, substitute 3 for t.

$$\begin{aligned} C(3) &= 120(3)^2 + 120(3) + 8030 \\ &= 1080 + 360 + 8030 \\ &= 9470 \end{aligned}$$

Thus, the cost is $9470.

71. Given that $h(t) = -16t^2 + 96t$ represents the height of a rocket t seconds after firing, note that the model is a quadratic function with graph a parabola opening down. The maximum height will be the second coordinate of the vertex. First find the first coordinate of the vertex using the formula.

$$-\frac{b}{2a} = -\frac{96}{2(-16)} = -\frac{96}{-32} = 3.$$

To find the second coordinate, the maximum height, evaluate the function when $t = 3$.

$$\begin{aligned} h(3) &= -16(3)^2 + 96(3) \\ &= -144 + 288 = 144 \end{aligned}$$

The maximum height is 144 ft after 3 seconds. Since the height is zero when the rocket is on the ground, to find the time when the rocket returns to the ground solve the following equation:

$$\begin{aligned} -16t^2 + 96t &= 0 \\ -16t(t - 6) &= 0 \\ t(t - 6) &= 0 \\ t = 0 \quad t - 6 &= 0 \\ t &= 6 \end{aligned}$$

Thus, the rocket is on the ground at time $t = 0$, just before it is fired, and returns to the ground 6 seconds later.

72. First translate *p varies directly as the square root of l* into the variation equation

$$p = c\sqrt{l}.$$

For p to be three times as long, that is $3p$, l would have to be 9 times as long since if we replace l with $9l$, then

$$c\sqrt{9l} = c3\sqrt{l} = 3(c\sqrt{l}) = 3p.$$

73. First plot the points $(0,10)$, $(2,25)$, $(4,55)$, $(6,75)$, and $(8,85)$ in a coordinate system as shown below.

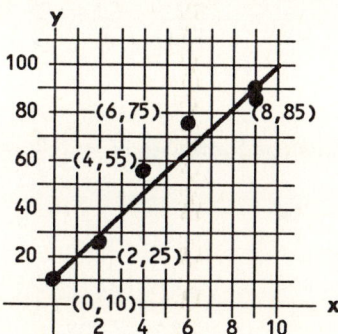

One line that appears to fit this data fairly closely can be drawn through $(0,10)$ and $(8,90)$. Use these two points to find the equation of the line. First find the slope of the line.

$$m = \frac{y_2 - y_1}{x_2 - x_1} = \frac{90 - 10}{8 - 0} = \frac{80}{8} = 10$$

Now use the point-slope form with $m = 10$ and $(x_1, y_1) = (0, 10)$.

$$\begin{aligned} y - y_1 &= m(x - x_1) \\ y - 10 &= 10(x - 0) \\ y - 10 &= 10x \\ y &= 10x + 10 \end{aligned}$$

Thus, one linear model that fits this data closely is

$$y = 10x + 10.$$

CHAPTER 3 RELATIONS, FUNCTIONS, AND GRAPHS

CHAPTER 3 Test

1. $d = \sqrt{(x_1 - x_2)^2 + (y_1 - y_2)^2}$
    $= \sqrt{(-2 - 1)^2 + (3 - 7)^2}$
    $= \sqrt{(-3)^2 + (-4)^2}$
    $= \sqrt{9 + 16} = \sqrt{25} = 5$

2. $(\bar{x}, \bar{y}) = \left(\dfrac{x_1 + x_2}{2}, \dfrac{y_1 + y_2}{2}\right)$
    $= \left(\dfrac{-2 + 1}{2}, \dfrac{7 + 3}{2}\right)$
    $= \left(\dfrac{-1}{2}, \dfrac{10}{2}\right)$
    $= \left(-\dfrac{1}{2}, 5\right)$

3. First find the slope of the line through $(-2,7)$ and $(1,3)$.

    $m = \dfrac{y_2 - y_1}{x_2 - x_1} = \dfrac{3 - 7}{1 - (-2)} = \dfrac{-4}{3} = -\dfrac{4}{3}$

    Use the point-slope form with $m = -\dfrac{4}{3}$ and $(x_1, y_1) = (-2, 7)$.

    $y - y_1 = m(x - x_1)$
    $y - 7 = -\dfrac{4}{3}(x - (-2))$
    $3(y - 7) = -4(x + 2)$
    $3y - 21 = -4x - 8$
    $4x + 3y - 13 = 0$

    Thus, the equation of the desired line is

    $4x + 3y - 13 = 0.$

4. To find the slope and y-intercept of the line with equation $x - 2y - 4 = 0$, write the equation in slope-intercept form by solving for y.

    $x - 2y - 4 = 0$
    $-2y = -x + 4$
    $y = \dfrac{1}{2}x - 2$

    Thus, the slope is $\dfrac{1}{2}$, and the y-intercept is $(0,-2)$.

5. First find the slope of the line with equation $x + 2y - 3 = 0$ by writing the equation in slope-intercept form.

    $x + 2y - 3 = 0$
    $2y = -x + 3$
    $y = -\dfrac{1}{2}x + \dfrac{3}{2}$

    Then this line has slope $-\dfrac{1}{2}$, which is also the slope of every parallel line. Use the point-slope form with $m = -\dfrac{1}{2}$ and $(x_1, y_1) = (1, -2)$.

    $y - y_1 = m(x - x_1)$
    $y - (-2) = -\dfrac{1}{2}(x - 1)$
    $2(y + 2) = -1(x - 1)$
    $2y + 4 = -x + 1$
    $x + 2y + 3 = 0$

    Thus, the equation of the desired line, in general form, is

    $x + 2y + 3 = 0.$

6. From the information given we have two data points with coordinates $(0, 80{,}000)$ and $(4, 96{,}000)$, where the x-coordinate represents the year and the y-coordinate represents the value of the house. Find the equation of the line passing through these two points. First find the slope.

    $m = \dfrac{y_2 - y_1}{x_2 - x_1} = \dfrac{96{,}000 - 80{,}000}{4 - 0}$
    $= \dfrac{16{,}000}{4} = 4000$

    Use 4000 for m and the point $(0, 80{,}000)$ in the point-slope form.

    $y - y_1 = m(x - x_1)$
    $y - 80{,}000 = 4000(x - 0)$
    $y = 4000x + 80{,}000$

To find the value of the house 10 years after it was purchased, substitute 10 for x.

$$y = 4000(10) + 80{,}000$$
$$= 40{,}000 + 80{,}000$$
$$= 120{,}000$$

Thus, the value of the house is $120,000.

7. The domain of $f(x) = \dfrac{1}{3x - 1}$ is all real numbers for which the fraction is defined, that is all real numbers for which the denominator is not zero. If $3x - 1 = 0$, then $x = \frac{1}{3}$. Thus, the domain is all reals $x \neq \frac{1}{3}$.

8. To find the range of $f(x) = 2x + 5$ with domain $\{1,2,3\}$, find the value of the function at each number in the domain.

$$f(1) = 2(1) + 5 = 2 + 5 = 7$$
$$f(2) = 2(2) + 5 = 4 + 5 = 9$$
$$f(3) = 2(3) + 5 = 6 + 5 = 11$$

Thus, the range is $\{7,9,11\}$.

9. Since every vertical line through the graph crosses the graph in only one point, the graph passes the vertical line test and is, therefore, the graph of a function.

10. As x takes on values from left to right, the values of $4x + 7$ get larger and larger. Thus, the function is increasing on the entire set of reals. That is, the values for which f is increasing are in the interval $(-\infty,\infty)$.

11. To obtain the graph of $y = h(x - 1) + 2$ from the graph of $y = h(x)$, slide the graph of $y = h(x)$ right 1 unit (from the -1) and then up 2 units (from the $+2$).

12. Given the function $f(x) = x^3 - 4x$. Then

$$f(-x) = (-x)^3 - 4(-x)$$
$$= -x^3 + 4x$$
$$= -(x^3 - 4x)$$
$$= -f(x)$$

Since $f(-x) = -f(x)$, f is an odd function.

13. Given the relation $y = 3x^2 - 1$. To test for symmetry with respect to the x-axis, substitute $-y$ for y to obtain $-y = 3x^2 - 1$, which is not the same as the original equation. Thus, the graph is not symmetric with respect to the x-axis. When x is replaced with $-x$, we obtain $y = 3(-x)^2 - 1 = 3x^2 - 1$, which is the same as the original equation. Thus, the graph is symmetric with respect to the y-axis. When x is replaced with $-x$ and y is replaced with $-y$ at the same time, the resulting equation is $-y = x^2 - 1$, which is not the same as the original. Thus, the graph is not symmetric with respect to the origin.

14. Graph $f(x) = 2$ for all values of $x \geq 1$, and then graph $f(x) = x - 1$ for all values of $x < 1$. The graph is given below.

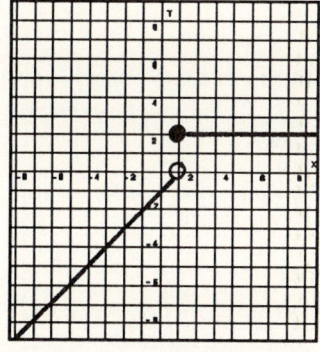

15. Given $f(x) = 3x - 5$ and $g(x) = 2x^2 - 1$.

$$g[f(x)] = g[3x - 5]$$
$$= 2(3x - 5)^2 - 1$$
$$= 2(9x^2 - 30x + 25) - 1$$
$$= 18x^2 - 60x + 50 - 1$$
$$= 18x^2 - 60x + 49$$

16. Since $f(x) = 5x + 6$ is an increasing function on $(-\infty,\infty)$, f is one-to-one.

17. Given $f(x) = x^2 + 1$ for $x \geq 0$. Interchange x and y and solve for y. Note that $y \geq 0$.

$$x = y^2 + 1$$
$$x - 1 = y^2$$
$$\pm\sqrt{x - 1} = y$$

But since $y \geq 0$, use the positive root only. Thus,

$$f^{-1}(x) = \sqrt{x - 1}.$$

18. Given $f(x) = x^2 + 4x$. First find the x-coordinate of the vertex using the formula

$$-\frac{b}{2a} = -\frac{4}{2(1)} = -\frac{4}{2} = -2.$$

Then substitute to find the y-coordinate.

$$f(2) = (-2)^2 + 4(-2)$$
$$= 4 - 8 = -4$$

Thus, the vertex is $(-2,-4)$. To find the x-intercepts, solve the following equation.

$$x^2 + 4x = 0$$
$$x(x + 4) = 0$$
$$x = 0 \qquad x + 4 = 0$$
$$x = -4$$

Thus, the intercepts are $(0,0)$ and $(-4,0)$. Note that since $a = 1 > 0$, the graph opens up. The graph is given below.

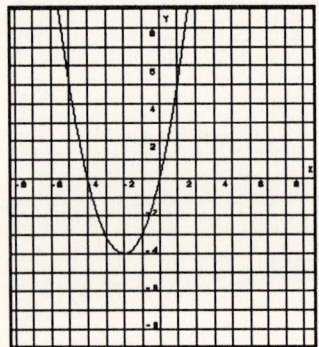

19. Let x = one number, and y = the other number. Then since their difference is 24,

$$y - x = 24,$$

so

$$y = x + 24.$$

The product of the two numbers is

$$P = xy$$
$$= x(x + 24)$$
$$= x^2 + 24x.$$

Since this is a quadratic function with graph a parabola that opens up, the minimum product will be the second coordinate of the vertex. The first coordinate of the vertex is

$$-\frac{b}{2a} = -\frac{24}{2(1)} = -\frac{24}{2} = -12.$$

Then since $x = -12$, $y = x + 24 = -12 + 24 = 12$. The two numbers are 12 and -12, and the minumum product is $(12)(-12) = -144$.

20. (a) Let n be the number of units. Since the fixed costs are $1200, and the cost per unit is $20, the cost function is given by

$$C(n) = 20n + 1200.$$

(b) Since the revenue produced on each unit is $30, the total revenue produced is $R(n) = 30n$. Then the model for the profit produced is

$$P(n) = R(n) - C(n)$$
$$= 30n - (20n + 1200)$$
$$= 10n - 1200$$

(c) To find the breakeven-point solve $P(n) = 0$.

$$10n - 1200 = 0$$
$$10n = 1200$$
$$n = 120$$

Thus, 120 units provides the break-even point.

21. First translate V varies directly as T and inversely as P to obtain the variation equation.

$$V = \frac{cT}{P}$$

Substitute 231 for V, 420 for T, and 20 for P to find the value of c.

$$231 = \frac{c(420)}{20}$$
$$231 = 21c$$
$$11 = c$$

Substitute this value for c, 300 for T, and 15 for P to find the value of V.

$$V = \frac{11(300)}{15} = 11(20) = 220$$

Thus, the volume is 220 in^3

PRACTICE EXERCISES

SECTION 4.1 155

CHAPTER 4 POLYNOMIAL AND RATIONAL FUNCTIONS

SECTION 4.1 Polynomials and Synthetic Division

1. (a) Given the polynomial $7 - x^2$. Since the term of highest degree, $-x^2$, has degree 2, the polynomial has degree 2. The leading coefficient is the coefficient of the term of highest degree. Thus, the leading coefficient is -1.

   (b) Given the polynomial 0. Since this is the zero polynomial, it has no degree and no leading coefficient.

2. Given the polynomial equation

   $$P(x) = x^3 - 3x^2 + 4x - 2 = 0.$$

   (a) Find $P(1)$.

   $$\begin{aligned} P(1) &= (1)^3 - 3(1)^2 + 4(1) - 2 \\ &= 1 - 3 + 4 - 2 \\ &= 0 \end{aligned}$$

   Thus, 1 is a solution to the equation.

   (b) Find $P(-2)$.

   $$\begin{aligned} P(-2) &= (-2)^3 - 3(-2)^2 + 4(-2) - 2 \\ &= -8 - 12 - 8 - 2 \\ &= -30 \end{aligned}$$

   Thus, -2 is not a solution to the equation.

   (c) Find $P(1 + i)$.

   $$\begin{aligned} P(1 + i) &= (1 + i)^3 - 3(1 + i)^2 + 4(1 + i) - 2 \\ &= 2i - 2 - 3(2i) + 4 + 4i - 2 \\ &= 2i - 2 - 6i + 4 + 4i - 2 \\ &= 0 \end{aligned}$$

   Thus, $1 + i$ is a solution to the equation.

3. Given the polynomial $F(y) = y^3 - 6y^2 + 13y - 10$.

   $$\begin{aligned} F(2) &= (2)^3 - 6(2)^2 + 13(2) - 10 \\ &= 8 - 24 + 26 - 10 \\ &= 0 \end{aligned}$$

   $$\begin{aligned} F(2 - i) &= (2 - i)^3 - 6(2 - i)^2 + 13(2 - i) - 10 \\ &= 2 - 11i - 6(3 - 4i) + 13(2 - i) - 10 \\ &= 2 - 11i - 18 + 24i + 26 - 13i - 10 \\ &= 0 \end{aligned}$$

   $$\begin{aligned} F(2 + i) &= (2 + i)^3 - 6(2 + i)^2 + 13(2 + i) - 10 \\ &= 2 + 11i - 6(3 + 4i) + 13(2 + i) - 10 \\ &= 2 + 11i - 18 - 24i + 26 + 13i - 10 \\ &= 0 \end{aligned}$$

   Thus, 2, $2 - i$, and $2 + i$ are all zeros of the polynomial.

4. (a) Find $(x^4 - 3x^3 + 2x - 5) \div (x - 2)$ using synthetic division. Be sure to write a 0 for the missing x^2 term.

   ```
   2|  1 - 3 + 0 + 2 - 5
         + 2 - 2 - 4 - 4
       _____
       1 - 1 - 2 - 2 - 9
   ```

   Thus, the quotient is $x^3 - x^2 - 2x - 2$, and the remainder is -9.

   (b) Find $(2x^2 - x^4 + x^5 - 1 - x) \div (x + 1)$ using synthetic division. Be sure to write a 0 for the missing x^3 term, and write the polynomial in descending order.

   ```
   -1|  1 - 1 + 0 + 2 - 1 - 1
          - 1 + 2 - 2 + 0 + 1
        _____
        1 - 2 + 2 + 0 - 1 + 0
   ```

   Thus, the quotient is $x^4 - 2x^3 + 2x^2 - 1$, and the remainder is 0.

5. Divide $5x^5 - 40x^3 + 410x^2 - 3350x - 5280$ by $x - 5$.

   ```
   5|  5 + 0 - 40 + 410 - 3350 - 5280
          + 25 + 125 + 425 + 4175 + 4125
        _____
        5 + 25 + 85 + 835 + 825 - 1155
   ```

   Thus, the quotient is $5x^4 + 25x^3 + 85x^2 + 835x + 825$, and the remainder is -1155.

CHAPTER 4 POLYNOMIALS AND RATIONAL FUNCTIONS

SECTION 4.1 Polynomials and Synthetic Division

1. Given the polynomial $P(x) = x^3 - 3x^2 + x - 5$.

   (a) Since the term of highest degree is x^3, and the degree of this term is 3, the degree of the polynomial is 3.

   (b) The leading coefficient is the coefficient of the term x^3, which is 1.

   (c) $\quad P(2) = (2)^3 - 3(2)^2 + (2) - 5$
   $= 8 - 12 + 2 - 5$
   $= -7$

   (d) $\quad P(-1) = (-1)^3 - 3(-1)^2 + (-1) - 5$
   $= -1 - 3 - 1 - 5$
   $= -10$

2. Given the polynomial $P(u) = -2x + \sqrt{2}$.

   (a) The term of highest degree is $-2x$, which has degree 1. The polynomial has degree 1.

   (b) The leading coefficient is the coefficient of the term of highest degree, $-2x$, which is -2.

   (c) $\quad P(2) = -2(2) + \sqrt{2} = -4 + \sqrt{2}$

   (d) $\quad P(-1) = -2(-1) + \sqrt{2} = 2 + \sqrt{2}$

3. Given the polynomial $P(y) = 2y^2 - y^4 - 1$.

   (a) The term of highest degree, $-y^4$, has degree 4. Thus the polynomial has degree 4.

   (b) The term of highest degree, $-y^4$, has coefficient -1, so this is the leading coefficient.

   (c) $\quad P(2) = 2(2)^2 - (2)^4 - 1$
   $= 2(4) - 16 - 1$
   $= 8 - 16 - 1$
   $= -9$

   (d) $\quad P(-1) = 2(-1)^2 - (-1)^4 - 1$
   $= 2(1) - (1) - 1$
   $= 2 - 1 - 1$
   $= 0$

4. Given the polynomial $P(t) = 8$. This is a constant polynomial.

   (a) Since this is a constant polynomial, it has degree 0 ($P(t) = 8t^0$).

   (b) Since this is a constant polynomial, leading coefficient is 8 ($P(t) = 8t^0$).

   (c) $\qquad P(2) = 8 \quad$ All values are 8.

   (d) $\qquad P(-1) = 8 \quad$ All values are 8.

5. Evaluate $f(x) = 3x^4 - 2x^2 - 1$ when $x = 1, -1,$ and 2.

   $f(1) = 3(1)^4 - 2(1)^2 - 1$
   $= 3 - 2 - 1 = 0$

   $f(-1) = 3(-1)^4 - 2(-1)^2 - 1$
   $= 3 - 2 - 1 = 0$

   $f(2) = 3(2)^4 - 2(2)^2 - 1$
   $= 48 - 8 - 1 = 39$

   Thus, 1 and -1 are zeros.

6. Evaluate $P(x) = 2x^4 - 5x^3 + 3x$ when $x = 0, 1,$ and -1.

   $P(0) = 2(0)^4 - 5(0)^3 + 3(0)$
   $= 0 - 0 + 0 = 0$

   $P(1) = 2(1)^4 - 5(1)^3 + 3(1)$
   $= 2 - 5 + 3 = 0$

   $P(-1) = 2(-1)^4 - 5(-1)^3 + 3(-1)$
   $= 2 + 5 - 3$
   $= 4$

   Thus, 0 and 1 are solutions to the equation $P(x) = 0$.

7. Substitute -1, $\dfrac{3 + i\sqrt{3}}{2}$, and $\dfrac{3 - i\sqrt{3}}{2}$ for x to see if each is a solution to the equation $x^3 - 3x^2 + 3x = 0$.

EXERCISES A

$(-1)^3 - 3(-1)^2 + 3(-1) = 0$
$-1 - 3 - 3 = 0$
$-7 = 0$

$\left(\dfrac{3+i\sqrt{3}}{2}\right)^3 - 3\left(\dfrac{3+i\sqrt{3}}{2}\right)^2 + 3\left(\dfrac{3+i\sqrt{3}}{2}\right) = 0$

$(3i\sqrt{3}) - 3\left(\dfrac{6+6i\sqrt{3}}{4}\right) + \dfrac{9+3i\sqrt{3}}{2} = 0$

$\dfrac{4(3i\sqrt{3})}{4} + \dfrac{-18-18i\sqrt{3}}{4} + \dfrac{18+6i\sqrt{3}}{4} = 0$

$\dfrac{12i\sqrt{3} - 18 - 18i\sqrt{3} + 18 + 6i\sqrt{3}}{4} = 0$

$\dfrac{0}{4} = 0$

$0 = 0$

$\left(\dfrac{3-i\sqrt{3}}{2}\right)^3 - 3\left(\dfrac{3-i\sqrt{3}}{2}\right)^2 + 3\left(\dfrac{3-i\sqrt{3}}{2}\right) = 0$

$(-3i\sqrt{3}) - 3\left(\dfrac{6-6i\sqrt{3}}{4}\right) + \dfrac{9-3i\sqrt{3}}{2} = 0$

$\dfrac{4(-3i\sqrt{3})}{4} + \dfrac{-18+18i\sqrt{3}}{4} + \dfrac{18-6i\sqrt{3}}{4} = 0$

$\dfrac{-12i\sqrt{3} - 18 + 18i\sqrt{3} + 18 - 6i\sqrt{3}}{4} = 0$

$\dfrac{0}{4} = 0$

$0 = 0$

Thus, $\dfrac{3+i\sqrt{3}}{2}$ and $\dfrac{3-i\sqrt{3}}{2}$ are solutions to the polynomial equation.

8. Given $h(t) = -16t^2 + 240t + 1600$. Substitute 20 for t and evaluate.

$h(20) = -16(20)^2 + 240(20) + 1600$
$= -16(400) + 4800 + 1600$
$= -6400 + 4800 + 1600$
$= 0$

Since $h(20) = 0$, the height of the object after 20 seconds is zero which means that it takes 20 seconds for the object to hit the ground.

9. Given the polynomial $M(t) = -0.05t^3 + 2t + 2$.

(a) Evaluate M when $t = 1$.

$M(1) = -0.05(1)^3 + 2(1) + 2$
$= -0.05(1) + 2 + 2$
$= -0.05 + 4 = 3.95$

Thus, in 1 hr, there are 3.95 parts per million.

(b) Evaluate M when $t = 3$.

$M(4) = -0.05(3)^3 + 2(3) + 2$
$= -0.05(27) + 6 + 2$
$= -1.35 + 8 = 6.65$

Thus, in 3 hr, there are 6.65 parts per million.

(c) Evaluate M when $t = 6$.

$M(5) = -0.05(6)^3 + 2(6) + 2$
$= -0.05(216) + 12 + 2$
$= -10.8 + 14 = 3.2$

Thus, in 6 hr, there are 3.2 parts per million.

10. Find $(3y^3 - 2y^2 + y - 5) \div (y - 2)$ using synthetic division.

```
 2|  3 - 2 + 1 -  5
  |    + 6 + 8 + 18
   ─────────────────
     3 + 4 + 9 + 13
```

Thus, the quotient is $3y^2 + 4y + 9$, and the remainder is 13.

11. Find $(3z^4 - 2z^2 + z - 3) \div (z - 3)$ using synthetic division. Be sure to write a 0 for the coefficient of the missing z^3 term.

```
 3|  3 + 0 -  2 +  1 -   3
  |    + 9 + 27 + 75 + 228
   ──────────────────────────
     3 + 9 + 25 + 76 + 225
```

Thus, the quotient is $3z^3 + 9z^2 + 25z + 76$, and the remainder is 225.

12. Find $(2x^5 - 3x^4 - 6x^3 + x + 8) \div (x - 3)$ using synthetic division. Remember $0x^2$.

```
 3|  2 - 3 - 6 + 0 +  1 +  8
  |    + 6 + 9 + 9 + 27 + 84
   ───────────────────────────
     2 + 3 + 3 + 9 + 28 + 92
```

Thus, the quotient is $2x^4 + 3x^3 + 3x^2 + 9x + 28$, and the remainder is 92.

13. Find $(y^3 - y + y^4 - 3) \div (y + 1)$ using synthetic division. Be sure to write the polynomial in descending order and supply a 0 for the coefficient of the missing y^2 term.

$$\begin{array}{r|rrrrr} -1 & 1 + 1 + 0 - 1 - 3 \\ & - 1 + 0 + 0 + 1 \\ \hline & 1 + 0 + 0 - 1 - 2 \end{array}$$

Thus, the quotient is $y^3 - 1$, and the remainder is -2.

14. Find $(6x^3 - x + 4) \div (x - \frac{1}{2})$ using synthetic division. Be sure to supply a 0 for the coefficient of the missing x^2 term.

$$\begin{array}{r|rrrr} \frac{1}{2} & 6 + 0 - 1 + 4 \\ & + 3 + \frac{3}{2} + \frac{1}{4} \\ \hline & 6 + 3 + \frac{1}{2} + \frac{17}{4} \end{array}$$

Thus, the quotient is $6x^2 + 3x + \frac{1}{2}$, and the remainder is $\frac{17}{4}$.

15. Divide $(7x^5 - 25x^4 + 80x^3 - 170x^2 + 305x + 2600)$ by $(x + 2)$ using a calculator and synthetic division.

$$\begin{array}{r|rrrrrr} -2 & 7 - 25 + 80 - 170 + 305 + 2600 \\ & - 14 + 78 - 316 + 972 - 2554 \\ \hline & 7 - 39 + 158 - 486 + 1277 + 46 \end{array}$$

Thus, the quotient is $7x^4 - 39x^3 + 158x^2 - 486x + 1277$ and the remainder is 46.

16. Yes, a constant polynomial, $P(x) = a_0$ ($a_0 \neq 0$) has no zeros.

17. The polynomial $P(x) = -x^7 + x^6 - x^5 + x^4 - x^3 + x^2 - x + 1$ cannot have a negative real number as a zero since all odd powers of x have a negative coefficient making each of these terms positive for a negative value of x, and since all even-powered terms would also be positive in this case, the polynomial will always be positive, never zero.

18. Divide $x^3 + x^2 - x + m$ by $x - 1$ using synthetic division, set the remainder equal to 0, and solve for m.

$$\begin{array}{r|rrrr} 1 & 1 + 1 - 1 + m \\ & + 1 + 2 + 1 \\ \hline & 1 + 2 + 1 + (m+1) \end{array}$$

Since the remainder is $m + 1$, set this equal to 0 and solve for m.

$$m + 1 = 0$$
$$m = -1$$

Thus, for the remainder to be 0, m must be -1.

19. Given $P(x) = x^3 - 2x^2 + 3x + 1$.

(a)
$$\begin{aligned} P(1) &= (1)^3 - 2(1)^2 + 3(1) + 1 \\ &= 1 - 2 + 3 + 1 \\ &= 3 \end{aligned}$$

(b) Divide $P(x)$ by $x - 1$ using synthetic division.

$$\begin{array}{r|rrrr} 1 & 1 - 2 + 3 + 1 \\ & + 1 - 1 + 2 \\ \hline & 1 - 1 + 2 + 3 \end{array}$$

The remainder is 3.

(c) The remainder, 3, is equal to $P(1)$.

20. When the two terms of highest degree are multiplied as part of the product of the two polynomials, that term will have degree $m + n$, the sum of the degrees of the two polynomials. Since this term cannot have a zero coefficient, and since it will be the term of highest degree in the product, the degree of the product is $m + n$, the sum of the degrees of the two polynomials.

21. The midpoint of the line segment joining $(6,-1)$ and $(3,5)$ is:

$$(\bar{x}, \bar{y}) = \left(\frac{x_1 + x_2}{2}, \frac{y_1 + y_2}{2}\right) = \left(\frac{6 + 3}{2}, \frac{-1 + 5}{2}\right)$$
$$= \left(\frac{9}{2}, \frac{4}{2}\right)$$
$$= \left(\frac{9}{2}, 2\right)$$

22. The slope of the line passing through $(4,-3)$ and $(7,-3)$ is:

$$m = \frac{y_2 - y_2}{x_2 - x_1} = \frac{-3 - (-3)}{4 - 7}$$
$$= \frac{0}{-3} = 0$$

EXERCISES A　　　　　　　　　　　　　　　　　　　　　　　SECTION 4.1　　159

23. First find the slope of the line with equation $3x + 6y - 2 = 0$ by writing the equation in slope-intercept form.

$$3x + 6y - 2 = 0$$
$$6y = -3x + 2$$
$$y = -\frac{3}{6}x + \frac{2}{6}$$
$$y = -\frac{1}{2}x + \frac{1}{3}$$

The slope of this line is $-\frac{1}{2}$, so the slope of a line perpendicular to it is 2. Use the point-slope form with the point $(6,1)$ and slope 2.

$$y - y_1 = m(x - x_1)$$
$$y - 1 = 2(x - 6)$$
$$y - 1 = 2x - 12)$$
$$0 = 2x - y - 11$$

Thus, the equation of the desired line is $2x - y - 11 = 0$.

24. (a) The conjugate of $3 - 2i$ is $3 + 2i$.

(b) The conjugate of $5 + i\sqrt{7}$ is $5 - i\sqrt{7}$.

CHAPTER 4 POLYNOMIAL AND RATIONAL FUNCTIONS

SECTION 4.2 The Remainder and Factor Theorems

1. Divide $P(x) = 2x^4 - 3x^2 + x - 1$ by $D(x) = x + 1$ using synthetic division.

   $$\begin{array}{r|rrrrr} -1 & 2 & +0 & -3 & +1 & -1 \\ & & -2 & +2 & +1 & -2 \\ \hline & 2 & -2 & -1 & +2 & -3 \end{array}$$

   Let $Q(x) = 2x^3 - 2x^2 - x + 2$ and $R(x) = -3$, then $P(x) = Q(x)D(x) + R(x)$.

2. Given $P(x) = x^3 - 3x^2 + x + 2$.

   (a) Find $P(-3)$.

   $$\begin{array}{r|rrrr} -3 & 1 & -3 & +1 & +2 \\ & & -3 & +18 & -57 \\ \hline & 1 & -6 & +19 & -55 \end{array}$$

   Thus, by the remainder theorem, $P(-3) = -55$.

   (b) Find $P(2)$.

   $$\begin{array}{r|rrrr} 2 & 1 & -3 & +1 & +2 \\ & & +2 & -2 & -2 \\ \hline & 1 & -1 & -1 & +0 \end{array}$$

   Thus, by the remainder theorem, $P(2) = 0$.

3. Divide $P(x) = 2x^3 + 8x^2 - x - 4$ by $x + 4$ using synthetic division. If the remainder is 0, then $x + 4$ is a factor of $P(x)$ by the factor theorem.

   $$\begin{array}{r|rrrr} -4 & 2 & +8 & -1 & -4 \\ & & -8 & +0 & +4 \\ \hline & 2 & +0 & -1 & +0 \end{array}$$

   Since the remainder is $P(-4) = 0$, $x + 4$ is a factor of $P(x)$.

4. Divide $x^4 - 2x^2 + 3x + m$ by $x + 1$ using synthetic division.

   $$\begin{array}{r|rrrrr} -1 & 1 & +0 & -2 & +3 & +m \\ & & -1 & +1 & +1 & -4 \\ \hline & 1 & -1 & -1 & +4 & +(m-4) \end{array}$$

   If $x + 1$ is a factor of the polynomial, then the remainder must be 0. That is, $m - 4 = 0$, or $m = 4$.

EXERCISES A   SECTION 4.2   161

CHAPTER 4 POLYNOMIALS AND RATIONAL FUNCTIONS

SECTION 4.2 The Remainder and Factor Theorems

1. Divide $x^3 - 3x^2 + x + 1$ by $x - 1$ using synthetic division. By the remainder theorem, the remainder is $P(1)$.

   $$\underline{1|}\begin{array}{r} 1 - 3 + 1 + 1 \\ + 1 - 2 - 1 \\ \hline 1 - 2 - 1 + 0 \end{array}$$

   Thus, the quotient is $x^2 - 2x - 1$, the remainder is 0, and $P(1) = 0$.

2. Divide $x^3 - 3x^2 + x + 1$ by $x - 5$ using synthetic division. By the remainder theorem, the remainder is $P(5)$.

   $$\underline{5|}\begin{array}{r} 1 - 3 + 1 + 1 \\ + 5 + 10 + 55 \\ \hline 1 + 2 + 11 + 56 \end{array}$$

   Thus, the quotient is $x^2 + 2x + 11$, the remainder is 56, and $P(5) = 56$.

3. Divide $x^5 + 3x^3 - x + 1$ by $x + 1$ using synthetic division. By the remainder theorem, the remainder is $P(-1)$.

   $$\underline{-1|}\begin{array}{r} 1 + 0 + 3 + 0 - 1 + 1 \\ - 1 + 1 - 4 + 4 - 3 \\ \hline 1 - 1 + 4 - 4 + 3 - 2 \end{array}$$

   Thus, the quotient is $x^4 - x^3 + 4x^2 - 4x + 3$, the remainder is -2, and $P(-1) = -2$.

4. Divide $x^5 + 3x^3 - x + 1$ by $x + 3$ using synthetic division. By the remainder theorem, the remainder is $P(-3)$.

   $$\underline{-3|}\begin{array}{r} 1 + 0 + 3 + 0 - 1 + 1 \\ - 3 + 9 - 36 + 108 - 321 \\ \hline 1 - 3 + 12 - 36 + 107 - 320 \end{array}$$

   Thus, the quotient is $x^4 - 3x^3 + 12x^2 - 36x + 107$, the remainder is -320, and $P(-3) = -320$.

5. Divide $P(x) = x^3 - 27$ by $x - 3$ using synthetic division and find the remainder.

   $$\underline{3|}\begin{array}{r} 1 + 0 + 0 - 27 \\ + 3 + 9 + 27 \\ \hline 1 + 3 + 9 + 0 \end{array}$$

   Thus, the quotient is $x^2 + 3x + 9$, and the remainder is 0.

6. Divide $P(x) = x^3 - 27$ by $x - 4$ using synthetic division and find the remainder.

   $$\underline{4|}\begin{array}{r} 1 + 0 + 0 - 27 \\ + 4 + 16 + 64 \\ \hline 1 + 4 + 16 + 37 \end{array}$$

   Thus, the quotient is $x^2 + 4x + 16$, and the remainder is 37.

7. Divide $P(x) = x^4 + x^2 - 3$ by $x + 2$ using synthetic division and find the remainder.

   $$\underline{-2|}\begin{array}{r} 1 + 0 + 1 + 0 - 3 \\ - 2 + 4 - 10 + 20 \\ \hline 1 - 2 + 5 - 10 + 17 \end{array}$$

   Thus, the quotient is $x^3 - 2x^2 + 5x - 10$, and the remainder is 17.

8. Divide $P(x) = x^4 + x^2 - 3$ by $x - 3$ using synthetic division and find the remainder.

   $$\underline{3|}\begin{array}{r} 1 + 0 + 1 + 0 - 3 \\ + 3 + 9 + 30 + 90 \\ \hline 1 + 3 + 10 + 30 + 87 \end{array}$$

   Thus, the quotient is $x^3 + 3x^2 + 10x + 30$, and the remainder is 87.

9. Divide $P(x) = x^3 - 6x^2 - 9x + 14$ by $x - 7$ using synthetic division and write
   $$P(x) = Q(x)D(x) + R(x).$$

   $$\underline{7|}\begin{array}{r} 1 - 6 - 9 + 14 \\ + 7 + 7 - 14 \\ \hline 1 + 1 - 2 + 0 \end{array}$$

   Thus, $Q(x) = x^2 + x - 2$, $R(x) = 0$, and
   $P(x) = (x^2 + x - 2)(x - 7) + 0$.

10. Divide $P(x) = x^3 - 6x^2 - 9x + 14$ by $x + 1$ using synthetic division and write
    $$P(x) = Q(x)D(x) + R(x).$$

    $$\underline{-1|}\begin{array}{r} 1 - 6 - 9 + 14 \\ - 1 + 7 + 2 \\ \hline 1 - 7 - 2 + 16 \end{array}$$

    Thus, $Q(x) = x^2 - 7x - 2$, $R(x) = 16$, and
    $P(x) = (x^2 - 7x - 2)(x + 1) + 16$.

11. Divide $P(x) = 2x^3 - x^2 - 15x + 18$ by $x + 3$ using synthetic division, and use the factor theorem to determine whether $x + 3$ is a factor of $P(x)$.

$$\begin{array}{r|rrrr} -3 & 2 & -1 & -15 & +18 \\ & & -6 & +21 & -18 \\ \hline & 2 & -7 & +6 & +0 \end{array}$$

Since the remainder is 0, $x + 3$ is a factor of $P(x)$.

12. Divide $P(x) = 2x^3 - x^2 - 15x + 18$ by $x - 1$ using synthetic division, and use the factor theorem to determine whether $x - 1$ is a factor of $P(x)$.

$$\begin{array}{r|rrrr} 1 & 2 & -1 & -15 & +18 \\ & & +2 & +1 & -14 \\ \hline & 2 & +1 & -14 & +4 \end{array}$$

Since the remainder is 4, $x - 1$ is not a factor of $P(x)$.

13. If $P(x)$ is divided by $x - r$, the remainder is equal to $P(r)$.

14. If $P(x)$ is divided by $x + r$, which is equal to $x - (-r)$, the remainder is equal to $P(-r)$.

15. If r is a zero of polynomial $P(x)$, then one factor of $P(x)$ is $x - r$.

16. If $-r$ is a zero of polynomial $P(x)$, then one factor of $P(x)$ is $x - (-r)$ which is $x + r$.

17. To show that 3 is a solution to the polynomial equation $x^3 - 3x^2 - x + 3 = 0$, use synthetic division to divide the polynomial by $x - 3$. If the remainder is 0, then 3 is a solution. To find the remaining solutions, solve the quadratic equation formed by setting $Q(x) = 0$.

$$\begin{array}{r|rrrr} 3 & 1 & -3 & -1 & +3 \\ & & +3 & +0 & -3 \\ \hline & 1 & +0 & -1 & +0 \end{array}$$

Since the remainder is 0, 3 is solution to the equation. The quotient is $x^2 - 1$, so solve the the equation $x^2 - 1 = 0$ to find the remaining solutions.

$$x^2 - 1 = 0$$
$$x^2 = 1$$
$$x = \pm 1$$

Thus, the solutions to the polynomial equation are 3, 1, and -1.

18. Given the polynomial $P(x) = x^8 - 3x^4 + x^2 - x^9 + 7$.

   (a) The polynomial has degree 9 since the term of highest degree, $-x^9$, has degree 9.

   (b) The leading coefficient is -1, the coefficient of the term of highest degree, $-x^9$.

   (c) $$P(-1) = (-1)^8 - 3(-1)^4 + (-1)^2 - (-1)^9 + 7$$
   $$= 1 - 3 + 1 + 1 + 7$$
   $$= 7$$

19. Given the revenue function $R(x) = 125x$ and the cost function $C(x) = 600 + 85x$, where x is the number of items sold and produced. The profit function is

$$P(x) = R(x) - C(x),$$

and the zeros of $P(x)$ are the break-even points. Thus, we must solve $P(x) = 0$.

$$125x - (600 + 85x) = 0$$
$$125x - 600 - 85x = 0$$
$$40x - 600 = 0$$
$$40x = 600$$
$$x = 15$$

The only break-even point is 15 items.

20. Given $P(x) = 15x^4 - 3x^3 + x^2 + 7x - 6$.

$$P(-x) = 15(-x)^4 - 3(-x)^3 + (-x)^2 + 7(-x) - 6$$
$$= 15x^4 - 3(-x^3) + x^2 - 7x - 6$$
$$= 15x^4 + 3x^3 + x^2 - 7x - 6$$

PRACTICE EXERCISES

CHAPTER 4 POLYNOMIAL AND RATIONAL FUNCTIONS

SECTION 4.3 More Theorems Involving Polynomials

1. (a) Given the polynomial

    $$F(x) = x^4 - 3x^2 + 2x - 5.$$

    Since $F(x)$ has degree 4, $F(x)$ has at least one and at most four zeros. Counting multiplicities, we can say that $F(x)$ has exactly four zeros.

    (b) Given the polynomial

    $$g(x) = (x - 2)^4(x + 6).$$

    Since the degree of $g(x)$ is five, $g(x)$ has at least one zero and at most five zeros. Counting multiplicities, we can say that $g(x)$ has exactly five zeros. In this case, since $g(x)$ is in factored form, we can actually identify the zeros as: 2, 2, 2, 2, and −6. Note that 2 is a zero of multiplicity four.

2. If $P(x)$ has 3 as a double zero, then $(x - 3)^2$ is a factor of $P(x)$. If −1 is a triple zero, then $(x + 1)^3$ is a factor of $P(x)$. If 5 is a single zero, then $(x - 5)$ is a factor of $P(x)$. Thus,

    $$\begin{aligned}P(x) &= (x - 3)^2(x + 1)^3(x - 5) \\ &= (x^2 - 6x + 9)(x^3 + 3x^2 + 3x + 1)(x - 5) \\ &= (x^5 - 3x^4 - 6x^3 + 10x^2 + 21x + 9)(x - 5) \\ &= x^6 - 8x^5 + 9x^4 + 40x^3 - 29x^2 - 96x - 45\end{aligned}$$

3. (a) Since $3 - i$ is a root of $P(x) = 0$ and $P(x)$ has real coefficients, $3 + i$ is also a root. Also, we have given that 2 is a root. Then by the factor theorem, $(x - (3 - i)) = (x - 3 + i)$ is a factor of $P(x)$, $(x - (3 + i)) = (x - 3 - i)$ is a factor of $P(x)$, and $(x - 2)$ is a factor of $P(x)$. Thus,

    $$\begin{aligned}P(x) &= (x - 3 + i)(x - 3 - i)(x - 2) \\ &= (x^2 - 6x + 10)(x - 2) \\ &= x^3 - 8x^2 + 22x - 20.\end{aligned}$$

    (b) If $T(x)$ has degree five, has real coefficients, and $1 - i$ is a double root of $T(x) = 0$, then $1 + i$ is also a double root. Then if −4 and 3 are single roots, the polynomial equation must have six roots, which is impossible since the degree of the polynomial is only five.

4. Suppose $T(x)$ is a polynomial of degree four with rational coefficients. If $3 - \sqrt{3}$ is a root of

    $T(x) = 0$, then $3 + \sqrt{3}$ is also a root. Also, if $1 + i$ is a root, then since rational coefficients are also real, $1 - i$ is also a root. Thus,

    $$\begin{aligned}P(x) &= (x - 3 + \sqrt{3})(x - 3 - \sqrt{3})(x - 1 - i)(x - 1 + i) \\ &= (x^2 - 6x + 6)(x^2 - 2x + 2) \\ &= x^4 - 8x^3 + 20x^2 - 24x + 12.\end{aligned}$$

5. (a) Given the polynomial equation

    $$P(x) = 3x^4 + x^3 - 2x - 5 = 0.$$

    Since $P(x)$ has degree four, there are four solutions to the equation. The number of sign changes in $P(x)$ is 1, so the number of positive real solutions to the equation is 1. Next find

    $$P(-x) = 3x^4 - x^3 + 2x - 5.$$

    Since the number of sign changes in $P(-x)$ is 3, the number of negative real solutions to the equation is either 3 or 1. Putting this information together, the equation has either 1 negative real solution, 1 positive real solution, and 2 nonreal solutions, or 3 negative real solutions, 1 positive real solution, and 0 nonreal solutions. This can be abbreviated to: Either 1,1,2 or 3,1,0.

    (b) Given the polynomial equation

    $$Q(x) = 2x^5 + 3x^3 + x + 1 = 0.$$

    Since $Q(x)$ has degree five, there are five solutions to the equation. The number of sign changes in $Q(x)$ is 0, so the number of positive real solutions to the equation is 0. Next find

    $$Q(-x) = -2x^5 - 3x^3 - x + 1.$$

    Since the number of sign changes in $Q(-x)$ is 1, the number of negative real solutions to the equation is 1. Putting this information together, the equation has 1 negative real solution, 0 positive real solutions, and 4 nonreal solutions. This can be abbreviated to: 1,0,4

(c) Given the polynomial equation

$$R(x) = 3x^6 + x^4 - x^3 + x = 0.$$

Since the degree of $R(x)$ is six, the equation has six solutions. In fact, since we can factor out an x, we obtain

$$x(3x^5 + x^3 - x^2 + 1) = 0,$$

and conclude that 0 is one solution. The remaining factor is of degree five so there will be five solutions to

$$3x^5 + x^3 - x^2 + 1 = 0.$$

There are 2 sign changes in this polynomial so the number of positive real solutions is either 2 or 0. Next substitute $-x$ for x and determine the number of sign changes.

$$-3x^5 - x^3 - x^2 + 1 = 0$$

Since this has 1 sign change, there is 1 negative real solution to the equation. Putting this information together we conclude that the equation has 0 as one solution and either 1 negative real solution, 2 positive real solutions, and 2 nonreal solutions, or else 1 negative real solution, 0 positive real solutions, and 4 nonreal solutions. This can be abbreviated to: 0 is one solution and either 1,2,2 or 1,0,4.

(d) Given the polynomial equation

$$W(x) = x^8 + 3x^6 + x^2 + 2 = 0.$$

Since the polynomial is of degree eight, there are eight solutions to the equation. There are 0 sign changes in $W(x)$ so there are 0 positive real solutions. Since $W(-x) = W(x)$, there are 0 sign changes in $W(x)$ so there are 0 negative real solutions. Thus, putting this information together, we conclude that there are 0 negative real solutions, 0 positive real solutions, and 8 nonreal solutions (four pairs of complex conjugates). In abbreviated form: 0,0,8.

EXERCISES A — SECTION 4.3

CHAPTER 4 POLYNOMIALS AND RATIONAL FUNCTIONS

SECTION 4.3 More Theorems Involving Polynomials

1. A polynomial of degree $n \geq 1$ has at least 1 distinct zero by the Fundamental Theorem of Algebra.

2. If 5 is a zero of multiplicity 2 of polynomial $P(x)$, then $(x - 5)^2$ is a quadratic factor of $P(x)$.

3. Counting multiplicities, a polynomial of degree 5 has exactly five real or complex zeros.

4. If $3 + 2i$ is a zero of polynomial $P(x)$, with real coefficients, then the congugate, $3 - 2i$, is also a zero of $P(x)$.

5. If $1 - 3\sqrt{2}$ is a zero of polynomial $P(x)$, with rational coefficients, then $1 + 3\sqrt{2}$ is also a zero of $P(x)$.

6. Since the given polynomial has degree 5, the greatest number of distinct solutions to the polynomial equation is five. The fewest number of distinct solutions is one. Counting multiplicities, the equation has exactly five solutions.

7. Since 3 is a double root of $P(x) = 0$, $(x - 3)^2$ is a factor of $P(x)$. Also, since $1 + 2i$, is a single root and the coefficients of $P(x)$ are rational, hence real, another root is $1 - 2i$. Thus, $[x - (1 + 2i)]$ and $[x - (1 - 2i)]$ are both factors. Then the polynomial is given by:

$$P(x) = (x - 3)^2[x - (1 + 2i)][x - (1 - 2i)]$$
$$= (x^2 - 6x + 9)(x^2 - 2x + 1 - 4i^2)$$
$$= (x^2 - 6x + 9)(x^2 - 2x + 5)$$
$$= x^4 - 2x^3 + 5x^2 - 6x^3 + 12x^2 - 30x + 9x^2 - 18x + 45$$
$$= x^4 - 8x^3 + 26x^2 - 48x + 45$$

8. Since 2 is a single root, $(x - 2)$ is a factor of $P(x) = 0$. Since $3 + \sqrt{2}$ is a single root and the coefficients are rational, $3 - \sqrt{2}$ is also a single root. Thus, $[x - (3 + \sqrt{2})][x - (3 - \sqrt{2})]$ are factors of $P(x) = 0$. Then $P(x)$ is given by the following:

$$P(x) = (x - 2)[x - (3 + \sqrt{2})][x - (3 - \sqrt{2})]$$
$$= (x - 2)[x^2 - 6x + 9 - 2]$$
$$= (x - 2)(x^2 - 6x + 7)$$
$$= x^3 - 6x^2 + 7x - 2x^2 + 12x - 14$$
$$= x^3 - 8x^2 + 19x - 14$$

9. Since 0 is a double root, $(x - 0)^2 = x^2$ is a factor of $P(x)$. Since $1 - \sqrt{3}$ is a single root, $1 + \sqrt{3}$ is also a single root. Thus,

$$[x - (1 - \sqrt{3})][x - (1 + \sqrt{3})]$$

is a factor of $P(x)$. Then $P(x)$ is given by the following:

$$P(x) = x^2[x - (1 - \sqrt{3})][x - (1 + \sqrt{3})]$$
$$= x^2(x^2 - 2x + 1 - 3)$$
$$= x^2(x^2 - 2x - 2)$$
$$= x^4 - 2x^3 - 2x^2$$

10. Since -2 is a single root, $(x + 2)$ is a factor of $P(x)$. Since $-\sqrt{2}$ is a single root, $\sqrt{2}$ is also a root making $(x + \sqrt{2})$ and $(x - \sqrt{2})$ factors of $P(x)$. And since $-i\sqrt{2}$ is a single root, $i\sqrt{2}$ is also a root making $(x + i\sqrt{2})$ and $(x - i\sqrt{2})$ factors of $P(x)$. Then $P(x)$ is given by the following:

$$P(x) = (x + 2)(x + \sqrt{2})(x - \sqrt{2})(x + i\sqrt{2})(x - i\sqrt{2})$$
$$= (x + 2)(x^2 - 2)(x^2 + 2)$$
$$= (x + 2)(x^4 - 4)$$
$$= x^5 + 2x^4 - 4x - 8$$

11. Since $x^3 - 1$ is of degree 3, $x^3 - 1 = 0$ has three roots so that there are three cube roots of 1. To find the roots, solve the equation. Begin by factoring.

$$x^3 - 1 = 0$$
$$(x - 1)(x^2 + x + 1) = 0$$
$$x - 1 = 0 \qquad x^2 + x + 1 = 0$$
$$x = 1$$

Then one root is 1. To find the other roots, solve the quadratic equation.

$$x = \frac{-b \pm \sqrt{b^2 - 4ac}}{2a}$$

$$= \frac{-(1) \pm \sqrt{(1)^2 - 4(1)(1)}}{2(1)}$$

$$= \frac{-1 \pm \sqrt{1-4}}{2}$$

$$= \frac{-1 \pm \sqrt{-3}}{2}$$

$$= \frac{-1 \pm i\sqrt{3}}{2}$$

Thus, the three cube roots of 1 are 1, $\frac{-1 + i\sqrt{3}}{2}$, and $\frac{-1 - i\sqrt{3}}{2}$.

12. Given $P(x) = x^2 - \sqrt{2}x - 1 - \sqrt{2}$.

Substitute $1 - \sqrt{2}$ for x and simplify.

$$P(1 + \sqrt{2}) = (1 + \sqrt{2})^2 - \sqrt{2}(1 + \sqrt{2}) - 1 - \sqrt{2}$$
$$= 1 + 2\sqrt{2} + 2 - \sqrt{2} - 2 - 1 - \sqrt{2}$$
$$= 0$$
$$P(1 - \sqrt{2}) = (1 - \sqrt{2})^2 - \sqrt{2}(1 - \sqrt{2}) - 1 - \sqrt{2}$$
$$= 1 - 2\sqrt{2} + 2 - \sqrt{2} + 2 - 1 - \sqrt{2}$$
$$= 4 - 4\sqrt{2} \neq 0$$

This does not contradict the theorem in this section since it is true for rational coefficients, not for real (irrational) ones.

13. Given $x^3 - x^2 + 2x + 1 = 0$. Since the polynomial has degree three, there are three solutions. The number of sign changes in $P(x)$ is 2, so the number of positive real solutions is either 2 or 0. Next find $P(-x) = -x^3 - x^2 - 2x + 1$. Since the number of sign changes here is 1, the number of negative real solutions is 1. Putting this information together, the equation either has 1 negative real solution, 2 positive real solutions, and 0 nonreal solutions, or else it has 1 negative real solution, 0 positive real solutions, and 2 nonreal solutions. In compact form: Either 1,2,0 or 1,0,2.

14. Given $x^3 + 3x^2 - x - 9 = 0$. Since the polynomial has degree 3, there are three solutions to the equation. Since the number of sign changes in $P(x)$ is 1, the number of positive real solutions is 1. Next find $P(-x) = -x^3 + 3x^2 + x - 9$. Since this polynomial has 2 sign changes, there are either 2 negative real solutions or else 0 negative real solutions. Putting all this information together, the equation either has 2 negative real solutions, 1 positive real solution, and 0 nonreal solutions, or else it has 0 negative real solutions, 1 positive real solution, and 2 nonreal solutions. In compact form: Either 2,1,0 or 0,1,2.

15. Given $-x^4 + x^3 - 2x^2 + x - 5 = 0$. Since the polynomial has degree 4, there are four solutions to the equation. Since the number of sign changes in $P(x)$ is 4, the number of positive real solutions is either 4, 2, or 0. Next find $P(-x) = -x^4 - x^3 - 2x^2 - x - 5$. Since this polynomial has 0 sign changes, there are 0 negative real solutions. Putting all this information together, the equation either has 0 negative real solutions, 4 positive real solutions, and 0 nonreal solutions, or 0 negative real solutions, 2 positive real solutions, and 2 nonreal solutions, or 0 negative real solutions, 0 positive real solutions, and 4 nonreal solutions. In compact form: Either 0,4,0; 0,2,2; or 0,0,4.

16. Given $3x^5 - x^4 + x^2 + x + 6 = 0$. Since the polynomial has degree 5, there are five solutions to the equation. Since the number of sign changes in $P(x)$ is 2, the number of positive real solutions is 2 or 0. Next find $P(-x) = -3x^5 - x^4 + x^2 - x + 6$. Since this polynomial has 3 sign changes, the number of negative real solutions is either 3 or 1. Putting all of this together, the equation either has 1 negative real solution, 0 positive real solutions, and 4 nonreal solutions, or 1 negative real solution, 2 positive real solutions, and 2 nonreal solutions, or 3 negative real solutions, 0 positive real solutions, and 2 nonreal solutions, or 3 negative real solutions, 2 positive real solutions, and 0 nonreal solutions. In compact form: Either 1,0,4; 1,2,2; 3,0,2; or 3,2,0.

17. Given $2x^6 - 4x^4 + x^2 - 3 = 0$. Since the polynomial has degree 6, there are six solutions to the equation. Since the number of sign changes in $P(x)$ is 3, the number of positive real solutions is either 3 or 1. Since $P(-x) = P(x)$, it too has 3 sign changes so there are either 3 or 1 negative real solutions. Putting this together, the equation either has 1 negative real solution, 1 positive real solution, and 4 nonreal solutions, or 1 negative real

EXERCISES A

solution, 3 positive real solutions, and 2 nonreal solutions, or 3 negative real solutions, 1 positive real solution, and 2 nonreal solutions, or 3 negative real solutions, 3 positive real solutions, and 0 nonreal solutions. In compact form: Either 1,1,4; 1,3,2; 3,1,2; or 3,3,0.

18. Given $8x^8 + 6x^6 + 4x^4 + 2x^2 + 1 = 0$. Since the polynomial has degree 8, there are eight solutions to the equation. Since the polynomial has 0 sign changes, there are 0 positive real solutions. Since $P(-x) = P(x)$, it too has 0 sign changes so there are 0 negative real solutions. Putting this together we have exactly 0 negative real solutions, 0 positive real solutions, and 8 nonreal solutions: 0,0,8.

19. To find the remainder when $3x^4 - x^2 + x - 5$ is divided by $x + 1$, use synthetic division. Don't forget to use 0 for the missing coefficient of the x^3 term.

    $$\begin{array}{r|rrrrr} -1 & 3 & +0 & -1 & +1 & -5 \\ & & -3 & +3 & -2 & +1 \\ \hline & 3 & -3 & +2 & -1 & -4 \end{array}$$

    Thus, the quotient is $3x^3 - 3x^2 + 2x - 1$, and the remainder is -4.

20. Divide $P(x) = x^3 - 2x^2 + x + 1$ by $x + 2$ using synthetic division. The remainder is $P(-2)$.

    $$\begin{array}{r|rrrr} -2 & 1 & -2 & +1 & +1 \\ & & -2 & +8 & -18 \\ \hline & 1 & -4 & +9 & -17 \end{array}$$

    Since the remainder is -17, $P(-2) = -17$.

21. If -4 is a zero of $P(x)$, then $x - (-4) = x + 4$ is a factor of $P(x)$.

22. Given the polynomial

    $$P(x) = 15x^4 - 3x^3 + x^2 - x - 6.$$

    (a) The leading coefficient is 15. Find all the factors of 15.

    $$15 = 3 \cdot 5$$

    Considering all possible combinations of these prime factors, both positive and negative, we obtain

    $$\pm 1, \pm 3, \pm 5, \pm 15$$

SECTION 4.3

as the list of all factors of 15.

(b) The constant term is -6. Find all factors of 6.

$$6 = 2 \cdot 3$$

Considering all possible combinations of these we obtain

$$\pm 1, \pm 2, \pm 3, \pm 6$$

as the list of all factors of -6.

CHAPTER 4 POLYNOMIAL AND RATIONAL FUNCTIONS

SECTION 4.4 Bounds and the Rational Root Theorem

1. First divide by −5 using synthetic division.

   ```
   -5 | 1 + 1 - 16 -  4 +  48
      |     - 5 + 20 - 20 + 120
      -----------------------------
        1 - 4 +  4 - 24 + 168
   ```

   Since the signs alternate in the third row, −5 is a lower bound for the real roots. Next divide by 4 using synthetic division.

   ```
   4 | 1 + 1 - 16 -  4 + 48
     |     + 4 + 20 + 16 + 48
     ---------------------------
       1 + 5 +  4 + 12 + 96
   ```

   Since the signs are all positive in the third row, 4 is an upper bound for the real roots.

2. Given $P(x) = 3x^4 + 11x^3 - 7x^2 - 11x + 4 = 0$. The possibilities for p are the divisors of 4, and the possibilities for q are the divisors of 3. Thus,

   $$p: \pm 1, \pm 2, \pm 4 \quad q: \pm 1, \pm 3$$

   and

   $$\frac{p}{q}: \pm 1, \pm 2, \pm 4, \pm \frac{1}{3}, \pm \frac{2}{3}, \pm \frac{4}{3}.$$

   Suppose we begin by trying 1.

   ```
   1 | 3 + 11 -  7 - 11 + 4
     |     +  3 + 14 +  7 - 4
     -------------------------
       3 + 14 +  7 -  4 + 0
   ```

   Thus, 1 is a solution making $x - 1$ a factor with the remaining factor $3x^3 + 14x^2 + 7x - 4$. The possibilities for zeros of this factor are the same as for the original polynomial. Suppose we try −1.

   ```
   -1 | 3 + 14 +  7 - 4
      |     -  3 - 11 + 4
      ---------------------
        3 + 11 -  4 + 0
   ```

   Thus, −1 is a solution making $x + 1$ a factor with the remaining factor $3x^2 + 11x - 4$. Since this factor is quadratic, it is best to find the remaining solutions by solving the following quadratic equation.

   $$3x^2 + 11x - 4 = 0$$
   $$(3x - 1)(x + 4) = 0$$
   $$3x - 1 = 0 \quad x + 4 = 0$$
   $$x = 1/3 \quad x = -4$$

   Thus, the four solutions to the original equation are −4, −1, 1, and 1/3.

3. Given the polynomial equation

   $$P(x) = x^3 - 2x^2 + 5x - 10 = 0.$$

   The possibilities for p are divisors of −10, and the possibilities for q are divisors of 1. Thus,

   $$p: \pm 1, \pm 2, \pm 5, \pm 10 \quad q: \pm 1$$

   and

   $$\frac{p}{q}: \pm 1, \pm 2, \pm 5, \pm 10.$$

   Suppose we try 2.

   ```
   2 | 1 - 2 + 5 - 10
     |     + 2 + 0 + 10
     -------------------
       1 + 0 + 5 +  0
   ```

   Thus, 2 is a solution making $x - 2$ a factor with the remaining factor $x^2 + 5$. Since this factor is quadratic, we can find the zeros of it by solving the quadratic following equation.

   $$x^2 + 5 = 0$$
   $$x^2 = -5$$
   $$x = \pm\sqrt{-5}$$
   $$x = \pm i\sqrt{5}$$

   Thus, the solutions to the original equation are 2, $i\sqrt{5}$, $-i\sqrt{5}$.

4. If x is the radius of the cylinder, then just as in the example, we obtain the equation

   $$\pi x^2 (30 - 2x) + \frac{4}{3}\pi x^3 = 936\pi.$$

   Simplifying we obtain
   $$x^3 - 45x^2 + 1404 = 0.$$
   Since the leading coefficient is 1, the rational solutions are integers that divide
   $$1404 = 2^2 \cdot 3^3 \cdot 13.$$
   With some trials, we can discover that 6 is a solution. Thus, the radius of the cylinder under these conditions is 6 ft.

EXERCISES A

CHAPTER 4 POLYNOMIALS AND RATIONAL FUNCTIONS

SECTION 4.4 Bounds and the Rational Root Theorem

1. Given the equation $4x^4 + 7x^2 - 2 = 0$.

   First try $r = 1$.

   ```
   1 | 4 + 0 + 7 + 0 - 2
     |   + 4 + 4 + 11 + 11
     -----------------------
       4 + 4 + 11 + 11 + 9
   ```

   Since the signs in the bottom row are all positive, 1 is the smallest positive integer upper bound for the real solutions to the equation.

   Next try $r = -1$.

   ```
   -1 | 4 + 0 + 7 + 0 - 2
      |   - 4 + 4 - 11 + 11
      -----------------------
        4 - 4 + 11 - 11 + 9
   ```

   Since the signs alternate in the bottom row, -1 is the largest negative integer lower bound for real solutions to the equation.

2. Given the equation $2x^3 - 3x^2 + 6x - 9 = 0$.

   First try $r = 1$.

   ```
   1 | 2 - 3 + 6 - 9
     |   + 2 - 1 + 5
     -----------------
       2 - 1 + 5 - 4
   ```

   Since the signs are not all positive in the bottom row, try $r = 2$.

   ```
   2 | 2 - 3 + 6 - 9
     |   + 4 + 2 + 16
     ------------------
       2 + 1 + 8 + 7
   ```

   Since the signs in the bottom row are all positive, 2 is the smallest positive integer upper bound for the real solutions to the equation.

   Next try $r = -1$.

   ```
   -1 | 2 - 3 + 6 - 9
      |   - 2 + 5 - 11
      ------------------
        2 - 5 + 11 - 20
   ```

   Since the signs alternate in the bottom row, -1 is the largest negative integer lower bound for real solutions to the equation.

3. Given the equation $x^4 + x^3 - 7x^2 - 5x + 10 = 0$.

   First try $r = 1$.

   ```
   1 | 1 + 1 - 7 - 5 + 10
     |   + 1 + 2 - 5 - 10
     ----------------------
       1 + 2 - 5 - 10 + 0
   ```

   Since the signs are not all positive, try $r = 2$.

   ```
   2 | 1 + 1 - 7 - 5 + 10
     |   + 2 + 6 - 2 - 14
     ----------------------
       1 + 3 - 1 - 7 - 4
   ```

   Since the signs are not all positive, try $r = 3$.

   ```
   3 | 1 + 1 - 7 - 5 + 10
     |   + 3 + 12 + 15 + 30
     ------------------------
       1 + 4 + 5 + 10 + 40
   ```

   Since the signs are now all positive, 3 is the smallest positive integer upper bound for the real solutions to the equation.

   Next try $r = -1$.

   ```
   -1 | 1 + 1 - 7 - 5 + 10
      |   - 1 + 0 + 7 - 2
      ----------------------
        1 + 0 - 7 + 2 + 8
   ```

   Since the signs do not alternate, try $r = -2$.

   ```
   -2 | 1 + 1 - 7 - 5 + 10
      |   - 2 + 2 + 10 - 10
      -----------------------
        1 - 1 - 5 + 5 + 0
   ```

   Since the signs do not alternate, try $r = -3$.

   ```
   -3 | 1 + 1 - 7 - 5 + 10
      |   - 3 + 6 + 3 + 6
      ----------------------
        1 - 2 - 1 - 2 + 16
   ```

   Since the signs do not alternate, try $r = -4$.

   ```
   -4 | 1 + 1 - 7 - 5 + 10
      |   - 4 + 12 - 20 + 100
      -------------------------
        1 - 3 + 5 - 25 + 110
   ```

   Since the signs now alternate, -4 is the largest negative integer lower bound for the real solutions to the equation.

4. Given $9x^4 - 6x^3 + 10x^2 - 30x - 165 = 0$.

   First try $r = 1$.

   $$\underline{1\rfloor\;\;\begin{array}{r}9 - 6 + 10 - 30 - 165 \\ +9 + 3 + 13 - 17\end{array}}$$
   $$\;\;\;\;9 + 3 + 13 - 17 - 182$$

   Since the signs are not all positive, try $r = 2$.

   $$\underline{2\rfloor\;\;\begin{array}{r}9 - 6 + 10 - 30 - 165 \\ +18 + 24 + 68 + 76\end{array}}$$
   $$\;\;\;\;9 + 12 + 34 + 38 - 89$$

   Since the signs are not all positive, try $r = 3$.

   $$\underline{3\rfloor\;\;\begin{array}{r}9 - 6 + 10 - 30 - 165 \\ +27 + 63 + 219 + 567\end{array}}$$
   $$\;\;\;\;9 + 21 + 73 + 189 + 402$$

   Since the signs are all positive, 3 is the smallest positive integer upper bound for the real solutions to the equation.

   Next try $r = -1$.

   $$\underline{-1\rfloor\;\;\begin{array}{r}9 - 6 + 10 - 30 - 165 \\ -9 + 15 - 25 + 55\end{array}}$$
   $$\;\;\;\;9 - 15 + 25 - 55 - 110$$

   Since the signs do not alternate, try $r = -2$.

   $$\underline{-2\rfloor\;\;\begin{array}{r}9 - 6 + 10 - 30 - 165 \\ -18 + 48 - 116 + 292\end{array}}$$
   $$\;\;\;\;9 - 24 + 58 - 146 + 127$$

   Since the signs now alternate, -2 is the largest negative integer lower bound for the real solutions to the equation.

5. Given $x^4 - 3x^3 + 9x^2 - 4x = 0$. Since x is a factor common to all terms, 0 is one solution to the equation. Thus, we can work with the equation $x^3 - 3x^2 + 9x - 4 = 0$. First try $r = 1$.

   $$\underline{1\rfloor\;\;\begin{array}{r}1 - 3 + 9 - 4 \\ +1 - 2 + 7\end{array}}$$
   $$\;\;\;\;1 - 2 + 7 + 3$$

   Since the signs are not all positive, try $r = 2$.

   $$\underline{2\rfloor\;\;\begin{array}{r}1 - 3 + 9 - 4 \\ +2 - 2 + 14\end{array}}$$
   $$\;\;\;\;1 - 1 + 7 + 10$$

   Since the signs are not all positive, try $r = 3$.

   $$\underline{3\rfloor\;\;\begin{array}{r}1 - 3 + 9 - 4 \\ +3 + 0 + 27\end{array}}$$
   $$\;\;\;\;1 + 0 + 9 + 23$$

   Since the signs are all positive, 3 is the smallest positive integer upper bound for the real solutions to the equation.

   Next try $r = -1$.

   $$\underline{-1\rfloor\;\;\begin{array}{r}1 - 3 + 9 - 4 \\ -1 + 4 - 13\end{array}}$$
   $$\;\;\;\;1 - 4 + 13 - 17$$

   Since the signs now alternate -1 is the largest negative integer lower bound for the real solutions to the equation.

6. Given $2x^4 - 15x^3 + 41x^2 - 48x + 20 = 0$. If we try $r = 1, 2, 3, 4, 5, 6,$ and 7, the signs will not all be positive. Suppose we try $r = 8$.

   $$\underline{8\rfloor\;\;\begin{array}{r}2 - 15 + 41 - 48 + 20 \\ +16 + 8 + 392 + 2752\end{array}}$$
   $$\;\;\;\;2 + 1 + 49 + 344 + 2772$$

   Since the signs are now all positive, 8 is the smallest positive integer upper bound for the real solutions to the equation.

   Next try $r = -1$.

   $$\underline{-1\rfloor\;\;\begin{array}{r}2 - 15 + 41 - 48 + 20 \\ -2 + 17 - 58 + 106\end{array}}$$
   $$\;\;\;\;2 - 17 + 58 - 106 + 126$$

   Since the signs alternate, -1 is the largest negative integer lower bound for the real solutions to the equation.

7. Given the equation $x^3 + 2x^2 - x - 2 = 0$. With $a_0 = -2$, the possible values of p are factors of -2, and with $a_n = 1$, the possible values of q are factors of 1. Then

   $$p: \pm 1, \pm 2, \qquad q: \pm 1$$

   and the possible rational solutions are

   $$\frac{p}{q}: \quad \pm 1, \pm 2$$

EXERCISES A

Suppose we try 1.

```
 1 | 1 + 2 - 1 - 2
   |     + 1 + 3 + 2
   |_____
     1 + 3 + 2 + 0
```

Then one solution to the equation is 1, one factor is $(x - 1)$, and the equation can be factored into $(x - 1)(x^2 + 3x + 2) = 0$. Since the remaining factor is quadratic, we can find the solutions to

$$x^2 + 3x + 2 = 0$$

by factoring and using the zero-product rule.

$$(x + 1)(x + 2) = 0$$
$$x + 1 = 0 \qquad x + 2 = 0$$
$$x = -1 \qquad x = -2$$

Thus, the solutions to the original equation are 1, -1, and -2. Notice that it is best to use previous techniques once the polynomial has been reduced to a factorization that has a quadratic factor.

8. Given the equation $x^3 - x^2 - 14x + 24 = 0$. With $a_0 = 24$, the possible values of p are factors of 24, and with $a_n = 1$, the possible values of q are factors of 1. Then

$$p: \pm 1, \pm 2, \pm 3, \pm 4, \pm 6, \pm 8, \pm 12, \pm 24 \quad q: \pm 1$$

and the possible rational solutions are

$$\frac{p}{q}: \pm 1, \pm 2, \pm 3, \pm 4, \pm 6, \pm 8, \pm 12, \pm 24.$$

Suppose we try 2.

```
 2 | 1 - 1 - 14 + 24
   |     + 2 +  2 - 24
   |_____
     1 + 1 - 12 +  0
```

Then one solution to the equation is 2, one factor is $(x - 2)$, and the equation can be factored into $(x - 2)(x^2 + x - 12) = 0$. Since the remaining factor is quadratic, we can find the solutions to

$$x^2 + x - 12 = 0$$

by factoring and using the zero-product rule.

$$(x + 4)(x - 3) = 0$$
$$x + 4 = 0 \qquad x - 3 = 0$$
$$x = -4 \qquad x = 3$$

SECTION 4.4   171

Thus, the solutions to the original equation are -4, 2, and 3. Notice that it is best to use previous techniques once the polynomial has been reduced to a factorization that has a quadratic factor.

9. Given the equation $x^3 - 5x^2 + x + 12 = 0$. With $a_0 = 12$, the possible values of p are factors of 12, and with $a_n = 1$, the possible values of q are factors of 1. Then

$$p: \pm 1, \pm 2, \pm 3, \pm 4, \pm 6, \pm 12 \quad q: \pm 1$$

and the possible rational solutions are

$$\frac{p}{q}: \pm 1, \pm 2, \pm 3, \pm 4, \pm 6, \pm 12.$$

It is easy to verify using synthetic division that 1, 2, and 3 are not solutions. Suppose we try 4.

```
 4 | 1 - 5 + 1 + 12
   |     + 4 - 4 - 12
   |_____
     1 - 1 - 3 +  0
```

Then one solution to the equation is 4, one factor is $(x - 4)$, and the equation can be factored into $(x - 4)(x^2 - x - 3) = 0$. Since the remaining factor is quadratic, we can find the solutions to

$$x^2 - x - 3 = 0$$

by using the quadratic formula.

$$x = \frac{-b \pm \sqrt{b^2 - 4ac}}{2a}$$
$$= \frac{-(-1) \pm \sqrt{(-1)^2 - 4(1)(-3)}}{2(1)}$$
$$= \frac{1 \pm \sqrt{1 + 12}}{2}$$
$$= \frac{1 \pm \sqrt{13}}{2}$$

Thus, the solutions to the original equation are

4, $\frac{1 + \sqrt{13}}{2}$, and $\frac{1 - \sqrt{13}}{2}$. Notice that the only way to obtain the irrational solutions is to use previous techniques once the polynomial has been reduced to a factorization that has a quadratic factor.

10. Given the equation $x^3 - 6x^2 + 4x - 24 = 0$. With $a_0 = 24$, the possible values of p are factors of 24, and with $a_n = 1$, the possible values of q are factors of 1. Then

p: $\pm 1, \pm 2, \pm 3, \pm 4, \pm 6, \pm 8, \pm 12, \pm 24$ q: ± 1

and the possible rational solutions are

$\dfrac{p}{q}$: $\pm 1, \pm 2, \pm 3, \pm 4, \pm 6, \pm 8, \pm 12, \pm 24$.

It is easy to verify using synthetic division that 1, 2, 3, and 4 are not solutions. Suppose we try 6.

```
6 | 1 - 6 + 4 - 24
    + 6 + 0 + 24
    ─────────────
    1 + 0 + 4 +  0
```

Then one solution to the equation is 6, one factor is $(x - 6)$, and the equation can be factored into $(x - 6)(x^2 + 4) = 0$. Since the remaining factor is quadratic, we can find the solutions to

$$x^2 + 4 = 0$$

by the method of taking roots.

$$x^2 = -4$$
$$x = \pm\sqrt{-4}$$
$$x = \pm 2i$$

Thus, the solutions to the original equation are 6, $2i$, and $-2i$.

11. Given the equation $3x^3 + x^2 + 12x + 4 = 0$. With $a_0 = 4$, the possible values of p are factors of 4, and with $a_n = 3$, the possible values of q are factors of 3. Then

p: $\pm 1, \pm 2, \pm 4$ q: $\pm 1, \pm 3$

and the possible rational solutions are

$\dfrac{p}{q}$: $\pm 1, \pm 2, \pm 4, \pm\dfrac{1}{3}, \pm\dfrac{2}{3}, \pm\dfrac{4}{3}$

Since there are no sign changes in the polynomial, there are no positive real solutions so we can eliminate all the positive numbers from the list. Suppose we try -1.

```
-1 | 3 + 1 + 12 +  4
     - 3 + 2 - 14
     ──────────────
     3 - 2 + 14 - 10
```

Since the signs alternate, -1 is a lower bound for the negative real solutions so we can eliminate all the remaining possibilities with the exception of $-\dfrac{1}{3}$, and $-\dfrac{2}{3}$. Suppose we try $-\dfrac{1}{3}$.

```
-1/3 | 3 + 1 + 12 + 4
       - 1 + 0 - 4
       ─────────────
       3 + 0 + 12 + 0
```

Thus, $-\dfrac{1}{3}$ is one solution and the remaining factor of the polynomial is the quadratic quotient $3x^2 + 12$. To find the remaining solutions solve

$$3x^2 + 12 = 0.$$
$$x^2 + 4 = 0$$
$$x^2 = -4$$
$$x = \pm 2i$$

Thus, the solutions to the original polynomial equation are $-\dfrac{1}{3}$, $2i$, and $-2i$.

12. Given the equation $x^4 - 4x^3 + x^2 + 8x - 6 = 0$. With $a_0 = -6$, the possible values of p are factors of -6, and with $a_n = 1$, the possible values of q are factors of 1. Then

p: $\pm 1, \pm 2, \pm 3, \pm 6$ q: ± 1

and the possible rational solutions are

$\dfrac{p}{q}$: $\pm 1, \pm 2, \pm 3, \pm 6$

Suppose we try 1.

```
1 | 1 - 4 + 1 + 8 - 6
    + 1 - 3 - 2 + 6
    ─────────────────
    1 - 3 - 2 + 6 + 0
```

Then one solution to the equation is 1, one factor is $(x - 1)$, and the equation can be factored into $(x - 1)(x^3 - 3x^2 - 2x + 6) = 0$. The remaining solutions will be solutions to the cubic equation, and

EXERCISES A

the possibilities are included in the same list as for the original equation. It is easy to show that 1 is not a double root, and that 2 is not a solution. Suppose we try 3.

$$\begin{array}{r|rrrr} 3 & 1 - 3 - 2 + 6 \\ & + 3 + 0 - 6 \\ \hline & 1 + 0 - 2 + 0 \end{array}$$

Then 3 is a second solution, and the remaining solutions must be solutions to

$$x^2 - 2 = 0.$$

Solve this equation by taking roots.

$$x^2 = 2$$
$$x = \pm\sqrt{2}$$

Thus, the solutions to the original equation are 1, 3, $\sqrt{2}$, and $-\sqrt{2}$.

13. Given the equation $9x^5 + 12x^4 + 10x^3 + x^2 - 2x = 0$. Then since x is a factor of each term, 0 is one solution and we can concentrate on finding the solutions of the equation $9x^4 + 12x^3 + 10x^2 + x - 2 = 0$. With $a_0 = -2$, the possible values of p are factors of 2, and with $a_n = 9$, the possible values of q are factors of 9. Then

$$p: \pm 1, \pm 2 \qquad q: \pm 1, \pm 3, \pm 9$$

and the possible rational solutions are

$$\frac{p}{q}: \pm 1, \pm 2, \pm\frac{1}{3}, \pm\frac{2}{3}, \pm\frac{1}{9}, \pm\frac{2}{9}$$

Suppose we try $\frac{1}{3}$.

$$\begin{array}{r|rrrrr} \frac{1}{3} & 9 + 12 + 10 + 1 - 2 \\ & + 3 + 5 + 5 + 2 \\ \hline & 9 + 15 + 15 + 6 + 0 \end{array}$$

Then $\frac{1}{3}$ is another solution, and we can concentrate on the equation $9x^3 + 15x^2 + 15x + 6 = 0$, or equivalently, $3x^3 + 5x^2 + 5x + 2 = 0$.

Suppose we try $-\frac{2}{3}$.

$$\begin{array}{r|rrrr} -\frac{2}{3} & 3 + 5 + 5 + 2 \\ & - 2 - 2 - 2 \\ \hline & 3 + 3 + 3 + 0 \end{array}$$

Then a third solution is $-\frac{2}{3}$, and the remaining solutions are solutions of the quadratic equation $3x^2 + 3x + 3 = 0$, or equivalently, $x^2 + x + 1 = 0$. Use the quadratic formula.

$$\begin{aligned} x &= \frac{-b \pm \sqrt{b^2 - 4ac}}{2a} \\ &= \frac{-1 \pm \sqrt{1^2 - 4(1)(1)}}{2(1)} \\ &= \frac{-1 \pm \sqrt{1 - 4}}{2} \\ &= \frac{-1 \pm \sqrt{-3}}{2} \\ &= \frac{-1 \pm i\sqrt{3}}{2} \end{aligned}$$

Thus, the solutions to the original equation are 0, $\frac{1}{3}$, $-\frac{2}{3}$, $\frac{-1+i\sqrt{3}}{2}$, $\frac{-1-i\sqrt{3}}{2}$.

14. Given the equation $2x^5 - x^4 - x^3 + 4x^2 - 2x = 0$. Since x is a factor of each term, 0 is one solution and we can concentrate on solving $2x^4 - x^3 - x^2 + 4x - 2 = 0$. With $a_0 = -2$, the possible values of p are factors of 2, and with $a_n = 2$, the possible values of q are factors of 2. Then

$$p: \pm 1, \pm 2 \qquad q: \pm 1, \pm 2$$

and the possible rational solutions are

$$\frac{p}{q}: \pm 1, \pm 2, \pm\frac{1}{2}$$

Suppose we try 1.

```
 1 | 2  - 1  - 1  + 4  - 2
   |    + 2  + 1  + 0  + 4
     -----------------------
     2  + 1  + 0  + 4  + 2
```

Since the remainder is not 0, 1 is not a solution, and since the signs are all positive, 1 is an upper bound for the solutions so that 2 can also be eliminated. Suppose we try −1.

```
-1 | 2  - 1  - 1  + 4  - 2
   |    - 2  + 3  - 2  - 2
     -----------------------
     2  - 3  + 2  + 2  - 4
```

Thus, −1 is not a solution. Try −2.

```
-2 | 2  - 1  - 1  + 4  - 2
   |    - 4  + 10 - 18 + 28
     ------------------------
     2  - 5  + 9  - 14 + 26
```

Thus, −2 is not a solution either. Suppose we try $\frac{1}{2}$.

```
 1/2 | 2  - 1  - 1  + 4  - 2
     |    + 1  + 0  - 1/2 + 7/4
       -------------------------
       2  + 0  - 1  + 7/2 - 1/4
```

Since the remainder is not 0, $\frac{1}{2}$ is not a solution.

This leaves only one possibility left, $-\frac{1}{2}$.

```
-1/2 | 2  - 1  - 1  + 4  - 2
     |    - 1  + 1  + 0  - 2
       -----------------------
       2  - 2  + 0  + 4  - 4
```

Thus, $-\frac{1}{2}$ is not a solution either. Since we have exhausted all the possibilities for rational solutions, 0 is the only rational solution to the original equation.

15. Let x = the width of the carton,
    $3x$ = the length of the carton,
    $x - 2$ = the height of the carton.
    Since the volume of the carton is the product of the three dimensions, we have the equation

$$3x(x)(x - 2) = 2400$$

which simplifies to

$$x^3 - 2x^2 - 800 = 0.$$

With $a_0 = -800$, the possible values of p are factors of 800, and with $a_n = 1$, the possible values of q are factors of 1. Then since q must be 1 or −1, the only rational solutions must be integers that divide 800. Since this is an extensive list, it is wise to try to select appropriate choices. Since the constant term is −800, it might be wise to try 10 first.

```
 10 | 1  - 2  + 0  - 800
    |    + 10 + 80 + 800
      ---------------------
      1  + 8  + 80 +  0
```

Thus, one solution is $x = 10$. Any other solutions would have to be solutions to the quadratic equation $x^2 + 8x + 80 = 0$. But since the discriminant is $b^2 - 4ac = (8)^2 - 4(1)(80) = 64 - 320 < 0$, the remaining solutions are complex (nonreal) solutions and will not fit the conditions of the problem. As a result, the width of the carton is $x = 10$, the length is $3x = 30$, and the height is $x - 2 = 8$. Thus the dimensions of the carton are 30 in by 10 in by 8 in.

16. Given $P(x) = 4x^4 - x^2 + 3x + 8$. Divide $P(x)$ by $x + 3$ using synthetic division. The remainder will be $P(-3)$.

```
-3 | 4  + 0  - 1  +  3  +  8
   |    - 12 + 36 - 105 + 306
     --------------------------
     4  - 12 + 35 - 102 + 314
```

Thus, $P(-3) = 314$.

17. Divide $P(x) = x^5 - 5x^4 + x^2 - 6x + 5$ by $x - 5$ using synthetic division. If the remainder is 0, then by the factor theorem, $x - 5$ is a factor of $P(x)$.

```
 5 | 1  - 5  + 0  + 1  - 6  + 5
   |    + 5  + 0  + 0  + 5  - 5
     ------------------------------
     1  + 0  + 0  + 1  - 1  + 0
```

Thus, $x - 5$ is a factor.

18. Given the polynomial equation

$$x^6 + 3x^4 - x^2 + 5x - 7 = 0.$$

Since the degree of the polynomial is 6, the greatest number of distinct solutions to the equation is six. Also, the fewest number of distinct solutions is one. Counting multiplicities, we can say that the equation has exactly six solutions.

EXERCISES A

19. Given polynomial $P(x)$ with real coefficients and such that $5 - 2i$ is a zero of $P(x)$. Then a second zero must be the conjugate of $5 - 2i$, which is $5 + 2i$.

20. Given polynomial $P(x)$ with rational coefficients and such that $1 + \sqrt{7}$ is a zero of $P(x)$. Then a second zero must be $1 - \sqrt{7}$.

21. If 2 is a double root of $P(x) = 0$, with rational coefficients, then $(x - 2)^2$ is a factor of $P(x)$. Also, if $1 + i$ is a single root of $P(x) = 0$, then $1 - i$ is also a root and $(x - (1 + i))$ and $(x - (1 - i))$ are factors of $P(x)$. Finally, if $1 + \sqrt{2}$ is a single root of $P(x) = 0$, then $1 - \sqrt{2}$ is also a root and $(x - (1 + \sqrt{2}))$ and $(x - (1 - \sqrt{2}))$ are factors of $P(x)$. Thus,

$$P(x)$$
$$= (x - 2)^2(x - (1 + i))(x - (1 - i))(x - (1 + \sqrt{2}))(x - (1 - \sqrt{2}))$$
$$= (x - 2)^2(x^2 - 2x + 2)(x^2 - 2x - 1)$$
$$= (x - 2)^2(x^4 - 4x^3 + 5x^2 - 2x - 2)$$
$$= (x^2 - 4x + 4)(x^4 - 4x^3 + 5x^2 - 2x - 2)$$
$$= x^6 - 8x^5 + 25x^4 - 38x^3 + 26x^2 - 8$$

CHAPTER 4 POLYNOMIAL AND RATIONAL FUNCTIONS

SECTION 4.5 Graphing Polynomial Functions

1. (a) Given $P(x) = x^5 + 2x^3 - 3$.

   Since $n = 5$ (n is odd) and $a_5 = 1 > 0$, the graph of $P(x)$ will eventually go up to the right and down to the left.

   (b) Given $P(x) = -x^5 + x^4 - 2x$.

   Since $n = 5$ (n is odd) and $a_5 = -1 < 0$, the graph of $P(x)$ will eventually go down to the right and up to the left.

   (c) Given $P(x) = -x^4 + 2x^2 + 1$.

   Since $n = 4$ (n is even) and $a_4 = -1 < 0$, the graph of $P(x)$ will eventually go down to the right and down to the left.

2. Given the function $y = P(x) = -x^3 - 2x^2 + x + 2$.

   Since the degree of $P(x)$ is 3, the graph has at most two turning points. Also, since 3 is odd and $a_3 < 0$, the graph must eventually go down to the right and up to the left. The y-intercept is $(0,2)$. Since $P(x)$ has one variation in sign, there is one positive real zero. Also, since $P(-x) = x^3 - 2x^2 - x + 2$ has two variations of sign there will be 2 or 0 negative real zeros. Since the leading coefficient is -1, the only rational zeros possible must be integers that divide 2. Thus the possibilities for rational zeros are ± 1 and ± 2. It is easy to verify using synthetic division that the zeros are 1, -1, and -2. Using this information and perhaps a few additional points we can obtain the graph shown below.

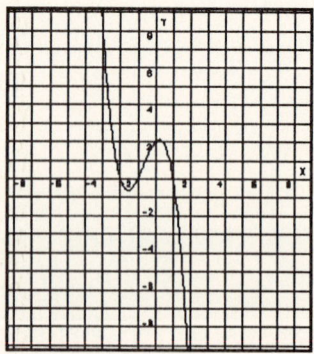

3. Given $P(x) = -x^4 + 4x^3 + 3x^2 - 14x + 8$.

   Since the degree of $P(x)$ is 4, the graph has at most three turning points. Also, since 4 is even and $a_4 < 0$, the graph must eventually go down to the right and down to the left. The y-intercept is $(0,8)$. Since there are three variations of sign, $P(x)$ must have 1 or 3 positive zeros, and since there is one variation of sign of $P(-x) = = -x^4 - 4x^3 + 3x^2 + 14x + 8$, there must be 1 negative zero. Since the leading coefficient is -1, the possible rational zeros are integers that divide 8. Using synthetic division, it is easy to verify that 1 is a zero of multiplicity two, and -2 and 4 are zeros of multiplicity one. Using this information and plotting several additional points we can obtain the graph of the function shown below.

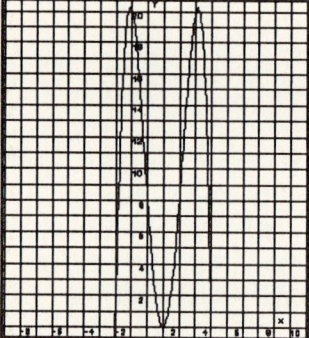

EXERCISES A

CHAPTER 4 POLYNOMIALS AND RATIONAL FUNCTIONS

SECTION 4.5 Graphing Polynomial Functions

1. Given $P(x) = x^2 - 5x + 4$. Since the degree of $P(x)$ is 2, the graph can have at most one turning point so the graph must be the one given in (d).

2. Given $P(x) = x^3 - x^2 - 4x + 4$. Since the graph has at most 2 turning points (the degree of $P(x)$ is 3), we suspect that the graph is given in either (a) or (c). Since $P(2) = 0$, the graph passes through (2,0), so we conclude that the graph is given in (a).

3. Given $P(x) = x^3 - 4x^2 - x + 4$. Since the graph has at most 2 turning points (the degree of $P(x)$ is 3), we suspect that the graph is given in either (a) or (c). Since $P(4) = 0$, the graph passes through (4,0), so we conclude that the graph is given in (c).

4. Given $P(x) = x^4 - 5x^2 + 4$. Since the graph has at most 3 turning points (the degree of $P(x)$ is 4), we suspect that the graph is given in (b). Since the y-intercept, (0,4), agrees with this, we conclude that the graph is indeed that given in (b).

5. Given $P(x) = x^3 + 3x^2 - x + 7$.

    (a) Since the degree of $P(x)$ is 3, the maximum number of zeros is 3.

    (b) Since the degree of $P(x)$ is $n = 3$, the maximum number of turning points is $n - 1 = 2$.

    (c) Since $P(0) = 7$, the y-intercept is (0,7).

    (d) Since the degree of $P(x) = n = 3$ is odd, and since $a_n = 1 > 0$, the graph eventually goes upward for large positive values of x.

    (e) Since the degree of $P(x) = n = 3$ is odd, and since $a_n = 1 > 0$, the graph eventually goes downward for small negative values of x.

6. Given $P(x) = -2x^3 + x^2 - 5x + 12$.

    (a) Since the degree of $P(x)$ is 3, the maximum number of zeros is 3.

    (b) Since the degree of $P(x)$ is $n = 3$, the maximum number of turning points is $n - 1 = 2$.

    (c) Since $P(0) = 12$, the y-intercept is (0,12).

    (d) Since the degree of $P(x) = n = 3$ is odd, and since $a_n = -2 < 0$, the graph eventually goes downward for large positive values of x.

    (e) Since the degree of $P(x) = n = 3$ is odd, and since $a_n = -2 < 0$, the graph eventually goes upward for small negative values of x.

7. Given $P(x) = 5x^4 + x^3 - 3x - 1$.

    (a) Since the degree of $P(x)$ is 4, the maximum number of zeros is 4.

    (b) Since the degree of $P(x)$ is $n = 4$, the maximum number of turning points is $n - 1 = 3$.

    (c) Since $P(0) = -1$, the y-intercept is (0,-1).

    (d) Since the degree of $P(x) = n = 4$ is even, and since $a_n = 5 > 0$, the graph eventually goes upward for large positive values of x.

    (e) Since the degree of $P(x) = n = 4$ is even, and since $a_n = 5 > 0$, the graph eventually goes upward for small negative values of x.

8. Given $P(x) = -3x^6 + 4x^4 + 2x^2 - x - 5$.

    (a) Since the degree of $P(x)$ is 6, the maximum number of zeros is 6.

    (b) Since the degree of $P(x)$ is $n = 6$, the maximum number of turning points is $n - 1 = 5$.

    (c) Since $P(0) = -5$, the y-intercept is (0,-5).

    (d) Since the degree of $P(x) = n = 6$ is even, and since $a_n = -3 < 0$, the graph eventually goes downward for large positive values of x.

    (e) Since the degree of $P(x) = n = 6$ is even, and since $a_n = -3 < 0$, the graph eventually goes downward for small negative values of x.

9. Given the function $y = P(x) = -x^3 + 4x$.

    Since the degree of $P(x)$ is 3, the graph has at most two turning points. Also, since 3 is odd and $a_3 < 0$, the graph must eventually go down to the right and up to the left. The y-intercept is (0,0). Since $P(x)$ has one variation in sign, there is one positive real zero. Also, since $P(-x) = x^3 - 4x$ has one variation of sign there will be 1 negative real zero. Since $P(x) = -x(x - 2)(x + 2)$, it is easy to see that the zeros of $P(x)$ are 0, 2, and -2 giving x-intercepts $(0,0)$, $(2,0)$, and $(-2,0)$. Using this information and perhaps a few additional points we can obtain the graph shown below.

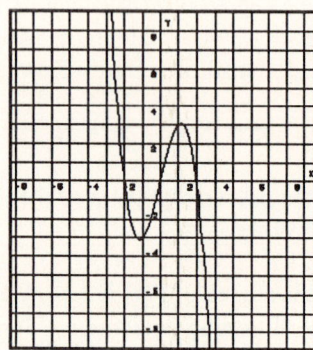

10. Given the function $y = P(x) = x^3 - 2x$.

Since the degree of $P(x)$ is 3, the graph has at most two turning points. Also, since 3 is odd and $a_3 > 0$, the graph must eventually go up to the right and down to the left. The y-intercept is $(0,0)$. Since $P(x)$ has one variation in sign, there is one positive real zero. Also, since $P(-x) = -x^3 + 2x$ has one variation of sign there will be 1 negative real zero. Since $P(x) = x(x^2 - 2)$, it is easy to see that the zeros of $P(x)$ are 0, $\sqrt{2}$, and $-\sqrt{2}$ (the latter two obtained by solving $x^2 - 2 = 0$ by taking roots) giving x-intercepts $(0,0)$, $(\sqrt{2},0)$, and $(-\sqrt{2},0)$. Using this information and perhaps a few additional points we can obtain the graph shown below.

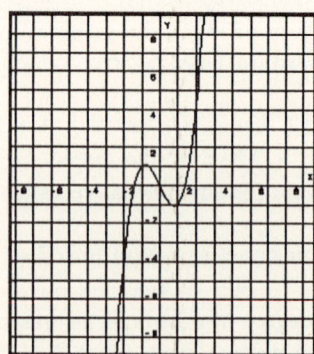

11. Given the function $y = P(x) = x^3 + x^2 + x + 1$.

Since the degree of $P(x)$ is 3, the graph has at most two turning points. Also, since 3 is odd and $a_3 > 0$, the graph must eventually go up to the right and down to the left. The y-intercept is $(0,1)$. Since $P(x)$ has no variation in signs, there is no positive real zero. Also, since $P(-x) = -x^3 + x^2 - x + 1$ has three variations of sign there will be 3 or 1 negative real zeros. Since a_0 and a_3 are both 1, the only possible rational zeros are 1 (which is impossible from above) or -1. It is easy to verify that $P(-1) = 0$ so the only x-intercept is $(-1,0)$. Using this information and perhaps a few additional points we can obtain the graph shown below.

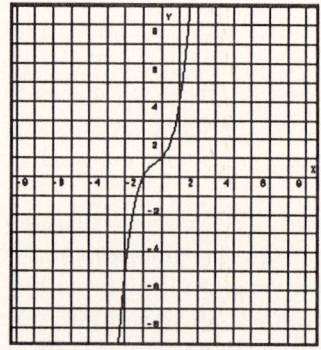

12. Given the function $y = P(x) = x^4 - 4x^2$.

Since the degree of $P(x)$ is 4, the graph has at most three turning points. Also, since 4 is even and $a_4 > 0$, the graph must eventually go up to the right and up to the left. The y-intercept is $(0,0)$. Since $P(x)$ has one variation in sign, there is one positive real zero. Also, since $P(-x) = P(x)$ also has one variation of sign there will be 1 negative real zero. Since $P(x) = x^2(x - 2)(x + 2)$, it is easy to see that the zeros of $P(x)$ are 0, 2, and -2 giving x-intercepts $(0,0)$, $(2,0)$, and $(-2,0)$. Using this information and perhaps a few additional points we can obtain the graph shown below.

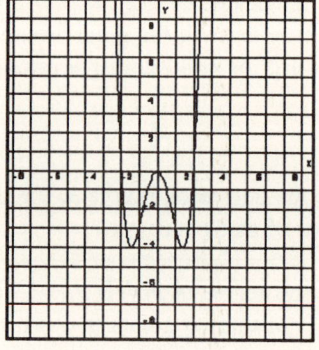

EXERCISES A

13. Given $P(x) = x^3 + 5x^2 - 2x - 10$. Since

$$P(1) = (1)^3 + 5(1)^2 - 2(1) - 10$$
$$= 1 + 5 - 2 - 10$$
$$= -6 < 0$$

and

$$P(2) = (2)^3 + 5(2)^2 - 2(2) - 10$$
$$= 8 + 20 - 4 - 10$$
$$= 14 > 0$$

by the intermediate value theorem, $P(x)$ takes on every value between $P(1) = -6$ and $P(2) = 14$ in the interval $[1,2]$. Since 0 is between -6 and 14, there must be a value of x between 1 and 2 for which $P(x) = 0$. That is, $P(x)$ has a zero between $a = 1$ and $b = 2$.

14. Given $P(x) = x^4 - 9x^2 + 8$. Since

$$P(-3) = (-3)^4 - 9(-3)^2 + 8$$
$$= 81 - 81 + 8$$
$$= 8 > 0$$

and

$$P(-2) = (-2)^4 - 9(-2)^2 + 8$$
$$= 16 - 36 + 8$$
$$= -12 < 0$$

by the intermediate value theorem, $P(x)$ takes on every value between $P(-2) = -12$ and $P(-3) = 8$ in the interval $[-3,-2]$. Since 0 is between -12 and 8, there must be a value of x between -3 and -2 for which $P(x) = 0$. That is, $P(x)$ has a zero between $a = -3$ and $b = -2$.

15. Given that $P(x) = x^3 - 2x^2 - 3x + 6$ has an irrational zero between $a = 1$ and $b = 2$.

By trying several values such as 1.1, 1.2, 1.3, 1.4, and so forth, we discover that

$$P(1.7) = 0.033 > 0$$

and

$$P(1.8) = -0.048 < 0.$$

Then by the intermediate value theorem, $P(x)$ has a zero between 1.7 and 1.8. Next try several values such as 1.71, 1.72, 1.73, 1.74, and so forth, we discover that

$$P(1.73) = 0.001917 > 0$$

and

$$P(1.74) = -0.007176 < 0.$$

Then by the intermediate value theorem, $P(x)$ has a zero between 1.73 and 1.74. In fact, since the value of $P(1.73)$ is closer to zero than the value of $P(1.74)$, we can conclude that to the nearest hundredth, 1.73 is a good approximation for the irrational zero we were looking for. You can actually find this zero by using the rational root theorem to show that 2 is a zero and the remaining factor is $x^2 - 3$, which has zeros $\pm\sqrt{3}$. We were approximating $\sqrt{3}$.

16. Given that $P(x) = x^3 + 2x^2 - 5x - 10$ has an irrational zero between $a = 2$ and $b = 3$.

By trying several values such as 2.1, 2.2, 2.3, 2.4, and so forth, we discover that

$$P(2.2) = -0.672 < 0$$

and

$$P(2.3) = 1.247 > 0.$$

Then by the intermediate value theorem, $P(x)$ has a zero between 2.2 and 2.3. Next try several values such as 2.21, 2.22, 2.23, 2.24, and so forth, we discover that

$$P(2.23) = -0.114633 < 0$$

and

$$P(2.24) = 0.074624 > 0.$$

Then by the intermediate value theorem, $P(x)$ has a zero between 2.23 and 2.24. In fact, since the value of $P(2.24)$ is closer to zero than the value of $P(2.23)$, we can conclude that to the nearest hundredth, 2.24 is a good approximation for the irrational zero.

17. The graph of $P(x) = x^3$ is given below.

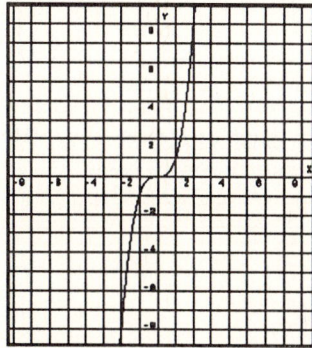

(a) The graph of $P(x) = x^3 + 2$ can be obtained from the graph of $P(x) = x^3$ by sliding the graph up 2 units as shown below.

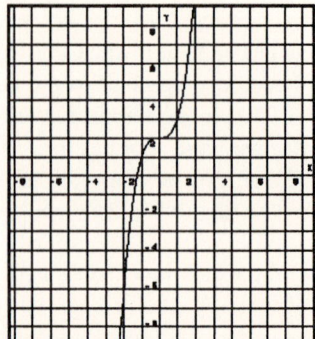

(b) The graph of $P(x) = x^3 - 3$ can be obtained from the graph of $P(x) = x^3$ by sliding the graph down 3 units as shown below.

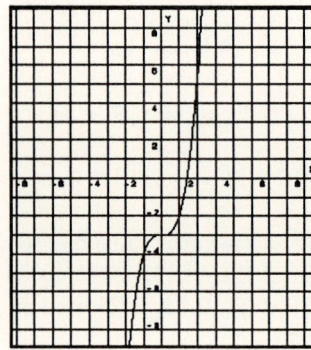

(c) The graph of $P(x) = (x + 1)^3$ can be obtained from the graph of $P(x) = x^3$ by sliding the graph left 1 unit as shown below.

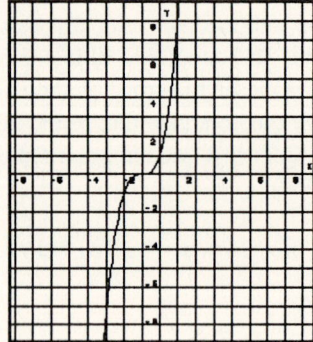

(d) The graph of $P(x) = (x - 3)^3$ can be obtained from the graph of $P(x) = x^3$ by slifing the graph right 3 units as shown below.

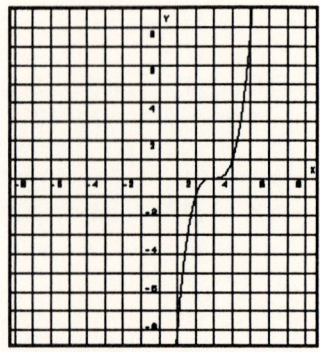

18. Given the equation $x^3 - 5x^2 + x - 5 = 0$. With $a_0 = -5$, the possible values of p are factors of 5, and with $a_n = 1$, the possible values of q are factors of 1. Then

$$p: \pm 1, \pm 5 \qquad q: \pm 1$$

and the possible rational solutions are

$$\frac{p}{q}: \pm 1, \pm 5$$

It is easy to verify using synthetic division that 1 and -1 are not solutions. Suppose we try 5.

```
5 |  1 - 5 + 1 - 5
        + 5 + 0 + 5
     ─────────────────
     1 + 0 + 1 + 0
```

Then one solution to the equation is 5, one factor is $(x - 5)$, and the equation can be factored into $(x - 5)(x^2 + 1) = 0$. Since the remaining factor is quadratic, we can find the solutions to

$$x^2 + 1 = 0$$

by taking roots.

$$x^2 = -1$$
$$x = \pm\sqrt{-1}$$
$$x = \pm i$$

Thus, the solutions to the original equation are 5, i, and $-i$. Notice that the only way to obtain the complex solutions is to use previous techniques once the polynomial has been reduced to a factorization that has a quadratic factor.

19. Given the equation $2x^4 - 5x^3 - 13x^2 + 25x + 15 = 0$. With $a_0 = 15$, the possible values of p are factors of 15, and with $a_n = 2$, the possible values

EXERCISES A   SECTION 4.5   181

of q are factors of 2. Then

$$p: \pm 1, \pm 3, \pm 5, \pm 15 \qquad q: \pm 1, \pm 2$$

and the possible rational solutions are

$$\frac{p}{q}: \pm 1, \pm 3, \pm 5, \pm 15, \pm \frac{1}{2}, \pm \frac{3}{2}, \pm \frac{5}{2}, \pm \frac{15}{2}$$

It is easy to verify using synthetic division that 1 and 2 are not solutions. Suppose we try 3.

```
 3 |  2 - 5 - 13 + 25 + 15
       + 6 + 3 - 30 - 15
      ─────────────────────
      2 + 1 - 10 -  5 +  0
```

Then one solution to the equation is 3, one factor is $(x - 3)$, and the equation can be factored into $(x - 3)(2x^3 + x^2 - 10x - 5) = 0$. Since the remaining factor has $a_0 = -5$, we can eliminate possible solutions that were listed that involve 15.

Suppose we try $-\frac{1}{2}$.

```
-1/2 |  2 + 1 - 10 - 5
         - 1 +  0 + 5
        ─────────────────
        2 + 0 - 10 + 0
```

Then $-\frac{1}{2}$ and the remaining quadratic factor is $2x^2 - 10 = 0$ which simplifies to $x^2 - 5 = 0$ and can be solved by taking roots.

$$x^2 - 5 = 0$$
$$x^2 = 5$$
$$x = \pm\sqrt{5}$$

Thus, the solutions to the original equation are 3, $-\frac{1}{2}$, $\sqrt{5}$, and $-\sqrt{5}$. Notice that the only way to obtain the irrational solutions is to use previous techniques once the polynomial has been reduced to a factorization that has a quadratic factor.

20. Divide $y^5 - 3y^4 + y^3 - 7y^2 + y - 2$ by $y + 1$ using synthetic division to find the quotient and remainder.

```
-1 |  1 - 3 + 1 -  7 +  1 -  2
        - 1 + 4 -  5 + 12 - 13
      ──────────────────────────
      1 - 4 + 5 - 12 + 13 - 15
```

Thus, the quotient is $y^4 - 4y^3 + 5y^2 - 12y + 13$ and the remainder is -15.

21. Use synthetic division and the remainder theorem to find $P(4.1)$ if $P(x) = 1.2x^3 - 3.7x^2 - 5.35x + 1.4268$.

```
4.1 |  1.2 - 3.7  - 5.35  + 1.4268
           + 4.92 + 5.002 - 1.4268
       ─────────────────────────────
       1.2 + 1.22 - 0.348 + 0
```

Since the remainder is the value of $P(x)$ when $x = 4.1$, we have $P(4.1) = 0$.

22. Given the equation $x^3 + 8 = 0$. The solutions are called cube roots of -8. Since the polynomial has degree 3, there are three cube roots of -8. Obviously one cube root of -8 is -2. When $x^3 + 8$ is divided by $x + 2$, the quotient is $x^2 - 2x + 4$. The remaining cube roots of -8 are solutions to $x^2 - 2x + 4 = 0$, found using the quadratic formula.

$$x = \frac{-b \pm \sqrt{b^2 - 4ac}}{2a}$$
$$= \frac{-(-2) \pm \sqrt{(-2)^2 - 4(1)(4)}}{2(1)}$$
$$= \frac{2 \pm \sqrt{4 - 16}}{2}$$
$$= \frac{2 \pm \sqrt{-12}}{2}$$
$$= \frac{2 \pm 2i\sqrt{3}}{2} = 1 \pm i\sqrt{3}$$

Thus, the three cube roots of -8 are -2, $1 + i\sqrt{3}$, and $1 - i\sqrt{3}$.

23. Given the function $f(x) = \frac{x^2 - 5x - 6}{x + 3}$. The domain of this function is all real numbers for which the equation is defined, that is, all real numbers except those that make the denominator 0. Since the denominator is 0 when $x = -3$, the domain is all reals except -3.

24. Given the function $g(x) = \frac{3x + 1}{x^2 + 5}$. Since the denominator is never 0, the domain is all real numbers.

CHAPTER 4 POLYNOMIAL AND RATIONAL FUNCTIONS

SECTION 4.6 Rational Functions

1. (a) Given $f(x) = \dfrac{x^2 - 4x + 3}{3x + 6}$. The vertical asymptotes are found by setting the denominator equal to 0 and solving for x.

$$3x + 6 = 0$$
$$3x = -6$$
$$x = -2$$

Thus, we obtain $x = -2$ as the only vertical asymptote.

(b) $g(x) = \dfrac{x - 2}{x^2 - 2x - 3}$. Set the denominator equal to 0 and solve for x.

$$x^2 - 2x - 3 = 0$$
$$(x + 1)(x - 3) = 0$$
$$x + 1 = 0 \qquad x - 3 = 0$$
$$x = -1 \qquad x = 3$$

Thus, we obtain $x = -1$ and $x = 3$ as the vertical asymptotes.

(c) $h(x) = \dfrac{x^2}{x^2 + 4}$. Since $x^2 + 4$ cannot be 0 for any real values of x, there are no vertical asymptotes.

2. (a) Given $f(x) = \dfrac{x^2 - 4x + 3}{3x + 6}$. Since the degree of the polynomial in the numerator is $n = 2$, the degree of the polynomial in the denominator is $m = 1$, and $n > m$, there is no horizontal asymptote.

(b) $g(x) = \dfrac{x - 2}{x^2 - 2x - 3}$. Since the degree of the polynomial in the numerator is $n = 1$, the degree of the polynomial in the denominator is $m = 2$, and $n < m$, the x-axis, $y = 0$, is a horizontal asymptote.

(c) $h(x) = \dfrac{x^2}{x^2 + 4}$. Since the degree of the polynomial in the numerator and the degree of the polynomial in the denominator are both 2 ($n = m$), the line

$$y = \frac{a_n}{b_m} = \frac{1}{1} = 1$$

is a horizontal asymptote.

3. Given $f(x) = \dfrac{x^2 - 4x + 3}{3x + 6}$. Since the degree of the numerator is 2, which is 1 more than the degree of the denominator which is 1, we divide $x^2 - 4x + 3$ by $3x + 6$ obtaining the quotient $\tfrac{1}{3}x - 2$ and remainder 15. Thus,

$$f(x) = \frac{x^2 - 4x + 3}{3x + 6} = \frac{1}{3}x - 2 + \frac{15}{3x + 6}.$$

As $|x|$ gets larger and larger, $15/(3x + 6)$ approaches zero so that $f(x)$ approaches $\tfrac{1}{3}x - 2$.

Thus, the line $y = \tfrac{1}{3}x - 2$ is an oblique asymptote of the graph.

4. Given the function $f(x) = \dfrac{x^2 - 4x + 3}{3x + 6}$. From Practice Exercise 1, $x = -2$ is a vertical asymptote. From Practice Exercise 2, the graph has no horizontal asymptote. From Practice Exercise 3, $y = \tfrac{1}{3}x - 2$ is an oblique asymptote. The numerator can be factored into $(x - 3)(x - 1)$, so the zeros of $f(x)$ are 3 and 1 making $(3,0)$ and $(1,0)$ the x-intercepts of the graph. Since $f(0) = \tfrac{1}{2}$, the

PRACTICE EXERCISES   SECTION 4.6 183

y-intercept of the graph is $\left(0,\frac{1}{2}\right)$. The intervals determined by the zeros of $f(x)$, 3 and 1, and the vertical asymptote, $x = -2$, are

$(-\infty,-2)$, $(-2,1)$, $(1,3)$, and $(3,\infty)$.

A test point from each interval shows that the graph is below the x-axis in $(-\infty,-2)$ and $(1,3)$, and above the x-axis in $(-2,1)$ and $(3,\infty)$. All three tests for symmetry fail so the graph is not symmetric with respect to the x-axis, the y-axis, or the origin. The graph is given below.

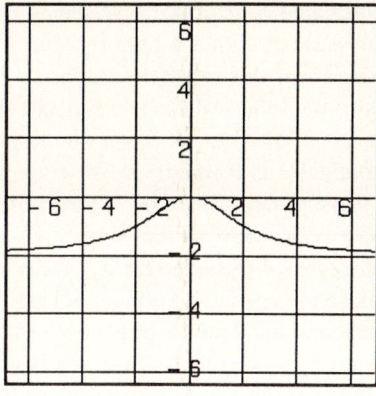

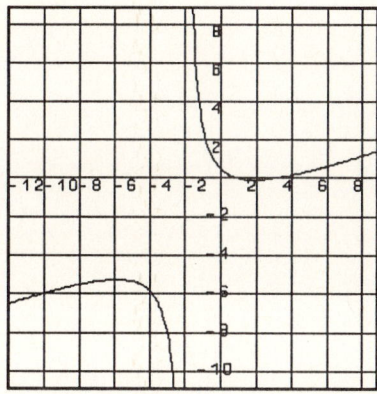

5. Given $g(x) = \dfrac{-2x^2}{x^2 + 4}$. Since the denominator is never 0 for real values of x, there are no vertical asymptotes. Since the degree of the numerator is equal to the degree of the denominator, the horizontal asymptote is $y = -2$ (-2 is a_n over b_m) and there are no oblique asymptotes. The only zero of the function (the only value of x that makes the numerator 0) is $x = 0$, so the x-intercept is $(0,0)$. The y-intercept is also $(0,0)$ since $g(0) = 0$. When x is replaced by $-x$, the equation remains the same so the graph is symmetric with respect to the y-axis (and only the y-axis). This time there are only two intervals to consider, $(-\infty,0)$ and $(0,\infty)$. Since the graph is symmetric with respect to the y-axis, the sign of $g(x)$ will be the same in both intervals. Using the test point 1 in $(0,\infty)$, we have $g(1) < 0$, so the graph is below the x-axis on both intervals. The graph is given below.

6. Given $G(x) = \dfrac{2x^2}{x^2 - 4}$. Since the denominator is 0 when x is 2 and -2, the vertical asymptotes are $x = 2$ and $x = -2$. Since the degree of the numerator and the degree of the denominator are both 2, $y = 2$ (2 is a_n over b_m) is the horizontal asymptote, and there is no oblique asymptote. Since $x = 0$ is a zero of $G(x)$, the x-intercept of the graph is $(0,0)$. Since $G(0) = 0$, $(0,0)$ is also the y-intercept of the graph. When x is replaced with $-x$, the equation remains the same so the graph is symmetric with respect to the y-axis. The intervals to consider are $(-\infty,-2)$, $(-2,0)$, $(0,2)$, and $(2,\infty)$. By symmetry with respect to the y-axis, the graph will behave the same on $(-\infty,-2)$ as on $(2,\infty)$, which is above the x-axis, and the same on $(-2,0)$ as on $(0,2)$, which is below the x-axis. The graph is given below.

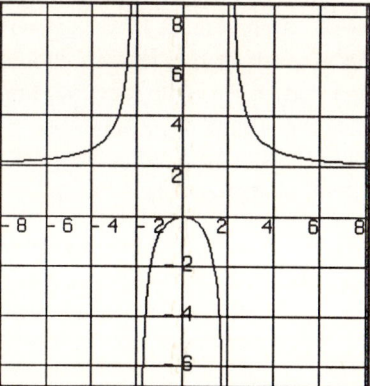

7. The total resistance of the circuit is given by the rational function

$$R = \frac{4r}{4+r},$$

when a variable resistor with resistance r ohms is placed in parallel with a 4-ohm resistor. We can use the graph of this rational function to see what happens to the total resistance when the variable resistance is allowed to get larger and larger. Since the denominator is 0 when r is -4, r = -4 is the vertical asymptote. Since the degree of the numerator is the same as the degree of the numerator, y = 4 (4 is a_n over b_m) is the horizontal asymptote and there are no oblique asymptotes. The x-intercept and the y-intercept are both (0,0). Since all tests for symmetry fail, the graph is not symmetric with respect to either axis or to the origin. Finally, the intervals to consider are: (-∞,-4), where the graph is above the x-axis, (-4,0), where the graph is below the x-axis, and (0,∞), where the graph is above the x-axis. The graph is given below.

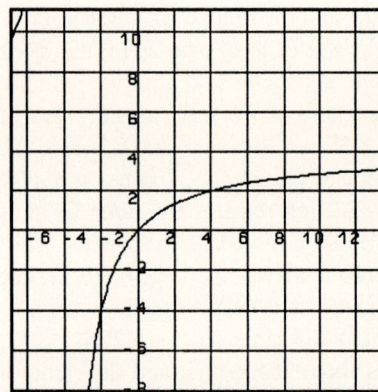

Actually, we are only interested in that portion of the graph where r ≥ 0 since the resistance must be nonnegative. Only a piece of the second branch of the graph in quadrant II is shown. Notice that as r gets larger and larger without bound, the curve approaches the horizontal asymptote y = 4, so we see that under these conditions, the total resistance of the circuit is approaching 4 ohms.

EXERCISES A

CHAPTER 4 POLYNOMIALS AND RATIONAL FUNCTIONS

SECTION 4.6 Rational Functions

1. Given $f(x) = \dfrac{5}{x+2}$.

   (a) The vertical asymptotes are found by setting the denominator equal to 0 and solving for x.
   $$x + 2 = 0$$
   $$x = -2$$
   Thus, the vertical asymptote is $x = -2$.

   (b) Since the degree of the polynomial in the numerator is $n = 0$, the degree of the polynomial in the denominator is $m = 1$, and $n < m$, the x-axis, $y = 0$, is a horizontal asymptote.

   (c) There are no oblique asymptotes since the degree of the numerator, 0, is not 1 more than the degree of the denominator, 1.

   (d) Since the numerator of $f(x)$ is 5, never 0, there are no zeros of $f(x)$. Thus, there are no x-intercepts.

   (e) The y-intercept is found by considering $f(0) = \dfrac{5}{2}$, so the y-intercept is $\left(0, \dfrac{5}{2}\right)$.

   (f) By replacing x with $-x$, replacing y by $-y$, or by replacing x with $-x$ and y with $-y$ at the same time, in the equation $y = \dfrac{5}{x+2}$, we do not obtain the same equation. Thus all tests for symmetry fail, and the graph is not symmetric with respect to either axis or the origin.

2. Given $F(x) = \dfrac{4x}{x-5}$.

   (a) The vertical asymptotes are found by setting the denominator equal to 0 and solving for x.
   $$x - 5 = 0$$
   $$x = 5$$
   Thus, the vertical asymptote is $x = 5$.

   (b) Since the degree of the polynomial in the numerator is $n = 1$, the degree of the polynomial in the denominator is $m = 1$, and $n = m$, the horizontal asymptote is $y = 4$ (4 is a_n over b_m).

   (c) There are no oblique asymptotes since the degree of the numerator, 1, is not 1 more than the degree of the denominator, 1.

   (d) Since the numerator of $F(x)$ is $4x$, and $4x = 0$ means that $x = 0$, 0 is the only zero of $F(x)$. Thus, the only x-intercept is $(0,0)$.

   (e) The y-intercept is found by considering $F(0) = 0$, so the y-intercept is also $(0,0)$.

   (f) By replacing x with $-x$, replacing y by $-y$, or by replacing x with $-x$ and y with $-y$ at the same time, in the equation $y = \dfrac{4x}{x-5}$, we do not obtain the same equation. Thus all tests for symmetry fail, and the graph is not symmetric with respect to either axis or the origin.

3. Given $f(x) = \dfrac{x+1}{x^2+x-2}$.

   (a) The vertical asymptotes are found by setting the denominator equal to 0 and solving for x.
   $$x^2 + x - 2 = 0$$
   $$(x - 1)(x + 2) = 0$$
   $$x - 1 = 0 \quad\quad x + 2 = 0$$
   $$x = 1 \quad\quad\quad x = -2$$
   Thus, the vertical asymptotes are $x = 1$ and $x = -2$.

   (b) Since the degree of the polynomial in the numerator is $n = 1$, the degree of the polynomial in the denominator is $m = 2$, and $n < m$, the x-axis, $y = 0$, is a horizontal asymptote.

   (c) There are no oblique asymptotes since the degree of the numerator, 1, is not 1 more than the degree of the denominator, 2.

(d) Since the numerator of $f(x)$ is 0 when $x = -1$, the only zero of $f(x)$ is -1. Thus, the only x-intercept is $(-1,0)$.

(e) The y-intercept is found by considering

$f(0) = -\frac{1}{2}$, so the y-intercept is $\left(0,-\frac{1}{2}\right)$.

(f) By replacing x with $-x$, replacing y by $-y$, or by replacing x with $-x$ and y with $-y$ at the same time, in the equation $y = \frac{x+1}{x^2 + x - 2}$, we do not obtain the same equation. Thus all tests for symmetry fail, and the graph is not symmetric with respect to either axis or the origin.

4. Given $F(x) = \frac{x^2 + x - 6}{x - 5}$.

(a) The vertical asymptotes are found by setting the denominator equal to 0 and solving for x.

$$x - 5 = 0$$
$$x = 5$$

Thus, the vertical asymptote is $x = 5$.

(b) Since the degree of the polynomial in the numerator is $n = 2$, the degree of the polynomial in the denominator is $m = 1$, and $n > m$, there are no horizontal asymptotes.

(c) Since the degree of the polynomial in the numerator is $n = 2$, the degree of the polynomial in the denominator is $m = 1$, and $n = m + 1$, the oblique asymptote is found by dividing the polynomial in the numerator by the one in the denominator to obtain

$$F(x) = \frac{x^2 + x - 6}{x - 5} = x + 6 + \frac{24}{x - 5}.$$

As $|x|$ becomes larger and larger, the fraction

$\frac{24}{x - 5}$ goes to 0 so the oblique asymptote is

$y = x + 6$.

(d) The zeros of $F(x)$ are the values of x that make the numerator 0, that is, the solutions to the quadratic equation

$$x^2 + x - 6 = 0$$
$$(x - 2)(x + 3) = 0$$
$$x - 2 = 0 \qquad x + 3 = 0$$
$$x = 2 \qquad x = -3$$

Thus, the x-intercepts of the graph are $(2,0)$ and $(-3,0)$.

(e) The y-intercept is found by considering

$f(0) = \frac{6}{5}$, so the y-intercept is $\left(0,\frac{6}{5}\right)$.

(f) By replacing x with $-x$, replacing y by $-y$, or by replacing x with $-x$ and y with $-y$ at the same time, in the equation $y = \frac{x^2 + x - 6}{x - 5}$, we do not obtain the same equation. Thus all tests for symmetry fail, and the graph is not symmetric with respect to either axis or the origin.

5. Given $f(x) = \frac{5x^4}{x^4 + 1}$.

(a) The vertical asymptotes are found by setting the denominator equal to 0 and solving for x. But since $x^4 + 1$ is never 0, for real values of x, there are no vertical asymptotes.

(b) Since the degree of the polynomial in the numerator is $n = 4$, the degree of the polynomial in the denominator is $m = 4$, and $n = m$, the horizontal asymptote is $y = 5$ (5 is a_n over b_m).

(c) There are no oblique asymptotes since the degree of the numerator, 4, is not 1 more than the degree of the denominator, 4.

(d) Since the numerator of $f(x)$ is $5x^4$, and $5x^4 = 0$ means that $x = 0$, 0 is the only zero of $f(x)$. Thus, the only x-intercept is $(0,0)$.

(e) The y-intercept is found by considering $f(0) = 0$, so the y-intercept is also $(0,0)$.

EXERCISES A

(f) By replacing x with $-x$ in the $y = \dfrac{5x^4}{x^4 + 1}$, we obtain the same equation so the graph is symmetric with respect to the y-axis. However, by replacing y by $-y$, or by replacing x with $-x$ and y with $-y$ at the same time, in this equation, we do not obtain the same equation. Thus, the tests for symmetry with respect to the x-axis and origin fail, and the graph is not symmetric with respect to the x-axis or the origin.

6. Given $g(x) = \dfrac{x^2 - 9}{x - 1}$.

(a) The vertical asymptotes are found by setting the denominator equal to 0 and solving for x.

$$x - 1 = 0$$
$$x = 1$$

Thus, the vertical asymptote is $x = 1$.

(b) Since the degree of the polynomial in the numerator is $n = 2$, the degree of the polynomial in the denominator is $m = 1$, and $n > m$, there are no horizontal asymptotes.

(c) Since the degree of the polynomial in the numerator is $n = 2$, the degree of the polynomial in the denominator is $m = 1$, and $n = m + 1$, the oblique asymptote is found by dividing the polynomial in the numerator by the one in the denominator to obtain

$$g(x) = \dfrac{x^2 - 9}{x - 1} = x + 1 - \dfrac{8}{x - 1}.$$

As $|x|$ becomes larger and larger, the fraction $\dfrac{8}{x - 1}$ goes to 0 so the oblique asymptote is $y = x + 1$.

(d) The zeros of $F(x)$ are the values of x that make the numerator 0, that is, the solutions to the quadratic equation

$$x^2 - 9 = 0$$
$$(x - 3)(x + 3) = 0$$
$$x - 3 = 0 \qquad x + 3 = 0$$
$$x = 3 \qquad x = -3$$

Thus, the x-intercepts of the graph are $(3,0)$ and $(-3,0)$.

(e) The y-intercept is found by considering $g(0) = 9$, so the y-intercept is $(0,9)$.

(f) By replacing x with $-x$, replacing y by $-y$, or by replacing x with $-x$ and y with $-y$ at the same time, in the equation $y = \dfrac{x^2 - 9}{x - 1}$, we do not obtain the same equation. Thus all tests for symmetry fail, and the graph is not symmetric with respect to either axis or the origin.

7. Given $f(x) = \dfrac{2}{x - 2}$. Since the denominator is 0 when x is 2, the vertical asymptote is $x = 2$. Since the degree of the numerator is $n = 0$, the degree of the denominator is $m = 1$, and $n < m$, the x-axis, $y = 0$, is the horizontal asymptote, and there is no oblique asymptote. Since $f(x)$ is never 0, there are no x-intercepts. Since $f(0) = -1$, $(0,-1)$ is the y-intercept of the graph. When x is replaced with $-x$, y is replaced with $-y$, or both are replaced at the same time, the equation is changed so the graph is not symmetric with respect to either axis or to the origin. The intervals to consider are $(-\infty,2)$ and $(2,\infty)$. The graph is below the x-axis in the first interval and above the x-axis in the second. The graph is given below.

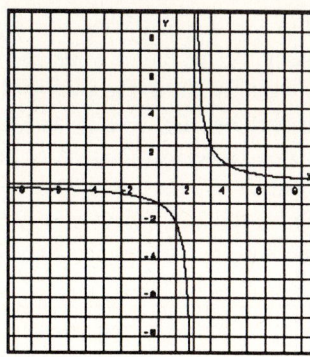

8. Given $g(x) = \dfrac{x}{x + 1}$. Since the denominator is 0 when x is -1, the vertical asymptote is $x = -1$. Since the degree of the numerator and the degree of the denominator are both 1, $y = 1$ (1 is a_n over b_m) is the horizontal asymptote, and there is no oblique asymptote. Since $x = 0$ is a zero of $g(x)$, the x-intercept of the graph is $(0,0)$. Since $g(0) = 0$,

(0,0) is also the y-intercept of the graph. When x is replaced with $-x$, y is replaced by $-y$, or both are replaced at the same time, the equation is changed, so the graph is not symmetric with respect to either axis or to the origin. The intervals to consider are $(-\infty,-1)$, $(-1,0)$, and $(0,\infty)$. Using test points we can see that the graph is below the x-axis on $(-1,0)$ and above the x-axis on the other two intervals. The graph is given below.

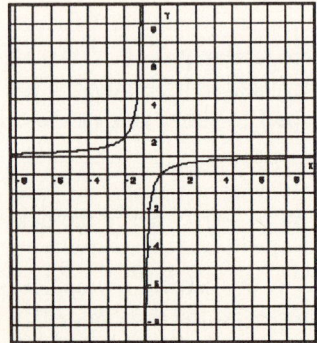

9. Given $F(x) = \dfrac{1}{x^2 - 4}$. Since the denominator is 0 when x is 2 and -2, the vertical asymptotes are $x = 2$ and $x = -2$. Since the degree of the numerator is 0 and the degree of the denominator is 2, the x-axis ($y = 0$) is the horizontal asymptote, and there is no oblique asymptote. Since there are no zeros of $F(x)$, there are no x-intercepts of the graph. Since $F(0) = -\dfrac{1}{4}$, $\left(0,-\dfrac{1}{4}\right)$ is the y-intercept of the graph. When x is replaced with $-x$, the equation remains the same so the graph is symmetric with respect to the y-axis. The intervals to consider are $(-\infty,-2)$, $(-2,2)$, and $(2,\infty)$. By symmetry with respect to the y-axis, the graph will behave the same on $(-\infty,-2)$ as on $(2,\infty)$, which is above the x-axis, and on $(-2,2)$ the graph is below the x-axis. The graph is given below.

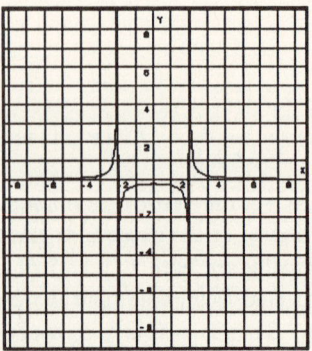

10. Given $G(x) = \dfrac{x}{x^2 - 4}$. Since the denominator is 0 when x is 2 and -2, the vertical asymptotes are $x = 2$ and $x = -2$. Since the degree of the numerator is less than the degree of the denominator the horizontal asymptote is the x-axis, $y = 0$, and there is no oblique asymptote. Since $x = 0$ is a zero of $G(x)$, the x-intercept of the graph is (0,0). Since $G(0) = 0$, (0,0) is also the y-intercept of the graph. When x is replaced with $-x$, y is replaced with $-y$, or both are replaced at the same time, the equation is changed so the graph is not symmetric with respect to either axis or to the origin. The intervals to consider are $(-\infty,-2)$, $(-2,0)$, $(0,2)$, and $(2,\infty)$. Using test points we can show that the graph is above the x-axis on $(-2,0)$ and $(2,\infty)$, and the graph is below the x-axis on $(-\infty,-2)$ and $(0,2)$. The graph is given below.

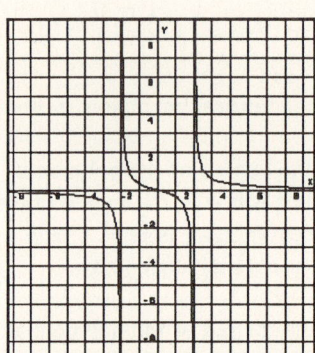

11. Given $f(x) = \dfrac{2x^2 + 1}{x^2 - 1}$. Since the denominator is 0 when x is 1 and -1, the vertical asymptotes are $x = 1$ and $x = -1$. Since the degree of the numerator and the degree of the denominator are both 2, $y = 2$ (2 is a_n over b_m) is the horizontal asymptote, and there is no oblique asymptote. Since $G(x)$ has no zeros, there are no x-intercepts. Since $G(0) = -1$, $(0,-1)$ is the y-intercept of the graph. When x is replaced with $-x$, the equation remains the same so the graph is symmetric with respect to the y-axis. The intervals to consider are $(-\infty,-1)$, $(-1,1)$, and $(1,\infty)$. By symmetry with respect to the y-axis, the graph will behave the same on $(-\infty,-1)$ as on $(1,\infty)$, which is above the x-axis, and on $(-1,1)$ the graph is below the x-axis. The graph is given below.

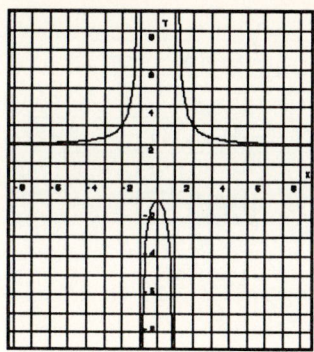

12. Given $g(x) = \dfrac{5}{x^2 + 5}$. Since the denominator is never 0, there are no vertical asymptotes. Since the degree of the numerator is 0 and the degree of the denominator is 2, the horizontal asymptote is the x-axis, $y = 0$, and there is no oblique asymptote. Since $g(x)$ has no zeros, there are no x-intercepts of the graph. Since $g(0) = 1$, $(0,1)$ is the y-intercept of the graph. When x is replaced with $-x$, the equation remains the same so the graph is symmetric with respect to the y-axis. There are no intervals to consider, the graph is above the x-axis for all values of x. The graph is given below.

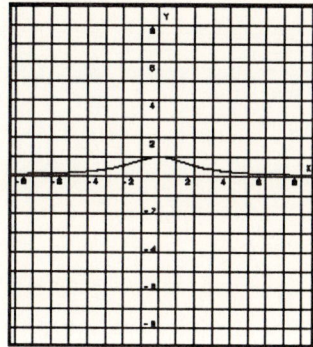

13. First sketch the graph of the rational function

$$a(t) = \frac{c(t)}{n(t)} = \frac{144t + 1728}{t^2 + 24t},$$

concentrating on the interval $0 \leq t \leq 12$, as given below.

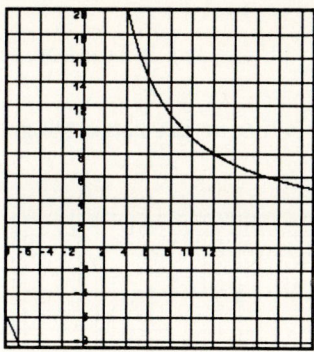

From the graph we can see that the average cost of production decreases to a minimum of $8 per item after 12 hours.

14. If a rational function has a vertical asymptote $x = 1$, then the denominator of the function has a factor of $x - 1$. If the function has a horizontal asymptote $y = 0$, then the degree of the polynomial in the numerator must be less than the degree of the polynomial in the denominator. One possibility for this function is

$$f(x) = \frac{c}{x - 1}$$

where c is a constant. If the y-intercept of the

so this forces c to be 3. Thus, one possible function is

$$f(x) = \frac{3}{x - 1}.$$

15. The graph of a rational function can cross a horizontal asymptote as shown in Example 7 and also in Exercise 10 above.

16. Given the function $y = P(x) = 2x^2 - x^4$.

Since the degree of $P(x)$ is 4, the graph has at most three turning points. Also, since 4 is even and $a_4 < 0$, the graph must eventually go down to the right and down to the left. The y-intercept is $(0,0)$. Since $P(x)$ has one variation in sign, there is one positive real zero. Also, since $P(-x) = P(x)$, $P(-x)$ also has one variation of sign so there will be one negative real zero. Since $P(x) = -x^2(x^2 - 2)$, it is easy to see that the zeros of $P(x)$ are 0, $\sqrt{2}$, and $-\sqrt{2}$. Using this information and perhaps a few additional points we can obtain the graph shown below.

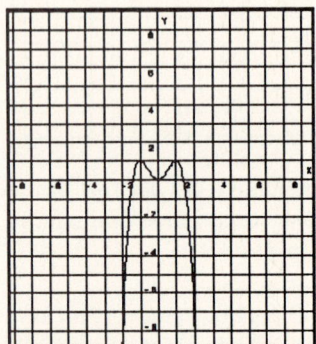

17. Given the polynomial equation $P(x) = 0$ where $P(x) = x^4 + 5x^3 + 2x^2 + x + 2$. Since the degree of $P(x)$ is 4, there are four solutions to the equation, counting multiplicities. Since $P(x)$ has 0 variations in signs, the equation has 0 positive real solutions. Next find $P(-x)$.

$$P(-x) = x^4 - 5x^3 + 2x^2 - x + 2$$

Since $P(-x)$ has 4 variations in sign, there are 4, 2, or 0 negative real solutions. Putting this information together, there are either 0 negative real solutions, 0 positive real solutions, and 4 nonreal solutions, or 2 negative real solutions, 0 positive real solutions, and 2 nonreal solutions, or 4 negative real solutions, 0 positive real solutions, and 0 nonreal solutions. More compactly, we have either 0,0,4; 2,0,2; or 4,0,0.

18. Solve $2x^2 + 15x - 8 = 0$.

$$\begin{aligned} 2x^2 + 15x - 8 &= 0 \\ (2x - 1)(x + 8) &= 0 \end{aligned}$$

$$\begin{array}{ll} 2x - 1 = 0 & x + 8 = 0 \\ 2x = 1 & x = -8 \\ x = \tfrac{1}{2} & \end{array}$$

Thus, the solutions are $\tfrac{1}{2}$ and -8.

19. Solve $3x^2 + x - 1 = 0$.

Since this quadratic expression cannot be factored, we use the quadratic formula.

$$\begin{aligned} x &= \frac{-b \pm \sqrt{b^2 - 4ac}}{2a} \\ &= \frac{-(1) \pm \sqrt{(1)^2 - 4(3)(-1)}}{2(3)} \\ &= \frac{-1 \pm \sqrt{1 + 12}}{6} \\ &= \frac{-1 \pm \sqrt{13}}{6} \end{aligned}$$

20. Solve $3x^2 + 9 = 0$.

$$\begin{aligned} 3x^2 + 9 &= 0 \\ 3x^2 &= -9 \\ x^2 &= -3 \\ x &= \pm\sqrt{-3} = \pm i\sqrt{3} \end{aligned}$$

PRACTICE EXERCISES

SECTION 4.7   191

CHAPTER 4 POLYNOMIAL AND RATIONAL FUNCTIONS

SECTION 4.7   Polynomial and Rational Inequalities

1. Solve $2x^2 + x - 6 \geq 0$.

   First solve the quadratic equation.

   $$2x^2 + x - 6 = 0$$
   $$(2x - 3)(x + 2) = 0$$
   $$2x - 3 = 0 \quad\quad x + 2 = 0$$
   $$2x = 3 \quad\quad\quad\quad x = -2$$
   $$x = \tfrac{3}{2}$$

   These solutions separate the number line into the three intervals

   $$(-\infty, -2), \quad \left(-2, \tfrac{3}{2}\right), \quad \left(\tfrac{3}{2}, \infty\right).$$

   Using 0 as a test point, we have

   $$2(0)^2 + (0) - 6 = -6 < 0.$$

   Thus, 0 is not a solution to the inequality so that none of the numbers in $\left(-2, \tfrac{3}{2}\right)$ is a solution. The solution is then

   $$(-\infty, -2] \quad \text{or} \quad \left[\tfrac{3}{2}, \infty\right),$$

   or using inequalities,

   $$x \leq -2 \quad \text{or} \quad x \geq \tfrac{3}{2}.$$

   Notice that brackets are used to include the endpoints of the intervals since the original inequality is $\geq$.

2. Solve $-x^2 + 4x - 4 > 0$.

   First solve the quadratic equation.

   $$-x^2 + 4x - 4 = 0$$
   $$x^2 - 4x + 4 = 0$$
   $$(x - 2)(x - 2) = 0$$
   $$x - 2 = 0 \quad\quad x - 2 = 0$$
   $$x = 2 \quad\quad\quad\quad x = 2$$

   Since there is only one solution, there are only two intervals, $(-\infty, 2)$ and $(2, \infty)$. Use the test point 0.

   $$-(0)^2 + 4(0) - 4 = -4 < 0$$

   Since 0 is not a solution, $(-\infty, 2)$ is not part of the solution, and therefore, $(2, \infty)$ is not either. Since 2 itself makes the expression equal to 0, not > 0, 0 is not part of the solution. As a result, we have eliminated all possible numbers as solutions so the inequality has no solution.

3. Solve $x^2 + 2x - 1 > 2(x + 1)$.

   First simplify the inequality to obtain

   $$x^2 + 2x - 1 > 2x + 2$$

   which simplifies to

   $$x^2 - 3 > 0.$$

   Solve the quadratic equation.

   $$x^2 - 3 = 0$$
   $$x^2 = 3$$
   $$x = \pm\sqrt{3}$$

   Use 0 as the test point.

   $$(0)^2 - 3 = -3 < 0$$

   Then 0 is not a solution, so the interval $(-\sqrt{3}, \sqrt{3})$ is not the solution. Thus, the solution is

   $$(-\infty, -\sqrt{3}) \quad \text{or} \quad (\sqrt{3}, \infty),$$

   or using inequalities,

   $$x < -\sqrt{3} \quad \text{or} \quad x > \sqrt{3}.$$

   Notice that parentheses (not brackets) are used since the original inequality was $>$.

4. Solve $(x + 2)(x + 5)(x - 1) > 0$.

   The zeros of the polynomial are -2, -5, and 1 which divide the number line into the intervals

$(-\infty,-5)$, $(-5,-2)$, $(-2,1)$, and $(1,\infty)$.

Choose a test point in each, say −6, −3, 0, and 2.

$x=-6$: $(-6+2)(-6+5)(-6-1) = (-4)(-1)(-7)$
$= -28 < 0$

$x=-3$: $(-3+2)(-3+5)(-3-1) = (-1)(2)(-4)$
$= 8 > 0$

$x=0$: $(0+2)(0+5)(0-1) = (2)(5)(-1)$
$= -10 < 0$

$x=2$: $(2+2)(2+5)(2-1) = (4)(7)(1)$
$= 28 > 0$

Since −3 and 2 are both solutions, the solutions include the intervals $(-5,-2)$ and $(1,\infty)$. Thus, the solution is

$(-5,-2)$ or $(1,\infty)$,

or using inequalities,

$-5 < x < -2$ or $x > 1$.

5. Solve $\dfrac{2x+1}{x-1} \le 3$.

First rewrite the inequality so that 0 is on the right side, and simplify the result.

$$\dfrac{2x+1}{x-1} - 3 \le 0$$
$$\dfrac{2x+1}{x-1} - \dfrac{3(x-1)}{x-1} \le 0$$
$$\dfrac{2x+1-3x+3}{x-1} \le 0$$
$$\dfrac{-x+4}{x-1} \le 0$$

The numerator is 0 when $x = 4$, and the denominator is 0 when $x = 1$. Thus, we consider the intervals

$(-\infty,1)$, $(1,4)$, and $(4,\infty)$.

Test points 0, 2, and 5 show that the solution includes the intervals $(-\infty,1)$ and $(4,\infty)$. Since the inequality is $\le$, include 4 in the solution (4 makes the numerator 0, hence the rational expression 0), but do not include 1 which makes the denominator 0 (1 makes the rational expression undefined). Thus, the solution is

$(-\infty,1)$ or $[4,\infty)$,

or using inequalities,

$x < 1$ or $x \ge 4$.

6. The height of the rocket is given by:

$$h = -16t^2 + 128t + 10.$$

The interval of time when the rocket is at a height less than 250 ft is the solution to the inequality

$$-16t^2 + 128t + 10 < 250.$$

This inequality can be simplified to

$$t^2 - 8t + 15 > 0,$$

or in factored form, to

$$(t-3)(t-5) > 0.$$

The values of t that make the expression 0 are 3 and 5, giving the intervals

$(-\infty,3)$, $(3,5)$, and $(5,\infty)$.

The test point 0 shows that the solution to the inequality includes $(-\infty,3)$, and as a result, also includes $(5,\infty)$. We are tempted to conclude that the solution is then $(-\infty,3)$ or $(5,\infty)$. However, since time must be nonnegative, the smallest t can be is 0. Thus, the solution is

$[0,3)$ or $(5,\infty)$.

That is, the rocket is below 250 ft prior to 3 sec ($0 \le t < 3$) or after 5 sec ($t > 5$).

EXERCISES A

CHAPTER 4 POLYNOMIALS AND RATIONAL FUNCTIONS

SECTION 4.7 Polynomial and Rational Inequalities

1. Solve $x^2 + 2x - 15 > 0$.

   First solve the quadratic equation.

   $$x^2 + 2x - 15 = 0$$
   $$(x - 3)(x + 5) = 0$$
   $$x - 3 = 0 \qquad x + 5 = 0$$
   $$x = 3 \qquad x = -5$$

   These solutions separate the number line into the three intervals

   $$(-\infty, -5), (-5, 3), \text{ and } (3, \infty).$$

   Using 0 as a test point, we have

   $$(0)^2 + 2(0) - 15 = -15 < 0.$$

   Thus, 0 is not a solution to the inequality so that none of the numbers in $(-5, 3)$ is a solution. The solution is then
   $$(-\infty, -5) \text{ or } (3, \infty)$$
   or using inequalities,
   $$x < -5 \text{ or } x > 3.$$

   Notice that parentheses are used since endpoints of the intervals are not included in the solution because the original inequality symbol is $>$.

2. Solve $2x^2 + 3x \geq 20$, which is equivalent to $2x^2 + 3x - 20 \geq 0$.

   First solve the quadratic equation.

   $$2x^2 + 3x - 20 = 0$$
   $$(2x - 5)(x + 4) = 0$$
   $$2x - 5 = 0 \qquad x + 4 = 0$$
   $$2x = 5 \qquad x = -4$$
   $$x = \tfrac{5}{2}$$

   These solutions separate the number line into the three intervals

   $$(-\infty, -4), \left(-4, \tfrac{5}{2}\right), \left(\tfrac{5}{2}, \infty\right).$$

   Using 0 as a test point, we have

   $$2(0)^2 + 3(0) - 20 = -20 < 0.$$

   Thus, 0 is not a solution to the inequality so that none of the numbers in $\left(-4, \tfrac{5}{2}\right)$ is a solution. The solution is then

   $$(-\infty, -4] \text{ or } \left[\tfrac{5}{2}, \infty\right),$$

   or using inequalities,

   $$x \leq -4 \text{ or } x \geq \tfrac{5}{2}.$$

   Notice that brackets are used to include the endpoints of the intervals since the original inequality is $\geq$.

3. Solve $x^2 - 2x - 2 > 0$.

   First solve the quadratic equation $x^2 - 2x - 2 = 0$.

   $$x = \frac{-b \pm \sqrt{b^2 - 4ac}}{2a}$$
   $$= \frac{-(-2) \pm \sqrt{(-2)^2 - 4(1)(-2)}}{2(1)}$$
   $$= \frac{2 \pm \sqrt{4 + 8}}{2}$$
   $$= \frac{2 \pm \sqrt{12}}{2}$$
   $$= \frac{2 \pm 2\sqrt{3}}{2}$$
   $$= 1 \pm \sqrt{3}$$

   These solutions separate the number line into the three intervals

   $$(-\infty, 1-\sqrt{3}), (1-\sqrt{3}, 1+\sqrt{3}), (1+\sqrt{3}, \infty).$$

   Using 0 as a test point, we have

   $$(0)^2 - 2(0) - 2 = -2 < 0.$$

   Thus, 0 is not a solution to the inequality so that none of the numbers in $(1-\sqrt{3}, 1+\sqrt{3})$ is a solution. The solution is then

$(-\infty, 1-\sqrt{3})$ or $(1+\sqrt{3}, \infty)$,

or using inequalities,

$$x < 1-\sqrt{3} \quad \text{or} \quad x > 1+\sqrt{3}.$$

4. Solve $3x^2 - x + 2 > 0$.

   First solve the quadratic equation

   $$3x^2 - x + 2 = 0.$$

   Notice that the discriminant is

   $$b^2 - 4ac = (-1)^2 - 4(3)(2) = -23 < 0,$$

   so the solutions are complex (nonreal) numbers. As a result, there are no real solutions and the number line is not divided into intervals. Thus, either every number is a solution or else there are no solutions to the inequality. The test point 0 shows that

   $$3(0)^2 - (0) + 2 = 2 > 0.$$

   Thus, 0 is a solution to the inequality so that every real number is a solution. Using interval notation, the solution is $(-\infty, \infty)$.

5. Solve $4x^2 - 20x + 25 \leq 0$.

   First solve the quadratic equation

   $$\begin{aligned} 4x^2 - 20x + 25 &= 0. \\ (2x-5)(2x-5) &= 0 \\ 2x - 5 = 0 \quad &\quad 2x - 5 = 0 \\ 2x = 5 \quad &\quad 2x = 5 \\ x = \tfrac{5}{2} \quad &\quad x = \tfrac{5}{2} \end{aligned}$$

   Since there is only one solution to the equation, the number line is divided into only two intervals,

   $$\left(-\infty, \tfrac{5}{2}\right) \quad \text{or} \quad \left(\tfrac{5}{2}, \infty\right).$$

   The test point 0 shows that

   $$4(0)^2 - 20(0) + 25 = 25 > 0$$

   so 0 is not a solution. Thus, no real number in either interval can be a solution. Since the original inequality is $\leq$, the only solution is $\tfrac{5}{2}$, which makes the expression equal to 0. Thus, the solution to the inequality is the one number $x = \tfrac{5}{2}$.

6. Solve $\dfrac{x-3}{x+1} \geq 0$.

   The numerator is 0 when $x = 3$, and the denominator is 0 when $x = -1$. Thus we consider the intervals

   $$(-\infty, -1), (-1, 3), \text{ and } (3, \infty).$$

   Test points -2, 0, and 4 show that the solution includes the intervals $(-\infty, -1)$ and $(3, \infty)$. Since the inequality is $\geq$, we include 3 in the solution (3 makes the numerator 0, hence the rational expression 0), but do not include -1 which makes the denominator 0 (-1 makes the rational expression undefined). Thus, the solution is

   $$(-\infty, -1) \quad \text{or} \quad [3, \infty),$$

   or using inequalities,

   $$x < -1 \quad \text{or} \quad x \geq 3.$$

7. Solve $\dfrac{2x-1}{x+3} < 1$.

   First rewrite the inequality so that 0 is on the right side, and simplify the result.

   $$\begin{aligned} \frac{2x-1}{x+3} &< 1 \\ \frac{2x-1}{x+3} - 1 &< 0 \\ \frac{2x-1}{x+3} - \frac{x+3}{x+3} &< 0 \\ \frac{2x-1-x-3}{x+3} &< 0 \\ \frac{x-4}{x+3} &< 0 \end{aligned}$$

   The numerator is 0 when $x = 4$, and the denominator is 0 when $x = -3$. Thus, we consider the intervals

   $$(-\infty, -3), (-3, 4), \text{ and } (4, \infty).$$

EXERCISES A   SECTION 4.7   195

Test points -4, 0, and 5 show that the solution includes the interval $(-3,4)$. Since the original inequality is $<$, we do not include either endpoint. Thus, the solution is

$$(-3,4)$$

or using inequalities

$$-3 < x < 4.$$

8. Solve $\dfrac{3x+2}{x-3} \geq 2$.

   First rewrite the inequality so that 0 is on the right side, and simplify the result.

   $$\dfrac{3x+2}{x-3} \geq 2$$
   $$\dfrac{3x+2}{x-3} - 2 \geq 0$$
   $$\dfrac{3x+2}{x-3} - \dfrac{2(x-3)}{x-3} \geq 0$$
   $$\dfrac{3x+2-2x+6}{x-3} \geq 0$$
   $$\dfrac{x+8}{x-3} \geq 0$$

   The numerator is 0 when $x = -8$, and the denominator is 0 when $x = 3$. Thus, we consider the intervals

   $$(-\infty,-8), \ (-8,3), \text{ and } (3,\infty).$$

   Test points -9, 0, and 4 show that the solution includes the intervals $(-\infty,-8)$ and $(3,\infty)$. Since the inequality is $\geq$, include -8 in the solution (-8 makes the numerator 0, hence the rational expression 0), but do not include 3 which makes the denominator 0 (3 makes the rational expression undefined). Thus, the solution is

   $$(-\infty,-8] \text{ or } (3,\infty),$$

   or using inequalities,

   $$x \leq -8 \text{ or } x > 3.$$

9. Solve $(x-2)(x+2)(x-5) \geq 0$.

   The values of x that make the polynomial 0 are 2, -2, and 5. Since these three values separate the number line into four intervals, the intervals to consider are

   $$(-\infty,-2), \ (-2,2), \ (2,5), \text{ and } (5,\infty).$$

   The test points -3, 0, 3, and 6 show that the solution includes the intervals $(-2,2)$ and $(5,\infty)$. Since the inequality is $\geq$, we also include the endpoints of these intervals where the expression is equal to 0. Thus, the solution to the inequality is

   $$[-2,2] \text{ or } [5,\infty),$$

   or using inequalities,

   $$-2 \leq x \leq 2 \text{ or } x \geq 5.$$

10. Solve $(x-2)(x+2) \geq 3x$.

    First simplify the inequality by clearing parentheses and subtracting $3x$ from both sides to obtain:

    $$x^2 - 3x - 4 \geq 0$$

    Then solve the related quadratic equation.

    $$x^2 - 3x - 4 = 0$$
    $$(x+1)(x-4) = 0$$
    $$x+1=0 \qquad x-4=0$$
    $$x = -1 \qquad x = 4$$

    The solutions -1 and 4 divide the number line into the intervals

    $$(-\infty,-1), \ (-1,4), \text{ and } (4,\infty).$$

    Test points -2, 0, and 5 show that the solution includes $(-\infty,-1)$ and $(4,\infty)$. Since the inequality is $\geq$, we also include the endpoints -1 and 4 which make the expression 0. Thus, the solution to the inequality is

    $$(-\infty,-1] \text{ or } [4,\infty),$$

    or using inequalities,

    $$x \leq -1 \text{ or } x \geq 4.$$

11. Solve $\dfrac{(x-3)(x+6)}{x-1} < 0$.

    The numerator is 0 when x is 3 and -6, and the denominator is 0 when x is 1. These values divide the number line into the four intervals

    $$(-\infty,-6), \ (-6,1), \ (1,3), \text{ and } (3,\infty).$$

Test points -7, 0, 2, and 4 show that the solution includes the intervals $(-\infty, -6)$ and $(1, 3)$. Since the inequality is $<$, we do not include values that make the numerator 0 (of course, we **never** include values that make the denominator 0). Thus, the solution is

$$(-\infty, -6) \text{ or } (1, 3),$$

or using inequalities,

$$x < -6 \text{ or } 1 < x < 3.$$

12. Solve $(x^2 - 2x)(x^2 + 8x + 15) \geq 0$.

    First factor the left side completely to obtain

    $$x(x - 2)(x + 3)(x + 5) \geq 0.$$

    The values of x that make the expression 0 are -5, -3, 0, and 2, which divide the number line into the five intervals

    $$(-\infty, -5), (-5, -3), (-3, 0), (0, 2), (2, \infty).$$

    Test points -6, -4, -1, 1, and 3 show that the solution includes the intervals $(-\infty, -5)$, $(-3, 0)$, and $(2, \infty)$. Since the inequality is $\geq$, we include the endpoints of these intervals which are values that make the expression 0. Thus the solution is

    $$(-\infty, -5] \text{ or } [-3, 0] \text{ or } [2, \infty),$$

    or using inequalities,

    $$x \leq -5 \text{ or } -3 \leq x \leq 0 \text{ or } x \geq 2.$$

13. Solve $x^3 \geq x$.

    First simplify the inequality

    $$x^3 - x \geq 0$$

    and factor the result.

    $$x(x^2 - 1) = x(x + 1)(x - 1) \geq 0.$$

    The values of x that make the polynomial 0 are 0, -1, and 1. These values divide the number line into four intervals,

    $$(-\infty, -1), (-1, 0), (0, 1), \text{ and } (1, \infty).$$

    Using test points -2, $-1/2$, $1/2$, and 2 we can show that the solution includes the intervals $(-1, 0)$ and $(1, \infty)$. Since the original inequality is $\geq$, the endpoints where the expression is 0 must be included. Thus, the solution to the inequality is

    $$[-1, 0] \text{ or } [1, \infty),$$

    or using inequalities,

    $$-1 \leq x \leq 0 \text{ or } x \geq 1.$$

14. Solve $x + \dfrac{1}{x} < 2$.

    First simplify the inequality.

    $$x + \frac{1}{x} < 2$$
    $$\frac{x^2 + 1}{x} < 2$$
    $$\frac{x^2 + 1}{x} - 2 < 0$$
    $$\frac{x^2 + 1}{x} - \frac{2x}{x} < 0$$
    $$\frac{x^2 - 2x + 1}{x} < 0$$
    $$\frac{(x - 1)(x - 1)}{x} < 0$$

    The values that make the numerator and denominator 0 are 1 and 0 which divide the number line into three intervals,

    $$(-\infty, 0), (0, 1), \text{ and } (1, \infty).$$

    Test points -1, $1/2$, and 2 show that the solution includes the interval $(-\infty, 0)$ but not the other two. Thus the solution is

    $$(-\infty, 0)$$

    or using inequalities,

    $$x < 0.$$

15. Since the radicand must be nonnegative for a radical expression to define a real number, the interval(s) where $\sqrt{x(x + 5)}$ defines a real number are found by solving $x(x + 5) \geq 0$. The solutions to the related equation are 0 and -5, which divide the number line into the three intervals

    $$(-\infty, -5), (-5, 0), \text{ and } (0, \infty).$$

EXERCISES A

Using test points −6, −1, and 1 we discover that the solution includes the intervals $(-\infty,-5)$ and $(0,\infty)$. Since the inequality is $\geq$, we include the endpoints where the expression is 0. Thus the solution is

$$(-\infty,-5] \text{ or } [0,\infty)$$

or using inequalities

$$x \leq -5 \text{ or } x \geq 0.$$

16. To find the interval(s) where $\sqrt{\dfrac{3-x}{x+8}}$ is defined, we must find the values of x that make the radicand nonnegative, that is solve $\dfrac{3-x}{x+8} \geq 0$. The values of x that make the numerator and denominator 0 are 3 and −8 which divide the number line into the intervals

$$(-\infty,-8), (-8,3), \text{ and } (3,\infty).$$

Test points −9, 0, and 4 show that the solution includes the interval $(-8,3)$. Since the inequality is $\geq$, we also include the endpoint that makes the numerator 0, 3, but of course do not include −8 which makes the denominator 0. Thus, the solution to the inequality, the values that make the expression a real number, are

$$(-8,3]$$

or using inequalities,

$$-8 < x \leq 3.$$

17. The interval(s) containing m in which the equation $x^2 + mx + 4 = 0$ has two complex (nonreal) solutions can be found by considering the discriminant and making sure the discriminant is negative. The discriminant is

$$b^2 - 4ac = m^2 - 4(1)(4) = m^2 - 16.$$

Thus, we must solve $m^2 - 16 < 0$. First solve the related equation.

$$m^2 - 16 = 0$$
$$(m+4)(m-4) = 0$$
$$m+4 = 0 \qquad m-4 = 0$$
$$m = -4 \qquad m = 4$$

Then the solutions 4 and −4 divide the number line into the intervals

$$(-\infty,-4), (-4,4), \text{ and } (4,\infty).$$

Using 0 as a test point, we see that the solution includes is the interval

$$(-4,4)$$

or using inequalities,

$$-4 < m < 4.$$

18. Given $R = 2t^2 + t$ and $C = t^2 + 6t$. A profit will be generated for values of t that solve $R > C$, that is that solve $2t^2 + t > t^2 + 6t$, which simplifies to

$$t^2 - 5t = t(t-5) > 0.$$

The solutions to the related equation are 0 and 5 which divide the number line into the intervals

$$(-\infty,0), (0,5), \text{ and } (5,\infty).$$

The test point 6 shows that the solution to this inequality includes the intervals $(-\infty,0)$ and $(5,\infty)$. However, since t must be nonnegative, the interval $(-\infty,0)$ must be discarded. Thus, the values of t that will realize a profit are given by

$$(5,\infty)$$

or using an inequality,

$$t > 5.$$

19. From Example 7, the height of an object after time t that is propelled upward from an initial height h_0 with initial velocity v_0 is given by

$$h = -16t^2 + v_0 t + h_0.$$

In this case, $h_0 = 200$ and $v_0 = 48$. To find the interval of time where the object is at least 40 ft high, we must solve

$$-16t^2 + 48t + 200 \geq 40.$$

Simplifying and dividing both sides by −16 (remember to reverse the inequality) we obtain

$$t^2 - 3t - 10 \leq 0,$$

which can be factored into

$$(t + 2)(t - 5) \leq 0.$$

The solutions to the related quadratic equation are -2 and 5 which divide the number line into the intervals

$$(-\infty, -2), (-2, 5), \text{ and } (5, \infty).$$

The test point 0 shows that the solution to the inequality is the interval $(-2, 5)$. However, since t represents time, t must be nonnegative. Also, since the inequality is $\leq$, 5 should be included in the solution. Thus, the period of time when the object is at least 40 ft high is given by the interval

$$[0, 5]$$

or using an inequality,

$$0 \text{ sec} \leq t \leq 5 \text{ sec}.$$

20. Since the height of the rocket is given by

$$h = -9.8t^2 + 147t,$$

to find the period of time when the rocket is higher than 529.2 m we must solve:

$$-9.8t^2 + 147t > 529.2$$

Simplifying and dividing both sides by -9.8 (remember to reverse the inequality) we obtain

$$t^2 - 15t + 54 < 0$$

which becomes

$$(t - 6)(t - 9) < 0.$$

The values of t that solve the related quadratic equation are 6 and 9, and these divide the number line into the intervals

$$(-\infty, 6), (6, 9), \text{ and } (9, \infty).$$

The test point 0 shows that we can eliminate $(-\infty, 6)$ and $(9, \infty)$. Thus the interval which contains the solutions is

$$(6, 9)$$

or using an inequality,

$$6 \text{ sec} < t < 9 \text{ sec}.$$

21. From Example 7, the height of an object after time t that is propelled upward from an initial height h_0 with initial velocity v_0 is given by

$$h = -16t^2 + v_0 t + h_0.$$

In this case, we can think of the bottom of the gorge as being height 0 giving $h_0 = 0$, and $v_0 = 176$. To find the interval of time where the object can be viewed from the rim, the height must be greater then 448 ft. Thus, we must solve

$$-16t^2 + 176t > 448.$$

Simplifying and dividing through by -16 (remember to reverse the inequality gives

$$t^2 - 11t + 28 < 0$$

or in factored form,

$$(t - 4)(t - 7) < 0.$$

The solutions to the related equation are 4 and 7 which divide the number line into the intervals

$$(-\infty, 4), (4, 7), \text{ and } (7, \infty).$$

The test point 0 shows that we can eliminate the intervals $(-\infty, 4)$ and $(7, \infty)$. Thus, the period of time when the flare can be viewed from the rim of the canyon is the interval

$$(4, 7)$$

or using an inequality,

$$4 \text{ sec} < t < 7 \text{ sec}.$$

22. Given $f(x) = \dfrac{-2}{x + 3}$.

(a) The vertical asymptotes are found by setting the denominator equal to 0 and solving for x. When $x + 3 = 0$, then $x = -3$, so $x = -3$ is the vertical asymptote.

(b) Since the degree of the numerator is less than the degree of the denominator, the x-axis, $y = 0$, is the horizontal asymptote.

(c) There is no oblique asymptote since the degree of the numerator is not 1 more than the degree of the denominator.

EXERCISES A

(d) Since the numerator is never 0, there are no zeros of $f(x)$ so there are no x-intercepts.

(e) Since $f(0) = -\frac{2}{3}$, the y-intercept is $\left(0, -\frac{2}{3}\right)$.

(f) When x is replaced with $-x$, y is replaced with $-y$, or both are replaced at the same time, the equation for the function is changed. Thus, the graph is not symmetric with respect to either axis or to the origin.

23. Given $g(x) = \dfrac{x^2 - 25}{x - 2}$.

    (a) The vertical asymptotes are found by setting the denominator equal to 0 and solving for x. When $x - 2 = 0$, then $x = 2$, so $x = 2$ is the vertical asymptote.

    (b) Since the degree of the numerator is greater than the degree of the denominator, there are no horizontal asymptotes.

    (c) Since the degree of the numerator is 1 more than the degree of the denominator, divide to obtain
    $$g(x) = \frac{x^2 - 25}{x - 2} = x + 2 - \frac{21}{x - 2}.$$
    As $|x|$ becomes larger and larger, the fraction goes to 0 so the oblique asymptote is $y = x + 2$.

    (d) Since the numerator is 0 when x is -5 or 5, the zeros of $g(x)$ are -5 and 5 so the x-intercepts are $(-5, 0)$ and $(5, 0)$.

    (e) Since $f(0) = \frac{25}{2}$, the y-intercept is $\left(0, \frac{25}{2}\right)$.

    (f) When x is replaced with $-x$, y is replaced with $-y$, or both are replaced at the same time, the equation for the function is changed. Thus, the graph is not symmetric with respect to either axis or to the origin.

CHAPTER 4 POLYNOMIAL AND RATIONAL FUNCTIONS

CHAPTER 4 Review Exercises

1. Given $P(x) = 3x^4 - 2x^2 + 7x - 8$.

   (a) Since the term of highest degree, $3x^4$, has degree 4, $P(x)$ has degree 4.

   (b) The leading coefficient is the coefficient of the term of highest degree, $3x^4$, which is 3.

   (c) $\quad P(1) = 3(1)^4 - 2(1)^2 + 7(1) - 8$
   $\qquad\quad = 3 - 2 + 7 - 8$
   $\qquad\quad = 0$

   (d) $\quad P(-1) = 3(-1)^4 - 2(-1)^2 + 7(-1) - 8$
   $\qquad\qquad = 3 - 2 - 7 - 8$
   $\qquad\qquad = -14$

2. Given $Q(x) = x^3 - 2x^2 + 9x - 18$.

   $Q(2) = (2)^3 - 2(2)^2 + 9(2) - 18$
   $\quad\quad = 8 - 8 + 18 - 18$
   $\quad\quad = 0$

   $Q(3i) = (3i)^3 - 2(3i)^2 + 9(3i) - 18$
   $\quad\quad\; = 27i^3 - 18i^2 + 27i - 18$
   $\quad\quad\; = 27(-i) - 18(-1) + 27i - 18$
   $\quad\quad\; = -27i + 18 + 27i - 18$
   $\quad\quad\; = 0$

   $Q(-3i) = (-3i)^3 - 2(-3i)^2 + 9(-3i) - 18$
   $\qquad\quad = -27i^3 - 18i^2 - 27i - 18$
   $\qquad\quad = -27(-i) - 18(-1) - 27i - 18$
   $\qquad\quad = 27i + 18 - 27i - 18$
   $\qquad\quad = 0$

   Thus, all three numbers are zeros of $Q(x)$.

3. Divide $x^4 - 3x^3 + x - 2$ by $x + 3$ using synthetic division. Remember to use a 0 for the coefficient of the missing x^2-term.

   $\underline{-3|}\;\; 1 - 3 + 0 + 1 - 2$
   $\qquad\quad\;\;\; - 3 + 18 - 54 + 159$
   $\qquad\quad\;\overline{1 - 6 + 18 - 53 + 157}$

   Thus the quotient is $x^3 - 6x^2 + 18x - 53$, and the remainder is 157.

4. If $P(x)$ is divided by $x - 5$, the remainder is $P(5)$.

5. The remainder theorem is used to obtain the answer in Exercise 4.

6. If $P(x)$ is a polynomial and $P(-3) = 0$, then $x + 3$ is a factor of $P(x)$.

7. The factor theorem is used to obtain the answer to Exercise 6.

8. Given $R(x) = 10.95x$ and $C(x) = 500 + 5.95x$. The zeros of $P(x) = R(x) - C(x)$ are the break-even points. To show that 100 is a break-even point find $P(100)$.

   $P(x) = 10.95x - (500 + 5.95x)$
   $\quad\;\; = 10.95x - 500 - 5.95x$
   $\quad\;\; = 5x - 500$

   $P(100) = 5(100) - 500$
   $\qquad\;\; = 500 - 500$
   $\qquad\;\; = 0$

   Thus, 100 is a break-even point.

9. Divide $P(x) = x^4 - 2x^3 + mx - 60$ by $x + 3$ using synthetic division, set the remainder equal to 0, and solve for m.

   $\underline{-3|}\;\; 1 - 2 + 0 + \quad m\;\; - \quad\quad 60$
   $\qquad\quad\;\;\; - 3 + 15 - \quad 45\; - \; 3(m-45)$
   $\qquad\quad\;\overline{1 - 5 + 15 + (m-45) + [-60-3(m-45)]}$

   Then
   $-60 - 3(m - 45) = 0$
   $-60 - 3m + 135 = 0$
   $-3m + 75 = 0$
   $-3m = -75$
   $m = 25$

   Thus, when $m = 25$, the remainder will be zero, and in fact, $P(-3) = 0$.

10. Given $P(x) = 2x^4 - 3x^2 + x - 8$.

    (a) Divide $P(x)$ by $x - 1$ using synthetic division.

    $\underline{1|}\;\; 2 + 0 - 3 + 1 - 8$
    $\qquad\quad\;\; + 2 + 2 - 1 + 0$
    $\qquad\;\overline{2 + 2 - 1 + 0 - 8}$

    Thus, since the remainder is -8, $P(1) = -8$.

(b) Divide $P(x)$ by $x + 1$ using synthetic division.

$$\begin{array}{r|rrrrr}
-1 & 2 & +0 & -3 & +1 & -8 \\
 & & -2 & +2 & +1 & -2 \\
\hline
 & 2 & -2 & -1 & +2 & -10
\end{array}$$

Thus, since the remainder is -10, $P(-1) = -10$.

(c) Divide $P(x)$ by $x - 2$ using synthetic division.

$$\begin{array}{r|rrrrr}
2 & 2 & +0 & -3 & +1 & -8 \\
 & & +4 & +8 & +10 & +22 \\
\hline
 & 2 & +4 & +5 & +11 & +14
\end{array}$$

Thus, since the remainder is 14, $P(2) = 14$.

11. Given $P(x) = x^4 - 4x^3 - 3x^2 + 10x + 8$.

(a) Divide $P(x)$ by $x - 4$. If the remainder is 0, then $x - 4$ is a factor of $P(x)$.

$$\begin{array}{r|rrrrr}
4 & 1 & -4 & -3 & +10 & +8 \\
 & & +4 & +0 & -12 & -8 \\
\hline
 & 1 & +0 & -3 & -2 & +0
\end{array}$$

Thus, since the remainder is 0, $x - 4$ is a factor of $P(x)$.

(b) Divide $P(x)$ by $x + 1$. If the remainder is 0, then $x + 1$ is a factor of $P(x)$.

$$\begin{array}{r|rrrrr}
-1 & 1 & -4 & -3 & +10 & +8 \\
 & & -1 & +5 & -2 & -8 \\
\hline
 & 1 & -5 & +2 & +8 & +0
\end{array}$$

Thus, since the remainder is 0, $x + 1$ is a factor of $P(x)$.

(c) Divide $P(x)$ by $x + 2$. If the remainder is 0, then $x + 2$ is a factor of $P(x)$.

$$\begin{array}{r|rrrrr}
-2 & 1 & -4 & -3 & +10 & +8 \\
 & & -2 & +12 & -18 & +16 \\
\hline
 & 1 & -6 & +9 & -8 & +24
\end{array}$$

Thus, since the remainder is 24, $x + 2$ is not a factor of $P(x)$.

12. If -4 is a zero of multiplicity three of $P(x)$, then $(x + 4)^3$ is a factor of $P(x)$.

13. If $2 - 7i$ is a zero of $P(x)$, and $P(x)$ has real coefficients, then the conjugate of $2 - 7i$, $2 + 7i$, is also a zero of $P(x)$.

14. If $3 + \sqrt{11}$ is a zero of $P(x)$, and $P(x)$ has rational coefficients, then $3 - \sqrt{11}$ is also a zero of $P(x)$.

15. Given $5x^7 + 3x^5 - 2x^4 + x^2 - 9 = 0$. Since the degree of this polynomial is 7, the fewest number of distinct solutions to the equation is 1, and the greatest number of solutions is 7. Counting multiplicities, the equation has exactly 7 solutions.

16. If -2 is a double zero of $P(x)$, then $(x + 2)^2$ is a factor of $P(x)$. Since $P(x)$ has rational coefficients, if $1 + \sqrt{2}$ is a single zero, then $1 - \sqrt{2}$ is also a single zero, and two factors of $P(x)$ are

$(x - (1 + \sqrt{2}))$ and $(x - (1 - \sqrt{2}))$. Since the coefficients are rational, they are also real, so if

$1 - i\sqrt{2}$ is a single zero then $1 + i\sqrt{2}$ is also a single zero, and two more factors of $P(x)$ are

$(x - (1 - i\sqrt{2}))$ and $(x - (1 + i\sqrt{2}))$. Then $P(x)$ is given by:

$$\begin{aligned}
P(x) &= (x+2)^2(x-(1+\sqrt{2}))(x-(1-\sqrt{2})) \\
&\quad (x-(1-i\sqrt{2}))(x-(1+i\sqrt{2})) \\
&= (x + 2)^2(x^2 - 2x - 1)(x^2 - 2x + 3) \\
&= (x^2 + 4x + 4)(x^4 - 4x^3 + 6x^2 - 4x - 3) \\
&= x^6 - 6x^4 + 4x^3 + 5x^2 - 28x - 12
\end{aligned}$$

17. Given $P(x) = 2x^6 - 4x^5 - 2x^3 + 3x^2 + x - 7 = 0$. Since $P(x)$ has degree 6, there are 6 solutions to the equation. Since $P(x)$ has three sign changes, there are either 3 or 1 positive real solutions to the equation. Consider

$$P(-x) = 2x^6 + 4x^5 + 2x^3 + 3x^2 - x - 7.$$

Since $P(-x)$ has one sign change, there is 1 negative real solution to the equation. Putting this information together we have either 1 negative real solution, 3 positive real solutions, and 2 nonreal solutions, or 1 negative real solution, 1 positive real solution, and 4 nonreal solutions. More compactly, either 1,3,2 or 1,1,4.

18. Given $8x^4 - 16x^3 - 2x^2 - 16x - 10 = 0$.

Since 2 is a common factor to all terms, we can

divide through by 2 to simplify the equation to

$$4x^4 - 8x^3 - x^2 - 8x - 5 = 0$$

Suppose we try 1.

```
 1 | 4 - 8 - 1 - 8 - 5
   |   + 4 - 4 - 5 + 13
   ---------------------
     4 - 4 - 5 - 13 + 8
```

Since the signs are not all positive, try 2.

```
 2 | 4 - 8 - 1 - 8 - 5
   |   + 8 + 0 - 2 - 20
   ---------------------
     4 + 0 - 1 - 10 - 25
```

Since the signs are not all positive, try 3.

```
 3 | 4 - 8 - 1 - 8 - 5
   |   + 12 + 12 + 33 + 75
   ------------------------
     4 + 4 + 11 + 25 + 70
```

Since the signs are all positive, 3 is the smallest positive integer upper bound for the real solutions to the equation.

Suppose we try -1.

```
-1 | 4 - 8 - 1 - 8 - 5
   |   - 4 + 12 - 11 + 19
   ----------------------
     4 - 12 + 11 - 19 + 14
```

Thus, -1 is the greatest negative integer lower bound for the real solutions to the equation.

19. Suppose that $\frac{a}{b}$ is a rational solution to the equation $4x^4 - 3x^3 + 2x^2 - x + 5 = 0$.

    (a) Then a is a factor of the constant term, 5.

    (b) Then b is a factor of the leading coefficient, 4.

20. Given $x^5 - 3x^3 + x^2 + x - 9 = 0$. Any possible rational solution to the equation, $\frac{p}{q}$, has the property that q must be a factor of the leading coefficient, 1. Since the only factors of 1 are 1 and -1, q is either 1 or -1, making $\frac{p}{q}$ equal to p or $-p$. In either case, the result is an integer.

21. Given $6x^3 - 11x^2 + 9x - 2 = 0$. Then the possibilities for p are divisors of -2: $\pm 1, \pm 2$.

The possibilities for q are divisors of 6: $\pm 1, \pm 2, \pm 3$, and ± 6. Thus, the possibilities for the rational solutions are

$$\frac{p}{q}: \pm 1, \pm 2, \pm \frac{1}{2}, \pm \frac{1}{3}, \pm \frac{1}{6}, \pm \frac{2}{3}$$

Since $P(-x) = -6x^3 - 11x^2 - 9x - 2$ has no sign changes, there are no negative real (hence rational) solutions so we do not need to try any of the negative possibilities. Suppose we try $\frac{1}{3}$.

```
 1/3 | 6 - 11 + 9 - 2
     |   + 2 - 3 + 2
     ------------------
       6 - 9 + 6 + 0
```

Thus, $\frac{1}{3}$ is one solution, and the remaining solutions can be found by solving the quadratic equation $6x^2 - 9x + 6 = 0$, which simplifies to $2x^2 - 3x + 2 = 0$. Use the quadratic formula.

$$x = \frac{-b \pm \sqrt{b^2 - 4ac}}{2a}$$

$$= \frac{-(-3) \pm \sqrt{(-3)^2 - 4(2)(2)}}{2(2)}$$

$$= \frac{3 \pm \sqrt{9 - 16}}{4}$$

$$= \frac{3 \pm \sqrt{-7}}{4}$$

$$= \frac{3 \pm i\sqrt{7}}{4}$$

Thus, the solutions to the original equation are

$$\frac{1}{3}, \frac{3 + i\sqrt{7}}{4}, \text{ and } \frac{3 - i\sqrt{7}}{4}.$$

22. The surface area of a sphere is given by

$$A = 4\pi r^2.$$

The volume of a sphere is given by

$$V = \frac{4}{3}\pi r^3.$$

Thus we must solve the equation

$$4\pi r^2 = \frac{4}{3}\pi r^3.$$

This equation simplifies to $r^3 - 3r^2 = 0$, or to $r^2(r - 3) = 0$. Then the solutions are $r = 0$ and $r = 3$. But $r = 0$ is impossible since the radius of a sphere must be positive. Thus the sphere has radius $r = 3$.

23. Given $P(x) = x^3 - 7x^2 + x - 5$.

    (a) Since the degree of $P(x)$ is 3, the maximum number of zeros of $P(x)$ is 3 so the graph has at most 3 x-intercepts.

    (b) Since the degree of $P(x)$ is 3, the maximum number of turning points for the graph is 2, 1 less than the degree.

    (c) Since $P(0) = -5$, the y-intercept is $(0,-5)$.

    (d) Since the degree of $P(x)$ is 3, and 3 is odd, and since the leading coefficient is 1, and 1 is positive, the graph must eventually go upward for large positive values of x.

    (e) Since the degree of $P(x)$ is 3, and 3 is odd, and since the leading coefficient is 1, and 1 is positive, the graph must eventually go downward for small negative values of x.

24. Given $P(x) = -x^4 + 3x^2 - x + 8$.

    (a) Since the degree of $P(x)$ is 4, the maximum number of zeros of $P(x)$ is 4 so the graph has at most 4 x-intercepts.

    (b) Since the degree of $P(x)$ is 4, the maximum number of turning points for the graph is 3, 1 less than the degree.

    (c) Since $P(0) = 8$, the y-intercept is $(0,8)$.

    (d) Since the degree of $P(x)$ is 4, and 4 is even, and since the leading coefficient is -1, and -1 is negative, the graph must eventually go downward for large positive values of x.

    (e) Since the degree of $P(x)$ is 4, and 4 is even, and since the leading coefficient is -1, and -1 is negative, the graph must eventually go downward for small negative values of x.

25. Given $y = P(x) = -x^3 - x^2 + 2x$.

    Since the degree of $P(x)$ is 3, the graph has at most two turning points. Also, since 3 is odd and $a_3 < 0$, the graph must eventually go down to the right and up to the left. The y-intercept is $(0,0)$. Since $P(x)$ has one variation in sign, there is one positive real zero. Also, since $P(-x) = x^3 - x^2 - 2x$ has one variation in sign, there is one negative real solution. In fact, $P(x) = -x(x^2 + x - 2) = -x(x + 2)(x - 1)$, so we can see that the zeros of $P(x)$ are 0, -2, and 1 giving us x-intercepts $(0,0)$, $(-2,0)$, and $(1,0)$. Using this information together with a few additional points we can obtain the graph of the function as shown below.

    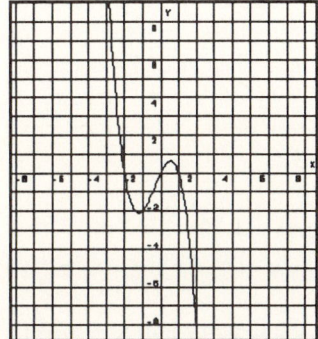

26. Given that $P(x) = x^3 + 5x^2 - 11x - 55$ has an irrational zero between 3 and 4. Try the numbers 3.0, 3.1, 3.2, 3.3, and so forth to discover that

    $$P(3.3) = -0.913$$

    and

    $$P(3.4) = 4.704.$$

    Then by the intermediate value theorem, the zero is between 3.3 and 3.4. Then try 3.31, 3.32, 3.33, 3.34, and so forth to discover that

    $$P(3.31) = -0.364809$$

    and

    $$P(3.32) = 0.186368.$$

    Then the zero is between 3.31 and 3.32. Since $P(3.32)$ is closer to 0 than $P(3.31)$, we conclude that the zero is closer to 3.32. Thus, to the nearest hundredth, the desired zero is about 3.32.

27. Given $f(x) = \dfrac{10}{x - 9}$.

    (a) Since the denominator is 0 when x is 9, the vertical asymptote is $x = 9$.

    (b) Since the degree of the numerator is less than the degree of the denominator, the x-axis, $y = 0$, is the horizontal asymptote.

(c) Since the degree of the numerator is not 1 more than the degree of the denominator, there is no oblique asymptote.

(d) Since the numerator is never 0, there are no zeros of the function, and therefore, there are no x-intercepts.

(e) Since $f(0) = -\frac{10}{9}$, the y-intercept is $\left(0, -\frac{10}{9}\right)$.

(f) If x is replaced by $-x$, y is replaced by $-y$, or both are replaced together, the equation for the function is changed. Thus, the graph is not symmetric with respect to either axis or to the origin.

28. Given $g(x) = \dfrac{6x^2}{9x^2 - 1}$.

(a) Since the denominator is 0 when x is $\frac{1}{3}$ or $-\frac{1}{3}$, the vertical asymptotes are $x = \frac{1}{3}$ and $x = -\frac{1}{3}$.

(b) Since the degree of the numerator is equal to the degree of the denominator, the horizontal asymptote is $y = \frac{2}{3}$ (this is a_n over b_m, reduced).

(c) Since the degree of the numerator is not 1 more than the degree of the denominator, there is no oblique asymptote.

(d) Since the numerator is 0, when x is 0, the zero of the function is 0, and therefore, the x-intercept is (0,0).

(e) Since $g(0) = 0$, the y-intercept is also (0,0).

(f) If x is replaced by $-x$, the equation remains unchanged. Thus, the graph is symmetric with respect to the y-axis. However, if y is replaced by $-y$, or both are replaced together, the equation for the function is changed. Thus, the graph is not symmetric with respect to the x-axis or to the origin.

29. Given $y = f(x) = \dfrac{1}{x^2 - 2x + 1}$. Since the denominator is 0 when x is 1, the vertical asymptote is $x = 1$. Since the degree of the numerator is less than the degree of the denominator the horizontal asymptote is the x-axis, $y = 0$, and there is no oblique asymptote. Since $f(x)$ has no zeros, the graph has no x-intercepts. Since $f(0) = 1$, (0,1) is the y-intercept of the graph. When x is replaced with $-x$, y is replaced with $-y$, or both are replaced at the same time, the equation is changed so the graph is not symmetric with respect to either axis or to the origin. The intervals to consider are $(-\infty, 1)$ and $(1, \infty)$, and the graph is above the x-axis on both. The graph is given below.

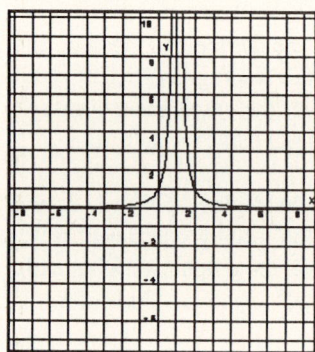

30. Solve $3x^2 + x \leq 10$.

First solve the quadratic equation.

$$3x^2 + x = 10$$
$$3x^2 + x - 10 = 0$$
$$(3x - 5)(x + 2) = 0$$
$$3x - 5 = 0 \qquad x + 2 = 0$$
$$3x = 5 \qquad x = -2$$
$$x = \frac{5}{3}$$

The solutions divide the number line into the intervals

$$(-\infty, -2), \left(-2, \frac{5}{3}\right), \left(\frac{5}{3}, \infty\right).$$

The test point 0 is a solution to the inequality, so all numbers in the interval $\left(-2, \frac{5}{3}\right)$ are part of the solution. Since the inequality is $\leq$, we include the endpoints where equality holds. Thus, the solution is

$$\left[-2, \frac{5}{3}\right]$$

CHAPTER 4 REVIEW

or using inequalities,

$$-2 \leq x \leq \tfrac{5}{3}.$$

31. Solve $x^2 - 3x + 1 \geq 0$.

First solve the quadratic equation

$$x^2 - 3x + 1 = 0$$

using the quadratic formula.

$$\begin{aligned} x &= \frac{-b \pm \sqrt{b^2 - 4ac}}{2a} \\ &= \frac{-(-3) \pm \sqrt{(-3)^2 - 4(1)(1)}}{2(1)} \\ &= \frac{3 \pm \sqrt{9-4}}{2} \\ &= \frac{3 \pm \sqrt{5}}{2} \end{aligned}$$

These solutions divide the number line into the three intervals

$$\left(-\infty, \tfrac{3-\sqrt{5}}{2}\right), \left(\tfrac{3-\sqrt{5}}{2}, \tfrac{3+\sqrt{5}}{2}\right), \left(\tfrac{3+\sqrt{5}}{2}, \infty\right).$$

The test point 0 is a solution to the inequality, and since 0 is in $\left(-\infty, \tfrac{3-\sqrt{5}}{2}\right)$, and the inequality is $\geq$, the solution to the inequality is

$$\left(-\infty, \tfrac{3-\sqrt{5}}{2}\right] \text{ or } \left[\tfrac{3+\sqrt{5}}{2}, \infty\right)$$

or using inequalities,

$$x \leq \tfrac{3-\sqrt{5}}{2} \text{ or } x \geq \tfrac{3+\sqrt{5}}{2}.$$

32. Solve $(x-3)(x^2 - 2x - 35) < 0$.

First factor the left side.

$$(x-3)(x-7)(x+5) < 0$$

Then the solutions to the related equation are 3, 7, and -5, which divide the number line into the four intervals

$$(-\infty, -5), (-5, 3), (3, 7), \text{ and } (7, \infty).$$

Test points -6, 0, 4, and 8 show that the solution includes the intervals $(-\infty, -5)$ and $(3, 7)$. Thus, the solution is

$$(-\infty, -5) \text{ or } (3, 7)$$

or using inequalities,

$$x < -5 \text{ or } 3 < x < 7.$$

33. Solve $\dfrac{x-5}{2x-3} \geq 1$.

First rewrite the inequality with 0 on one side.

$$\begin{aligned} \frac{x-5}{2x-3} - 1 &\geq 0 \\ \frac{x-5}{2x-3} - \frac{2x-3}{2x-3} &\geq 0 \\ \frac{x-5-2x+3}{2x-3} &\geq 0 \\ \frac{-x-2}{2x-3} &\geq 0 \end{aligned}$$

Then the values that make the numerator and denominator 0 are -2 and $\tfrac{3}{2}$. These divide the number line into the three intervals

$$(-\infty, -2) \left(-2, \tfrac{3}{2}\right), \left(\tfrac{3}{2}, \infty\right).$$

Test points -3, 0, and 2 show that the solution includes the interval $\left(-2, \tfrac{3}{2}\right)$. Since the inequality is $\geq$, we include the value that makes the numerator 0, -2, but, of course, do not include the value that makes the denominator 0. Thus, the solution to the inequality is

$$\left[-2, \tfrac{3}{2}\right)$$

or using inequalities,

$$-2 \leq x < \tfrac{3}{2}.$$

34. Only the graph in (b) could be the graph of a polynomial. The graph in (a) has a "sharp corner" and the graph in (c) is not a "continuous" curve.

35. false (The remainder is $P(5)$ not $P(-5)$.)

36. true

37. true

38. true

39. false   ($P(x)$ must have at least degree 5.)

40. true

41. false   ($x = r$ is a vertical asymptote.)

42. true

43. true

44. Given $y = f(x) = \dfrac{x^3 + 1}{x^2}$.   Since the denominator is 0 when x is 0, the vertical asymptote is $x = 0$. Since the degree of the numerator is 1 more than the degree of the denominator, there is an oblique asymptote $y = x$, found by dividing the denominator into the numerator. Since $x = -1$ is a zero of $f(x)$, the x-intercept of the graph is $(-1,0)$. Since $f(0)$ is undefined, there is no y-intercept of the graph. When x is replaced with $-x$, y is replaced with $-y$, or both are replaced at the same time, the equation is changed, so the graph is not symmetric with respect to either axis or to the origin. The intervals to consider are $(-\infty,-1)$, $(-1,0)$, and $(0,\infty)$. The graph is below the x-axis on $(-\infty,-1)$, and above the x-axis on the other two intervals. The graph is given below.

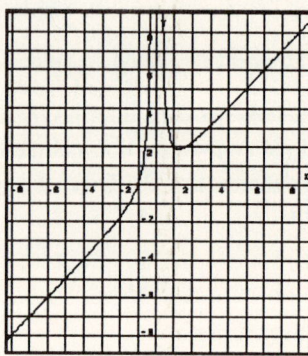

45. Given the function $y = P(x) = 5x^4 - 5$.

Since the degree of $P(x)$ is 4, the graph has at most three turning points. Also, since 4 is even and $a_4 > 0$, the graph must eventually go up to the right and up to the left. The y-intercept is $(0,-5)$. Since $P(x)$ has one variation in sign, there is one positive real zero. Also, since $P(-x) = P(x)$ has one variation of sign there will be 1 negative real zero. Since $P(x) = 5(x^2 + 1)(x - 1)(x + 1)$, the zeros of $P(x)$ are easy to determine and are 1 and

-1. Thus, the x-intercepts of the graph are $(-1,0)$, $(1,0)$. Using perhaps a few additional points we can obtain the graph shown below.

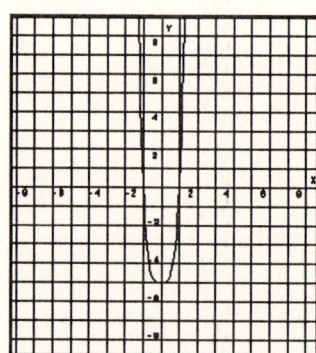

46. Given $P(x) = 3x^5 + 7x^3 + 10x + 1 = 0$. Since the degree of $P(x)$ is 5, the equation has 5 solutions. Since $P(x)$ has no variation in signs, there are 0 positive real solutions. Since $P(-x) = -3x^5 - 7x^3 - 10x + 1$ has one variation in signs, there is 1 negative real solution. Thus, there is 1 negative real solution, 0 positive real solutions, and 4 nonreal solutions, or more compactly, the answer is 1,0,4.

47. Given $x^3 - 3x^2 - 9x - 5 = 0$. Since p must be a factor of -5, the possibilities for p are ± 1 and ± 5. Since q must be a factor of 1, the possibilities for q are ± 1. Thus, the possible rational solutions are ± 1 and ± 5. Suppose we try -1.

```
-1 | 1 - 3 - 9 - 5
   |   - 1 + 4 + 5
   -------------------
     1 - 4 - 5 + 0
```

Thus, -1 is one solution and the remaining solutions can be found by solving the quadratic equation

$$x^2 - 4x - 5 = 0.$$
$$(x + 1)(x - 5) = 0$$
$$x + 1 = 0 \qquad x - 5 = 0$$
$$x = -1 \qquad x = 5$$

Thus, the solutions are -1, -1, and 5. Note that -1 is a double root.

48. Given $3x^6 + 13x^5 + 14x^4 + 6x^2 + 26x + 28 = 0$.

Suppose we try 1.

```
1 | 3 + 13 + 14 +  0 +  6 + 26 + 28
  |    +  3 + 16 + 30 + 30 + 36 + 62
  ------------------------------------
    3 + 16 + 30 + 30 + 36 + 62 + 90
```

Since the signs are all positive, 1 is the smallest positive integer upper bound for the real solutions.

Suppose we try −4.

```
-4 | 3 + 13 + 14 +   0 +   6 +  26 +    28
   |    - 12 -  4 - 40 + 160 - 664 + 2552
     ─────────────────────────────────────
     3 +  1 + 10 - 40 + 166 - 638 + 2580
```

Since the signs do not alternate, suppose we try −5.

```
-5 | 3 + 13 + 14 +   0 +   6 +   26 +    28
   |    - 15 + 10 - 120 + 600 - 3030 + 15020
     ──────────────────────────────────────
     3 -  2 + 24 - 120 + 606 - 3004 + 15048
```

Since the signs now alternate, the largest negative integer lower bound for the real solutions is −5.

49. If $P(x)$ is a polynomial and all the coefficients are positive, then for any positive number r, $P(r) > 0$ (not 0) since every term will be positive. Thus no real number r can be a zero of $P(x)$.

50. Solve $3x^2 - 2x + 1 < 0$.

Since the quadratic equation $3x^2 - 2x + 1 = 0$ has complex (nonreal) solutions (note that the discriminant is $b^2 - 4ac = (-2)^2 - 4(3)(1) = 4 - 12 = -8 < 0$), either every real number is a solution, or else there is no solution. The test point 0 shows that 0 is not a solution since $1 < 0$ is false, thus, there is no solution to the inequality.

51. Solve $3x^2 + x - 10 > 0$.

First solve the quadratic equation.

$$3x^2 + x - 10 = 0$$
$$(3x - 5)(x + 2) = 0$$
$$3x - 5 = 0 \qquad x + 2 = 0$$
$$3x = 5 \qquad\quad x = -2$$
$$x = \tfrac{5}{3}$$

The solutions divide the number line into the intervals

$$(-\infty, -2), \left(-2, \tfrac{5}{3}\right), \left(\tfrac{5}{3}, \infty\right).$$

The test point 0 is not a solution to the inequality, so all numbers in the interval $\left(-2, \tfrac{5}{3}\right)$ are not part of the solution. Thus, the solution is

$$(-\infty, -2) \text{ or } \left(\tfrac{5}{3}, \infty\right)$$

or using inequalities,

$$x < -2 \text{ or } x > \tfrac{5}{3}.$$

52. Solve $\dfrac{x-7}{x+3} \leq 0$.

The values of x that make the numerator 0 and that make the denominator 0 are −3 and 7. These values divide the number line into three intervals,

$$(-\infty, -3), (-3, 7), \text{ and } (7, \infty).$$

Test points −4, 0, and 8 show that the solution includes the points in the interval $(-3, 7)$. Since the inequality is $\leq$, we also include the value that makes the numerator 0, namely 7. (Do not include −3, the value that makes the denominator 0.) Thus, the solution is

$$(-3, 7]$$

or using inequalities,

$$-3 < x \leq 7.$$

53. The volume of a hemisphere is $\tfrac{2}{3}\pi r^3$, and the volume of a cylinder is $\pi r^2 h$, so the total volume of the nose section is the sum of these. Since the overall height of the section is 40 inches, $r + h = 40$, so that $h = 40 - r$. Substituting we obtain the following expression for the volume:

$$V(r) = \pi r^2 h + \tfrac{2}{3}\pi r^3 = \pi r^2(40 - r) + \tfrac{2}{3}\pi r^3$$
$$= 40\pi r^2 - \tfrac{1}{3}\pi r^3$$

Since the volume is to be 5019 in³, and

$$V(6) = 2163.90375$$

and

$$V(7) = 5798.332841,$$

there must be a value of r between 6 and 7 for which $V(r) = 5019$. By trying 6.1, 6.2, 6.3, and so forth, we can discover that

$$V(6.5) = 5021.704957.$$

Since this is the closest to 5019 that we obtain using values in the list above, we conclude that the radius of the sphere must be about 6.5 in.

CHAPTER 4 POLYNOMIAL AND RATIONAL FUNCTIONS

CHAPTER 4 Test

1. Divide $x^4 - 2x^3 - x^2 + 3x - 2$ by $x - 2$ using synthetic division.

   $$\underline{2\rfloor} \begin{array}{r} 1 - 2 - 1 + 3 - 2 \\ + 2 + 0 - 2 + 2 \\ \hline 1 + 0 - 1 + 1 + 0 \end{array}$$

   Thus, the quotient is $x^3 - x + 1$ and the remainder is 0.

2. Given the revenue function $R(x) = 40x$ and the cost function $C(x) = 80 + 30x$. The profit function is given by

   $$\begin{aligned} P(x) &= R(x) - C(x) \\ &= 40x - (80 + 30x) \\ &= 40x - 80 - 30x \\ &= 10x - 80. \end{aligned}$$

   Then
   $$P(8) = 10(8) - 80 = 0.$$

   Thus, 8 is a break-even point since $P(8) = 0$, that is, since 8 is a zero of $P(x)$.

3. Given $P(x) = 3x^4 - 2x^3 + x - 8$. Divide $P(x)$ by $x - 2$ using synthetic division. The remainder is $P(2)$. Do not forget to use a 0 for the missing x^2 term.

   $$\underline{2\rfloor} \begin{array}{r} 3 - 2 + 0 + 1 - 8 \\ + 6 + 8 + 16 + 34 \\ \hline 3 + 4 + 8 + 17 + 26 \end{array}$$

   Then the quotient is $3x^3 + 4x^2 + 8x + 17$, and the remainder is 26. Thus, $P(2) = 26$.

4. Given $P(x) = 2x^3 + 9x^2 + 9x - 2$. Divide $P(x)$ by $x + 2$ using synthetic division. If the remainder is 0, then $x + 2$ is a factor of $P(x)$.

   $$\underline{-2\rfloor} \begin{array}{r} 2 + 9 + 9 - 2 \\ - 4 - 10 + 2 \\ \hline 2 + 5 - 1 + 0 \end{array}$$

   Since the remainder is 0, $x + 2$ is a factor of $P(x)$.

5. If $P(x)$ is a polynomial with real coefficients and $4 + 3i$ is a zero of $P(x)$, then the conjugate of $4 + 3i$, $4 - 3i$, is also a zero of $P(x)$.

6. If 3 is a double root of $P(x) = 0$, then $(x - 3)^2$ is a factor of $P(x)$. Since $P(x)$ has rational coefficients, the coefficients are also real, and if $1 - i$ is a root, then $1 + i$ is also a root. Then $(x - (1 - i))$ and $(x - (1 + i))$ are also factors of $P(x)$. Thus,

   $$\begin{aligned} P(x) &= (x - 3)^2(x - (1 - i))(x - (1 + i)) \\ &= (x - 3)^2(x^2 - 2x + 2) \\ &= (x^2 - 6x + 9)(x^2 - 2x + 2) \\ &= x^4 - 8x^3 + 23x^2 - 30x + 18. \end{aligned}$$

7. Given $P(x) = 3x^6 + 2x^4 - 9x + 17 = 0$. Since $P(x)$ has degree 6, there are six solutions to the equation. Since $P(x)$ has 2 variations in sign, there are either 2 or 0 positive real solutions. Since $P(-x) = 3x^6 + 2x^4 + 9x + 17$ has 0 variations in sign, there are 0 negative real solutions. Putting this information together we have that there are either 0 negative real solutions, 0 positive real solutions, and 6 nonreal solutions, or 0 negative real solutions, 2 positive real solutions, and 4 nonreal solutions. More compactly: Either 0,0,6 or 0,2,4.

8. Given $8x^3 + 2x^2 - 15x = 0$. Use synthetic division to determine the bounds for the roots. Suppose we try 1.

   $$\underline{1\rfloor} \begin{array}{r} 8 + 2 - 15 + 0 \\ + 8 + 10 - 5 \\ \hline 8 + 10 - 5 - 5 \end{array}$$

   Since the signs are not all positive, try 2.

   $$\underline{2\rfloor} \begin{array}{r} 8 + 2 - 15 + 0 \\ + 16 + 36 + 42 \\ \hline 8 + 18 + 21 + 42 \end{array}$$

   Since the signs are now all positive, 2 is the smallest positive integer upper bound for the solutions.

   Suppose we try -1.

   $$\underline{-1\rfloor} \begin{array}{r} 8 + 2 - 15 + 0 \\ - 8 + 6 + 9 \\ \hline 8 - 6 - 9 + 9 \end{array}$$

   Since the signs do not alternate, try -2.

   $$\underline{-2\rfloor} \begin{array}{r} 8 + 2 - 15 + 0 \\ - 16 + 28 - 26 \\ \hline 8 - 14 + 13 - 26 \end{array}$$

   Since the signs now alternate, -2 is the largest negative integer lower bound for the solutions.

9. Given $2x^3 + 3x^2 - 3x - 2 = 0$. Since p must be a factor of $a_0 = -2$, the possibilities for p are: ± 1, ± 2. Since q must be a factor of $a_3 = 2$, the possibilities for q are: ± 1, ± 2. Then the possible rational solutions are

$$\frac{p}{q}: \pm 1, \pm 2, \pm \frac{1}{2}.$$

Suppose we try 1.

```
1 | 2 + 3 - 3 - 2
  |   + 2 + 5 + 2
  -----------------
    2 + 5 + 2 + 0
```

Thus, since the remainder is 0, one solution is 1. To find the remaining solutions solve the quadratic equation

$$2x^2 + 5x + 2 = 0$$
$$(2x + 1)(x + 2) = 0$$
$$2x + 1 = 0 \qquad x + 2 = 0$$
$$2x = -1 \qquad x = -2$$
$$x = -\frac{1}{2}$$

Thus, the solutions are 1, -2, and $-\frac{1}{2}$.

10. Let x = the width of the crate,
    $2x$ = the length of the crate,
    $x - 3$ = the height of the crate.

Since the volume of the crate is the product of the three dimensions, we must solve

$$x(2x)(x - 3) = 1400.$$

This equation can be simplified to

$$x^3 - 3x^2 - 700 = 0.$$

Since $a_3 = 1$, any possible rational solutions must be integers that are factors of 700. We try ones that seem the most likely, for example, try 10. Since

```
10 | 1 - 3 +  0 - 700
   |   + 10 + 70 + 700
   ---------------------
     1 + 7 + 70 +   0
```

we see that 10 is one solution. Since the remaining factor is $x^2 + 7x + 70$, and
$$b^2 - 4ac = 49 - 280 < 0,$$
10 is the only real solution to the equation. Thus, the width is 10 in, the length is $2(10) = 20$ in, and the height is $10 - 3 = 7$ in. That is, the dimensions of the crate are 20 in by 10 in by 7 in.

11. Given the function $y = P(x) = x^4 - 16$.

Since the degree of $P(x)$ is 4, the graph has at most three turning points. Also, since 4 is even and $a_4 > 0$, the graph must eventually go up to the right and up to the left. The y-intercept is $(0, -16)$. Since $P(x)$ has one variation in sign, there is one positive real zero. Also, since $P(-x) = P(x)$ has one variation of sign there will be one negative real zero, and the function is even making the graph symmetric with respect to the y-axis. Since $P(x) = (x^2 - 4)(x^2 + 4) = (x - 2)(x + 2)(x^2 + 4)$, the real zeros of $P(x)$ are 2 and -2 so the x-intercepts of the graph are $(2,0)$ and $(-2,0)$. Using this information and perhaps a few additional points we can obtain the graph shown below.

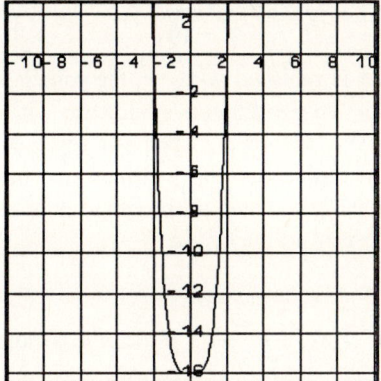

12. Given $f(x) = \dfrac{x}{x^2 + x - 2} = \dfrac{x}{(x + 2)(x - 1)}$.

The vertical asymptotes of the graph are found by setting the denominator equal to 0. Since the denominator is 0 when x is 1 and when x is -2, the vertical asymptotes are $x = 1$ and $x = -2$.

13. Given $f(x) = \dfrac{x}{x^2 - 3x - 4}$.

Since the degree of the numerator is $n = 1$, the degree of the denominator is $m = 2$, and $n < m$, the x-axis, $y = 0$, is the horizontal asymptote.

14. Given $f(x) = \dfrac{x^2 + 2x - 1}{x - 5}$.

Since the degree of the numerator is $n = 2$, the degree of the denominator is $m = 1$, and $n = m + 1$, the graph has an oblique asymptote found by dividing the numerator by the denominator. Doing this we can write

$$f(x) = \dfrac{x^2 + 2x - 1}{x - 5} = x + 7 + \dfrac{34}{x - 5}.$$

Since $\dfrac{34}{x - 5}$ goes to 0 as $|x|$ gets larger and larger, the oblique asymptote is $y = x + 7$.

15. Given $y = f(x) = \dfrac{x^2}{x^4 - 16}$.

When x is replaced with $-x$, the equation remains the same so the graph is symmetric with respect to the y-axis. However, when x is replaced with $-x$ and y is replaced with $-y$ at the same time, the equation is changed, so the graph is not symmetric with respect to the origin.

16. Given $f(x) = \dfrac{-4}{x^2 + 4}$. Since the denominator is never 0 for real values of x, there are no vertical asymptotes. Since the degree of the numerator is less than the degree of the denominator, the x-axis, $y = 0$, is the horizontal asymptote, and there is no oblique asymptote. Since there are no zeros of $f(x)$, there are no x-intercepts of the graph. Since $f(0) = -1$, $(0, -1)$ is the y-intercept of the graph. When x is replaced with $-x$, the equation remains the same so the graph is symmetric with respect to the y-axis. Notice that $f(x) < 0$ for all values of x, so the graph is entirely below the x-axis. The graph is given below.

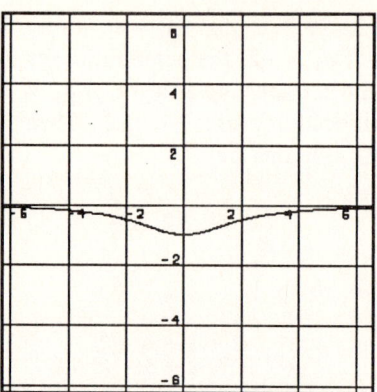

17. Solve $\dfrac{(x - 2)(x + 5)}{x + 1} \leq 0$.

The values of x that make the numerator and that make the denominator 0 are 2, -5, and -1. These values separate the number line into the four intervals

$(-\infty, -5), (-5, -1), (-1, 2),$ and $(2, \infty)$.

Test points -6, -2, 0, and 3 show that the solution includes the intervals $(-\infty, -5)$ and $(-1, 2)$. Since the inequality is $\leq$, we include the endpoints that make the numerator 0, -5 and 2, but of course do not include -1, the value that makes the denominator 0. Thus the solution is

$(-\infty, -5]$ or $(-1, 2]$.

PRACTICE EXERCISES SECTION 5.1

CHAPTER 5 EXPONENTIAL AND LOGARITHMIC FUNCTIONS

SECTION 5.1 Introduction to Logarithms

1. (a) $u^p = 5$

    The equivalent logarithmic form is $p = \log_u 5$.

    (b) $4 = \log_5 q$

    The equivalent exponential form is $5^4 = q$.

    (c) $s = \log_{-5} z$

    There is no equivalent form since the base for logarithms must be a positive real number (other than 1). That is, $\log_{-5} z$ is not defined.

2. (a) $x = \log_2 32$

    First convert to exponential form.

    $$2^x = 32$$

    Since $2^5 = 32$, it is clear that $x = 5$.

    (b) $\log_x \frac{1}{2} = -1$

    First convert to exponential form.

    $$x^{-1} = \frac{1}{2}$$
    $$x^{-1} = 2^{-1}$$

    Now it is clear that $x = 2$ since the exponents on both sides of the equation are equal forcing the bases to be equal.

3. (a) $5^{2x} = 25$

    Since $25 = 5^2$, $5^{2x} = 5^2$, so by equating exponents we have $2x = 2$ making $x = 1$.

    (b) $(x + 2)^3 = 8$

    Since $8 = 2^3$, $(x + 2)^3 = 2^3$, so by equating bases we have $x + 2 = 2$ making $x = 0$.

    (c) $\log_x 81 = 4$

    First convert to exponential form.

$x^4 = 81$

Since $81 = 3^4$, $x^4 = 3^4$ so by equating bases we have $x = 3$. Notice that $(-3)^4$ is also equal to 81, but we cannot have a negative number as a base for a logarithm. Thus, the only answer is $x = 3$.

(d) $\log_\pi x = 1$

Convert to exponential form.

$$\pi^1 = x$$

Thus, $x = \pi$.

(e) $\log_\pi 1 = x$

Convert to exponential form.

$$\pi^x = 1$$

Since $\pi^0 = 1$, $\pi^0 = \pi^x$, so equating exponents, we have $x = 0$.

CHAPTER 5 EXPONENTIAL AND LOGARITHMIC FUNCTIONS

SECTION 5.1 Introduction to Logarithms

1. Given $2^3 = 8$.

   Converting to logarithmic form we obtain:
   $$\log_2 8 = 3$$

2. Given $5^v = 9$.

   Converting to logarithmic form we obtain:
   $$\log_5 9 = v$$

3. Given $a^{-3} = c$.

   Converting to logarithmic form we obtain:
   $$\log_a c = -3$$

4. Given $u^{-v} = w$.

   Converting to logarithmic form we obtain:
   $$\log_u w = -v$$

5. Given $\log_3 9 = 2$

   Converting to exponential form we obtain:
   $$3^2 = 9$$

6. Given $\log_a b = 7$

   Converting to exponential form we obtain:
   $$a^7 = b$$

7. Given $\log_3 \frac{1}{27} = b$

   Converting to exponential form we obtain:
   $$3^b = \frac{1}{27}$$

8. Given $\log_a 6 = c$

   Converting to exponential form we obtain:
   $$a^c = 6$$

9. Solve $3^y = 243$.

   Since $243 = 3^5$, we have $3^y = 3^5$. Equating exponents, $y = 5$. Thus, the solution is 5.

10. Solve $b^{-3} = \frac{1}{125}$.

    Since $\frac{1}{125} = \frac{1}{5^3} = 5^{-3}$, $b^{-3} = 5^{-3}$, the exponents are equal so we equate the bases to obtain $b = 5$. Thus, the solution is 5.

11. Solve $25^x = 5$.

    Since $25 = 5^2$, we have $(5^2)^x = 5$, or $5^{2x} = 5^1$.

    Equating exponents, $2x = 1$, or $x = \frac{1}{2}$. Thus, the solution is $\frac{1}{2}$.

12. Solve $a^{1/2} = 7$.

    Since $a^{1/2} = \sqrt{a}$, the equation becomes $\sqrt{a} = 7$. Squaring both sides gives $a = 49$. Thus, the solution is 49.

13. Solve $c^{-1/2} = 6$.

    Since $c^{-1/2} = \frac{1}{c^{1/2}} = \frac{1}{\sqrt{c}}$, the equation becomes
    $$\frac{1}{\sqrt{c}} = 6.$$
    Squaring both sides we obtain
    $$\frac{1}{c} = 36, \text{ or } c = \frac{1}{36}.$$
    Thus, the solution is $\frac{1}{36}$.

14. Solve $\log_2 8 = y$.

    First convert to exponential form.

EXERCISES A

$$2^y = 8$$

Since $8 = 2^3$, we have $2^y = 2^3$. Equating exponents, $y = 3$. Thus, the solution is 3.

15. Solve $\log_3 x = 4$.

    First convert to exponential form.

    $$3^4 = x$$

    Then $x = 3^4 = 81$. The solution is 81.

16. Solve $\log_a \frac{1}{27} = -3$.

    First convert to exponential form.

    $$a^{-3} = \frac{1}{27}$$

    Since $\frac{1}{27} = \frac{1}{3^3} = 3^{-3}$, $a^{-3} = 3^{-3}$, so equating bases we obtain $a = 3$. The solution is 3.

17. Solve $\log_8 0.125 = w$.

    First convert to exponential form.

    $$8^w = 0.125 = \frac{1}{8} = 8^{-1}$$

    Then equating exponents, $w = -1$. The solution is -1.

18. Solve $2^{x^2} = 16$.

    Since $16 = 2^4$, we have $2^{x^2} = 2^4$. Equating exponents, $x^2 = 4$, which gives $x = \pm\sqrt{4} = \pm 2$. Thus, the solutions are 2 and -2.

19. Given that $D = 10 \log_{10} \frac{S}{S_0}$.

    (a) Substitute 100 for S and 10 for S_0.

$$D = 10 \log_{10} \frac{100}{10}$$
$$= 10 \log_{10} 10$$
$$= 10(1) \qquad \log_a a = 1$$
$$= 10$$

(b) Substitute 600 for S and 6 for S_0.

$$D = 10 \log_{10} \frac{600}{6}$$
$$= 10 \log_{10} 100$$
$$= 10 \log_{10} 10^2$$
$$= 10(2) \qquad \log_a a^x = x$$
$$= 20$$

20. Given that $A = 100(1 + 0.05)^n$, substitute 2 for n and evaluate.

$$A = 100(1 + 0.05)^2$$
$$= 100(1.05)^2$$
$$= 100(1.1025)$$
$$= 110.25$$

Thus, the value of the account is $110.25.

21. Given that $P = 14.7(2.7)^{-0.2x}$. Substitute 10 for x and evaluate.

$$P = 14.7(2.7)^{-0.2(10)}$$
$$= 14.7(2.7)^{-2}$$
$$= \frac{14.7}{(2.7)^2} = \frac{14.7}{7.29} = 2.016460905$$

Thus, to the nearest tenth, the the pressure is about 2.0 lb/in².

22. Given $P(x) = -3x^5 + x^3 + x^2 - 4x - 9$.

    (a) Since the degree of $P(x)$ is $n = 5$, there are at most 5 real zeros of $P(x)$.

    (b) Since the degree of $P(x)$ is $n = 5$, there are at most $n - 1 = 4$ turning points of the graph.

    (c) Since $P(0) = -9$, the y-intercept is $(0, -9)$.

    (d) Since the degree of $P(x)$ is 5, which is odd, and since $a_5 = -3 < 0$, the graph of $P(x)$ will eventually go downward to the right for large positive values of x.

(e) Since the degree of $P(x)$ is 5, which is odd, and since $a_5 = -3 < 0$, the graph of $P(x)$ will eventually go upward to the left for small negative values of x.

23. Given the rational function $C(x) = \dfrac{1000x}{100 - x}$, which approximates the cost of removing pollutants from the smoke of a power plant where x is the percent of pollutants to be removed. By making a table of values, noting that $x = 100$ is a vertical asymptote, noting that $y = -1$ is a horizontal asymptote, and concentrating on the interval $0 \le x < 100$, we can obtain the graph of the function given below.

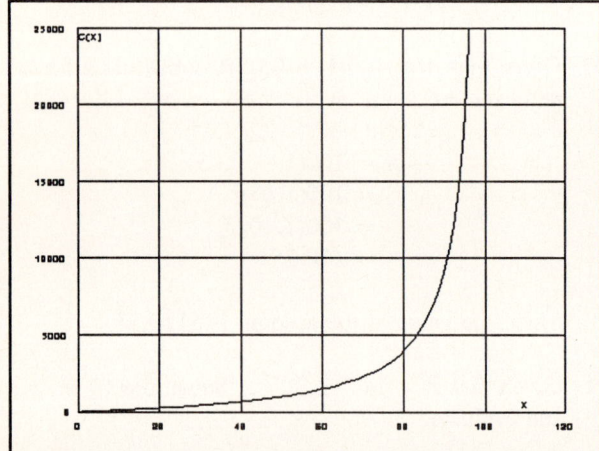

The cost of removing 95% of the pollutants is given by

$$C(95) = \dfrac{1000(95)}{100 - 95}$$
$$= \dfrac{95{,}000}{5} = 19{,}000$$

Thus, the cost of removing 95% of the pollutants is about \$19,000.

24. Graph $y = f(x) = 3x + 1$.

This is a linear function, the graph is a straight line.

When $x = 0$, $y = 1$, and when $y = 0$, $x = -\dfrac{1}{3}$.

Thus the intercepts are $(0,1)$ and $\left(-\dfrac{1}{3}, 0\right)$. The graph is given below.

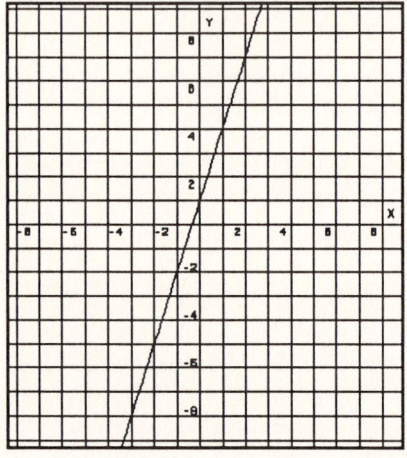

25. Graph $y = g(x) = x^2 + 2$.

This is a quadratic function, the graph is a parabola that opens up with vertex $(0,2)$. There are no x-intercepts since the equation $x^2 + 2 = 0$ has no real solutions. Two additional points on the graph are $(1,3)$ and $(-1,3)$. The graph is given below.

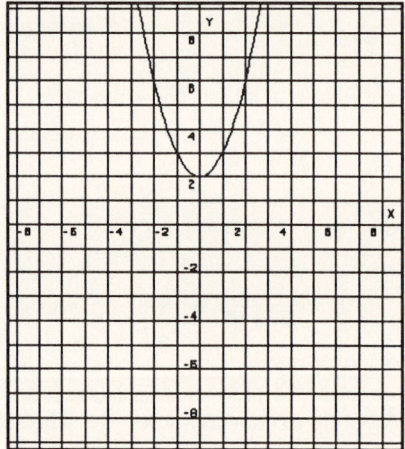

PRACTICE EXERCISES — SECTION 5.2   215

CHAPTER 5 EXPONENTIAL AND LOGARITHMIC FUNCTIONS

SECTION 5.2 Exponential and Logarithmic Functions

1. Graph the function $y = 10^x$.

   Make a table of values and plot the points. Connect them with a smooth curve to obtain the graph given below.

   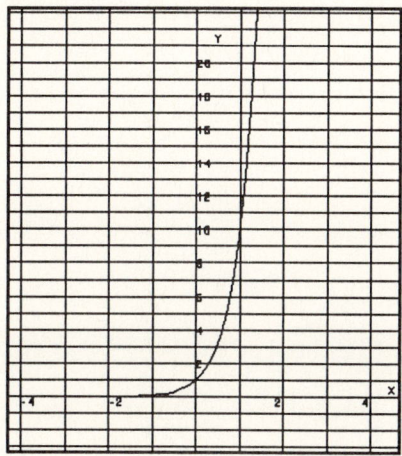

2. (a) $(1.09)^{-40}$

   Enter 1.09 and press the y^x button. This places 1.09 in for y. Then enter 40 and use the change sign button to make the value -40 for x. Press the $=$ button to obtain the result.

   $$(1.09)^{-40} = 0.031837582$$

   Thus, to four decimal places,

   $$(1.09)^{-40} = 0.0318.$$

   (b) $(\sqrt{2})^\pi$

   Enter 2, press the square root button, then press y^x to enter the value as y. Then enter π and press the $=$ button. The result is

   $$(\sqrt{2})^\pi = 2.970686424$$

   Thus, to four decimal places, $(\sqrt{2})^\pi = 2.9707$.

3. Graph $f(x) = 2^{x^2} - 4$.

   Make a table of values, plot the points and join them with a smooth curve to obtain the graph.

   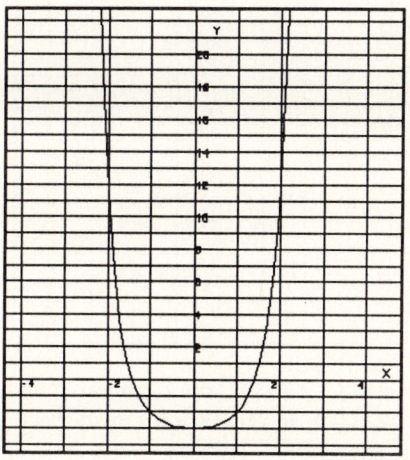

4. Graph $y = 3^x$ and $y = \log_3 x$ in the same coordinate system. Make a table of values for $y = 3^x$. Convert $y = \log_3 x$ to exponential form, $3^y = x$, and then make a table of values for it. The graphs of the two functions are given below.

   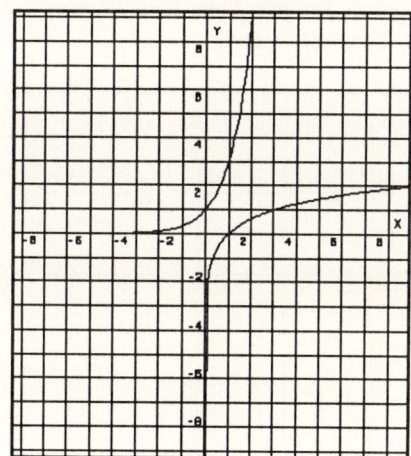

5. (a) $\log_\pi \pi^5 = 5$ since $\log_a a^x = x$ for any x.

   (b) $\log_{100} 10^4 = \log_{100} 100^2 = 2$.

   (c) $4^{\log_4 (5x+2)} = 5x + 2$ since $a^{\log_a x} = x$ for any x.

6. Graph $f(x) = 3 + \log_2 x^2$. The graph can be obtained by making a table of values. Alternatively, since the graph of the function $f(x) = \log_2 x^2$ is given in Figure 5.6, the graph of $f(x) = 3 + \log_2 x^2$ can be obtained by sliding the graph up 3 units. The graph is shown below.

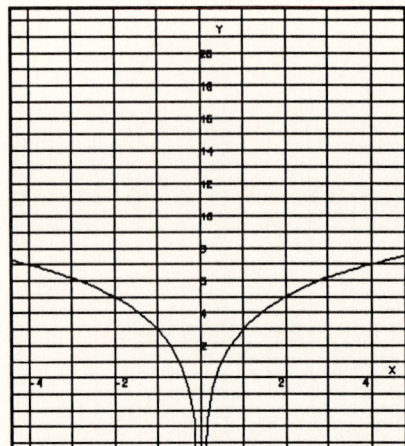

7. The value of the rare coin originally purchased for $100, and whose value triples every year, is given by the exponential function

$$V(t) = (100)3^t.$$

Graph this function for $t \geq 0$. Substituting values for t, we obtain the ordered pairs (0,100), (1,300), (2,900), (3,2700), (4,8100) and (5, 24,300). Plot these points and join them with a smooth curve to obtain the graph below.

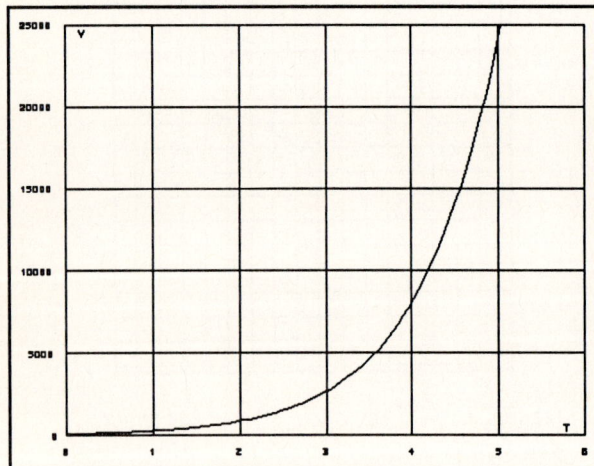

After 5 years (when $t = 5$), the value of the coin is $24,300.

EXERCISES A   SECTION 5.2   217

CHAPTER 5 EXPONENTIAL AND LOGARITHMIC FUNCTIONS

SECTION 5.2 Exponential and Logarithmic Functions

1. Graph $f(x) = 5^x$.

   Make a table of values, plot the points, and join them with a smooth curve to obtain the graph shown below.

   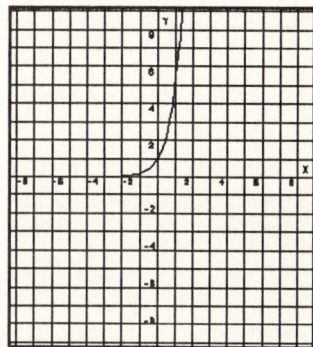

2. Graph $f(x) = \left(\frac{1}{5}\right)^x$.

   Make a table of values, plot the points, and join them with a smooth curve to obtain the graph shown below.

   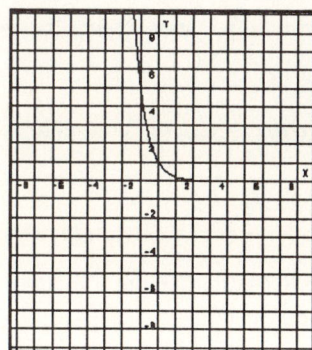

3. Graph $f(x) = -5^x$.

   Make a table of values (be careful of the minus sign, all the values are negative!), plot the points, and join them with a smooth curve to obtain the graph shown below.

   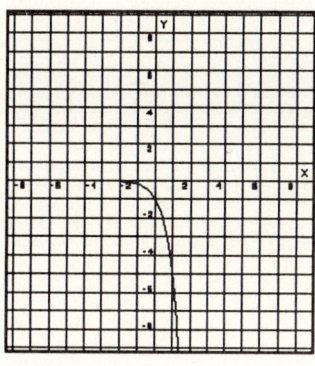

4. Graph $f(x) = 2 + 5^x$.

   Note that the graph can be obtained from the graph in Exercise 1 by sliding that graph up 2 units.

   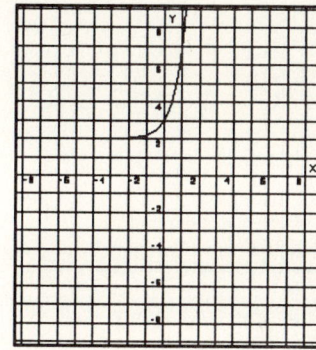

5. Graph $f(x) = 2^{x^2+1}$.

   Make a table of values, plot the points, and join them with a smooth curve to obtain the graph shown below.

   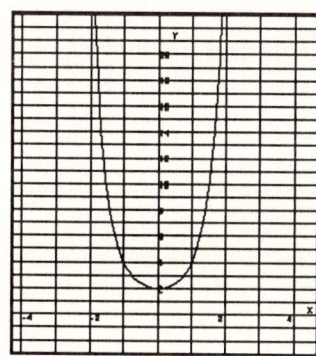

6. Graph $f(x) = 2^x - 2^{-x}$.

   Make a table of values, plot the points, and join them with a smooth curve to obtain the graph below.

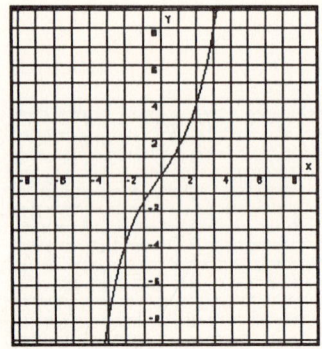

7. Graph $f(x) = \log_8 x$.

Make a table of values, plot the points, and join them with a smooth curve to obtain the graph shown below.

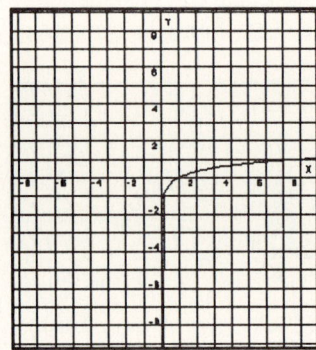

8. Graph $f(x) = \log_8(-x)$, for $x < 0$.

Make a table of values, plot the points, and join them with a smooth curve to obtain the graph shown below.

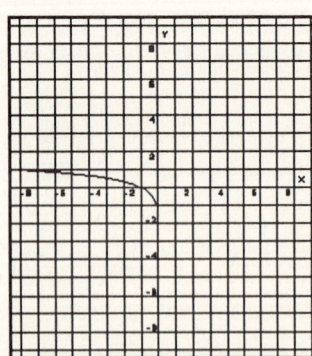

9. Graph $f(x) = \log_2 |x|$, for $x \neq 0$.

Make a table of values, plot the points, and join them with a smooth curve to obtain the graph shown below.

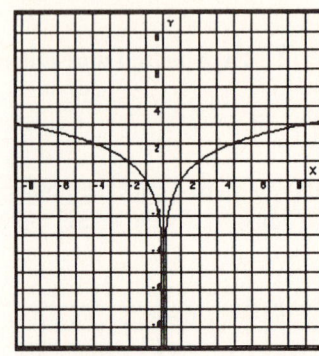

10. $\log_5 5^3 = 3$ since $\log_a a^x = x$ for any x.

11. $\log_{10} 0.001 = \log_{10} 10^{-3} = -3$ since $\log_a a^x = x$ for any x.

12. $e^{\log_e(x+7)} = x + 7$ since $a^{\log_a u} = u$ for any u.

13. $5^{\sqrt{3}} = 16.2425$ Use the y^x button with y equal to 5 and $x = \sqrt{3}$.

14. $-\pi^{\sqrt{3}} = -7.2625$ Use the y^x button with y equal to π and $x = \sqrt{3}$, then negate the result.

15. $6^{-\sqrt{2}} = 0.0793$ Use the y^x button with y equal to 6 and $x = -\sqrt{2}$.

16. $(1.02)^{60} = 3.2810$ Use the y^x button with y equal to 1.02 and $x = 60$.

17. The graph of every exponential function, $f(x) = a^x$, passes through the point $(0,1)$ since $a^0 = 1$ for every base a.

18. The domain of every exponential function, $f(x) = a^x$, is the set of all real numbers since a^x is defined for every real number x.

19. The domain of every logarithmic function, $f(x) = \log_a x$, is the set of positive real numbers since logarithms are only defined for positive numbers.

EXERCISES A   SECTION 5.2   219

20. If the logarithmic function $y = \log_a x$ passes through the point $(3,1)$, then $1 = \log_a 3$ and changing to exponential form, $a^1 = 3$, which means that the base a is 3.

21. The number 1 is omitted as a base for logarithms since if $a = 1$, then $\log_a x = \log_1 x = y$ is equivalent to $1^y = x$, or $x = 1$, for all values of y. As a result, this would not define a function.

22. Given $y = f(x) = 2^{x+1} + 1$.

    To find the inverse, interchange x and y and solve for y. Notice that $y > 1$.

    $$x = 2^{y+1} + 1$$
    $$x - 1 = 2^{y+1}$$
    $$\log_2(x - 1) = \log_2 2^{y+1}$$
    $$\log_2(x - 1) = y + 1$$
    $$-1 + \log_2(x - 1) = y$$

    Thus, $f^{-1}(x) = -1 + \log_2(x - 1)$, for all $x > 1$.

23. Given $y = f(x) = 2\log_3(x - 4)$.

    To find the inverse, interchange x and y and solve for y. Notice that y can be any real number, but $x > 4$.

    $$x = 2\log_3(y - 4)$$
    $$\frac{x}{2} = \log_3(y - 4)$$
    $$3^{\frac{x}{2}} = 3^{\log_3(y-4)}$$
    $$3^{\frac{x}{2}} = y - 4$$
    $$3^{\frac{x}{2}} + 4 = y$$

    Thus, $f^{-1}(x) = 3^{x/2} + 4$, for all real numbers x.

24. The formula $N = (5000)2^t$ gives the number of bacteria N in terms of time t, in hours.

    (a) At the start of the experiment, $t = 0$. Thus,
    $$N = (5000)2^0 = (5000)(1) = 5000.$$

    (b) After 5 hours, the number of bacteria is
    $$N = (5000)2^5 = (5000)(32) = 160,000.$$

    (c) After 3.25 hours, the number of bacteria is
    $$N = (5000)2^{3.25}$$
    $$= 47,568.2846 \quad \text{Use } y^x \text{ button}$$
    $$\approx 47,568.$$

25. The amount of a radioactive isotope that remains after t years when 50 grams of the isotope were present initially is given by
    $$A = 50\left(\frac{1}{2}\right)^{t/4}.$$

    (a) The number of grams remaining after 2 years (when $t = 2$) is given by
    $$A = 50\left(\frac{1}{2}\right)^{2/4}$$
    $$= 50(0.5)^{1/2}$$
    $$= 50(0.5)^{0.5}$$
    $$= 35.35533906 \quad \text{Use } y^x \text{ button}$$

    Thus, about 35.36 grams remain.

    (b) The number of grams remaining after 100 years (when $t = 100$) is given by
    $$A = 50\left(\frac{1}{2}\right)^{100/4}$$
    $$= 50(0.5)^{25}$$
    $$= 0.00000149 \quad \text{Use } y^x \text{ button}$$

    Thus, about 0.0000015 grams remain.

    (c) The number of grams remaining after 5.5 years (when $t = 5.5$) is given by
    $$A = 50\left(\frac{1}{2}\right)^{5.5/4}$$
    $$= 50(0.5)^{1.375}$$
    $$= 19.27763532 \quad \text{Use } y^x \text{ button}$$

    Thus, about 19.28 grams remain.

26. Solve $\log_a 32 = 5$.

    First convert to exponential form.
    $$a^5 = 32 = 2^5$$

    Since the exponents are equal, equate the bases to obtain $a = 2$. The solution is 2.

27. Solve $\log_{10} x = -2$.

First convert to exponential form.

$$10^{-2} = x$$

Then $x = 10^{-2} = \dfrac{1}{10^2} = \dfrac{1}{100} = 0.01$.

28. Solve $10^{\log_{10}(2x+5)} = 7$.

Use the fact that $a^{\log_a u} = u$ for any real number u to obtain the equation

$$\begin{aligned} 2x + 5 &= 7 \\ 2x &= 2 \\ x &= 1 \end{aligned}$$

Thus, the solution is 1.

29. true

30. false (actually $(a^m)^n = a^{mn}$)

31. true

PRACTICE EXERCISES   SECTION 5.3   221

CHAPTER 5 EXPONENTIAL AND LOGARITHMIC FUNCTIONS

SECTION 5.3 Properties of Logarithms

1. Solve $\log_5(2x + 3) = \log_5(x + 7)$.

   Since the bases are both 5, we have

   $$2x + 3 = x + 7$$
   $$2x - x = 7 - 3$$
   $$x = 4$$

   Thus, the solution is 4.

2. (a) $\log_3 3u = \log_3 3 + \log_3 u = 1 + \log_3 u$

   (b) $\log_2(x + y) + \log_2(x - y)$

   $= \log_2(x + y)(x - y)$

   $= \log_2(x^2 - y^2)$

3. (a) $\log_4 \frac{1}{4} = \log_4 1 - \log_4 4 = 0 - 1 = -1$

   (b) $\log_5 15 - \log_5 3 = \log_5 \frac{15}{3} = \log_5 5 = 1$

4. (a) $\log_3 9^2 = 2\log_3 9 = 2\log_3 3^2 = 2(2) = 4$

   (b) $\log_3 \sqrt{27} = \log_3 27^{1/2}$

   $= \log_3 (3^3)^{1/2}$

   $= \log_3 3^{3/2}$

   $= \frac{3}{2}$

5. (a) $\log_a \frac{ax}{y} = \log_a ax - \log_a y$

   $= \log_a a + \log_a x - \log_a y$

   $= 1 + \log_a x - \log_a y$

   (b) $\log_a \frac{x\sqrt{y}}{z^3} = \log_a x\sqrt{y} - \log_a z^3$

   $= \log_a x + \log_a \sqrt{y} - 3\log_a z$

   $= \log_a x + \log_a y^{1/2} - 3\log_a z$

   $= \log_a x + \frac{1}{2}\log_a y - 3\log_a z$

6. $3\log_a u - \log_a v - \frac{1}{2}\log_a w$

   $= \log_a u^3 - \log_a v - \log_a w^{1/2}$

   $= \log_a u^3 - (\log_a v + \log_a w^{1/2})$

   $= \log_a u^3 - \log_a vw^{1/2}$

   $= \log_a \frac{u^3}{vw^{1/2}} = \log_a \frac{u^3}{v\sqrt{w}}$

7. (a) $\log_a 6 = \log_a 2 \cdot 3 = \log_a 2 + \log_a 3$

   $= (0.3010) + (0.4771) = 0.7781$

   (b) $\log_a \frac{\sqrt{3}}{2a} = \log_a \sqrt{3} - \log_a 2a$

   $= \frac{1}{2}\log_a 3 - \log_a 2 - \log_a a$

   $= \frac{1}{2}(0.4771) - (0.3010) - 1$

   $= -1.06245 \approx -1.0625$

8. (a) $\log_{27} 9 = \frac{\log_3 9}{\log_3 27} = \frac{\log_3 3^2}{\log_3 3^3} = \frac{2}{3}$

   (b) $\log_{16} 4 = \frac{\log_4 4}{\log_4 16} = \frac{1}{\log_4 4^2} = \frac{1}{2}$

CHAPTER 5 EXPONENTIAL AND LOGARITHMIC FUNCTIONS

SECTION 5.3 Properties of Logarithms

1. Solve $\log_2(3x + 1) = \log_2(5x - 7)$.

   Since the bases are equal, solve the following:

   $$3x + 1 = 5x - 7$$
   $$-2x = -8$$
   $$x = 4$$

   Since 4 does check (note that 4 does not force us to try and evaluate the logarithm of a negative number), the solution is 4.

2. Solve $\log_5 x^2 = \log_5(3x + 4)$.

   Since both bases are 5, solve the following:

   $$x^2 = 3x + 4$$
   $$x^2 - 3x - 4 = 0$$
   $$(x - 4)(x + 1) = 0$$
   $$x - 4 = 0 \qquad x + 1 = 0$$
   $$x = 4 \qquad x = -1$$

   Neither 4 nor -1 force us to try and evaluate the logarithm of a negative number. Since both check, the solutions are 4 and -1.

3. Solve $\frac{1}{3}\log_2 x^2 = \log_8 2x$.

   First use the power rule to write the left side as

   $$\log_2 (x^2)^{1/3}.$$

   Then use the base conversion formula to write the right side using base 2.

   $$\log_8 2x = \frac{\log_2 2x}{\log_2 8} = \frac{\log_2 2x}{3}$$
   $$= \frac{1}{3}\log_2 2x = \log_2 (2x)^{1/3}$$

   Then the equation becomes

   $$\log_2 (x^2)^{1/3} = \log_2 (2x)^{1/3}.$$

   Since the bases are the same, solve the following:

   $$(x^2)^{1/3} = (2x)^{1/3}$$
   $$x^2 = 2x \qquad \text{Cube both sides}$$
   $$x^2 - 2x = 0$$
   $$x(x - 2) = 0$$
   $$x = 0 \qquad x - 2 = 0$$
   $$\qquad\qquad x = 2$$

   But 0 will not check in the original equation since the logarithm of 0 is not defined. However, 2 does check. Thus, the solution is 2.

4. $\log_{10} 500 = \log_{10} 5 \cdot 10^2$
   $= \log_{10} 5 + \log_{10} 10^2$
   $= 0.6990 + 2$
   $= 2.6990$

5. $\log_{10} 27 = \log_{10} 3^3$
   $= 3 \log_{10} 3$
   $= 3(0.4771)$
   $= 1.4313$

6. $\log_{10} \frac{2}{5} = \log_{10} 2 - \log_{10} 5$
   $= (0.3010) - (0.6990)$
   $= -0.3980$

7. $\log_{10} \frac{1}{9} = \log_{10} 1 - \log_{10} 9$
   $= 0 - \log_{10} 3^2$
   $= -2 \log_{10} 3$
   $= -2(0.4771)$
   $= -0.9542$

8. $\log_{10} \sqrt{6} = \log_{10} 6^{1/2}$
   $= \frac{1}{2} \log_{10} 6$
   $= \frac{1}{2} \log_{10} 2 \cdot 3$
   $= \frac{1}{2} (\log_{10} 2 + \log_{10} 3)$
   $= \frac{1}{2} (0.3010 + 0.4771)$
   $= \frac{1}{2} (0.7781)$
   $= 0.3891$

9. $\log_{10} \sqrt[7]{3} = \log_{10} 3^{1/7}$
   $= \frac{1}{7} \log_{10} 3$
   $= \frac{1}{7} (0.4771)$
   $= 0.0682$

EXERCISES A

10. $\log_5 3 = \dfrac{\log_{10} 3}{\log_{10} 5}$
$= \dfrac{0.4771}{0.6990}$
$= 0.6825$

11. $\dfrac{\log_{10} 2^7}{\log_{10} 5} = \dfrac{7 \log_{10} 2}{\log_{10} 5}$
$= \dfrac{7(0.3010)}{0.6990}$
$= 3.0143$

12. $\log_{10} 1.8 = \log_{10} \dfrac{9}{5}$
$= \log_{10} 9 - \log_{10} 5$
$= \log_{10} 3^2 - \log_{10} 5$
$= 2 \log_{10} 3 - \log_{10} 5$
$= 2(0.4771) - (0.6990)$
$= 0.2552$

13. $\log_a xyz = \log_a x + \log_a y + \log_a z$

14. $\log_a \dfrac{xz^2}{y} = \log_a xz^2 - \log_a y$
$= \log_a x + \log_a z^2 - \log_a y$
$= \log_a x + 2\log_a z - \log_a y$

15. $\log_a \dfrac{y\sqrt{x}}{z^3} = \log_a y\sqrt{x} - \log_a z^3$
$= \log_a y + \log_a x^{1/2} - \log_a z^3$
$= \log_a y + \dfrac{1}{2}\log_a x - 3\log_a z$

16. $\log_a z(x+1)^3 = \log_a z + \log_a (x+1)^3$
$= \log_a z + 3\log_a (x+1)$

17. $\log_a x^2 \sqrt{\dfrac{y}{z}} = \log_a x^2 + \log_a \left(\dfrac{y}{z}\right)^{1/2}$
$= 2\log_a x + \dfrac{1}{2}\log_a \dfrac{y}{z}$
$= 2\log_a x + \dfrac{1}{2}(\log_a y - \log_a z)$
$= 2\log_a x + \dfrac{1}{2}\log_a y - \dfrac{1}{2}\log_a z$

18. $\log_a \sqrt{x\sqrt{y}} = \log_a (x\sqrt{y})^{1/2}$
$= \dfrac{1}{2} \log_a xy^{1/2}$
$= \dfrac{1}{2} \log_a x + \dfrac{1}{2} \log_a y^{1/2}$
$= \dfrac{1}{2} \log_a x + \dfrac{1}{4} \log_a y$

19. $\log_a x + \dfrac{1}{2}\log_a y = \log_a x + \log_a y^{1/2}$
$= \log_a x + \log_a \sqrt{y}$
$= \log_a x\sqrt{y}$

20. $3\log_a x - 2\log_a xy = \log_a x^3 - \log_a (xy)^2$
$= \log_a \dfrac{x^3}{(xy)^2}$
$= \log_a \dfrac{x^3}{x^2 y^2}$
$= \log_a \dfrac{x}{y^2}$

21. $\dfrac{1}{2}\log_a x - 3\log_a yz + 6\log_a z$
$= \log_a x^{1/2} - \log_a (yz)^3 + \log_a z^6$
$= \log_a \dfrac{x^{1/2} z^6}{(yz)^3}$
$= \log_a \dfrac{\sqrt{x}\, z^6}{y^3 z^3}$
$= \log_a \dfrac{z^3 \sqrt{x}}{y^3}$

22. $x\log_a y - 3\log_a z = \log_a y^x - \log_a z^3$
$= \log_a \dfrac{y^x}{z^3}$

23. $-\log_a (x-1) + \log_a (x^2 - 1) = \log_a \dfrac{(x^2-1)}{(x-1)}$
$= \log_a \dfrac{(x-1)(x+1)}{(x-1)}$
$= \log_a (x+1)$

24. $\dfrac{1}{2}\left[\log_a x - \log_a y - \dfrac{1}{2}\log_a z\right]$
$= \dfrac{1}{2}\log_a x - \dfrac{1}{2}\log_a y - \dfrac{1}{4}\log_a z$
$= \log_a x^{1/2} - \log_a y^{1/2} - \log_a z^{1/4}$
$= \log_a \dfrac{x^{1/2}}{y^{1/2} z^{1/4}}$
$= \log_a \sqrt{\dfrac{x}{y\sqrt{z}}}$

25. true

26. false $\quad \left(\log_a \dfrac{u}{v} = \log_a u - \log_a v \neq \dfrac{\log_a u}{\log_a v}\right)$

27. true

28. true

29. false $\left(\log_a \sqrt{u} = \frac{1}{2}\log_a u \neq (\log_a u)^{1/2}\right)$

30. true

31. $\log_{32} 8 = \dfrac{\log_2 8}{\log_2 32} = \dfrac{\log_2 2^3}{\log_2 2^5} = \dfrac{3}{5}$

32. $\log_{27} 3 = \dfrac{\log_3 3}{\log_3 27} = \dfrac{\log_3 3}{\log_3 3^3} = \dfrac{1}{3}$

33. $\log_9 243 = \dfrac{\log_3 243}{\log_3 9} = \dfrac{\log_3 3^5}{\log_3 3^2} = \dfrac{5}{2}$

34. Given that $P = \log_2(m + n)$, and let $m = 8$ and $n = 8$.

$P = \log_2(8 + 8) = \log_2 16 = \log_2 2^4 = 4,$

but

$\log_2 8 + \log_2 8 = \log_2 2^3 + \log_2 2^3 = 3 + 3 = 6.$

Thus, $P = \log_2(m + n) \neq \log_2 m + \log_2 n.$

35. Given that $R = \log_3 mn$, and $m = 1$ and $n = 9$.

$R = \log_3(1 \cdot 9) = \log_3 9 = \log_3 3^2 = 2,$

but

$(\log_3 m)(\log_3 n) = (\log_3 1)(\log_3 9) = (0)(2) = 0.$

Thus, $R = \log_3 mn \neq (\log_3 m)(\log_3 n).$

36. Graph $f(x) = 2^{x^2+2}$.

Make a table of values, plot the points and connect them with a smooth curve to obtain the graph given below.

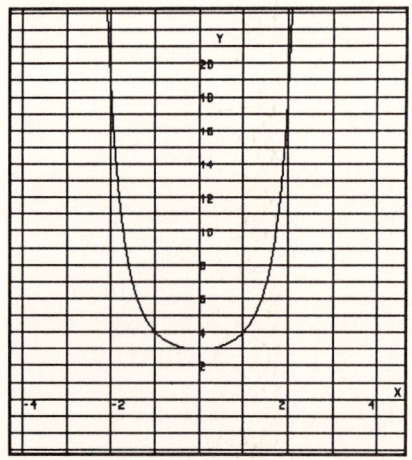

37. $\log_\pi \pi^{3x+2} = 3x + 2$ since $\log_a a^u = u$ for any real number u.

38. If $m = \log_{10} 35$, then

$$10^m = 10^{\log_{10} 35} = 35$$

since $a^{\log_a u} = u$ for any real number u.

39. When 64 milligrams of a drug are absorbed in the bloodstream, the number of milligrams D remaining after t hours is given by the function

$$D = (64)2^{-t/8}.$$

The graph of this function is given below. It was obtained by finding several points and connecting them to form a smooth curve.

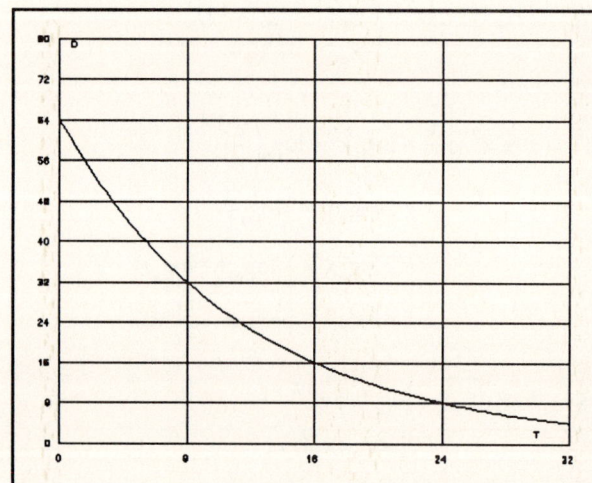

EXERCISES A

(a) The amount of the drug remaining after 0 hours (when $t = 0$) is equal to

$$D = (64)2^{-0/8} = (64)2^0 = (64)(1) = 64 mg.$$

(b) The amount of the drug remaining after 8 hours (when $t = 8$) is equal to

$$D = (64)2^{-8/8} = (64)2^{-1} = (64)\left(\frac{1}{2}\right) = 32 mg.$$

(c) The amount of the drug remaining after 16 hours (when $t = 16$) is equal to

$$D = (64)2^{-16/8} = (64)2^{-2} = (64)\left(\frac{1}{4}\right) = 16 mg.$$

(d) The amount of the drug remaining after 24 hours (when $t = 24$) is equal to

$$D = (64)2^{-24/8} = (64)2^{-3} = (64)\left(\frac{1}{8}\right) = 8 mg.$$

CHAPTER 5 EXPONENTIAL AND LOGARITHMIC FUNCTIONS

SECTION 5.4 Common and Natural Logarithms

1. (a) $\log 9254 = 3.9663295$

    Enter 9254 and press the LOG button.

    (b) $\log 0.00328 = -2.4841262$

    (c) $\log(-5.5)$ is not defined since logarithms of negative numbers are not defined. An error message shows on the calculator if you try to find this logarithm.

    (d) $\log 1{,}298{,}000{,}000{,}000 = \log 1.298 \times 10^{12}$

    $\qquad\qquad\qquad\qquad\quad = 12.1132747$

    For very large or very small numbers, scientific notation must be used for calculator entry.

2. (a) $\log_3 26.5 = \dfrac{\log 26.5}{\log 3} = 2.9830$

    Be sure to divide the two logarithms, do not subtract. The answer is given correct to four decimal places.

    (b) $\log_{15} 0.00096 = \dfrac{\log 0.00096}{\log 15} = -2.5659$

3. (a) $\log x = 2.3597$

    To find x, enter 2.3597 and press INV and LOG to find the antilog. We obtain $x = 228.9285724$, which is 228, correct to three significant digits.

    (b) $\log x = -0.8665$

    Then $x = 0.135987816$, which is 0.136, correct to three significant digits.

4. (a) $\ln 2395 = 7.7811$

    Enter 2395 and press the ln$x$ button. The answer is given correct to four decimal places.

    (b) $\ln 0.000000554 = \ln 5.54 \times 10^{-7} = -14.4061$

5. (a) $\ln x = 3.8441$ Then $x = 46.71662047$.

    Enter 3.8441, press the INV and ln$x$ buttons, then $x = 46.7$, correct to three significant digits.

    (b) $\ln x = -0.5459$

    Then $x = 0.579$, correct to three significant digits.

6. Graph $g(x) = -2\ln x$. The ordered pairs that complete the table are: $(0.1, 4.6)$, $(0.5, 1.4)$, $(0.8, 0.4)$, $(1, 0)$, $(2, -1.4)$, $(3, -2.2)$, $(4, -2.8)$, $(5, -3.2)$, and $(6, -3.6)$. The graph is given below. Notice that the graph could also be obtained by reflecting the graph of $f(x) = e^{-x/2}$, given in Figure 5.9, in the line $y = x$, since the two functions are inverses.

    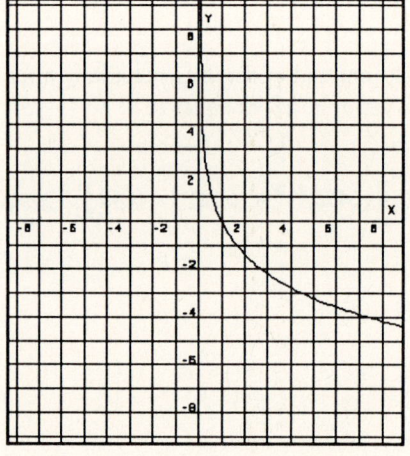

EXERCISES A SECTION 5.4 227

CHAPTER 5 EXPONENTIAL AND LOGARITHMIC FUNCTIONS

SECTION 5.4 Common and Natural Logarithms

1. $\log 625 = 2.7959$

   Enter 625 then press the LOG button.

2. $\log 0.0059 = -2.2291$

3. $\log 2.00046 = 0.3011$

4. $\log 0.0000669 = -4.1746$

5. $\log (5.12 \times 10^{11}) = 11.7093$

6. $\log 0.00000000741 = -8.1302$

   Use scientific notation.

7. $\log_4 15.6 = \dfrac{\log 15.6}{\log 4} = 1.9817$

   Be sure to divide the two logarithms, do not subtract.

8. $\log_{3.5} 468 = \dfrac{\log 468}{\log 3.5} = 4.9079$

9. $\log_{\sqrt{2}} 348.2 = \dfrac{\log 348.2}{\log \sqrt{2}} = 16.8875$

10. $\log x = 1.3565$

    Then $x = 22.7$, correct to three significant digits. Enter 1.3565 then press INV followed by LOG buttons.

11. $\log x = -3.2117$

    Then $x = 0.000614$, correct to three significant digits.

12. $\log x = 14.6531$

    Then $x = 4.50 \times 10^{14}$, correct to three significant digits. Notice that the calculator gives the antilog in scientific notation.

13. $\log x = -0.0035$

    Then $x = 0.992$, correct to three significant digits.

14. $\log x = -35.2264$

    Then $x = 5.94 \times 10^{-36}$, correct to three significant digits.

15. $\log x = 51.3225$

    Then $x = 2.10 \times 10^{51}$, correct to three significant digits.

16. $\ln 846 = 6.7405$

    Enter 846 and press the ln$x$ button.

17. $\ln 2.003 = 0.6946$

18. $\ln 0.000525 = -7.5521$

19. $\ln 0.0000387 = -10.1597$

20. $\ln (2.66 \times 10^{15} = 35.5171$

21. $\ln \pi = 1.1447$

    Use the π button then press ln$x$.

22. $\ln x = 2.4308$

    Then $x = 11.4$, correct to three significant digits. Enter 2.4308 then press INV and ln$x$ buttons.

23. $\ln x = -4.1553$

    Then $x = 0.0157$, correct to three significant digits.

24. $\ln x = 21.7439$

    Then $x = 2.77 \times 10^9$, correct to three significant digits.

25. $\ln x = -0.0070$

    Then $x = 0.993$, correct to three significant digits.

26. $\ln x = -41.4239$

    Then $x = 1.02 \times 10^{-18}$, correct to three significant digits.

27. $\ln x = 17.3591$

Then $x = 34{,}600{,}000$, correct to three significant digits.

28. Graph $y = e^{2x}$.

Make a table of values. To find the value of y when x is 2, for example, enter $2(2) = 4$, and press INV then $\ln x$ on your calculator. The result is 54.59815003, or about 54.6. The graph is given below.

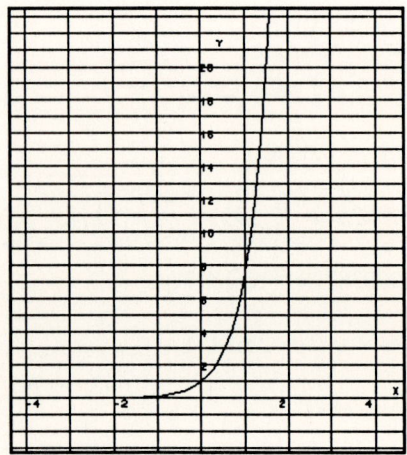

29. Graph $y = e^{-x}$.

Make a table of values. To find the value of y when x is 2, for example, enter $-(2) = -2$, and press INV then $\ln x$ on your calculator. The result is 0.1353353, or about 0.1. The graph is given below.

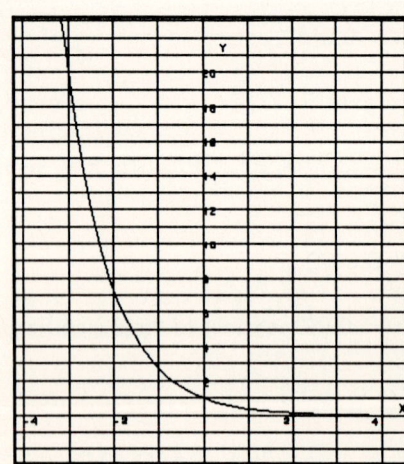

30. Graph $y = \ln x^2$.

Make a table of values. To find the value of y when x is 2, for example, enter $(2)^2 = 4$, and press $\ln x$ on your calculator. The result is 1.386294361, or about 1.4. The graph is given below.

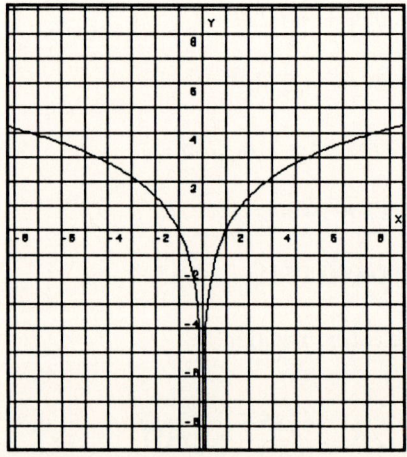

31. The demand equation for x units of a product relative to the price per unit P is given by

$$P = 100 - 0.2e^{0.005x}.$$

(a) To find the price when the demand is 100 units, substitute 100 for x and evaluate.

$$\begin{aligned}P &= 100 - 0.2e^{(0.005)(100)} \\ &= 100 - 0.2e^{0.5} \\ &= 99.67025575\end{aligned}$$

Thus the price is about \$99.67.

(b) To find the price when the demand is 1000 units, substitute 1000 for x and evaluate.

$$\begin{aligned}P &= 100 - 0.2e^{(0.005)(1000)} \\ &= 100 - 0.2e^{5} \\ &= 70.31736818\end{aligned}$$

Thus the price is about \$70.32.

32. The atmospheric pressure P, in pounds per square inch, at an altitude x, in miles above sea level, is given by

$$P = 14.7e^{-0.2x}.$$

(a) To find the pressure at sea level, substitute 0 for x and evaluate.

EXERCISES A

$P = 14.7e^{(-0.2)(0)}$
$ = 14.7e^0$
$ = 14.7(1)$
$ = 14.7$

Thus, the pressure at sea level is 14.7 lb/in².

(b) To find the pressure on the airplane at an altitude of 6.25 miles, substitute 6.25 for x and evaluate.

$P = 14.7e^{(-0.2)(6.25)}$
$ = 14.7e^{-1.25}$
$ = 14.7(0.286504796)$
$ = 4.211620514$

Thus, the pressure on the airplane is about 4.2 lb/in².

33. If 15 mg of a drug is taken orally, the amount A remaining in the system after x hours can be approximated by

$$A = 15e^{-0.3x}.$$

(a) To find the amount remaining after 3 hours, substitute 3 for x and evaluate.

$A = 15e^{(-0.3)(3)}$
$ = 15e^{-0.9}$
$ = 6.098544896$

Thus, there are about 6.10 mg remaining in the system after 3 hours.

(b) To find the amount remaining after 1 day, which is 24 hours, substitute 24 for x and evaluate.

$A = 15e^{(-0.3)(24)}$
$ = 15e^{-7.2}$
$ = 0.011198787$

Thus, there are about 0.01 mg remaining in the system after 1 day.

34. $\ln \dfrac{e\sqrt{z}}{\sqrt[3]{x}} = \ln e + \ln \sqrt{z} - \ln \sqrt[3]{x}$

$\phantom{\ln \dfrac{e\sqrt{z}}{\sqrt[3]{x}}} = 1 + \ln z^{1/2} - \ln x^{1/3}$

$\phantom{\ln \dfrac{e\sqrt{z}}{\sqrt[3]{x}}} = 1 + \dfrac{1}{2}\ln z - \dfrac{1}{3}\ln x$

35. $\ln \sqrt{e\sqrt{y}} = \ln (e\sqrt{y})^{1/2}$

$\phantom{\ln \sqrt{e\sqrt{y}}} = \dfrac{1}{2} \ln e y^{1/2}$

$\phantom{\ln \sqrt{e\sqrt{y}}} = \dfrac{1}{2}[\ln e + \ln y^{1/2}]$

$\phantom{\ln \sqrt{e\sqrt{y}}} = \dfrac{1}{2}\ln e + \dfrac{1}{2}\ln y^{1/2}$

$\phantom{\ln \sqrt{e\sqrt{y}}} = \dfrac{1}{2}(1) + \dfrac{1}{4}\ln y$

$\phantom{\ln \sqrt{e\sqrt{y}}} = \dfrac{1}{2} + \dfrac{1}{4}\ln y$

36. $3\ln xy + \dfrac{1}{2}\ln z = \ln (xy)^3 + \ln z^{1/2}$

$\phantom{3\ln xy + \dfrac{1}{2}\ln z} = \ln (xy)^3 z^{1/2}$

$\phantom{3\ln xy + \dfrac{1}{2}\ln z} = \ln x^3 y^3 \sqrt{z}$

37. $3\ln x - \dfrac{1}{2}\ln y - 5\ln z = \ln x^3 - \ln y^{1/2} - \ln z^5$

$\phantom{3\ln x - \dfrac{1}{2}\ln y - 5\ln z} = \ln \dfrac{x^3}{y^{1/2}z^5}$

$\phantom{3\ln x - \dfrac{1}{2}\ln y - 5\ln z} = \ln \dfrac{x^3}{z^5 \sqrt{y}}$

38. Solve $7^x = 7^{2x+3}$.

Since the bases are equal, we can equate the exponents and solve the following equation.

$x = 2x + 3$
$-x = 3$
$x = -3$

Since -3 checks in the original equation, the solution is -3.

39. Solve $\log_3 x = \log_3 (x^2 - 2)$.

Since the bases are the same, solve the equation:

$x = x^2 - 2$
$0 = x^2 - x - 2$
$0 = (x - 2)(x + 1)$
$x - 2 = 0 \qquad x + 1 = 0$
$x = 2 \qquad x = -1$

But since -1 forces us to take the logarithm of a negative number in the original equation, we must discard it. Since 2 does check, the solution is 2.

CHAPTER 5 EXPONENTIAL AND LOGARITHMIC FUNCTIONS

SECTION 5.5 Exponential and Logarithmic Equations

1. Solve $10^{2x+1} = 100^{x/2}$.

   Write both sides of the equation as powers with the same base 10.

   $$10^{2x+1} = (10^2)^{x/2} = 10^x$$

   Then equate the exponents.

   $$2x + 1 = x$$
   $$2x - x = -1$$
   $$x = -1$$

   Thus, the solution is -1.

2. Solve $5^{2x} = 8$.

   Since we cannot express both sides as powers with the same base, take the logarithm base 5 of both sides.

   $$\log_5 5^{2x} = \log_5 8$$

   Then the left side reduces to $2x$.

   $$2x = \log_5 8$$

   Divide both sides by 2 and evaluate using the base conversion formula.

   $$x = \frac{\log_5 8}{2}$$
   $$= \frac{\frac{\log 8}{\log 5}}{2} = 0.646014837$$

   Thus, correct to three significant digits, the solution is 0.646.

3. *Analysis:* The atmospheric pressure P, in lb/in^2, at an altitude x, in miles, is approximated by

   $$P = 14.7e^{-0.2x}.$$

   Tabulation: The pressure at sea level is found by letting $x = 0$.

   $$P = 14.7e^{(-0.2)(0)} = 14.7e^0 = 14.7$$

   We want to find the altitude x when the pressure is half the pressure at sea level, that is, when the pressure is $\frac{14.7}{2} = 7.35$ lb/in^2.

   Translation: Solve the equation:

   $$7.35 = 14.7e^{-0.2x}$$

   Divide both sides by 14.7, and then take the natural logarithm of both sides.

   $$\frac{7.35}{14.7} = e^{-0.2x}$$
   $$\ln\left(\frac{7.35}{14.7}\right) = \ln e^{-0.2x}$$
   $$\ln\left(\frac{7.35}{14.7}\right) = -0.2x$$
   $$\frac{\ln\left(\frac{7.35}{14.7}\right)}{-0.2} = x$$
   $$3.465735903 = x$$

   Approximation: 3.47 mi is a reasonable number for the altitide in this situation.

   Check: Substitute 3.47 for x and evaluate.

   $$P = 14.7e^{(-0.2)(3.47)} \approx 7.343734449 \approx 7.35$$

   Thus, we conclude that the altitude at which the pressure is half that at sea level is about 3.47 mi.

4. Solve $\log(x + 10) - \log(x + 1) = 1$.

   Since the difference of two logs comes from the log of a quotient, we first rewrite the left side.

   $$\log\left(\frac{x + 10}{x + 1}\right) = 1$$

   Convert the result to exponential form.

   $$\frac{x + 10}{x + 1} = 10^1 = 10$$

   Multiply both sides by the denominator $x + 1$.

PRACTICE EXERCISES

$$x + 10 = 10(x + 1)$$
$$x + 10 = 10x + 10$$
$$0 = 9x$$
$$0 = x$$

The solution is 0, which does check in the original equation.

5. Solve $\ln x + \ln(x + 1) = \ln 2$.

Since the sum of two logs comes from the log of a product, first rewrite the left side.

$$\ln x(x + 1) = \ln 2$$

Since $\ln u = \ln v$ means that $u = v$, we have:

$$x(x + 1) = 2$$
$$x^2 + x = 2$$
$$x^2 + x - 2 = 0$$
$$(x + 2)(x - 1) = 0$$
$$x + 2 = 0 \quad x - 1 = 0$$
$$x = -2 \quad x = 1$$

But since -2 does not check, remember that the log of a negative number is undefined, and if we substitute -2 for x in the original equation, we get

$$\ln(-2) + \ln(-1),$$

it must be discarded. Thus, the only solution is 1, which does check.

6. Solve $\log_5 x - \log_2 x = 3$.

Since the bases are different, we cannot combine the two logarithms as before. Begin by changing all logarithms to base 10 logs.

$$\frac{\log x}{\log 5} - \frac{\log x}{\log 2} = 3$$

Then factor out $\log x$.

$$\log x \left[\frac{1}{\log 5} - \frac{1}{\log 2} \right] = 3$$

Solve for $\log x$.

$$\log x = \frac{3}{\dfrac{1}{\log 5} - \dfrac{1}{\log 2}}$$
$$\log x = -1.58625119$$
$$x = 0.025926793$$

Evaluation has been made using a calculator with no rounding. Thus, the solution is about 0.0259, correct to three significant digits. Be careful to find the value of x, it is the antilog of $\log x$ found in the next to the last step.

CHAPTER 5 EXPONENTIAL AND LOGARITHMIC FUNCTIONS

SECTION 5.5 Exponential and Logarithmic Equations

1. Solve $4^x = 8$.

   First rewrite both sides using the same base, 2.

   $$(2^2)^x = 2^3$$
   $$2^{2x} = 2^3$$

   Since the bases are now equal, equate the exponents.

   $$2x = 3$$
   $$x = \frac{3}{2}$$

   Since $\frac{3}{2}$ checks in the original equation, $\frac{3}{2}$ is the solution.

2. Solve $2^{5x} = 8^{x+5}$.

   First rewrite both sides using the same base, 2.

   $$2^{5x} = (2^3)^{x+5}$$
   $$2^{5x} = 2^{3x+15}$$

   Since the bases are now equal, equate the exponents.

   $$5x = 3x + 15$$
   $$2x = 15$$
   $$x = \frac{15}{2}$$

   Since $\frac{15}{2}$ checks in the original equation, $\frac{15}{2}$ is the solution.

3. Solve $2^{x^2} = 16^x$.

   First rewrite both sides using the same base, 2.

   $$2^{x^2} = (2^4)^x$$
   $$2^{x^2} = 2^{4x}$$

   Since the bases are now equal, equate the exponents.

   $$x^2 = 4x$$
   $$x^2 - 4x = 0$$
   $$x(x - 4) = 0$$

   $x = 0 \qquad x - 4 = 0$
   $\qquad\qquad x = 4$

   Since 0 and 4 both check in the original equation, 0 and 4 are the solutions.

4. Solve $2^{3x} = 7$.

   This time we cannot write both sides using the same base. To remove the variable from the exponent, take the common log of both sides and use the power rule.

   $$\log 2^{3x} = \log 7$$
   $$3x \log 2 = \log 7$$
   $$x = \frac{\log 7}{3 \log 2}$$
   $$x = 0.935784974$$

   Thus, correct to three significant digits, the solution is 0.936.

5. Solve $5^{x+1} = 3^{2x}$.

   This time we cannot write both sides using the same base. To remove the variable from the exponents, take the common log of both sides and use the power rule.

   $$\log 5^{x+1} = \log 3^{2x}$$
   $$(x+1) \log 5 = 2x \log 3$$
   $$x \log 5 + \log 5 = 2x \log 3$$
   $$x \log 5 - 2x \log 3 = -\log 5$$
   $$x(\log 5 - 2 \log 3) = -\log 5$$
   $$x = \frac{-\log 5}{\log 5 - 2 \log 3}$$
   $$x = 2.738132742$$

   Thus, correct to three significant digits, the solution is 2.74.

6. Solve $(1.08)^x = 100$.

   Take the common log of both sides.

   $$\log(1.08)^x = \log 100$$
   $$x \log(1.08) = 2 \qquad \log 100 = \log 10^2 = 2$$
   $$x = \frac{2}{\log(1.08)}$$
   $$x = 59.83768044$$

EXERCISES A
SECTION 5.5   233

Thus, correct to three significant digits, the solution is 59.8.

7. Solve $8^{\log_2 x} = 125$.

   First simplify the left side.

   $$8^{\log_2 x} = (2^3)^{\log_2 x} = 2^{3\log_2 x} = 2^{\log_2 x^3} = x^3$$

   Thus, the equation becomes $x^3 = 125$. Since $125 = 5^3$, we have $x^3 = 5^3$. Equating bases (the exponents are equal), $x = 5$.

8. Solve $2^{x^2} 3^{x^2} = 6^{5x-6}$.

   Since the exponents are the same on the left side, the left side can be simplified to

   $$(2 \cdot 3)^{x^2} = 6^{x^2}.$$

   Then $6^{x^2} = 6^{5x-6}$, so equating exponents we have

   $$x^2 = 5x - 6.$$
   $$x^2 - 5x + 6 = 0$$
   $$(x - 2)(x - 3) = 0$$
   $$x - 2 = 0 \qquad x - 3 = 0$$
   $$x = 2 \qquad x = 3$$

   Since both 2 and 3 check in the original equation, the solutions are 2 and 3.

9. Solve $7^{2x} 3^x = 10$.

   First rewrite the left side.

   $$7^{2x} 3^x = (7^2)^x 3^x = [(7^2) \cdot 3]^x = [49 \cdot 3]^x = 147^x$$

   Then the equation becomes

   $$147^x = 10.$$

   Take the common log of both sides.

   $$\log 147^x = \log 10$$
   $$x \log 147 = 1 \qquad \log 10 = \log_{10} 10 = 1$$
   $$x = \frac{1}{\log 147}$$
   $$x = 0.461399899$$

   Thus, to three significant digits, the solution is 0.461.

10. Solve $\log_2 (x + 2) = 3$.

    Convert to exponential form.

    $$2^3 = x + 2$$
    $$8 = x + 2$$
    $$6 = x$$

    Since 6 checks in the original equation, the solution is 6.

11. Solve $\log_3 x + \log_3 9 = 5$.

    Since $\log_3 9 = \log_3 3^2 = 2$, the equation reduces to

    $$\log_3 x + 2 = 5$$
    $$\log_3 x = 3$$

    Convert to exponential form.

    $$3^3 = x$$
    $$27 = x$$

    Since 27 checks in the original equation, the solution is 27.

12. Solve $\ln(x + 1) + \ln(x - 1) = \ln(4x + 4)$.

    First rewrite the left side using the product rule.

    $$\ln(x + 1)(x - 1) = \ln(4x + 4)$$

    Since the two logarithms have the same base, we can solve

    $$(x + 1)(x - 1) = 4x + 4.$$

    Clear parentheses and collect like terms.

    $$x^2 - 1 = 4x + 4$$
    $$x^2 - 4x - 5 = 0$$
    $$(x - 5)(x + 1) = 0$$
    $$x - 5 = 0 \qquad x + 1 = 0$$
    $$x = 5 \qquad x = -1$$

    But -1 does not check since using -1 forces us to try and take the natural logarithm of a negative number which is undefined. Since 5 does check, the only solution is 5.

13. Solve $\log_5 (2x - 1) - \log_5 (x - 5) = 1$.

    First simplify the left side using the quotient rule.

$$\log_5(2x-1) - \log_5(x-5) = \log_5 \frac{2x-1}{x-5}$$

Then the equation becomes

$$\log_5 \frac{2x-1}{x-5} = 1.$$

Convert to exponential form.

$$\frac{2x-1}{x-5} = 5^1$$
$$2x - 1 = 5(x - 5)$$
$$2x - 1 = 5x - 25$$
$$24 = 3x$$
$$8 = x$$

Since 8 does check in the original equation, the solution is 8.

14. Solve $\log_5 x + \log_7 x = 2$.

Since the bases are different, we begin by converting all logarithms to the same base 10 using the base conversion formula.

$$\frac{\log x}{\log 5} + \frac{\log x}{\log 7} = 2$$
$$\log x \left[\frac{1}{\log 5} + \frac{1}{\log 7} \right] = 2$$
$$\log x = \frac{2}{\left[\frac{1}{\log 5} + \frac{1}{\log 7} \right]}$$
$$\log x = 0.765119364$$
$$x = 5.822632295$$

Do not forget to take the antilog in the last step to obtain x from $\log x$. Thus, correct to three significant digits, the solution is 5.82.

15. Solve $\log_4 4x = \log_2(x + 1)$.

First convert the right side to base 2.

$$\log_4 4x = \frac{\log_2 4x}{\log_2 4} = \frac{\log_2 4x}{2}$$

Then the equation to solve is:

$$\frac{\log_2 4x}{2} = \log_2(x+1)$$
$$\log_2 4x = 2\log_2(x+1)$$
$$\log_2 4x = \log_2(x+1)^2$$

Since the bases are equal, we solve the following equation.

$$4x = (x+1)^2$$
$$4x = x^2 + 2x + 1$$
$$0 = x^2 - 2x + 1$$
$$0 = (x-1)(x-1)$$
$$x - 1 = 0 \qquad x - 1 = 0$$
$$x = 1 \qquad x = 1$$

Since 1 checks in the original equation, the solution is 1.

16. Solve $(\log_5 x)^2 = 2 \log_5 x$.

This equation is quadratic in form. Substitute u for $\log_5 x$ and solve for u.

$$u^2 = 2u$$
$$u^2 - 2u = 0$$
$$u(u - 2) = 0$$
$$u = 0 \qquad u - 2 = 0$$
$$u = 2$$

Backsubstitute to find the value of x.

$$\log_5 x = 0 \qquad \log_5 x = 2$$
$$5^0 = x \qquad 5^2 = x$$
$$1 = x \qquad 25 = x$$

Since both 1 and 25 check in the original equation, the solutions are 1 and 25.

17. Solve $\ln x^3 = (\ln x)^3$.

First use the power rule on the left side.

$$3 \ln x = (\ln x)^3$$

Substitute u for $\ln x$.

$$3u = u^3$$
$$0 = u^3 - 3u$$
$$0 = u(u^2 - 3)$$

One solution is $u = 0$, and the other two come from solving $u^2 - 3 = 0$, which gives $u = \pm\sqrt{3}$.

Backsubstitute to find the value of x.

$$\ln x = 0 \qquad \ln x = \sqrt{3} \qquad \ln x = -\sqrt{3}$$
$$x = 1 \qquad x = 5.65 \qquad x = 0.177$$

Since all three check in the original equation, to three significant digits, the solutions are 1, 5.65, and 0.177.

EXERCISES A

SECTION 5.5   235

18. Solve $x^{\ln x} = e^2 x$.

Take the natural log of both sides.

$$\ln x^{\ln x} = \ln e^2 x$$
$$(\ln x)(\ln x) = \ln e^2 + \ln x$$
$$(\ln x)^2 = 2 + \ln x$$
$$(\ln x)^2 - \ln x - 2 = 0$$

Substitute u for $\ln x$.

$$u^2 - u - 2 = 0$$
$$(u - 2)(u + 1) = 0$$
$$u - 2 = 0 \qquad u + 1 = 0$$
$$u = 2 \qquad u = -1$$
$$\ln x = 2 \qquad \ln x = -1$$
$$x = 7.39 \qquad x = 0.368$$

Since both numbers check in the original equation, to three significant digits, the solutions are 7.39 and 0.368.

19. Solve $\log(\log x) = 1$.

First convert to exponential form.

$$\log x = 10^1 = 10$$

Then convert to exponential form once more.

$$x = 10^{10}$$

The solution is 10^{10}, which does check.

20. Solve $\log \sqrt{x} = \sqrt{\log x}$.

Use the power rule on the left side, then square both sides to eliminate the radical.

$$\log x^{1/2} = \sqrt{\log x}$$
$$\tfrac{1}{2} \log x = \sqrt{\log x}$$
$$\tfrac{1}{4} (\log x)^2 = \log x$$
$$(\log x)^2 = 4 \log x$$
$$(\log x)^2 - 4 \log x = 0$$
$$u^2 - 4u = 0$$
$$u(u - 4) = 0$$
$$u = 0 \qquad u = 4$$

Then backsubstitute to find the value of x.

$$\log x = 0 \qquad \log x = 4$$
$$x = 1 \qquad x = 10{,}000$$

Since both check, the solutions are 1 and 10,000.

21. Solve $|\log x| = 2$.

First translate to the two related equations.

$$\log x = 2 \qquad \log x = -2$$
$$x = 100 \qquad x = 0.01$$

Since both check in the original equation, the solutions are 100 and 0.01.

22. The demand equation for x units of a product relative to the price per unit P is given by

$$P = 100 - 0.2 e^{0.005x}.$$

To find the demand when the price is $60.00, substitute 60 for P and solve for x.

$$60 = 100 - 0.2 e^{0.005x}$$
$$-40 = -0.2 e^{0.005x}$$
$$200 = e^{0.005x}$$
$$\ln 200 = \ln e^{0.005x}$$
$$\ln 200 = 0.005x$$
$$\frac{\ln 200}{0.005} = x$$
$$1059.663473 = x$$

Thus, the number of units is about 1060.

23. The atmospheric pressure P, in lb/in², at an altitude x, in miles, is approximated by

$$P = 14.7 e^{-0.2x}.$$

The altitude at sea level is found by substituting 0 for x.

$$P = 14.7 e^{(-0.2)(0)} = 14.7 e^0 = 14.7$$

Substitute $\dfrac{14.7}{4} = 3.675$ for P and solve for x to find the altitude at which the pressure is one-fourth that at sea level.

$$3.675 = 14.7 e^{-0.2x}$$
$$0.25 = e^{-0.2x}$$
$$\ln 0.25 = -0.2x$$
$$\frac{\ln 0.25}{-0.2} = x$$
$$6.931471806 = x$$

Thus, the desired altitude is about 6.93 mi.

24. When 15 mg of a drug is injected into the bloodstream, the amount A remaining in the system after x hours can be approximated by

$$A = 15e^{-0.3x}.$$

To be effective, there must be at least 5 mg remaining in the system. Substitute 5 for A and solve for x.

$$5 = 15e^{-0.3x}$$
$$\frac{1}{3} = e^{-0.3x}$$
$$\ln \frac{1}{3} = -0.3x$$
$$\frac{\ln \frac{1}{3}}{-0.3} = x$$
$$3.662040962 = x$$

Thus, the drug will be effective for about 3.66 hr.

Newton's law of cooling relates the time t, in minutes, that it takes for an object heated to a temperature T_0 to cool to a temperature T when placed in an area of constant temperature T_c. The formula is given by

$$T = T_c + (T_0 - T_c)e^{-kt},$$

where the constant k depends on the nature of the object considered. Use this information in Exercises 25-26.

25. Given that $k = 0.025$, $T_0 = 250$, $T_c = 70$, and $t = 10$. Substitute these values to find T.

$$T = 70 + (250 - 70)e^{-0.025(10)}$$
$$= 70 + 180e^{-0.25}$$
$$= 70 + 180(0.778800783)$$
$$= 210.184141$$

Thus, the temperature of the rod is about 210°F.

26. Given that $T_c = 80$, $T_0 = 400$, $T = 200$, and $t = 15$. First substitute to find the value of k.

$$200 = 80 + (400 - 80)e^{-k(15)}$$
$$120 = 320e^{-15k}$$
$$0.375 = e^{-15k}$$
$$\ln 0.375 = -15k$$
$$\frac{\ln 0.375}{-15} = k$$
$$0.065388616 = k$$

Thus, $k = 0.0654$. Use this value of k together with the given values and $t = 30$ to find T.

$$T = 80 + (400 - 80)e^{-0.0654(30)}$$
$$= 80 + 320e^{-1.962}$$
$$= 80 + 320(0.140576985)$$
$$= 124.9846354$$

Thus, after 30 minutes, the temperature of the object will be about 125°F.

27. When a beam of light passes through water, the intensity of the light I is diminished by the number of feet x from the surface of the water as given by

$$I = I_0 e^{-kx},$$

where I_0 is the intensity of the light at the surface of the water, and k is the coefficient of extinction depending on the particular water under consideration. Given that $k = 0.05$, substitute $0.5I_0$ for I to find the value of x.

$$0.5I_0 = I_0 e^{-0.05x}$$
$$0.5 = e^{-0.05x}$$
$$\ln 0.5 = -0.05x$$
$$\frac{\ln 0.5}{-0.05} = x$$
$$13.86294361 = x$$

Thus, to the nearest foot, the depth at which the intensity is 50% that at the surface is 14 ft.

28. Graph $y = \ln(x^2 + 1)$.

Use a calculator to make a table of values, plot the points, and connect them with a smooth curve to obtain the graph shown below.

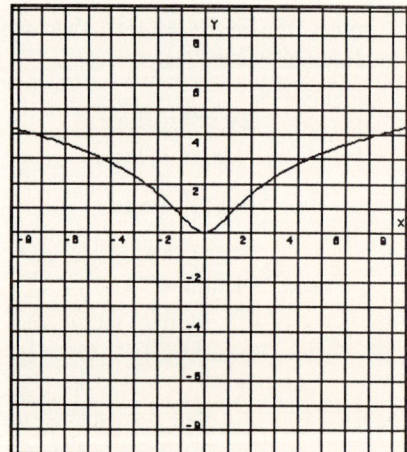

EXERCISES A

29. Graph $y = e^{-0.5x}$.

Use a calculator to make a table of values, plot the points, and connect them with a smooth curve to obtain the graph shown below.

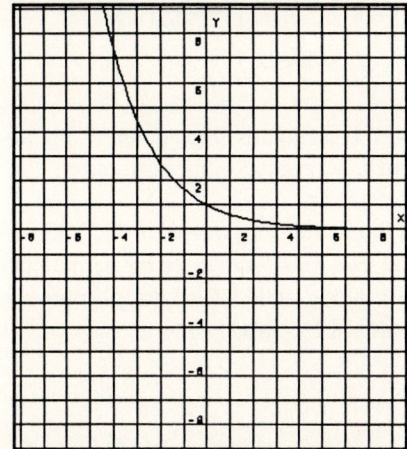

30. Use the pH equation, $pH = -\log[H^+]$ and substitute 3.25×10^{-7} to find the pH of the soft drink.

$$\begin{aligned}
pH &= -\log[3.25 \times 10^{-7}] \\
&= -[\log 3.25 + \log 10^{-7}] \\
&= -[\log 3.25 + (-7)] \\
&= -\log 3.25 + 7 \\
&= 6.488116639
\end{aligned}$$

Thus, the pH of the soft drink is about 6.5.

CHAPTER 5 EXPONENTIAL AND LOGARITHMIC FUNCTIONS

SECTION 5.6 More Applications of Exponentials and Logarithms

1. Use the formula

$$A = P\left(1 + \frac{r}{k}\right)^{kt}$$

with $P = 10{,}000$, $r = 0.12$, $k = 12$, and $t = 5$.

$$A = 10{,}000\left(1 + \frac{0.12}{12}\right)^{(12)(5)}$$
$$= 10{,}000(1 + 0.01)^{60}$$
$$= 10{,}000(1.01)^{60}$$
$$\approx \$18{,}166.97$$

Notice that the amount when the interest is compounded monthly is $105.87 more than when interest is compounded quarterly as was done in Example 1.

2. Use the formula

$$A = P\left(1 + \frac{r}{k}\right)^{kt}$$

with $A = 5000$, $r = 0.08$, $k = 4$, and $t = 10$, and solve for P.

$$5000 = P\left(1 + \frac{0.08}{4}\right)^{(4)(10)}$$
$$5000 = P(1 + 0.02)^{40}$$
$$5000 = P(1.02)^{40}$$
$$\frac{5000}{(1.02)^{40}} = P$$
$$\$2264.45 \approx P$$

Thus, about $2264.45 would need to be deposited to have the desired $5000 in 10 years.

3. To find the monthly payment, substitute 60,000 for P, 12 for k, 0.12 for r and 20 for t in the amortization formula

$$p = \frac{Pr}{k\left[1 - \left(1 + \frac{r}{k}\right)^{-kt}\right]}$$

$$p = \frac{(60{,}000)(0.12)}{12\left[1 - \left(1 + \frac{0.12}{12}\right)^{-12(20)}\right]}$$
$$= \$660.65$$

Follow the calculator steps given in the text, if necessary, to evaluate this expression.

Using the monthly payment of $660.65, to find the total interest paid, substitute the same values for k, r, and t, along with 660.65 for p in the second amortization formula:

$$I = kp\left[t - \frac{1 - \left(1 + \frac{r}{k}\right)^{-kt}}{r}\right]$$

$$= (12)(660.65)\left[20 - \frac{1 - \left(1 + \frac{0.12}{12}\right)^{-(12)(20)}}{0.12}\right]$$
$$= \$98{,}556.15$$

Thus, the couple will pay $98,556.15 in interest over the 20-year period.

4. Use the Malthusian model

$$N = Ie^{kt}$$

with $I = 120{,}000$, $N = 100{,}000$, and $k = -0.02$. Notice that k is negative since the population is decreasing rather than increasing. Solve for t.

$$100{,}000 = 120{,}000e^{-0.02t}$$
$$\frac{100{,}000}{120{,}000} = e^{-0.02t}$$
$$\frac{5}{6} = e^{-0.02t}$$
$$\ln \frac{5}{6} = \ln e^{-0.02t}$$
$$\ln \frac{5}{6} = -0.02t$$
$$\frac{\ln \frac{5}{6}}{-0.02} = t$$
$$9.11607784 = t$$

Thus, it will take about 9.1 yr for the population to decrease to 100,000.

PRACTICE EXERCISES

5. We can assume that the population can be approximated using the Malthusian model

$$N = Ie^{kt}.$$

First we must find the value of k. When $t = 0$, the initial population was 300 (in 1980), giving $I = 300$. Ten years later, in 1990, the population was 380, so substitute 380 for N and 10 for t and solve for k.

$$380 = 300e^{k(10)}$$
$$\frac{380}{300} = e^{10k}$$
$$\ln\left(\frac{380}{300}\right) = \ln e^{10k}$$
$$\ln\left(\frac{380}{300}\right) = 10k$$
$$\frac{\ln\left(\frac{380}{300}\right)}{10} = k$$
$$0.024 \approx k$$

Using this value of k we can find the population in the year 2000, when $t = 20$.

$$N = 300e^{(0.024)(20)}$$
$$= 300e^{0.48}$$
$$= 484.8223207$$

Thus, the antelope population will be about 485 in the year 2000.

6. The formula used for a model of radioactive decay is

$$A = Ie^{kt}.$$

If the half-life of a lead isotope is 25 years, substitute 25 for t, and $0.5I$ for A to find the value of k.

$$0.5I = Ie^{k(25)}$$
$$0.5 = e^{25k}$$
$$\ln 0.5 = \ln e^{25k}$$
$$\ln 0.5 = 25k$$
$$\frac{\ln 0.5}{25} = k$$
$$-0.028 \approx k$$

Use this value of k together with 500 for I and 100 for t to find A, the amount remaining.

$$A = 500e^{(-0.028)(100)}$$
$$= 500e^{-2.8}$$
$$\approx 30.40503131$$

Thus, about 30.4 grams remain after 100 years.

7. Use the model

$$A = Ie^{-0.000125t}$$

for carbon-14 dating. Since the mummy had lost 40% of its carbon-14, this means that 60% remained. Use $0.60I$ for A and solve for t.

$$0.60I = Ie^{-0.000125t}$$
$$0.60 = e^{-0.000125y}$$
$$\ln 0.60 = \ln e^{-0.000125t}$$
$$\ln 0.60 = -0.000125t$$
$$\frac{\ln 0.60}{-0.000125} = t$$
$$4086.60499 = t$$

Thus, the person died about 4087 yr ago.

8. The magnitude of an earthquake is given by

$$M = \log \frac{A}{A_0},$$

where the ratio $\frac{A}{A_0}$ is called the intensity of the earthquake. Given that the California earthquake measured 7.1 on the Richter scale, substitute 7.1 for M and solve for $\frac{A}{A_0}$.

$$7.1 = \log \frac{A}{A_0}$$
$$10^{7.1} = 10^{\log \frac{A}{A_0}}$$
$$10^{7.1} = \frac{A}{A_0}$$
$$12{,}589{,}254.12 = \frac{A}{A_0}$$

Thus, the the quake was about 12,589,254 times more intense than a reference level zero earthquake.

9. The loudness of a sound in decibels is given by

$$D = 10 \log \frac{S}{S_0},$$

Where $S_0 = 10^{-12}$ watt/m² is the measure of a sound at the threshold of human hearing, that is, the zero level for decibels. To find the intensity in watt/m² of a gunshot with a decibel reading measured at 100 decibels, substitute 100 for D, 10^{-12} for S_0, and solve for S.

$$100 = 10 \log \frac{S}{10^{-12}}$$
$$\frac{100}{10} = \log \frac{S}{10^{-12}}$$
$$10 = \log S - \log 10^{-12}$$
$$10 = \log S - (-12)$$
$$10 = \log S + 12$$
$$-2 = \log S$$
$$10^{-2} = 10^{\log S}$$
$$10^{-2} = S$$

Thus, the intensity of the sound of the gunshot is about 10^{-2} watt/m².

EXERCISES A SECTION 5.6 241

CHAPTER 5 EXPONENTIAL AND LOGARITHMIC FUNCTIONS

SECTION 5.6 More Applications of Exponentials and Logarithms

1. Use the compound interest formula
$$A = P\left(1 + \frac{r}{k}\right)^{kt}.$$

   (a) Substitute $P = 1000$, $r = 0.09$, $k = 1$, and $t = 4$.
   $$A = 1000\left(1 + \frac{0.09}{1}\right)^{(1)(4)}$$
   $$= 1000(1 + 0.09)^4$$
   $$= 1000(1.09)^4$$
   $$\approx \$1411.58$$

   (b) Substitute $P = 1000$, $r = 0.09$, $k = 2$, and $t = 4$.
   $$A = 1000\left(1 + \frac{0.09}{2}\right)^{(2)(4)}$$
   $$= 1000(1 + 0.045)^8$$
   $$= 1000(1.045)^8$$
   $$\approx \$1422.10$$

   (c) Substitute $P = 1000$, $r = 0.09$, $k = 4$, and $t = 4$.
   $$A = 1000\left(1 + \frac{0.09}{4}\right)^{(4)(4)}$$
   $$= 1000(1 + 0.0225)^{16}$$
   $$= 1000(1.0225)^{16}$$
   $$\approx \$1427.62$$

   (d) Substitute $P = 1000$, $r = 0.09$, $k = 12$, and $t = 4$.
   $$A = 1000\left(1 + \frac{0.09}{12}\right)^{(12)(4)}$$
   $$= 1000(1 + 0.0075)^{48}$$
   $$= 1000(1.0075)^{48}$$
   $$\approx \$1431.41$$

   (e) Substitute $P = 1000$, $r = 0.09$, $k = 52$, and $t = 4$.
   $$A = 1000\left(1 + \frac{0.09}{52}\right)^{(52)(4)}$$
   $$= 1000(1 + 0.001730769)^{208}$$
   $$= 1000(1.001730769)^{208}$$
   $$\approx \$1432.88$$

   (f) Substitute $P = 1000$, $r = 0.09$, $k = 365$, and $t = 4$.
   $$A = 1000\left(1 + \frac{0.09}{365}\right)^{(365)(4)}$$
   $$= 1000(1 + 0.000246575)^{1460}$$
   $$= 1000(1.000246575)^{1460}$$
   $$\approx \$1433.27$$

2. Use the compound interest formula
$$A = P\left(1 + \frac{r}{k}\right)^{kt}.$$

   Substitute $2P$ for A, 0.12 for r, and 1 for k, and solve for t.
   $$2P = P\left(1 + \frac{0.12}{1}\right)^{(1)t}$$
   $$2 = (1.12)^t$$
   $$\log 2 = \log(1.12)^t$$
   $$\log 2 = t \log(1.12)$$
   $$\frac{\log 2}{\log(1.12)} = t$$
   $$6.116255374 = t$$

   Thus, it will take about 6.12 years for any amount to double.

3. Use the compound interest formula
$$A = P\left(1 + \frac{r}{k}\right)^{kt}.$$

   Substitute $2P$ for A, 0.12 for r, and 2 for k, and solve for t.
   $$2P = P\left(1 + \frac{0.12}{2}\right)^{(2)t}$$
   $$2 = (1.06)^{2t}$$
   $$\log 2 = \log(1.06)^{2t}$$
   $$\log 2 = 2t \log(1.06)$$
   $$\frac{\log 2}{2\log(1.06)} = t$$
   $$5.947830523 = t$$

   Thus, it will take about 5.95 years for any amount to double.

4. To find the monthly payment, substitute 18,000 for P, 12 for k, 0.12 for r and 5 for t in the amortization formula

$$p = \frac{Pr}{k\left[1 - \left(1 + \frac{r}{k}\right)^{-kt}\right]}$$

$$p = \frac{(18{,}000)(0.12)}{12\left[1 - \left(1 + \frac{0.12}{12}\right)^{-12(5)}\right]}$$

$$= \$400.40$$

Follow the calculator steps given in the text, if necessary, to evaluate this expression.

Using the monthly payment of $400.40, to find the total interest paid, substitute the same values for k, r, and t, along with 400.40 for p in the second amortization formula.

$$I = kp\left[t - \frac{1 - \left(1 + \frac{r}{k}\right)^{-kt}}{r}\right]$$

$$= (12)(400.40)\left[5 - \frac{1 - \left(1 + \frac{0.12}{12}\right)^{-(12)(5)}}{0.12}\right]$$

$$= \$6024.00$$

Thus, Jim Kirk will pay $6024.00 in interest over the 5-year period.

5. Use the Malthusian model

$$N = Ie^{kt}$$

with $N = 2I$, $k = 0.021$, and solve for t. Notice that k is positive since the population is increasing.

$$2I = Ie^{0.021t}$$
$$2 = e^{0.021t}$$
$$\ln 2 = \ln e^{0.021t}$$
$$\ln 2 = 0.021t$$
$$\frac{\ln 2}{0.021} = t$$
$$33.0070086 = t$$

Thus, it will take about 33 yr for the population to double.

6. Use the Malthusian model

$$N = Ie^{kt}$$

with $I = 237{,}000{,}000$, $k = 0.009$, and $t = 15$. Notice that k is positive since the population is increasing.

$$N = 237{,}000{,}000\,e^{0.009(15)}$$
$$= 237{,}000{,}000\,e^{0.135}$$
$$= 237{,}000{,}000(1.144536784)$$
$$= 271{,}255{,}217.9$$

Thus, the population will be about 271 million in the year 2000.

7. Use the Malthusian model

$$N = Ie^{kt}$$

with $I = 8000$, $k = -0.062$, and $N = 2000$. Notice that k is negative since the population is decreasing. Solve for t.

$$2000 = 8000\,e^{-0.062t}$$
$$0.25 = e^{-0.062t}$$
$$\ln 0.25 = -0.062t$$
$$\frac{\ln 0.25}{-0.062} = t$$
$$22.35958647 = t$$

Thus, it will take about 22.4 yr for the antelope population to decline to 2000.

8. We can assume that the population can be approximated using the Malthusian model

$$N = Ie^{kt}.$$

First we must find the value of k. When $t = 0$, the initial population was 800 (in 1960), giving $I = 800$. Twenty years later, in 1980, the population was 1150, so substitute 1150 for N and 20 for t and solve for k.

$$1150 = 800\,e^{k(20)}$$
$$\frac{1150}{800} = e^{20k}$$
$$\ln\left(\frac{1150}{800}\right) = \ln e^{20k}$$
$$\ln\left(\frac{1150}{800}\right) = 20k$$
$$\frac{\ln\left(\frac{1150}{800}\right)}{20} = k$$
$$0.018 \approx k$$

Using this value of k we can find the population in the year 2000, when $t = 40$.

$$N = 800\,e^{(0.018)(40)}$$
$$= 800\,e^{0.72}$$
$$= 1643.5465569$$

EXERCISES A

Thus, the population of Alpine will be about 1644 in the year 2000.

9. A radioactive isotope decays according to the equation $A = Ie^{-0.03t}$ where t is measured in years. To find the half-life of this isotope, substitute $0.5I$ for A and solve for t.

$$0.5I = Ie^{-0.03t}$$
$$0.5 = e^{-0.03t}$$
$$\ln 0.5 = \ln e^{-0.03t}$$
$$\ln 0.5 = -0.03t$$
$$\frac{\ln 0.5}{-0.03} = t$$
$$23.10490602 = t$$

Thus, the half-life is about 23.1 years.

10. First use the equation $A = Ie^{kt}$ with $A = 0.5I$ and $t = 1650$ to find the value of k.

$$0.5I = Ie^{k(1650)}$$
$$0.5 = e^{1650k}$$
$$\ln 0.5 = 1650k$$
$$\frac{\ln 0.5}{1650} = k$$
$$-0.000420089 = k$$

Thus, $k \approx -0.00042$. Use this value for k and substitute $0.75I$ for A and solve for t.

$$0.75I = Ie^{-0.00042t}$$
$$0.75 = e^{-0.00042t}$$
$$\ln 0.75 = -0.00042t$$
$$\frac{\ln 0.75}{-0.00042} = t$$
$$684.9573154 = t$$

Thus, it will take about 685 years for the original amount to decay to three-fourths that amount.

11. Use the equation $A = Ie^{kt}$ with $k = -0.000125$ and $0.55I$ for A, and solve for t.

$$0.55I = Ie^{-0.000125t}$$
$$0.55 = e^{-0.000125t}$$
$$\ln 0.55 = -0.000125t$$
$$\frac{\ln 0.55}{-0.000125} = t$$
$$4782.696006 = t$$

Thus, the mask is about 4783 years old.

12. First use the equation $A = Ie^{kt}$ with $A = 0.5I$ and $t = 3.2$ to find the value of k.

$$0.5I = Ie^{k(3.2)}$$
$$0.5 = e^{3.2k}$$
$$\ln 0.5 = 3.2k$$
$$\frac{\ln 0.5}{3.2} = k$$
$$-0.216608493 = k$$

Thus, $k \approx -0.2166$. Use this value for k and substitute 30 for I and 9 for t and evaluate.

$$A = 30e^{-0.2166(9)}$$
$$= 30e^{-1.9494}$$
$$= 30(0.142359461)$$
$$= 4.270783849$$

Thus, about 4.27 mg remain in the system at 6:00 P.M. that evening.

13. Use the Richter scale formula

$$M = \log \frac{A}{A_0}$$

with $A = 2{,}511{,}890 A_0$ and find the value of M.

$$M = \log \frac{2{,}511{,}890 A_0}{A_0}$$
$$= \log 2{,}511{,}890$$
$$= 6.400000617$$

Thus, the magnitude was about 6.4 on the Richter scale.

14. The magnitude of an earthquake is given by

$$M = \log \frac{A}{A_0},$$

where the ratio $\frac{A}{A_0}$ is called the intensity of the earthquake. Given that the earthquake measured 7.2 on the Richter scale, substitute 7.2 for M and solve for $\frac{A}{A_0}$.

$$7.2 = \log \frac{A}{A_0}$$
$$10^{7.2} = 10^{\log \frac{A}{A_0}}$$
$$10^{7.2} = \frac{A}{A_0}$$

Thus, the the quake was about $10^{7.2}$ times more intense than a reference level zero earthquake.

15. The loudness of a sound in decibels is given by

$$D = 10 \log \frac{S}{S_0},$$

Where $S_0 = 10^{-12}$ watt/m^2 is the measure of a sound at the threshold of human hearing, that is, the zero level for decibels. To find the decibel level of the conversation in a restaurant that produces 4.15×10^{-6} watt/m^2 of power, substitute this value for S and evaluate D.

$$\begin{aligned} D &= 10 \log \frac{4.15 \times 10^{-6}}{10^{-12}} \\ &= 10 \log (4.15 \times 10^6) \\ &= 10(6.618048097) \\ &= 66.18048097 \end{aligned}$$

Thus, the decibel level of the conversation is about 66 decibels.

16. The loudness of a sound in decibels is given by

$$D = 10 \log \frac{S}{S_0},$$

Where $S_0 = 10^{-12}$ watt/m^2 is the measure of a sound at the threshold of human hearing, that is, the zero level for decibels. To find the intensity in watt/m^2 of a noise with a decibel reading measured at 60 decibels, substitute 100 for D, 10^{-12} for S_0, and solve for S.

$$\begin{aligned} 60 &= 10 \log \frac{S}{10^{-12}} \\ \frac{60}{10} &= \log \frac{S}{10^{-12}} \\ 6 &= \log S - \log 10^{-12} \\ 6 &= \log S - (-12) \\ 6 &= \log S + 12 \\ -6 &= \log S \\ 10^{-6} &= 10^{\log S} \\ 10^{-6} &= S \end{aligned}$$

Thus, the intensity of the sound of the noise is about 10^{-6} watt/m^2.

17. If the intensity of a sound is 4.6×10^{-7} watt/m^2, the decibel level is given by:

$$\begin{aligned} D &= 10 \log \frac{4.6 \times 10^{-7}}{10^{-12}} \\ &= 10 \log (4.6 \times 10^5) \\ &= 10(5.662757832) \\ &= 56.62757832 \end{aligned}$$

If the intensity is multiplied by 100 to obtain an intensity of 4.6×10^{-5}, then the decibel reading is:

$$\begin{aligned} D &= 10 \log \frac{4.6 \times 10^{-5}}{10^{-12}} \\ &= 10 \log (4.6 \times 10^7) \\ &= 10(7.662757832) \\ &= 76.62757832 \end{aligned}$$

Thus, the decibel is increased by about 20 decibels, the decibel rate is **not** multiplied by 100.

18. Solve $5^{2x} = 7^{x+1}$.

Since we cannot express both sides using the same base, take the common log of both sides.

$$\begin{aligned} \log 5^{2x} &= \log 7^{x+1} \\ 2x \log 5 &= (x+1) \log 7 \\ 2x \log 5 &= x \log 7 + \log 7 \\ 2x \log 5 - x \log 7 &= \log 7 \\ x(2 \log 5 - \log 7) &= \log 7 \\ x &= \frac{\log 7}{2 \log 5 - \log 7} \\ x &= 1.528643062 \end{aligned}$$

Thus, the solution is 1.53, correct to three significant digits.

19. Solve $(\log x)^2 = \log x^5$.

First use the power rule on the right side.

$$\begin{aligned} (\log x)^2 &= 5 \log x \\ u^2 &= 5u \\ u^2 - 5u &= 0 \\ u(u - 5) &= 0 \\ u = 0 & \quad u = 5 \end{aligned}$$

Backsubstitute to find the values of x.

$$\begin{aligned} \log x &= 0 & \log x &= 5 \\ x &= 1 & x &= 10^5 = 100,000 \end{aligned}$$

Since both check in the original equation, the solutions are 1 and 100,000.

20. $x = \log 4{,}820{,}000 = 6.6830$

21. $x = \ln(-0.0045)$ is undefined.

22. $\log x = -9.4994$ Then $x = 3.17 \times 10^{-10}$.

23. $\ln x = 13.2317$ Then $x = 558{,}000$.

24. $\ln \frac{x\sqrt{y}}{z^4} = \ln x + \ln \sqrt{y} - \ln z^4 = \ln x + \frac{1}{2} \ln y - 4 \ln z$

CHAPTER 5 EXPONENTIAL AND LOGARITHMIC FUNCTIONS

CHAPTER 5 Review Exercises

1. Suppose that a is any base for logarithms.

   (a) $\log_a a = 1$ (Since $a^1 = a$.)

   (b) $\log_a 1 = 0$ (Since $a^0 = 1$.)

2. Converting $2^x = 5$ to logarithmic form gives $\log_2 5 = x$.

3. Converting $\log_3 a = w$ to exponential form gives $3^w = a$.

4. Solve $\log_a 64 = 3$.

   First convert to exponential form, then express 64 as a power.

   $$a^3 = 64 = 4^3$$

   Since the exponents are equal, equate the bases to obtain $a = 4$.

5. Solve $125^x = 5$.

   Since $125^x = (5^3)^x = 5^{3x}$, we must solve $5^{3x} = 5$ or $5^{3x} = 5^1$. Since the bases are equal, equate the exponents.

   $$3x = 1$$
   $$x = \tfrac{1}{3}$$

6. If a is any base and x is any positive real number, then

   $$a^{\log_a x} = x.$$

7. If a is any base and x is any real number, then

   $$\log_a a^x = x.$$

8. Graph $f(x) = -3^x$.

   Make a table of values. Be sure to watch the minus sign. Remember that the exponent is on 3 only, then the power is negated. For example, when $x = 2$, then $f(2) = -3^2 = -9$ (*not* $(-3)^2 = 9$). Plot the points in your table and connect them with a smooth curve to obtain the graph shown below.

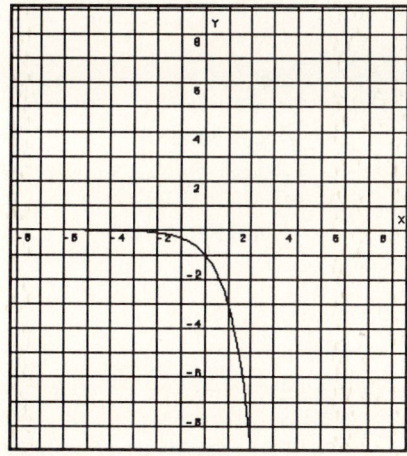

9. Graph $f(x) = \log_5(-x)$ for $x < 0$.

   Remember that we cannot take the logarithm of a negative number. This is why the restriction $x < 0$ is imposed, since under this condition, $-x > 0$. Make a table of values using the base conversion formula or by converting to the equivalent exponential form. Plot the points and connect them with a smooth curve to obtain the graph given below.

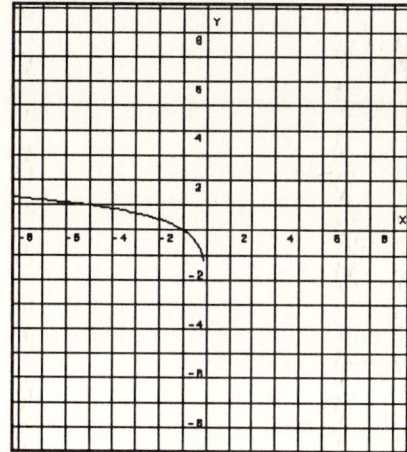

10. To approximate $(\sqrt{5})^\pi$ using a calculator, use the y^x button with $y = \sqrt{5}$ and $x = \pi$.

    $$(\sqrt{5})^\pi = 12.5297$$

11. $\log_a 21 = \log_a 3 \cdot 7$
    $= \log_a 3 + \log_a 7$
    $= 0.5283 + 0.9358 = 1.4641$

12. $\log_a \dfrac{14}{6} = \log_a \dfrac{7}{3}$
    $= \log_a 7 - \log_a 3$
    $= 0.9358 - 0.5283 = 0.4075$

13. $\log_a \left(\dfrac{7}{3}\right)^{3/2} = \dfrac{3}{2} \log_a \dfrac{7}{3}$
    $= \dfrac{3}{2}(\log_a 7 - \log_a 3)$
    $= \dfrac{3}{2}(0.9358 - 0.5283)$
    $= \dfrac{3}{2}(0.4075) = 0.6113$

14. $\log_3 7 = \dfrac{\log_a 7}{\log_a 3}$
    $= \dfrac{0.9358}{0.5283}$
    $= 1.7713$ *Divide, don't subtract*

15. $\log_a \left(\dfrac{xy}{z}\right)^{2/3} = \dfrac{2}{3} \log_a \dfrac{xy}{z}$
    $= \dfrac{2}{3}[\log_a xy - \log_a z]$
    $= \dfrac{2}{3}[\log_a x + \log_a y - \log_a z]$
    $= \dfrac{2}{3} \log_a x + \dfrac{2}{3} \log_a y - \dfrac{2}{3} \log_a z$

16. $3\log_a xy + \dfrac{1}{2} \log_a z = \log_a (xy)^3 + \log_a z^{1/2}$
    $= \log_a (xy)^3 z^{1/2}$
    $= \log_a x^3 y^3 \sqrt{z}$

17. false $(\log_a u + \log_a v = \log_a uv \neq \log_a(u+v))$

18. true (Remember that $\log_a a = 1$.)

19. $\ln(2.58 \times 10^{-7}) = -15.1703$

20. $\log_2 0.0425 = \dfrac{\log 0.0425}{\log 2} = -4.5564$ *Divide!*

21. If $\log x = -8.6219$, then $x = 2.39 \times 10^{-9}$. Some calculators will give $x = 0.000000002$.

22. If $\ln x = 41.6219$, then $x = 1.19 \times 10^{18}$.

23. Solve $81^{x-2} = 9^{x+5}$.

    First write each side using 3 as the base.

    $(3^4)^{x-2} = (3^2)^{x+5}$
    $3^{4(x-2)} = 3^{2(x+5)}$

Then equate the exponents.

$4(x-2) = 2(x+5)$
$4x - 8 = 2x + 10$
$2x = 18$
$x = 9$

Since 9 does check in the original equation, the solution is 9.

24. Solve the equation:

    $\log_2(3x+1) + \log_2(x-1) = \log_2(10x+14)$

    First use the product rule to write the left side as a single logarithm.

    $\log_2(3x+1)(x-1) = \log_2(10x+14)$

    Since the bases are equal, we solve:

    $(3x+1)(x-1) = 10x + 14$
    $3x^2 - 2x - 1 = 10x + 14$
    $3x^2 - 12x - 15 = 0$
    $x^2 - 4x - 5 = 0$
    $(x-5)(x+1) = 0$
    $x - 5 = 0 \quad x + 1 = 0$
    $x = 5 \quad x = -1$

    But -1 does not check in the original equation since using -1 forces us to find the logarithm of a negative number, which is undefined. Since 5 does check, the only solution is 5.

25. Solve $7^{x-5} = 3^{x+7}$.

    Since the bases cannot be made the same, take the common log of each side.

    $\log 7^{x-5} = \log 3^{x+7}$
    $(x-5)\log 7 = (x+7)\log 3$
    $x \log 7 - 5 \log 7 = x \log 3 + 7 \log 3$
    $x \log 7 - x \log 3 = 7 \log 3 + 5 \log 7$
    $x[\log 7 - \log 3] = 7 \log 3 + 5 \log 7$
    $x = \dfrac{7 \log 3 + 5 \log 7}{\log 7 - \log 3}$
    $x = 20.55928332$

    Thus, to three significant digits, the solution is 20.6.

26. Solve $(1.05)^x (1.02)^{2x} = 1000$.

    First simplify the right side.

    $[(1.05)(1.02)^2]^x = 1000$
    $[1.09242]^x = 1000$

CHAPTER 5 REVIEW

Then take the common log of both sides.

$$\log[1.09242]^x = \log 1000$$
$$x \log[1.09242] = 3 \quad \log 1000 = \log 10^3 = 3$$
$$x = \frac{3}{\log[1.09242]}$$
$$x = 78.14607788$$

Thus, to three significant digits, the solution is 78.1.

27. Solve $(\log_5 x)(\log_2 x) = \log x$.

First convert all logs to base 10.

$$\left(\frac{\log x}{\log 5}\right)\left(\frac{\log x}{\log 2}\right) = \log x$$
$$(\log x)^2 \left(\frac{1}{(\log 5)(\log 2)}\right) - \log x = 0$$
$$\log x \left[\left(\frac{1}{(\log 5)(\log 2)}\right) \log x - 1\right] = 0$$
$$\log x = 0 \quad \left(\frac{1}{(\log 5)(\log 2)}\right) \log x - 1 = 0$$
$$\log x = (\log 5)(\log 2)$$

Then since $\log x = 0$, $x = 1$. And since $\log x = (\log 5)(\log 2)$, $x = 1.62334541$, which to three significant digits, is 1.62. The solutions are 1 and 1.62.

28. Solve $\ln(\ln x) = 0$

First convert to exponential form, remember that the base is e.

$$e^0 = \ln x$$

Then
$$1 = \ln x.$$

Then convert to exponential form again.

$$e^1 = x$$

Thus, $x = e$, and the solution is e.

29. The intensity I of a light is diminished as it is directed through water according to

$$I = I_0 e^{-kx},$$

where I_0 is the intensity at the surface of the water and x is the depth in feet below the surface. When $x = 20$, the intensity is reduced by 50%, so substitute $0.5 I_0$ for I and solve for k.

$$0.5 I_0 = I_0 e^{-k(20)}$$
$$0.5 = e^{-20k}$$
$$\ln 0.5 = \ln e^{-20k}$$
$$\ln 0.5 = -20k$$
$$\frac{\ln 0.5}{-20} = k$$
$$0.034657359 = k$$

Thus, the coefficient of extinction of the water is about $k = 0.0347$.

30. Use the compound interest formula

$$A = P\left(1 + \frac{r}{k}\right)^{kt}.$$

(a) Substitute $P = 2000$, $r = 0.14$, $k = 2$, and $t = 2$.

$$A = 2000\left(1 + \frac{0.14}{2}\right)^{(2)(2)}$$
$$= 2000(1 + 0.07)^4$$
$$= 2000(1.07)^4$$
$$\approx \$2621.59$$

(b) Substitute $P = 2000$, $r = 0.14$, $k = 12$, and $t = 2$.

$$A = 2000\left(1 + \frac{0.14}{12}\right)^{(12)(2)}$$
$$= 2000(1 + 0.011\overline{6})^{24}$$
$$= 2000(1.011\overline{6})^{24}$$
$$\approx \$2641.97$$

31. Use the compound interest formula

$$A = P\left(1 + \frac{r}{k}\right)^{kt}.$$

Substitute $4P$ for A, 0.10 for r, and 4 for k, and solve for t.

$$4P = P\left(1 + \frac{0.10}{4}\right)^{(4)t}$$
$$4 = (1.025)^{4t}$$
$$\log 4 = \log(1.025)^{4t}$$
$$\log 4 = 4t \log(1.025)$$
$$\frac{\log 4}{4 \log(1.025)} = t$$
$$14.03551726 = t$$

Thus, it will take about 14 years for any amount to quadruple.

32. To find the monthly payment, substitute 30,000 for P, 12 for k, 0.12 for r and 15 for t in the amortization formula

$$p = \frac{Pr}{k\left[1 - \left(1 + \frac{r}{k}\right)^{-kt}\right]}.$$

$$p = \frac{(30{,}000)(0.12)}{12\left[1 - \left(1 + \frac{0.12}{12}\right)^{-12(15)}\right]}$$

$$= \$360.05$$

Follow the calculator steps given in the text, if necessary, to evaluate this expression.

Using the monthly payment of $360.05, to find the total interest paid, substitute the same values for k, r, and t, along with 360.05 for p in the second amortization formula.

$$I = kp\left[t - \frac{1 - \left(1 + \frac{r}{k}\right)^{-kt}}{r}\right]$$

$$= (12)(360.05)\left[15 - \frac{1 - \left(1 + \frac{0.12}{12}\right)^{-(12)(15)}}{0.12}\right]$$

$$= \$34{,}809.03$$

Thus, a total of $34,809.03 in interest will be paid over the 15-year period.

33. We can assume that the population can be approximated using the Malthusian model

$$N = Ie^{kt}.$$

First we must find the value of k. When $t = 0$, the initial population was 25,000 (in 1980), giving $I = 25{,}000$. Ten years later, in 1990, the population was 28,000, so substitute 28,000 for N and 10 for t and solve for k.

$$28{,}000 = 25{,}000e^{k(10)}$$
$$\frac{28{,}000}{25{,}000} = e^{10k}$$
$$\ln\left(\frac{28{,}000}{25{,}000}\right) = \ln e^{10k}$$
$$\ln\left(\frac{28{,}000}{25{,}000}\right) = 10k$$
$$\frac{\ln\left(\frac{28{,}000}{25{,}000}\right)}{10} = k$$
$$0.011 \approx k$$

Using this value of k we can find the population in the year 2010, when $t = 30$.

$$N = 25{,}000e^{(0.011)(30)}$$
$$= 25{,}000e^{0.33}$$
$$= 34{,}774.20321$$

Thus, the population of Winter will be about 34,774 in the year 2010.

34. A radioactive isotope decays according to the equation $A = Ie^{-0.0046t}$ where t is measured in years. To find the half-life of this isotope, substitute $0.5I$ for A and solve for t.

$$0.5I = Ie^{-0.0046t}$$
$$0.5 = e^{-0.0046t}$$
$$\ln 0.5 = \ln e^{-0.0046t}$$
$$\ln 0.5 = -0.0046t$$
$$\frac{\ln 0.5}{-0.0046} = t$$
$$150.6841697 = t$$

Thus, the half-life is about 150.7 years.

35. The magnitude of an earthquake is given by

$$M = \log \frac{A}{A_0},$$

where the ratio $\frac{A}{A_0}$ is called the intensity of the earthquake. Given that the intensity of an earthquake was 2.65×10^5, substitute and find the magnitude on the Richter scale.

$$M = \log(2.65 \times 10^5)$$
$$= 5.423245874$$

Thus, the magnitude of the quake was about 5.4 on the Richter scale.

36. The loudness of a sound in decibels is given by

$$D = 10 \log \frac{S}{S_0},$$

Where $S_0 = 10^{-12}$ watt/m² is the measure of a sound at the threshold of human hearing, that is, the zero level for decibels. To find the decibel level of the band's performance that produces 5.5×10^{-2} watt/m² of power, substitute this value for S and evaluate D.

$$D = 10 \log \frac{5.5 \times 10^{-2}}{10^{-12}}$$
$$= 10 \log(5.5 \times 10^{10})$$
$$= 10(10.74036269)$$
$$= 107.4036269$$

Since the decibel level is about 107.4 decibels, which exceeds 90 decibels by 17.4, there is a potential danger of hearing loss to the members of the band.

37. The number of bacteria in a culture that originally starts with 100 and doubles each hour is given by

$$N = (100)2^t, \ t \geq 0.$$

(a) When $t = 0$, $N = (100)2^0 = (100)(1) = 100$.

(b) When $t = 2$, $N = (100)2^2 = (100)(4) = 400$.

(c) When $t = 4$, $N = (100)2^4 = (100)(16) = 1600$.

(d) When $t = 6$, $N = (100)2^6 = (100)(64) = 6400$.

Plot the points $(0,100)$, $(2,400)$, $(4,1600)$, and $(6,6400)$ and join them with a smooth curve to obtain the graph shown below.

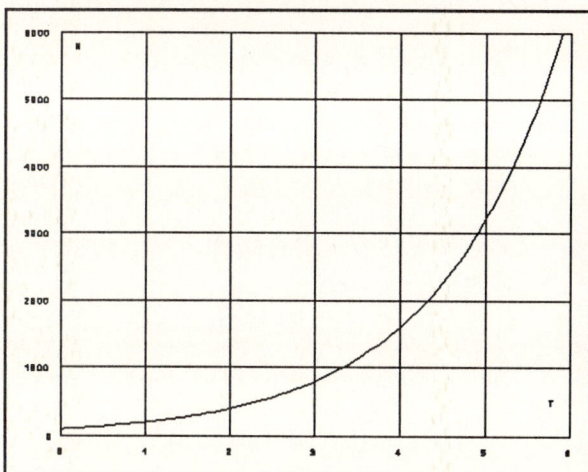

38. Graph $f(x) = e^{-x^2}$.

Make a table of values, plot the points, and connect them with a smooth curve. Be careful with the minus sign. Remember that $-x^2$ is the same as $-(x^2)$ **not** $(-x)^2$. The graph is given below.

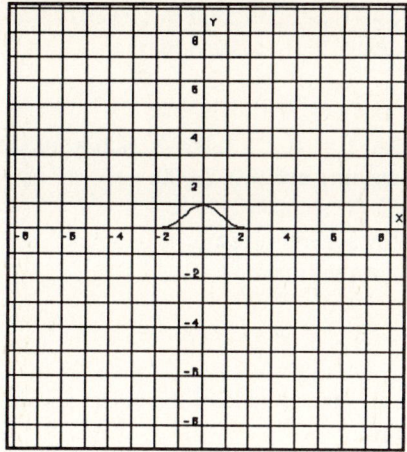

39. $\log_\pi \pi^{2x+1} = 2x + 1$ since $\log_a a^u = u$ for any positive real number u.

40. $(1.02)^{-50} = 0.3715$ Use the y^x button with $y = 1.02$ and $x = -50$.

41. $2\log_a(x^2 - y^2) - \log_a(x + y)$
$$= \log_a(x^2 - y^2)^2 - \log_a(x + y)$$
$$= \log_a \frac{(x^2 - y^2)^2}{x + y}$$
$$= \log_a \frac{(x - y)^2(x + y)^2}{(x + y)}$$
$$= \log_a (x - y)^2(x + y)$$

42. $\log_a y^5 \sqrt[3]{\frac{x^2}{z^2}} = \log_a y^5 + \log_a \sqrt[3]{\frac{x^2}{z^2}}$
$$= \log_a y^5 + \frac{1}{3}\log_a \frac{x^2}{z^2}$$
$$= \log_a y^5 + \frac{1}{3}[\log_a x^2 - \log_a z^2]$$
$$= \log_a y^5 + \frac{1}{3}\log_a x^2 - \frac{1}{3}\log_a z^2$$
$$= 5\log_a y + \frac{2}{3}\log_a x - \frac{2}{3}\log_a z$$

43. The number of bacteria N in a culture is given in terms of time t, in hours, by $N = (3000)2^t$.

(a) To find the number of bacteria in 6 hours, substitute 6 fot t and evaluate.

$$N = (3000)2^6 = (3000)(64) = 192{,}000$$

(b) To find the number of hours for the bacteria count to reach 3,072,000, substitute 3,072,000 for N and solve for t.

$$3{,}072{,}000 = (3000)2^t$$
$$1024 = 2^t$$
$$2^{10} = 2^t$$

Thus, $t = 10$. It takes 10 hours for the culture count to reach 3,072,000.

44. The atmospheric pressure P in pounds per square inch is approximated by

$$P = 14.7e^{-0.2x},$$

where x is the altitude of the object above sea level in miles.

(a) To find the pressure on an object 6 miles high, substitute 6 for x and evaluate.

$$P = 14.7e^{-0.2(6)}$$
$$= 14.7e^{-1.2}$$
$$\approx 4.43 \text{ lb/in}^2$$

(b) To find the height of a plane when the pressure reading on the surface of the plane is 4.89 lb/in², substitute 4.89 for P and solve for x.

$$4.89 = 14.7e^{-0.2x}$$
$$\frac{4.89}{14.7} = e^{-0.2x}$$
$$\ln\frac{4.89}{14.7} = -0.2x$$
$$\frac{\ln\frac{4.89}{14.7}}{-0.2} = x$$
$$5.503275952 = x$$

Thus, the altitude of the plane is about 5.5 mi.

45. The demand for x units of a product relative to the price per unit P is given by

$$P = 1000 - 0.5e^{0.003x}.$$

(a) To find the price if the demand is 500 units, substitute 500 for x and evaluate.

$$P = 1000 - 0.5e^{0.003(500)}$$
$$= 1000 - 0.5e^{1.5}$$
$$= 997.7591555$$

Thus the price is $997.76.

(b) To find the demand when the price is $850, substitute 850 for P and solve for x.

$$850 = 1000 - 0.5e^{0.003x}$$
$$-150 = -0.5e^{0.003x}$$
$$300 = e^{0.003x}$$
$$\ln 300 = 0.003x$$
$$\frac{\ln 300}{0.003} = x$$
$$1901.260825 = x$$

Thus, the demand is for 1901 units when the price per unit is $850.

46. The pH of a substance is given by

$$\text{pH} = -\log[H^+],$$

where $[H^+]$ is the concentration of hydrogen ions in the solution.

(a) To find the pH of a mixed drink with hydrogen ion concentration 2.75×10^{-7} moles per liter, substitute and evaluate.

$$\text{pH} = -\log(2.75 \times 10^{-7}) = 6.560667306$$

Thus, the pH is about 6.6.

(b) To find the hydrogen ion concentration of the water in a lake with a pH of 5.1, substitute and solve for $[H^+]$.

$$5.1 = -\log[H^+]$$

Then

$$-5.1 = \log[H^+],$$

and by changing to exponential form we obtain

$$[H^+] = 10^{-5.1} = 7.94 \times 10^{-6}.$$

47. The limiting magnitude L of a telescope is related to the diameter d of the lens by

$$L = 8.8 + 2.2 \ln d.$$

(a) To find the limiting magnitude of a telescope with a 3.5-inch lens, substitute 3.5 for d and evaluate.

$$L = 8.8 + 2.2 \ln 3.5 = 11.55607853$$

Thus the limiting magnitude is about 11.6.

(b) To find the diameter of the lens of a telescope with a limiting magnitude of 13, substitute and solve for d.

$$13 = 8.8 + 2.2\ln d$$
$$4.2 = 2.2\ln d$$
$$\frac{4.2}{2.2} = \ln d$$
$$e^{\frac{4.2}{2.2}} = d$$
$$6.746952417 = d$$

Thus, the diameter of the lens is about 6.7 in.

48. The amount of a drug remaining in the circulatory system x hours after a 15 mg dose is taken is approximated by

$$A = 15e^{0.25x}.$$

(a) To find the amount remaining in the system after 4 hours, substitute 4 for x and evaluate.

$$A = 15e^{-0.25(4)} = 15e^{-1} = 5.518191618$$

Thus, about 5.5 mg remain after 4 hours.

(b) Substitute 10 for A and solve for x to find the number of hours for the drug to be effective.

$$10 = 15e^{-0.25x}$$
$$\frac{10}{15} = e^{-0.25x}$$
$$\ln\frac{2}{3} = -0.25x$$
$$\frac{\ln\frac{2}{3}}{-0.25} = x$$
$$1.621860432 = x$$

Thus, the drug will be effective about 1.62 hr.

49. The average decrease in a test score when the test was repeated on monthly intervals, where x is the number months after testing, is given by

$$D = 75 - 15\ln(x + 1).$$

(a) To find the average score 6 months after testing, substitute 6 for x and evaluate.

$$D = 75 - 15\ln(6 + 1) = 75 - 15\ln 7 \approx 45.8$$

(b) To find the number of months for the score to fall to 30 substitute 30 for D and solve for x.

$$30 = 75 - 15\ln(x+1)$$
$$-45 = -15\ln(x+1)$$
$$3 = \ln(x+1)$$
$$e^3 = x + 1$$
$$e^3 - 1 = x$$
$$19.08553692 = x$$

Thus, it took about 19 months for the average score to fall below 30 points.

50. Use Newton's law of cooling,

$$T = T_c + (T_0 - T_c)e^{-kt}$$

and substitute 72 for T_c, 130 for T_0, 105 for T, and 10 for t to find the value of k.

$$105 = 72 + (130 - 72)e^{-10k}$$
$$33 = 58e^{-10k}$$
$$\frac{33}{58} = e^{-10k}$$
$$\ln\frac{33}{58} = -10k$$
$$\frac{\ln\frac{33}{58}}{-10} = k$$
$$0.056 \approx k$$

Use this value of k and substitute 30 for t to find the temperature of the object after 30 minutes.

$$T = 72 + (130 - 72)e^{-0.056(30)}$$
$$= 72 + 58e^{-1.68}$$
$$= 82.80969061$$

Thus, in 30 minutes, the temperature of the object will be about 82.8°F.

51. If $x = \ln 42.5$, then $x = 3.7495$.

52. If $\ln x = -0.5138$, then $x = 0.5982$.

53. $x = \log_{3.5} 27.8 = \dfrac{\log 27.8}{\log 3.5} = 2.6542$

54. Since $e^{\ln x} = x$, we have $x = 5$.

55. Solve $\log_3(8x + 1) - \log_3(x - 7) = 3$.

First rewrite the left side as a single logarithm.

$$\log_3 \frac{8x+1}{x-7} = 3$$

Convert to exponential form.

$$\frac{8x+1}{x-7} = 3^3$$
$$8x + 1 = 27(x - 7)$$
$$8x + 1 = 27x - 189$$
$$190 = 19x$$
$$10 = x$$

Since 10 checks in the original equation, the solution is 10.

56. Solve $\log_2 16^{2x+1} = 8$.

First convert to exponential form.

$$2^8 = 16^{2x+1}$$

Then write each side using the same base.

$$2^8 = (2^4)^{2x+1}$$
$$2^8 = 2^{4(2x+1)}$$

Then equate the exponents.

$$8 = 4(2x + 1)$$
$$8 = 8x + 4$$
$$4 = 8x$$
$$\frac{1}{2} = x$$

Since $\frac{1}{2}$ checks in the original equation, the solution is $\frac{1}{2}$.

57. Solve $\log_a a^{x^2} = x$.

Since the left side reduces to x^2, we must solve

$$x^2 = x$$
$$x^2 - x = 0$$
$$x(x - 1) = 0$$
$$x = 0 \quad x - 1 = 0$$
$$x = 1$$

Since both 0 and 1 check in the original equation, the solutions are 0 and 1.

58. Solve $e^{-x} = 5$.

Take the natural log of both sides.

$$\ln e^{-x} = \ln 5$$

$$-x = \ln 5$$
$$x = -\ln 5 \approx -1.6094$$

59. false $\quad \left(\log_a \frac{u}{v} = \log_a u - \log_a v \right)$

60. false (The log of a product is the difference of the logs, but the log of a difference is *not* the product of the logs.)

CHAPTER 5 EXPONENTIAL AND LOGARITHMIC FUNCTIONS

CHAPTER 5 Test

1. Solve $\log_2 x = 3$.

   First convert to exponential form.

   $$2^3 = x$$

   Then $x = 8$.

2. Solve $2^a = \frac{1}{4}$.

   Write both sides using 2 as a base.

   $$2^a = \frac{1}{2^2}$$
   $$2^a = 2^{-2}$$

   Since the bases are the same, equate the exponents. Then $a = -2$.

3. $\log_6 6^{1.2} = 1.2$ since $\log_a a^u = u$ for every real number u.

4. Given that the number of bacteria N in a culture is given in terms of time t, in hours, by

   $$N = (8000)2^t,$$

   to find the number of bacteria present when $t = 2.5$ hours, substitute and evaluate using the y^x button.

   $$N = (8000)2^{2.5} = 45,254.834$$

   Thus, the number of bacteria is about 45,255.

5. $\log_a 25 = \log_a 5^2 = 2\log_a 5 = 2(0.8982) = 1.7964$

6. $\log_a \frac{xy^2}{\sqrt[3]{z}} = \log_a xy^2 - \log_a \sqrt[3]{z}$

   $= \log_a x + \log_a y^2 - \log_a z^{1/3}$

   $= \log_a x + 2\log_a y - \frac{1}{3}\log_a z$

7. $\frac{1}{2}\log_a x - 2\log_a z + 3\log_a y$

   $= \log_a x^{1/2} - \log_a z^2 + \log_a y^3$

   $= \log_a \frac{y^3 x^{1/2}}{z^2}$

   $= \log_a \frac{y^3 \sqrt{x}}{z^2}$

8. $\log_5 423 = \frac{\log 423}{\log 5}$

   $= 3.7574$ *Divide, don't subtract!*

9. Graph $y = e^{3x}$.

   Make a table of values, plot the points, and connect them with a smooth curve. To find values, enter the number $3x$ then press the INV and ln$x$ buttons. For example, if $x = 1$, enter $3(1) = 3$, and press INV followed by ln$x$ to obtain 20.1, correct to the nearest tenth. The graph is given below.

   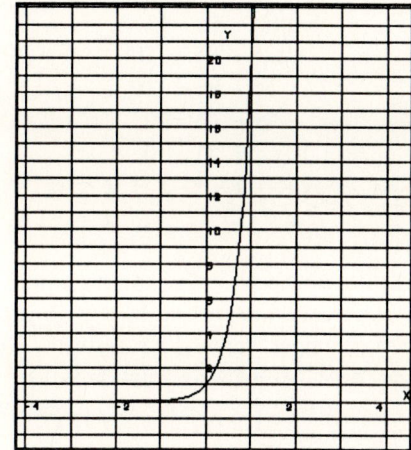

10. Solve $4^{2x} = 8^{3x-20}$.

    Write both sides using 2 as a base.

    $$(2^2)^{2x} = (2^3)^{3x-20}$$
    $$2^{4x} = 2^{3(3x-20)}$$

    Then equate the exponents.

    $$4x = 3(3x - 20)$$
    $$4x = 9x - 60$$
    $$-5x = -60$$
    $$x = 12$$

    Since 12 does check in the original equation, the solution is 12.

11. Solve $5^x = 3^{3x+1}$.

    Since we cannot write both sides using the same base, take the common log of both sides.

$$\log 5^x = \log 3^{2x+1}$$
$$x \log 5 = (2x + 1)\log 3$$
$$x \log 5 = 2x \log 3 + \log 3$$
$$x \log 5 - 2x \log 3 = \log 3$$
$$x(\log 5 - 2\log 3) = \log 3$$
$$x = \frac{\log 3}{\log 5 - 2\log 3}$$
$$x = -1.869066371$$

Thus, correct to three significant digits, the solution is -1.87.

12. Solve $\log x + \log(x - 9) = 1$.

First write the left side as a single logarithm using the quotient rule, then convert to exponential form.

$$\log x(x - 9) = 1$$
$$x(x - 9) = 10^1$$
$$x^2 - 9x = 10$$
$$x^2 - 9x - 10 = 0$$
$$(x - 10)(x + 1) = 0$$
$$x - 10 = 0 \quad x + 1 = 0$$
$$x = 10 \quad x = -1$$

But -1 does not check in the original equation since using -1 forces us to try and take the logarithm of a negative number which is not defined. However, 10 does check, so the only solution is 10.

13. Solve $\log_3 x + \log_5 x = 4$.

First convert all logarithms to the same base.

$$\frac{\log x}{\log 3} + \frac{\log x}{\log 5} = 4$$
$$\log x \left[\frac{1}{\log 3} + \frac{1}{\log 5}\right] = 4$$
$$\log x = \frac{4}{\left[\frac{1}{\log 3} + \frac{1}{\log 5}\right]}$$
$$\log x = 1.134243429$$
$$x = 13.62208006$$

Don't forget to take the antilog in the last step to obtain the value of x. Thus, correct to three significant digits, the solution is 13.6.

14. Substitute 0.03 for k, 200 for T_0, 68 for T_c, and 10 for t in Newton's law of cooling,

$$T = T_c + (T_0 - T_c)e^{-kt},$$

and evaluate.

$$T = 68 + (200 - 68)e^{-0.03(10)}$$
$$= 68 + 132e^{-0.3}$$
$$= 165.7880051$$

Thus, to the nearest degree, the temperature of the object will be about 166°F after 10 minutes.

15. Use the formula

$$A = P\left(1 + \frac{r}{k}\right)^{kt}$$

with $P = 5000$, $r = 0.08$, $k = 4$, and $t = 6$.

$$A = 5000\left(1 + \frac{0.08}{4}\right)^{(4)(6)}$$
$$= 5000(1 + 0.02)^{24}$$
$$= 5000(1.02)^{24}$$
$$\approx \$8042.19$$

Thus, the value of the account will be $8042.19.

16. The loudness of a sound in decibels is given by

$$D = 10 \log \frac{S}{S_0},$$

Where $S_0 = 10^{-12}$ watt/m² is the measure of a sound at the threshold of human hearing, that is, the zero level for decibels. To find the decibel level of a sound that produces an intensity of 2.5×10^{-3} watt/m², substitute this value for S and evaluate D.

$$D = 10 \log \frac{2.5 \times 10^{-3}}{10^{-12}}$$
$$= 10 \log(2.5 \times 10^9)$$
$$= 10(9.397940009) = 93.97940009$$

Thus, the decibel level of the sound is about 94 decibels.

17. Use the Malthusian model

$$N = Ie^{kt}$$

with $N = 2I$, $k = 0.015$, and solve for t. Notice that k is positive since the population is increasing.

$$2I = Ie^{0.015t}$$
$$2 = e^{0.015t}$$
$$\ln 2 = \ln e^{0.015t}$$
$$\ln 2 = 0.015t$$
$$\frac{\ln 2}{0.015} = t$$
$$46.20981204 = t$$

Thus, it will take about 46 yr for the population to double.

PRACTICE EXERCISES SECTION 6.1

CHAPTER 6 SYSTEMS OF EQUATIONS AND INEQUALITIES

SECTION 6.1 Linear Systems in Two Variables

1. Without solving, determine the number of solutions.

   (a) $3x - 2y = -1$
   $6x - 4y = -2$

   Write each equation in slope-intercept form.

   $3x - 2y = -1$ $6x - 4y = -2$
   $-2y = -3x - 1$ $-4y = -6x - 2$
   $y = \frac{3}{2}x + \frac{1}{2}$ $y = \frac{3}{2}x + \frac{1}{2}$

   Since both slopes are $\frac{3}{2}$ and both y-intercepts are $\left(0, \frac{1}{2}\right)$ the lines coincide, there are infinitely many solutions, and the system is dependent.

   (b) $3x - 5y = -1$
   $-6x + 15y = 1$

   Write each equation in slope-intercept form.

   $2x - 5y = -1$ $-6x + 15y = 1$
   $-5y = -2x - 1$ $15y = 6x + 1$
   $y = \frac{2}{5}x + \frac{1}{5}$ $y = \frac{2}{5}x + \frac{1}{15}$

   Since both slopes are $\frac{2}{5}$ but the y-intercepts are different, the lines are parallel, there is no solution, and the system is inconsistent.

   (c) $2x - y + 5 = 0$
   $3y + 6 = 0$

   Write each equation in slope-intercept form.

   $2x - y = -5$ $3y = 6$
   $-y = -2x - 5$ $y = 2$
   $y = 2x + 5$ $y = 0x + 2$

   Since the slopes are 2 and 0, the slopes are different so the lines intersect, there is exactly one solution, and the system is consistent and independent.

2. Solve by the substitution method.

   $3x + 5y = 6$
   $7x - y = 14$

   Solve the second equation for y,

   $-y = -7x + 14$
   $y = 7x - 14,$

   and substitute into the first.

   $3x + 5(7x - 14) = 6$
   $3x + 35x - 70 = 6$
   $38x = 76$
   $x = 2$

   Use this value for x and find y.

   $y = 7x - 14$
   $y = 7(2) - 14$
   $= 14 - 14$
   $= 0$

   Thus, the solution os (2,0).

3. Solve by the elimination method.

   $2x + 7y = 17$
   $5x - 8y = -34$

   Multiply the first equation by -5, and multiply the second by 2.

   $-10x - 35y = -85$
   $10x - 16y = -68$

   Add to eliminate x.

   $-51y = -153$
   $y = 3$

   Substitute 3 for y in the first original equation to obtain the numerical value of x.

   $2x + 7(3) = 17$
   $2x + 21 = 17$
   $2x = -4$
   $x = -2$

   Thus, the solution is (−2,3).

4. Solve by either method.

$$4x - 2y = -6$$
$$-2x + y = 3$$

Suppose we solve the second equation for y,

$$y = 2x + 3,$$

and substitute into the first.

$$4x - 2(2x + 3) = -6$$
$$4x - 4x - 6 = -6$$
$$-6 = -6$$

Since we obtain an identity, the system is dependent and has infinitely many solutions. Since $y = 2x + 3$, the solutions take the form $(x, 2x+3)$ for x any real number.

5. Solve by either method.

$$4x - 2y = 3$$
$$-2x + y = 3$$

Suppose we solve the second equation for y,

$$y = 2x + 3,$$

and substitute into the first.

$$4x - 2(2x + 3) = 3$$
$$4x - 4x - 6 = 3$$
$$-6 = 3$$

Since we obtain a contradiction, the system is inconsistent and has no solution.

EXERCISES A SECTION 6.1 257

CHAPTER 6 SYSTEMS OF EQUATIONS AND INEQUALITIES

SECTION 6.1 Linear Systems in Two Variables

1. Determine whether $(2,-3)$ is a solution to the given system.

   (a) $x + y = -1$
   $3x - y = 9$

   Substitute 2 for x and -3 for y in each equation.

   $2 + (-3) = -1 \qquad 3(2) - (-3) = 9$
   $\qquad -1 = -1 \qquad\qquad\quad 9 = 9$

   Since $(2,-3)$ solves both equations, $(2,-3)$ is a solution to the system.

   (b) $2x - 4 = 0$
   $y + 3 = 0$

   Substitute 2 for x and -3 for y in each equation.

   $2(2) - 4 = 0 \qquad (-3) + 3 = 0$
   $\qquad 0 = 0 \qquad\qquad\quad 0 = 0$

   Since $(2,-3)$ solves both equations, $(2,-3)$ is a solution to the system.

2. Given the system $3x - y = 2$
   $\qquad\qquad\qquad\quad x - 3y = 2$.

   (a) To determine the number of solutions to the system, write each equation in slope-intercept form.

   $3x - y = 2 \qquad\qquad x - 3y = 2$
   $-y = -3x + 2 \qquad -3y = -x + 2$
   $y = 3x - 2 \qquad\qquad y = \frac{1}{3}x - \frac{2}{3}$

   Since the slopes are different, the lines intersect so there is exactly one solution to the system.

   (b) The lines intersect.

   (c) The system is consistent and independent.

3. Given the system $3x + 2y = -5$
   $\qquad\qquad\qquad\quad 6x + 4y = 5$.

   (a) To determine the number of solutions to the system, write each equation in slope-intercept form.

   $3x + 2y = -5 \qquad\qquad 6x + 4y = 5$
   $2y = -3x - 5 \qquad\qquad 4y = -6x + 5$
   $y = -\frac{3}{2}x - \frac{5}{2} \qquad\qquad y = -\frac{3}{2}x + \frac{5}{4}$

   Since the slopes are equal and the y-intercepts are unequal, the lines are parallel and the system has no solution.

   (b) The lines are parallel.

   (c) The system is inconsistent.

4. Given the system $5x - 3y + 2 = 0$
   $\qquad\qquad\qquad\quad -5x + 3y - 2 = 0$.

   (a) To determine the number of solutions to the system, write each equation in slope-intercept form.

   $5x - 3y + 2 = 0 \qquad\qquad -5x + 3y - 2 = 0$
   $5x + 2 = 3y \qquad\qquad\quad -5x - 2 = -3y$
   $\frac{5}{3}x + \frac{2}{3} = y \qquad\qquad\qquad \frac{5}{3}x + \frac{2}{3} = y$

   Since the slopes are equal and the y-intercepts are equal, the lines coincide and the system has infinitely many solutions.

   (b) The lines coincide.

   (c) The system is dependent.

5. Given the system $x + 5 = 0$
   $\qquad\qquad\qquad\quad 5 + y = 0$.

   (a) To determine the number of solutions to the system, notice that the first equation has as its graph a vertical line, and the second equation has a horizontal line for its graph. Since a vertical line and a horizontal lint must intersect, the system has exactly one solution.

   (b) The lines intersect.

   (c) The system is consistent and independent.

6. Solve using the substitution method.

   $x + 2y = 7$
   $2x + y = 2$

Solve the first equation for x, $x = 7 - 2y$, and substitute this expression for x in the second equation.

$$2(7 - 2y) + y = 2$$
$$14 - 4y + y = 2$$
$$-3y = -12$$
$$y = 4$$

Substitute 4 for y in $x = 7 - 2y$ to find the value of x.

$$x = 7 - 2(4) = 7 - 8 = -1$$

Thus, the solution is $(-1, 4)$.

7. Solve using the substitution method.

$$5x - 2y = 3$$
$$-10x + 4y = -6$$

Solve the first equation for y,

$$-2y = -5x + 3$$
$$y = \frac{5}{2}x - \frac{3}{2},$$

and substitute this expression for y in the second equation.

$$-10x + 4\left(\frac{5}{2}x - \frac{3}{2}\right) = -6$$
$$-10x + 10x - 6 = -6$$
$$0 = 0$$

Since we obtain an identity, the system is dependent and has infinitely many solutions. The solutions can be expressed as

$$\left(x, \frac{5}{2}x - \frac{3}{2}\right) \quad \text{for } x \text{ any real number.}$$

8. Solve using the substitution method.

$$3x - y = 2$$
$$2y - 8 = 0$$

Solve the second equation for y.

$$2y - 8 = 0$$
$$2y = 8$$
$$y = 4$$

Then we obtain the numerical value of y immediately. Substitute 4 for y in the first equation to find the value of x.

$$3x - 4 = 2$$
$$3x = 6$$
$$x = 2$$

Thus, the solution is $(2, 4)$.

9. Solve by the elimination method.

$$2x - 5y = 6$$
$$4x + 3y = 12$$

Suppose we try to eliminate x. Multiply the first equation by -2.

$$-4x + 10y = -12$$
$$4x + 3y = 12$$

Adding these two equations we obtain

$$13y = 0$$
$$y = 0$$

Substitute 0 for y in the first original equation and solve for x.

$$2x - 5(0) = 6$$
$$2x = 6$$
$$x = 3$$

Thus, the solution is $(3, 0)$.

10. Solve by the elimination method.

$$2s - 11t = 3$$
$$-4s + 22t = -1$$

Multiply the first equation by 2.

$$4s - 22t = 6$$
$$-4s + 22t = -1$$

Adding these two equations we obtain the contradiction

$$0 = 5.$$

This means that the system is inconsistent and there is no solution.

11. Solve by the elimination method.

$$0.3x - 0.7y = 1.6$$
$$0.5x + 0.6y = 0.9$$

First eliminate the decimals by multiplying both equations by 10.

$$3x - 7y = 16$$
$$5x + 6y = 9$$

Now multiply the first equation by -5, and multiply the second by 3.

$$-15x + 35y = -80$$
$$15x + 18y = 27$$

EXERCISES A

Adding we obtain

$$53y = -53$$
$$y = -1.$$

Substitute -1 for y in the equation $3x - 7y = 16$.

$$3x - 7(-1) = 16$$
$$3x + 7 = 16$$
$$3x = 9$$
$$x = 3$$

Thus, the solution is $(3,-1)$.

12. Solve by either method.

$$5x - 7y = -2$$
$$7x - 5y = 2$$

Multiply the first equation by -7, and multiply the second by 5.

$$-35x + 49y = 14$$
$$35x - 25y = 10$$

Adding we obtain

$$24y = 24$$
$$y = 1$$

Substitute 1 for y in the first original equation.

$$5x - 7(1) = -2$$
$$5x - 7 = -2$$
$$5x = 5$$
$$x = 1$$

Thus, the solution is $(1,1)$.

13. Solve by either method.

$$5u - 3v - 2 = 0$$
$$-5u + 3v - 1 = 0$$

If we add the two equations like they are, we eliminate both variables and obtain the contradiction

$$-3 = 0.$$

Thus, the system is inconsistent and there is no solution.

14. Solve by either method.

$$\frac{3}{2}x - \frac{1}{3}y = \frac{9}{20}$$
$$\frac{3}{4}x + \frac{2}{9}y = \frac{11}{10}$$

It is often best to clear all fractions first, but in this case, if we multiply the second equation by -2, we can add to eliminate x fairly easily.

$$\frac{3}{2}x - \frac{1}{3}y = \frac{9}{20}$$
$$-\frac{3}{2}x - \frac{4}{9}y = -\frac{22}{10}$$

Adding we obtain

$$-\frac{7}{9}y = -\frac{35}{20}$$
$$y = \left(-\frac{35}{20}\right)\left(-\frac{9}{7}\right)$$
$$y = \frac{9}{4}$$

Substitute $\frac{9}{4}$ for y in the first equation and solve for x.

$$\frac{3}{2}x - \left(\frac{1}{3}\right)\left(\frac{9}{4}\right) = \frac{9}{20}$$
$$\frac{3}{2}x - \frac{3}{4} = \frac{9}{20}$$
$$\frac{3}{2}x = \frac{9}{20} + \frac{3}{4}$$
$$\frac{3}{2}x = \frac{24}{20}$$
$$x = \left(\frac{6}{5}\right)\left(\frac{2}{3}\right)$$
$$x = \frac{4}{5}$$

Thus, the solution is $\left(\frac{4}{5}, \frac{9}{4}\right)$.

15. Solve by either method.

$$0.02x + 1.05y = -1.07$$
$$0.1x - 0.6y = 0.5$$

First clear the decimals by multiplying the first equation by 100 and the second by 10.

$$2x + 105y = -107$$
$$x - 6y = 5$$

Solve the second equation for x, $x = 6y + 5$, and substitute this expression into the first equation.

$$2(5 + 6y) + 105y = -107$$
$$10 + 12y + 105y = -107$$
$$117y = -117$$
$$y = -1$$

Then since $x = 6y + 5$, substitute -1 for y to find the value of x.

$$x = 6(-1) + 5 = -6 + 5 = -1$$

Thus, the solution is $(-1,-1)$.

16. Solve for x and y assuming that a and b are nonzero constants.

$$ax + by = 1$$
$$3ax - by = -5$$

If we add the two equations, we can eliminate y and obtain:

$$4ax = -4$$
$$x = \frac{-4}{4a} = -\frac{1}{a}$$

Substitute $-\frac{1}{a}$ for x in the first equation and solve for y.

$$a\left(-\frac{1}{a}\right) + by = 1$$
$$-1 + by = 1$$
$$by = 2$$
$$y = \frac{2}{b}$$

Thus, the solution is $\left(-\frac{1}{a}, \frac{2}{b}\right)$.

17. To find a and b so that $(-1,3)$ is a solution to the system

$$ax - by = 11$$
$$-2ax - by = 14$$

substitute -1 for x and 3 for y to obtain a system of two equations in a and b.

$$-a - 3b = 11$$
$$2a - 3b = 14$$

By subtracting the top equation from the bottom equation we can eliminate b and obtain:

$$3a = 3$$
$$a = 1$$

Substitute 1 for a in the top equation, $-a - 3b = 11$, and solve for b.

$$-1 - 3b = 11$$
$$-3b = 12$$
$$b = -4$$

Thus, when $a = 1$ and $b = -4$, $(-1,3)$ will solve the given system.

18. To find the value(s) of m so that the system

$$x - 4y = m$$
$$-2x + 8y = 4$$

has no solution, we begin by trying to solve the system. Multiply the first equation by 2.

$$2x - 8y = 2m$$
$$-2x + 8y = 4$$

Adding we obtain:

$$0 = 2m + 4$$
$$-4 = 2m$$
$$-2 = m$$

If the system is to have no solution, then we should obtain a contradiction when we try to solve it. As a result, if m is any number **except** -2, we would have a contradiction and the system will have no solution.

19. To show that the system

$$2x + y = m$$
$$x - 4y = 3$$

has exactly one solution for every real number m, suppose we write both equations in slope-intercept form.

$$2x + y = m \qquad x - 4y = 3$$
$$y = -2x + m \qquad -4y = -x + 3$$
$$y = \tfrac{1}{4}x - \tfrac{3}{4}$$

Since the two lines have different slopes, regardless of the value of m, the two lines will always intersect and the system will always have exactly one solution.

20. To find the value(s) of m so that the system

$$5x + 2y = m$$
$$-15x - 6y = 9$$

has infinitely many solutions, suppose we try to solve the system. Multiply the first equation by 3.

$$15x + 6y = 3m$$
$$-15x - 6y = 9$$

Adding we obtain:

$$0 = 3m + 9$$
$$-9 = 3m$$
$$-3 = m$$

For the system to have infinitely many solutions,

EXERCISES A

when we try to solve it, we should obtain an identity. This will happen whenever m is -3, since we would then obtain $0 = 0$ when we solve the system. Thus, if m is -3, the system has infinitely many solutions.

21. Substitute -1 for x, 3 for y, and -2 for z in

$$2x + 4y - z.$$

$$2(-1) + 4(3) - (-2) = -2 + 12 + 2 = 12$$

22. Substitute -1 for x, 3 for y, and -2 for z in

$$-3x + y + 5z.$$

$$-3(-1) + (3) + 5(-2) = 3 + 3 - 10 = -4$$

23. Substitute 2 for x and -3 for z in

$$2x - y + 3z = -3$$

and solve for y.

$$\begin{aligned} 2(2) - y + 3(-3) &= -3 \\ 4 - y - 9 &= -3 \\ -y &= 2 \\ y &= -2 \end{aligned}$$

24. Substitute 2 for x and -3 for z in

$$7x + 2y - z = 17$$

and solve for y.

$$\begin{aligned} 7(2) - y - (-3) &= 17 \\ 14 - y + 3 &= 17 \\ -y &= 0 \\ y &= 0 \end{aligned}$$

CHAPTER 6 SYSTEMS OF EQUATIONS AND INEQUALITIES

SECTION 6.2 Linear Systems in More Than Two Variables

1. Solve the system.

$$3x + 2y + 3z = 3$$
$$4x - 5y + 7z = 1$$
$$2x + 3y - 2z = 6$$

Notice that no one variable is really easier to eliminate than any other. Suppose we eliminate x. Multiply the first equation by -2, and multiply the third equation by 3.

$$-6x - 4y - 6z = -6$$
$$6x + 9y - 6z = 18$$

Add to eliminate x.

$$5y - 12z = 12$$

To obtain a second equation is y and z, multiply the third equation by -2 and pair the result with the second equation.

$$4x - 5y + 7z = 1$$
$$-4x - 6y + 4z = -12$$

Add to eliminate x.

$$-11y + 11z = -11$$
$$y - z = 1 \quad \text{Divide by } -11$$

Then we obtain the following system of two equations in y and z.

$$5y - 12z = 12$$
$$y - z = 1$$

Solve the second for y,

$$y = z + 1$$

and substitute into the first.

$$5(z + 1) - 12z = 12$$
$$5z + 5 - 12z = 12$$
$$-7z = 7$$
$$z = -1$$

Substitute -1 for z in $y = z + 1$ to obtain the value of y.

$$y = -1 + 1 = 0$$

Substitute 0 for y and -1 for z in the first original equation to find the value of x.

$$3x + 2(0) + 3(-1) = 3$$
$$3x - 3 = 3$$
$$3x = 6$$
$$x = 2$$

Thus, the solution is $(2, 0, -1)$.

2. Solve the system.

$$2x - y + 3z = 4$$
$$-4x + 2y - 6z = 1$$
$$8x - 3y + 5z = 0$$

Suppose we try to eliminate x. Multiply the first equation by 2 and pair the result with the second equation.

$$4x - 2y + 6z = 8$$
$$-4x + 2y - 6z = 1$$

If we add, not only is x eliminated, but all three variables are eliminated and we obtain the contradiction

$$0 = 1.$$

As a result, the system is inconsistent, and there is no solution.

3. Solve the system.

$$x - 3y + z = 4$$
$$x + 5y - z = 2$$
$$-2x + 2y - z = -7$$

It is easiest to eliminate z since we will not need to multiply by constants before adding. Add the first two equations to obtain

$$2x + 2y = 6,$$

which simplifies to

$$x + y = 3.$$

Then add the first and third equations to obtain

$$-x - y = -3.$$

When we add the two equations that result, we obtain the identity $0 = 0$. Thus, the system is dependent (note that all three original equations have been used) and there are infinitely many solutions. To represent the solutions, solve $x + y = 3$ for y,

$$y = 3 - x,$$

PRACTICE EXERCISES

to write y in terms of x. Then substitute this expression for y in any of the original equations, say the first, and solve for z in terms of x.

$$x - 3(3 - x) + z = 4$$
$$x - 9 + 3x + z = 4$$
$$z = 13 - 4x$$

Then the solution to the system is $(x, 3-x, 13-4x)$, for x any real number.

NOTE: There are actually two other forms in which the system can be written, one in terms of y and the other in terms of z. If you obtained one of these, compare it with the solution above to see that both actually represent the same set of solutions.

4. Solve the nonsquare systems.

(a) $\begin{aligned} 2x + y - z &= 5 \\ x - y + 4z &= 1 \end{aligned}$

Adding the two equations we can eliminate y to obtain

$$3x + 3z = 6,$$

which simplifies to

$$x + z = 2.$$

Solve for z in terms of x.

$$z = 2 - x$$

Substitute this expression for z in the first original equation and solve for y to obtain y in terms of x.

$$2x + y - (2 - x) = 5$$
$$2x + y - 2 + x = 5$$
$$y = 7 - 3x$$

Then the solutions to the system are $(x, 7-3x, 2-x)$, for x any real number.

(b) $\begin{aligned} 5x - y + 2z &= 1 \\ -10x + 2y - 4z &= 1 \end{aligned}$

Multiply the first equation by 2.

$$10x - 2y + 4z = 2$$
$$-10x + 2y - 4z = 1$$

Adding we obtain the contradiction $0 = 3$. Thus, the system has no solution.

5. Solve the system.

$$3x + y - z = 0$$
$$2x - 3y + z = 0$$
$$-x + 5y - 2z = 0$$

Add the first two equations to eliminate z and obtain

$$5x - 2y = 0.$$

Multiply the second equation by 2 and pair the result with the third equation.

$$4x - 6y + 2z = 0$$
$$-x + 5y - 2z = 0$$

Add to eliminate z and obtain

$$3x - y = 0$$

Then the resulting sustem of two equations in x and y is

$$5x - 2y = 0$$
$$3x - y = 0$$

Solve the second for y, $y = 3x$, and substitute into the first.

$$5x - 2(3x) = 0$$
$$5x - 6x = 0$$
$$x = 0$$
$$x = 0$$

Then $y = 3x = 3(0) = 0$. Substitute 0 for x and 0 for y in the first original equation.

$$3(0) + (0) - z = 0$$
$$-z = 0$$
$$z = 0$$

Thus, the only solution is the trivial solution, $(0,0,0)$.

CHAPTER 6 SYSTEMS OF EQUATIONS AND INEQUALITIES

SECTION 6.2 Linear Systems in More Than Two Variables

1. Solve the system.

$$x + y + z = 3$$
$$-x + 2y - z = 0$$
$$3x - y + 2z = 2$$

Adding the first two equations will eliminate x (and also z) and give

$$3y = 3$$
$$y = 1$$

Multiply the second equation by 3 and pair the result with the third equation.

$$-3x + 6y - 3z = 0$$
$$3x - y + 2z = 2$$

Adding we obtain:

$$5y - z = 2$$

Substitute 1 for y in this equation to obtain the value of z.

$$5(1) - z = 2$$
$$-z = -3$$
$$z = 3$$

Substitute 1 for y and 3 for z in the first original equation to find the value of x.

$$x + 1 + 3 = 3$$
$$x = -1$$

Thus, the solution is $(-1,1,3)$.

2. Solve the system.

$$x + y + z = 6$$
$$x \quad - z = -2$$
$$\quad y + 3z = 11$$

Suppose we subtract the second equation from the first to eliminate x.

$$y + 2z = 8$$

Pair this equation with the third original equation to obtain the system of two equations in y and z.

$$y + 2z = 8$$
$$y + 3z = 11$$

Subtract the top equation from the bottom equation to obtain:

$$z = 3$$

Substitute 3 for z in $y + 2z = 8$ to obtain the value of y.

$$y + 2(3) = 8$$
$$y = 2$$

Substitute 3 for z in the second original equation to find the value of x.

$$x - 3 = -2$$
$$x = 1$$

Thus, the solution is $(1,2,3)$.

3. Solve the system.

$$x + 5y - z = 2$$
$$4x - y + 3z = 3$$
$$8x - 2y + 6z = 7$$

Multiply the second equation by -2 and pair the result with the third equation.

$$-8x + 2y - 6z = -6$$
$$8x - 2y + 6z = 7$$

Adding we obtain the contradiction

$$0 = 1,$$

so we know that the system is inconsistent and has no solution.

4. Solve the system.

$$3x + y + z = 0$$
$$-5x + 5y + z = 0$$
$$x + 2y + z = 0$$

Since the system is a homogeneous system, we know that one solution is $(0,0,0)$. To determine whether there are more solutions, we proceed as with solving any system. It is easy to eliminate z. subtract the second equation from the first,

$$8x - 4y = 0$$

which can be simplified to

$$2x - y = 0.$$

Then subtract the third equation from the first.

EXERCISES A

$$2x - y = 0$$

When these two equations are used together to form the system

$$2x - y = 0$$
$$2x - y = 0$$

and we subtract, we obtain the identity

$$0 = 0$$

Which means that the system is dependent and there are infinitely many solutions. Solve $2x - y = 0$ for y to obtain $y = 2x$. Then substitute $2x$ for y in the first original equation and solve for z in terms of x.

$$3x + 2x + z = 0$$
$$5x + z = 0$$
$$z = -5x$$

Thus, we can write the solution as $(x, 2x, -5x)$ for x any real number. Notice that $(0,0,0)$ is one of these solutions resulting when $x = 0$.

5. Solve the system.

$$2x + y = 0$$
$$x - 3y + z = 0$$
$$3x + y - z = 0$$

Since this is a homogeneous system we know that at least one solution is $(0,0,0)$. To determine whether there are more solutions proceed as in solving any system. If we add the second and third equations we will eliminate z and obtain

$$4x - 2y = 0$$

which simplifies to

$$2x - y = 0.$$

Pair this equation with the first original equation to obtain the following system.

$$2x + y = 0$$
$$2x - y = 0$$

Adding these two equations we eliminate y and obtain

$$4x = 0$$
$$x = 0.$$

When x is 0, substituting in $2x + y = 0$ we obtain $y = 0$. And, of course, substituting 0 for both x and y in the second original equation we obtain

SECTION 6.2 265

$z = 0$. Thus, the only solution to this homogeneous system is $(0,0,0)$.

6. Solve the system.

$$x - 3y + 2z = -1$$
$$4x + 3y + 3z = 6$$

Since this system has only two equations in three variables, it has either no solution or infinitely many solutions. Adding the two equations will eliminate y and result in

$$5x + 5z = 5$$

which can be simplified to

$$x + z = 1.$$

Since we do not obtain a contradiction, the system has infinitely many solutions. Solving for z in terms of x we obtain

$$z = 1 - x.$$

Substitute $1 - x$ for z in the first original equation and solve for y in terms of x.

$$x - 3y + 2(1 - x) = -1$$
$$x - 3y + 2 - 2x = -1$$
$$-3y - x = -3$$
$$-3y = x - 3$$
$$y = -\frac{1}{3}x + 1$$

Thus, the infinitely many solutions can be represented by $\left(x, -\frac{1}{3}x + 1, 1 - x\right)$ for x any real number.

7. Solve the system.

$$x - 3y + 5z = 2$$
$$-2x + 6y - 10z = 7$$

Since this system has only two equations in three variables, it has either no solution or infinitely many solutions. Multiply the first equation by 2.

$$2x - 6y + 10z = 4$$
$$-2x + 6y - 10z = 7$$

Adding we obtain the contradiction

$$0 = 11$$

which means that the system has no solution.

8. Solve the system.

$$\frac{1}{4}x - \frac{1}{3}y - \frac{1}{2}z = -2$$
$$\frac{1}{2}x - \frac{1}{2}y + \frac{1}{4}z = 2$$
$$-\frac{1}{4}x + \frac{1}{2}y - \frac{1}{2}z = -1$$

It might be best to eliminate the fractions first. Multiply the first equation by 12, the second by 4, and the third by 4 to obtain the following system.

$$3x - 4y - 6z = -24$$
$$2x - 2y + z = 8$$
$$-x + 2y - 2z = -4$$

Add the second and third equations to eliminate y.

$$x - z = 4$$

Multiply the third equation by 2 and pair the result with the first equation.

$$3x - 4y - 6z = -24$$
$$-2x + 4y - 4z = -8$$

Add to eliminate y.

$$x - 10z = -32$$

Then we obtain the system of two equations in x and z given below.

$$x - z = 4$$
$$x - 10z = -32$$

Subtract the first equation from the second to eliminate x.

$$-9z = -36$$
$$z = 4$$

Substitute 4 for z in $x - z = 4$ to find the value of x.

$$x - 4 = 4$$
$$x = 8$$

Substitute 8 for x and 4 for z in

$$-x + 2y - 2z = -4$$

to find the value of y.

$$-8 + 2y - 2(4) = -4$$
$$-8 + 2y - 8 = -4$$
$$2y = 12$$
$$y = 6$$

Thus, the solution is (8,6,4).

9. Solve the system.

$$\frac{1}{x} - \frac{2}{y} - \frac{1}{z} = 2$$
$$\frac{2}{x} - \frac{1}{y} + \frac{1}{z} = 7$$
$$\frac{3}{x} + \frac{2}{y} + \frac{1}{z} = 2$$

First make a change of variables to eliminate the fractions by substituting

$$u = \frac{1}{x}, \ v = \frac{1}{y}, \ w = \frac{1}{z}.$$

Then the system becomes:

$$u - 2v - w = 2$$
$$2u - v + w = 7$$
$$3u + 2v + w = 2$$

Add the first two equations to eliminate w,

$$3u - 3v = 9$$

which simplifies to

$$u - v = 3.$$

Then add the first and third equations to obtain:

$$4u = 4$$
$$u = 1$$

Substitute 1 for u in $u - v = 3$ to find the value of v.

$$1 - v = 3$$
$$-v = 2$$
$$v = -2$$

Then substitute 1 for u and -2 for v in

$$u - 2v - w = 2$$

to find the value of w.

$$1 - 2(-2) - w = 2$$
$$1 + 4 - w = 2$$
$$-w = -3$$
$$w = 3$$

Then backsubstitute to find the values of x, y, and z.

EXERCISES A

$$x = \frac{1}{u} = 1, \quad y = \frac{1}{v} = -\frac{1}{2}, \quad z = \frac{1}{w} = \frac{1}{3}$$

Thus, the solution to the original system is

$$\left(1, -\frac{1}{2}, \frac{1}{3}\right).$$

10. Solve the system.

$$\begin{aligned} x - y + z + w &= 2 \\ x + y - z + w &= 4 \\ x + y + z - w &= -2 \\ x - y - z - w &= 0 \end{aligned}$$

Suppose we eliminate w. Subtract the second equation from the first to obtain

$$-2y + 2z = -2$$

which simplifies to

$$y - z = 1.$$

Next add the first and the fourth equations to obtain

$$2x + 2z = 0$$

which simplifies to

$$x + z = 0.$$

Finally, add the first and the third equations to obtain

$$2x - 2y = 2$$

which simplifies to

$$x - y = 1.$$

Putting these three simplified equations in x, y, and z together we form the system:

$$\begin{aligned} y - z &= 1 \\ x \quad + z &= 0 \\ x - y \quad &= 1 \end{aligned}$$

Add the first two equations to eliminate z and pair the result with the third equation.

$$\begin{aligned} x + y &= 1 \\ x - y &= 1 \end{aligned}$$

Add these two equations to obtain:

$$2x = 2$$

$$x = 1$$

Substitute 1 for x in $x + y = 1$ to obtain the value of y.

$$\begin{aligned} 1 + y &= 1 \\ y &= 0 \end{aligned}$$

Substitute 1 for x in $x + z = 0$ to obtain the value of z.

$$\begin{aligned} 1 + z &= 0 \\ z &= -1 \end{aligned}$$

Finally, substitute 1 for x, 0 for y, and -1 for z in the first original equation,

$$x - y + z + w = 2,$$

to obtain the value of w.

$$\begin{aligned} 1 - 0 + (-1) + w &= 2 \\ w &= 2 \end{aligned}$$

Thus, the solution is $(1, 0, -1, 2)$, in the order (x, y, z, w).

11. Substitute $(1,0)$, $(-2,-12)$, and $(2,0)$ for x and y in the equation $y = ax^2 + bx + c$ to obtain, respectively, the following system of three equations in a, b, and c.

$$\begin{aligned} a + b + c &= 0 \\ 4a - 2b + c &= -12 \\ 4a + 2b + c &= 0 \end{aligned}$$

Suppose we eliminate c. Subtract the first equation from the second to obtain

$$3a - 3b = -12$$

which simplifies to

$$a - b = -4.$$

Then subtract the first equation from the third to obtain

$$3a + b = 0.$$

Then we have the following system of two equations in the two variables a and b.

$$\begin{aligned} a - b &= -4 \\ 3a + b &= 0 \end{aligned}$$

Adding we eliminate b and obtain

$$4a = -4$$
$$a = -1.$$

Substitute -1 for a in $a - b = -4$ to find the value of b.

$$-1 - b = -4$$
$$-b = -3$$
$$b = 3$$

Then substitute -1 for a and 3 for b in the first original equation, $a + b + c = 0$, to find the value of c.

$$-1 + 3 + c = 0$$
$$c = -2$$

Then with $a = -1$, $b = 3$, and $c = -2$, the equation of the parabola becomes

$$y = ax^2 + bx + c$$
$$= -x^2 + 3x - 2.$$

12. Substitute $(1,2)$, $(4,-1)$, and $(-2,-1)$ for x and y in the equation $x^2 + y^2 + ax + by + c = 0$ to obtain, respectively, the following system of three equations in a, b, and c.

$$a + 2b + c = -5$$
$$4a - b + c = -17$$
$$-2a - b + c = -5$$

Suppose we eliminate c. Subtract the first equation from the second to obtain

$$3a - 3b = -12$$

which simplifies to

$$a - b = -4.$$

Then subtract the first equation from the third to obtain

$$-3a - 3b = 0$$

which simplifies to

$$a + b = 0.$$

Then we obtain the system of two equations in a and b given below.

$$a - b = -4$$
$$a + b = 0$$

Adding we eliminate b and obtain

$$2a = -4$$
$$a = -2.$$

Substitute -2 for a in $a + b = 0$ to find the value of b.

$$-2 + b = 0$$
$$b = 2$$

Finally, substitute -2 for a and 2 for b in the first original equation, $a + 2b + c = -5$, to find the value of c.

$$-2 + 2(2) + c = -5$$
$$-2 + 4 + c = -5$$
$$c = -7$$

Then with $a = -2$, $b = 2$, and $c = -7$, the equation of the circle becomes

$$x^2 + y^2 - 2x + 2y - 7 = 0.$$

If the three given points are collinear, the resulting system of equations will be inconsistent and have no solution. Certainly three collinear points cannot all be on the same circle. You might try three points such as $(1,1)$, $(2,2)$, and $(-1,-1)$ to verify that the system is indeed inconsistent.

13. Solve the system for x and y in terms of the nonzero constants a and b.

$$ax - by = 2$$
$$2ax + by = 4$$

Add the two equations to eliminate b.

$$3ax = 6$$
$$x = \frac{6}{3a}$$
$$x = \frac{2}{a}$$

Substitute this value for x in the first equation to find the value of y.

$$a\left(\frac{2}{a}\right) - by = 2$$
$$2 - by = 2$$
$$-by = 0$$
$$y = \frac{0}{-b}$$
$$y = 0$$

Thus, the solution is $\left(\frac{2}{a}, 0\right)$.

EXERCISES A

14. To find the value(s) for m so that the system

$$2x - 5y = 3$$
$$6x - 15y = m$$

is dependent, suppose we solve each equation for y, that is, write each in slope-intercept form. For the system to be dependent, the slopes must be the same and the y-intercepts must also be the same, that is, the lines must coincide.

$$2x - 5y = 3 \qquad\qquad 6x - 15y = m$$
$$-5y = -2x + 3 \qquad -15y = -6x + m$$
$$y = \tfrac{2}{5}x - \tfrac{3}{5} \qquad\quad y = \tfrac{2}{5}x - \tfrac{m}{15}$$

Then the slopes are indeed equal, both are $\tfrac{2}{5}$, regardless of the value of m. However, if the y-intercepts are to be equal, then we must solve:

$$-\tfrac{3}{5} = -\tfrac{m}{15}$$
$$-9 = -m$$
$$9 = m$$

Thus, for the system to be dependent, $m = 9$.

15. To find the value(s) for m so that the system

$$3x + 7y = 2m$$
$$-6x - 14y = 5$$

is inconsistent, suppose we solve each equation for y, that is, write each in slope-intercept form. For the system to be inconsistent, the slopes must be the same and the y-intercepts must different, that is, the lines must be parallel.

$$3x + 7y = 2m \qquad\quad -6x - 14y = 5$$
$$7y = -3x + 2m \qquad -14y = 6x + 5$$
$$y = -\tfrac{3}{7}x + \tfrac{2m}{7} \qquad y = -\tfrac{3}{7}x - \tfrac{5}{14}$$

Then the slopes are indeed equal, both are $-\tfrac{3}{7}$, regardless of the value of m. However, if the y-intercepts are to be equal, then we must solve:

$$\tfrac{2m}{7} = -\tfrac{5}{14}$$
$$4m = -5$$
$$m = -\tfrac{5}{4}$$

Thus, for the system to be inconsistent, m can be any real number **except** $-\tfrac{5}{4}$.

CHAPTER 6 SYSTEMS OF EQUATIONS AND INEQUALITIES

SECTION 6.3 Problem Solving Using Systems of Equations

1. Let x = the price of one shirt,
   y = price of one pair of socks.

   In this case there are two value equations. Since 5 shirts and 4 pairs of socks were purchased for $87.00, one equation is

   $$5x + 4y = 87.$$

   Since 2 shirts and 6 pairs of socks were purchased for $48.00, the second equation is

   $$2x + 6y = 48,$$

   which simplifies to

   $$x + 3y = 24.$$

   Solve the second equation for x, $x = 24 - 3y$, and substitute this expression in the first equation.

   $$5(24 - 3y) + 4y = 87$$
   $$120 - 15y + 4y = 87$$
   $$-11y = -33$$
   $$y = 3$$

   Then since $x = 24 - 3y$,

   $$x = 24 - 3(3) = 24 - 9 = 15.$$

   Thus, the price of one shirt was $15.00, and the price of one pair of socks was $3.00.

2. Let x = number of quarts of antifreeze,
   y = number of quarts of 20% antifreeze solution.

   Since the radiator holds 14 quarts, the quantity equation is

   $$x + y = 14.$$

   The "value equation" in this case becomes the amount of antifreeze in the two expressions. Since the x quarts are pure antifreeze, the percent of antifreeze is 100% which means that the amount of antifreeze in x quarts is 100% of x which is

   $$(1.00)x = x \text{ quarts}.$$

   The amount of antifreeze in the 20% solution is 20% of y which is

   $$(0.20)y = 0.2y \text{ quarts}.$$

   The amount of antifreeze in the mixture of 14 quarts is 40% of 14, which is

   $$(0.40)14 = 5.6 \text{ qt}.$$

   Thus, the two expressions for the amount of antifreeze are equal giving us the second equation

   $$x + 0.2y = 5.6.$$

   This can be simplified by multiplying both sides by 10 to clear the decimals

   $$10x + 2y = 56$$

   or

   $$5x + y = 28.$$

   Thus, the system to solve is

   $$5x + y = 28$$
   $$x + y = 14$$

   Solve the second equation for y, $y = 14 - x$, and substitute this expression in the first equation.

   $$5x + (14 - x) = 28$$
   $$5x + 14 - x = 28$$
   $$4x = 14$$
   $$x = 3.5$$

   Then since $y = 14 - x$,

   $$y = 14 - 3.5 = 10.5.$$

   Thus, 3.5 qt of antifreeze should be mixed with 10.5 qt of the 20% antifreeze solution to fill the radiator with a 40% antifreeze solution.

3. Let x = speed of boat in still water,
   y = speed of the stream.

   Then $x + y$ = speed of boat downstream,
   $x - y$ = speed of boat upstream.

   Use the distance formula $d = rt$ twice to obtain two equations. Since she sailed 210 miles upstream in

PRACTICE EXERCISES

7 hours (and at a rate of $x - y$), one equation is

$$210 = (x - y)7$$

which can be simplified to

$$x - y = 30.$$

Since the distance downstream was also 210 miles and the time downstream was 6 hours (the rate was $x + y$), a second equation is

$$210 = (x + y)6$$

which can be simplified to

$$x + y = 35.$$

Thus, the system to solve is

$$x - y = 30$$
$$x + y = 35$$

Add to eliminate y and obtain

$$2x = 65$$
$$x = 32.5$$

Substitute 32.5 for x in the equation $x - y = 30$ to find the value of y.

$$32.5 - y = 30$$
$$-y = -2.5$$
$$y = 2.5$$

Thus, the speed of the boat in still water was 32.5 mph, and the speed of the stream was 2.5 mph.

4. Let x = score on the first test,
   y = score on the second test,
   z = score on the third test.

Since the average of the three scores was 88, this translates to the equation

$$\frac{x + y + z}{3} = 88$$
$$x + y + z = 264.$$

Since the first score was 13 points higher than the second, this translates to

$$x = 13 + y$$
$$x - y = 13.$$

Since the third was 5 points higher than the second, this translates to

$$z = 5 + y$$
$$-y + z = 5.$$

Thus, the system to solve is

$$x + y + z = 264$$
$$x - y = 13$$
$$-y + z = 5$$

If we eliminate x from the first two equations, we can pair the result with the third equation obtaining two equations in the two variables y and z. Subtract the second equation from the first which gives

$$2y + z = 251.$$

Then we must solve

$$2y + z = 251$$
$$-y + z = 5$$

Subtract the second from the first to eliminate z and obtain:

$$3y = 246$$
$$y = 82$$

Substitute 82 for y in $-y + z = 5$ to find the value of z.

$$-82 + z = 5$$
$$z = 87$$

Substitute 82 for y in $x - y = 13$ to find the value of x.

$$x - 82 = 13$$
$$x = 95$$

Thus the three scores are 95, 82, and 87.

5. A profit of $50 the first week, $100 the second week, and $300 the third week can be translated into the three data points

$$(1, 50), (2, 100), \text{ and } (3, 300).$$

If x represents the week and y the profit that week, we can assume that the equation

$$y = ax^2 + bx + c$$

can be used as a model for this situation. Substitute the three data points into the equation to obtain the following system.

$$a + b + c = 50$$
$$4a + 2b + c = 100$$
$$9a + 3b + c = 300$$

It is easiest to eliminate c. Suppose we subtract the first equation from the second to obtain

$$3a + b = 50.$$

Then subtract the second equation from the third to obtain

$$5a + b = 200.$$

Then the reduced system of two equations in two variables is

$$3a + b = 50$$
$$5a + b = 200$$

Subtract the first equation from the second to eliminate b.

$$2a = 150$$
$$a = 75$$

Substitute 75 for a in $3a + b = 50$ to find the value of b.

$$3(75) + b = 50$$
$$225 + b = 50$$
$$b = -175$$

Substitute 75 for a and -175 for b in the first original equation, $a + b + c = 50$, to find the value of c.

$$75 + (-175) + c = 50$$
$$-100 + c = 50$$
$$c = 150$$

Then since $a = 75$, $b = -175$, and $c = 150$, the equation describing the data points,

$$y = ax^2 + bx + c,$$

becomes

$$y = 75x^2 - 175x + 150.$$

To find the estimated profit in week 5, substitute 5 for x and evaluate.

$$y = 75(5)^2 - 175(5) + 150$$
$$= 75(25) - 875 + 150$$
$$= 1875 - 725$$
$$= 1150$$

Thus, we would estimate that the profit will be $1150 the fifth week.

EXERCISES A SECTION 6.3

CHAPTER 6 SYSTEMS OF EQUATIONS AND INEQUALITIES

SECTION 6.3 Problem Solving Using Systems of Equations

1. Let x = the first number,
    y = the second number.

    Since the sum of the two numbers is 44, this translates to

    $$x + y = 44.$$

    Since the difference of the two numbers is 20, this translates to

    $$x - y = 20.$$

    Thus we obtain the following system:

    $$x + y = 44$$
    $$x - y = 20$$

    Adding we can eliminate y and obtain

    $$2x = 64$$
    $$x = 32.$$

    Substitute 32 for x in $x + y = 44$ to find the value of y.

    $$32 + y = 44$$
    $$y = 12$$

    Thus, the two numbers are 12 and 32.

2. Let x = the cost of one book,
    y = the cost of one pen.

    Then four books would cost $4x$, six pens would cost $6y$, the the total cost of these items would be $4x + 6y$. Since the total cost is given as $9.00, we obtain the first equation

    $$4x + 6y = 9.$$

    In a similar manner, three books and nine pens that cost a total of $9.00 gives us the second equation

    $$3x + 9y = 9,$$

    which can be simplified to

    $$x + 3y = 3.$$

    Solve this equaiton for x, $x = 3 - 3y$, and substitute into the first.

    $$4(3 - 3y) + 6y = 9$$
    $$12 - 12y + 6y = 9$$
    $$-6y = -3$$
    $$y = 0.5$$

    Then since $x = 3 - 3y$, substitute 0.5 for y to find x.

    $$x = 3 - 3(0.5)$$
    $$= 3 - 1.5$$
    $$= 1.5$$

    Then the cost of one book is $1.50, and the cost of one pen is $0.50.

3. Let x = number of adults at the play,
    y = number of children at the play.

    Since there were 600 total people at the play, the first equation is

    $$x + y = 600.$$

    Since the admission price for adults was $2.00, the x adults would have paid $2x$ dollars. Also, since the price for children was $1.00, the y children would have paid $(1)y = y$ dollars. The total receipts were then $2x + y$ which was given to be $980. Thus, the second equation is

    $$2x + y = 980.$$

    Solve the first equation for x, $x = 600 - y$, and substitute into the second equation.

    $$2(600 - y) + y = 980$$
    $$1200 - 2y + y = 980$$
    $$-y = -220$$
    $$y = 220$$

    Then since $x = 600 - y$, substitute 220 for y to find x.

    $$x = 600 - 220$$
    $$= 380$$

    Thus, the number of adults in attendance was 380, and the number of children was 220.

4. Let x = number pounds of $0.90/lb candy,
   y = number pounds of $1.50/lb candy.

Since the mixture is to contain 60 pounds, the quantity equation is

$$x + y = 60.$$

The value of x pounds of candy worth $0.90 per pound is $0.90x$, and the value of y pounds of candy worth $1.50 per pound is $1.50y$. Adding these gives one expression for the value of the mixture. But since the mixture contains 60 pounds of candy worth $1.10 per pound, a second expression for the value of the mixture is $1.10(60)$. Equating the two expressions we obtain the value equation

$$0.90x + 1.50y = 1.10(60)$$

which can be simplified to

$$3x + 5y = 220.$$

Solve the quantity equation for x, $x = 60 - y$, and substitute this into the simplified value equation.

$$\begin{aligned} 3(60 - y) + 5y &= 220 \\ 180 - 3y + 5y &= 220 \\ 2y &= 40 \\ y &= 20 \end{aligned}$$

Since $x = 60 - y$, substitute 20 for y to find x.

$$x = 60 - 20 = 40$$

Thus, the mix requires 40 lb of the candy worth $0.90/lb and 20 lb of candy worth $1.50/lb.

5. Let x = number of dimes in the collection,
   y = number of quarters in the collection.

Since the total number of coins in the collection is 30, we obtain the quantity equation

$$x + y = 30.$$

The value of x dimes is $10x$ (in cents) and the value of y quarters is $25y$ (also in cents). Since the value of the collection is given to be $4.20, or 420 cents, we obtain the value equation

$$10x + 25y = 420$$

which simplifies to

$$2x + 5y = 84.$$

Solve the quantity equation for x, $x = 30 - y$, and substitute into the equation above.

$$\begin{aligned} 2(30 - y) + 5y &= 84 \\ 60 - 2y + 5y &= 84 \\ 3y &= 24 \\ y &= 8 \end{aligned}$$

Then since $x = 30 - y$, $x = 30 - 8 = 22$.

Thus, there are 22 dimes and 8 quarters in the collection.

6. Let x = number of liters of 25% acid solution,
   y = number of liters of 50% acid solution.

Since the mixture is to consist of a total of 25 L, the quantity equation is

$$x + y = 25.$$

The amount of acid in x liters of a 25% acid solution is $0.25x$, and the amount of acid in y liters of a 50% acid solution is $0.50y$. Adding these we obtain the amount of acid in the mixture, which can also be represented as $0.40(25)$. Thus we obtain

$$0.25x + 0.50y = 0.40(25)$$

which simplifies to

$$x + 2y = 40.$$

Solve the first equation for x, $x = 25 - y$, and substitute into the second.

$$\begin{aligned} 25 - y + 2y &= 40 \\ y &= 15 \end{aligned}$$

Then since $x = 25 - y$, $x = 25 - 15 = 10$.

Thus, the mixture contains 10 L of the 25% solution and 15 L of the 50% solution.

7. Let x = measure of one angle,
   y = measure of second angle.

Since the angles are supplementary, the sum of their measures totals 180°, so one equation is

$$x + y = 180.$$

EXERCISES A

The phrase "one measures 4° more than seven times the other" translates to:

$$x = 7y + 4.$$

Since this equation is already solved for x, substitute this value for x in the first equation.

$$7y + 4 + y = 180$$
$$8y = 176$$
$$y = 22$$

Then since $x = 7y + 4$,

$$x = 7(22) + 4 = 154 + 4 = 158.$$

Thus, the two angles measure 22° and 158°.

8. Let x = time traveled at 40 km/hr,
y = time traveled at 50 km/hr.

Then the distance traveled at 40 km/hr is $40x$, and the distance traveled at 50 km/hr is $50y$ (using the formula $d = rt$). The total of these two is given to be 370 km, so the first equation is

$$40x + 50y = 370$$

which simplifies to

$$4x + 5y = 37.$$

The phrase "had he gone 10 km/hr *faster* throughout" means "had he gone 50 km/hr the first part and 60 km/hr the second" so the second equation is

$$50x + 60y = 450$$

which simplifies to

$$5x + 6y = 45.$$

Multiply the first equation by 5 and the second by −4 to obtain the following system.

$$20x + 25y = 185$$
$$-20x - 24y = -180$$

Adding we eliminate x and obtain

$$y = 5.$$

Substitute 5 for y in $4x + 5y = 37$ to find the value of x.

$$4x + 5(5) = 37$$
$$4x = 12$$
$$x = 3$$

Thus, he traveled 3 hr at 40 km/hr and 5 hr and 50 km/hr.

9. Let x = number times he jogs each month,
y = number times he plays tennis each month.

Since the total number of times for these two activities is 15, the first equation is

$$x + y = 15.$$

Each time he jogs he spends 1.2 hr, so the total time spent jogging is $1.2x$ hr. Also, each time he plays tennis, he spends 2.8 hr for a total time of $2.8y$ hr. Together he can spend a total of 26 hr each month so the second equation is

$$1.2x + 2.8y = 26$$

which can be simplified to

$$3x + 7y = 65.$$

Solve the first equation for x, $x = 15 - y$, and substitute into the second.

$$3(15 - y) + 7y = 65$$
$$45 - 3y + 7y = 65$$
$$4y = 20$$
$$y = 5$$

Since $x = 15 - y$, $x = 15 - 5 = 10$. Thus, he should jog 10 times each month and play tennis 5 times each month.

10. Let x = Milt's present age,
y = Lew's present age,
z = Jenny's present age.

Since the sum of their ages is 53, the first equation is

$$x + y + z = 53.$$

The phrase "Jenny is 5 years younger than Lew" translates to the second equation

$$z = y - 5,$$

or equivalently,

$$y - z = 5.$$

The phrase "in two years, Milt will be the same age as Lew is now" translates to

$$x + 2 = y,$$

or equivalently,

$$x - y = -2.$$

Thus, we have the following system:

$$\begin{aligned} x + y + z &= 53 \\ y - z &= 5 \\ x - y &= -2 \end{aligned}$$

Add the first two equations to eliminate z,

$$x + 2y = 58,$$

and pair this with the third equation to obtain the system

$$\begin{aligned} x + 2y &= 58 \\ x - y &= -2 \end{aligned}$$

Subtract the second equation from the first to obtain:

$$\begin{aligned} 3y &= 60 \\ y &= 20 \end{aligned}$$

Substitute 20 for y in $x - y = -2$ to find the value of x.

$$\begin{aligned} x - 20 &= -2 \\ x &= 18 \end{aligned}$$

Substitute 20 for y in $y - z = 5$ to find the value of z.

$$\begin{aligned} 20 - z &= 5 \\ -z &= -15 \\ z &= 15 \end{aligned}$$

Thus, Milt is 18 years old, Lew is 20 years old, and Jenny is 15 years old.

11. Let x = the number of courtside seats,
    y = the number of endzone seats,
    z = the number of balcony seats.

Since the total number of seats in the arena is 12,000, the first equation is

$$x + y + z = 12,000.$$

The phrase "there are twice as many balcony seats as courtside seats" translates into

$$z = 2x,$$

or equivalently,

$$2x - z = 0.$$

Since each courtside seat sold for $10.00, the revenue on the sale of x courtside seats was $10x$ dollars. Similarly, the revenue for the balcony seats was $8z$ dollars, and the revenue for the endzone seats was $7y$ dollars. The total revenue for the sold-out arena was $99,000, so we have the third equation,

$$10x + 7y + 8z = 99,000.$$

Thus, the system to solve is:

$$\begin{aligned} x + y + z &= 12,000 \\ 2x - z &= 0 \\ 10x + 7y + 8z &= 99,000 \end{aligned}$$

Multiply the first equation by -7,

$$-7x - 7y - 7z = -84,000,$$

and add to the third equation to eliminate y and obtain:

$$3x + z = 15,000.$$

Put this equation with the second equation to obtain the system:

$$\begin{aligned} 3x + z &= 15,000 \\ 2x - z &= 0 \end{aligned}$$

Adding we eliminate z and obtain

$$\begin{aligned} 5x &= 15,000 \\ x &= 3000. \end{aligned}$$

Since the problem only asked for the number of courtside seats, we are finished at this point. Thus, the number of courtside seats was 3000.

12. Let x = the number of nickels in the collection,
    y = the number of dimes in the collection,
    z = the number of quarters in the collection.

Since the total number of coins is 40, the first equation is

$$x + y + z = 40.$$

EXERCISES A

The phrase "twice the number of nickels is the same as three times the number of quarters" translates to the second equation

$$2x = 3z,$$

or equivalently,

$$2x - 3z = 0$$

The value of the nickels is $5x$ cents, the value of the dimes is $10y$ cents, and the value of the quarters is $25z$ cents. Since the total value of the collection is $4.60, or 460 cents, we obtain the third equation

$$5x + 10y + 25z = 460,$$

or equivalently

$$x + 2y + 5z = 92.$$

Thus, the system to solve is:

$$\begin{aligned} x + y + z &= 40 \\ 2x \quad\quad - 3z &= 0 \\ x + 2y + 5z &= 92 \end{aligned}$$

Multiply the top equation by -2,

$$-2x - 2y - 2z = -80,$$

and add to the bottom equation to eliminate y and obtain

$$-x + 3z = 12$$

or equivalently,

$$x - 3z = -12.$$

Then the system reduces to

$$\begin{aligned} x - 3z &= -12 \\ 2x - 3z &= 0 \end{aligned}$$

Subtract the top equation from the bottom equation to obtain

$$x = 12.$$

Substitute 12 for x in $x - 3z = -12$ to obtain

$$\begin{aligned} 12 - 3z &= -12 \\ -3z &= -24 \\ z &= 8 \end{aligned}$$

Then substitute 12 for x and 8 for z in the first original equation to obtain

$$\begin{aligned} 12 + y + 8 &= 40 \\ y &= 20. \end{aligned}$$

Thus, the collection consists of 12 nickels, 20 dimes, and 8 quarters.

13. Let $x =$ the amount invested in bonds,
$y =$ the amount invested in certificates,
$z =$ the amount invested in the mutual fund.

Since the total amount to be invested is $5000, the first equation is

$$x + y + z = 5000.$$

Since the bonds earn 8%, the amount earned is $0.08x$. Since the certificates earn 7%, the amount earned is $0.07y$. If the fund does well, it earns 6% for an amount $0.06z$. Under these conditions the total amount earned will be $345, so the second equation is

$$0.08x + 0.07y + 0.06z = 345,$$

or equivalently

$$8x + 7y + 6z = 34,500.$$

If the fund does not do well, it will lose 3% which means that is will lose $0.03z$ dollars. This translates to $-0.03z$ dollars. Under these conditions, the total earned will be $165, so we obtain the third equation

$$0.08x + 0.07y - 0.03z = 165,$$

or equivalently,

$$8x + 7y - 3z = 16,500.$$

Thus the system to solve is:

$$\begin{aligned} x + y + z &= 5000 \\ 8x + 7y + 6z &= 34,500 \\ 8x + 7y - 3z &= 16,500 \end{aligned}$$

Notice if we subtract the third equation from the second, we will eliminate both x and y and obtain

$$\begin{aligned} 9z &= 18,000 \\ z &= 2000 \end{aligned}$$

Knowing the value of z already, we could substitute this value into the first equation (it has not yet been used) and either of the other two, say the third, to obtain a system of two equations in x and y,

$$x + y + 2000 = 5000$$
$$x + y = 3000$$

and

$$8x + 7y - 3(2000) = 16{,}500$$
$$8x + 7y - 6000 = 16{,}500$$
$$8x + 7y = 22{,}500.$$

Solve for x in the first equation, $x = 3000 - y$, and substitute into the second.

$$8(3000 - y) + 7y = 22{,}500$$
$$24{,}000 - 8y + 7y = 22{,}500$$
$$-y = -1500$$
$$y = 1500$$

Substitute 1500 for y in $x = 3000 - y$ to obtain

$$x = 3000 - 1500$$
$$x = 1500$$

Thus, $1500 were invested in bonds, $1500 were invested in certificates, and $2000 were invested in the mutual fund.

14. The cost of operating a home air conditioner can be approximated by a quadratic model

$$y = ax^2 + bx + c,$$

where y is the cost/day when the thermostat setting is at a temperature of x degrees. The table actually gives us three data points:

$$(70,6), (80,5), \text{ and } (90,3).$$

Substitute these values for x and y in the model above to obtain, respectively, the following system of three equations in a, b, and c.

$$4900a + 70b + c = 6$$
$$6400a + 80b + c = 5$$
$$8100a + 90b + c = 3$$

Subtract the first equation from the second to obtain

$$1500a + 10b = -1.$$

Then subtract the second from the third to obtain

$$1700a + 10b = -2.$$

We have eliminated c and have the system:

$$1500a + 10b = -1$$
$$1700a + 10b = -2$$

Subtract the top equation from the bottom one.

$$200a = -1$$
$$a = -\frac{1}{200}$$

Substitute this value for a in $1500a + 10b = -1$ to find the value of b.

$$1500\left(-\frac{1}{200}\right) + 10b = -1$$
$$-\frac{15}{2} + 10b = -1$$
$$-15 + 20b = -2$$
$$20b = 13$$
$$b = \frac{13}{20}$$

Finally, substitute these values for a and b in

$$4900a + 70b + c = 6$$

to find the value of c.

$$4900\left(-\frac{1}{200}\right) + 70\left(\frac{13}{20}\right) + c = 6$$
$$-\frac{49}{2} + \frac{91}{2} + c = 6$$
$$-49 + 91 + 2c = 12$$
$$2c = -30$$
$$c = -15$$

Thus, the quadratic function that fits this model is

$$y = -\frac{1}{200}x^2 + \frac{13}{20}x - 15.$$

To estimate the operating costs on a day when the thermostat is set at 100°, substitute 100 for x and simplify.

$$y = -\frac{1}{200}(100)^2 + \frac{13}{20}(100) - 15$$
$$= -50 + 65 - 15$$
$$= 0$$

Thus the operating costs are $0.00 when the thermostat is set at 100°. This does seem reasonable since at this setting the air conditioner probably would not come on if the exterior temperature stays below 100°.

EXERCISES A

SECTION 6.3   279

15. Solve the system.

$$3x + y - z = 0$$
$$x - y + 2z = 0$$
$$7x + y = 0$$

Since this is a homogeneous system, one solution is (0,0,0). To determine whether there are more solutions, suppose we eliminate y. Add the first two equations obtaining

$$4x + z = 0.$$

Then add the last two equations obtaining

$$8x + 2z = 0,$$

or equivalently

$$4x + z = 0.$$

Since we obtain exactly the same equations, subtracting them would give the identity $0 = 0$ so there are infinitely many solutions. Solve for z.

$$z = -4x.$$

Substitute $-4x$ for z in the first original equation to find y in terms of x.

$$3x + y - (-4x) = 0$$
$$3x + y + 4x = 0$$
$$y = -7x$$

Thus, the solutions can be written as $(x, -7x, -4x)$ for x any real number.

16. Solve the system.

$$5x + 2y - z = 2$$
$$3x - 3y + z = -1$$

Since the system has only two equations and there are three variables, there will be no solution or else infinitely many solutions. Add to eliminate z.

$$8x - y = 1$$

Since we do not obtain a contradiction, there are infinitely many solutions. Solve this equation for y.

$$-y = -8x + 1$$
$$y = 8x - 1$$

Substitute this expression for y in the equation

$$3x - 3y + z = -1$$

$$3x - 3(8x - 1) + z = -1$$
$$3x - 24x + 3 + z = -1$$
$$-21x + 3 + z = -1$$
$$z = 21x - 4$$

Thus, the solution can be written as $(x, 8x-1, 21x-4)$ for x any real number.

17. Graph the linear equation. $4x + 2y = -6$

Use the intercepts, $(0,-3)$ and $\left(-\frac{3}{2}, 0\right)$, to obtain the graph of the line shown below.

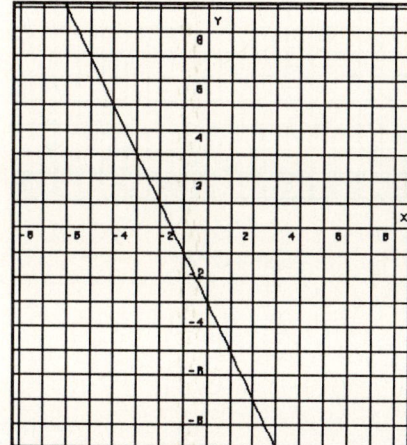

18. Graph the linear equation. $3x + 3y = 0$

Since both intercepts are $(0,0)$, we find another point, such as $(2,-2)$, to obtain the graph of the line shown below.

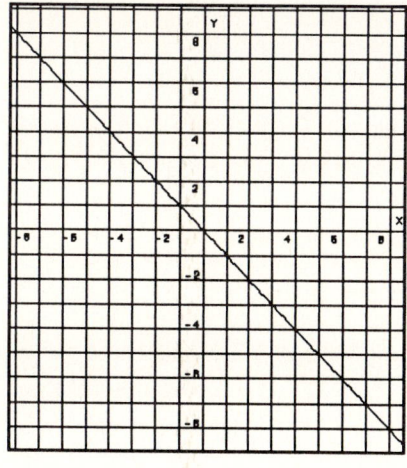

19. Graph the linear equation. $2y - 2 = 0$

Since x is missing, the graph is a horizontal line. The equation can be simplified to $y = 1$. The y-intercept is $(0,1)$, and the graph of the line is shown below.

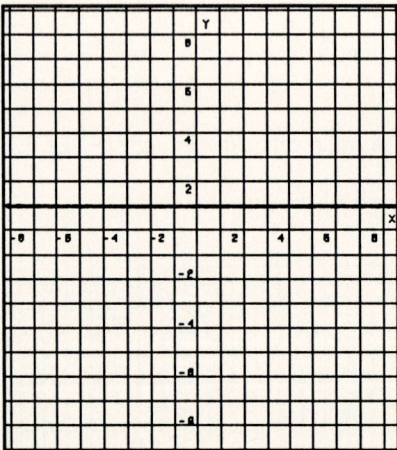

PRACTICE EXERCISES — SECTION 6.4

CHAPTER 6 SYSTEMS OF EQUATIONS AND INEQUALITIES

SECTION 6.4 Linear Systems of Inequalities

1. Graph $3x - 2y \leq 6$.

   First graph the line $3x - 2y = 6$ using intercepts $(0,-3)$ and $(2,0)$ and a solid line (the inequality is $\leq$). Use the test point $(0,0)$.

   $$3(0) - 2(0) \leq 6$$
   $$0 \leq 6 \quad \text{This is true}$$

   Since we obtain a true inequality, shade the half-plane containing $(0,0)$ to obtain the graph shown below.

   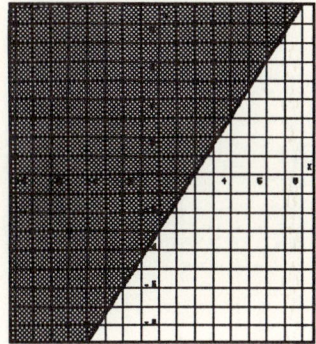

2. (a) Graph $5y + 2 \geq 0$.

   First graph the line with equation $5y + 2 = 0$, or equivalently, $y = -\frac{2}{5}$. This is a horizontal line with y-intercept $\left(0, -\frac{2}{5}\right)$. Use a solid line since the inequality is $\geq$. Use the test point $(0,0)$.

   $$5(0) + 2 \geq 0$$
   $$2 \geq 0 \quad \text{This is true}$$

   Since we obtain a true inequality, shade the half-plane that contains $(0,0)$ as shown in the graph below.

   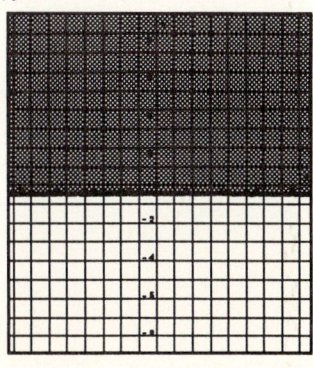

   (b) Graph $6x + 3 < 0$.

   First graph the line with equation $6x + 3 = 0$, or equivalently, $x = -\frac{1}{2}$. Use a dashed line since the inequality is $<$. The graph is a vertical line with x-intercept $\left(-\frac{1}{2}, 0\right)$. Use the test point $(0,0)$.

   $$6(0) + 3 < 0$$
   $$3 < 0 \quad \text{This is false}$$

   Since we obtain a false inequality, shade the half-plane that does not contain $(0,0)$ as shown below.

   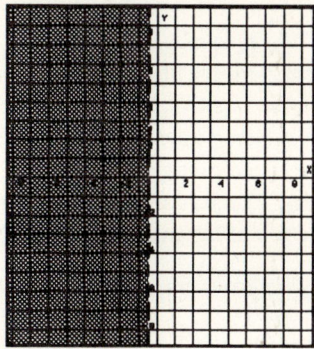

3. Graph the system.

   $$3x - y > 6$$
   $$3x - 3 \geq 0$$

   First graph the equation $3x - y = 6$ using intercepts $(0,-6)$ and $(2,0)$ and a dashed line since the inequality is $>$. Use the test point $(0,0)$.

   $$3(0) - (0) > 6$$
   $$0 > 6 \quad \text{This is false}$$

   Since we obtain a false inequality, the graph of this inequality is all points in the half-plane that does not contain $(0,0)$. Keep this in mind and graph the second inequality by first graphing the line with equation $3x - 3 = 0$, or equivalently, $x = 1$. The graph is a solid vertical line with x-intercept $(1,0)$. The test point $(0,0)$ shows that

   $$3(0) - 3 \geq 0$$
   $$-3 \geq 0 \quad \text{This is false}$$

we shade the region that does not contain (0,0). Putting these two together, the graph of the system is shown below.

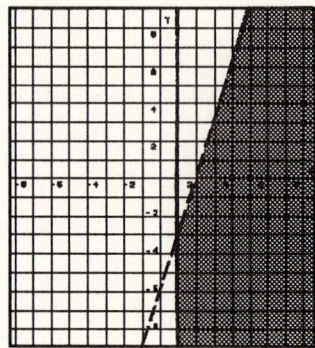

4. Graph the system.

$$x \geq 0$$
$$y \geq 0$$
$$2x + y \leq 4$$
$$x + 2y \leq 4$$

The first two inequalities describe all the points in the first quadrant together with those on the positive x-axis and the positive y-axis. Next graph the line $2x + y = 4$ using intercepts (0,4) and (2,0) and a solid line. Then graph the line $x + 2y = 4$ using intercepts (0,2) and (4,0) and a solid line. The point of intersection of the two lines, found by solving the system, is $\left(\frac{4}{3}, \frac{4}{3}\right)$. The test point (0,0) shows that the region "below" both lines, but still in quadrant I, is the graph. The graph is given below.

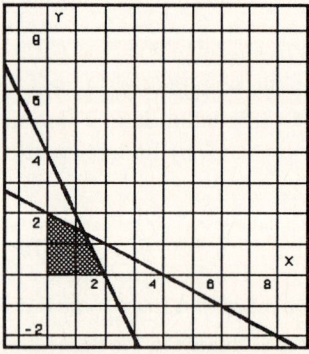

5. Let x = number of Model 20 computers,
   y = number of Model 30 computers.

The phrase "at least two of each model must be kept in inventory" translates to the two inequalities $x \geq 2$ and $y \geq 2$. The phrase "the number of Model 30 computers plus four times the number of Model 20 computers must not exceed 14" translates to the inequality $4x + y \leq 14$. Thus, the system we must graph is

$$x \geq 2$$
$$y \geq 2$$
$$4x + y \leq 14$$

First graph the solid vertical line $x = 2$ with intercept (2,0), and then graph the solid horizontal line $y = 2$ with intercept (0,2). The two inequalities $x \geq 2$ and $y \geq 2$ describe all points to the right and above of these lines, respectively. Finally graph $4x + y = 14$ using intercepts (0,14) and $\left(\frac{7}{2}, 0\right)$ and a solid line. The test point (0,0) shows that the region "below" this line should be shaded. Putting this information together, we obtain the graph given below.

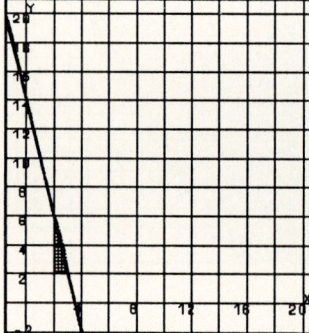

EXERCISES A — SECTION 6.4

CHAPTER 6 SYSTEMS OF EQUATIONS AND INEQUALITIES

SECTION 6.4 Linear Systems of Inequalities

1. Graph the inequality. $3x - 2y \leq 6$

   First graph the line $3x - 2y = 6$ using the intercepts $(0,-3)$ and $(2,0)$, and a solid line (the inequality is $\leq$). The test point $(0,0)$ gives

   $$3(0) - 2(0) \leq 6$$
   $$0 \leq 6 \quad \text{This is true}$$

   which shows that we shade the region containing $(0,0)$, the half-plane "above" the line. The graph is given below.

   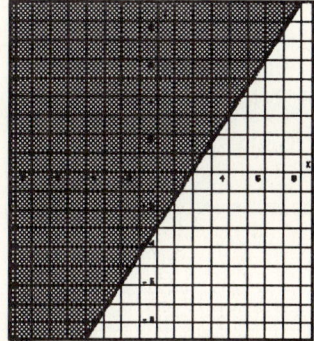

2. Graph the inequality. $2x + y \geq 3$

   First graph the line $2x + y = 3$ using the intercepts $(0,3)$ and $\left(\frac{3}{2},0\right)$, and a solid line (the inequality is $\geq$). The test point $(0,0)$ gives

   $$2(0) + (0) \geq 3$$
   $$0 \geq 3 \quad \text{This is false}$$

   which shows that we shade the region that does not contain $(0,0)$, the half-plane "above" the line. The graph is given below.

   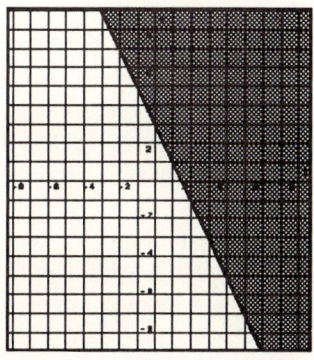

3. Graph the inequality. $x - y < -2$

   First graph the line $x - y = -2$ using the intercepts $(0,2)$ and $(-2,0)$, and a dashed line (the inequality is $<$). The test point $(0,0)$ gives

   $$(0) - (0) < -2$$
   $$0 < -2 \quad \text{This is false}$$

   which shows that we shade the region that does not contain $(0,0)$, the half-plane "above" the line. The graph is given below.

   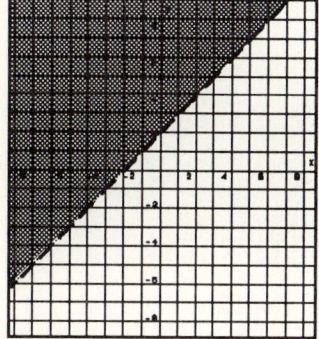

4. Graph the inequality. $3x + 2y < 0$

   First graph the line $3x + 2y = 0$. Since both intercepts are $(0,0)$, find an additional point such as $(2,-3)$, and use a dashed line (the inequality is $<$). The test point $(1,0)$ (we cannot use $(0,0)$ since it is on the line) gives

   $$3(1) + 2(0) < 0$$
   $$3 < 0 \quad \text{This is false}$$

   which shows that we shade the region that does not contain $(1,0)$, the half-plane "below" the line. The graph is given below.

   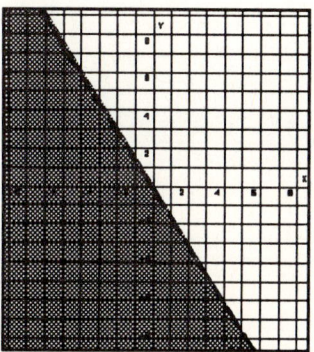

5. Graph the inequality. $4y - 8 \leq 0$

First graph the line $4y - 8 = 0$, or equivalently, $y = 2$. Since the x term is missing, the graph is a horizontal line with y-intercept $(0,2)$. Use a solid line (the inequality is $\leq$). The test point $(0,0)$ gives

$$4(0) - 8 \leq 0$$
$$-8 \leq 0 \quad \text{This is true}$$

which shows that we shade the region containing $(0,0)$, the half-plane "below" the line. The graph is given below.

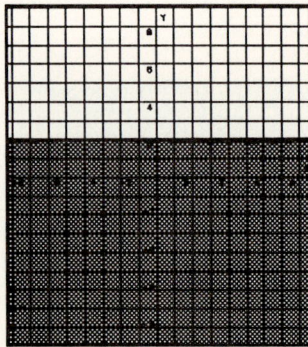

6. Graph the inequality. $4y + 12 > 0$

First graph the line $4y + 12 = 0$, or equivalently, $y = -3$. Since the x term is missing, the graph is a horizontal line with y-intercept $(0,-3)$. Use a dashed line (the inequality is $>$). The test point $(0,0)$ gives

$$4(0) + 12 > 0$$
$$12 > 0 \quad \text{This is true}$$

which shows that we shade the region containing $(0,0)$, the half-plane "above" the line. The graph is given below.

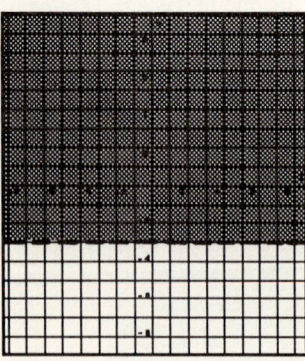

7. Graph the system. $\quad 2x - 3y \geq 6$
$\quad\quad\quad\quad\quad\quad\quad\; x + y < -1$

First graph the line $2x - 3y = 6$ using intercepts $(0,-2)$ and $(3,0)$, and a solid line (the inequality is $\geq$). The test point $(0,0)$ shows

$$2(0) - 3(0) \geq 6$$
$$0 \geq 6 \quad \text{This is false}$$

that we should shade the region that does not contain $(0,0)$, the region "below" the line.

Now graph the line $x + y = -1$ using intercepts $(0,-1)$ and $(-1,0)$, and a dashed line (the inequality is $<$). The test point $(0,0)$ shows

$$(0) + (0) < -1$$
$$0 < -1 \quad \text{This is false}$$

that we should shade the region that does not contain $(0,0)$, the region "below" the line.

Putting this information together, we obtain the graph of the system as shown below.

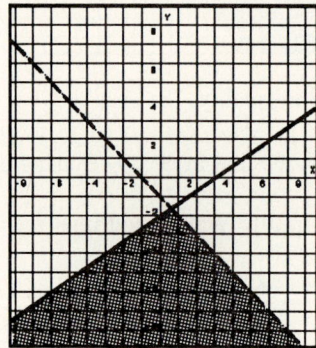

8. Graph the system. $\quad 2x - 3y < 6$
$\quad\quad\quad\quad\quad\quad\quad\; x + y \geq -1$

First graph the line $2x - 3y = 6$ using intercepts $(0,-2)$ and $(3,0)$, and a dashed line (the inequality is $<$). The test point $(0,0)$ shows

$$2(0) - 3(0) < 6$$
$$0 < 6 \quad \text{This is true}$$

that we should shade the region that contains $(0,0)$, the region "above" the line.

Now graph the line $x + y = -1$ using intercepts $(0,-1)$ and $(-1,0)$, and a solid line (the inequality is $\geq$). The test point $(0,0)$ shows

EXERCISES A   SECTION 6.4   285

$(0) + (0) \geq -1$

$0 \geq -1$ *This is true*

that we should shade the region that contains (0,0), the region "above" the line.

Putting this information together, we obtain the graph of the system as shown below.

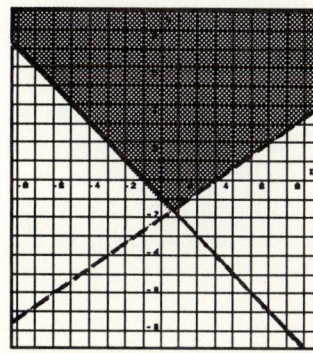

9. Graph the system. $2x - 2y \geq -4$
$x - y < 1$

First graph the line $2x - 2y = -4$ using intercepts (0,2) and (-2,0), and a solid line (the inequality is $\geq$). The test point (0,0) shows

$2(0) - 2(0) \geq -4$

$0 \geq -4$ *This is true*

that we should shade the region that contains (0,0), the region "below" the line.

Now graph the line $x - y = 1$ using intercepts (0,-1) and (1,0), and a dashed line (the inequality is $<$). The test point (0,0) shows

$(0) - (0) < 1$

$0 < 1$ *This is true*

that we should shade the region that contains (0,0), the region "above" the line.

Putting this information together, we obtain the graph of the system as shown below.

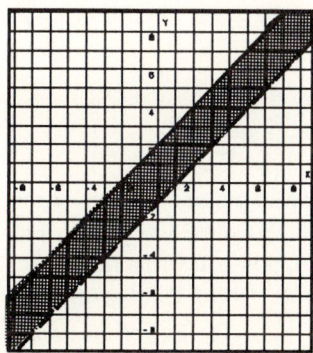

10. Graph the system. $-2x + 3y \leq 6$
$x + 1 > 0$

First graph the line $-2x + 3y = 6$ using intercepts (0,2) and (-3,0), and a solid line (the inequality is $\leq$). The test point (0,0) shows

$-2(0) + 3(0) \leq 6$

$0 \leq 6$ *This is true*

that we should shade the region that contains (0,0), the region "below" the line.

Now graph the line $x + 1 = 0$. Since y is missing, the graph is a vertical line through x-intercept (-1,0). Use and a dashed line (the inequality is $>$). The test point (0,0) shows

$(0) + 1 > 0$

$1 > 0$ *This is true*

that we should shade the region that contains (0,0), the region "right" of the line.

Putting this information together, we obtain the graph of the system as shown below.

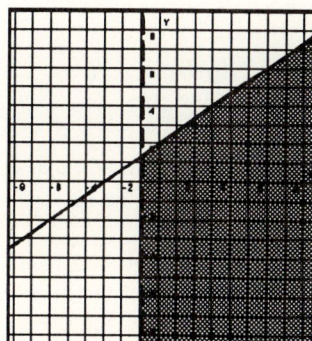

11. Graph the system. $x \geq 0$
$y \geq 0$
$3x + 7y < 21$

First note that the first two inequalities describe all points in the first quadrant together with the positive x-axis and the positive y-axis. Keeping this in mind, we can turn attention to the third inequality. First graph the line $3x + 7y = 21$ using intercepts (0,3) and (7,0), and a dashed line (the inequality is $<$). The test point (0,0) shows

$3(0) + 7(0) < 21$

$0 < 21$ *This is true*

that we should shade the region that contains (0,0), the region "below" the line.

Putting this information together, we shade the region below the graphed dashed line but stay within the first quadrant as shown below.

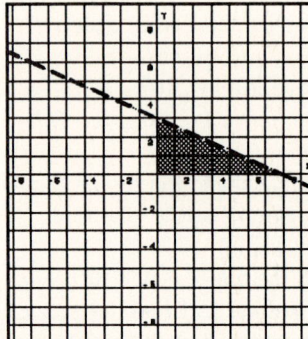

12. Graph the system.
$$x \leq 0$$
$$y \geq 0$$
$$2x - y \geq -6$$
$$x + 2y \leq 2$$

First note that the first two inequalities describe all points in the second quadrant together with the negative x-axis and the positive y-axis. Keeping this in mind, we can turn attention to the third inequality. First graph the line $2x - y = -6$ using intercepts $(0,6)$ and $(-3,0)$, and a solid line (the inequality is $\geq$). The test point $(0,0)$ shows

$$2(0) - (0) \geq -6$$
$$0 \geq -6 \quad \text{This is true}$$

that we should shade the region that contains $(0,0)$, the region "below" the line.

Next graph the line $x + 2y = 2$ using intercepts $(0,1)$ and $(2,0)$, and a solid line (the inequality is $\leq$). The test point $(0,0)$ shows

$$(0) + 2(0) \leq 2$$
$$0 \leq 2 \quad \text{This is true}$$

that we should shade the region that contains $(0,0)$, the region "below" the line.

Putting this information together, we shade the region below the graphed lines but stay within the second quadrant as shown below.

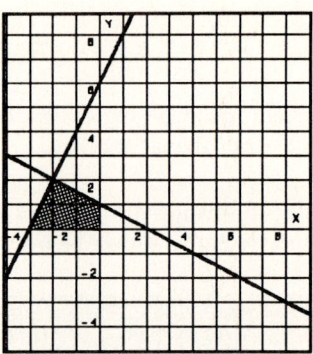

13. Let x = the number of Princess homes,
    y = the number of Knight homes.

The phrase "he must have at least three times as many Princess homes available as Knights" translates to

$$x \geq 3y,$$

or equivalently,

$$x - 3y \geq 0.$$

The phrase "he wants at least 6 Princess homes and 2 Knight homes available and ready for occupancy" translates to the two inequalities

$$x \geq 6 \quad \text{and} \quad y \geq 2.$$

Since each Princess model costs $30,000, the value of x such homes is $30,000x$. Also, since each Knight home costs $20,000, the value of y such homes is $20,000y$. If the inventory costs are to be kept at $600,000 or less, we have the fourth inequality

$$30,000x + 20,000y \leq 600,000$$

which simplifies to

$$3x + 2y \leq 60.$$

Thus, we must graph the following system.
$$x \geq 6$$
$$y \geq 2$$
$$x - 3y \geq 0$$
$$3x + 2y \leq 60$$

The graph consists of all points to the right of the line $x = 6$, above the line $y = 2$, below the line $x - 3y = 0$, and below the line $3x + 2y = 60$ as shown below. Notice that the region is a "bounded" region.

EXERCISES A

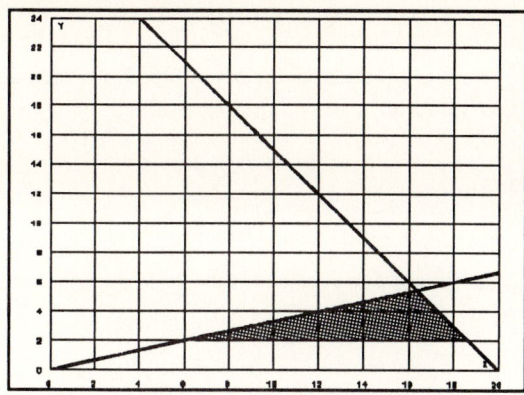

14. Let $x =$ the number of Standard tables,
    $y =$ the number of Deluxe tables.

Since the number of each cannot be negative, two inequalities in the system are

$$x \geq 0 \quad \text{and} \quad y \geq 0.$$

From the table, using the row describing the construction stage, we obtain the third inequality

$$3x + 6y \leq 96.$$

Also from the table, using the row describing the finishing stage, we obtain the fourth inequality

$$x + 4y \leq 36.$$

The graph of this system includes all points in the first quadrant that are below the two lines $3x + 6y = 96$ and $x + 4y = 36$. As shown below, the graph is a "bounded" region.

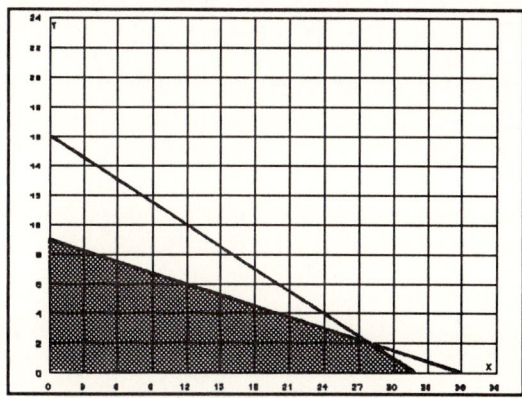

15. Let $x =$ the price of 1 lb of candy in cents,
    $y =$ the price of 1 lb of nuts in cents.

Since "2 lb of candy and 3 lb of nuts cost a total of $4.90 (490 cents)" we have the first equation

$$2x + 3y = 490.$$

Since 3 lb of candy and 5 lb of nuts cost a total of $7.90 (790 cents)" we have the second equation

$$3x + 5y = 790.$$

Multiply the first equation by -3,

$$-6x - 9y = -1470,$$

and the second by 2,

$$6x + 10y = 1580.$$

Adding these resulting equations will eliminate x and give

$$y = 110.$$

Substitute 110 for y in $2x + 3y = 490$ to find the value of x.

$$\begin{aligned} 2x + 3(110) &= 490 \\ 2x + 330 &= 490 \\ 2x &= 160 \\ x &= 80 \end{aligned}$$

Thus, the candy costs $0.80 per pound and the nuts cost $1.10 per pound.

16. Let $x =$ the measure of the smallest angle,
    $y =$ the measure of the middle-sized angle,
    $z =$ the measure of the largest angle.

Since the measures of the angles of a triangle add up to $180°$, the first equation is

$$x + y + z = 180.$$

Since "the smallest angle measures $28°$ less than the largest angle," we obtain the second equation,

$$x = z - 28,$$

or equivalently,

$$x - z = -28.$$

Since "the measure of the largest angle less the measure of the middle-sized angle is $2°$," we obtain the third equation

$$z - y = 2,$$

or equivalently,

$$y - z = -2.$$

Thus, the system we must solve is:

$$x + y + z = 180$$
$$x \quad - z = -28$$
$$y - z = -2$$

Subtract the second equation from the first to eliminate x,

$$y + 2z = 208,$$

and pair the result with the third equation giving a system of two equations in two variables y and z.

$$y + 2z = 208$$
$$y - z = -2$$

Subtract the second equation from the first to obtain

$$3z = 210$$
$$z = 70.$$

Substitute 70 for z in $y - z = -2$ to find the value of y.

$$y - 70 = -2$$
$$y = 68$$

Substitute 70 for z in $x - z = -28$ to find the value of x.

$$x - 70 = -28$$
$$x = 42$$

Thus the measures of the angles are 42°, 68°, and 70°.

17. Let x = the length of the pasture,
y = the width of the pasture.

The perimeter of a rectangle is given by $P = 2l + 2w$, and since the perimeter of the pasture is 50 mi, we obtain the first equation

$$2x + 2y = 50$$

which simplifies to

$$x + y = 25.$$

Since "the length is 1 mi more than twice the width," we obtain the second equation

$$x = 1 + 2y.$$

Since this equation is already solved for x, substitute this value for x in the first equation.

$$1 + 2y + y = 25$$
$$3y = 24$$
$$y = 8$$

Substitute 8 for y in $x = 1 + 2y$ to find the value of x.

$$x = 1 + 2(8)$$
$$x = 17$$

Thus, the dimensions of the pasture are

17 mi by 8 mi.

18. Given the function $f(x) = ae^x - be^{-x} + 1$. Since $f(0) = 2$, substitute 0 for x to obtain one equation in a and b.

$$ae^0 - be^{-0} + 1 = 2$$
$$a(1) - b(1) = 1$$
$$a - b = 1$$

Now substitute ln 3 for x to obtain a second equation in a and b.

$$ae^{\ln 3} - be^{-\ln 3} + 1 = 4$$
$$ae^{\ln 3} - be^{\ln 3^{-1}} + 1 = 4$$
$$a(3) - b(3^{-1}) + 1 = 4$$
$$3a - \frac{1}{3}b = 3$$
$$9a - b = 9$$

Then the system to solve is:

$$9a - b = 9$$
$$a - b = 1$$

Subtract the second equation from the first to obtain

$$8a = 8$$
$$a = 1$$

Substitute 1 for a in $a - b = 1$ to find the value of b.

$$1 - b = 1$$
$$-b = 0$$
$$b = 0$$

Thus, the desired function is

$$f(x) = ae^x - be^{-x} + 1$$
$$= (1)e^x - (0)e^{-x} + 1$$
$$= e^x + 1.$$

CHAPTER 6 SYSTEMS OF EQUATIONS AND INEQUALITIES

SECTION 6.5 Linear Programming

1. The system of inequalities

$$x \geq 0$$
$$y \geq 0$$
$$x + y \leq 18$$
$$5x + 3y \leq 60$$

is exactly the same as that given in Example 1, and the graph of the system is given in Figure 6.14. The only difference is in the objective function. This time, a heavy-duty hitch (the number of which is x) returns a profit of $30 when sold, and a standard model hitch (the number of whch is y) returns a profit of $70. Thus the objective function is

$$\text{profit} = 30x + 70y.$$

The vertices of the feasible region, the graph of the system, are $(0,0)$, $(0,18)$, $(3,15)$, and $(12,0)$. The optimal solution must be found using these pairs when substituted into the objective function. The possibilities are summarized in the table below.

Vertex	$30x + 70y$	
(0,0)	0	
(0,18)	1260	OPTIMAL SOLUTION
(3,15)	1140	
(12,0)	360	

Thus, a maximum profit will be realized in this case when 18 standard hitches and no heavy-duty hitches are made.

2. The system of inequalities

$$x \geq 0$$
$$y \geq 0$$
$$x + 3y \leq 33$$
$$4x + 3y \leq 42$$
$$6x + y \leq 42$$

is exactly the same as that given in Example 2, and the graph of the system is given in Figure 6.15. The only difference is in the objective function. This time, the profits are reversed. That is, the profit on the sale of skunks (of which there are x) is $3.00, and the profit on the sale of weasels (of which there are y) is $2.00. Thus, the objective function to be maximized is

$$\text{profit} = 3x + 2y.$$

The vertices of the feasible region, the graph of the system, are $(0,0)$, $(0,11)$, $(3,10)$, $(6,6)$, and $(7,0)$. The optimal solution must be found using these pairs when substituted into the objective function. The possibilities are summarized in the table below.

Vertex	$3x + 2y$	
(0,0)	0	
(0,11)	22	
(3,10)	29	
(6,6)	30	OPTIMAL SOLUTION
(7,0)	21	

Thus, a maximum profit will be realized in this case when six weasels and six skunks are produced.

CHAPTER 6 SYSTEMS OF EQUATIONS AND INEQUALITIES

SECTION 6.5 Linear Programming

1. Find the maximum and minimum values of the objective function

    $$P = x + 5y$$

    subject to the constraints graphed in the given feasible region in the text. Since the vertices of the feasible region are (1,1), (2,6), (6,8), and (7,2), the maximum and minimum values will occur at one of these vertices using the fundamental theorem of linear programming. We list the various values in the table below.

Vertex	x + 5y	
(1,1)	6	OPTIMAL SOLUTION
(2,6)	32	
(6,8)	46	OPTIMAL SOLUTION
(7,2)	17	

    Thus, the minimum value of the objective function is 6, and the maximum value is 46.

2. Find the maximum and minimum values of the objective function

    $$P = 75x + 100y$$

    subject to the constraints graphed in the given feasible region in the text. Notice that the feasible region is unbounded so it is possible to make the objective function as large as we please by choosing points farther and farther out to the right and up. Thus, there is no maximum value. Since the vertices of the feasible region are (1,4), (2,2), and (5,1), the minimum value will occur at one of these vertices using the fundamental theorem of linear programming. We list the various values in the table below.

Vertex	75x + 100y	
(1,4)	475	
(2,2)	350	OPTIMAL SOLUTION
(5,1)	475	

    Thus, the minimum value of the objective function is 350.

3. Find the maximum and minimum values of the objective function

    $$P = 10x + 50y$$

    subject to the constraints:

    $$x \geq 0$$
    $$y \geq 0$$
    $$x + 2y \leq 6$$
    $$x + y \leq 4$$

    The feasible region is shown below with vertices (0,0), (0,3), (2,2), and (4,0).

    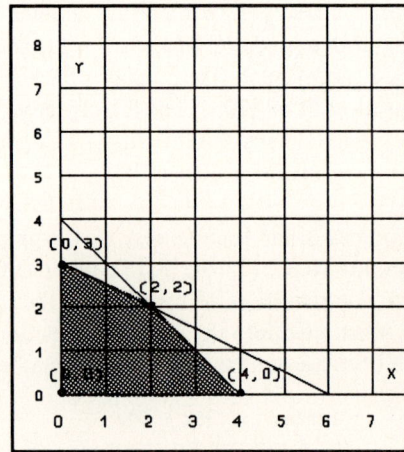

    The vertex (2,2) was obtained by solving the system

    $$x + 2y = 6$$
    $$x + y = 4.$$

    We list the various values in the table below.

Vertex	10x + 50y	
(0,0)	0	OPTIMAL SOLUTION
(0,3)	150	OPTIMAL SOLUTION
(2,2)	120	
(4,0)	40	

    Thus, the minimum value of the objective function is 0, and the maximum value is 150.

4. Find the maximum and minimum values of the objective function

    $$P = 10x + 50y$$

    subject to the constraints:

    $$x \geq 2$$
    $$2y \geq 1$$
    $$5x + 6y \geq 28$$

    The feasible region is shown below with vertices (2,3) and (5, 0.5).

EXERCISES A   SECTION 6.5   291

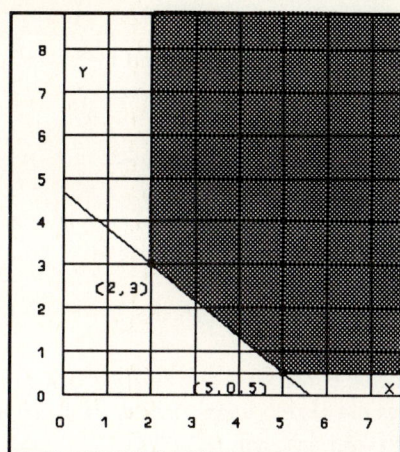

The vertex (2,3) was obtained substituting 2 for x in the equation $5x + 6y = 28$ and solving for y. Similarly, (5, 0.5) was obtained by substituting 0.5 for y in the equation $5x + 6y = 28$ and solving for x. Notice that since the feasible region is unbounded, we can make the objective function as large as we please so there is no maximum value. The minimum value will occur at one of the two vertices. We list the various values in the table below.

Vertex	$10x + 50y$	
(2,3)	170	
(5,0.5)	75	OPTIMAL SOLUTION

Thus, the minimum value of the objective function is 75.

5. Let x = the number of True-Shot balls produced daily,
    y = the number of True-Bounce balls produced daily.

Since the profit is $20 on each True-Shot model and $13 on each True-Bounce model, the objective function to be maximized is

$$P = 20x + 13y.$$

Since the number of True-Shot models produced daily must be between 20 and 100, inclusive, we have

$$20 \leq x \leq 100.$$

Since the number of True-Bounce models produced daily must be between 10 and 70, inclusive, we have

$$10 \leq y \leq 70.$$

The total number of balls produced daily should not exceed 150 gives

$$x + y \leq 150.$$

The graph of these constraints, the feasible region, is given below, and the vertices are easy to determine either by inspection or by substitution.

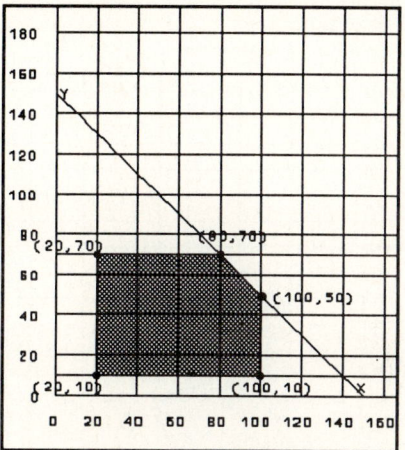

The various possibilities are summarized in the following table.

Vertex	$20x + 13y$	
(20,10)	530	
(20,70)	1310	
(80,70)	2510	
(100,50)	2650	OPTIMAL SOLUTION
(100,10)	2130	

Thus, the maximum profit will be realized when 100 True-Shot models and 50 True-Bounce models are produced each day.

6. Let x = the number of barrels of type G oil produced per day,
    y = the number of barrels of type H oil produced per day.

Since the refinery can produce up to 5000 barrels of oil daily,

$$x + y \leq 5000.$$

Since at least 1000 barrels and at most 3500 barrels of type G must be produced daily,

$$1000 \leq x \leq 3500.$$

Also, clearly the number of barrels of type H must be greater than or equal to 0, so

$y \geq 0$.

With a profit of $7 a barrel for type G and $3 a barrel for type H, the objective function to maximize subject to the constraints given above is

$$P = 7x + 3y.$$

The constraints are graphed below showing the feasible region.

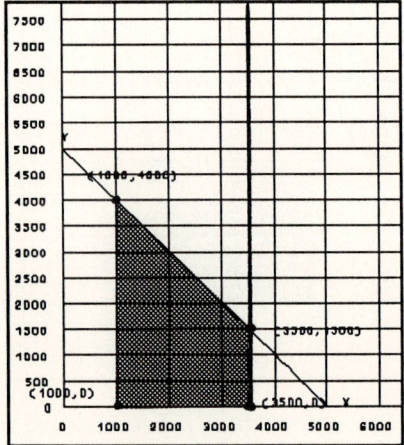

The vertices of the region obtained by substitution or inspection are (1000,0), (1000,4000), (3500,1500), and (3500,0).

Vertex	$7x + 3y$	
(1000,0)	7000	
(1000,4000)	19,000	
(3500,1500)	29,000	OPTIMAL SOLUTION
(3500,0)	24,500	

Thus, the maximum profit will be realized when 3500 barrels of type G and 1500 barrels of type H oil are produced daily.

7. Let x = the number of acres planted in corn,
   y = the number of acres planted in wheat.

Since there are only 100 acres available,

$$x + y \leq 100.$$

The cost is $5 per acre for seed to plant corn, so $5x$ is the cost of the seed for corn. The cost per acre for seed to plant wheat is $8, so the $8y$ is the cost of seed for the wheat. Since he can spend up to $704 for seed,

$$5x + 8y \leq 704.$$

Similarly, the cost for labor and fuel, which must be less than or equal to $1640 gives

$$20x + 12y \leq 1640.$$

Finally, the number of acres of corn and the number of acres of wheat cannot be negative, so

$$x \geq 0 \quad \text{and} \quad y \geq 0.$$

These five constraints are graphed below showing the feasible region with vertices (0,0), (0,88), (32,68), (55,45), and (82,0).

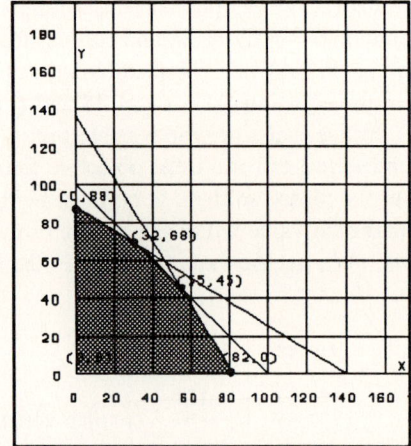

The vertex (55,45) comes from solving the system

$$20x + 12y = 1640$$
$$x + y = 100,$$

and the vertex (32,68) comes from solving the system

$$5x + 8y = 704$$
$$x + y = 100.$$

Since the profit on an acre of corn is $220 and the profit on an acre of wheat is $250, the objective function to maximize is

$$P = 220x + 250y.$$

We summarize the possibilities in the table below.

Vertex	$220x + 250y$	
(0,0)	0	
(0,88)	22,000	
(32,68)	24,040	OPTIMAL SOLUTION
(55,45)	23,350	
(82,0)	18,040	

EXERCISES A SECTION 6.5   293

Thus, the farmer should plant 32 acres of corn and 68 acres of wheat to realize a maximum profit of $24,040.

8. Let x = the number of units of product A to produce daily,
    y = the number of units of product B to produce daily.

Since there is a profit of $275 on each unit of A and $180 on each unit of B the objective function to maximize is

$$P = 275x + 180y.$$

There are three different machines used in the manufacturing process, X, Y, and Z. Since X is used 2 hours to make each unit of A and 1 hour to make each unit of B, and X can be used at most 18 hours each day, we have

$$2x + y \le 18.$$

Since Y is used 1 hour to make each unit of A and 2 hours to make each unit of B, and Y can be used at most 20 hours each day, we have

$$x + 2y \le 20.$$

Since Z is used 1 hour to make each unit of A and 1 hour to make each unit of B, and Z can be used at most 11 hours daily, we have

$$x + y \le 11.$$

Then obviously the number of units of each cannot be negative so we have

$$x \ge 0 \quad \text{and} \quad y \ge 0.$$

The graph of these constraints showing the feasible region with vertices (0,0), (0,10), (2,9), (7,4), and (9,0) is given below. The vertex (2,9) comes from solving

$$x + 2y = 20$$
$$x + y = 11$$

and the vertex (7,4) comes from solving

$$2x + y = 18$$
$$x + y = 11.$$

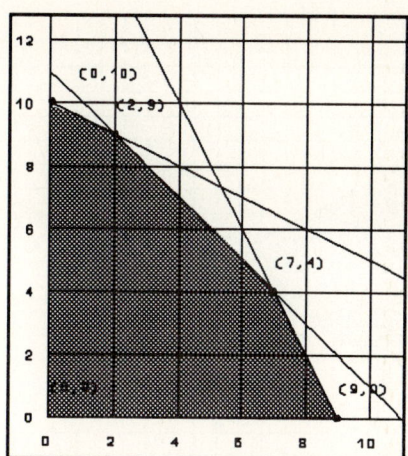

We summarize the possibilities in the following table.

Vertex	275x + 180y	
(0,0)	0	
(0,10)	1800	
(2,9)	2170	
(7,4)	2645	OPTIMAL SOLUTION
(9,0)	2475	

Thus, the maximum profit will be realized when 7 units of A and 4 units of B are produced daily.

9. Graph the system. $2x - 5y \le 10$
    $x + 2y < -2$

First graph the equation $2x - 5y = 10$ using intercepts $(0,-2)$ and $(5,0)$ and a solid line (the inequality is $\le$). The test point $(0,0)$ shows

$$2(0) - 5(0) \le 10 \quad \textit{This is true}$$

that we shade the region containing (0,0), that is the region "above" the line.

Then graph the equation $x + 2y = -2$ using intercepts $(0,-1)$ and $(-2,0)$ and a dashed line (the inequality is $<$). The test point $(0,0)$ shows

$$(0) - 5(0) < -2 \quad \textit{This is false}$$

that we shade the region that does not contain (0,0), that is, the region "below" the line. Putting this information together, we obtain the graph of the system shown below.

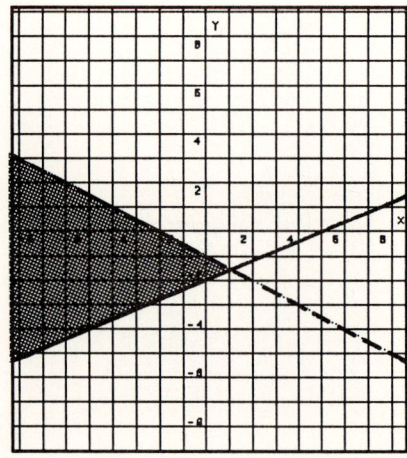

10. Graph the system. $y \geq x$
$y \geq -x$

First graph the equations $y = x$ and $y = -x$ using intercepts $(0,0)$, one additional point for each, and a solid line (the inequalities are $\geq$). The test point $(0,1)$ shows that we shade the region "above" both lines to obtain the graph below.

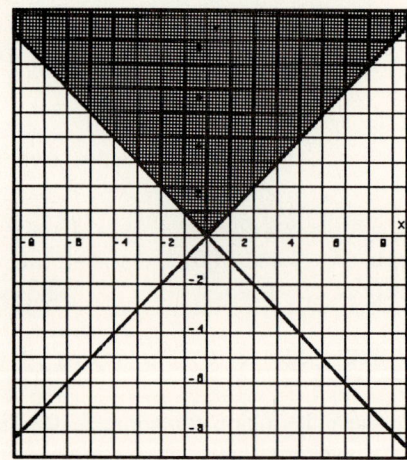

11. Let x = the number of $20 seats,
y = the number of $10 seats.

Since the total number of seats available is 1000, we have

$$x + y \leq 1000.$$

Since the revenue produced on the $20 seats is $20x$ and the revenue produced on the $10 seats is $10y$, and since the promoters want to make at least $16,000, we have

$$20x + 10y \geq 16,000.$$

If at least 300 seats are to be sold for $10 each, then

$$y \geq 300.$$

Finally, the number of $20 seats cannot be negative, so

$$x \geq 0.$$

The graph of this system of four inequalities is given below, where the vertices of the region, $(650,300)$, $(700,300)$, and $(600,400)$ come from solving the various systems of equations determined by the system of inequalities.

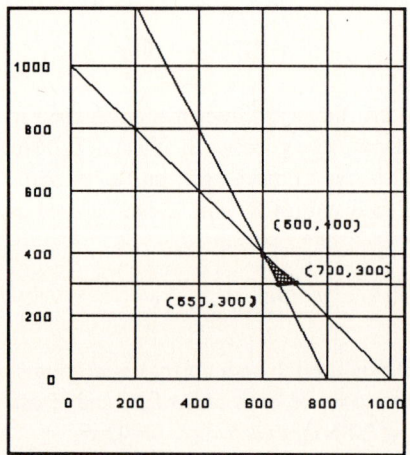

12. The degree of $P(x) = (x - 2)(x + 3)^2(x^2 - 5)$ is 5 since if these factors are multiplied, the term of highest degree will be the x^5-term.

13. If $2 + 3i$ is a zero of polynomial $P(x)$, with real coefficients, then the conjugate of $2 + 3i$, $2 - 3i$, will also be a zero of $P(x)$.

14. If $x - 2$ is one factor of the polynomial $P(x) = x^3 - x^2 - x - 2$, then by dividing $P(x)$ by $x - 2$ we obtain another factor. Use synthetic division.

$$\underline{2|}\ \begin{array}{r} 1 - 1 - 1 - 2 \\ + 2 + 2 + 2 \\ \hline 1 + 1 + 1 + 0 \end{array}$$

Then since the remainder is 0, $x - 2$ is indeed one factor of $P(x)$, and another factor is $x^2 + x + 1$.

15. Given $f(x) = \dfrac{1}{x - 2} + \dfrac{2}{x + 2}$.

To find another representation of $f(x)$ that has only one term, simply add the two expressions on the right side. The least common denominator is $(x - 2)(x + 2)$.

EXERCISES A

$$\frac{1}{x-2} + \frac{2}{x+2} = \frac{x+2}{(x-2)(x+2)} + \frac{2(x-2)}{(x-2)(x+2)}$$
$$= \frac{x+2+2(x-2)}{(x-2)(x+2)}$$
$$= \frac{x+2+2x-4}{(x-2)(x+2)}$$
$$= \frac{3x-2}{(x-2)(x+2)}$$
$$= \frac{3x-2}{x^2-4}$$

Thus, $f(x) = \dfrac{3x-2}{x^2-4}$.

16. Solve the system.

$$A + B = 3$$
$$2A - 2B = -2$$

First simplify the second equation by dividing through by 2.

$$A - B = -1$$

Then add the result to the first equation to eliminate B.

$$2A = 2$$
$$A = 1$$

Substitute 1 for A in $A + B = 3$ to find the value of B.

$$1 + B = 3$$
$$B = 2$$

Thus, $A = 1$ and $B = 2$.

17. Solve the system.

$$A + C = 4$$
$$ B - 2C = -4$$
$$-A + B + C = 4$$

Add the first and third equations to eliminate A and obtain

$$B + 2C = 8.$$

Pair this equation with the second equation in the original system to obtain the system of two equations in the two variables B and C given below.

$$B - 2C = -4$$
$$B + 2C = 8$$

Add these two equations to eliminate C.

$$2B = 4$$
$$B = 2$$

Substitute 2 for B in $B + 2C = 8$ to find the value of C.

$$2 + 2C = 8$$
$$2C = 6$$
$$C = 3$$

Substitute 3 for C in the first original equation,

$$A + C = 4,$$

to find the value of A.

$$A + 3 = 4$$
$$A = 1$$

Thus, $A = 1$, $B = 2$, and $C = 3$.

CHAPTER 6 SYSTEMS OF EQUATIONS AND INEQUALITIES

SECTION 6.6 Partial Fractions

1. Find the partial fraction decomposition of
$$f(x) = \frac{x-3}{x^2-1}.$$

Since the degree of the numerator is 1 and the degree of the denominator is 2, we proceed by factoring the denominator.
$$f(x) = \frac{x-3}{(x-1)(x+1)}$$

Using the partial fraction decomposition theorem, the decomposition must have two terms
$$\frac{A}{x-1} \quad \text{and} \quad \frac{B}{x+1}.$$

Thus,
$$\frac{x-3}{(x-1)(x+1)} = \frac{A}{x-1} + \frac{B}{x+1},$$

Where A and B are constant real numbers, yet to be determined. Multiply both sides of the equation by the LCD, $(x-1)(x+1)$.
$$x - 3 = A(x+1) + B(x-1)$$
$$x - 3 = Ax + A + Bx - B$$
$$x - 3 = (A+B)x + (A-B)$$

Since the polynomials on both sides are equal, the coefficients of like terms must be equal. Equating the coefficients of x and equating the constant terms gives the following system of two linear equations in the two variables, A and B.
$$A + B = 1$$
$$A - B = -3$$

Adding will eliminate B and give
$$2A = -2$$
$$A = -1$$

Substitute -1 for A in $A + B = 1$, $-1 + B = 1$, or $B = 2$. Then
$$f(x) = \frac{A}{x-1} + \frac{B}{x+1}$$
$$= \frac{-1}{x-1} + \frac{2}{x+1}.$$

2. Find the partial fraction decomposition of
$$f(x) = \frac{-2x^2 + 11x - 12}{(x-1)(x-2)^2}.$$

Using the partial fraction decomposition theorem,
$$\frac{-2x^2 + 11x - 12}{(x-1)(x-2)^2} = \frac{A}{x-1} + \frac{B}{x-2} + \frac{C}{(x-2)^2}.$$

Then A, B, and C are constant real numbers, yet to be determined. Multiply both sides of the equation by the LCD, $(x-1)(x-2)^2$.
$$-2x^2 + 11x - 12 = A(x-2)^2 + B(x-1)(x-2) + C(x-1)$$

Clear parentheses and collect like terms.
$$-2x^2 + 11x - 12 = (A+B)x^2 + (-4A - 3B + C)x + (4A + 2B - C)$$

Equate coefficients of like terms to obtain the following system.
$$A + B = -2$$
$$-4A - 3B + C = 11$$
$$4A + 2B - C = -12$$

Adding the second equation to the third gives
$$-B = -1$$
$$B = 1$$

Substitute 1 for B in the first equation, $A + 1 = -2$, to obtain $A = -3$. Then substitute 1 for B and -3 for A in the second equation to find the value of C.
$$-4(-3) - 3(1) + C = 11$$
$$12 - 3 + C = 11$$
$$C = 2$$

Then the partial fraction decomposition of the function is
$$f(x) = \frac{A}{x-1} + \frac{B}{x-2} + \frac{C}{(x-2)^2}$$
$$= \frac{-3}{x-1} + \frac{1}{x-2} + \frac{2}{(x-2)^2}.$$

PRACTICE EXERCISES

3. Find the partial fraction decomposition of

$$f(x) = \frac{x^2 + x + 10}{(x^2 + 5)(x - 1)}.$$

Using the partial fraction decomposition theorem,

$$\frac{x^2 + x + 10}{(x^2 + 5)(x - 1)} = \frac{Ax + B}{x^2 + 5} + \frac{C}{x - 1}.$$

Then A, B, and C are constant real numbers, yet to be determined. Multiply both sides of the equation by the LCD, $(x^2 + 5)(x - 1)$.

$$x^2 + x + 10 = (Ax + B)(x - 1) + C(x^2 + 5)$$

Clear parentheses and collect like terms.

$$x^2 + x + 10 = (A + C)x^2 + (-A + B)x + (-B + 5C)$$

Equate coefficients of like terms to obtain the following system.

$$\begin{aligned} A + + C &= 1 \\ -A + B &= 1 \\ -B + 5C &= 10 \end{aligned}$$

Adding the second equation to the third gives

$$-A + 5C = 11,$$

which can be paired with the first to obtain two equations in A and C. Adding these two equations gives

$$6C = 12$$
$$C = 2$$

Substitute 2 for C in the first equation, $A + 2 = 1$, to obtain $A = -1$. Then substitute -1 for A in the second equation, $-(-1) + B = 1$, to obtain $B = 0$.

Then the partial fraction decomposition of the function is

$$f(x) = \frac{Ax + B}{x^2 + 5} + \frac{C}{x - 1}$$
$$= \frac{-x}{x^2 + 5} + \frac{2}{x - 1}.$$

4. Find the partial fraction decomposition of

$$f(x) = \frac{x^5 + 6x^3 + x^2 + 9x + 1}{x^4 + 4x^2 + 4}.$$

Since the degree of the numerator is greater than the degree of the denominator, first divide the numerator by the denominator to obtain the following form.

$$f(x) = x + \frac{2x^3 + x^2 + 5x + 1}{x^4 + 4x^2 + 4}$$
$$= x + \frac{2x^3 + x^2 + 5x + 1}{(x^2 + 2)^2}$$

Then we can concentrate of the fraction in which the denominator has now been factored. Using the partial fraction decomposition theorem,

$$\frac{2x^3 + x^2 + 5x + 1}{(x^2 + 2)^2} = \frac{Ax + B}{x^2 + 2} + \frac{Cx + D}{(x^2 + 2)^2}.$$

Multiply both sides of the equation by the LCD, $(x^2 + 2)^2$.

$$2x^3 + x^2 + 5x + 1 = (Ax + B)(x^2 + 2) + Cx + D$$

Clear parentheses and collect like terms.

$$2x^3 + x^2 + 5x + 1 = Ax^3 + Bx^2 + (2A + C)x + (2B + D)$$

Equate coefficients of like terms to obtain $A = 2$, $B = 1$, and

$$\begin{aligned} 2A + C &= 5 & 2B + D &= 1 \\ 2(2) + C &= 5 & 2(1) + D &= 1 \\ C &= 1 & D &= -1 \end{aligned}$$

Then the partial fraction decomposition of the function is

$$f(x) = x + \frac{Ax + B}{x^2 + 2} + \frac{Cx + D}{(x^2 + 2)^2}$$
$$= x + \frac{2x + 1}{x^2 + 2} + \frac{x - 1}{(x^2 + 2)^2}.$$

CHAPTER 6 SYSTEMS OF EQUATIONS AND INEQUALITIES

SECTION 6.6 Partial Fractions

1. Given the function $f(x) = \dfrac{3x+2}{(x-1)(x+5)}$.

   Using the partial fraction decomposition theorem, the decomposition must have two terms,

   $$\dfrac{A}{x-1} \text{ and } \dfrac{B}{x+5}.$$

   Thus,

   $$f(x) = \dfrac{A}{x-1} + \dfrac{B}{x+5},$$

   where A and B are constant real numbers.

2. Given the function $f(x) = \dfrac{6x^2 - 7}{(x+3)^2(x+5)}$.

   Using the partial fraction decomposition theorem, the decomposition must have three terms,

   $$\dfrac{A}{x+3}, \dfrac{B}{(x+3)^2}, \text{ and } \dfrac{C}{x+5}.$$

   Thus,

   $$f(x) = \dfrac{A}{x+3} + \dfrac{B}{(x+3)^2} + \dfrac{C}{x+5},$$

   where A, B, and C are constant real numbers.

3. Given the function $f(x) = \dfrac{x^3 - 5}{(x^2 + 2x + 10)^2}$.

   Using the partial fraction decompoaition theorem, the decomposition must have two terms,

   $$\dfrac{Ax+B}{x^2+2x+10} \text{ and } \dfrac{Cx+D}{(x^2+2x+10)^2}.$$

   Thus,

   $$f(x) = \dfrac{Ax+B}{x^2+2x+10} + \dfrac{Cx+D}{(x^2+2x+10)^2},$$

   where A, B, C, and D are constant real numbers.

4. Given the function $f(x) = \dfrac{x+1}{x^3 - 2x^2 - 3x}$.

   First factor the denominator of $f(x)$ and reduce the fraction.

   $$f(x) = \dfrac{x+1}{x(x+1)(x-3)} = \dfrac{1}{x(x-3)}$$

   Using the partial fraction decomposition theorem, the decomposition must have two terms,

   $$\dfrac{A}{x} \text{ and } \dfrac{B}{x-3}.$$

   Thus,

   $$f(x) = \dfrac{A}{x} + \dfrac{B}{x-3},$$

   where A and B are constant real numbers.

5. Given the function

   $$f(x) = \dfrac{x-3}{x^2(x+2) - 2x(x+2) - 3(x+2)}.$$

   First factor the denominator of $f(x)$ and reduce the fraction.

   $$f(x) = \dfrac{x-3}{(x+2)(x^2 - 2x - 3)}$$
   $$= \dfrac{x-3}{(x+2)(x+1)(x-3)}$$
   $$= \dfrac{1}{(x+2)(x+1)}$$

   Using the partial fraction decomposition theorem, the decomposition must have two terms,

   $$\dfrac{A}{x+2} \text{ and } \dfrac{B}{x+1}.$$

   Thus,

   $$f(x) = \dfrac{A}{x+2} + \dfrac{B}{x+1},$$

   where A and B are constant real numbers.

6. Given the function $f(x) = \dfrac{x^2 + 5x + 5}{x^4 + 5x^2 + 4}$.

   First factor the denominator.

   $$f(x) = \dfrac{x^2 + 5x + 5}{(x^2 + 1)(x^2 + 4)}$$

   Then by the partial fraction decomposition theorem, the decomposition must have two terms,

   $$\dfrac{Ax+B}{x^2+1} \text{ and } \dfrac{Cx+D}{x^2+4}.$$

   Thus,

EXERCISES A

$$f(x) = \frac{Ax + B}{x^2 + 1} + \frac{Cx + D}{x^2 + 4},$$

where A, B, C, and D are constant real numbers.

7. Given $f(x) = \dfrac{5x - 1}{x^2 - 1} = \dfrac{A}{x - 1} + \dfrac{B}{x + 1}$,

   find A and B.

   Multiplying both sides of the representation for $f(x)$ by the LCD, $(x - 1)(x + 1)$, we obtain

   $$5x - 1 = A(x + 1) + B(x - 1).$$

   Clearing parentheses, we can simplify this equation to the form

   $$5x - 1 = (A + B)x + (A - B).$$

   Then equating coefficients of like terms, we obtain the following system:

   $$A + B = 5$$
   $$A - B = -1$$

   Adding these two equations we eliminate B and obtain

   $$2A = 4$$
   $$A = 2.$$

   Substitute 2 for A in $A + B = 5$ to find the value of B.

   $$2 + B = 5$$
   $$B = 3$$

   Thus, $A = 2$ and $B = 3$.

8. Given $f(x) = \dfrac{1 - x}{(x + 1)^2} = \dfrac{A}{x + 1} + \dfrac{B}{(x + 1)^2}$,

   find A and B.

   Multiplying both sides of the representation for $f(x)$ by the LCD, $(x + 1)^2$, we obtain

   $$1 - x = A(x + 1) + B.$$

   Clearing parentheses, we can simplify this equation to the form

   $$-x + 1 = Ax + (A + B).$$

   Then equating coefficients of like terms, we obtain the following system:

   $$A = -1$$
   $$A + B = 1$$

   Then $A = -1$, and substituting -1 for A in the second equation gives

   $$-1 + B = 1$$
   $$B = 2.$$

   Thus, $A = -1$ and $B = 2$.

9. Given $f(x) = \dfrac{6x^2 + 1}{(x^2 + 1)(x - 2)} = \dfrac{Ax + B}{x^2 + 1} + \dfrac{C}{x - 2}$,

   find A, B, and C.

   Multiplying both sides of the representation for $f(x)$ by the LCD, $(x^2 + 1)(x - 2)$, we obtain

   $$6x^2 + 1 = (Ax + B)(x - 2) + C(x^2 + 1).$$

   Clearing parentheses, we can simplify this equation to the form

   $$6x^2 + 1 = (A + C)x^2 + (-2A + B)x + (-2B + C).$$

   Then equating coefficients of like terms, we obtain the following system:

   $$A + C = 6$$
   $$-2A + B = 0$$
   $$ - 2B + C = 1$$

   Subtract the third from the first to eliminate C.

   $$A + 2B = 5$$

   Multiply the second equation by -2 and add to this equation to eliminate B.

   $$5A = 5$$
   $$A = 1$$

   Substitute 1 for A in $A + C = 6$ to find the value of C.

   $$1 + C = 6$$
   $$C = 5$$

   Substitute 1 for A in $-2A + B = 0$ to find the value of B.

$$-2(1) + B = 0$$
$$B = 2$$

Thus, $A = 1$, $B = 2$, and $C = 5$.

10. Given

$$f(x) = \frac{x^3 + x^2 + 3x + 1}{(x^2 + 2)^2} = \frac{Ax + B}{x^2 + 2} + \frac{Cx + D}{(x^2 + 2)^2},$$

find A, B, C, and D.

Multiplying both sides of the representation for $f(x)$ by the LCD, $(x^2 + 2)^2$, we obtain:

$$x^3 + x^2 + 3x + 1 = (Ax + B)(x^2 + 2) + (Cx + D)$$

Clearing parentheses, we can simplify this equation to the form:

$$x^3 + x^2 + 3x + 1 = Ax^3 + Bx^2 + (2A + C)x + (2B + D)$$

Then equating coefficients of like terms, we obtain

$A = 1$, $B = 1$, $2A + C = 3$, and $2B + D = 1$.

Substitute 1 for A in $2A + C = 3$ to obtain $C = 1$. Then substitute 1 for B in $2B + D = 1$ to obtain $D = -1$.

Thus, $A = 1$, $B = 1$, $C = 1$, and $D = -1$.

11. Given

$$f(x) = \frac{-2x^2 + 15x + 13}{(x-2)^2(x+3)} = \frac{A}{x-2} + \frac{B}{(x-2)^2} + \frac{C}{x+3},$$

find A, B, and C.

Multiplying both sides of the representation for $f(x)$ by the LCD, $(x - 2)^2(x + 3)$, we obtain:

$$-2x^2 + 15x + 13 =$$
$$A(x - 2)(x + 3) + B(x + 3) + C(x - 2)^2$$

Clearing parentheses, we can simplify this equation to the form:

$$-2x^2 + 15x + 13 =$$
$$(A + C)x^2 + (A + B - 4C)x + (-6A + 3B + 4C)$$

Then equating coefficients of like terms, we obtain the following system:

$$A + C = -2$$
$$A + B - 4C = 15$$
$$-6A + 3B + 4C = 13$$

Multiply the second equation by -3 and add the result to the third equation to eliminate B and obtain

$$-9A + 16C = -32.$$

Add this equation to 9 times the first original equation to eliminate A and obtain

$$25C = -50$$
$$C = -2.$$

Substitute -2 for C in $A + C = -2$ to find the value of A.

$$A + (-2) = -2$$
$$A = 0$$

Substitute 0 for A and -2 for C in

$$A + B - 4C = 15$$

to find the value of B.

$$0 + B - 4(-2) = 15$$
$$B + 8 = 15$$
$$B = 7$$

Thus, $A = 0$, $B = 7$, and $C = -2$.

12. Given

$$f(x) = \frac{4x}{(x^2 + 1)^2(x - 1)} = \frac{Ax + B}{x^2 + 1} + \frac{Cx + D}{(x^2 + 1)^2} + \frac{E}{x - 1},$$

find A, B, C, D and E.

Multiplying both sides of the representation for $f(x)$ by the LCD, $(x^2 + 1)^2(x - 1)$, we obtain

$$4x = (Ax + B)(x^2 + 1)(x - 1) + (Cx + D)(x - 1) + E(x^2 + 1)^2.$$

Clearing parentheses, we can simplify this equation to the form

$$4x = (A + E)x^4 + (-A + B)x^3 + (A - B + C + 2E)x^2 + (-A + B - C + D)x + (-B - D + E).$$

Substitute 1 for x and simplify to obtain

EXERCISES A

$$4 = 4E$$
$$1 = E.$$

Equating coefficients of like terms, we have $A + E = 0$, and with $E = 1$, $A = -1$. Then since $-A + B = 0$, and $A = -1$, $B = -1$. Substitute these values into $A - B + C + 2E = 0$ to obtain $C = -2$. Then substitute -1 for B and 1 for E in $-B - D + E = 0$ to obtain $D = 2$.

Thus, $A = 1$, $B = -1$, $C = -2$, $D = 2$, and $E = 1$.

13. Find the partial fraction decomposition of

$$f(x) = \frac{3x - 3}{x^2 - 9}.$$

First factor the denominator.

$$f(x) = \frac{3x - 3}{(x - 3)(x + 3)}$$

Then by the partial fraction decomposition theorem, we know that

$$f(x) = \frac{3x - 3}{x^2 - 9} = \frac{A}{x - 3} + \frac{B}{x + 3}.$$

Multiply both sides of the two expressions for $f(x)$ by the LCD, $(x - 3)(x + 3)$.

$$3x - 3 = A(x + 3) + B(x - 3)$$

Clearing parentheses and simplifying, we obtain:

$$3x - 3 = (A + B)x + (3A - 3B)$$

Then equating coefficients and simplifying, we obtain the following system:

$$A + B = 3$$
$$A - B = -1$$

Adding the two equations gives

$$2A = 2$$
$$A = 1.$$

Substitute 1 for A in the first equation to obtain

$$1 + B = 3$$
$$B = 2.$$

Thus, $A = 1$ and $B = 2$, and

$$f(x) = \frac{1}{x - 3} + \frac{2}{x + 3}.$$

14. Find the partial fraction decomposition of

$$f(x) = \frac{6x^2 + 20x + 19}{(x + 2)^2(x + 1)}.$$

Then by the partial fraction decomposition theorem, we know that

$$f(x) = \frac{6x^2 + 20x + 19}{(x + 2)^2(x + 1)} = \frac{A}{x + 2} + \frac{B}{(x + 2)^2} + \frac{C}{x + 1}.$$

Multiply both sides of the two expressions for $f(x)$ by the LCD, $(x + 2)^2(x + 1)$.

$$6x^2 + 20x + 19 = A(x + 2)(x + 1) + B(x + 1) + C(x + 2)^2$$

Clearing parentheses and simplifying, we obtain:

$$6x^2 + 20x + 19 = (A + C)x^2 + (3A + B + 4C)x + (2A + B + 4C)$$

Then equating coefficients and simplifying, we obtain the following system:

$$A + C = 6$$
$$3A + B + 4C = 20$$
$$2A + B + 4C = 19$$

Subtract the third equation from the second to eliminate B and C and obtain $A = 1$. Substitute 1 for A in the first equation to obtain $C = 5$. Then substitute 1 for A and 5 for C in the second equaiton to obtain $B = -3$. Thus,

$$f(x) = \frac{1}{x + 2} + \frac{-3}{(x + 2)^2} + \frac{5}{x + 1}.$$

15. Find the partial fraction decomposition of

$$f(x) = \frac{-9x - 7}{(x^2 + 1)(x - 5)}.$$

Then by the partial fraction decomposition theorem, we know that

$$f(x) = \frac{-9x - 7}{(x^2 + 1)(x - 5)} = \frac{Ax + B}{x^2 + 1} + \frac{C}{x - 5}.$$

Multiply both sides of the two expressions for $f(x)$ by the LCD, $(x^2 + 1)(x - 5)$.

$$-9x - 7 = (Ax + B)(x - 5) + C(x^2 + 1)$$

Clearing parentheses and simplifying, we obtain:

$$-9x - 7 = (A + C)x^2 + (-5A + B)x + (-5B + C)$$

Then equating coefficients, we obtain the following system:

$$A + C = 0$$
$$-5A + B = -9$$
$$ - 5B + C = -7$$

Subtract the third equation from the first equation to obtain

$$A + 5B = 7.$$

Multiply this equation by 5 and add to the second original equation to eliminate A and obtain $B = 1$. Substitute 1 for B in the second original equation to obtain $A = 2$. Then substitute 2 for A in the first original equation to obtain $C = -2$. Thus,

$$f(x) = \frac{2x+1}{x^2+1} + \frac{-2}{x-5}.$$

16. Find the partial fraction decomposition of

$$f(x) = \frac{2x^3 - x^2 + 3x - 1}{(x^2+1)^2}.$$

Then by the partial fraction decomposition theorem, we know that

$$f(x) = \frac{2x^3 - x^2 + 3x - 1}{(x^2+1)^2} = \frac{Ax+B}{x^2+1} + \frac{Cx+D}{(x^2+1)^2}.$$

Multiply both sides of the two expressions for $f(x)$ by the LCD, $(x^2 + 1)^2$.

$$2x^3 - x^2 + 3x - 1 = (Ax + B)(x^2 + 1) + Cx + D$$

Clearing parentheses and simplifying, we obtain:

$$2x^3 - x^2 + 3x - 1 = Ax^3 + Bx^2 + (A + C)x + (B + D)$$

Then equating coefficients, we obtain $A = 2$, $B = -1$, $A + C = 3$, and $B + D = -1$. Substituting the known values for A and B into the other two equations we obtain $C = 1$ and $D = 0$. Thus

$$f(x) = \frac{2x-1}{x^2+1} + \frac{x}{(x^2+1)^2}.$$

17. The vertices of the feasible region shown in the text are (2,1), (4,5), (7,6), and (8,2). By the fundamental theorem of linear programming, the maximum and minimum values of the objective function

$$P = 2x - 5y$$

will occur at one of these vertices. The table below shows the various possibilities.

Vertex	$2x - 5y$	
(2,1)	-1	
(4,5)	-17	OPTIMAL SOLUTION
(7,6)	-16	
(8,2)	6	OPTIMAL SOLUTION

Thus, the maximum value is 6 and the minimum value is -17.

18. Let x = the number of Pro-Model racquets made daily,
y = the number of Hacker-Model racquets daily.

Since a profit of $15 is made on each Pro-Model racquet and a profit of $10 is made on each Hacker-Model racquet, the objective function we must maximize is

$$P = 15x + 10y.$$

The daily production of the Pro-model should be between 10 and 50, inclusive, gives us the inequality

$$10 \leq x \leq 50.$$

The daily production of the Hacker-Model should be between 20 and 40, inclusive, gives us the inequality

$$20 \leq y \leq 40.$$

Since the total number of racquets made daily should not exceed 80, a third constraint is

$$x + y \leq 80.$$

The graph of these constraints, the feasible region, is given below where the vertices, (10,20), (10,40), (40,40), (50,30), and (50,20), are determined by substitution and inspection.

EXERCISES A

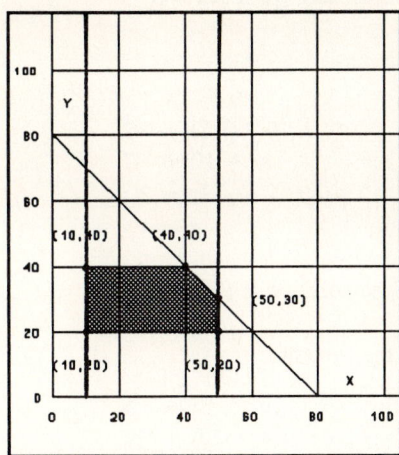

The table below summarizes the various possibilities.

Vertex	15x + 10y	
(10,20)	350	
(10,40)	550	
(40,40)	1000	
(50,30)	1050	OPTIMAL SOLUTION
(50,20)	950	

Thus, the maximum profit of $1050 will be realized when 50 Pro-Model and 30 Hacker-Model racquets are made daily.

CHAPTER 6 SYSTEMS OF EQUATIONS AND INEQUALITIES

CHAPTER 6 Review Exercises

1. To determine whether $(2,-3)$ is a solution to the system
$$3x + y = 3$$
$$-x + 2y = -8$$
Substitute 2 for x and -3 for y in each equation.

$$3(2) + (-3) = 3 \qquad -(2) + 2(-3) = -8$$
$$6 - 3 = 3 \qquad\qquad -2 - 6 = -8$$
$$3 = 3 \qquad\qquad\qquad -8 = -8$$

Thus, $(2,-3)$ is a solution to the system.

2. Solve by the substitution method.
$$3x + y = -1$$
$$-2x - 3y = -11$$
Solve the first equation for y, $y = -3x - 1$, and substitute this expression for y in the second equation.

$$-2x - 3(-3x - 1) = -11$$
$$-2x + 9x + 3 = -11$$
$$7x = -14$$
$$x = -2$$

Substitute -2 for x in $y = -3x - 1$ to find the value of y.

$$y = -3(-2) - 1$$
$$y = 6 - 1$$
$$y = 5$$

Thus, the solution is $(-2, 5)$.

3. Solve by the elimination method.
$$3x + 5y = -7$$
$$-4x + 2y = -8$$
Multiply the first equation by 4 and the second equation by 3 to obtain:
$$12x + 20y = -28$$
$$-12x + 6y = -24$$
Add to eliminate x ant obtain
$$26y = -52$$
$$y = -2.$$
Substitute -2 for y in the first original equation to find the value of x.

$$3x + 5(-2) = -7$$
$$3x - 10 = -7$$
$$3x = 3$$
$$x = 1$$

Thus, the solution is $(1,-2)$.

4. Solve the system.
$$3x - 5y = 2$$
$$-6x + 10y = -1$$
Multiply the top equation by 2.
$$6x - 10y = 4$$
$$-6x + 10y = -1$$
Adding we eliminate both x and y and obtain the contradiction $0 = 3$. Thus, the system is inconsistent and has no solution.

5. Solve the system.
$$\frac{3}{x} + \frac{1}{y} = 2$$
$$\frac{5}{x} - \frac{3}{y} = 8$$
Substitute u for $\frac{1}{x}$ and v for $\frac{1}{y}$.
$$3u + v = 2$$
$$5u - 3v = 8$$
Solve the first equation for v, $v = -3u + 2$, and substitute this expression for v in the second equation.

$$5u - 3(-3u + 2) = 8$$
$$5u + 9u - 6 = 8$$
$$14u = 14$$
$$u = 1$$

Substitute 1 for u in $v = -3u + 2$ to find the value of v.

$$v = -3(1) + 2$$
$$v = -3 + 2$$
$$v = -1$$

Backsubstitute to obtain the values of x and y.

$$x = \frac{1}{u} = \frac{1}{1} = 1 \quad \text{and} \quad y = \frac{1}{v} = \frac{1}{-1} = -1$$

Thus, the solution is $(1,-1)$.

CHAPTER 6 REVIEW

7. Solve the system.

$$2x + 3y + 4z = 0$$
$$x - 5y + 3z = -1$$
$$3x + y + z = 5$$

Multiply the second equation by -2,

$$-2x + 10y - 6z = 2,$$

and add the result to the first equation to obtain:

$$13y - 2z = 2.$$

Then multiply the second equation by -3,

$$-3x + 15y - 9z = 3,$$

and add the result to the third equation to obtain:

$$16y - 8z = 8,$$

which simplifies to

$$2y - z = 1.$$

Then we obtain the system of two equations in y and z given below.

$$13y - 2z = 2$$
$$2y - z = 1$$

Multiply the second equation by -2,

$$-4y + 2z = -2,$$

and add to the first to eliminate z and obtain

$$9y = 0$$
$$y = 0.$$

Substitute 0 for y in $2y - z = 1$ to find the value of z.

$$2(0) - z = 1$$
$$-z = 1$$
$$z = -1$$

Substitute 0 for y and -1 for z in the second original equation to find the value of x.

$$x - 5(0) + 3(-1) = -1$$
$$x - 3 = -1$$
$$x = 2$$

Thus, the solution is $(2, 0, -1)$.

REVIEW EXERCISES 305

8. Solve the system.

$$x - 2y + 4z = 0$$
$$x + 2y - 5z = 0$$
$$2x - z = 0$$

Since this is a homogeneous system, one solution is $(0,0,0)$. To determine whether there are infinitely many solutions, we try to solve by elimination. If we add the first two equations, we can eliminate y and obtain

$$2x - z = 0.$$

Since the result is the same as the original third equation, subtracting these two gives the identity $0 = 0$ which tells us that there are indeed infinitely many solutions. Solving for z, we obtain $z = 2x$. Substitute $2x$ for z in the first original equation and solve for y in terms of x.

$$x - 2y + 4(2x) = 0$$
$$x - 2y + 8x = 0$$
$$-2y = -9x$$
$$y = \frac{9}{2}x$$

Thus, the solutions are $\left(x, \frac{9}{2}x, 2x\right)$ for x any real number. Notice that $(0,0,0)$ is one of these solutions, found by letting $x = 0$.

9. Solve the system.

$$3x + 2y = 12$$
$$-x + y - 5z = 1$$

Since there are only two equations and three unknowns, the system has either no solution or infinitely many solutions. Multiply the second equation by 3,

$$-3x + 3y - 15z = 3,$$

and add the result to the first to eliminate x,.

$$5y - 15z = 15$$

which simplifies to

$$y - 3z = 3.$$

Since we do not obtain a contradiction, there are infinitely many solutions. Solve for y in terms of z.

$$y = 3z + 3$$

Substitute this expression for y in the first original equation and solve for x in terms of z.

$$3x + 2(3z + 3) = 12$$
$$3x + 6z + 6 = 12$$
$$3x = 6 - 6z$$
$$x = 2 - 2z$$

Thus, the solution is $(2-2z, 3z+3, z)$ for z any real number.

10. Since the parabola with equation

$$y = ax^2 + bx + c$$

passes through $(1,4)$, $(-1,10)$, and $(3,14)$, substitute these values for x and y into the equation to obtain the following system of equations in a, b, and c.

$$a + b + c = 4$$
$$a - b + c = 10$$
$$9a + 3b + c = 14$$

Add the first two equations to eliminate b, and obtain

$$2a + 2c = 14$$

which simplifies to

$$a + c = 7.$$

Multiply the second equation by 3,

$$3a - 3b + 3c = 30,$$

and add the result to the third equation to eliminate b and obtain

$$12a + 4c = 44,$$

which simplifies to

$$3a + c = 11.$$

Then we have reduced the system to

$$a + c = 7$$
$$3a + c = 11.$$

Subtract the top equation from the bottom equation to eliminate c and obtain

$$2a = 4$$
$$a = 2.$$

Substitute 2 for a in $a + c = 7$ to find the value of c.

$$2 + c = 7$$
$$c = 5$$

Substitute 2 for a and 5 for c in the first original equation to find the value of b.

$$2 + b + 5 = 4$$
$$b = -3$$

Thus, $a = 2$, $b = -3$, and $c = 5$, and the equation of the parabola through the three given points is

$$y = 2x^2 - 3x + 5.$$

11. Let $x =$ the time traveled at 40 mph,
    $y =$ the time traveled at 50 mph.

Then using the distance formula, $d = rt$, the distance traveled at 40 mph is $40x$ and the distance traveled at 50 mph is $50y$. Since the total distance traveled was 370 miles, we obtain

$$40x + 50y = 370$$

which simplifies to

$$4x + 5y = 37.$$

Had he gone 55 mph over the same period of time, which is a total time of $x + y$, he would have gone 440 miles. Thus

$$55(x + y) = 440$$

which simplifies to

$$x + y = 8.$$

Solve this equation for y, $y = 8 - x$, and substitute into the first equation.

$$4x + 5(8 - x) = 37$$
$$4x + 40 - 5x = 37$$
$$-x = -3$$
$$x = 3$$

Then $y = 8 - x = 8 - 3 = 5$. Thus, he traveled 3 hr at 40 mph and 5 hr at 50 mph.

12. Let x = the measure of one angle,
y = the measure of the other angle.

Since the two angles are complementary, the sum of their measures is 90°, and we obtain

$$x + y = 90.$$

The phrase "one measures 6° less than seven times the measure of the other" translates to

$$x = 7y - 6.$$

Substitute this expression for x in the first equation.

$$(7y - 6) + y = 90$$
$$8y - 6 = 90$$
$$8y = 96$$
$$y = 12$$

Then $x = 7y - 6 = 7(12) - 6 = 84 - 6 = 78$. Thus, the angles measure 12° and 78°.

13. Let x = number lb of candy worth \$1.60/lb,
y = number lb of candy worth \$1.20/lb.

Since the total number of pounds in the mixture is 80, the quantity equation is

$$x + y = 80.$$

The value of x pounds of candy worth \$1.60/lb is $1.60x$, the value of y pounds of candy worth \$1.20/lb is $1.20y$, and the value of the 80 pounds of candy in the mixture is $(1.36)(80)$. Thus, the value equation is

$$1.60x + 1.20y = (1.36)(80),$$

which simplifies to

$$4x + 3y = 272.$$

Solve the first equation for y, $y = 80 - x$, and substitute this expression for y in the second equation.

$$4x + 3(80 - x) = 272$$
$$4x + 240 - 3x = 272$$
$$x = 32$$

Then $y = 80 - 32 = 48$. Thus, the mix contains 32 lb of candy worth \$1.60/lb and 48 lb of candy worth \$1.20/lb.

14. Let x = the amount invested in bonds,
y = the amount invested in certificates,
z = the amount invested in stocks.

Since the total amount invested is \$9000, we have

$$x + y + z = 9000.$$

Since the bonds earn 9%, the amount earned in bonds is $0.09x$. Similarly the amount earned in certificates is $0.08y$. If the stocks do well they will earn 10% which gives $0.10z$ as the amount earned. Thus, a second equation is

$$0.09x + 0.08y + 0.10z = 780,$$

which simplifies to

$$9x + 8y + 10z = 78,000.$$

If the stocks lose 3%, the amount lost is $0.03z$, so this amount must be subtracted from the earnings to give us the third equation

$$0.09x + 0.08y - 0.03z = 520,$$

or equivalently

$$9x + 8y - 3z = 52,000.$$

Thus we obtain the following system:

$$\begin{array}{rcr} x + y + z &=& 9000 \\ 9x + 8y + 10z &=& 78,000 \\ 9x + 8y - 3z &=& 52,000 \end{array}$$

If we subtract the third equation from the second, we will eliminate x and y and obtain the value of z.

$$13z = 26,000$$
$$z = 2000$$

Substitute 2000 for z in the first equation and in the second equation to obtain two equations in x and y.

$$x + y = 7000$$
$$9x + 8y = 58,000$$

Solve the first equation for x, $x = 7000 - y$, and substitute into the second.

$$9(7000 - y) + 8y = 58,000$$
$$63,000 - 9y + 8y = 58,000$$
$$y = 5000$$

Then $x = 7000 - y = 7000 - 5000 = 2000$. Thus, \$2000 are in bonds, \$5000 in certificates, and \$2000 in stocks.

15. Given that the polynomial function

$$P(x) = x^3 + ax + b$$

has the property that $P(1) = 4$ and $P(-1) = 12$, substitute to obtain a system of two equations in the unknowns a and b.

$$P(1) = 4 = (1)^3 + a(1) + b$$
$$3 = a + b$$

$$P(-1) = 12 = (-1)^3 + a(-1) + b$$
$$13 = -a + b$$

Then the system of equations is:

$$a + b = 3$$
$$-a + b = 13$$

Adding we have

$$2b = 16$$
$$b = 8.$$

Substitute 8 for b in $a + b = 3$ to find the value of a.

$$a + 8 = 3$$
$$a = -5$$

Thus, $a = -5$ and $b = 8$, and the polynomial function is

$$P(x) = x^3 - 5x + 8.$$

16. Graph the system.

$$2x + y \leq 2$$
$$y - 1 > 0$$

First graph the line with equation $2x + y = 2$ using intercepts $(0,2)$ and $(1,0)$ and a solid line (the inequality is $\leq$). The test point $(0,0)$ shows

$$2(0) + (0) \leq 2 \quad \text{This is true}$$

that the solution is the region containing the test point, the region "below" the line.

The line $y - 1 = 0$, or $y = 1$, is a horizontal line (dashed) through y-intercept $(0,1)$. The test point $(0,0)$ shows

$$(0) - 1 > 0 \quad \text{This is false}$$

that the solution is the region not containing the test point, the region "above" the line. Putting this information together, we obtain the graph of the system given below.

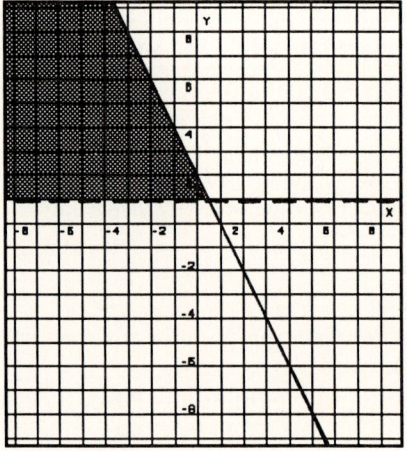

17. Graph the system.

$$x \geq 0$$
$$y \geq 0$$
$$3x + 2y < 6$$

The first two inequalities describe the region in the first quadrant, together with the positive x-axis and the positive y-axis. Thus, we can concentrate on the third inequality. Graph the line $3x + 2y = 6$ using intercepts $(0,3)$ and $(2,0)$ and a dashed line (the inequality is $<$). The test point $(0,0)$ shows

$$3(0) + 2(0) < 6 \quad \text{This is true}$$

that the solution is the region that contains the test point, the region "below" the line. The graph of the system is then the region below the line but still in the first quadrant as shown below.

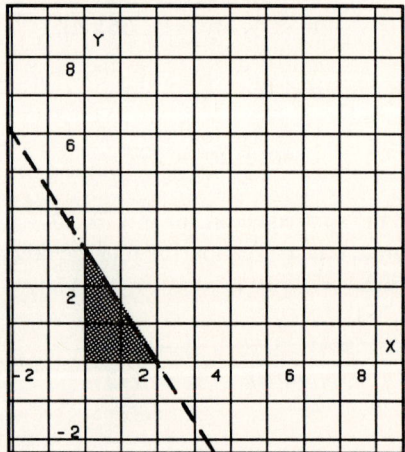

18. The maximum and minimum values of the objective function $P = 18x + 3y$ will occur at one of the vertices of the feasible region shown in the text.

We summarize the possibilities using a table.

Vertex	18x + 3y	
(2,1)	39	
(1,5)	33	OPTIMAL SOLUTION
(4,6)	90	
(8,5)	159	OPTIMAL SOLUTION
(6,1)	111	

Thus, the minimum value is 33 and the maximum value is 159.

19. The maximum and minimum values of the objective function

$$P = 6x - 10y$$

subject to the constraints

$$1 \leq x \leq 5$$
$$y \geq 2$$
$$y - 3x \leq 0$$
$$2x + 3y \leq 22$$

will occur at one of the vertices of the feasible region (the graph of the system) given below. The vertices are obtained by solving the various systems of equations determined by the constraints or by direct substitution.

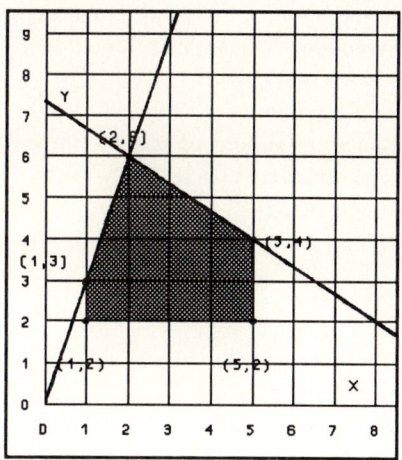

We summarize the possibilities using a table.

Vertex	6x - 10y	
(1,2)	-14	
(1,3)	-24	
(2,6)	-48	OPTIMAL SOLUTION
(5,4)	-10	
(5,2)	10	OPTIMAL SOLUTION

Thus, the maximum value is 10 and the minimum value is -48.

20. Let $x =$ the number of hamburgers to make daily,
$y =$ the number of tacos to make daily.

The constraint inequality for the number of pounds of ground beef available is

$$\tfrac{1}{3}x + \tfrac{1}{6}y \leq 90,$$

which simplifies to

$$2x + y \leq 540.$$

The constraint inequality for the total labor costs is

$$0.18x + 0.06y \leq 36,$$

which simplifies to

$$3x + y \leq 600.$$

Also, since x and y must be nonnegative, we have

$$x \geq 0 \text{ and } y \geq 0.$$

The graph of these four constraints, the feasible region is given below.

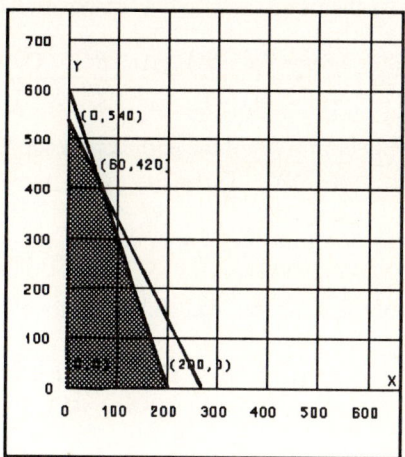

The objective function, which is the profit to be maximized, is $P = 42x + 20y$ (using cents). The possibilities are summarized in the following table.

Vertex	42x + 20y	
(0,0)	0	
(0,540)	10,800	
(60,420)	10,920	OPTIMAL SOLUTION
(200,0)	8400	

Thus, the maximum profit of 10,920 cents, or $109.20, occurs when 60 hamburgers and 420 taco fillers are made.

21. Given the partial fraction decomposition of

$$f(x) = \frac{-2x-5}{(x+4)^2} = \frac{A}{x+4} + \frac{B}{(x+4)^2}.$$

Multiply both sides of the expressions for $f(x)$ by the LCD, $(x+4)^2$.

$$-2x - 5 = A(x+4) + B$$
$$-2x - 5 = Ax + (4A + B)$$

Equating coefficients, we obtain $A = -2$ and $4A + B = -5$. Substitute -2 for A in the second equation to find the value of B.

$$4(-2) + B = -5$$
$$-8 + B = -5$$
$$B = 3$$

Thus, $A = -2$ and $B = 3$, and

$$f(x) = \frac{-2}{x+4} + \frac{3}{(x+4)^2}.$$

22. Given $f(x) = \dfrac{8x^2 - 4x + 16}{(x^2+3)(x-1)}$.

Then by the partial fraction decomposition theorem,

$$f(x) = \frac{8x^2 - 4x + 16}{(x^2+3)(x-1)} = \frac{Ax+B}{x^2+3} + \frac{C}{x-1}.$$

Multiply both sides of the expressions for $f(x)$ by the LCD, $(x^2 + 3)(x - 1)$.

$$8x^2 - 4x + 16 = (Ax + B)(x-1) + C(x^2 + 3)$$

Clearing parentheses and simplifying we have:

$$8x^2 - 4x + 16 = (A+C)x^2 + (-A+B)x + (-B+3C)$$

Equating coefficients we obtain the system:

$$\begin{array}{rcr} A\phantom{{}+B} + C &=& 8 \\ -A + B \phantom{{}+ 3C} &=& -4 \\ -B + 3C &=& 16 \end{array}$$

Add the first two equations to eliminate A and obtain

$$B + C = 4.$$

Add this result to the third original equation to eliminate B and obtain

$$4C = 20$$
$$C = 5$$

Substitute 5 for C in $A + C = 8$ to find the value of A.

$$A + 5 = 8$$
$$A = 3$$

Substitute 3 for A in $-A + B = -4$ to find B.

$$-3 + B = -4$$
$$B = -1$$

Thus, $A = 3$, $B = -1$, $C = 5$, and

$$f(x) = \frac{3x-1}{x^2+3} + \frac{5}{x-1}.$$

23. Let $x =$ number of gallons of 15% acid solution, $y =$ number of gallons of 20% acid solution.

Since the total number of gallons in the mixture is 100, the quantity equation is

$$x + y = 100.$$

The "value equation" this time is the amount of acid in the solution. The amount of acid in x gallons of a 15% acid solution is $0.15x$. Similarly the amount of acid in y gallons of a 20% acid solution is $0.20y$. The number of gallons of acid in 100 gallons of an 18% acid solution is $(0.18)(100) = 18$. Thus the second equation is

$$0.15x + 0.20y = 18,$$

which can be simplified to

$$3x + 4y = 360.$$

Solve the first equation for x, $x = 100 - y$, and substitute into the second equation.

$$3(100 - y) + 4y = 360$$
$$300 - 3y + 4y = 360$$
$$y = 60$$

Then $x = 100 - y = 100 - 60 = 40$. Thus, there are 40 gallons of the 15% acid solution and 60 gallons of the 20% acid solution in the mixture.

24. Let x = number of nickels in collection,
y = number of dimes in collection,
z = number of quarters in collection.

Since there is a total of 70 coins in the collection,

$$x + y + z = 70.$$

The value equation, in cents, is

$$5x + 10y + 25z = 800,$$

which reduces to

$$x + 2y + 5z = 160.$$

Since the number of dimes is twice the number of quarters, we have $y = 2z$, or

$$y - 2z = 0.$$

Thus, the system to solve is:

$$\begin{array}{rcl} x + y + z &=& 70 \\ x + 2y + 5z &=& 160 \\ y - 2z &=& 0 \end{array}$$

Subtract the second equation from the first equation to eliminate x and obtain

$$-y - 4z = -90.$$

Pair this equation with the third original equation and add to eliminate y and obtain

$$\begin{array}{rcl} -6z &=& -90 \\ z &=& 15 \end{array}$$

Substitute 15 for z in $y - 2z = 0$ to find the value of y.

$$\begin{array}{rcl} y - 2(15) &=& 0 \\ y &=& 30 \end{array}$$

Then substitute 30 for y and 15 for z in the first original equation to find the value of x.

$$\begin{array}{rcl} x + 30 + 15 &=& 70 \\ x &=& 25 \end{array}$$

Thus, there are 25 nickels, 30 dimes, and 15 quarters in the collection.

25. Let x = the measure of the smallest angle,
y = the measure of the middle-sized angle,
z = the measure of the largest angle.

Since the sum of the measures of the angles of a triangle is 180°, the first equation is

$$x + y + z = 180.$$

Since the smallest angle has measure one-third the measure of the middle-sized angle,

$$x = \tfrac{1}{3}y,$$

which simplifies to

$$3x - y = 0.$$

Since the largest angle measures 5° more than the middle sized angle,

$$z = 5 + y,$$

which is equivalent to

$$y - z = -5.$$

Thus, the system to solve is:

$$\begin{array}{rcl} x + y + z &=& 180 \\ 3x - y &=& 0 \\ y - z &=& -5 \end{array}$$

Add the first and third equations to eliminate z and obtain

$$x + 2y = 175.$$

Pair this equation with the second original equation to obtain a system of two equations in x and y:

$$\begin{array}{rcl} x + 2y &=& 175 \\ 3x - y &=& 0 \end{array}$$

Solve the second equation for y, $y = 3x$, and substitute to find the value of x.

$$\begin{array}{rcl} x + 2(3x) &=& 175 \\ 7x &=& 175 \\ x &=& 25 \end{array}$$

Substitute 25 for x in $y = 3x$ to obtain $y = 75$. Then substitute 75 for y in $y - z = -5$ to obtain $z = 80$. Thus, the three angles have measure 25°, 75°, and 80°.

26. Assume that the gasoline costs y required to operate the car can be approximated by a quadratic function of the speed x at which the car is driven, of the form

$$y = ax^2 + bx + c.$$

The data in the table can be interpreted as three data points (30,6), (40,5), and (50,8). Substitute these values for x and y and simplify to obtain the following system of three equations in the three unknowns a, b, and c.

$$900a + 30b + c = 6$$
$$1600a + 40b + c = 5$$
$$2500a + 50b + c = 8$$

Subtract the top equation from the second equation to eliminate c and obtain

$$700a + 10b = -1.$$

Then subtract the second equation from the third equation to eliminate c and obtain

$$900a + 10b = 3.$$

Then we have a system of two equations in a and b.

$$700a + 10b = -1$$
$$900a + 10b = 3$$

Subtract the top equation from the bottom equation to eliminate b.

$$200a = 4$$
$$a = \frac{4}{200} = \frac{1}{50}$$

Substitute this value for a into $700a + 10b = -1$ to find the value of b.

$$700\left(\frac{1}{50}\right) + 10b = -1$$
$$14 + 10b = -1$$
$$10b = -15$$
$$b = \frac{-15}{10} = -\frac{3}{2}$$

Then substitute these values of a and b into $900a + 30b + c = 6$ to find the value of c.

$$900\left(\frac{1}{50}\right) + 30\left(-\frac{3}{2}\right) + c = 6$$
$$18 - 45 + c = 6$$
$$c = 33$$

Thus, the quadratic model is

$$y = \frac{1}{50}x^2 - \frac{3}{2}x + 33.$$

To find the cost of operating the car at 65 mph, we substitute 65 for x and evaluate.

$$y = \frac{1}{50}(65)^2 - \frac{3}{2}(65) + 33$$
$$= 84.5 - 97.5 + 33$$
$$= 20$$

Thus, it would cost about 20 cents per mile to operate the car at 65 mph.

The model does not make sense when x is 0, that is at 0 mph, since substituting we obtain $y = 33$, which means that the gasoline costs are 33 cents per mile when the car is not moving.

27. Solve the system.

$$2x + 2y = 6$$
$$-3x - 3y = -9$$

If the first equation is multiplied by $\frac{1}{2}$, and the second by $\frac{1}{3}$, and the results added, we obtain the identity $0 = 0$. Thus the system is dependent and has infinitely many solutions. Solve the first equation for y in terms of x.

$$2x + 2y = 6$$
$$2y = 6 - 2x$$
$$y = 3 - x$$

Thus, the solutions are $(x, 3-x)$ for x any real number.

28. Solve the system.

$$2(x+3y) + (x+y) = -28$$
$$3(x+3y) - (x+y) = -32$$

It is probably easiest to clear the parentheses in each equation and simplify to obtain the following system.

$$3x + 7y = -28$$
$$x + 4y = -16$$

Solve the second equation for x, $x = -16 - 4y$, and substitute into the first equation.

$$3(-16 - 4y) + 7y = -28$$
$$-48 - 12y + 7y = -28$$
$$-5y = 20$$
$$y = -4$$

Then $x = -16 - 4y = -16 - 4(-4) = -16 + 16 = 0$. Thus, the solution is $(0, -4)$.

29. Solve the system.

$$2x + y - z = 0$$
$$-x - y + 5z = 0$$
$$3x - 2y + z = 0$$

Since this is a homogeneous system, we know that there is at least one solution, (0,0,0). To determine whether there are infinitely many solutions, we try to solve the system by elimination. Adding the first two equations eliminates y and gives

$$x + 4z = 0.$$

Multiply the first equation by 2,

$$4x + 2y - 2z = 0,$$

and add the result to the third to eliminate y and obtain

$$7x - z = 0.$$

Solve this equation for z, $z = 7x$, and substitute into $x + 4z = 0$.

$$x + 4(7x) = 0$$
$$29x = 0$$
$$x = 0$$

Since $x = 0$, substituting we find that $z = 0$ and $y = 0$. Thus, the only solution is the trivial solution (0,0,0).

30. Solve the system.

$$3x - y + z = 4$$
$$-6x + 2y - 2z = 4$$

If we multiply the first equation by 2 and add the result to the second equation, we obtain the contradiction $0 = 12$. Thus, the system has no solution.

31. To find the value(s) of m so that the system

$$2x - 3y = m$$
$$-4x + 6y = 4$$

is inconsistent, that is has no solution (the graphs of the lines are parallel), solve each equation for y (write each in slope-intercept form) and compare the slopes and y-intercepts.

$$2x - 3y = m \qquad -4x + 6y = 4$$
$$-3y = -2x + m \qquad 6y = 4x + 4$$
$$y = \tfrac{2}{3}x - \tfrac{m}{3} \qquad y = \tfrac{2}{3}x + \tfrac{2}{3}$$

Since the slopes are the same, the lines will always be parallel or else coincide. The only time they will coincide is when $m = -2$, for then the y-intercepts are equal also. Thus, the system will be inconsistent for m any real number except -2.

32. Given that

$$f(x) = \frac{x^3 + 2x^2 + 10x + 7}{(x^2 + x + 7)^2}$$
$$= \frac{Ax + B}{x^2 + x + 7} + \frac{Cx + D}{(x^2 + x + 7)^2}.$$

Then multiply both sides of the representation for $f(x)$ by the LCD, $(x^2 + x + 7)^2$.

$$x^3 + 2x^2 + 10x + 7 = (Ax + B)(x^2 + x + 7) + Cx + D$$

Clear parentheses and collect like terms.

$$x^3 + 2x^2 + 10x + 7 = Ax^3 + (A + B)x^2 + (7A + B + C)x + (7B + D)$$

Equating coefficients of like terms, we obtain $A = 1$, $A + B = 2$, $7A + B + C = 10$, and $7B + D = 7$.

Substitute 1 for A in $A + B = 2$ to obtain $B = 1$.

Substitute 1 for A and 1 for B in $7A + B + C = 10$ to obtain $C = 2$.

Substitute 1 for B in $7B + D = 7$ to obtain $D = 0$.

Thus, $A = 1$, $B = 1$, $C = 2$, $D = 0$, and

$$f(x) = \frac{x + 1}{x^2 + x + 7} + \frac{2}{(x^2 + x + 7)^2}.$$

33. Given $f(x) = \dfrac{2x^3 - 16x^2 + 28x - 12}{x^2 - 8x + 12}$.

Since the degree of the numerator is greater than the degree of the denominator, we must first divide the denominator into the numerator to write $f(x)$ in the following form.

$$f(x) = 2x + \frac{4x - 12}{x^2 - 8x + 12}$$

We can then concentrate on the fraction on the right. First factor the denominator.

$$\frac{4x - 12}{x^2 - 8x + 12} = \frac{4x - 12}{(x - 2)(x - 6)}$$

Then by the partial fraction decomposition theorem,

$$\frac{4x - 12}{(x - 2)(x - 6)} = \frac{A}{x - 2} + \frac{B}{x - 6}.$$

Multiply both sides by the LCD, $(x - 2)(x - 6)$.

$$4x - 12 = A(x - 6) + B(x - 2)$$

Clear parentheses and collect like terms.

$$4x - 12 = (A + B)x + (-6A - 2B)$$

Equating coefficients and simplifying, we obtain the following system.

$$\begin{aligned} A + B &= 4 \\ -3A - B &= -6 \end{aligned}$$

Adding we obtain

$$\begin{aligned} -2A &= -2 \\ A &= 1. \end{aligned}$$

Substitute 1 for A in $A + B = 4$ to obtain $B = 3$.

Thus,

$$f(x) = 2x + \frac{1}{x - 2} + \frac{3}{x - 6}.$$

Don't forget the term $2x$!

34. The minimum value of the objective function

$$P = 24x + 5y$$

will be obtained at one of the vertices of the feasible region graphed in the text. There is no maximum value since the region is unbounded. Use a table to summarize the possibilities.

Vertex	$24x + 5y$
(1,5)	49 OPTIMAL SOLUTION
(2,2)	58
(4,1)	101

Thus, the minimum value is 49.

35. Graph the system.

$$\begin{aligned} 1 \leq x &\leq 4 \\ y &\geq 1 \\ 3x - 4y &\geq -12 \end{aligned}$$

The first inequality places the graph between the vertical parallel lines $x = 1$ and $x = 4$. The second inequality places the graph above the horizontal line $y = 1$. The third inequality places the graph "below" the line with intercepts $(0,3)$ and $(-4,0)$. The graph is the bounded region shown below.

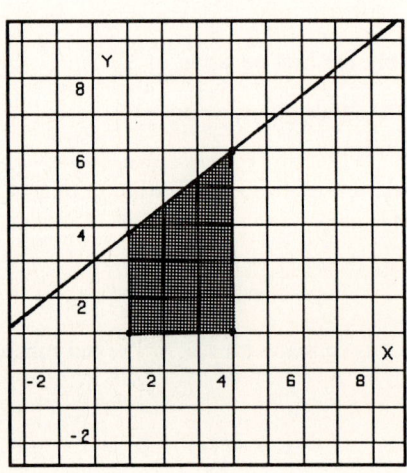

CHAPTER 6 SYSTEMS OF EQUATIONS AND INEQUALITIES

CHAPTER 6 Test

1. Solve the system.

$$2x - 5y = -16$$
$$x + 2y = 19$$

   Solve the second equation for x, $x = -2y + 19$, and substitute this expression for x in the first equation.

$$2(-2y + 19) - 5y = -16$$
$$-4y + 38 - 5y = -16$$
$$-9y = -54$$
$$y = 6$$

   Then $x = -2y + 19 = -2(6) + 19 = 7$. Thus the solution is $(7,6)$.

2. Solve the system.

$$2(x+y) - 3(x-y) = -2$$
$$3(x+y) + (x-y) = 8$$

   It is best to clear the parentheses and simplify each equation first. This results in the following system.

$$-x + 5y = -2$$
$$2x + y = 4$$

   Solve the second equation for y, $y = -2x + 4$, and substitute this expression for y in the first equation.

$$-x + 5(-2x + 4) = -2$$
$$-x - 10x + 20 = -2$$
$$-11x = -22$$
$$x = 2$$

   Then $y = -2x + 4 = -2(2) + 4 = -4 + 4 = 0$. Thus, the solution is $(2,0)$.

3. Let x = the speed of the plane in still air,
   y = the speed of the wind.

   When the plane flies with the wind, its speed relative to the ground is *increased* by the wind. Thus,

   $x + y$ = the speed of the plane with the wind.

   When the plane flies into the wind, its speed relative to the ground is *decreased* by the wind. Thus,

   $x - y$ = the speed of the plane against the wind.

   Use the formula, $d = rt$, twice. Since the distance traveled with the wind is 3000 miles, the rate with the wind is $x + y$, and the time with the wind is 5 hours,

$$3000 = (x + y)5$$

   which simplifies to

$$x + y = 600.$$

   Similarly, we use the information given when the situation is into the wind to obtain

$$3000 = (x - y)6$$

   which simplifies to

$$x - y = 500.$$

   Thus, the system to solve is:

$$x + y = 600$$
$$x - y = 500$$

   Adding these we eliminate y and obtain

$$2x = 1100$$
$$x = 550.$$

   Substitute 550 for x in $x + y = 600$ to find the value of y.

$$550 + y = 600$$
$$y = 50$$

   Thus, the speed of the plane is 550 mph, and the speed of the wind is 50 mph.

4. Let x = the number of pounds of candy,
   y = the number of pounds of nuts.

   Since the number of pounds in the mixture is 20, we obtain the quantity equation

$$x + y = 20.$$

   Since x lb of candy at 80 cents per pound has value $80x$, y lb of nuts at 70 cents per pound has value $70y$, and 20 pounds of the mix at 77 cents per pound has value $77(20) = 1540$, the value equation is

$$80x + 70y = 1540,$$

which simplifies to

$$8x + 7y = 154.$$

Thus, the system to solve is:

$$x + y = 20$$
$$8x + 7y = 154$$

Since we are only asked to find the number of pounds of candy (x), solve the first equation for y, $y = 20 - x$, and substitute this expression for y into the second equation to obtain an equation in x.

$$8x + 7(20 - x) = 154$$
$$8x + 140 - 7x = 154$$
$$x = 14$$

Thus, 14 lb of candy are used in the mix.

5. Solve the system.

$$x - 2y + z = 1$$
$$2x + y - 3z = -8$$
$$3x - y - 5z = -13$$

Suppose we eliminate y. Multiply the second equation by 2,

$$4x + 2y - 6z = -16,$$

and add the result to the first equation to obtain

$$5x - 5z = -15,$$

which simplifies to

$$x - z = -3.$$

Then add the second and third equations to obtain

$$5x - 8z = -21.$$

Then we have the following system:

$$x - z = -3$$
$$5x - 8z = -21$$

Solve the first equation for x, $x = z - 3$, and substitute into the second equation.

$$5(z - 3) - 8z = -21$$
$$5z - 15 - 8z = -21$$
$$-3z = -6$$
$$z = 2$$

Substitute 2 for z in $x = z - 3$, $x = 2 - 3 = -1$.

Then substitute -1 for x and 2 for z in the first original equation to find the value of y.

$$-1 - 2y + 2 = 1$$
$$-2y = 0$$
$$y = 0$$

Thus, the solution is $(-1, 0, 2)$.

6. Solve the system.

$$2x + 3y + 8z = 0$$
$$x - 2y - 3z = 0$$
$$-x + 5y + 9z = 0$$

Since this is a homogeneous system, one solution is the trivial solution $(0,0,0)$. To determine whether there are more solutions, we try to solve using elimination. If we add the last two equations, we eliminate x and obtain

$$3y + 6z = 0,$$

which simplifies to

$$y + 2z = 0.$$

Multiply the second equation by -2,

$$-2x + 4y + 6z = 0,$$

and add the result to the first equation to eliminate x and obtain a second equation in y and z,

$$7y + 14z = 0,$$

which reduces to

$$y + 2z = 0.$$

Since both equations are the same, when we subtract them we obtain the identity $0 = 0$, so we know that there are infinitely many solutions to the system. Solve $y + 2z = 0$ for y, $y = -2z$. Substitute $-2z$ for y in the middle original equation and solve for x in terms of z.

$$x - 2(-2z) - 3z = 0$$
$$x + 4z - 3z = 0$$
$$x + z = 0$$
$$x = -z$$

Thus, the solution is $(-z, -2z, z)$ for z any real number. [*NOTE:* The answer might also be given as $(x, 2x, -x)$ for x any real number, or as $(y/2, y, -y/2)$ for y any real number.]

CHAPTER 6 REVIEW

7. Let x = the amount invested in bonds,
   y = the amount invested in certificates,
   z = the amount invested in stocks.

   Since the total amount invested is $5000, the first equation is

   $$x + y + z = 5000.$$

   The earnings on the investments when the stocks do well is given by

   $$0.09x + 0.08y + 0.10z = 450,$$

   which simplifies to

   $$9x + 8y + 10z = 45,000.$$

   The earnings on the investments when the stocks do not do well, that is when the stocks *lose* money, is given by

   $$0.09x + 0.08y - 0.03z = 190,$$

   which simplifies to

   $$9x + 8y - 3z = 19,000.$$

   Thus, the system to solve is:

   $$\begin{aligned} x + y + z &= 5000 \\ 9x + 8y + 10z &= 45,000 \\ 9x + 8y - 3z &= 19,000 \end{aligned}$$

   If we subtract the last two equations, we can eliminate both x and y and obtain the value of z.

   $$\begin{aligned} 13z &= 26,000 \\ z &= 2000 \end{aligned}$$

   Substitute 2000 for z in the first equation,

   $$\begin{aligned} x + y + 2000 &= 5000 \\ x + y &= 3000, \end{aligned}$$

   and in the second equation,

   $$\begin{aligned} 9x + 8y + 10(2000) &= 45,000 \\ 9x + 8y &= 25,000. \end{aligned}$$

   Since all we are asked to find is the amount invested in bonds, x, solve $x + y = 3000$ for y, $y = 3000 - x$, and substitute into the equation $9x + 8y = 25,000$.

CHAPTER TEST

$$\begin{aligned} 9x + 8(3000 - x) &= 25,000 \\ 9x + 24,000 - 8x &= 25,000 \\ x &= 1000 \end{aligned}$$

Thus, $1000 was invested in bonds.

8. Graph the system.

   $$\begin{aligned} 2x + y &\geq 4 \\ x - 1 &> 0 \end{aligned}$$

   First graph $2x + y = 4$ using intercepts $(0,4)$ and $(2,0)$ and a solid line (the inequality is $\geq$). The test point $(0,0)$ shows

   $$2(0) + (0) \geq 4 \quad \text{This is false}$$

   that we shade the region that does not contain the test point, the region "above" the line. Then graph $x - 1 = 0$, or $x = 1$, using a vertical dashed line through x-intercept $(1,0)$. The test point $(0,0)$ shows that we shade the region to the right of the line. Putting these two together, we obtain the graph of the system shown below.

   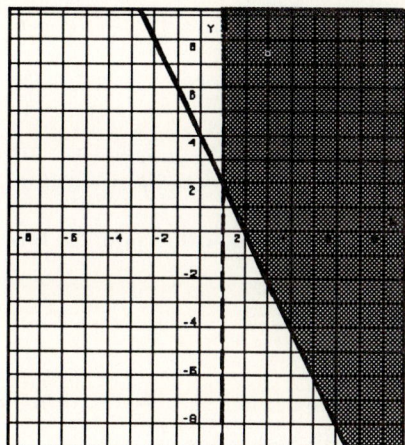

9. Given the polynomial function

   $$P(x) = x^3 + ax^2 + bx + c$$

   with $P(1) = 5$, $P(-1) = 9$, and $P(2) = 15$. Substitute 1 for x to obtain an equation in a, b, and c,

   $$5 = (1)^3 + a(1)^2 + b(1) + c$$

   which simplifies to

   $$a + b + c = 4.$$

   Similarly we can substitute -1 for x and 2 for x to obtain the other two equations in the following system.

$$a + b + c = 4$$
$$a - b + c = 10$$
$$4a + 2b + c = 7$$

Add the first two equations to eliminate b and obtain

$$2a + 2c = 14,$$

which simplifies to

$$a + c = 7.$$

Multiply the middle equation by 2,

$$2a - 2b + 2c = 20,$$

and add the result to the third equation to eliminate b and obtain

$$6a + 3c = 27,$$

which simplifies to

$$2a + c = 9.$$

Then we have the following system in the two variables a and c.

$$2a + c = 9$$
$$a + c = 7$$

Subtract the second equation from the first to obtain the value of a,

$$a = 2.$$

Substitute 2 for a in $a + c = 7$ to obtain the value of c.

$$2 + c = 7$$
$$c = 5$$

Substitute 2 for a and 5 for c in the first original equation to find the value of b.

$$2 + b + 5 = 4$$
$$b = -3$$

Thus, $a = 2$, $b = -3$, $c = 5$, and

$$P(x) = x^3 + 2x^2 - 3x + 5.$$

10. The maximum and minimum values of the objective function $P = 5x + 10y$ subject to the constraints graphed in the feasible region in the text must occur at the vertices of the region: (2,1), (1,4), (3,7), (7,5), and (6,1). The possibilities are summarized in the following table.

Vertex	$5x + 10y$	
(2,1)	20	OPTIMAL SOLUTION
(1,4)	45	
(3,7)	85	OPTIMAL SOLUTION
(7,5)	85	OPTIMAL SOLUTION
(6,1)	40	

Thus, the minimum value is 20 and the maximum value is 85. Note that the maximum value is assumed at two different points.

11. Let x = the number of Valedictorian rings made,
    y = the number of Salutatorian rings made.

Since up to a total of 24 rings can be made each day, we have one constraint,

$$x + y \le 24.$$

Since it takes 3 hours to make one Valedictorian ring, 2 hours to make one Salutatorian ring, and up to a total of 60 man-hours are available daily, we have a second constraint,

$$3x + 2y \le 60.$$

Clearly, the number of each made daily cannot be negative, so we also have

$$x \ge 0 \quad \text{and} \quad y \ge 0.$$

These four constraints are graphed in the feasible region below. The vertices, (0,0), (0,24), (12,12), and (20,0) are obtained by solving the various systems of equations determined by the inequalities.

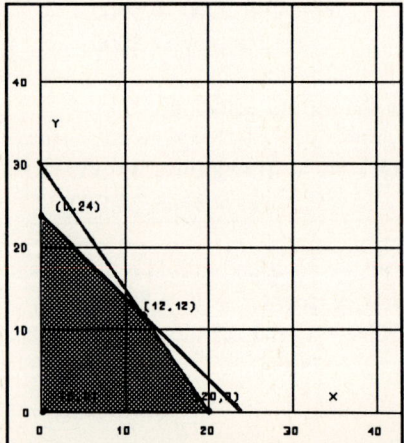

Since each Valedictorian ring returns a profit of $30 and each Salutatorian ring returns a profit of $40, the objective function we are to maximize is

$$P = 30x + 40y.$$

The various possibilities are summarized in the table below.

Vertex	30x + 40y	
(0,0)	0	
(0,24)	960	OPTIMAL SOLUTION
(12,12)	840	
(20,0)	600	

Thus, the profit will be maximized at $960 when no Valedictorian-model rings and 24 Salutatorian-model rings are made daily.

12. Given the function

$$f(x) = \frac{x + 18}{x^2 + x - 12}.$$

First factor the denominator.

$$f(x) = \frac{x + 18}{(x - 3)(x + 4)}$$

Then by the partial fraction decomposition theorem, $f(x)$ can be expressed in the following form.

$$f(x) = \frac{x + 18}{(x - 3)(x + 4)} = \frac{A}{x - 3} + \frac{B}{x + 4}$$

Multiply both expressions for $f(x)$ by the LCD, $(x - 3)(x + 4)$.

$$x + 18 = A(x + 4) + B(x - 3)$$

Clearing parentheses and collecting like terms we have:

$$x + 18 = (A + B)x + (4A - 3B)$$

Equating coefficients of like terms gives the following system of equations.

$$A + B = 1$$
$$4A - 3B = 18$$

Solve the first equation for A, $A = 1 - B$, and substitute into the second equation.

$$4(1 - B) - 3B = 18$$
$$4 - 4B - 3B = 18$$
$$-7B = 14$$
$$B = -2$$

Since $A = 1 - B$, $A = 1 - (-2) = 1 + 2 = 3$.

Thus, $A = 3$, $B = -2$, and

$$f(x) = \frac{3}{x - 3} + \frac{-2}{x + 4}.$$

CHAPTER 7 MATRICES AND DETERMINANTS

SECTION 7.1 Matrices

1. Given the following matrices.

$$U = \begin{bmatrix} 3 & -4 \\ 2 & 5 \end{bmatrix} \quad V = \begin{bmatrix} a & b \\ c & d \end{bmatrix}$$

$$W = \begin{bmatrix} -3 & 5 & 0 \end{bmatrix} \quad X = \begin{bmatrix} -3 \\ 5 \\ 0 \end{bmatrix}$$

(a) If $U = V$, then the corresponding elements are equal which means that $a = 3$, $b = -4$, $c = 2$, and $d = 5$.

(b) The matrices W and X are **not** equal since they do not have the same order.

(c) The order of matrix U is 2×2 since U has 2 rows and 2 columns.

2. Given the following matrices.

$$A = \begin{bmatrix} 1 & 2 \\ -3 & 4 \end{bmatrix} \quad B = \begin{bmatrix} 0 & -3 \\ 2 & 5 \end{bmatrix} \quad C = \begin{bmatrix} -4 & -2 \\ 3 & 1 \end{bmatrix}$$

$$D = \begin{bmatrix} 0 & -2 & 5 \\ 3 & 7 & -1 \end{bmatrix} \quad E = \begin{bmatrix} -5 & 7 & 2 \\ 4 & -1 & 3 \end{bmatrix} \quad F = \begin{bmatrix} 5 & 0 & -1 \\ 2 & -1 & 0 \end{bmatrix}$$

(a) $B + C = \begin{bmatrix} 0 & -3 \\ 2 & 5 \end{bmatrix} + \begin{bmatrix} -4 & -2 \\ 3 & 1 \end{bmatrix}$

$= \begin{bmatrix} 0+(-4) & -3+(-2) \\ 2+3 & 5+1 \end{bmatrix}$

$= \begin{bmatrix} -4 & -5 \\ 5 & 6 \end{bmatrix}$

(b) $C + B = \begin{bmatrix} -4 & -2 \\ 3 & 1 \end{bmatrix} + \begin{bmatrix} 0 & -3 \\ 2 & 5 \end{bmatrix}$

$= \begin{bmatrix} -4+0 & -2+(-3) \\ 3+2 & 1+5 \end{bmatrix}$

$= \begin{bmatrix} -4 & -5 \\ 5 & 6 \end{bmatrix}$

Notice that $B + C = C + B$.

(c) $A + E = \begin{bmatrix} 1 & 2 \\ -3 & 4 \end{bmatrix} + \begin{bmatrix} -5 & 7 & 2 \\ 4 & -1 & 3 \end{bmatrix}$

is not defined since the orders of A and E are different.

(d) $(D + E) + F$

$= \left(\begin{bmatrix} 0 & -2 & 5 \\ 3 & 7 & -1 \end{bmatrix} + \begin{bmatrix} -5 & 7 & 2 \\ 4 & -1 & 3 \end{bmatrix} \right) + \begin{bmatrix} 5 & 0 & -1 \\ 2 & -1 & 0 \end{bmatrix}$

$= \begin{bmatrix} -5 & 5 & 7 \\ 7 & 6 & 2 \end{bmatrix} + \begin{bmatrix} 5 & 0 & -1 \\ 2 & -1 & 0 \end{bmatrix}$

$= \begin{bmatrix} 0 & 5 & 6 \\ 9 & 5 & 2 \end{bmatrix}$

(e) $D + (E + F)$

$= \begin{bmatrix} 0 & -2 & 5 \\ 3 & 7 & -1 \end{bmatrix} + \left(\begin{bmatrix} -5 & 7 & 2 \\ 4 & -1 & 3 \end{bmatrix} + \begin{bmatrix} 5 & 0 & -1 \\ 2 & -1 & 0 \end{bmatrix} \right)$

$= \begin{bmatrix} 0 & -2 & 5 \\ 3 & 7 & -1 \end{bmatrix} + \begin{bmatrix} 0 & 7 & 1 \\ 6 & -2 & 3 \end{bmatrix}$

$= \begin{bmatrix} 0 & 5 & 6 \\ 9 & 5 & 2 \end{bmatrix}$

Note that $(D + E) + F = D + (E + F)$.

3. Subtract the given matrices.

(a) $\begin{bmatrix} 3 & 0 \\ -2 & 1 \end{bmatrix} - \begin{bmatrix} 4 & -2 \\ -3 & 5 \end{bmatrix} = \begin{bmatrix} 3-4 & 0-(-2) \\ -2-(-3) & 1-5 \end{bmatrix}$

$= \begin{bmatrix} -1 & 2 \\ 1 & -4 \end{bmatrix}$

(b) $\begin{bmatrix} 1 & 2 & 3 \end{bmatrix} - \begin{bmatrix} 3 \\ 5 \\ 8 \end{bmatrix}$

is undefined since the matrices are of different orders.

4. Given the following matrices.

$$A = \begin{bmatrix} 2 & 5 & -1 \\ 0 & 3 & 7 \end{bmatrix} \quad B = \begin{bmatrix} -1 & 2 & 4 \\ 3 & -3 & 0 \end{bmatrix}$$

(a) $-2B = -2 \begin{bmatrix} -1 & 2 & 4 \\ 3 & -3 & 0 \end{bmatrix}$

$= \begin{bmatrix} -2(-1) & -2(2) & -2(4) \\ -2(3) & -2(-3) & -2(0) \end{bmatrix}$

$= \begin{bmatrix} 2 & -4 & -8 \\ -6 & 6 & 0 \end{bmatrix}$

PRACTICE EXERCISES

(b) $A - B = \begin{bmatrix} 2 & 5 & -1 \\ 0 & 3 & 7 \end{bmatrix} - \begin{bmatrix} -1 & 2 & 4 \\ 3 & -3 & 0 \end{bmatrix}$

$= \begin{bmatrix} 2-(-1) & 5-2 & -1-4 \\ 0-3 & 3-(-3) & 7-0 \end{bmatrix}$

$= \begin{bmatrix} 3 & 3 & -5 \\ -3 & 6 & 7 \end{bmatrix}$

(c) $3A - 4B = 3\begin{bmatrix} 2 & 5 & -1 \\ 0 & 3 & 7 \end{bmatrix} - 4\begin{bmatrix} -1 & 2 & 4 \\ 3 & -3 & 0 \end{bmatrix}$

$= \begin{bmatrix} 6 & 15 & -3 \\ 0 & 9 & 21 \end{bmatrix} - \begin{bmatrix} -4 & 8 & 16 \\ 12 & -12 & 0 \end{bmatrix}$

$= \begin{bmatrix} 10 & 7 & -19 \\ -12 & 21 & 21 \end{bmatrix}$

5. Use the data base matrices in Example 5 to answer the following.

(a) The matrix $M + J$ is given below with the entry corresponding to Baker's automobile insurance sales in the two months of May and June given in brackets.

$M + J = \begin{bmatrix} 90{,}000 & 10{,}000 & 19{,}000 & 130{,}000 \\ 42{,}000 & 25{,}000 & [25{,}000] & 139{,}000 \\ 120{,}000 & 45{,}000 & 19{,}000 & 77{,}000 \end{bmatrix}$

Thus, Baker's automobile insurance sales for the two-month period were $25,000.

(b) The matrix $M - J$, given below, shows the increase or decrease in sales in each category for each salesperson from May to June and Abbott's entry is placed in brackets.

$J - M = \begin{bmatrix} [50{,}000] & -10{,}000 & 9000 & 50{,}000 \\ -18{,}000 & 15{,}000 & -9000 & 9000 \\ 120{,}000 & -9000 & 3000 & 13{,}000 \end{bmatrix}$

Thus, Abbott's life insurance sales increased $50,000 from May to June.

(c) The matrix $0.08(M + J)$, given below, shows the commission earned by each salesperson during the two-month period with the entry corresponding to Chance relative to sales on health insurance given in brackets.

$0.08(M + J) = \begin{bmatrix} 7200 & 800 & 1520 & 10{,}400 \\ 3360 & 2000 & 2000 & 11{,}120 \\ 9600 & [3600] & 1520 & 6160 \end{bmatrix}$

(d) To determine which salesperson earned the greatest total commission on sales during the two-month period, consider again the matrix $0.08(M + J)$ given below.

$0.08(M + J) = \begin{bmatrix} 7200 & 800 & 1520 & 10{,}400 \\ 3360 & 2000 & 2000 & 11{,}120 \\ 9600 & 3600 & 1520 & 6160 \end{bmatrix}$

The total of the entries in each row gives the total commission earned by each sales person. Thus, the total commission earned by Abbott is:

$7200 + 800 + 1520 + 10{,}400 = \$19{,}920$

The total commission earned by Baker is:

$3360 + 2000 + 2000 + 11{,}120 = \$18{,}480$

The total commission earned by Chance is:

$9600 + 3600 + 1520 + 6160 = \$20{,}880$

Thus, Chance had the largest total commission over the two-month period, $20,880.

CHAPTER 7 MATRICES AND DETERMINANTS

SECTION 7.1 Matrices

Exercises 1-16 refer to the following matrices.

$$A = \begin{bmatrix} 1 & 2 \\ 3 & 4 \end{bmatrix} \quad B = \begin{bmatrix} 5 & -2 \\ 9 & -4 \end{bmatrix} \quad C = \begin{bmatrix} 5^0 & \sqrt{4} \\ |-3| & 2^2 \end{bmatrix}$$

$$D = \begin{bmatrix} 3 & -1 & 4 \\ 2 & 0 & 5 \end{bmatrix} \quad E = \begin{bmatrix} -2 & 3 & 0 \\ 5 & 4 & 7 \end{bmatrix}$$

$$F = \begin{bmatrix} 3 & 2 & 0 & -1 \end{bmatrix} \quad G = \begin{bmatrix} 3 \\ 2 \\ 0 \\ -1 \end{bmatrix}$$

1. The order of matrix A is 2×2 since A has 2 rows and 2 columns.

2. The order of matrix F is 1×4 since F has 1 row and 4 columns.

3. Since square matrices have the same number of rows as columns, the only square matrices listed above are A, B, and C.

4. Since a column vector has only one column, the only matrix listed above that is a column vector is G.

5. Since A and C have the same order, and since $5^0 = 1$, $\sqrt{4} = 2$, $|-3| = 3$, and $2^2 = 4$, the corresponding elements in A and C are equal, thus $A = C$.

6. The element in the first row and second column of D, d_{12}, is -1.

7. Since A has dimension 2×2, the zero matrix of the same order is
$$\begin{bmatrix} 0 & 0 \\ 0 & 0 \end{bmatrix}.$$

8. The element a_{21} in matrix A is the element in the second row and first column of A. Thus, $a_{21} = 3$.

9. Since $A = \begin{bmatrix} 1 & 2 \\ 3 & 4 \end{bmatrix}$, $-A = \begin{bmatrix} -1 & -2 \\ -3 & -4 \end{bmatrix}$.

10. $A + B = \begin{bmatrix} 1 & 2 \\ 3 & 4 \end{bmatrix} + \begin{bmatrix} 5 & -2 \\ 9 & -4 \end{bmatrix}$
$$= \begin{bmatrix} 1+5 & 2+(-2) \\ 3+9 & 4+(-4) \end{bmatrix}$$
$$= \begin{bmatrix} 6 & 0 \\ 12 & 0 \end{bmatrix}$$

11. Since A and D have different dimensions, $A + D$ is not defined.

12. $D - E = \begin{bmatrix} 3 & -1 & 4 \\ 2 & 0 & 5 \end{bmatrix} - \begin{bmatrix} -2 & 3 & 0 \\ 5 & 4 & 7 \end{bmatrix}$
$$= \begin{bmatrix} 3-(-2) & -1-3 & 4-0 \\ 2-5 & 0-4 & 5-7 \end{bmatrix}$$
$$= \begin{bmatrix} 5 & -4 & 4 \\ -3 & -4 & -2 \end{bmatrix}.$$

13. Since B and E have different dimensions, $B - E$ is not defined.

14. $-2A = -2 \begin{bmatrix} 1 & 2 \\ 3 & 4 \end{bmatrix} = \begin{bmatrix} (-2)(1) & (-2)(2) \\ (-2)(3) & (-2)(4) \end{bmatrix} = \begin{bmatrix} -2 & -4 \\ -6 & -8 \end{bmatrix}$

15. Since a_{12} is the element in the first row and second column of A, $a_{12} = 2$. Thus,
$$a_{12}B = 2 \begin{bmatrix} 5 & -2 \\ 9 & -4 \end{bmatrix} = \begin{bmatrix} (2)(5) & (2)(-2) \\ (2)(9) & (2)(-4) \end{bmatrix} = \begin{bmatrix} 10 & -4 \\ 18 & -8 \end{bmatrix}.$$

16. $2A - 3B = 2 \begin{bmatrix} 1 & 2 \\ 3 & 4 \end{bmatrix} - 3 \begin{bmatrix} 5 & -2 \\ 9 & -4 \end{bmatrix}$
$$= \begin{bmatrix} 2 & 4 \\ 6 & 8 \end{bmatrix} - \begin{bmatrix} 15 & -6 \\ 27 & -12 \end{bmatrix}$$
$$= \begin{bmatrix} 2-15 & 4-(-6) \\ 6-27 & 8-(-12) \end{bmatrix}$$
$$= \begin{bmatrix} -13 & 10 \\ -21 & 20 \end{bmatrix}$$

17. For any matrix A, $A + 0 = 0$, where 0 is the zero matrix with the same dimension as A.

18. For any matrix E, $0 - E = -E$, where 0 is the zero matrix with the same dimension as E.

EXERCISES A

19. Begin by adding the two matrices on the left, then equate corresponding elements.

$$\begin{bmatrix} a & b \\ c & d \end{bmatrix} + \begin{bmatrix} -1 & 3 \\ 0 & 2 \end{bmatrix} = \begin{bmatrix} 6 & -2 \\ 5 & 1 \end{bmatrix}$$

$$\begin{bmatrix} a-1 & b+3 \\ c+0 & d+2 \end{bmatrix} = \begin{bmatrix} 6 & -2 \\ 5 & 1 \end{bmatrix}$$

Then

$$\begin{aligned} a - 1 &= 6 & b + 3 &= -2 \\ a &= 7 & b &= -5 \end{aligned}$$

$$\begin{aligned} c + 0 &= 5 & d + 2 &= 1 \\ c &= 5 & d &= -1. \end{aligned}$$

20. Begin by subtracting the two matrices on the left, then equate corresponding elements.

$$\begin{bmatrix} a & 1 \\ 0 & b \end{bmatrix} - \begin{bmatrix} b & -2 \\ 3 & 2a \end{bmatrix} = \begin{bmatrix} -5 & 3 \\ -3 & 8 \end{bmatrix}$$

$$\begin{bmatrix} a-b & 1-(-2) \\ 0-3 & b-2a \end{bmatrix} = \begin{bmatrix} -5 & 3 \\ -3 & 8 \end{bmatrix}$$

$$\begin{bmatrix} a-b & 3 \\ -3 & -2a+b \end{bmatrix} = \begin{bmatrix} -5 & 3 \\ -3 & 8 \end{bmatrix}$$

Then since $3 = 3$ and $-3 = -3$, we find the values of a and b by solving the system

$$\begin{aligned} a - b &= -5 \\ -2a + b &= 8 \end{aligned}$$

Adding the two equations eliminates b.

$$\begin{aligned} -a &= 3 \\ a &= -3 \end{aligned}$$

Substitute -3 for a in the first equation to find the value of b.

$$\begin{aligned} -3 - b &= -5 \\ -b &= -2 \\ b &= 2 \end{aligned}$$

Thus, $a = -3$ and $b = 2$.

21. (a) The matrix that gives the totals for the tournament in each category is the sum of F and S.

$$F + S = \begin{bmatrix} 21 & 5 & 35 \\ 18 & 9 & 32 \\ 10 & 7 & 28 \\ 8 & 3 & 30 \\ 15 & 12 & 32 \end{bmatrix} + \begin{bmatrix} 19 & 4 & 33 \\ 26 & 8 & 30 \\ 8 & 10 & 26 \\ 8 & 5 & 32 \\ 17 & 12 & 28 \end{bmatrix}$$

$$= \begin{bmatrix} 40 & 9 & 68 \\ 44 & 17 & 62 \\ 18 & 17 & 54 \\ 16 & 8 & 62 \\ 32 & 24 & 60 \end{bmatrix}$$

(b) The matrix

$$\frac{1}{2}(F + S) = \frac{1}{2}\begin{bmatrix} 40 & 9 & 68 \\ 44 & 17 & 62 \\ 18 & 17 & 54 \\ 16 & 8 & 62 \\ 32 & 24 & 60 \end{bmatrix}$$

$$= \begin{bmatrix} 20 & 4.5 & 34 \\ 22 & 8.5 & 31 \\ 9 & 8.5 & 27 \\ 8 & 4 & 31 \\ 16 & 12 & 30 \end{bmatrix}$$

represents the average points, rebounds, and minutes played for each player in the tournament.

22. Solve the system.

$$\begin{aligned} 3x + 5y &= 9 \\ 2x - 7y &= -25 \end{aligned}$$

Multiply the first equation by 2, the second equation by -3, and add the results to eliminate x.

$$\begin{aligned} 31y &= 93 \\ y &= 3 \end{aligned}$$

Substitute 3 for y in the first original equation to find the value of x.

$$\begin{aligned} 3x + 5(3) &= 9 \\ 3x + 15 &= 9 \\ 3x &= -6 \\ x &= -2 \end{aligned}$$

Thus, the solution is $(-2, 3)$.

23. Solve the system.

$$4x - y = -3$$
$$-2y - 5z = -1$$
$$3x + 2z = -2$$

Multiply the first equation by -2,

$$-8x + 2y = 6,$$

and add the result to the second equation to eliminate y and obtain

$$-8x - 5z = 5.$$

Pair this equation with the original third equation to obtain the following system:

$$-8x - 5z = 5$$
$$3x + 2z = -2$$

Multiply the first equation by 2, the second by 5, and add to eliminate z.

$$-x = 0$$
$$x = 0$$

Substitute 0 for x in $3x + 2z = -2$ to find the value of z.

$$3(0) + 2z = -2$$
$$2z = -2$$
$$z = -1$$

Substitute 0 for x in the first original equation,

$$4x - y = -3,$$

to find the value of y.

$$4(0) - y = -3$$
$$-y = -3$$
$$y = 3$$

Thus, the solution is $(0, 3, -1)$.

PRACTICE EXERCISES

SECTION 7.2   **325**

CHAPTER 7   MATRICES AND DETERMINANTS

SECTION 7.2   Matrix Multiplication

1. Given the following matrices.

$$A = \begin{bmatrix} 1 & -3 & 4 \end{bmatrix} \quad B = \begin{bmatrix} 10 \\ 20 \\ 30 \end{bmatrix}$$

$$\begin{aligned} A \cdot B &= \begin{bmatrix} 1 & -3 & 4 \end{bmatrix} \cdot \begin{bmatrix} 10 \\ 20 \\ 30 \end{bmatrix} \\ &= (1)(10) + (-3)(20) + (4)(30) \\ &= 10 - 60 + 120 \\ &= 70 \end{aligned}$$

2. Given the following matrices.

$$A = \begin{bmatrix} 2 & -1 \\ 0 & 3 \end{bmatrix} \quad B = \begin{bmatrix} 3 & -1 & 0 \\ 4 & 2 & 1 \end{bmatrix}$$

$$AB = \begin{bmatrix} 2 & -1 \\ 0 & 3 \end{bmatrix} \begin{bmatrix} 3 & -1 & 0 \\ 4 & 2 & 1 \end{bmatrix}$$

$$= \begin{bmatrix} [2\ -1] \cdot \begin{bmatrix} 3 \\ 4 \end{bmatrix} & [2\ -1] \cdot \begin{bmatrix} -1 \\ 2 \end{bmatrix} & [2\ -1] \cdot \begin{bmatrix} 0 \\ 1 \end{bmatrix} \\ [0\ 3] \cdot \begin{bmatrix} 3 \\ 4 \end{bmatrix} & [0\ 3] \cdot \begin{bmatrix} -1 \\ 2 \end{bmatrix} & [0\ 3] \cdot \begin{bmatrix} 0 \\ 1 \end{bmatrix} \end{bmatrix}$$

$$= \begin{bmatrix} (2)(3)+(-1)(4) & (2)(-1)+(-1)(2) & (2)(0)+(-1)(1) \\ (0)(3)+(3)(4) & (0)(-1)+(3)(2) & (0)(0)+(3)(1) \end{bmatrix}$$

$$= \begin{bmatrix} 2 & -4 & -1 \\ 12 & 6 & 3 \end{bmatrix}$$

3. Find each product.

(a)

$$\begin{bmatrix} -1 & 0 & 2 \\ 2 & 4 & -1 \\ 3 & -1 & 1 \end{bmatrix} \begin{bmatrix} 1 & -2 \\ 0 & 3 \\ 4 & -1 \end{bmatrix}$$

$$= \begin{bmatrix} (-1)(1)+(0)(0)+(2)(4) & (-1)(-2)+(0)(3)+(2)(-1) \\ (2)(1)+(4)(0)+(-1)(4) & (2)(-2)+(4)(3)+(-1)(-1) \\ (3)(1)+(-1)(0)+(1)(4) & (3)(-2)+(-1)(3)+(1)(-1) \end{bmatrix}$$

$$= \begin{bmatrix} 7 & 0 \\ -2 & 9 \\ 7 & -10 \end{bmatrix}$$

Notice that the matrix on the left has order 3×3, the one on the right has order 3×2, so the product is defined and has order 3×2.

(b) The product

$$\begin{bmatrix} 1 & -2 \\ 0 & 3 \\ 4 & -1 \end{bmatrix} \begin{bmatrix} -1 & 0 & 2 \\ 2 & 4 & -1 \\ 3 & -1 & 1 \end{bmatrix}$$

is not defined since the matrix on the left has order 3×2, the one on the right has order 3×3, and the number of columns in the left matrix (2) is not the same as the number of rows in the right matrix (3).

(c) $\begin{bmatrix} 3 & -1 \end{bmatrix} \begin{bmatrix} 2 & 0 \\ 4 & -5 \end{bmatrix}$

$$= \begin{bmatrix} (3)(2)+(-1)(4) & (3)(0)+(-1)(-5) \end{bmatrix}$$
$$= \begin{bmatrix} 2 & 5 \end{bmatrix}$$

(d) $\begin{bmatrix} 5 & -1 \\ 3 & 0 \end{bmatrix} \begin{bmatrix} 2 & 4 \\ -1 & 3 \end{bmatrix}$

$$= \begin{bmatrix} (5)(2)+(-1)(-1) & (5)(4)+(-1)(3) \\ (3)(2)+(0)(-1) & (3)(4)+(0)(3) \end{bmatrix}$$

$$= \begin{bmatrix} 11 & 17 \\ 6 & 12 \end{bmatrix}$$

(e) $\begin{bmatrix} 2 & 4 \\ -1 & 3 \end{bmatrix} \begin{bmatrix} 5 & -1 \\ 3 & 0 \end{bmatrix}$

$$= \begin{bmatrix} (2)(5)+(4)(3) & (2)(-1)+(4)(0) \\ (-1)(5)+(3)(3) & (-1)(-1)+(3)(0) \end{bmatrix}$$

$$= \begin{bmatrix} 22 & -2 \\ 4 & 1 \end{bmatrix}$$

4. If the estimated costs are \$0.60 by telephone, \$0.25 by mail, and \$0.85 for a personal contact, then the matrix C becomes:

$$C = \begin{bmatrix} 0.60 \\ 0.25 \\ 0.85 \end{bmatrix}.$$

First find the matrix NC, with N as given in Example 4.

$$NC = \begin{bmatrix} 5000 & 10{,}000 & 1000 \\ 8000 & 20{,}000 & 6000 \\ 3000 & 5000 & 700 \end{bmatrix} \begin{bmatrix} 0.60 \\ 0.25 \\ 0.85 \end{bmatrix} = \begin{bmatrix} 6350 \\ 14{,}900 \\ 3645 \end{bmatrix}$$

Then using $T = \begin{bmatrix} 1 & 1 & 1 \end{bmatrix}$, the total cost is given by:

$$T \cdot (NC) = \begin{bmatrix} 1 & 1 & 1 \end{bmatrix} \cdot \begin{bmatrix} 6350 \\ 14{,}900 \\ 3645 \end{bmatrix} = \$24{,}895$$

CHAPTER 7 MATRICES AND DETERMINANTS

SECTION 7.2 Matrix Multiplication

1. Find the scalar product.

$$[1 \quad -2] \cdot \begin{bmatrix} 3 \\ 5 \end{bmatrix} = (1)(3) + (-2)(5)$$
$$= 3 - 10$$
$$= -7$$

2. Find the scalar product.

$$[2 \quad 3 \quad 5] \cdot \begin{bmatrix} -1 \\ 0 \\ 2 \end{bmatrix} = (2)(-1) + (3)(0) + (5)(2)$$
$$= -2 + 0 + 10$$
$$= 8$$

3. Determine m and n so that the product

$$A_{m \times n} B_{3 \times 2} = C_{4 \times 2}$$

is defined.

For the product to be defined, the number of columns in A must equal the number of rows in B, that is, $n = 3$. Also, for the product to equal C, since C has 4 rows, A must also have 4 rows making $m = 4$.

4. Determine m and n so that the product

$$A_{4 \times m} B_{2 \times n} = C_{4 \times 3}$$

is defined.

For the product to be defined, the number of columns in A must equal the number of rows in B, that is, $m = 2$. Also, for the product to equal C, since C has 3 columns, B must also have 3 columns making $n = 3$.

Exercises 5-14 refer to the following matrices.

$$A = \begin{bmatrix} 2 & 5 \\ -1 & 3 \end{bmatrix} \quad B = \begin{bmatrix} 0 & -2 \\ 1 & 4 \end{bmatrix} \quad C = \begin{bmatrix} 1 & 1 \\ 1 & 1 \end{bmatrix}$$

$$D = \begin{bmatrix} 1 & 3 & -2 \\ 4 & 0 & 5 \end{bmatrix} \quad E = \begin{bmatrix} 0 & 4 \\ 4 & 0 \end{bmatrix}$$

$$F = \begin{bmatrix} 3 & 0 \\ -2 & 1 \\ 0 & 5 \end{bmatrix} \quad G = [3 \quad -2] \quad H = \begin{bmatrix} -2 \\ 4 \\ 1 \end{bmatrix}$$

5. $AB = \begin{bmatrix} 2 & 5 \\ -1 & 3 \end{bmatrix} \begin{bmatrix} 0 & -2 \\ 1 & 4 \end{bmatrix}$

$$= \begin{bmatrix} [2 \quad 5] \cdot \begin{bmatrix} 0 \\ 1 \end{bmatrix} & [2 \quad 5] \cdot \begin{bmatrix} -2 \\ 4 \end{bmatrix} \\ [-1 \quad 3] \cdot \begin{bmatrix} 0 \\ 1 \end{bmatrix} & [-1 \quad 3] \cdot \begin{bmatrix} -2 \\ 4 \end{bmatrix} \end{bmatrix}$$

$$= \begin{bmatrix} (2)(0)+(5)(1) & (2)(-2)+(5)(4) \\ (-1)(0)+(3)(1) & (-1)(-2)+(3)(4) \end{bmatrix}$$

$$= \begin{bmatrix} 5 & 16 \\ 3 & 14 \end{bmatrix}$$

6. $CD = \begin{bmatrix} 1 & 1 \\ 1 & 1 \end{bmatrix} \begin{bmatrix} 1 & 3 & -2 \\ 4 & 0 & 5 \end{bmatrix}$

$$= \begin{bmatrix} [1 \quad 1] \cdot \begin{bmatrix} 1 \\ 4 \end{bmatrix} & [1 \quad 1] \cdot \begin{bmatrix} 3 \\ 0 \end{bmatrix} & [1 \quad 1] \cdot \begin{bmatrix} -2 \\ 5 \end{bmatrix} \\ [1 \quad 1] \cdot \begin{bmatrix} 1 \\ 4 \end{bmatrix} & [1 \quad 1] \cdot \begin{bmatrix} 3 \\ 0 \end{bmatrix} & [1 \quad 1] \cdot \begin{bmatrix} -2 \\ 5 \end{bmatrix} \end{bmatrix}$$

$$= \begin{bmatrix} (1)(1)+(1)(4) & (1)(3)+(1)(0) & (1)(-2)+(1)(5) \\ (1)(1)+(1)(4) & (1)(3)+(1)(0) & (1)(-2)+(1)(5) \end{bmatrix}$$

$$= \begin{bmatrix} 5 & 3 & 3 \\ 5 & 3 & 3 \end{bmatrix}$$

7. $FD = \begin{bmatrix} 3 & 0 \\ -2 & 1 \\ 0 & 5 \end{bmatrix} \begin{bmatrix} 1 & 3 & -2 \\ 4 & 0 & 5 \end{bmatrix}$

$$= \begin{bmatrix} (3)(1)+(0)(4) & (3)(3)+(0)(0) & (3)(-2)+(0)(5) \\ (-2)(1)+(1)(4) & (-2)(3)+(1)(0) & (-2)(-2)+(1)(5) \\ (0)(1)+(5)(4) & (0)(3)+(5)(0) & (0)(-2)+(5)(5) \end{bmatrix}$$

$$= \begin{bmatrix} 3 & 9 & -6 \\ 2 & -6 & 9 \\ 20 & 0 & 25 \end{bmatrix}$$

8. $DH = \begin{bmatrix} 1 & 3 & -2 \\ 4 & 0 & 5 \end{bmatrix} \begin{bmatrix} -2 \\ 4 \\ 1 \end{bmatrix}$

$$= \begin{bmatrix} (1)(-2)+(3)(4)+(-2)(1) \\ (4)(-2)+(0)(4)+(5)(1) \end{bmatrix}$$

$$= \begin{bmatrix} 8 \\ -3 \end{bmatrix}$$

EXERCISES A SECTION 7.2 327

9. To find B^T, interchange the rows and columns of matrix B. That is, the first row of B is the first column of B^T, and the second row of B is the second column of B^T. Thus,

$$B^T = \begin{bmatrix} 0 & 1 \\ -2 & 4 \end{bmatrix}.$$

10. To find B^2, multiply B times itself. Thus,

$$B^2 = \begin{bmatrix} 0 & -2 \\ 1 & 4 \end{bmatrix}\begin{bmatrix} 0 & -2 \\ 1 & 4 \end{bmatrix}$$
$$= \begin{bmatrix} (0)(0)+(-2)(1) & (0)(-2)+(-2)(4) \\ (1)(0)+(4)(1) & (1)(-2)+(4)(4) \end{bmatrix}$$
$$= \begin{bmatrix} -2 & -8 \\ 4 & 14 \end{bmatrix}$$

11. To find F^T, interchange the rows and columns of F.

$$F^T = \begin{bmatrix} 3 & -2 & 0 \\ 0 & 1 & 5 \end{bmatrix}$$

12. Since $(A^T)^T = A$, finding $(A^T)^T E$ is the same as finding AE.

$$AE = \begin{bmatrix} 2 & 5 \\ -1 & 3 \end{bmatrix}\begin{bmatrix} 0 & 4 \\ 4 & 0 \end{bmatrix}$$
$$= \begin{bmatrix} (2)(0)+(5)(4) & (2)(4)+(5)(0) \\ (-1)(0)+(3)(4) & (-1)(4)+(3)(0) \end{bmatrix}$$
$$= \begin{bmatrix} 20 & 8 \\ 12 & -4 \end{bmatrix}$$

13. Since $A^T = \begin{bmatrix} 2 & -1 \\ 5 & 3 \end{bmatrix}$, and $B^T = \begin{bmatrix} 0 & 1 \\ -2 & 4 \end{bmatrix}$,

$$A^T + B^T = \begin{bmatrix} 2 & -1 \\ 5 & 3 \end{bmatrix} + \begin{bmatrix} 0 & 1 \\ -2 & 4 \end{bmatrix}$$
$$= \begin{bmatrix} 2+0 & -1+1 \\ 5+(-2) & 3+4 \end{bmatrix}$$
$$= \begin{bmatrix} 2 & 0 \\ 3 & 7 \end{bmatrix}.$$

14. First find $A + B$.

$$A + B = \begin{bmatrix} 2 & 5 \\ -1 & 3 \end{bmatrix} + \begin{bmatrix} 0 & -2 \\ 1 & 4 \end{bmatrix}$$
$$= \begin{bmatrix} 2+0 & 5+(-2) \\ -1+1 & 3+4 \end{bmatrix}$$
$$= \begin{bmatrix} 2 & 3 \\ 0 & 7 \end{bmatrix}$$

Then interchange the rows and columns of this matrix to obtain

$$(A + B)^T = \begin{bmatrix} 2 & 0 \\ 3 & 7 \end{bmatrix}.$$

Notice from Exercises 13 and 14 that $(A + B)^T = A^T + B^T$.

Use the following matrices to verify each statement in Exercises 15-22.

$$A = \begin{bmatrix} 2 & 1 \\ 0 & 5 \end{bmatrix} \quad B = \begin{bmatrix} 3 & 1 \\ 1 & -2 \end{bmatrix} \quad C = \begin{bmatrix} 0 & 4 \\ -2 & 1 \end{bmatrix}$$

15. $AB = \begin{bmatrix} 2 & 1 \\ 0 & 5 \end{bmatrix}\begin{bmatrix} 3 & 1 \\ 1 & -2 \end{bmatrix} = \begin{bmatrix} 7 & 0 \\ 5 & -10 \end{bmatrix}$

$BA = \begin{bmatrix} 3 & 1 \\ 1 & -2 \end{bmatrix}\begin{bmatrix} 2 & 1 \\ 0 & 5 \end{bmatrix} = \begin{bmatrix} 6 & 8 \\ 2 & -9 \end{bmatrix}$

Thus, $AB \neq BA$.

16. $A(BC) = \begin{bmatrix} 2 & 1 \\ 0 & 5 \end{bmatrix}\left(\begin{bmatrix} 3 & 1 \\ 1 & -2 \end{bmatrix}\begin{bmatrix} 0 & 4 \\ -2 & 1 \end{bmatrix}\right)$
$= \begin{bmatrix} 2 & 1 \\ 0 & 5 \end{bmatrix}\begin{bmatrix} -2 & 13 \\ 4 & 2 \end{bmatrix}$
$= \begin{bmatrix} 0 & 28 \\ 20 & 10 \end{bmatrix}$

$$(AB)C = \left(\begin{bmatrix} 2 & 1 \\ 0 & 5 \end{bmatrix}\begin{bmatrix} 3 & 1 \\ 1 & -2 \end{bmatrix}\right)\begin{bmatrix} 0 & 4 \\ -2 & 1 \end{bmatrix}$$

$$= \begin{bmatrix} 7 & 0 \\ 5 & -10 \end{bmatrix}\begin{bmatrix} 0 & 4 \\ -2 & 1 \end{bmatrix}$$

$$= \begin{bmatrix} 0 & 28 \\ 20 & 10 \end{bmatrix}$$

Thus, $A(BC) = (AB)C$.

17. Remember that $I = \begin{bmatrix} 1 & 0 \\ 0 & 1 \end{bmatrix}$, thus,

$$AI = \begin{bmatrix} 2 & 1 \\ 0 & 5 \end{bmatrix}\begin{bmatrix} 1 & 0 \\ 0 & 1 \end{bmatrix} = \begin{bmatrix} 2 & 1 \\ 0 & 5 \end{bmatrix} = A.$$

18. $A + (-A) = \begin{bmatrix} 2 & 1 \\ 0 & 5 \end{bmatrix} + \begin{bmatrix} -2 & -1 \\ 0 & -5 \end{bmatrix} = \begin{bmatrix} 0 & 0 \\ 0 & 0 \end{bmatrix} = 0$

19. $A + 0 = \begin{bmatrix} 2 & 1 \\ 0 & 5 \end{bmatrix} + \begin{bmatrix} 0 & 0 \\ 0 & 0 \end{bmatrix} = \begin{bmatrix} 2 & 1 \\ 0 & 5 \end{bmatrix} = A$

20. $A(B + C) = \begin{bmatrix} 2 & 1 \\ 0 & 5 \end{bmatrix}\left(\begin{bmatrix} 3 & 1 \\ 1 & -2 \end{bmatrix} + \begin{bmatrix} 0 & 4 \\ -2 & 1 \end{bmatrix}\right)$

$$= \begin{bmatrix} 2 & 1 \\ 0 & 5 \end{bmatrix}\begin{bmatrix} 3 & 5 \\ -1 & -1 \end{bmatrix}$$

$$= \begin{bmatrix} 5 & 9 \\ -5 & -5 \end{bmatrix}$$

$$AB + AC = \begin{bmatrix} 2 & 1 \\ 0 & 5 \end{bmatrix}\begin{bmatrix} 3 & 1 \\ 1 & -2 \end{bmatrix} + \begin{bmatrix} 2 & 1 \\ 0 & 5 \end{bmatrix}\begin{bmatrix} 0 & 4 \\ -2 & 1 \end{bmatrix}$$

$$= \begin{bmatrix} 7 & 0 \\ 5 & -10 \end{bmatrix} + \begin{bmatrix} -2 & 9 \\ -10 & 5 \end{bmatrix}$$

$$= \begin{bmatrix} 5 & 9 \\ -5 & -5 \end{bmatrix}$$

Thus, $A(B + C) = AB + AC$.

21. $(B + C)A = \left(\begin{bmatrix} 3 & 1 \\ 1 & -2 \end{bmatrix} + \begin{bmatrix} 0 & 4 \\ -2 & 1 \end{bmatrix}\right)\begin{bmatrix} 2 & 1 \\ 0 & 5 \end{bmatrix}$

$$= \begin{bmatrix} 3 & 5 \\ -1 & -1 \end{bmatrix}\begin{bmatrix} 2 & 1 \\ 0 & 5 \end{bmatrix}$$

$$= \begin{bmatrix} 6 & 28 \\ -2 & -6 \end{bmatrix}$$

$$BA + CA = \begin{bmatrix} 3 & 1 \\ 1 & -2 \end{bmatrix}\begin{bmatrix} 2 & 1 \\ 0 & 5 \end{bmatrix} + \begin{bmatrix} 0 & 4 \\ -2 & 1 \end{bmatrix}\begin{bmatrix} 2 & 1 \\ 0 & 5 \end{bmatrix}$$

$$= \begin{bmatrix} 6 & 8 \\ 2 & -9 \end{bmatrix} + \begin{bmatrix} 0 & 20 \\ -4 & 3 \end{bmatrix}$$

$$= \begin{bmatrix} 6 & 28 \\ -2 & -6 \end{bmatrix}$$

Thus, $(B + C)A = BA + CA$.

22. In Exercise 20, we found

$$A(B + C) = \begin{bmatrix} 5 & 9 \\ -5 & -5 \end{bmatrix},$$

and in Exercise 21 we found

$$(B + C)A = \begin{bmatrix} 6 & 28 \\ -2 & -6 \end{bmatrix}.$$

Thus, $A(B + C) \neq (B + C)A$.

23. (a) First find $[1\ 1\ 1\ 1\ 1]F$.

$$[1\ 1\ 1\ 1\ 1]F = [1\ 1\ 1\ 1\ 1]\begin{bmatrix} 21 & 5 & 35 \\ 18 & 9 & 32 \\ 10 & 7 & 28 \\ 8 & 3 & 30 \\ 15 & 12 & 32 \end{bmatrix}$$

$= [21+18+10+8+15\ \ 5+9+7+3+12\ \ 35+32+28+30+32]$
$= [72\ \ 36\ \ 157]$

This matrix represents the total points, rebounds, and minutes played by the starters in the first game.

(b) First find $F + S$.

$$F + S = \begin{bmatrix} 40 & 9 & 68 \\ 44 & 17 & 62 \\ 18 & 17 & 54 \\ 16 & 8 & 62 \\ 32 & 24 & 60 \end{bmatrix}$$

Then

EXERCISES A

$$[1\ 1\ 1\ 1\ 1](F + S) = [1\ 1\ 1\ 1\ 1]\begin{bmatrix} 40 & 9 & 68 \\ 44 & 17 & 62 \\ 18 & 17 & 54 \\ 16 & 8 & 62 \\ 32 & 24 & 60 \end{bmatrix}$$

$$= [150\ 75\ 306]$$

Then this matrix represents the total points, rebounds, and minutes played by the starters in the tournament.

(c) Since $[1\ 1\ 1\ 1\ 1](F + S) = [150\ 75\ 306]$,

$$\frac{1}{2}[1\ 1\ 1\ 1\ 1](F + S) = \frac{1}{2}[150\ 75\ 306]$$
$$= [75\ 37.5\ 153].$$

This matrix represents the average number of points, rebounds, and minutes played by the starters in the tournament.

24. (a) $JC = \begin{bmatrix} 2100 & 400 & 55 \\ 3500 & 700 & 60 \end{bmatrix}\begin{bmatrix} 3 \\ 4 \\ 10 \end{bmatrix} = \begin{bmatrix} 8450 \\ 13{,}900 \end{bmatrix}$

This matrix represents the total revenue in rentals at each store in the month of January.

(b) $FC = \begin{bmatrix} 3200 & 750 & 84 \\ 4300 & 900 & 105 \end{bmatrix}\begin{bmatrix} 3 \\ 4 \\ 10 \end{bmatrix} = \begin{bmatrix} 13{,}440 \\ 17{,}550 \end{bmatrix}$

This matrix represents the total revenue in rentals at each store in the month of February.

(c) $J + F = \begin{bmatrix} 2100 & 400 & 55 \\ 3500 & 700 & 60 \end{bmatrix} + \begin{bmatrix} 3200 & 750 & 84 \\ 4300 & 900 & 105 \end{bmatrix}$
$= \begin{bmatrix} 5300 & 1150 & 139 \\ 7800 & 1600 & 165 \end{bmatrix}$

Then

$$(J + F)C = \begin{bmatrix} 5300 & 1150 & 139 \\ 7800 & 1600 & 165 \end{bmatrix}\begin{bmatrix} 3 \\ 4 \\ 10 \end{bmatrix} = \begin{bmatrix} 21{,}890 \\ 31{,}450 \end{bmatrix}$$

represents the total revenue in rentals at each store during the two-month period.

(d) Using the result from part (c),

$$[1\ 1] \cdot ((J + F)C) = [1\ 1] \cdot \begin{bmatrix} 21{,}890 \\ 31{,}450 \end{bmatrix}$$
$$= 21{,}890 + 31{,}450$$
$$= 53{,}340$$

Notice that the result here is a number, not a matrix, since we found the dot product above. This number represents the total revenue of $53,340 at both stores during the two-month period.

Use the following matrices in Exercises 25-28.

$$A = \begin{bmatrix} -5 & 0 \\ 2 & 7 \end{bmatrix} \quad B = \begin{bmatrix} 0 & 3 \\ -2 & 1 \end{bmatrix} \quad C = \begin{bmatrix} u & v \\ w & z \end{bmatrix}$$

25. If $A = C$, then we can equate corresponding elements to obtain $u = -5$, $v = 0$, $w = 2$, and $z = 7$.

26. The element b_{21} in matrix B is the element in the 2nd row and 1st column, so $b_{21} = -2$.

27. Since 0 represents the zero matrix of order 2×2, $A + 0 = A$.

28. $A - B = \begin{bmatrix} -5 & 0 \\ 2 & 7 \end{bmatrix} - \begin{bmatrix} 0 & 3 \\ -2 & 1 \end{bmatrix} = \begin{bmatrix} -5 & -3 \\ 4 & 6 \end{bmatrix}$

29. Solve the following system.

$$2x + y = 1$$
$$x - 3y = 11$$

Solve the first equation for y, $y = 1 - 2x$, and substutute this expression for y in the second equation.

$$x - 3(1 - 2x) = 11$$
$$x - 3 + 6x = 11$$
$$7x = 14$$
$$x = 2$$

Then since $y = 1 - 2x$, $y = 1 - 2(2) = 1 - 4 = -3$. Thus, the solution is $(2, -3)$.

30. Solve the following system.

$$2x - y + z = 1$$
$$x + 3y - z = -4$$
$$3x + 2y + z = 0$$

Add the first two equations to eliminate z and obtain

$$3x + 2y = -3.$$

Then add the last two equations to eliminate z and obtain

$$4x + 5y = -4.$$

We now must solve the following system:

$$3x + 2y = -3$$
$$4x + 5y = -4$$

Multiply the first equation by -4, the second by 3, and add to eliminate x and obtain:

$$7y = 0$$
$$y = 0$$

Substitute 0 for y in $3x + 2y = -3$ to find the value of x.

$$3x + 2(0) = -3$$
$$3x = -3$$
$$x = -1$$

Substitute -1 for x and 0 for y in the first original equation,

$$2x - y + z = 1,$$

to find the value of z.

$$2(-1) - 0 + z = 1$$
$$-2 + z = 1$$
$$z = 3$$

Thus, the solution is $(-1, 0, 3)$.

PRACTICE EXERCISES

CHAPTER 7 MATRICES AND DETERMINANTS

SECTION 7.3 Solving Systems of Equations Using Matrices

1. Solve using the Gaussian method.

$$2x - y = -5$$
$$3x + 4y = 9$$

The augmented matrix is:

$$\begin{bmatrix} 2 & -1 & -5 \\ 3 & 4 & 9 \end{bmatrix}$$

Multiply the first row by -1 and add the result to the second row. This will give a 1 in the first column.

$$\begin{bmatrix} 2 & -1 & -5 \\ 1 & 5 & 14 \end{bmatrix}$$

Interchange the two rows to place the 1 in the upper left corner of the matrix.

$$\begin{bmatrix} 1 & 5 & 14 \\ 2 & -1 & -5 \end{bmatrix}$$

Keep the first row, multiply it by -2, and add the result to the second row. This will place a 0 in the lower left corner of the matrix.

$$\begin{bmatrix} 1 & 5 & 14 \\ 0 & -11 & -33 \end{bmatrix}$$

Multiply the second row by $-\frac{1}{11}$ to obtain 1 in the second column.

$$\begin{bmatrix} 1 & 5 & 14 \\ 0 & 1 & 3 \end{bmatrix}$$

Keep the second row, multiply it by -5, and add the result to the first row. This will place a 0 in the middle of the first row.

$$\begin{bmatrix} 1 & 0 & -1 \\ 0 & 1 & 3 \end{bmatrix}$$

Then since $x = -1$ and $y = 3$, the solution to the system is (-1,3).

2. Solve using the Gaussian method.

$$x - 2y + z = 0$$
$$2x + y - 3z = 5$$
$$3x - y - z = 5$$

Then the augmented matrix is:

$$\begin{bmatrix} 1 & -2 & 1 & 0 \\ 2 & 1 & -3 & 5 \\ 3 & -1 & -1 & 5 \end{bmatrix}$$

Since we already have a 1 in the upper left corner, we begin by obtaining 0's in the rest of the column. Keep the first row, multiply it by -2, and add the result to the second row. Then multiply the first row by -3, and add the result to the third row.

$$\begin{bmatrix} 1 & -2 & 1 & 0 \\ 0 & 5 & -5 & 5 \\ 0 & 5 & -4 & 5 \end{bmatrix}$$

Next we obtain a 1 in the middle of the second column by multiplying the second row by $\frac{1}{5}$.

$$\begin{bmatrix} 1 & -2 & 1 & 0 \\ 0 & 1 & -1 & 1 \\ 0 & 5 & -4 & 5 \end{bmatrix}$$

Keep the middle row, multiply it by 2, and add the result to the top row. Then multiply the middle row by -5, and add the result to the bottom row. This will give the two 0's in the second column.

$$\begin{bmatrix} 1 & 0 & -1 & 2 \\ 0 & 1 & -1 & 1 \\ 0 & 0 & 1 & 0 \end{bmatrix}$$

The above step also gave us the desired 1 in the third row. Keep the third row, multiply it by 1 and add the result to the second row and to the first row. This will give the desired 0's in the third column and finish the process.

$$\begin{bmatrix} 1 & 0 & 0 & 2 \\ 0 & 1 & 0 & 1 \\ 0 & 0 & 1 & 0 \end{bmatrix}$$

Since $x = 2$, $y = 1$, and $z = 0$, the solution to the system is (2,1,0).

3. Solve using the Gaussian method.

$$w + x - y + z = 2$$
$$w - x + 2y + z = 4$$
$$w - y - z = -1$$
$$2w + x + z = 3$$

We already have a 1 in the upper left corner, so we concentrate on obtaining 0's in the rest of the first column. Keep the first row, multiply it by -1, and add the result to the second row and to the third row. Then multiply the first row by -2, and add the result to the fourth row.

$$\begin{bmatrix} 1 & 1 & -1 & 1 & 2 \\ 0 & -2 & 3 & 0 & 2 \\ 0 & -1 & 0 & -2 & -3 \\ 0 & -1 & 2 & -1 & -1 \end{bmatrix}$$

Multiply the third row by -1 and interchange the result with the second row to obtain the desired 1 in the second row, second column.

$$\begin{bmatrix} 1 & 1 & -1 & 1 & 2 \\ 0 & 1 & 0 & 2 & 3 \\ 0 & -2 & 3 & 0 & 2 \\ 0 & -1 & 2 & -1 & -1 \end{bmatrix}$$

Keep the second row, multiply it by -1 and add the result to the first row. Then multiply the second row by 2, and add the result to the third row. Finally, multiply the second row by 1 and add the result to the fourth row (that is, simply add the second row to the fourth row).

$$\begin{bmatrix} 1 & 0 & -1 & -1 & -1 \\ 0 & 1 & 0 & 2 & 3 \\ 0 & 0 & 3 & 4 & 8 \\ 0 & 0 & 2 & 1 & 2 \end{bmatrix}$$

To obtain a 1 in the third row, third column, we could multiply the elements in the third row by $\frac{1}{3}$, but we would then need to work with fractions. Keep the fourth row, multiply it by -1, and add the result to the third row. (Also leave the first rows just as they are.)

$$\begin{bmatrix} 1 & 0 & -1 & -1 & -1 \\ 0 & 1 & 0 & 2 & 3 \\ 0 & 0 & 1 & 3 & 6 \\ 0 & 0 & 2 & 1 & 2 \end{bmatrix}$$

Keep the third row, multiply it by 1 and add the result to the first row (simply add the third row to the first row). Then multiply the third row by -2, and add the result to the fourth row. This will give the remaining two zeros in the third column (we were lucky since one 0 was already there).

$$\begin{bmatrix} 1 & 0 & 0 & 2 & 5 \\ 0 & 1 & 0 & 2 & 3 \\ 0 & 0 & 1 & 3 & 6 \\ 0 & 0 & 0 & -5 & -10 \end{bmatrix}$$

We can now obtain the desired 1 in the fourth row by multiplying the fourth row by $-\frac{1}{5}$.

$$\begin{bmatrix} 1 & 0 & 0 & 2 & 5 \\ 0 & 1 & 0 & 2 & 3 \\ 0 & 0 & 1 & 3 & 6 \\ 0 & 0 & 0 & 1 & 2 \end{bmatrix}$$

Keep the fourth row, multiply it by -2, and add the result to the first row and then to the second row. Then multiply the fourth row by -3, and add the result to the third row. This will give the desired 0's in the fourth column and complete the process.

$$\begin{bmatrix} 1 & 0 & 0 & 0 & 1 \\ 0 & 1 & 0 & 0 & -1 \\ 0 & 0 & 1 & 0 & 0 \\ 0 & 0 & 0 & 1 & 2 \end{bmatrix}$$

Since $w = 1$, $x = -1$, $y = 0$, and $z = 2$, the solution to the system is $(1,-1,0,2)$.

4. Solve using the Gaussian method.

$$\begin{aligned} x - y + z &= 3 \\ 2x - 2y + 2z &= -1 \\ -x + y - z &= -3 \end{aligned}$$

Since we already have a 1 in the upper left corner, we try to obtain 0's in the rest of the first column. Keep the first row, multiply it by -2, and add the result to the second row. Then multiply the first row by 1 and add the result to the third row (simply add the first row to the third row).

$$\begin{bmatrix} 1 & -1 & 1 & 3 \\ 0 & 0 & 0 & -7 \\ 0 & 0 & 0 & 0 \end{bmatrix}$$

Notice that the third row is equivalent to the identity $0 = 0$, but the second row is equivalent to the contradiction $0 = -7$. Since a contradiction is obtained, we know that the system is inconsistent and has no solution. Had the second row not been a contradiction, and we obtained all 0's in the third row, then there would have been infinitely many solutions to the system as was the case in Example 4.

EXERCISES A

CHAPTER 7 MATRICES AND DETERMINANTS

SECTION 7.3 Solving Systems of Equations Using Matrices

1. Given the matrix

$$\begin{matrix} (A) \\ (B) \end{matrix} \begin{bmatrix} 2 & 3 & -4 \\ 1 & -2 & 5 \end{bmatrix}.$$

The following steps will write the matrix in reduced echelon form.

(a) Interchange rows (A) and (B).

$$\begin{matrix} (B) \\ (A) \end{matrix} \begin{bmatrix} 1 & -2 & 5 \\ 2 & 3 & -4 \end{bmatrix}$$

(b) Retain (B) and replace (A) with (C) = -2(B) + (A).

$$\begin{matrix} (B) \\ (C) \end{matrix} \begin{bmatrix} 1 & -2 & 5 \\ 0 & 7 & -14 \end{bmatrix}$$

(c) Replace (C) with (D) = $\frac{1}{7}$(C).

$$\begin{matrix} (B) \\ (D) \end{matrix} \begin{bmatrix} 1 & -2 & 5 \\ 0 & 1 & -2 \end{bmatrix}$$

(d) Retain (D) and replace (B) with (E) = 2(D) + (B).

$$\begin{matrix} (E) \\ (D) \end{matrix} \begin{bmatrix} 1 & 0 & 1 \\ 0 & 1 & -2 \end{bmatrix}$$

2. Given the matrix

$$\begin{matrix} (A) \\ (B) \\ (C) \end{matrix} \begin{bmatrix} 2 & 1 & -1 & 4 \\ 3 & -1 & 0 & 3 \\ -4 & 2 & 3 & -10 \end{bmatrix}.$$

The following steps will write the matrix in reduced echelon form.

(a) Replace (A) with (D) = (B) - (A).

$$\begin{matrix} (D) \\ (B) \\ (C) \end{matrix} \begin{bmatrix} 1 & -2 & 1 & -1 \\ 3 & -1 & 0 & 3 \\ -4 & 2 & 3 & -10 \end{bmatrix}$$

(b) Retain (D), replace (B) with (E) = -3(D) + (B), and replace (C) with (F) = 4(D) + (C).

$$\begin{matrix} (D) \\ (E) \\ (F) \end{matrix} \begin{bmatrix} 1 & -2 & 1 & -1 \\ 0 & 5 & -3 & 6 \\ 0 & -6 & 7 & -14 \end{bmatrix}$$

(c) Retain (D) and (F) and replace (E) with (G) = (-1)[(E) + (F)].

$$\begin{matrix} (D) \\ (G) \\ (F) \end{matrix} \begin{bmatrix} 1 & -2 & 1 & -1 \\ 0 & 1 & -4 & 8 \\ 0 & -6 & 7 & -14 \end{bmatrix}$$

(d) Retain (G), replace (D) with (H) = 2(G) + (D), and replace (F) with (I) = 6(G) + (F).

$$\begin{matrix} (H) \\ (G) \\ (I) \end{matrix} \begin{bmatrix} 1 & 0 & -7 & 15 \\ 0 & 1 & -4 & 8 \\ 0 & 0 & -17 & 34 \end{bmatrix}$$

(e) Retain (H) and (G) and replace (I) with (J) = $-\frac{1}{17}$(I).

$$\begin{matrix} (H) \\ (G) \\ (J) \end{matrix} \begin{bmatrix} 1 & 0 & -7 & 15 \\ 0 & 1 & -4 & 8 \\ 0 & 0 & 1 & -2 \end{bmatrix}$$

(f) Retain (J), replace (G) with (K) = 4(J) + (G), and replace (H) with (L) = 7(J) + (H).

$$\begin{matrix} (L) \\ (K) \\ (J) \end{matrix} \begin{bmatrix} 1 & 0 & 0 & 1 \\ 0 & 1 & 0 & 0 \\ 0 & 0 & 1 & -2 \end{bmatrix}$$

3. Solve using the Gaussian method.

$$x + 3y = 1$$
$$3x + 7y = 5$$

First write the augmented matrix of the system.

$$\begin{bmatrix} 1 & 3 & 1 \\ 3 & 7 & 5 \end{bmatrix}$$

Since the first row already has a 1 in the desired position, we begin by obtaining the necessary 0 in the first column. Keep the first row, multiply it by -3, and add the result to the second row.

$$\begin{bmatrix} 1 & 3 & 1 \\ 0 & -2 & 2 \end{bmatrix}$$

Keep the first row and multiply the second row by $-\frac{1}{2}$ to obtain the desired 1 in the second row, second column.

$$\begin{bmatrix} 1 & 3 & 1 \\ 0 & 1 & -1 \end{bmatrix}$$

Keep the second row, multiply it by -3, and add the result to the first row.

$$\begin{bmatrix} 1 & 0 & 4 \\ 0 & 1 & -1 \end{bmatrix}$$

The matrix now corresponds to $x = 4$ and $y = -1$. Thus, the solution to the system is $(4, -1)$.

4. Solve using the Gaussian method.

$$\begin{aligned} 2x - y &= 1 \\ -4x + 2y &= 1 \end{aligned}$$

First write the augmented matrix of the system.

$$\begin{bmatrix} 2 & -1 & 1 \\ -4 & 2 & 1 \end{bmatrix}$$

Notice that if we keep the first row, multiply it by 2, and add the result to the second row, we obtain:

$$\begin{bmatrix} 2 & -1 & 1 \\ 0 & 0 & 3 \end{bmatrix}$$

The second row corresponds to the contradiction $0 = 3$, so we know that the system has no solution.

5. Solve using the Gaussian method.

$$\begin{aligned} 4x - 2y &= -3 \\ x + y &= 3 \end{aligned}$$

First write the augmented matrix of the system.

$$\begin{bmatrix} 4 & -2 & -3 \\ 1 & 1 & 3 \end{bmatrix}$$

Interchange the two rows to obtain the desired 1 in the first row, first column.

$$\begin{bmatrix} 1 & 1 & 3 \\ 4 & -2 & -3 \end{bmatrix}$$

Keep the first row, multiply it by -4, and add the result to the the second row.

$$\begin{bmatrix} 1 & 1 & 3 \\ 0 & -6 & -15 \end{bmatrix}$$

Multiply the second row by $-\frac{1}{6}$ to obtain the desired 1 in the second row, second column.

$$\begin{bmatrix} 1 & 1 & 3 \\ 0 & 1 & \frac{5}{2} \end{bmatrix}$$

Finally, keep the second row, multiply it by -1, and add the result to the first row.

$$\begin{bmatrix} 1 & 0 & \frac{1}{2} \\ 0 & 1 & \frac{5}{2} \end{bmatrix}$$

Then this corresponds to $x = \frac{1}{2}$ and $y = \frac{5}{2}$.

Thus, the solution to the system is $\left(\frac{1}{2}, \frac{5}{2}\right)$.

6. Solve using the Gaussian method.

$$\begin{aligned} 5x - 3y &= -2 \\ 2x + 7y &= -9 \end{aligned}$$

Write the augmented matrix of the system.

$$\begin{bmatrix} 5 & -3 & -2 \\ 2 & 7 & -9 \end{bmatrix}$$

Keep the second row, multiply it by -2, and add the result to the first row to obtain the desired 1 in the first row, first column.

$$\begin{bmatrix} 1 & -17 & 16 \\ 2 & 7 & -9 \end{bmatrix}$$

Keep the first row, multiply it by -2, and add the result to the second row.

$$\begin{bmatrix} 1 & -17 & 16 \\ 0 & 41 & -41 \end{bmatrix}$$

Multiply the second row by $\frac{1}{41}$.

EXERCISES A

$$\begin{bmatrix} 1 & -17 & 16 \\ 0 & 1 & -1 \end{bmatrix}$$

Keep the second row, multiply it by 17, and add the result to the first row.

$$\begin{bmatrix} 1 & 0 & -1 \\ 0 & 1 & -1 \end{bmatrix}$$

This corresponds to $x = -1$ and $y = -1$. Thus, the solution to the system is $(-1,-1)$.

7. Solve using the Gaussian method.

$$\begin{aligned} 3x + 2y &= 1 \\ -9x - 5y &= -3 \end{aligned}$$

Write the augmented matrix of the system.

$$\begin{bmatrix} 3 & 2 & 1 \\ -9 & -5 & -3 \end{bmatrix}$$

Multiply the first row by $\frac{1}{3}$ to obtain the desired 1 in the first row, first column.

$$\begin{bmatrix} 1 & \frac{2}{3} & \frac{1}{3} \\ -9 & -5 & -3 \end{bmatrix}$$

Keep the first row, multiply it by 9, and add the result to the second row.

$$\begin{bmatrix} 1 & \frac{2}{3} & \frac{1}{3} \\ 0 & 1 & 0 \end{bmatrix}$$

Keep the second row, multiply it by $-\frac{2}{3}$, and add the result to the first row.

$$\begin{bmatrix} 1 & 0 & \frac{1}{3} \\ 0 & 1 & 0 \end{bmatrix}$$

This corresponds to $x = \frac{1}{3}$ and $y = 0$. Thus, the solution to the system is $\left(\frac{1}{3},0\right)$.

8. Solve using the Gaussian method.

$$\begin{aligned} x + 2y &= 1 \\ -2x - 4y &= -2 \end{aligned}$$

Write the augmented matrix of the system.

$$\begin{bmatrix} 1 & 2 & 1 \\ -2 & -4 & -2 \end{bmatrix}$$

Notice that if we multiply the first equation by 2, and add the result to the second equation, we obtain:

$$\begin{bmatrix} 1 & 2 & 1 \\ 0 & 0 & 0 \end{bmatrix}$$

Since the second row is all 0's, this corresponds to the identity $0 = 0$, and we know that the system has infinitely many solutions. When this occurs, it is best to solve the system using previous techniques.

Solve the first equation for x,

$$x = 1 - 2y.$$

Then the solution can be written as $(1-2y,y)$ for y any real number.

9. Solve using the Gaussian method.

$$\begin{aligned} x + y + z &= 2 \\ -x - y + 3z &= 6 \\ 2x + y - z &= -1 \end{aligned}$$

Write the augmented matrix of the system.

$$\begin{bmatrix} 1 & 1 & 1 & 2 \\ -1 & -1 & 3 & 6 \\ 2 & 1 & -1 & -1 \end{bmatrix}$$

Since we already have the desired 1 in the first row, first column, we begin by obtaining the 0's in the first column. Keep the first row, multiply it by 1, and add the result to the second row (simply add the first row to the second). Then multiply the first row by -2, and add the result to the third row.

$$\begin{bmatrix} 1 & 1 & 1 & 2 \\ 0 & 0 & 4 & 8 \\ 0 & -1 & -3 & -5 \end{bmatrix}$$

We can obtain the desired 1 in the second row, second column by multiplying the third row by -1 and interchanging the result with the second row.

$$\begin{bmatrix} 1 & 1 & 1 & 2 \\ 0 & 1 & 3 & 5 \\ 0 & 0 & 4 & 8 \end{bmatrix}$$

Keep the second row, multiply it by -1, and add the result to the first row.

$$\begin{bmatrix} 1 & 0 & -2 & -3 \\ 0 & 1 & 3 & 5 \\ 0 & 0 & 4 & 8 \end{bmatrix}$$

Multiply the third row by $\frac{1}{4}$.

$$\begin{bmatrix} 1 & 0 & -2 & -3 \\ 0 & 1 & 3 & 5 \\ 0 & 0 & 1 & 2 \end{bmatrix}$$

Finally, keep the third row, multiply it by 2, and add the result to the first row. Then multiply the third row by -3, and add the result to the second row.

$$\begin{bmatrix} 1 & 0 & 0 & 1 \\ 0 & 1 & 0 & -1 \\ 0 & 0 & 1 & 2 \end{bmatrix}$$

This corresponds to $x = 1$, $y = -1$, and $z = 2$. Thus, the solution to the system is $(1, -1, 2)$.

10. Solve using the Gaussian method.

$$\begin{aligned} x \phantom{{}+y} + z &= 1 \\ y + z &= 4 \\ 2x + y \phantom{{}+z} &= -3 \end{aligned}$$

Write the augmented matrix of the system.

$$\begin{bmatrix} 1 & 0 & 1 & 1 \\ 0 & 1 & 1 & 4 \\ 2 & 1 & 0 & -3 \end{bmatrix}$$

Keep the first and second rows. Multiply the first row by -2, and add the result to the third row.

$$\begin{bmatrix} 1 & 0 & 1 & 1 \\ 0 & 1 & 1 & 4 \\ 0 & 1 & -2 & -5 \end{bmatrix}$$

Keep the first and second rows. Multiply the second row by -1, and add the result to the third row.

$$\begin{bmatrix} 1 & 0 & 1 & 1 \\ 0 & 1 & 1 & 4 \\ 0 & 0 & -3 & -9 \end{bmatrix}$$

Multiply the third row by $-\frac{1}{3}$.

$$\begin{bmatrix} 1 & 0 & 1 & 1 \\ 0 & 1 & 1 & 4 \\ 0 & 0 & 1 & 3 \end{bmatrix}$$

Finally, keep the third row, multiply it by -1, and add the result to the first row, then also to the second row.

$$\begin{bmatrix} 1 & 0 & 0 & -2 \\ 0 & 1 & 0 & 1 \\ 0 & 0 & 1 & 3 \end{bmatrix}$$

Then this corresponds to $x = -2$, $y = 1$, and $z = 3$. Thus, the solution to the system is $(-2, 1, 3)$.

11. Solve using the Gaussian method.

$$\begin{aligned} 2x - y + 3z &= 0 \\ x + y - z &= 4 \\ 2y + 5z &= -3 \end{aligned}$$

Write the augmented matrix of the system.

$$\begin{bmatrix} 2 & -1 & 3 & 0 \\ 1 & 1 & -1 & 4 \\ 0 & 2 & 5 & -3 \end{bmatrix}$$

Interchange the first and second rows.

$$\begin{bmatrix} 1 & 1 & -1 & 4 \\ 2 & -1 & 3 & 0 \\ 0 & 2 & 5 & -3 \end{bmatrix}$$

Keep the first row, multiply it by -2, and add the result to the second row.

$$\begin{bmatrix} 1 & 1 & -1 & 4 \\ 0 & -3 & 5 & -8 \\ 0 & 2 & 5 & -3 \end{bmatrix}$$

Keep the first and third rows, add the third row to the second, and multiply the result by -1.

$$\begin{bmatrix} 1 & 1 & -1 & 4 \\ 0 & 1 & -10 & 11 \\ 0 & 2 & 5 & -3 \end{bmatrix}$$

Keep the second row, multiply it by -1, and add the result to the first row. Then multiply the second row by -2, and add the result to the third row.

$$\begin{bmatrix} 1 & 0 & 9 & -7 \\ 0 & 1 & -10 & 11 \\ 0 & 0 & 25 & -25 \end{bmatrix}$$

EXERCISES A

Multiply the third row by $\frac{1}{25}$.

$$\begin{bmatrix} 1 & 0 & 9 & -7 \\ 0 & 1 & -10 & 11 \\ 0 & 0 & 1 & -1 \end{bmatrix}$$

Finally, keep the third row, multiply it by -9, and add the result to the first row. Then multiply the third row by 10, and add the result to the second row.

$$\begin{bmatrix} 1 & 0 & 0 & 2 \\ 0 & 1 & 0 & 1 \\ 0 & 0 & 1 & -1 \end{bmatrix}$$

Then this corresponds to $x = 2$, $y = 1$, and $z = -1$. Thus, the solution to the system is $(2,1,-1)$.

12. Solve using the Gaussian method.

$$\begin{aligned} 3x - y + 2z &= 2 \\ x - z &= 1 \\ 2x - y + 3z &= 1 \end{aligned}$$

Write the augmented matrix of the system.

$$\begin{bmatrix} 3 & -1 & 2 & 2 \\ 1 & 0 & -1 & 1 \\ 2 & -1 & 3 & 1 \end{bmatrix}$$

Interchange the first and second rows.

$$\begin{bmatrix} 1 & 0 & -1 & 1 \\ 3 & -1 & 2 & 2 \\ 2 & -1 & 3 & 1 \end{bmatrix}$$

Keep the first row, multiply it by -3, and add the result to the second row. Then multiply the first row by -2, and add the result to the third row.

$$\begin{bmatrix} 1 & 0 & -1 & 1 \\ 0 & -1 & 5 & -1 \\ 0 & -1 & 5 & -1 \end{bmatrix}$$

Multiply the second row by -1.

$$\begin{bmatrix} 1 & 0 & -1 & 1 \\ 0 & 1 & -5 & 1 \\ 0 & -1 & 5 & -1 \end{bmatrix}$$

Keep the first and second rows. Multiply the second row by 1, and add the result to the third row (simply add the second row to the third row).

$$\begin{bmatrix} 1 & 0 & -1 & 1 \\ 0 & 1 & -5 & 1 \\ 0 & 0 & 0 & 0 \end{bmatrix}$$

Since the third row is all 0's, this corresponds to the identity $0 = 0$, and we know that the system has infinitely many solutions. It is best to solve it using previous methods.

Returning to the original system of equations, solve the second equation for z in terms of x,

$$z = x - 1.$$

Substitute this expression for z in the first original equation,

$$3x - y + 2z = 2,$$

and solve for y in terms of x.

$$\begin{aligned} 3x - y + 2(x - 1) &= 2 \\ 3x - y + 2x - 2 &= 2 \\ -y &= -5x + 4 \\ y &= 5x - 4 \end{aligned}$$

Thus, the solution can be written as $(x, 5x-4, x-1)$ for x any real number.

Exercises 13-20 refer to the following matrices.

$$A = \begin{bmatrix} 1 & 0 \\ -2 & 5 \end{bmatrix} \quad B = \begin{bmatrix} -2 & 5 \\ 1 & 4 \end{bmatrix} \quad C = \begin{bmatrix} 3 & -1 & 2 \\ 0 & -2 & 4 \end{bmatrix}$$

$$D = \begin{bmatrix} 2 & 1 & -1 \end{bmatrix} \quad E = \begin{bmatrix} 3 \\ 0 \\ 5 \end{bmatrix} \quad F = \begin{bmatrix} 4 & -3 \end{bmatrix}$$

13. The order of matrix C is 2×3 since C has 2 rows and 3 columns.

14. The element c_{13} in matrix C is the element in the 1st row and 3rd column of C. Thus, $c_{13} = 2$.

15. Since $B = \begin{bmatrix} -2 & 5 \\ 1 & 4 \end{bmatrix}$, $-B = \begin{bmatrix} 2 & -5 \\ -1 & -4 \end{bmatrix}$.

16. $A + B = \begin{bmatrix} 1 & 0 \\ -2 & 5 \end{bmatrix} + \begin{bmatrix} -2 & 5 \\ 1 & 4 \end{bmatrix}$

$= \begin{bmatrix} 1+(-2) & 0+5 \\ -2+1 & 5+4 \end{bmatrix}$

$= \begin{bmatrix} -1 & 5 \\ -1 & 9 \end{bmatrix}$

17. $2A - 4B = 2\begin{bmatrix} 1 & 0 \\ -2 & 5 \end{bmatrix} - 4\begin{bmatrix} -2 & 5 \\ 1 & 4 \end{bmatrix}$

$= \begin{bmatrix} 2 & 0 \\ -4 & 10 \end{bmatrix} - \begin{bmatrix} -8 & 20 \\ 4 & 16 \end{bmatrix}$

$= \begin{bmatrix} 2-(-8) & 0-20 \\ -4-4 & 10-16 \end{bmatrix}$

$= \begin{bmatrix} 10 & -20 \\ -8 & -6 \end{bmatrix}$

18. First find AB.

$AB = \begin{bmatrix} 1 & 0 \\ -2 & 5 \end{bmatrix}\begin{bmatrix} -2 & 5 \\ 1 & 4 \end{bmatrix}$

$= \begin{bmatrix} (1)(-2)+(0)(1) & (1)(5)+(0)(4) \\ (-2)(-2)+(5)(1) & (-2)(5)+(5)(4) \end{bmatrix}$

$= \begin{bmatrix} -2 & 5 \\ 9 & 10 \end{bmatrix}$

Interchange the rows and columns of this matrix to obtain

$(AB)^T = \begin{bmatrix} -2 & 9 \\ 5 & 10 \end{bmatrix}.$

19. $AI = \begin{bmatrix} 1 & 0 \\ -2 & 5 \end{bmatrix}\begin{bmatrix} 1 & 0 \\ 0 & 1 \end{bmatrix} = \begin{bmatrix} 1 & 0 \\ -2 & 5 \end{bmatrix} = A$

20. $ED = \begin{bmatrix} 3 \\ 0 \\ 5 \end{bmatrix}\begin{bmatrix} 2 & 1 & -1 \end{bmatrix}$

$= \begin{bmatrix} (3)(2) & (3)(1) & (3)(-1) \\ (0)(2) & (0)(1) & (0)(-1) \\ (5)(2) & (5)(1) & (5)(-1) \end{bmatrix}$

$= \begin{bmatrix} 6 & 3 & -3 \\ 0 & 0 & 0 \\ 10 & 5 & -5 \end{bmatrix}$

21. If $C = [200\ 300\ 400\ 550]$ gives the selling price of four models of television sets, and

$$N = \begin{bmatrix} 5 \\ 3 \\ 4 \\ 2 \end{bmatrix}$$

gives the number of models sold on Saturday, then

$C \cdot N = [200\ 300\ 400\ 550] \cdot \begin{bmatrix} 5 \\ 3 \\ 4 \\ 2 \end{bmatrix}$

$= (200)(5)+(300)(3)+(400)(4)+(550)(2)$
$= 1000+900+1600+1100$
$= 4600$

represents the total sales of the four models on Saturday. Thus, the total sales amounted to $4600.

PRACTICE EXERCISES SECTION 7.4

CHAPTER 7 MATRICES AND DETERMINANTS

SECTION 7.4 The Inverse of a Square Matrix

1. Find the inverse of

$$A = \begin{bmatrix} -2 & -1 \\ 3 & 1 \end{bmatrix}.$$

First form the augmented matrix $[A|I]$.

$$\begin{bmatrix} -2 & -1 & | & 1 & 0 \\ 3 & 1 & | & 0 & 1 \end{bmatrix}$$

To obtain a 1 in the first column, keep the first row, multiply it by 1, and add it to the second row (simply add the first row to the second row).

$$\begin{bmatrix} -2 & -1 & | & 1 & 0 \\ 1 & 0 & | & 1 & 1 \end{bmatrix}$$

Interchange the two rows.

$$\begin{bmatrix} 1 & 0 & | & 1 & 1 \\ -2 & -1 & | & 1 & 0 \end{bmatrix}$$

Keep the first row, multiply it by 2, and add the result to the second row.

$$\begin{bmatrix} 1 & 0 & | & 1 & 1 \\ 0 & -1 & | & 3 & 2 \end{bmatrix}$$

Multiply the second row by -1 to obtain the desired 1 in the second row, second column. Notice that we already have the desired 0 in the first row, second column, so this completes the process.

$$\begin{bmatrix} 1 & 0 & | & 1 & 1 \\ 0 & 1 & | & -3 & -2 \end{bmatrix}$$

Then

$$A^{-1} = \begin{bmatrix} 1 & 1 \\ -3 & -2 \end{bmatrix}.$$

This can be checked by showing that

$$AA^{-1} = A^{-1}A = I.$$

2. Find the inverse of

$$A = \begin{bmatrix} 1 & 1 & -1 \\ 2 & 1 & 1 \\ 3 & -2 & -1 \end{bmatrix}.$$

First form the augmented matrix $[A|I]$.

$$\begin{bmatrix} 1 & 1 & -1 & | & 1 & 0 & 0 \\ 2 & 1 & 1 & | & 0 & 1 & 0 \\ 3 & -2 & -1 & | & 0 & 0 & 1 \end{bmatrix}$$

Since we already have a 1 in the upper left corner, concentrate on obtaining the two 0's in the first column. Keep the first row, multiply it by -2, and add the result to the second row. Then multiply the first row by -3, and add the result to the third row.

$$\begin{bmatrix} 1 & 1 & -1 & | & 1 & 0 & 0 \\ 0 & -1 & 3 & | & -2 & 1 & 0 \\ 0 & -5 & 2 & | & -3 & 0 & 1 \end{bmatrix}$$

Multiply the second row by -1 to obtain the desired 1 in the second row, second column.

$$\begin{bmatrix} 1 & 1 & -1 & | & 1 & 0 & 0 \\ 0 & 1 & -3 & | & 2 & -1 & 0 \\ 0 & -5 & 2 & | & -3 & 0 & 1 \end{bmatrix}$$

Keep the second row, multiply it by -1, and add the result to the first row. Then multiply the second row by 5, and add the result to the third row. This will obtain the desired 0's in the second column.

$$\begin{bmatrix} 1 & 0 & 2 & | & -1 & 1 & 0 \\ 0 & 1 & -3 & | & 2 & -1 & 0 \\ 0 & 0 & -13 & | & 7 & -5 & 1 \end{bmatrix}$$

Multiply the third row by $-\frac{1}{13}$ to obtain the desired 1 in the third row, third column.

$$\begin{bmatrix} 1 & 0 & 2 & | & -1 & 1 & 0 \\ 0 & 1 & -3 & | & 2 & -1 & 0 \\ 0 & 0 & 1 & | & -\frac{7}{13} & \frac{5}{13} & -\frac{1}{13} \end{bmatrix}$$

Keep the third row, multiply it by -2, and add the result to the first row. Then multiply the third row by 3, and add the result to the second row. This will give the two 0's necessary to complete the third column and finish the process.

$$\begin{bmatrix} 1 & 0 & 0 & | & \frac{1}{13} & \frac{3}{13} & \frac{2}{13} \\ 0 & 1 & 0 & | & \frac{5}{13} & \frac{2}{13} & -\frac{3}{13} \\ 0 & 0 & 1 & | & -\frac{7}{13} & \frac{5}{13} & -\frac{1}{13} \end{bmatrix}$$

Thus, the inverse is

$$A^{-1} = \begin{bmatrix} \frac{1}{13} & \frac{3}{13} & \frac{2}{13} \\ \frac{5}{13} & \frac{2}{13} & -\frac{3}{13} \\ -\frac{7}{13} & \frac{5}{13} & -\frac{1}{13} \end{bmatrix}.$$

You can check this by showing that

$$AA^{-1} = A^{-1}A = I.$$

3. Solve the system using the inverse method.

$$\begin{aligned} -2x - y &= 0 \\ 3x + y &= 2 \end{aligned}$$

This system is equivalent to the matrix equation $AX = B$, where

$$A = \begin{bmatrix} -2 & -1 \\ 3 & 1 \end{bmatrix} \quad X = \begin{bmatrix} x \\ y \end{bmatrix} \quad B = \begin{bmatrix} 0 \\ 2 \end{bmatrix}.$$

In Practice Exercise 1, we found the inverse of the matrix A to be

$$A^{-1} = \begin{bmatrix} 1 & 1 \\ -3 & -2 \end{bmatrix}.$$

The solution to the matrix equation is

$$\begin{bmatrix} x \\ y \end{bmatrix} = X = A^{-1}B = \begin{bmatrix} 1 & 1 \\ -3 & -2 \end{bmatrix}\begin{bmatrix} 0 \\ 2 \end{bmatrix} = \begin{bmatrix} 2 \\ -4 \end{bmatrix},$$

making $x = 2$ and $y = -4$. Thus, the solution to the system is $(2,-4)$. This can be checked in the system of equations.

4. Solve the system using the inverse method.

$$\begin{aligned} x + y - z &= -2 \\ 2x + y + z &= 3 \\ 3x - 2y - z &= -10 \end{aligned}$$

This system is equivalent to the matrix equation $AX = B$, where

$$A = \begin{bmatrix} 1 & 1 & -1 \\ 2 & 1 & 1 \\ 3 & -2 & -1 \end{bmatrix} \quad X = \begin{bmatrix} x \\ y \\ z \end{bmatrix} \quad B = \begin{bmatrix} -2 \\ 3 \\ -10 \end{bmatrix}.$$

In Practice Exercise 2, we found the inverse of the matrix A to be

$$A^{-1} = \begin{bmatrix} \frac{1}{13} & \frac{3}{13} & \frac{2}{13} \\ \frac{5}{13} & \frac{2}{13} & -\frac{3}{13} \\ -\frac{7}{13} & \frac{5}{13} & -\frac{1}{13} \end{bmatrix}.$$

The solution to the matrix equation is

$$\begin{bmatrix} x \\ y \\ z \end{bmatrix} = X = A^{-1}B = \begin{bmatrix} \frac{1}{13} & \frac{3}{13} & \frac{2}{13} \\ \frac{5}{13} & \frac{2}{13} & -\frac{3}{13} \\ -\frac{7}{13} & \frac{5}{13} & -\frac{1}{13} \end{bmatrix}\begin{bmatrix} -2 \\ 3 \\ -10 \end{bmatrix} = \begin{bmatrix} -1 \\ 2 \\ 3 \end{bmatrix},$$

making $x = -1$, $y = 2$, and $z = 3$. Thus, the solution to the system is $(-1,2,3)$. This can be checked in the system of equations.

5. Most of the work has already been done in Example 5 when the inverse of the coefficient matrix was obtained,

$$A^{-1} = \frac{1}{3}\begin{bmatrix} 8 & -1 \\ -5 & 1 \end{bmatrix}.$$

In the additional game, the stadium seats 35,000, so $n = 35,000$, and the revenue produced was \$220,000, so $r = 220,000$, giving the matrix

$$\begin{bmatrix} n \\ r \end{bmatrix} = \begin{bmatrix} 35{,}000 \\ 220{,}000 \end{bmatrix}.$$

To find the number of tickets at each price, x and y, all we must do is multiply A^{-1} times the above matrix.

$$\begin{bmatrix} x \\ y \end{bmatrix} = \frac{1}{3}\begin{bmatrix} 8 & -1 \\ -5 & 1 \end{bmatrix}\begin{bmatrix} 35{,}000 \\ 220{,}000 \end{bmatrix}$$
$$= \frac{1}{3}\begin{bmatrix} 60{,}000 \\ 45{,}000 \end{bmatrix}$$
$$= \begin{bmatrix} 20{,}000 \\ 15{,}000 \end{bmatrix}$$

Thus, $x = 20,000$ and $y = 15,000$. That is, 20,000 \$5 seats were sold and 15,000 \$8 seats were sold.

EXERCISES A SECTION 7.4 341

CHAPTER 7 MATRICES AND DETERMINANTS

SECTION 7.4 The Inverse of a Square Matrix

1. First find AB.

$$AB = \begin{bmatrix} 2 & 3 \\ -3 & -5 \end{bmatrix} \begin{bmatrix} 5 & 3 \\ -3 & -2 \end{bmatrix}$$
$$= \begin{bmatrix} (2)(5)+(3)(-3) & (2)(3)+(3)(-2) \\ (-3)(5)+(-5)(-3) & (-3)(3)+(-5)(-2) \end{bmatrix}$$
$$= \begin{bmatrix} 1 & 0 \\ 0 & 1 \end{bmatrix}$$
$$= I$$

Then find BA.

$$BA = \begin{bmatrix} 5 & 3 \\ -3 & -2 \end{bmatrix} \begin{bmatrix} 2 & 3 \\ -3 & -5 \end{bmatrix}$$
$$= \begin{bmatrix} (5)(2)+(3)(-3) & (5)(3)+(3)(-5) \\ (-3)(2)+(-2)(-3) & (-3)(3)+(-2)(-5) \end{bmatrix}$$
$$= \begin{bmatrix} 1 & 0 \\ 0 & 1 \end{bmatrix}$$
$$= I$$

Thus, since $AB = BA = I$, $B = A^{-1}$.

2. First find AB.

$$AB = \begin{bmatrix} 1 & 0 & 1 \\ 2 & 1 & 1 \\ 3 & 2 & 2 \end{bmatrix} \begin{bmatrix} 0 & 2 & -1 \\ -1 & -1 & 1 \\ 1 & -2 & 1 \end{bmatrix}$$
$$= \begin{bmatrix} 0+0+1 & 2+0+(-2) & (-1)+0+1 \\ 0+(-1)+1 & 4+(-1)+(-2) & (-2)+1+1 \\ 0+(-2)+2 & 6+(-2)+(-4) & (-3)+2+2 \end{bmatrix}$$
$$= \begin{bmatrix} 1 & 0 & 0 \\ 0 & 1 & 0 \\ 0 & 0 & 1 \end{bmatrix}$$
$$= I$$

Then find BA.

$$BA = \begin{bmatrix} 0 & 2 & -1 \\ -1 & -1 & 1 \\ 1 & -2 & 1 \end{bmatrix} \begin{bmatrix} 1 & 0 & 1 \\ 2 & 1 & 1 \\ 3 & 2 & 2 \end{bmatrix}$$
$$= \begin{bmatrix} 0+4+(-3) & 0+2+(-2) & 0+2+(-2) \\ (-1)+(-2)+3 & 0+(-1)+2 & (-1)+(-1)+2 \\ 1+(-4)+3 & 0+(-2)+2 & 1+(-2)+2 \end{bmatrix}$$
$$= \begin{bmatrix} 1 & 0 & 0 \\ 0 & 1 & 0 \\ 0 & 0 & 1 \end{bmatrix}$$
$$= I$$

Thus, $AB = BA = I$, so $B = A^{-1}$.

3. Find the inverse of the matrix.

$$\begin{bmatrix} 3 & -4 \\ 4 & -5 \end{bmatrix}$$

Form the augmented matrix $[A|I]$.

$$\begin{bmatrix} 3 & -4 & | & 1 & 0 \\ 4 & -5 & | & 0 & 1 \end{bmatrix}$$

Keep the first row, multiply it by -1, and add the result to the second row.

$$\begin{bmatrix} 3 & -4 & | & 1 & 0 \\ 1 & -1 & | & -1 & 1 \end{bmatrix}$$

Interchange the two rows.

$$\begin{bmatrix} 1 & -1 & | & -1 & 1 \\ 3 & -4 & | & 1 & 0 \end{bmatrix}$$

Keep the first row, multiply it by -3, and add the result to the second row.

$$\begin{bmatrix} 1 & -1 & | & -1 & 1 \\ 0 & -1 & | & 4 & -3 \end{bmatrix}$$

Multiply the second row by -1.

$$\begin{bmatrix} 1 & -1 & | & -1 & 1 \\ 0 & 1 & | & -4 & 3 \end{bmatrix}$$

Keep the second row, add the second row to the first.

$$\begin{bmatrix} 1 & 0 & | & -5 & 4 \\ 0 & 1 & | & -4 & 3 \end{bmatrix}$$

Thus, $A^{-1} = \begin{bmatrix} -5 & 4 \\ -4 & 3 \end{bmatrix}$.

4. Find the inverse of the matrix.

$$\begin{bmatrix} 2 & -1 \\ 4 & -3 \end{bmatrix}$$

Form the augmented matrix $[A|I]$.

$$\begin{bmatrix} 2 & -1 & | & 1 & 0 \\ 4 & -3 & | & 0 & 1 \end{bmatrix}$$

Multiply the first row by $\frac{1}{2}$.

$$\begin{bmatrix} 1 & -\frac{1}{2} & | & \frac{1}{2} & 0 \\ 4 & -3 & | & 0 & 1 \end{bmatrix}$$

Keep the first row, multiply it by −4, and add the result to the second row.

$$\begin{bmatrix} 1 & -\frac{1}{2} & | & \frac{1}{2} & 0 \\ 0 & -1 & | & -2 & 1 \end{bmatrix}$$

Multiply the second row by −1.

$$\begin{bmatrix} 1 & -\frac{1}{2} & | & \frac{1}{2} & 0 \\ 0 & 1 & | & 2 & -1 \end{bmatrix}$$

Keep the second row, multiply it by $\frac{1}{2}$ and add the result to the first row.

$$\begin{bmatrix} 1 & 0 & | & \frac{3}{2} & -\frac{1}{2} \\ 0 & 1 & | & 2 & -1 \end{bmatrix}$$

Thus, $A^{-1} = \begin{bmatrix} \frac{3}{2} & -\frac{1}{2} \\ 2 & -1 \end{bmatrix}$.

5. Find the inverse of the matrix.

$$\begin{bmatrix} 2 & 2 \\ 2 & 2 \end{bmatrix}$$

Form the augmented matrix $[A|I]$.

$$\begin{bmatrix} 2 & 2 & | & 1 & 0 \\ 2 & 2 & | & 0 & 1 \end{bmatrix}$$

Multiply the first row by $\frac{1}{2}$.

$$\begin{bmatrix} 1 & 1 & | & \frac{1}{2} & 0 \\ 2 & 2 & | & 0 & 1 \end{bmatrix}$$

Keep the first row, multiply it by −2, and add the result to the second row.

$$\begin{bmatrix} 1 & 1 & | & \frac{1}{2} & 0 \\ 0 & 0 & | & -1 & 1 \end{bmatrix}$$

Notice that we obtain 0's in the first two positions in the second row. As a result, it is impossible to make the element in the second row, second column into a 1, and the matrix has no inverse.

6. Find the inverse of the matrix.

$$\begin{bmatrix} 2 & -2 & 3 \\ 1 & 0 & -1 \\ -2 & 1 & 0 \end{bmatrix}$$

Form the augmented matrix $[A|I]$.

$$\begin{bmatrix} 2 & -2 & 3 & | & 1 & 0 & 0 \\ 1 & 0 & -1 & | & 0 & 1 & 0 \\ -2 & 1 & 0 & | & 0 & 0 & 1 \end{bmatrix}$$

Interchange the first two rows.

$$\begin{bmatrix} 1 & 0 & -1 & | & 0 & 1 & 0 \\ 2 & -2 & 3 & | & 1 & 0 & 0 \\ -2 & 1 & 0 & | & 0 & 0 & 1 \end{bmatrix}$$

Keep the first row, multiply it by −2, and add the result to the second row. Then multiply the first row by 2, and add the result to the third row.

$$\begin{bmatrix} 1 & 0 & -1 & | & 0 & 1 & 0 \\ 0 & -2 & 5 & | & 1 & -2 & 0 \\ 0 & 1 & -2 & | & 0 & 2 & 1 \end{bmatrix}$$

Interchange the second and third rows.

$$\begin{bmatrix} 1 & 0 & -1 & | & 0 & 1 & 0 \\ 0 & 1 & -2 & | & 0 & 2 & 1 \\ 0 & -2 & 5 & | & 1 & -2 & 0 \end{bmatrix}$$

Keep the first two rows. Multiply the second row by 2, and add the result to the third row.

$$\begin{bmatrix} 1 & 0 & -1 & | & 0 & 1 & 0 \\ 0 & 1 & -2 & | & 0 & 2 & 1 \\ 0 & 0 & 1 & | & 1 & 2 & 2 \end{bmatrix}$$

Keep the third row, add it to the first row. Then multiply the third row by 2, and add the result to the second row.

$$\begin{bmatrix} 1 & 0 & 0 & | & 1 & 3 & 2 \\ 0 & 1 & 0 & | & 2 & 6 & 5 \\ 0 & 0 & 1 & | & 1 & 2 & 2 \end{bmatrix}$$

Thus, $A^{-1} = \begin{bmatrix} 1 & 3 & 2 \\ 2 & 6 & 5 \\ 1 & 2 & 2 \end{bmatrix}$.

EXERCISES A

7. Find the inverse of the matrix.
$$\begin{bmatrix} 1 & 2 & 0 \\ 0 & 1 & 1 \\ -1 & 1 & 1 \end{bmatrix}$$

Form the augmented matrix $[A|I]$.
$$\begin{bmatrix} 1 & 2 & 0 & | & 1 & 0 & 0 \\ 0 & 1 & 1 & | & 0 & 1 & 0 \\ -1 & 1 & 1 & | & 0 & 0 & 1 \end{bmatrix}$$

Keep the first and second rows. Add the first row to the third row.
$$\begin{bmatrix} 1 & 2 & 0 & | & 1 & 0 & 0 \\ 0 & 1 & 1 & | & 0 & 1 & 0 \\ 0 & 3 & 1 & | & 1 & 0 & 1 \end{bmatrix}$$

Keep the second row, multiply it by -2 and add to the first row. Then multiply it by -3 and add to the third row.
$$\begin{bmatrix} 1 & 0 & -2 & | & 1 & -2 & 0 \\ 0 & 1 & 1 & | & 0 & 1 & 0 \\ 0 & 0 & -2 & | & 1 & -3 & 1 \end{bmatrix}$$

Multiply the third row by $-\frac{1}{2}$.
$$\begin{bmatrix} 1 & 0 & -2 & | & 1 & -2 & 0 \\ 0 & 1 & 1 & | & 0 & 1 & 0 \\ 0 & 0 & 1 & | & -\frac{1}{2} & \frac{3}{2} & -\frac{1}{2} \end{bmatrix}$$

Keep the third row, multiply it by 2 and add to the first row. Then multiply the third row by -1 and add to the second row.
$$\begin{bmatrix} 1 & 0 & 0 & | & 0 & 1 & -1 \\ 0 & 1 & 0 & | & \frac{1}{2} & -\frac{1}{2} & \frac{1}{2} \\ 0 & 0 & 1 & | & -\frac{1}{2} & \frac{3}{2} & -\frac{1}{2} \end{bmatrix}$$

Thus, $A^{-1} = \begin{bmatrix} 0 & 1 & -1 \\ \frac{1}{2} & -\frac{1}{2} & \frac{1}{2} \\ -\frac{1}{2} & \frac{3}{2} & -\frac{1}{2} \end{bmatrix}$.

8. Find the inverse of the matrix.
$$\begin{bmatrix} 1 & 3 & -1 \\ 3 & 1 & 0 \\ -2 & 0 & 1 \end{bmatrix}$$

Form the augmented matrix $[A|I]$.
$$\begin{bmatrix} 1 & 3 & -1 & | & 1 & 0 & 0 \\ 3 & 1 & 0 & | & 0 & 1 & 0 \\ -2 & 0 & 1 & | & 0 & 0 & 1 \end{bmatrix}$$

Keep the first row, multiply it by -3 and add to the second row. Then multiply the first row by 2 and add to the third row.
$$\begin{bmatrix} 1 & 3 & -1 & | & 1 & 0 & 0 \\ 0 & -8 & 3 & | & -3 & 1 & 0 \\ 0 & 6 & -1 & | & 2 & 0 & 1 \end{bmatrix}$$

Multiply the second row by $-\frac{1}{8}$.
$$\begin{bmatrix} 1 & 3 & -1 & | & 1 & 0 & 0 \\ 0 & 1 & -\frac{3}{8} & | & \frac{3}{8} & -\frac{1}{8} & 0 \\ 0 & 6 & -1 & | & 2 & 0 & 1 \end{bmatrix}$$

Keep the second row, multiply it by -3 and add to the first row. Then multiply the second row by -6 and add to the third row.
$$\begin{bmatrix} 1 & 0 & \frac{1}{8} & | & -\frac{1}{8} & \frac{3}{8} & 0 \\ 0 & 1 & -\frac{3}{8} & | & \frac{3}{8} & -\frac{1}{8} & 0 \\ 0 & 0 & \frac{5}{4} & | & -\frac{1}{4} & \frac{3}{4} & 1 \end{bmatrix}$$

Multiply the third row by $\frac{4}{5}$.
$$\begin{bmatrix} 1 & 0 & \frac{1}{8} & | & -\frac{1}{8} & \frac{3}{8} & 0 \\ 0 & 1 & -\frac{3}{8} & | & \frac{3}{8} & -\frac{1}{8} & 0 \\ 0 & 0 & 1 & | & -\frac{1}{5} & \frac{3}{5} & \frac{4}{5} \end{bmatrix}$$

Keep the third row, multiply it by $\frac{3}{8}$ and add to the second row. Then multiply the third row by $-\frac{1}{8}$ and add to the first row.

$$\begin{bmatrix} 1 & 0 & 0 & | & -\frac{1}{10} & \frac{3}{10} & -\frac{1}{10} \\ 0 & 1 & 0 & | & \frac{3}{10} & \frac{1}{10} & \frac{3}{10} \\ 0 & 0 & 1 & | & -\frac{1}{5} & \frac{3}{5} & \frac{4}{5} \end{bmatrix}$$

Thus, $A^{-1} = \begin{bmatrix} -\frac{1}{10} & \frac{3}{10} & -\frac{1}{10} \\ \frac{3}{10} & \frac{1}{10} & \frac{3}{10} \\ -\frac{1}{5} & \frac{3}{5} & \frac{4}{5} \end{bmatrix}$.

9. Find the values of x and y.

$$\begin{bmatrix} x \\ y \end{bmatrix} = \begin{bmatrix} 2 & -1 \\ 3 & 5 \end{bmatrix} \begin{bmatrix} -2 \\ 4 \end{bmatrix}$$
$$= \begin{bmatrix} (2)(-2)+(-1)(4) \\ (3)(-2)+(5)(4) \end{bmatrix}$$
$$= \begin{bmatrix} -8 \\ 14 \end{bmatrix}$$

Thus, $x = -8$ and $y = 14$.

10. Find the values of x, y, and z.

$$\begin{bmatrix} x \\ y \\ z \end{bmatrix} = \begin{bmatrix} 2 & -1 & 3 \\ 4 & 0 & 1 \\ 2 & -2 & 0 \end{bmatrix} \begin{bmatrix} 1 \\ -1 \\ 3 \end{bmatrix}$$
$$= \begin{bmatrix} (2)(1)+(-1)(-1)+(3)(3) \\ (4)(1)+(0)(-1)+(1)(3) \\ (2)(1)+(-2)(-1)+(0)(3) \end{bmatrix}$$
$$= \begin{bmatrix} 12 \\ 7 \\ 4 \end{bmatrix}$$

Thus, $x = 12$, $y = 7$, and $z = 4$.

11. Use the inverse method to solve the following system of equations.

$$3x - 4y = 2$$
$$4x - 5y = 3$$

This system is equivalent to the matrix equation $AX = B$, where

$$A = \begin{bmatrix} 3 & -4 \\ 4 & -5 \end{bmatrix} \quad X = \begin{bmatrix} x \\ y \end{bmatrix} \quad B = \begin{bmatrix} 2 \\ 3 \end{bmatrix}.$$

In Exercise 3 we found that $A^{-1} = \begin{bmatrix} -5 & 4 \\ -4 & 3 \end{bmatrix}$. Then

$$\begin{bmatrix} x \\ y \end{bmatrix} = X = A^{-1}B = \begin{bmatrix} -5 & 4 \\ -4 & 3 \end{bmatrix} \begin{bmatrix} 2 \\ 3 \end{bmatrix} = \begin{bmatrix} 2 \\ 1 \end{bmatrix}.$$

Thus, $x = 2$ and $y = 1$, and the solution to the system is $(2,1)$.

12. Use the inverse method to solve the following system of equations.

$$2x - y = 11$$
$$4x - 3y = 25$$

This system is equivalent to the matrix equation $AX = B$, where

$$A = \begin{bmatrix} 2 & -1 \\ 4 & -3 \end{bmatrix} \quad X = \begin{bmatrix} x \\ y \end{bmatrix} \quad B = \begin{bmatrix} 11 \\ 25 \end{bmatrix}.$$

In Exercise 4 we found that $A^{-1} = \begin{bmatrix} \frac{3}{2} & -\frac{1}{2} \\ 2 & -1 \end{bmatrix}$. Then

$$\begin{bmatrix} x \\ y \end{bmatrix} = X = A^{-1}B = \begin{bmatrix} \frac{3}{2} & -\frac{1}{2} \\ 2 & -1 \end{bmatrix} \begin{bmatrix} 11 \\ 25 \end{bmatrix} = \begin{bmatrix} 4 \\ -3 \end{bmatrix}.$$

Thus, $x = 4$ and $y = -3$, and the solution to the system is $(4,-3)$.

13. Use the inverse method to solve the following system of equations.

$$2x - 2y + 3z = -5$$
$$x - z = 5$$
$$-2x + y = -4$$

This system is equivalent to the matrix equation $AX = B$, where

$$A = \begin{bmatrix} 2 & -2 & 3 \\ 1 & 0 & -1 \\ -2 & 1 & 0 \end{bmatrix} \quad X = \begin{bmatrix} x \\ y \\ z \end{bmatrix} \quad B = \begin{bmatrix} -5 \\ 5 \\ -4 \end{bmatrix}.$$

EXERCISES A

In Exercise 6 we found that $A^{-1} = \begin{bmatrix} 1 & 3 & 2 \\ 2 & 6 & 5 \\ 1 & 2 & 2 \end{bmatrix}$. Then

$$\begin{bmatrix} x \\ y \\ z \end{bmatrix} = X = A^{-1}B = \begin{bmatrix} 1 & 3 & 2 \\ 2 & 6 & 5 \\ 1 & 2 & 2 \end{bmatrix} \begin{bmatrix} -5 \\ 5 \\ -4 \end{bmatrix} = \begin{bmatrix} 2 \\ 0 \\ -3 \end{bmatrix}.$$

Thus, $x = 2$, $y = 0$, $z = -3$, and the solution to the system is $(2, 0, -3)$.

14. Use the inverse method to solve the following system of equations.

$$\begin{aligned} x + 2y &= 4 \\ y + z &= 6 \\ -x + y + z &= 2 \end{aligned}$$

This system is equivalent to the matrix equation $AX = B$, where

$$A = \begin{bmatrix} 1 & 2 & 0 \\ 0 & 1 & 1 \\ -1 & 1 & 1 \end{bmatrix} \quad X = \begin{bmatrix} x \\ y \\ z \end{bmatrix} \quad B = \begin{bmatrix} 4 \\ 6 \\ 2 \end{bmatrix}.$$

In Exercise 7 we found that $A^{-1} = \begin{bmatrix} 0 & 1 & -1 \\ \frac{1}{2} & -\frac{1}{2} & \frac{1}{2} \\ -\frac{1}{2} & \frac{3}{2} & -\frac{1}{2} \end{bmatrix}$.

Then

$$\begin{bmatrix} x \\ y \\ z \end{bmatrix} = X = A^{-1}B = \begin{bmatrix} 0 & 1 & -1 \\ \frac{1}{2} & -\frac{1}{2} & \frac{1}{2} \\ -\frac{1}{2} & \frac{3}{2} & -\frac{1}{2} \end{bmatrix} \begin{bmatrix} 4 \\ 6 \\ 2 \end{bmatrix} = \begin{bmatrix} 4 \\ 0 \\ 6 \end{bmatrix}.$$

Thus, $x = 4$, $y = 0$, $z = 6$, and the solution to the system is $(4, 0, 6)$.

15. Use the inverse method to solve the following system of equations.

$$\begin{aligned} x + 3y - z &= 13 \\ 3x + y &= 12 \\ -2x + z &= -7 \end{aligned}$$

This system is equivalent to the matrix equation $AX = B$, where

$$A = \begin{bmatrix} 1 & 3 & -1 \\ 3 & 1 & 0 \\ -2 & 0 & 1 \end{bmatrix} \quad X = \begin{bmatrix} x \\ y \\ z \end{bmatrix} \quad B = \begin{bmatrix} 13 \\ 12 \\ -7 \end{bmatrix}.$$

In Exercise 8 we found that $A^{-1} = \begin{bmatrix} -\frac{1}{10} & \frac{3}{10} & -\frac{1}{10} \\ \frac{3}{10} & \frac{1}{10} & \frac{3}{10} \\ -\frac{1}{5} & \frac{3}{5} & \frac{4}{5} \end{bmatrix}$.

Then

$$\begin{bmatrix} x \\ y \\ z \end{bmatrix} = X = A^{-1}B = \begin{bmatrix} -\frac{1}{10} & \frac{3}{10} & -\frac{1}{10} \\ \frac{3}{10} & \frac{1}{10} & \frac{3}{10} \\ -\frac{1}{5} & \frac{3}{5} & \frac{4}{5} \end{bmatrix} \begin{bmatrix} 13 \\ 12 \\ -7 \end{bmatrix} = \begin{bmatrix} 3 \\ 3 \\ -1 \end{bmatrix}.$$

Thus, $x = 3$, $y = 3$, $z = -1$, and the solution to the system is $(3, 3, -1)$.

16. Given the system

$$\begin{aligned} 2x + y &= a \\ 5x + 3y &= b. \end{aligned}$$

Since we are to solve the system for various values of a and b, we can find the inverse of the coefficient matrix and use the inverse method. Form the augmented matrix $[A|I]$.

$$\begin{bmatrix} 2 & 1 & | & 1 & 0 \\ 5 & 3 & | & 0 & 1 \end{bmatrix}$$

Multiply the first row by -2, and add the result to the second row.

$$\begin{bmatrix} 2 & 1 & | & 1 & 0 \\ 1 & 1 & | & -2 & 1 \end{bmatrix}$$

Interchange the two rows.

$$\begin{bmatrix} 1 & 1 & | & -2 & 1 \\ 2 & 1 & | & 1 & 0 \end{bmatrix}$$

Keep the first row, multiply it by -2, and add the result to the second row.

$$\begin{bmatrix} 1 & 1 & | & -2 & 1 \\ 0 & -1 & | & 5 & -2 \end{bmatrix}$$

Multiply the second row by -1.

$$\begin{bmatrix} 1 & 1 & | & -2 & 1 \\ 0 & 1 & | & -5 & 2 \end{bmatrix}$$

Keep the second row, multiply it by -1, and add the result to the first row.

$$\begin{bmatrix} 1 & 0 & | & 3 & -1 \\ 0 & 1 & | & -5 & 2 \end{bmatrix}$$

Thus, $A^{-1} = \begin{bmatrix} 3 & -1 \\ -5 & 2 \end{bmatrix}$. With this work accomplished, we can solve the systems for the given values of a and b very quickly using the inverse method.

(a) Since $a = 1$ and $b = 1$, $B = \begin{bmatrix} 1 \\ 1 \end{bmatrix}$, and

$$\begin{bmatrix} x \\ y \end{bmatrix} = X = A^{-1}B = \begin{bmatrix} 3 & -1 \\ -5 & 2 \end{bmatrix} \begin{bmatrix} 1 \\ 1 \end{bmatrix} = \begin{bmatrix} 2 \\ -3 \end{bmatrix}.$$

Thus, $x = 2$, $y = -3$, and the solution to the system is $(2,-3)$.

(b) Since $a = 6$ and $b = 18$, $B = \begin{bmatrix} 6 \\ 18 \end{bmatrix}$, and

$$\begin{bmatrix} x \\ y \end{bmatrix} = X = A^{-1}B = \begin{bmatrix} 3 & -1 \\ -5 & 2 \end{bmatrix} \begin{bmatrix} 6 \\ 18 \end{bmatrix} = \begin{bmatrix} 0 \\ 6 \end{bmatrix}.$$

Thus, $x = 0$, $y = 6$, and the solution to the system is $(0,6)$.

(c) Since $a = -1$ and $b = 0$, $B = \begin{bmatrix} -1 \\ 0 \end{bmatrix}$, and

$$\begin{bmatrix} x \\ y \end{bmatrix} = X = A^{-1}B = \begin{bmatrix} 3 & -1 \\ -5 & 2 \end{bmatrix} \begin{bmatrix} -1 \\ 0 \end{bmatrix} = \begin{bmatrix} -3 \\ 5 \end{bmatrix}.$$

Thus, $x = -3$, $y = 5$, and the solution to the system is $(-3,5)$.

17. Given the system

$$\begin{aligned} x \quad\quad - z &= a \\ x + y \quad\quad &= b \\ y + 2z &= c. \end{aligned}$$

Since we are to solve the system for various values of a, b, and c, we can find the inverse of the coefficient matrix and use the inverse method. Form the augmented matrix $[A|I]$.

$$\begin{bmatrix} 1 & 0 & -1 & | & 1 & 0 & 0 \\ 1 & 1 & 0 & | & 0 & 1 & 0 \\ 0 & 1 & 2 & | & 0 & 0 & 1 \end{bmatrix}$$

Multiply the first row by -1, and add the result to the second row.

$$\begin{bmatrix} 1 & 0 & -1 & | & 1 & 0 & 0 \\ 0 & 1 & 1 & | & -1 & 1 & 0 \\ 0 & 1 & 2 & | & 0 & 0 & 1 \end{bmatrix}$$

Multiply the second row by -1 and add to the third row.

$$\begin{bmatrix} 1 & 0 & -1 & | & 1 & 0 & 0 \\ 0 & 1 & 1 & | & -1 & 1 & 0 \\ 0 & 0 & 1 & | & 1 & -1 & 1 \end{bmatrix}$$

Keep the third row, and add it to the first row. Then multiply the third row by -1, and add the result to the second row.

$$\begin{bmatrix} 1 & 0 & 0 & | & 2 & -1 & 1 \\ 0 & 1 & 0 & | & -2 & 2 & -1 \\ 0 & 0 & 1 & | & 1 & -1 & 1 \end{bmatrix}$$

Thus, $A^{-1} = \begin{bmatrix} 2 & -1 & 1 \\ -2 & 2 & -1 \\ 1 & -1 & 1 \end{bmatrix}$. With this work accomplished, we can solve the systems for the given values of a, b, and c very quickly using the inverse method.

(a) Since $a = 5$, $b = 2$, and $c = -6$, $B = \begin{bmatrix} 5 \\ 2 \\ -6 \end{bmatrix}$, and

$$\begin{bmatrix} x \\ y \\ z \end{bmatrix} = X = A^{-1}B = \begin{bmatrix} 2 & -1 & 1 \\ -2 & 2 & -1 \\ 1 & -1 & 1 \end{bmatrix} \begin{bmatrix} 5 \\ 2 \\ -6 \end{bmatrix} = \begin{bmatrix} 2 \\ 0 \\ -3 \end{bmatrix}.$$

Thus, $x = 2$, $y = 0$, $z = -3$, and the solution to the system is $(2,0,-3)$.

(b) Since $a = 0$, $b = 5$, and $c = 6$, $B = \begin{bmatrix} 0 \\ 5 \\ 6 \end{bmatrix}$,

and

EXERCISES A

$$\begin{bmatrix} x \\ y \\ z \end{bmatrix} = X = A^{-1}B = \begin{bmatrix} 2 & -1 & 1 \\ -2 & 2 & -1 \\ 1 & -1 & 1 \end{bmatrix} \begin{bmatrix} 0 \\ 5 \\ 6 \end{bmatrix} = \begin{bmatrix} 1 \\ 4 \\ 1 \end{bmatrix}.$$

Thus, $x = 1$, $y = 4$, $z = 1$, and the solution to the system is $(1,4,1)$.

(c) Since $a = -3$, $b = -4$, and $c = 0$, $B = \begin{bmatrix} -3 \\ -4 \\ 0 \end{bmatrix}$,

and

$$\begin{bmatrix} x \\ y \\ z \end{bmatrix} = X = A^{-1}B = \begin{bmatrix} 2 & -1 & 1 \\ -2 & 2 & -1 \\ 1 & -1 & 1 \end{bmatrix} \begin{bmatrix} -3 \\ -4 \\ 0 \end{bmatrix} = \begin{bmatrix} -2 \\ -2 \\ 1 \end{bmatrix}.$$

Thus, $x = -2$, $y = -2$, $z = 1$, and the solution to the system is $(-2,-2,1)$.

18. Given the matrices

$$A = \begin{bmatrix} 1 & 3 \\ 2 & 7 \end{bmatrix} \text{ and } B = \begin{bmatrix} -3 & 2 \\ 2 & -1 \end{bmatrix}.$$

(a) $AB = \begin{bmatrix} 1 & 3 \\ 2 & 7 \end{bmatrix} \begin{bmatrix} -3 & 2 \\ 2 & -1 \end{bmatrix} = \begin{bmatrix} 3 & -1 \\ 8 & -3 \end{bmatrix}$

(b) To find the inverse of AB, $(AB)^{-1}$, form the augmented matrix.

$$\begin{bmatrix} 3 & -1 & | & 1 & 0 \\ 8 & -3 & | & 0 & 1 \end{bmatrix}$$

Multiply the first row by $\frac{1}{3}$.

$$\begin{bmatrix} 1 & -\frac{1}{3} & | & \frac{1}{3} & 0 \\ 8 & -3 & | & 0 & 1 \end{bmatrix}$$

Multiply the first row by -8 and add to the second row.

$$\begin{bmatrix} 1 & -\frac{1}{3} & | & \frac{1}{3} & 0 \\ 0 & -\frac{1}{3} & | & -\frac{8}{3} & 1 \end{bmatrix}$$

Multiply the third row by -1 and add to the first row.

SECTION 7.4

$$\begin{bmatrix} 1 & 0 & | & 3 & -1 \\ 0 & -\frac{1}{3} & | & -\frac{8}{3} & 1 \end{bmatrix}$$

Finally, multiply the second row by -3.

$$\begin{bmatrix} 1 & 0 & | & 3 & -1 \\ 0 & 1 & | & 8 & -3 \end{bmatrix}$$

Thus, $(AB)^{-1} = \begin{bmatrix} 3 & -1 \\ 8 & -3 \end{bmatrix}$.

(c) Find A^{-1} by starting with the augmented matrix.

$$\begin{bmatrix} 1 & 3 & | & 1 & 0 \\ 2 & 7 & | & 0 & 1 \end{bmatrix}$$

Multiply the first row by -2 and add to the second row.

$$\begin{bmatrix} 1 & 3 & | & 1 & 0 \\ 0 & 1 & | & -2 & 1 \end{bmatrix}$$

Keep the third row, multiply it by -3, and add the result to the first row.

$$\begin{bmatrix} 1 & 0 & | & 7 & -3 \\ 0 & 1 & | & -2 & 1 \end{bmatrix}$$

Thus, $A^{-1} = \begin{bmatrix} 7 & -3 \\ -2 & 1 \end{bmatrix}$.

(d) Find B^{-1} by starting with the augmented matrix.

$$\begin{bmatrix} -3 & 2 & | & 1 & 0 \\ 2 & -1 & | & 0 & 1 \end{bmatrix}$$

Keep the second row, add it to the first row, and multiply the result by -1.

$$\begin{bmatrix} 1 & -1 & | & -1 & -1 \\ 2 & -1 & | & 0 & 1 \end{bmatrix}$$

Multiply the first row by -2 and add to the second row.

$$\begin{bmatrix} 1 & -1 & | & -1 & -1 \\ 0 & 1 & | & 2 & 3 \end{bmatrix}$$

Keep the second row and add it to the first row.

$$\begin{bmatrix} 1 & 0 & | & 1 & 2 \\ 0 & 1 & | & 2 & 3 \end{bmatrix}$$

Thus, $B^{-1} = \begin{bmatrix} 1 & 2 \\ 2 & 3 \end{bmatrix}$.

(e) $A^{-1}B^{-1} = \begin{bmatrix} 7 & -3 \\ -2 & 1 \end{bmatrix}\begin{bmatrix} 1 & 2 \\ 2 & 3 \end{bmatrix} = \begin{bmatrix} 1 & 5 \\ 0 & -1 \end{bmatrix}$

(f) $B^{-1}A^{-1} = \begin{bmatrix} 1 & 2 \\ 2 & 3 \end{bmatrix}\begin{bmatrix} 7 & -3 \\ -2 & 1 \end{bmatrix} = \begin{bmatrix} 3 & -1 \\ 8 & -3 \end{bmatrix}$

(g) It would appear that $(AB)^{-1} = B^{-1}A^{-1}$ and $(AB)^{-1} \neq A^{-1}B^{-1}$.

19. Let x = the number of ounces of Supergro,
    y = the number of ounces of Healthy Mix.

Since Supergro conatins 30% protein and 6% fat, and Healthy Mix contains 20% protein and 2% fat, the amount of protein in the mix is

$$0.30x + 0.20y,$$

and the amount of fat in the mix is

$$0.06x + 0.02y.$$

Since each of the three mixes contains different amounts of protein, say a ounces, and fat, say b ounces, we really have the following system

$$0.30x + 0.20y = a$$
$$0.06x + 0.02y = b$$

with three sets of values for a and b. Suppose we find the inverse of the coefficient matrix, and use the inverse method to solve the three systems. First write the augmented matrix.

$$\begin{bmatrix} 0.30 & 0.20 & | & 1 & 0 \\ 0.06 & 0.02 & | & 0 & 1 \end{bmatrix}$$

To clear the matrix of decimals, multiply the first row by 10 and the second row by 100.

$$\begin{bmatrix} 3 & 2 & | & 10 & 0 \\ 6 & 2 & | & 0 & 100 \end{bmatrix}$$

Multiply the first row by $\frac{1}{3}$.

$$\begin{bmatrix} 1 & \frac{2}{3} & | & \frac{10}{3} & 0 \\ 6 & 2 & | & 0 & 100 \end{bmatrix}$$

Multiply the first row by -6 and add to the second row.

$$\begin{bmatrix} 1 & \frac{2}{3} & | & \frac{10}{3} & 0 \\ 0 & -2 & | & -20 & 100 \end{bmatrix}$$

Multiply the second row by $-\frac{1}{2}$.

$$\begin{bmatrix} 1 & \frac{2}{3} & | & \frac{10}{3} & 0 \\ 0 & 1 & | & 10 & -50 \end{bmatrix}$$

Finally, multiply the second row by $-\frac{2}{3}$ and add to the first row.

$$\begin{bmatrix} 1 & 0 & | & -\frac{10}{3} & \frac{100}{3} \\ 0 & 1 & | & 10 & -50 \end{bmatrix}$$

Thus the inverse of the coefficient matrix is

$$A^{-1} = \begin{bmatrix} -\frac{10}{3} & \frac{100}{3} \\ 10 & -50 \end{bmatrix}.$$

For Mix 1, $a = 13.5$ and $b = 2.4$. Thus,

$$\begin{bmatrix} x \\ y \end{bmatrix} = X = A^{-1}B = \begin{bmatrix} -\frac{10}{3} & \frac{100}{3} \\ 10 & -50 \end{bmatrix}\begin{bmatrix} 13.5 \\ 2.4 \end{bmatrix} = \begin{bmatrix} 35 \\ 15 \end{bmatrix}$$

Thus, $x = 35$ and $y = 15$, which means that 35 oz of Supergrow and 15 oz of Healthy Mix should be used for Mix 1.

For Mix 2, $a = 14$ and $b = 2.6$. Thus,

$$\begin{bmatrix} x \\ y \end{bmatrix} = X = A^{-1}B = \begin{bmatrix} -\frac{10}{3} & \frac{100}{3} \\ 10 & -50 \end{bmatrix}\begin{bmatrix} 14 \\ 2.6 \end{bmatrix} = \begin{bmatrix} 40 \\ 10 \end{bmatrix}$$

Thus, $x = 40$ and $y = 10$, which means that 40 oz of Supergrow and 10 oz of Healthy Mix should be used for Mix 2.

For Mix 3, $a = 16$ and $b = 3.1$. Thus,

$$\begin{bmatrix} x \\ y \end{bmatrix} = X = A^{-1}B = \begin{bmatrix} -\frac{10}{3} & \frac{100}{3} \\ 10 & -50 \end{bmatrix}\begin{bmatrix} 16 \\ 3.1 \end{bmatrix} = \begin{bmatrix} 50 \\ 5 \end{bmatrix}$$

Thus, $x = 50$ and $y = 5$, which means that 50 oz of Supergrow and 5 oz of Healthy Mix should be used for Mix 3.

EXERCISES A

20. Solve by the Gaussian method.

$$4x - 5y = -38$$
$$2x + 3y = 14$$

First form the augmented matrix.

$$\begin{bmatrix} 4 & -5 & -38 \\ 2 & 3 & 14 \end{bmatrix}$$

Multiply the first equation by $\frac{1}{4}$.

$$\begin{bmatrix} 1 & -\frac{5}{4} & -\frac{38}{4} \\ 2 & 3 & 14 \end{bmatrix}$$

Multiply the first equation by -2, and add the result to the second equation.

$$\begin{bmatrix} 1 & -\frac{5}{4} & -\frac{38}{4} \\ 0 & \frac{11}{2} & 33 \end{bmatrix}$$

Multiply the second equation by $\frac{2}{11}$.

$$\begin{bmatrix} 1 & -\frac{5}{4} & -\frac{38}{4} \\ 0 & 1 & 6 \end{bmatrix}$$

Finally, multiply the second equation by $\frac{5}{4}$, and add the result to the first equation.

$$\begin{bmatrix} 1 & 0 & -2 \\ 0 & 1 & 6 \end{bmatrix}$$

This corresponds to $x = -2$ and $y = 6$. Thus, the solution to the system is $(-2,6)$.

21. Solve using the Gaussian method.

$$x - 2y + z = -4$$
$$2x \quad\quad - z = 7$$
$$-3x + 4y + z = -8$$

First write the augmented matrix.

$$\begin{bmatrix} 1 & -2 & 1 & -4 \\ 2 & 0 & -1 & 7 \\ -3 & 4 & 1 & -8 \end{bmatrix}$$

Multiply the first row by -2 and add to the second. Then multiply the first row by 3 and add to the third.

$$\begin{bmatrix} 1 & -2 & 1 & -4 \\ 0 & 4 & -3 & 15 \\ 0 & -2 & 4 & -20 \end{bmatrix}$$

Multiply the third row by $-\frac{1}{2}$, and interchange the result with the second row.

$$\begin{bmatrix} 1 & -2 & 1 & -4 \\ 0 & 1 & -2 & 10 \\ 0 & 4 & -3 & 15 \end{bmatrix}$$

Multiply the second row by 2 and add to the first, Then multiply the second row by -4 and add to the third.

$$\begin{bmatrix} 1 & 0 & -3 & 16 \\ 0 & 1 & -2 & 10 \\ 0 & 0 & 5 & -25 \end{bmatrix}$$

Multiply the third row by $-\frac{1}{5}$.

$$\begin{bmatrix} 1 & 0 & -3 & 16 \\ 0 & 1 & -2 & 10 \\ 0 & 0 & 1 & -5 \end{bmatrix}$$

Finally, multiply the third row by 3 and add to the first. Then multiply the third row by 2 and add to the second.

$$\begin{bmatrix} 1 & 0 & 0 & 1 \\ 0 & 1 & 0 & 0 \\ 0 & 0 & 1 & -5 \end{bmatrix}$$

Then this corresponds to $x = 1$, $y = 0$, and $z = -5$. Thus, the solution to the system is $(1,0,-5)$.

CHAPTER 7 MATRICES AND DETERMINANTS

SECTION 7.5 Determinants and Cramer's Rule

1. Evaluate the determinants.

   (a) $\begin{vmatrix} -1 & 3 \\ 5 & 2 \end{vmatrix} = (-1)(2) - (3)(5) = -2 - 15 = -17$

   (b) $\begin{vmatrix} 6 & 6 \\ 6 & 6 \end{vmatrix} = (6)(6) - (6)(6) = 36 - 36 = 0$

   (c) $\begin{vmatrix} c_1 & b_1 \\ c_2 & b_2 \end{vmatrix} = (c_1)(b_2) - (c_2)(b_1)$

2. Solve the system using Cramer's rule.

   $$2x - y = -8$$
   $$3x + 7y = 22$$

   $|A| = \begin{vmatrix} 2 & -1 \\ 3 & 7 \end{vmatrix} = (2)(7) - (-1)(3) = 17$

   $|A_x| = \begin{vmatrix} -8 & -1 \\ 22 & 7 \end{vmatrix} = (-8)(7) - (-1)(22) = -34$

   $|A_y| = \begin{vmatrix} 2 & -8 \\ 3 & 22 \end{vmatrix} = (2)(22) - (-8)(3) = 68$

   Then

   $$x = \frac{|A_x|}{|A|} = \frac{-34}{17} = -2$$

   and

   $$y = \frac{|A_y|}{|A|} = \frac{68}{17} = 4.$$

   Thus, the solution to the system is $(-2, 4)$, which checks in both equations.

3. Given the following matrix:

   $$A = \begin{bmatrix} 1 & 2 & 3 \\ 4 & 5 & 6 \\ 7 & 8 & 9 \end{bmatrix}$$

   (a) The minor, M_{23}, of the element $a_{23} = 6$ is the determinant formed by eliminating the 2nd row and 3rd column of A.

   $M_{23} = \begin{vmatrix} 1 & 2 \\ 7 & 8 \end{vmatrix} = (1)(8) - (2)(7) = -6$

   (b) The minor, M_{33}, of the element $a_{33} = 9$ is the determinant formed by eliminating the 3rd row and 3rd column of A.

   $M_{33} = \begin{vmatrix} 1 & 2 \\ 4 & 5 \end{vmatrix} = (1)(5) - (2)(4) = -3$

4. Given the following matrix:

   $$A = \begin{bmatrix} 1 & 2 & 3 \\ 4 & 5 & 6 \\ 7 & 8 & 9 \end{bmatrix}$$

   (a) A_{23} is the cofactor of the element $a_{23} = 6$. First find the minor, M_{23}, of the element $a_{23} = 6$ which is the determinant formed by eliminating the 2nd row and 3rd column of A. Then multiply the minor by $(-1)^{2+3} = (-1)^5 = -1$. Alternatively, we could have obtained the factor of -1 using the "checkerboard of signs" shown over the elements of A in the text. The element $a_{23} = 6$ has a minus sign over it so we use -1 as a factor to form the cofactor.

   $A_{23} = (-1)^{2+3} M_{23} = (-1)^{2+3} \begin{vmatrix} 1 & 2 \\ 7 & 8 \end{vmatrix}$
   $= (-1)^5 [(1)(8) - (2)(7)]$
   $= (-1)(-6) = 6$

   (b) A_{33} is the cofactor of the element $a_{33} = 9$. First find the minor, M_{33}, of the element $a_{33} = 9$ which is the determinant formed by eliminating the 3rd row and 3rd column of A. Then multiply the minor by $(-1)^{3+3} = (-1)^6 = +1$. Alternatively, we could have obtained the factor of $+1$ using the "checkerboard of signs" shown over the elements of A in the text. The element $a_{33} = 9$ has a plus sign over it so we use $+1$ as a factor to form the cofactor.

PRACTICE EXERCISES

$$A_{33} = (-1)^{3+3}M_{33} = (-1)^{3+3}\begin{vmatrix} 1 & 2 \\ 4 & 5 \end{vmatrix}$$
$$= (-1)^6[(1)(5) - (2)(8)]$$
$$= (+1)(-3) = -3$$

5. Given the following matrix:

$$A = \begin{bmatrix} 2 & 1 & 0 \\ -1 & 4 & 0 \\ 3 & 2 & -1 \end{bmatrix}$$

(a) Find $|A|$ using elements in the third row.

$$|A| = a_{31}A_{31} + a_{32}A_{32} + a_{33}A_{33}$$
$$= (3)(-1)^{3+1}\begin{vmatrix} 1 & 0 \\ 4 & 0 \end{vmatrix} + (2)(-1)^{3+2}\begin{vmatrix} 2 & 0 \\ -1 & 0 \end{vmatrix}$$
$$+ (-1)(-1)^{3+3}\begin{vmatrix} 2 & 1 \\ -1 & 4 \end{vmatrix}$$
$$= (3)(+1)[0 - 0] + (2)(-1)[0 - 0]$$
$$+ (-1)(+1)[8 - (-1)]$$
$$= (3)(0) + (-2)(0) + (-1)(9)$$
$$= 0 + 0 + (-9)$$
$$= -9$$

(b) Find $|A|$ using elements in the third column.

$$|A| = a_{13}A_{13} + a_{23}A_{23} + a_{33}A_{33}$$
$$= (0)(-1)^{1+3}\begin{vmatrix} -1 & 4 \\ 3 & 2 \end{vmatrix} + (0)(-1)^{2+3}\begin{vmatrix} 2 & 1 \\ 3 & 2 \end{vmatrix}$$
$$+ (-1)(-1)^{3+3}\begin{vmatrix} 2 & 1 \\ -1 & 4 \end{vmatrix}$$
$$= (0)(+1)[-2 - 12] + (0)(-1)[4 - 3]$$
$$+ (-1)(+1)[8 - (-1)]$$
$$= (0)(-14) + (0)(1) + (-1)(9)$$
$$= 0 + 0 + (-9)$$
$$= -9$$

(c) It is probably easier to expand down the third column taking advantage of the 0's. In fact, the first two terms of the expansion will automatically be 0 so we would not even need to find the value of the first two minors in this case.

6. Find

$$|A| = \begin{vmatrix} -1 & 2 & 0 & 3 \\ 2 & 4 & 3 & -1 \\ 0 & 1 & 0 & -2 \\ -3 & -1 & 1 & 2 \end{vmatrix}$$

by expanding down the third column.

$$|A| = 0 + (3)(-1)\begin{vmatrix} -1 & 2 & 3 \\ 0 & 1 & -2 \\ -3 & -1 & 2 \end{vmatrix} +$$
$$0 + (1)(-1)\begin{vmatrix} -1 & 2 & 3 \\ 2 & 4 & -1 \\ 0 & 1 & -2 \end{vmatrix}$$
$$= (-3)\left[(-1)(+1)\begin{vmatrix} 1 & -2 \\ -1 & 2 \end{vmatrix} + (-3)(+1)\begin{vmatrix} 2 & 3 \\ 1 & -2 \end{vmatrix}\right]$$
$$+ (-1)\left[(-1)(+1)\begin{vmatrix} 4 & -1 \\ 1 & -2 \end{vmatrix} + (2)(-1)\begin{vmatrix} 2 & 3 \\ 1 & -2 \end{vmatrix}\right]$$
$$= (-3)[(-1)[2-2] + (-3)[-4-3]]$$
$$+ (-1)[(-1)[-8+1] + (-2)[-4-3]]$$
$$= (-3)[(-1)[0] + (-3)[-7]]$$
$$+ (-1)[(-1)[-7] + (-2)[-7]]$$
$$= (-3)[0 + 21] + (-1)[7 + 14]$$
$$= (-3)[21] + (-1)[21]$$
$$= -63 - 21 = -84$$

7. Solve the system using Cramer's rule.

$$\begin{aligned} y + 4z &= 6 \\ 3x \quad + z &= 7 \\ 5y - z &= 9 \end{aligned}$$

$$|A| = \begin{vmatrix} 0 & 1 & 4 \\ 3 & 0 & 1 \\ 0 & 5 & -1 \end{vmatrix}$$
$$= 0 + (3)(-1)\begin{vmatrix} 1 & 4 \\ 5 & -1 \end{vmatrix} + 0$$
$$= (-3)[-1 - 20]$$
$$= (-3)[-21]$$
$$= 63$$

$$|A_x| = \begin{vmatrix} 6 & 1 & 4 \\ 7 & 0 & 1 \\ 9 & 5 & -1 \end{vmatrix}$$

$$= (1)(-1)\begin{vmatrix} 7 & 1 \\ 9 & -1 \end{vmatrix} + 0 + (5)(-1)\begin{vmatrix} 6 & 4 \\ 7 & 1 \end{vmatrix}$$
$$= (-1)[-7-9] + (-5)[6-28]$$
$$= (-1)[-16] + (-5)[-22]$$
$$= 16 + 110$$
$$= 126$$

$$|A_y| = \begin{vmatrix} 0 & 6 & 4 \\ 3 & 7 & 1 \\ 0 & 9 & -1 \end{vmatrix}$$

$$= 0 + (3)(-1)\begin{vmatrix} 6 & 4 \\ 9 & -1 \end{vmatrix} + 0$$
$$= (-3)[-6-36]$$
$$= (-3)[-42]$$
$$= 126$$

$$|A_z| = \begin{vmatrix} 0 & 1 & 6 \\ 3 & 0 & 7 \\ 0 & 5 & 9 \end{vmatrix}$$

$$= 0 + (3)(-1)\begin{vmatrix} 1 & 6 \\ 5 & 9 \end{vmatrix} + 0$$
$$= (-3)[9-30]$$
$$= (-3)[-21]$$
$$= 63$$

Then

$$x = \frac{|A_x|}{|A|} = \frac{126}{63} = 2,$$

$$y = \frac{|A_y|}{|A|} = \frac{126}{63} = 2,$$

and

$$z = \frac{|A_z|}{|A|} = \frac{63}{63} = 1.$$

Thus, the solution to the system is $(2,2,1)$, which can be checked in each equation.

EXERCISES A   SECTION 7.5

CHAPTER 7 MATRICES AND DETERMINANTS

SECTION 7.5 Determinants and Cramer's Rule

1. Evaluate the determinant.

    $$\begin{vmatrix} 3 & 5 \\ 2 & 4 \end{vmatrix} = (3)(4) - (5)(2) = 12 - 10 = 2$$

2. Evaluate the determinant.

    $$\begin{vmatrix} -1 & 0 \\ 3 & 4 \end{vmatrix} = (-1)(4) - (0)(3) = -4 - 0 = -4$$

3. Evaluate the determinant.

    $$\begin{vmatrix} 2 & 5 \\ -3 & -1 \end{vmatrix} = (2)(-1) - (5)(-3) = -2 - (-15) = 13$$

4. Evaluate the determinant.

    $$\begin{vmatrix} -2 & -3 \\ b & a \end{vmatrix} = (-2)(a) - (-3)(b) = -2a + 3b$$

Use the following matrix in Exercises 5-8.

$$A = \begin{bmatrix} 2 & -1 & 3 \\ 1 & -2 & 4 \\ -3 & 5 & -4 \end{bmatrix}$$

5. The minor of the element 3, which is a_{13}, is the determinant formed by deleting the 1st row and 3rd column of A.

    $$M_{13} = \begin{vmatrix} 1 & -2 \\ -3 & 5 \end{vmatrix} = (1)(5) - (-2)(-3) = 5 - 6 = -1$$

6. The cofactor of the element 3, which is a_{13}, is $(-1)^{1+3}$ times the minor of 3. Since the minor of 3 was found to be -1 in Exercise 5, the cofactor of 3 is

    $$A_{13} = (-1)^{1+3}(-1) = (+1)(-1) = -1.$$

7. Find $|A|$ by expanding along the first row.

    $$|A| = \begin{vmatrix} 2 & -1 & 3 \\ 1 & -2 & 4 \\ -3 & 5 & -4 \end{vmatrix}$$
    $$= 2(-1)^{1+1}\begin{vmatrix} -2 & 4 \\ 5 & -4 \end{vmatrix} + (-1)(-1)^{1+2}\begin{vmatrix} 1 & 4 \\ -3 & -4 \end{vmatrix}$$
    $$\quad + 3(-1)^{1+3}\begin{vmatrix} 1 & -2 \\ -3 & 5 \end{vmatrix}$$
    $$= 2[(-2)(-4)-(4)(5)] + 1[(1)(-4)-(4)(-3)]$$
    $$\quad + 3[(1)(5)-(-2)(-3)]$$
    $$= 2[8-20] + 1[-4+12] + 3[5-6]$$
    $$= 2[-12] + 1[8] + 3[-1]$$
    $$= -24 + 8 - 3$$
    $$= -19$$

8. Find $|A|$ by expanding along the first column.

    $$|A| = \begin{vmatrix} 2 & -1 & 3 \\ 1 & -2 & 4 \\ -3 & 5 & -4 \end{vmatrix}$$
    $$= 2(-1)^{1+1}\begin{vmatrix} -2 & 4 \\ 5 & -4 \end{vmatrix} + (1)(-1)^{2+1}\begin{vmatrix} -1 & 3 \\ 5 & -4 \end{vmatrix}$$
    $$\quad + (-3)(-1)^{3+1}\begin{vmatrix} -1 & 3 \\ -2 & 4 \end{vmatrix}$$
    $$= 2[(-2)(-4)-(4)(5)] + (-1)[(-1)(-4)-(3)(5)]$$
    $$\quad + (-3)[(-1)(4)-(3)(-2)]$$
    $$= 2[8-20] - 1[4-15] - 3[-4+6]$$
    $$= 2[-12] - 1[-11] - 3[2]$$
    $$= -24 + 11 - 6$$
    $$= -19$$

9. To evaluate the determinant we expand along the first column taking advantage of the 0's.

    $$\begin{vmatrix} 0 & 2 & -1 \\ 1 & -3 & 0 \\ 0 & 1 & -2 \end{vmatrix}$$
    $$= 0 + (1)(-1)^{2+1}\begin{vmatrix} 2 & -1 \\ 1 & -2 \end{vmatrix} + 0$$
    $$= (-1)[(2)(-2)-(-1)(1)]$$
    $$= (-1)[-4+1]$$
    $$= (-1)[-3]$$
    $$= 3$$

10. To evaluate the determinant, expand down the second column to take advantage of the 0 and 1's.

$$\begin{vmatrix} 3 & 1 & 2 \\ 4 & 0 & -2 \\ -2 & 1 & 3 \end{vmatrix}$$

$= (1)(-1)^{1+2} \begin{vmatrix} 4 & -2 \\ -2 & 3 \end{vmatrix} + 0 + (1)(-1)^{3+2} \begin{vmatrix} 3 & 2 \\ 4 & -2 \end{vmatrix}$
$= (-1)[(4)(3)-(-2)(-2)] + (-1)[(3)(-2)-(2)(4)]$
$= (-1)[12-4] + (-1)[-6-8]$
$= (-1)[8] + (-1)[-14]$
$= -8 + 14$
$= 6$

11. To evaluate the determinant, we expand along the first row.

$$\begin{vmatrix} 1 & -2 & 1 \\ 2 & 3 & 2 \\ 3 & 1 & 3 \end{vmatrix}$$

$= (1)(-1)^{1+1} \begin{vmatrix} 3 & 2 \\ 1 & 3 \end{vmatrix} + (-2)(-1)^{1+2} \begin{vmatrix} 2 & 2 \\ 3 & 3 \end{vmatrix}$
$\quad + (1)(-1)^{1+3} \begin{vmatrix} 2 & 3 \\ 3 & 1 \end{vmatrix}$
$= 1[(3)(3)-(2)(1)] + 2[(2)(3)-(2)(3)]$
$\quad + 1[(2)(1)-(3)(3)]$
$= 1[9-2] + 2[6-6] + 1[2-9]$
$= 1[7] + 2[0] + 1[-7]$
$= 7 + 0 - 7$
$= 0$

12. Solve using Cramer's rule.

$$x + 2y = -1$$
$$2x - 3y = 12$$

The determinant of the coefficient matrix is:

$$|A| = \begin{vmatrix} 1 & 2 \\ 2 & -3 \end{vmatrix} = (1)(-3) - (2)(2) = -3 - 4 = -7$$

Replace the coefficients of x in the coefficient matrix with the constants to obtain:

$$|A_x| = \begin{vmatrix} -1 & 2 \\ 12 & -3 \end{vmatrix} = (-1)(-3) - (2)(12) = 3 - 24 = -21$$

Replace the coefficients of y in the coefficient matrix with the constants to obtain:

$$|A_y| = \begin{vmatrix} 1 & -1 \\ 2 & 12 \end{vmatrix} = (1)(12) - (-1)(2) = 12 + 2 = 14$$

Then

$$x = \frac{|A_x|}{|A|} = \frac{-21}{-7} = 3 \qquad y = \frac{|A_y|}{|A|} = \frac{14}{-7} = -2$$

Thus, the solution to the system is $(3, -2)$.

13. Solve using Cramer's rule.

$$3x + 2y = -3$$
$$6x - 3y = 8$$

The determinant of the coefficient matrix is:

$$|A| = \begin{vmatrix} 3 & 2 \\ 6 & -3 \end{vmatrix} = (3)(-3) - (2)(6) = -9 - 12 = -21$$

Replace the coefficients of x in the coefficient matrix with the constants to obtain:

$$|A_x| = \begin{vmatrix} -3 & 2 \\ 8 & -3 \end{vmatrix} = (-3)(-3) - (2)(8) = 9 - 16 = -7$$

Replace the coefficients of y in the coefficient matrix with the constants to obtain:

$$|A_y| = \begin{vmatrix} 3 & -3 \\ 6 & 8 \end{vmatrix} = (3)(8) - (-3)(6) = 24 + 18 = 42$$

Then

$$x = \frac{|A_x|}{|A|} = \frac{-7}{-21} = \frac{1}{3} \qquad y = \frac{|A_y|}{|A|} = \frac{42}{-21} = -2$$

Thus, the solution to the system is $\left(\frac{1}{3}, -2\right)$.

14. Solve using Cramer's rule.

$$4x - y = -3$$
$$2x - 3y = 1$$

EXERCISES A

The determinant of the coefficient matrix is:

$$|A| = \begin{vmatrix} 4 & -1 \\ 2 & -3 \end{vmatrix} = (4)(-3) - (-1)(2) = -12 + 2 = -10$$

Replace the coefficients of x in the coefficient matrix with the constants to obtain:

$$|A_x| = \begin{vmatrix} -3 & -1 \\ 1 & -3 \end{vmatrix} = (-3)(-3) - (-1)(1) = 9 + 1 = 10$$

Replace the coefficients of y in the coefficient matrix with the constants to obtain:

$$|A_y| = \begin{vmatrix} 4 & -3 \\ 2 & 1 \end{vmatrix} = (4)(1) - (-3)(2) = 4 + 6 = 10$$

Then

$$x = \frac{|A_x|}{|A|} = \frac{10}{-10} = -1 \quad y = \frac{|A_y|}{|A|} = \frac{10}{-10} = -1$$

Thus, the solution to the system is $(-1, -1)$.

15. Solve using Cramer's rule.

$$\begin{aligned} 3x - y &= -11 \\ 2x + 4z &= -2 \\ -3y + 5z &= -1 \end{aligned}$$

The determinant of the coefficient matrix, expanding along the first row, is:

$$|A| = \begin{vmatrix} 3 & -1 & 0 \\ 2 & 0 & 4 \\ 0 & -3 & 5 \end{vmatrix}$$

$$= (3)(-1)^{1+1} \begin{vmatrix} 0 & 4 \\ -3 & 5 \end{vmatrix} + (-1)(-1)^{1+2} \begin{vmatrix} 2 & 4 \\ 0 & 5 \end{vmatrix} + 0$$

$$= (3)[(0)(5) - (4)(-3)] + (1)[(2)(5) - (4)(0)]$$
$$= (3)[12] + (1)[10]$$
$$= 36 + 10$$
$$= 46$$

Replace the coefficients of x in the coefficient matrix with the constants, and expand along the first row, to obtain:

$$|A_x| = \begin{vmatrix} -11 & -1 & 0 \\ -2 & 0 & 4 \\ -1 & -3 & 5 \end{vmatrix}$$

$$= (-11)(-1)^{1+1} \begin{vmatrix} 0 & 4 \\ -3 & 5 \end{vmatrix} + (-1)(-1)^{1+2} \begin{vmatrix} -2 & 4 \\ -1 & 5 \end{vmatrix} + 0$$

$$= (-11)[(0)(5) - (4)(-3)] + (1)[(-2)(5) - (4)(-1)]$$
$$= (-11)[12] + (1)[-6]$$
$$= -132 - 6$$
$$= -138$$

Replace the coefficients of y in the coefficient matrix with the constants, and expand along the first column, to obtain:

$$|A_y| = \begin{vmatrix} 3 & -11 & 0 \\ 2 & -2 & 4 \\ 0 & -1 & 5 \end{vmatrix}$$

$$= (3)(-1)^{1+1} \begin{vmatrix} -2 & 4 \\ -1 & 5 \end{vmatrix} + (2)(-1)^{2+1} \begin{vmatrix} -11 & 0 \\ -1 & 5 \end{vmatrix} + 0$$

$$= (3)[(-2)(5) - (4)(-1)] + (-2)[(-11)(5) - (0)(-1)]$$
$$= (3)[-6] + (-2)[-55]$$
$$= -18 + 110$$
$$= 92$$

Replace the coefficients of z in the coefficient matrix with the constants, and expand along the first column, to obtain:

$$|A_z| = \begin{vmatrix} 3 & -1 & -11 \\ 2 & 0 & -2 \\ 0 & -3 & -1 \end{vmatrix}$$

$$= (3)(-1)^{1+1} \begin{vmatrix} 0 & -2 \\ -3 & -1 \end{vmatrix} + (2)(-1)^{2+1} \begin{vmatrix} -1 & -11 \\ -3 & -1 \end{vmatrix} + 0$$

$$= (3)[(0)(-1) - (-2)(-3)]$$
$$\quad + (-2)[(-1)(-1) - (-11)(-3)]$$
$$= (3)[-6] + (-2)[-32]$$
$$= -18 + 64$$
$$= 46$$

Then

$$x = \frac{|A_x|}{|A|} = \frac{-138}{46} = -3 \quad y = \frac{|A_y|}{|A|} = \frac{92}{46} = 2$$

$$z = \frac{|A_z|}{|A|} = \frac{46}{46} = 1$$

Thus, the solution to the system is $(-3, 2, 1)$.

16. Solve using Cramer's rule.

$$3x + y - z = -3$$
$$x - y + z = 0$$
$$-3x + 2y - 4z = -3$$

The determinant of the coefficient matrix, expanding along the first row, is:

$$|A| = \begin{vmatrix} 2 & 1 & -1 \\ 1 & -1 & 1 \\ -3 & 2 & -4 \end{vmatrix}$$

$$= (2)(-1)^{1+1}\begin{vmatrix} -1 & 1 \\ 2 & -4 \end{vmatrix} + (1)(-1)^{1+2}\begin{vmatrix} 1 & 1 \\ -3 & -4 \end{vmatrix}$$
$$\quad + (-1)(-1)^{1+3}\begin{vmatrix} 1 & -1 \\ -3 & 2 \end{vmatrix}$$
$$= (2)[(-1)(-4)-(1)(2)] + (-1)[(1)(-4)-(1)(-3)]$$
$$\quad + (-1)[(1)(2)-(-1)(-3)]$$
$$= (2)[2] + (-1)[-1] + (-1)[-1]$$
$$= 4 + 1 + 1$$
$$= 6$$

Replace the coefficients of x in the coefficient matrix with the constants, and expand along the first column, to obtain:

$$|A_x| = \begin{vmatrix} -3 & 1 & -1 \\ 0 & -1 & 1 \\ -3 & 2 & -4 \end{vmatrix}$$
$$= (-3)(-1)^{1+1}\begin{vmatrix} -1 & 1 \\ 2 & -4 \end{vmatrix} + 0 + (-3)(-1)^{3+1}\begin{vmatrix} 1 & -1 \\ -1 & 1 \end{vmatrix}$$
$$= (-3)[(-1)(-4)-(1)(2)] + (-3)[(1)(1)-(-1)(-1)]$$
$$= (-3)[2] + (-3)[0]$$
$$= -6 - 0$$
$$= -6$$

Replace the coefficients of y in the coefficient matrix with the constants, and expand along the second column, to obtain:

$$|A_y| = \begin{vmatrix} 2 & -3 & -1 \\ 1 & 0 & 1 \\ -3 & -3 & -4 \end{vmatrix}$$
$$= (-3)(-1)^{1+2}\begin{vmatrix} 1 & 1 \\ -3 & -4 \end{vmatrix} + 0 + (-3)(-1)^{3+2}\begin{vmatrix} 2 & -1 \\ 1 & 1 \end{vmatrix}$$
$$= (3)[(1)(-4)-(1)(-3)] + (3)[(2)(1)-(-1)(1)]$$
$$= (3)[-1] + (3)[3]$$
$$= -3 + 9$$
$$= 6$$

Replace the coefficients of z in the coefficient matrix with the constants, and expand along the third column, to obtain:

$$|A_z| = \begin{vmatrix} 2 & 1 & -3 \\ 1 & -1 & 0 \\ -3 & 2 & -3 \end{vmatrix}$$
$$= (-3)(-1)^{1+3}\begin{vmatrix} 1 & -1 \\ -3 & 2 \end{vmatrix} + 0 + (-3)(-1)^{3+3}\begin{vmatrix} 2 & 1 \\ 1 & -1 \end{vmatrix}$$
$$= (-3)[(1)(2)-(-1)(-3)] + (-3)[(2)(-1)-(1)(1)]$$
$$= (-3)[-1] + (-3)[-3]$$
$$= 3 + 9$$
$$= 12$$

Then

$$x = \frac{|A_x|}{|A|} = \frac{-6}{6} = -1 \quad y = \frac{|A_y|}{|A|} = \frac{6}{6} = 1$$
$$z = \frac{|A_z|}{|A|} = \frac{12}{6} = 2$$

Thus, the solution to the system is $(-1, 1, 2)$.

17. Solve for x.

$$\begin{vmatrix} x & 2 \\ 3 & 1 \end{vmatrix} = 3$$

Expand the determinant on the left side.

$$\begin{vmatrix} x & 2 \\ 3 & 1 \end{vmatrix} = (x)(1) - (2)(3) = x - 6$$

Then the equation becomes:

$$x - 6 = 3$$
$$x = 9$$

Thus, the solution is 9.

18. Solve for x.

$$\begin{vmatrix} x & 1 \\ 1 & x \end{vmatrix} = 3$$

Expand the determinant on the left side.

EXERCISES A

$$\begin{vmatrix} x & 1 \\ 1 & x \end{vmatrix} = (x)(x) - (1)(1) = x^2 - 1$$

Then the equation becomes:

$$x^2 - 1 = 3$$
$$x^2 = 4$$
$$x = \pm\sqrt{4}$$
$$x = \pm 2$$

Thus, the solutions are 2 and −2.

19. Solve for x.

$$\begin{vmatrix} x & 0 & 0 \\ 2 & x & -1 \\ -3 & 2 & 1 \end{vmatrix} = 3$$

Expand the determinant on the left side along the first row.

$$\begin{vmatrix} x & 0 & 0 \\ 2 & x & -1 \\ -3 & 2 & 1 \end{vmatrix} = x(-1)^{1+1}\begin{vmatrix} x & -1 \\ 2 & 1 \end{vmatrix} + 0 + 0$$
$$= x[(x)(1) - (-1)(2)]$$
$$= x(x + 2)$$

Then the equation to solve is:

$$x(x + 2) = 3$$
$$x^2 + 2x = 3$$
$$x^2 + 2x - 3 = 0$$
$$(x - 1)(x + 3) = 0$$
$$x - 1 = 0 \qquad x + 3 = 0$$
$$x = 1 \qquad x = -3$$

Thus, the solutions are 1 and −3.

20. Evaluate the 4 × 4 determinant.

$$\begin{vmatrix} 2 & 3 & 0 & -2 \\ 1 & -2 & -1 & 0 \\ 0 & 1 & 0 & 0 \\ -1 & 4 & -2 & 3 \end{vmatrix}$$

We can take advantage of the 0's in the third row. Expanding along the third row gives:

$$0 + (1)(-1)^{3+2}\begin{vmatrix} 2 & 0 & -2 \\ 1 & -1 & 0 \\ -1 & -2 & 3 \end{vmatrix} + 0 + 0$$

Then expand the resulting 3 × 3 determinant along the first row. Do not forget the factor of (−1).

$$(-1)\begin{vmatrix} 2 & 0 & -2 \\ 1 & -1 & 0 \\ -1 & -2 & 3 \end{vmatrix}$$
$$= (-1)\left((2)(-1)^{1+1}\begin{vmatrix} -1 & 0 \\ -2 & 3 \end{vmatrix} + 0 + (-2)(-1)^{1+3}\begin{vmatrix} 1 & -1 \\ -1 & -2 \end{vmatrix}\right)$$
$$= (-1)((2)[(-1)(3) - (0)(-2)]$$
$$\qquad + (-2)[(1)(-2) - (-1)(-1)])$$
$$= (-1)((2)[-3] + (-2)[-3])$$
$$= (-1)(-6 + 6)$$
$$= (-1)(0)$$
$$= 0$$

Thus, the 4 × 4 determinant is 0.

21. (a) Evaluate $\begin{vmatrix} 0 & 0 \\ 0 & 0 \end{vmatrix}$.

$$\begin{vmatrix} 0 & 0 \\ 0 & 0 \end{vmatrix} = (0)(0) - (0)(0) = 0$$

(b) Evaluate $\begin{vmatrix} 0 & 0 & 0 \\ 0 & 0 & 0 \\ 0 & 0 & 0 \end{vmatrix}$.

Since we can expand along the first row (in fact any row or column), and all the entries will be multiplied by 0, the determinant is 0.

(c) Evaluate $\begin{vmatrix} 0 & 0 & 0 & 0 \\ 0 & 0 & 0 & 0 \\ 0 & 0 & 0 & 0 \\ 0 & 0 & 0 & 0 \end{vmatrix}$.

Since we can expand along the first row (in fact any row or column), and all the entries will be multiplied by 0, the determinant is 0.

(d) Using an argument similar to those above, the determinant of any $n \times n$ zero matrix is 0.

22. (a) Evaluate $\begin{vmatrix} a & 0 \\ 0 & b \end{vmatrix}$.

$$\begin{vmatrix} a & 0 \\ 0 & b \end{vmatrix} = (a)(b) - (0)(0) = ab$$

(b) Evaluate $\begin{vmatrix} a & 0 & 0 \\ 0 & b & 0 \\ 0 & 0 & c \end{vmatrix}$.

$$\begin{vmatrix} a & 0 & 0 \\ 0 & b & 0 \\ 0 & 0 & c \end{vmatrix} = a(-1)^{1+1} \begin{vmatrix} b & 0 \\ 0 & c \end{vmatrix} + 0 + 0$$
$$= a[(b)(c) - (0)(0)]$$
$$= a[bc - 0]$$
$$= abc$$

(c) Evaluate $\begin{vmatrix} a & 0 & 0 & 0 \\ 0 & b & 0 & 0 \\ 0 & 0 & c & 0 \\ 0 & 0 & 0 & d \end{vmatrix}$.

Expand along the first row. The resulting 3×3 determinant can be expanded using the results of part (b).

$$\begin{vmatrix} a & 0 & 0 & 0 \\ 0 & b & 0 & 0 \\ 0 & 0 & c & 0 \\ 0 & 0 & 0 & d \end{vmatrix} = a(-1)^{1+1} \begin{vmatrix} b & 0 & 0 \\ 0 & c & 0 \\ 0 & 0 & d \end{vmatrix}$$
$$= a(bcd) \quad \text{Using part (b)}$$
$$= abcd$$

(d) The determinant of any diagonal matrix is the product of the elements along the main diagonal, that is, the product of $a_{11}, a_{22}, a_{33}, \ldots,$ and a_{nn}.

23. (a) Evaluate $\begin{vmatrix} a & 0 \\ b & 0 \end{vmatrix}$.

Since we can evaluate this determinant along the second column, and since all the elements there are 0, all the terms in the expansion will be 0. Thus, the determinant is 0.

(b) Evaluate $\begin{vmatrix} a & b & 0 \\ c & d & 0 \\ e & f & 0 \end{vmatrix}$.

Since we can evaluate this determinant along the third column, and since all the elements there are 0, all the terms in the expansion will be 0. Thus, the determinant is 0.

(c) Evaluate $\begin{vmatrix} a & b & c & 0 \\ d & e & f & 0 \\ g & h & i & 0 \\ j & k & l & 0 \end{vmatrix}$.

Since we can evaluate this determinant along the fourth column, and since all the elements there are 0, all the terms in the expansion will be 0. Thus, the determinant is 0.

(d) The determinant of any matrix that has a column of 0's can be evaluated along that column giving 0 for the determinant.

24. Given the matrix $A = \begin{bmatrix} a & b \\ c & d \end{bmatrix}$.

$$|A| = \begin{vmatrix} a & b \\ c & d \end{vmatrix} = ad - bc$$

Then the transpose of A is:

$$A^T = \begin{bmatrix} a & c \\ b & d \end{bmatrix}$$

The determinant of A^T is:

EXERCISES A

$$|A^T| = \begin{vmatrix} a & c \\ b & d \end{vmatrix} = ad - cb = ad - bc$$

Thus, $|A| = |A^T|$.

25. Given the matrix $A = \begin{bmatrix} a & b \\ c & d \end{bmatrix}$.

First find the inverse of A. Form the augmented matrix $[A|I]$.

$$\begin{bmatrix} a & b & | & 1 & 0 \\ c & d & | & 0 & 1 \end{bmatrix}$$

Multiply the first row by $\frac{1}{a}$.

$$\begin{bmatrix} 1 & \frac{b}{a} & | & \frac{1}{a} & 0 \\ c & d & | & 0 & 1 \end{bmatrix}$$

Multiply the first row by $-c$ and add the result to the second row.

$$\begin{bmatrix} 1 & \frac{b}{a} & | & \frac{1}{a} & 0 \\ 0 & \frac{ad-bc}{a} & | & -\frac{c}{a} & 1 \end{bmatrix}$$

Multiply the second row by $\frac{a}{ad-bc}$.

$$\begin{bmatrix} 1 & \frac{b}{a} & | & \frac{1}{a} & 0 \\ 0 & 1 & | & \frac{-c}{ad-bc} & \frac{a}{ad-bc} \end{bmatrix}$$

Multiply the second row by $-\frac{b}{a}$ and add the result to the first row.

$$\begin{bmatrix} 1 & 0 & | & \frac{d}{ad-bc} & \frac{-b}{ad-bc} \\ 0 & 1 & | & \frac{-c}{ad-bc} & \frac{a}{ad-bc} \end{bmatrix}$$

Thus, $A^{-1} = \begin{bmatrix} \frac{d}{ad-bc} & \frac{-b}{ad-bc} \\ \frac{-c}{ad-bc} & \frac{a}{ad-bc} \end{bmatrix}$. Then the determinant of A^{-1} is given by:

$$|A^{-1}| = \begin{vmatrix} \frac{d}{ad-bc} & \frac{-b}{ad-bc} \\ \frac{-c}{ad-bc} & \frac{a}{ad-bc} \end{vmatrix}$$
$$= \left(\frac{d}{ad-bc}\right)\left(\frac{a}{ad-bc}\right) - \left(\frac{-b}{ad-bc}\right)\left(\frac{-c}{ad-bc}\right)$$
$$= \frac{ad-bc}{(ad-bc)^2}$$
$$= \frac{1}{ad-bc}$$

Since $|A| = ad - bc$, substituting in the above we have the desired result,

$$|A^{-1}| = \frac{1}{|A|},$$

which, of course, is true only if $|A| \neq 0$.

26. $\begin{vmatrix} a & b \\ na & nb \end{vmatrix} = (a)(nb) - (b)(na) = abn - abn = 0$

27. Since

$$\begin{vmatrix} a & b \\ c & d \end{vmatrix} = ad - bc,$$

and

$$\begin{vmatrix} b & a \\ d & c \end{vmatrix} = bc - ad,$$

we have that

$$\begin{vmatrix} a & b \\ c & d \end{vmatrix} = ad - bc = -(bc - ad) = -\begin{vmatrix} b & a \\ d & c \end{vmatrix}.$$

28. Find the inverse of the matrix.

$$\begin{bmatrix} 2 & -4 \\ -1 & 3 \end{bmatrix}$$

First form the augmented matrix $[A|I]$.

$$\begin{bmatrix} 2 & -4 & | & 1 & 0 \\ -1 & 3 & | & 0 & 1 \end{bmatrix}$$

Multiply the second row by -1, and interchange the result with the first row.

$$\begin{bmatrix} 1 & -3 & | & 0 & -1 \\ 2 & -4 & | & 1 & 0 \end{bmatrix}$$

Multiply the first row by -2, and add the result to the second row.

$$\begin{bmatrix} 1 & -3 & | & 0 & -1 \\ 0 & 2 & | & 1 & 2 \end{bmatrix}$$

Multiply the second row by $\frac{1}{2}$.

$$\begin{bmatrix} 1 & -3 & | & 0 & -1 \\ 0 & 1 & | & \frac{1}{2} & 1 \end{bmatrix}$$

Multiply the second row by 3, and add the result to the first row.

$$\begin{bmatrix} 1 & 0 & | & \frac{3}{2} & 2 \\ 0 & 1 & | & \frac{1}{2} & 1 \end{bmatrix}$$

Thus the inverse of the matrix is

$$\begin{bmatrix} \frac{3}{2} & 2 \\ \frac{1}{2} & 1 \end{bmatrix}.$$

29. Solve the system using the inverse method.

$$\begin{aligned} 2x - 4y &= -20 \\ -x + 3y &= 15 \end{aligned}$$

This system is equivalent to the matrix equation $AX = B$, where

$$A = \begin{bmatrix} 2 & -4 \\ -1 & 3 \end{bmatrix} \quad X = \begin{bmatrix} x \\ y \end{bmatrix} \quad B = \begin{bmatrix} -20 \\ 15 \end{bmatrix}.$$

In Exercise 28, we found that

$$A^{-1} = \begin{bmatrix} \frac{3}{2} & 2 \\ \frac{1}{2} & 1 \end{bmatrix},$$

so using the inverse method, we have

$$\begin{bmatrix} x \\ y \end{bmatrix} = X = A^{-1}B = \begin{bmatrix} \frac{3}{2} & 2 \\ \frac{1}{2} & 1 \end{bmatrix} \begin{bmatrix} -20 \\ 15 \end{bmatrix} = \begin{bmatrix} 0 \\ 5 \end{bmatrix}.$$

Thus, $x = 0$, $y = 5$, and the solution to the system of equations is $(0,5)$.

PRACTICE EXERCISES SECTION 7.6

CHAPTER 7 MATRICES AND DETERMINANTS

SECTION 7.6 More on Determinants

1. Given that
$$A = \begin{bmatrix} 1 & 2 & -3 \\ -1 & 0 & 4 \\ 5 & 2 & 3 \end{bmatrix}$$

with $|A| = 44$, find $|C|$ if

$$C = \begin{bmatrix} 1 & -1 & 5 \\ 2 & 0 & 2 \\ -3 & 4 & 3 \end{bmatrix}.$$

Notice that $C = A^T$, that is, the first row of A is the first column of C, the second row of A is the second column of C, and the third row of A is the third column of C. Thus, by Theorem 1, since the determinant of a matrix is the same as the determinant of its transpose,

$$|C| = |A^T| = |A| = 44.$$

2. Given that
$$B = \begin{bmatrix} 6 & -4 & 8 \\ 1 & 3 & -5 \\ 0 & 0 & 0 \end{bmatrix}.$$

Then by Theorem 3, since every element in the third row of B is 0, $|B| = 0$. This is easy to see if we were to expand the determinant of B along the third row since every term would be 0.

3. Given that
$$A = \begin{bmatrix} 1 & 2 & -3 \\ -1 & 0 & 4 \\ 5 & 2 & 3 \end{bmatrix}$$

with $|A| = 44$, find $|C|$ if

$$C = \begin{bmatrix} 2 & 1 & -3 \\ 0 & -1 & 4 \\ 2 & 5 & 3 \end{bmatrix}.$$

Since C is formed from A by interchanging the first two columns, by Theorem 4, the determinant of C is the negative of the determinant of A. That is,

$$|C| = -|A| = -44.$$

4. Since the first and third rows in

$$\begin{vmatrix} -3 & -3 & -3 \\ 1 & 4 & -6 \\ -3 & -3 & -3 \end{vmatrix}$$

are the same, by Theorem 5, this determinant is 0.

5. Since the elements in the third row are 3 times those in the first row in

$$\begin{vmatrix} 1 & -2 & 0 \\ 3 & 7 & -4 \\ 3 & -6 & 0 \end{vmatrix}$$

by Theorem 6, this determinant is 0.

6. Use Theorem 7 to find

$$\begin{vmatrix} 2 & 1 & 1 \\ 1 & -1 & 1 \\ 1 & -2 & -1 \end{vmatrix}.$$

Remember that if a determinant has several 0's in a row or column, it is easy to find the determinant using that row or column. This is the substance of Theorem 7 which gives a way to force 0's in a row or column without changing the value of the determinant. Suppose we use the 1 in the first row, third column to force 0's in the other two positions in the third column. Keep the first row, multiply it by -1, and add the result to the second row. Then multiply the first row by 1, and add the result to the third row (simply add the first row to the third). The resulting determinant is:

$$\begin{vmatrix} 2 & 1 & 1 \\ -1 & -2 & 0 \\ 3 & -1 & 0 \end{vmatrix}$$

Expand along the third column.

$$\begin{vmatrix} 2 & 1 & 1 \\ -1 & -2 & 0 \\ 3 & -1 & 0 \end{vmatrix} = (1)(+1)\begin{vmatrix} -1 & -2 \\ 3 & -1 \end{vmatrix} + 0 + 0$$
$$= (1)[1 - (-6)]$$
$$= (1)[7]$$
$$= 7$$

7. Find A^{-1} if

$$A = \begin{bmatrix} 1 & 0 & -1 \\ -2 & 1 & 0 \\ 2 & -2 & 3 \end{bmatrix}.$$

First find $|A|$ since if $|A| = 0$, we can stop, A^{-1} does not exist. Expand along the first row.

$$|A| = (1)(+1)\begin{vmatrix} 1 & 0 \\ -2 & 3 \end{vmatrix} + 0 + (-1)(+1)\begin{vmatrix} -2 & 1 \\ 2 & -2 \end{vmatrix}$$
$$= (1)[3 - 0] + (-1)[4 - 2]$$
$$= (1)[3] + (-1)[2]$$
$$= 3 - 2$$
$$= 1$$

Since $|A| = 1 \neq 0$, we find the matrix of cofactors.

$$cofA = \begin{bmatrix} A_{11} & A_{12} & A_{13} \\ A_{21} & A_{22} & A_{23} \\ A_{31} & A_{32} & A_{33} \end{bmatrix}$$

$$= \begin{bmatrix} +\begin{vmatrix} 1 & 0 \\ -2 & 3 \end{vmatrix} & -\begin{vmatrix} -2 & 0 \\ 2 & 3 \end{vmatrix} & +\begin{vmatrix} -2 & 1 \\ 2 & -2 \end{vmatrix} \\ -\begin{vmatrix} 0 & -1 \\ -2 & 3 \end{vmatrix} & +\begin{vmatrix} 1 & -1 \\ 2 & 3 \end{vmatrix} & -\begin{vmatrix} 1 & 0 \\ 2 & -2 \end{vmatrix} \\ +\begin{vmatrix} 0 & -1 \\ 1 & 0 \end{vmatrix} & -\begin{vmatrix} 1 & -1 \\ -2 & 0 \end{vmatrix} & +\begin{vmatrix} 1 & 0 \\ -2 & 1 \end{vmatrix} \end{bmatrix}$$

$$= \begin{bmatrix} 3 & 6 & 2 \\ 2 & 5 & 2 \\ 1 & 2 & 1 \end{bmatrix}$$

Then

$$(cofA)^T = \begin{bmatrix} 3 & 2 & 1 \\ 6 & 5 & 2 \\ 2 & 2 & 1 \end{bmatrix}$$

and

$$A^{-1} = \frac{1}{|A|}(cofA)^T = \frac{1}{1}\begin{bmatrix} 3 & 2 & 1 \\ 6 & 5 & 2 \\ 2 & 2 & 1 \end{bmatrix}$$

so finally,

$$A^{-1} = \begin{bmatrix} 3 & 2 & 1 \\ 6 & 5 & 2 \\ 2 & 2 & 1 \end{bmatrix}.$$

Notice that we obtained the same result using this method as we obtained in Example 2 of Section 7.4 using the first method.

8. Find A^{-1} if

$$A = \begin{bmatrix} -2 & -1 \\ 3 & 1 \end{bmatrix}.$$

Since $|A| = -2 - (-3) = -2 + 3 = 1 \neq 0$, A^{-1} exists.

$$cofA = \begin{bmatrix} A_{11} & A_{12} \\ A_{21} & A_{22} \end{bmatrix}$$
$$= \begin{bmatrix} +[[1]] & -[[3]] \\ -[[-1]] & +[[-2]] \end{bmatrix}$$
$$= \begin{bmatrix} 1 & -3 \\ 1 & -2 \end{bmatrix}$$

Then

$$(cofA)^T = \begin{bmatrix} 1 & 1 \\ -3 & -2 \end{bmatrix},$$

and

$$A^{-1} = \frac{1}{|A|}(cofA)^T = \frac{1}{1}\begin{bmatrix} 1 & 1 \\ -3 & -2 \end{bmatrix} = \begin{bmatrix} 1 & 1 \\ -3 & -2 \end{bmatrix}.$$

Compare this result with the result in Practice Exercise 1, Section 7.4.

EXERCISES A — SECTION 7.6

CHAPTER 7 MATRICES AND DETERMINANTS

SECTION 7.6 More on Determinants

1. Given that $\begin{vmatrix} 1 & 3 \\ -1 & 5 \end{vmatrix} = \begin{vmatrix} 1 & -1 \\ 3 & 5 \end{vmatrix}$.

   This is true by Theorem 1 since the matrix on the right is the transpose of the one on the left.

2. Given that $\begin{vmatrix} 0 & 7 \\ 0 & -8 \end{vmatrix} = 0$.

   This is true by Theorem 3 since the matrix has a column of 0's making its determinant 0.

3. Given that $\begin{vmatrix} 4 & -4 \\ -1 & 3 \end{vmatrix} = 4\begin{vmatrix} 1 & -1 \\ -1 & 3 \end{vmatrix}$.

   This is true by Theorem 2 since the matrix on the left is formed from the matrix on the right by multiplying all the elements in the first row by 4.

4. Given that $\begin{vmatrix} -5 & 0 \\ 1 & 6 \end{vmatrix} = -\begin{vmatrix} 0 & -5 \\ 6 & 1 \end{vmatrix}$.

   This is true by Theorem 4 since the matrix on the left is obtained from the one on the right by interchanging the two columns. This process changes the sign of the determinant.

5. Given that $\begin{vmatrix} 1 & 1 & -6 \\ 2 & 2 & 7 \\ 3 & 3 & -8 \end{vmatrix} = 0$.

   This is true by Theorem 5 since the matrix has two columns (the first and second) with equal corresponding elements. This makes the determinant 0.

6. Given that $\begin{vmatrix} 1 & 3 \\ 4 & -5 \end{vmatrix} = \begin{vmatrix} 1 & 3 \\ 4-4 & -5-12 \end{vmatrix}$.

   This is true by Theorem 7 since the elements in the second row of the matrix on the right are obtained by adding -4 times the first row to the second row.

7. Given that $\begin{vmatrix} -2 & -6 \\ 1 & 3 \end{vmatrix} = 0$.

   This is true by Theorem 6 since the elements in the first row are a multiple of (-2 times) the elements in the second row. This makes the determinant 0.

8. Given that $3\begin{vmatrix} -1 & 2 & 0 \\ 2 & -1 & 5 \\ -1 & 2 & 0 \end{vmatrix} = \begin{vmatrix} -1 & 2 & 0 \\ 6 & -3 & 15 \\ -1 & 2 & 0 \end{vmatrix}$.

   This is true by Theorem 2 since the matrix on the right is obtained from the one on the left by multiplying all the elements in the second row by 3. This changes the determinant by a multiple of 3.

9. Given that $\begin{vmatrix} 5 & 0 & 7 \\ -6 & 0 & 5 \\ 1 & 0 & -4 \end{vmatrix} = 0$.

   This is true by Theorem 3 since the matrix has a column with all elements 0. Any matrix with a row or column of 0's has 0 for its determinant.

10. The determinant on the left was transformed to the determinant on the right using Theorem 7.

    $\begin{vmatrix} 1 & 4 \\ -3 & 5 \end{vmatrix} = \begin{vmatrix} 1 & 0 \\ -3 & x \end{vmatrix}$

    Since both determinants have the same first column, the first column must have been used in the transformation. Since 0 in the determinant on the right could be obtained from 4 in the determinant on the left by multiplying 1 (in the first column) by -4 and adding the result to 4 (in the second column), we assume that this was the procedure. Then x must equal (-4) times (-3) plus 5. That is,

    $x = (-4)(-3) + 5 = 12 + 5 = 17$.

11. The determinant on the left was transformed to the determinant on the right using Theorem 7.

$$\begin{vmatrix} 1 & -1 & 3 \\ 0 & 2 & -2 \\ 4 & -3 & 5 \end{vmatrix} = \begin{vmatrix} 1 & -1 & 3 \\ 0 & 2 & -2 \\ x & 0 & -4 \end{vmatrix}$$

If the first row is multiplied by −3, and the result added to the third row, we obtain the 0 and −4 in the third row. Thus, we assume this was the procedure used to transform the determinant. This means that

$$x = (-3)(1) + 4 = -3 + 4 = 1.$$

12. Use Theorem 7 to introduce two 0's in row 1 of

$$\begin{vmatrix} 2 & 1 & -5 \\ -3 & 4 & -2 \\ 1 & 3 & -1 \end{vmatrix}.$$

Use the 1 in the first row, second column to obtain 0's in the positions a_{11} and a_{13}. Keep the second column, multiply the elements by −2, and add the results to the first column. This will transform the determinant to the form:

$$\begin{vmatrix} 0 & 1 & -5 \\ -11 & 4 & -2 \\ -5 & 3 & -1 \end{vmatrix}$$

Then in this determinant, keep the second column, multiply the elements by 5, and add the results to the third column. This will transform the determinant to the form:

$$\begin{vmatrix} 0 & 1 & 0 \\ -11 & 4 & 18 \\ -5 & 3 & 14 \end{vmatrix}$$

Of course, these two steps could be combined in one writing of the determinant to save time and space.

13. Use Theorem 7 to introduce two 0's in column 2 of

$$\begin{vmatrix} 2 & -3 & 1 \\ -1 & -2 & 3 \\ 4 & 1 & -2 \end{vmatrix}.$$

Use the 1 in the third row, second column to obtain 0's in the positions a_{12} and a_{22}. Keep the third row, multiply the elements by 3, and add the results to the first row. This will transform the determinant to the form:

$$\begin{vmatrix} 14 & 0 & -5 \\ -1 & -2 & 3 \\ 4 & 1 & -2 \end{vmatrix}$$

Then in this determinant, keep the third row, multiply the elements by 2, and add the results to the second row. This will transform the determinant to the form:

$$\begin{vmatrix} 14 & 0 & -5 \\ 7 & 0 & -1 \\ 4 & 1 & -2 \end{vmatrix}$$

Of course, these two steps could be combined in one writing of the determinant to save time and space.

14. Use Theorem 7 to introduce two 0's in a row or column of the determinant

$$\begin{vmatrix} 0 & 1 & 2 \\ -1 & 2 & 5 \\ 2 & -3 & 1 \end{vmatrix}$$

and evaluate the determinant.

Since we already have a 0 in the first row, and a 1 in the first row, second column, use this 1 to obtain another 0 in the first row in the position of a_{13}. Keep the second column, multiply the elements by −2, and add the results to the third column to obtain:

$$\begin{vmatrix} 0 & 1 & 0 \\ -1 & 2 & 1 \\ 2 & -3 & 7 \end{vmatrix}$$

Expand the determinant along the first row taking advantage of the two zeros. Then

$$\begin{vmatrix} 0 & 1 & 0 \\ -1 & 2 & 1 \\ 2 & -3 & 7 \end{vmatrix} = (1)(-1)^{1+2} \begin{vmatrix} -1 & 1 \\ 2 & 7 \end{vmatrix}$$

$$= (-1)[(-1)(7)-(1)(2)]$$
$$= (-1)[-7-2]$$
$$= (-1)[-9]$$
$$= 9$$

EXERCISES A

15. Use Theorem 7 to introduce two 0's in a row or column of the determinant

$$\begin{vmatrix} 2 & -1 & 5 \\ 3 & 1 & -2 \\ 4 & -3 & 2 \end{vmatrix}$$

and evaluate the determinant.

Suppose we keep the 1 in the second row, second column and use it to obtain 0's in the remaining positions in the second column. Keep the second row, multiply it by 1 and add the results to the first row (simply add the second row to the first row). Then multiply the second row by 3 and add the results to the third row.

$$\begin{vmatrix} 5 & 0 & 3 \\ 3 & 1 & -2 \\ 13 & 0 & -4 \end{vmatrix}$$

Expand the determinant along the second column taking advantage of the two zeros. Then

$$\begin{vmatrix} 5 & 0 & 3 \\ 3 & 1 & -2 \\ 13 & 0 & -4 \end{vmatrix} = (1)(-1)^{2+2} \begin{vmatrix} 5 & 3 \\ 13 & -4 \end{vmatrix}$$
$$= (1)[(5)(-4)-(3)(13)]$$
$$= (1)[-20-39]$$
$$= (1)[-59]$$
$$= -59$$

16. Use Theorem 7 to introduce two 0's in a row or column of the determinant

$$\begin{vmatrix} 4 & -5 & 7 \\ 2 & 3 & -3 \\ -6 & -2 & -4 \end{vmatrix}$$

and evaluate the determinant.

Notice that the elements in the first column are all multiples of 2. As a result, it is easy to use the 2 in the second row, first column, to force 0's in the remaining positions in the first column. Keep the second row, multiply it by -2, and add the results to the first row. Then keeping the second row, multiply the elements by 3, and add the results to the third row.

SECTION 7.6   365

$$\begin{vmatrix} 0 & -11 & 13 \\ 2 & 3 & -3 \\ 0 & 7 & -13 \end{vmatrix}$$

Expand the determinant along the first column taking advantage of the two zeros. Then

$$\begin{vmatrix} 0 & -11 & 13 \\ 2 & 3 & -3 \\ 0 & 7 & -13 \end{vmatrix} = (2)(-1)^{2+1} \begin{vmatrix} -11 & 13 \\ 7 & -13 \end{vmatrix}$$
$$= (-2)[(-11)(-13)-(13)(7)]$$
$$= (-2)[143-91]$$
$$= (-2)[52]$$
$$= -104$$

17. Use Theorem 7 to introduce two 0's in a row or column of the determinant

$$\begin{vmatrix} 2 & 3 & 0 & -2 \\ 1 & 4 & -1 & 0 \\ 3 & -2 & 2 & 5 \\ -2 & 0 & 1 & -3 \end{vmatrix}$$

and evaluate the determinant.

Suppose we use the 1 in the second row to obtain 0's in the remaining two positions (we already have one 0 in the second row, fourth column). Keep the first column, multiply is by -4, and add the results to the second column. Then multiply the first column by 1 and add the results to the third column (simply add the first column to the third column).

$$\begin{vmatrix} 2 & -5 & 2 & -2 \\ 1 & 0 & 0 & 0 \\ 3 & -14 & 5 & 5 \\ -2 & 8 & -1 & -3 \end{vmatrix}$$

Expand the determinant along the second row taking advantage of the three zeros to obtain:

$$(1)(-1)^{2+1} \begin{vmatrix} -5 & 2 & -2 \\ -14 & 5 & 5 \\ 8 & -1 & -3 \end{vmatrix} = (-1) \begin{vmatrix} -5 & 2 & -2 \\ -14 & 5 & 5 \\ 8 & -1 & -3 \end{vmatrix}$$

We can then concentrate on the 3 × 3 determinant. Suppose we introduce two 0's in the third row of it using the element −1. Keep the second column, multiply it by 8, and add the results to the first column. Then multiply the second column by −3, and add the results to the third column. Remember that we still have a factor of (−1) from above.

$$(-1)\begin{vmatrix} 11 & 2 & -8 \\ 26 & 5 & -10 \\ 0 & -1 & 0 \end{vmatrix}$$

Then expanding along the third row we have:

$$(-1)(-1)(-1)^{3+2}\begin{vmatrix} 11 & -8 \\ 26 & -10 \end{vmatrix} = (-1)[-110 + 208] = -98$$

18. Use the optional method to find the inverse of the matrix.

$$\begin{bmatrix} 2 & 1 \\ 5 & -3 \end{bmatrix}$$

Since $|A| = -6 - 5 = -11 \neq 0$, the inverse exists.

$$cofA = \begin{bmatrix} A_{11} & A_{12} \\ A_{21} & A_{22} \end{bmatrix}$$
$$= \begin{bmatrix} +|[-3]| & -|[5]| \\ -|[1]| & +|[2]| \end{bmatrix}$$
$$= \begin{bmatrix} -3 & -5 \\ -1 & 2 \end{bmatrix}$$

Then

$$(cofA)^T = \begin{bmatrix} -3 & -1 \\ -5 & 2 \end{bmatrix},$$

and

$$A^{-1} = \frac{1}{|A|}(cofA)^T = -\frac{1}{11}\begin{bmatrix} -3 & -1 \\ -5 & 2 \end{bmatrix}.$$

19. Use the optional method to find the inverse of the matrix.

$$\begin{bmatrix} 2 & 2 \\ 2 & 2 \end{bmatrix}$$

Since $|A| = 4 - 4 = 0$, the inverse does not exist.

20. Find A^{-1} if

$$A = \begin{bmatrix} 1 & 2 & 0 \\ 0 & 1 & 1 \\ -1 & 1 & 1 \end{bmatrix}.$$

First find $|A|$ since if $|A| = 0$, we can stop, A^{-1} does not exist. Expand along the first row.

$$|A| = (1)(+1)\begin{vmatrix} 1 & 1 \\ 1 & 1 \end{vmatrix} + (2)(-1)\begin{vmatrix} 0 & 1 \\ -1 & 1 \end{vmatrix}$$
$$= (1)[1 - 1] + (-2)[0 - (-1)]$$
$$= (1)[0] + (-2)[1]$$
$$= 0 - 2$$
$$= -2$$

Since $|A| = -2 \neq 0$, we find the matrix of cofactors.

$$cofA = \begin{bmatrix} A_{11} & A_{12} & A_{13} \\ A_{21} & A_{22} & A_{23} \\ A_{31} & A_{32} & A_{33} \end{bmatrix}$$

$$= \begin{bmatrix} +\begin{vmatrix}1 & 1\\1 & 1\end{vmatrix} & -\begin{vmatrix}0 & 1\\-1 & 1\end{vmatrix} & +\begin{vmatrix}0 & 1\\-1 & 1\end{vmatrix} \\ -\begin{vmatrix}2 & 0\\1 & 1\end{vmatrix} & +\begin{vmatrix}1 & 0\\-1 & 1\end{vmatrix} & -\begin{vmatrix}1 & 2\\-1 & 1\end{vmatrix} \\ +\begin{vmatrix}2 & 0\\1 & 1\end{vmatrix} & -\begin{vmatrix}1 & 0\\0 & 1\end{vmatrix} & +\begin{vmatrix}1 & 2\\0 & 1\end{vmatrix} \end{bmatrix}$$

$$= \begin{bmatrix} 0 & -1 & 1 \\ -2 & 1 & -3 \\ 2 & -1 & 1 \end{bmatrix}$$

Then

$$(cofA)^T = \begin{bmatrix} 0 & -2 & 2 \\ -1 & 1 & -1 \\ 1 & -3 & 1 \end{bmatrix}$$

and

$$A^{-1} = \frac{1}{|A|}(cofA)^T = -\frac{1}{2}\begin{bmatrix} 0 & -2 & 2 \\ -1 & 1 & -1 \\ 1 & -3 & 1 \end{bmatrix}.$$

EXERCISES A

21. Find A^{-1} if

$$A = \begin{bmatrix} 1 & 3 & -1 \\ 3 & 1 & 0 \\ -2 & 0 & 1 \end{bmatrix}.$$

First find $|A|$ since if $|A| = 0$, we can stop, A^{-1} does not exist. Expand along the third column.

$$|A| = (-1)(+1)\begin{vmatrix} 3 & 1 \\ -2 & 0 \end{vmatrix} + (1)(+1)\begin{vmatrix} 1 & 3 \\ 3 & 1 \end{vmatrix}$$
$$= (-1)[0 + 2] + (1)[1 - 9]$$
$$= (-1)[2] + (1)[-8]$$
$$= -2 - 8$$
$$= -10$$

Since $|A| = -10 \neq 0$, we find the matrix of cofactors.

$$cof A = \begin{bmatrix} A_{11} & A_{12} & A_{13} \\ A_{21} & A_{22} & A_{23} \\ A_{31} & A_{32} & A_{33} \end{bmatrix}$$

$$= \begin{bmatrix} +\begin{vmatrix} 1 & 0 \\ 0 & 1 \end{vmatrix} & -\begin{vmatrix} 3 & 0 \\ -2 & 1 \end{vmatrix} & +\begin{vmatrix} 3 & 1 \\ -2 & 0 \end{vmatrix} \\ -\begin{vmatrix} 3 & -1 \\ 0 & 1 \end{vmatrix} & +\begin{vmatrix} 1 & -1 \\ -2 & 1 \end{vmatrix} & -\begin{vmatrix} 1 & 3 \\ -2 & 0 \end{vmatrix} \\ +\begin{vmatrix} 3 & -1 \\ 1 & 0 \end{vmatrix} & -\begin{vmatrix} 1 & -1 \\ 3 & 0 \end{vmatrix} & +\begin{vmatrix} 1 & 3 \\ 3 & 1 \end{vmatrix} \end{bmatrix}$$

$$= \begin{bmatrix} 1 & -3 & 2 \\ -3 & -1 & -6 \\ 1 & -3 & -8 \end{bmatrix}$$

Then

$$(cof A)^T = \begin{bmatrix} 1 & -3 & 1 \\ -3 & -1 & -3 \\ 2 & -6 & -8 \end{bmatrix}$$

and

$$A^{-1} = \frac{1}{|A|}(cof A)^T = -\frac{1}{10}\begin{bmatrix} 1 & -3 & 1 \\ -3 & -1 & -3 \\ 2 & -6 & -8 \end{bmatrix}.$$

22. Solve the system using Cramer's rule.

$$\begin{aligned} 2x \quad - z &= -6 \\ 3y + 5z &= 29 \\ x - y + z &= 0 \end{aligned}$$

$$|A| = \begin{vmatrix} 2 & 0 & -1 \\ 0 & 3 & 5 \\ 1 & -1 & 1 \end{vmatrix}$$
$$= (2)(+1)\begin{vmatrix} 3 & 5 \\ -1 & 1 \end{vmatrix} + 0 + (-1)(+1)\begin{vmatrix} 0 & 3 \\ 1 & -1 \end{vmatrix}$$
$$= (2)[3 + 5] + (-1)[0 - 3]$$
$$= (2)[8] + (-1)[-3]$$
$$= 16 + 3$$
$$= 19$$

$$|A_x| = \begin{vmatrix} -6 & 0 & -1 \\ 29 & 3 & 5 \\ 0 & -1 & 1 \end{vmatrix}$$
$$= 0 + (-1)(-1)\begin{vmatrix} -6 & -1 \\ 29 & 5 \end{vmatrix} + (1)(+1)\begin{vmatrix} -6 & 0 \\ 29 & 3 \end{vmatrix}$$
$$= (1)[-30 + 29] + (1)[-18 - 0]$$
$$= (1)[-1] + (1)[-18]$$
$$= -1 - 18$$
$$= -19$$

$$|A_y| = \begin{vmatrix} 2 & -6 & -1 \\ 0 & 29 & 5 \\ 1 & 0 & 1 \end{vmatrix}$$
$$= (1)(+1)\begin{vmatrix} -6 & -1 \\ 29 & 5 \end{vmatrix} + 0 + (1)(+1)\begin{vmatrix} 2 & -6 \\ 0 & 29 \end{vmatrix}$$
$$= (1)[-30 + 29] + (1)[58 - 0]$$
$$= (1)[-1] + (1)[58]$$
$$= -1 + 58$$
$$= 57$$

$$|A_z| = \begin{vmatrix} 2 & 0 & -6 \\ 0 & 3 & 29 \\ 1 & -1 & 0 \end{vmatrix}$$
$$= (1)(+1)\begin{vmatrix} 0 & -6 \\ 3 & 29 \end{vmatrix} + (-1)(-1)\begin{vmatrix} 2 & -6 \\ 0 & 29 \end{vmatrix} + 0$$
$$= (1)[0 + 18] + (1)[58 - 0]$$
$$= (1)[18] + (1)[58]$$
$$= 18 + 58$$
$$= 76$$

$$x = \frac{|A_x|}{|A|} = \frac{-19}{19} = -1 \qquad y = \frac{|A_y|}{|A|} = \frac{57}{19} = 3$$

$$z = \frac{|A_z|}{|A|} = \frac{76}{19} = 4$$

The solution is $(-1, 3, 4)$.

CHAPTER 7 MATRICES AND DETERMINANTS

CHAPTER 7 Review Exercises

Exercises 1-14 refer to the following matrices.

$$A = \begin{bmatrix} -1 & 2 \\ 3 & 4 \end{bmatrix} \quad B = \begin{bmatrix} 0 & -3 \\ 1 & 5 \end{bmatrix} \quad C = \begin{bmatrix} -2 & 4 & 5 \\ 0 & -1 & 2 \end{bmatrix}$$

$$D = \begin{bmatrix} 1 & -3 & 4 \end{bmatrix} \quad E = \begin{bmatrix} 2 \\ -1 \\ 0 \end{bmatrix} \quad F = \begin{bmatrix} 2 & 0 & -1 \\ 4 & 1 & 3 \\ -2 & -1 & 5 \end{bmatrix}$$

1. Since C has 2 rows and 3 columns, the order of C is 2×3.

2. Since A has order 2×2, the zero matrix with the same order is

$$\begin{bmatrix} 0 & 0 \\ 0 & 0 \end{bmatrix}.$$

3. The element a_{21} in matrix A is the element in the 2nd row and 1st column of A. Thus, $a_{21} = 3$.

4. Since $B = \begin{bmatrix} 0 & -3 \\ 1 & 5 \end{bmatrix}$, $-B = \begin{bmatrix} 0 & 3 \\ -1 & -5 \end{bmatrix}$.

5. To find the transpose of matrix C, C^T, interchange the rows and columns of C. Thus,

$$C^T = \begin{bmatrix} -2 & 0 \\ 4 & -1 \\ 5 & 2 \end{bmatrix}.$$

6. $A + B = \begin{bmatrix} -1 & 2 \\ 3 & 4 \end{bmatrix} + \begin{bmatrix} 0 & -3 \\ 1 & 5 \end{bmatrix}$

$$= \begin{bmatrix} -1+0 & 2+(-3) \\ 3+1 & 4+5 \end{bmatrix}$$

$$= \begin{bmatrix} -1 & -1 \\ 4 & 9 \end{bmatrix}$$

7. $A - B = \begin{bmatrix} -1 & 2 \\ 3 & 4 \end{bmatrix} - \begin{bmatrix} 0 & -3 \\ 1 & 5 \end{bmatrix}$

$$= \begin{bmatrix} -1-0 & 2-(-3) \\ 3-1 & 4-5 \end{bmatrix}$$

$$= \begin{bmatrix} -1 & 5 \\ 2 & -1 \end{bmatrix}$$

8. $-3C = -3 \begin{bmatrix} -2 & 4 & 5 \\ 0 & -1 & 2 \end{bmatrix}$

$$= \begin{bmatrix} (-3)(-2) & (-3)(4) & (-3)(5) \\ (-3)(0) & (-3)(-1) & (-3)(2) \end{bmatrix}$$

$$= \begin{bmatrix} 6 & -12 & -15 \\ 0 & 3 & -6 \end{bmatrix}$$

9. $2A - 3B = 2 \begin{bmatrix} -1 & 2 \\ 3 & 4 \end{bmatrix} - 3 \begin{bmatrix} 0 & -3 \\ 1 & 5 \end{bmatrix}$

$$= \begin{bmatrix} -2 & 4 \\ 6 & 8 \end{bmatrix} - \begin{bmatrix} 0 & -9 \\ 3 & 15 \end{bmatrix}$$

$$= \begin{bmatrix} -2 & 13 \\ 3 & -7 \end{bmatrix}$$

10. $D \cdot E = \begin{bmatrix} 1 & -3 & 4 \end{bmatrix} \cdot \begin{bmatrix} 2 \\ -1 \\ 0 \end{bmatrix}$

$= (1)(2) + (-3)(-1) + (4)(0)$
$= 2 + 3 + 0$
$= 5$

11. $AC = \begin{bmatrix} -1 & 2 \\ 3 & 4 \end{bmatrix} \begin{bmatrix} -2 & 4 & 5 \\ 0 & -1 & 2 \end{bmatrix}$

$$= \begin{bmatrix} (-1)(-2)+(2)(0) & (-1)(4)+(2)(-1) & (-1)(5)+(2)(2) \\ (3)(-2)+(4)(0) & (3)(4)+(4)(-1) & (3)(5)+(4)(2) \end{bmatrix}$$

$$= \begin{bmatrix} 2 & -6 & -1 \\ -6 & 8 & 23 \end{bmatrix}$$

12. The product CA is undefined since C has 3 columns and A only has 2 rows.

13. $CF = \begin{bmatrix} -2 & 4 & 5 \\ 0 & -1 & 2 \end{bmatrix} \begin{bmatrix} 2 & 0 & -1 \\ 4 & 1 & 3 \\ -2 & -1 & 5 \end{bmatrix}$

$$= \begin{bmatrix} -4+16-10 & 0+4-5 & 2+12+25 \\ 0-4-4 & 0-1-2 & 0-3+10 \end{bmatrix}$$

$$= \begin{bmatrix} 2 & -1 & 39 \\ -8 & -3 & 7 \end{bmatrix}$$

14. Since only square matrices have determinants, and since C is of order 2×3, $|C|$ is undefined.

CHAPTER 7 REVIEW

15. (a) The matrix that fives the two-month totals in each category is

$$J + A = \begin{bmatrix} 200 & 180 & 20 \\ 300 & 120 & 40 \end{bmatrix} + \begin{bmatrix} 150 & 120 & 30 \\ 220 & 100 & 10 \end{bmatrix}$$

$$= \begin{bmatrix} 200+150 & 180+120 & 20+30 \\ 300+220 & 120+100 & 40+10 \end{bmatrix}$$

$$= \begin{bmatrix} 350 & 300 & 50 \\ 520 & 220 & 50 \end{bmatrix}$$

(b) The matrix $\frac{1}{2}(J+A)$ represents the average number of rentals per agency in each category over the two-month period. Using $J + A$ from part (a), we have:

$$\frac{1}{2}(J+A) = \frac{1}{2}\begin{bmatrix} 350 & 300 & 50 \\ 520 & 220 & 50 \end{bmatrix}$$

$$= \begin{bmatrix} \frac{1}{2}(350) & \frac{1}{2}(300) & \frac{1}{2}(50) \\ \frac{1}{2}(520) & \frac{1}{2}(220) & \frac{1}{2}(50) \end{bmatrix}$$

$$= \begin{bmatrix} 175 & 150 & 25 \\ 260 & 110 & 25 \end{bmatrix}$$

(c) The matrix

$$[1 \ 1]J = [1 \ 1]\begin{bmatrix} 200 & 180 & 20 \\ 300 & 120 & 40 \end{bmatrix}$$

$$= [200+300 \ \ 180+120 \ \ 20+40]$$

$$= [500 \ \ 300 \ \ 60]$$

represents the total number of rentals in each category at both agencies during July.

(d) The matrix

$$[1 \ 1](J+A) = [1 \ 1]\begin{bmatrix} 350 & 300 & 50 \\ 520 & 220 & 50 \end{bmatrix}$$

$$= [350+520 \ \ 300+220 \ \ 50+50]$$

$$= [870 \ \ 520 \ \ 100]$$

represents the total number of rentals in each category at both agencies during the two-month period.

(e) Using the results of part (d),

$$([1 \ 1](J+A))\cdot\begin{bmatrix} 1 \\ 1 \\ 1 \end{bmatrix}$$

$$= [870 \ \ 520 \ \ 100]\cdot\begin{bmatrix} 1 \\ 1 \\ 1 \end{bmatrix}$$

$$= (870)(1)+(520)(1)+(100)(1)$$

$$= 1490$$

represents the total number of rentals in each category at both agencies during the two-month period.

(f) The matrix

$$JC = \begin{bmatrix} 200 & 180 & 20 \\ 300 & 120 & 40 \end{bmatrix}\begin{bmatrix} 80 \\ 120 \\ 150 \end{bmatrix}$$

$$= \begin{bmatrix} (200)(80)+(180)(120)+(20)(150) \\ (300)(80)+(120)(120)+(40)(150) \end{bmatrix}$$

$$= \begin{bmatrix} 40,600 \\ 44,400 \end{bmatrix}$$

represents the total revenue produced by each agency during July.

(g) Using $J + A$ from part (a), the matrix

$$(J+A)C = \begin{bmatrix} 350 & 300 & 50 \\ 520 & 220 & 50 \end{bmatrix}\begin{bmatrix} 80 \\ 120 \\ 150 \end{bmatrix}$$

$$= \begin{bmatrix} (350)(80)+(300)(120)+(50)(150) \\ (520)(80)+(220)(120)+(50)(150) \end{bmatrix}$$

$$= \begin{bmatrix} 71,500 \\ 75,500 \end{bmatrix}$$

represents the total revenue by each agency during the two-month period.

(h) Using $(J + A)C$ from part (g),

$$[1\ 1] \cdot ((J+A)C) = [1\ 1] \cdot \begin{bmatrix} 71{,}500 \\ 75{,}500 \end{bmatrix}$$
$$= (1)(71{,}500) + (1)(75{,}500)$$
$$= 147{,}000$$

represents $147,000, the total revenue produced by both agencies during the two-month period.

16. Solve using the Gaussian method.

$$x - 3y = 4$$
$$4x + 5y = -1$$

Write the augmented matrix of the system.

$$\begin{bmatrix} 1 & -3 & 4 \\ 4 & 5 & -1 \end{bmatrix}$$

Since we already have a 1 in the upper left corner, use this 1 to obtain the 0 in the first column. Keep the first row, multiply it by −4, and add the results to the second row.

$$\begin{bmatrix} 1 & -3 & 4 \\ 0 & 17 & -17 \end{bmatrix}$$

Multiply the second row by $\frac{1}{17}$.

$$\begin{bmatrix} 1 & -3 & 4 \\ 0 & 1 & -1 \end{bmatrix}$$

Keep the second row, multiply it by 3, and add the results to the first row.

$$\begin{bmatrix} 1 & 0 & 1 \\ 0 & 1 & -1 \end{bmatrix}$$

Since this corresponds to $x = 1$ and $y = -1$, the solution to the system is $(1,-1)$.

17. Solve using the Gaussian method.

$$x - 2y + 3z = 7$$
$$-x + 3y + 2z = 8$$
$$3x - 4y - z = -9$$

Keep the first row, add it to the second row. Then multiply the first row by −3, and add the results to the third row.

$$\begin{bmatrix} 1 & -2 & 3 & 7 \\ 0 & 1 & 5 & 15 \\ 0 & 2 & -10 & -30 \end{bmatrix}$$

Keep the second row, multiply it by 2, and add the results to the first row. Then multiply the second row by −2, and add the results to the third row.

$$\begin{bmatrix} 1 & 0 & 13 & 37 \\ 0 & 1 & 5 & 15 \\ 0 & 0 & -20 & -60 \end{bmatrix}$$

Multiply the third row by $-\frac{1}{20}$.

$$\begin{bmatrix} 1 & 0 & 13 & 37 \\ 0 & 1 & 5 & 15 \\ 0 & 0 & 1 & 3 \end{bmatrix}$$

Keep the third row, multiply it by −13, and add the results to the first row. Then multiply the third row by −5, and add the results to the second row.

$$\begin{bmatrix} 1 & 0 & 0 & -2 \\ 0 & 1 & 0 & 0 \\ 0 & 0 & 1 & 3 \end{bmatrix}$$

Since this corresponds to $x = -2$, $y = 0$, and $z = 3$, the solution to the system is $(-2,0,3)$.

18. Find the inverse of the matrix.

$$\begin{bmatrix} -5 & -2 \\ 3 & 1 \end{bmatrix}$$

First form the augmented matrix $[A|I]$.

$$\begin{bmatrix} -5 & -2 & | & 1 & 0 \\ 3 & 1 & | & 0 & 1 \end{bmatrix}$$

Multiply the first row by $-\frac{1}{5}$.

$$\begin{bmatrix} 1 & \frac{2}{5} & | & -\frac{1}{5} & 0 \\ 3 & 1 & | & 0 & 1 \end{bmatrix}$$

Keep the first row, multiply it by −3, and add the results to the second row.

$$\begin{bmatrix} 1 & \frac{2}{5} & | & -\frac{1}{5} & 0 \\ 0 & -\frac{1}{5} & | & \frac{3}{5} & 1 \end{bmatrix}$$

Multiply the second row by −5.

$$\begin{bmatrix} 1 & \frac{2}{5} & | & -\frac{1}{5} & 0 \\ 0 & 1 & | & -3 & -5 \end{bmatrix}$$

Keep the second row, multiply it by $-\frac{2}{5}$, and add the results to the first row.

$$\begin{bmatrix} 1 & 0 & | & 1 & 2 \\ 0 & 1 & | & -3 & -5 \end{bmatrix}$$

Thus, $A^{-1} = \begin{bmatrix} 1 & 2 \\ -3 & -5 \end{bmatrix}$.

19. Find the inverse of the matrix.

$$\begin{bmatrix} 3 & 0 & 1 \\ 1 & -1 & 0 \\ 0 & 1 & 2 \end{bmatrix}$$

Form the augmented matrix.

$$\begin{bmatrix} 3 & 0 & 1 & | & 1 & 0 & 0 \\ 1 & -1 & 0 & | & 0 & 1 & 0 \\ 0 & 1 & 2 & | & 0 & 0 & 1 \end{bmatrix}$$

Interchange the first two rows.

$$\begin{bmatrix} 1 & -1 & 0 & | & 0 & 1 & 0 \\ 3 & 0 & 1 & | & 1 & 0 & 0 \\ 0 & 1 & 2 & | & 0 & 0 & 1 \end{bmatrix}$$

Keep the first row (and the third row), multiply the first row by -3, and add the results to the second row.

$$\begin{bmatrix} 1 & -1 & 0 & | & 0 & 1 & 0 \\ 0 & 3 & 1 & | & 1 & -3 & 0 \\ 0 & 1 & 2 & | & 0 & 0 & 1 \end{bmatrix}$$

Interchange the second and third rows.

$$\begin{bmatrix} 1 & -1 & 0 & | & 0 & 1 & 0 \\ 0 & 1 & 2 & | & 0 & 0 & 1 \\ 0 & 3 & 1 & | & 1 & -3 & 0 \end{bmatrix}$$

Keep the second row, multiply it by -3 and add to the third row. Then add the second row to the first.

$$\begin{bmatrix} 1 & 0 & 2 & | & 0 & 1 & 1 \\ 0 & 1 & 2 & | & 0 & 0 & 1 \\ 0 & 0 & -5 & | & 1 & -3 & -3 \end{bmatrix}$$

Multiply the third row by $-\frac{1}{5}$.

$$\begin{bmatrix} 1 & 0 & 2 & | & 0 & 1 & 1 \\ 0 & 1 & 2 & | & 0 & 0 & 1 \\ 0 & 0 & 1 & | & -\frac{1}{5} & \frac{3}{5} & \frac{3}{5} \end{bmatrix}$$

Keep the third row, multiply it by -2, and add the results to the second row and then to the first row.

$$\begin{bmatrix} 1 & 0 & 0 & | & \frac{2}{5} & -\frac{1}{5} & -\frac{1}{5} \\ 0 & 1 & 0 & | & \frac{2}{5} & -\frac{6}{5} & -\frac{1}{5} \\ 0 & 0 & 1 & | & -\frac{1}{5} & \frac{3}{5} & \frac{3}{5} \end{bmatrix}$$

Thus, $A^{-1} = \begin{bmatrix} \frac{2}{5} & -\frac{1}{5} & -\frac{1}{5} \\ \frac{2}{5} & -\frac{6}{5} & -\frac{1}{5} \\ -\frac{1}{5} & \frac{3}{5} & \frac{3}{5} \end{bmatrix}$.

20. Solve using the inverse method.

$$\begin{aligned} -5x - 2y &= 7 \\ 3x + y &= -5 \end{aligned}$$

Then this system is equivalent to the matrix equation $AX = B$, where

$$A = \begin{bmatrix} -5 & -2 \\ 3 & 1 \end{bmatrix} \quad X = \begin{bmatrix} x \\ y \end{bmatrix} \quad B = \begin{bmatrix} 7 \\ -5 \end{bmatrix}.$$

Using A^{-1} found in Exercise 18, we have:

$$\begin{bmatrix} x \\ y \end{bmatrix} = X = A^{-1}B = \begin{bmatrix} 1 & 2 \\ -3 & -5 \end{bmatrix}\begin{bmatrix} 7 \\ -5 \end{bmatrix} = \begin{bmatrix} -3 \\ 4 \end{bmatrix}$$

Thus, $x = -3$, $y = 4$, and the solution to the system is $(-3, 4)$.

21. Solve using the inverse method.

$$\begin{aligned} 3x + z &= 9 \\ x - y &= 4 \\ y + 2z &= 4 \end{aligned}$$

Then this system is equivalent to the matrix equation $AX = B$, where

$$A = \begin{bmatrix} 3 & 0 & 1 \\ 1 & -1 & 0 \\ 0 & 1 & 2 \end{bmatrix} \quad X = \begin{bmatrix} x \\ y \\ z \end{bmatrix} \quad B = \begin{bmatrix} 9 \\ 4 \\ 4 \end{bmatrix}.$$

Using A^{-1} found in Exercise 19, we have:

$$\begin{bmatrix} x \\ y \\ z \end{bmatrix} = X = A^{-1}B = \begin{bmatrix} \frac{2}{5} & -\frac{1}{5} & -\frac{1}{5} \\ \frac{2}{5} & -\frac{6}{5} & -\frac{1}{5} \\ -\frac{1}{5} & \frac{3}{5} & \frac{3}{5} \end{bmatrix} \begin{bmatrix} 9 \\ 4 \\ 4 \end{bmatrix} = \begin{bmatrix} 2 \\ -2 \\ 3 \end{bmatrix}$$

Thus, $x = 2$, $y = -2$, $z = 3$, and the solution to the system is $(2, -2, 3)$.

22. Let $x =$ the number of individual memberships,
$y =$ the number of family memberships.

Then the total number of memberships is represented by:

$$x + y$$

The total revenue produced by these memberships is given by:

$$300x + 400y$$

The table gives three sets of values, a and b, where a is the number of memberships and b is the revenue produced. These values give rise to the three systems

$$\begin{aligned} x + y &= a \\ 300x + 400y &= b \end{aligned}$$

formed by replacing a and b by the values in the table. As a result, we can solve these systems by finding the inverse of the coefficient matrix and using the inverse method. It is easy to show that the inverse matrix is:

$$\begin{bmatrix} 4 & -\frac{1}{100} \\ -3 & \frac{1}{100} \end{bmatrix} = \begin{bmatrix} 4 & -0.01 \\ -3 & 0.01 \end{bmatrix}$$

Thus, the solutions to each system can be found by multiplying this inverse matrix times the matrix of constants from the table. Thus, for January, we have:

$$\begin{bmatrix} x \\ y \end{bmatrix} = \begin{bmatrix} 4 & -0.01 \\ -3 & 0.01 \end{bmatrix} \begin{bmatrix} 30 \\ 10,000 \end{bmatrix} = \begin{bmatrix} 20 \\ 10 \end{bmatrix}$$

This means that in January, 20 individual and 10 family memberships must be sold.

For February, we have:

$$\begin{bmatrix} x \\ y \end{bmatrix} = \begin{bmatrix} 4 & -0.01 \\ -3 & 0.01 \end{bmatrix} \begin{bmatrix} 50 \\ 17,000 \end{bmatrix} = \begin{bmatrix} 30 \\ 20 \end{bmatrix}$$

This means that in February, 30 individual and 20 family memberships must be sold.

For March, we have:

$$\begin{bmatrix} x \\ y \end{bmatrix} = \begin{bmatrix} 4 & -0.01 \\ -3 & 0.01 \end{bmatrix} \begin{bmatrix} 65 \\ 23,000 \end{bmatrix} = \begin{bmatrix} 30 \\ 35 \end{bmatrix}$$

This means that in March, 30 individual and 35 family memberships must be sold.

23. Evaluate the determinant.

$$\begin{vmatrix} -3 & 1 \\ 5 & 7 \end{vmatrix} = (-3)(7) - (1)(5) = -21 - 5 = -26$$

24. We expand along the first row in the evaluation of the following determinant.

$$\begin{vmatrix} 2 & -1 & 3 \\ -2 & 2 & 4 \\ 1 & -3 & 5 \end{vmatrix}$$

$$= (2)(-1)^{1+1}\begin{vmatrix} 2 & 4 \\ -3 & 5 \end{vmatrix} + (-1)(-1)^{1+2}\begin{vmatrix} -2 & 4 \\ 1 & 5 \end{vmatrix}$$
$$+ (3)(-1)^{1+3}\begin{vmatrix} -2 & 2 \\ 1 & -3 \end{vmatrix}$$
$$= (2)[(2)(5)-(4)(-3)] + (1)[(-2)(5)-(4)(1)]$$
$$\quad (3)[(-2)(-3)-(2)(1)]$$
$$= (2)[22] + (1)[-14] + (3)[4]$$
$$= 44 - 14 + 12$$
$$= 42$$

25. Since the element 4, in the matrix whose determinant is given in Exercise 24, is in the 2nd row, 3rd column, the cofactor of 4 is $(-1)^{2+3}$ times the minor of 4, the determinant formed by deleting the 2nd row and 3rd column of the matrix. Thus,

$$A_{23} = (-1)^{2+3}\begin{vmatrix} 2 & -1 \\ 1 & -3 \end{vmatrix} = (-1)[-6+1] = 5.$$

26. Solve using Cramer's rule.

$$\begin{aligned} -2x + y &= 9 \\ 3x + 4y &= 14 \end{aligned}$$

The determinant of the coefficient matrix, $|A|$, is:

$$|A| = \begin{vmatrix} -2 & 1 \\ 3 & 4 \end{vmatrix} = (-2)(4) - (1)(3) = -8 - 3 = -11$$

To find $|A_x|$, replace the coefficients of x with the constants in $|A|$.

$$|A_x| = \begin{vmatrix} 9 & 1 \\ 14 & 4 \end{vmatrix} = (9)(4) - (1)(14) = 36 - 14 = 22$$

To find $|A_y|$, replace the coefficients of y with the constants in $|A|$.

$$|A_y| = \begin{vmatrix} -2 & 9 \\ 3 & 14 \end{vmatrix} = (-2)(14) - (9)(3) = -28 - 27 = -55$$

Then

$$x = \frac{|A_x|}{|A|} = \frac{22}{-11} = -2 \quad y = \frac{|A_y|}{|A|} = \frac{-55}{-11} = 5.$$

Thus, the solution to the system is $(-2, 5)$.

27. Solve using Cramer's rule.

$$\begin{aligned} 3x - y + z &= -12 \\ -2x + y + 3z &= 9 \\ x - y - 2z &= -6 \end{aligned}$$

The determinant of the coefficient matrix, found by expanding along the first row, is:

$$|A| = \begin{vmatrix} 3 & -1 & 1 \\ -2 & 1 & 3 \\ 1 & -1 & -2 \end{vmatrix}$$

$$= (3)(+1)\begin{vmatrix} 1 & 3 \\ -1 & -2 \end{vmatrix} + (-1)(-1)\begin{vmatrix} -2 & 3 \\ 1 & -2 \end{vmatrix}$$

$$+ (1)(+1)\begin{vmatrix} -2 & 1 \\ 1 & -1 \end{vmatrix}$$

$$= (3)[-2+3] + (1)[4-3] + (1)[2-1]$$
$$= (3)[1] + (1)[1] + (1)[1]$$
$$= 3 + 1 + 1$$
$$= 5$$

Replace the coefficients of x with the constants, and evaluate along the first row.

$$|A_x| = \begin{vmatrix} -12 & -1 & 1 \\ 9 & 1 & 3 \\ -6 & -1 & -2 \end{vmatrix}$$

$$= (-12)(+1)\begin{vmatrix} 1 & 3 \\ -1 & -2 \end{vmatrix} + (-1)(-1)\begin{vmatrix} 9 & 3 \\ -6 & -2 \end{vmatrix}$$

$$+ (1)(+1)\begin{vmatrix} 9 & 1 \\ -6 & -1 \end{vmatrix}$$

$$= (-12)[-2+3] + (1)[-18+18] + (1)[-9+6]$$
$$= (-12)[1] + (1)[0] + (1)[-3]$$
$$= -12 + 0 - 3$$
$$= -15$$

Replace the coefficients of y with the constants, and evaluate along the first row.

$$|A_y| = \begin{vmatrix} 3 & -12 & 1 \\ -2 & 9 & 3 \\ 1 & -6 & -2 \end{vmatrix}$$

$$= (3)(+1)\begin{vmatrix} 9 & 3 \\ -6 & -2 \end{vmatrix} + (-12)(-1)\begin{vmatrix} -2 & 3 \\ 1 & -2 \end{vmatrix}$$

$$+ (1)(+1)\begin{vmatrix} -2 & 9 \\ 1 & -6 \end{vmatrix}$$

$$= (3)[-18+18] + (12)[4-3] + (1)[12-9]$$
$$= (3)[0] + (12)[1] + (1)[3]$$
$$= 0 + 12 + 3$$
$$= 15$$

Replace the coefficients of z with the constants, and evaluate along the first row.

$$|A_z| = \begin{vmatrix} 3 & -1 & -12 \\ -2 & 1 & 9 \\ 1 & -1 & -6 \end{vmatrix}$$

$$= (3)(+1)\begin{vmatrix} 1 & 9 \\ -1 & -6 \end{vmatrix} + (-1)(-1)\begin{vmatrix} -2 & 9 \\ 1 & -6 \end{vmatrix}$$

$$+ (-12)(+1)\begin{vmatrix} -2 & 1 \\ 1 & -1 \end{vmatrix}$$

$$= (3)[-6+9] + (1)[12-9] + (-12)[2-1]$$
$$= (3)[3] + (1)[3] + (-12)[1]$$
$$= 9 + 3 - 12$$
$$= 0$$

Thus,

$$x = \frac{|A_x|}{|A|} = \frac{-15}{5} = -3 \quad y = \frac{|A_y|}{|A|} = \frac{15}{5} = 3$$

$$z = \frac{|A_z|}{|A|} = \frac{0}{5} = 0.$$

The solution to the system is $(-3, 3, 0)$.

28. Solve for x.

$$\begin{vmatrix} 2 & x \\ x & 5 \end{vmatrix} = 6$$
$$(2)(5) - x^2 = 6$$
$$10 - x^2 = 6$$
$$-x^2 = -4$$
$$x^2 = 4$$
$$x = \pm\sqrt{4}$$
$$x = \pm 2$$

Thus, the solutions are 2 and -2.

29. To solve for x, we expand the determinant on the left along the first column taking advantage of the 0's.

$$\begin{vmatrix} x & 1 & 3 \\ 0 & 2 & x \\ 0 & 1 & -1 \end{vmatrix} = -3$$

$$x \begin{vmatrix} 2 & x \\ 1 & -1 \end{vmatrix} = -3$$

$$x(-2 - x) = -3$$
$$-2x - x^2 = -3$$
$$0 = x^2 + 2x - 3$$

Thus we obtain the quadratic equation:

$$x^2 + 2x - 3 = 0$$
$$(x + 3)(x - 1) = 0$$
$$x + 3 = 0 \qquad x - 1 = 0$$
$$x = -3 \qquad x = 1$$

Thus, the solutions are -3 and 1.

30. We know that

$$\begin{vmatrix} 0 & 0 \\ 5 & -7 \end{vmatrix} = 0$$

since the first row contains all 0's. Any determinant that has a row or column that is all zeros is 0.

31. We know that

$$\begin{vmatrix} 3 & -2 & 3 \\ 5 & 0 & 5 \\ 1 & 6 & 1 \end{vmatrix} = 0$$

since the first and third columns are equal. Any determinant that has two rows or two columns the same is 0.

32. We know that

$$\begin{vmatrix} 2 & -3 \\ 1 & 7 \end{vmatrix} = - \begin{vmatrix} -3 & 2 \\ 7 & 1 \end{vmatrix}$$

since the two columns were interchanged. Any determinant formed from another determinant by interchanging two rows or columns is the negative of the first determinant.

33. We know that

$$\begin{vmatrix} 3 & -3 \\ 0 & 4 \end{vmatrix} = 3 \begin{vmatrix} 1 & -1 \\ 0 & 4 \end{vmatrix}$$

since the determinant on the left was obtained by multiplying the first row of the determinant on the right by the constant 3. Any determinant formed from another determinant by multiplying a row or column by a constant has value equal to that constant times the original determinant.

34. Suppose that $\begin{vmatrix} 0 & 1 \\ x & 5 \end{vmatrix}$ was obtained from $\begin{vmatrix} 2 & 1 \\ -3 & 5 \end{vmatrix}$ by using Theorem 7 of Section 7.6. Notice that both determinants have the same second column. If the second column of the second is multiplied by -2 and the results added to the first column, we obtain the 0 in the first row, first column. Thus, x comes from multiplying 5 by -2 and adding the result to -3. That is

$$x = (5)(-2) + (-3)$$
$$x = -13.$$

35. Evaluate the determinant by introducing two 0's in column 2.

$$\begin{vmatrix} -2 & 5 & 2 \\ 4 & 1 & 7 \\ 1 & -3 & -1 \end{vmatrix}$$

Use the 1 in the second column to obtain the desired 0's. Keep the second row, multiply it by -5, and add the results to the first row. Then multiply the second row by 3, and add the results to the third row.

$$\begin{vmatrix} -22 & 0 & -33 \\ 4 & 1 & 7 \\ 13 & 0 & 20 \end{vmatrix}$$

Then expanding along the second column, we have:

$$\begin{vmatrix} -22 & 0 & -33 \\ 4 & 1 & 7 \\ 13 & 0 & 20 \end{vmatrix} = 0 + (1)(+1)\begin{vmatrix} -22 & -33 \\ 13 & 20 \end{vmatrix} + 0$$

$$= (1)[(-22)(20) - (-33)(13)]$$
$$= (1)[-440 + 429]$$
$$= -11$$

36.
$$|A| = \begin{vmatrix} a_1 & a_2 & 0 & 0 \\ a_3 & a_4 & 0 & 0 \\ 0 & 0 & b_1 & b_2 \\ 0 & 0 & b_3 & b_4 \end{vmatrix}$$

$$= a_1 \begin{vmatrix} a_4 & 0 & 0 \\ 0 & b_1 & b_2 \\ 0 & b_3 & b_4 \end{vmatrix} - a_3 \begin{vmatrix} a_2 & 0 & 0 \\ 0 & b_1 & b_2 \\ 0 & b_3 & b_4 \end{vmatrix}$$

$$= a_1 a_4 \begin{vmatrix} b_1 & b_2 \\ b_3 & b_4 \end{vmatrix} - a_3 a_2 \begin{vmatrix} b_1 & b_2 \\ b_3 & b_4 \end{vmatrix}$$

$$= (a_1 a_4 - a_3 a_2) \begin{vmatrix} b_1 & b_2 \\ b_3 & b_4 \end{vmatrix}$$

$$= \begin{vmatrix} a_1 & a_2 \\ a_3 & a_4 \end{vmatrix} \begin{vmatrix} b_1 & b_2 \\ b_3 & b_4 \end{vmatrix}$$

37. Let x = the number of standard models,
    y = the number of deluxe models.

Since the labor costs for one standard table are $20, and the labor costs for for one deluxe table are $40, the total labor costs amount to

$$20x + 40y.$$

Similarly, the total costs for materials for the two types of tables are

$$25x + 80y.$$

Since the table gives the available labor and materials costs for three different weeks, letting a be the total labor costs and b the total materials costs, we have three systems,

$$20x + 40y = a$$
$$25x + 80y = b$$

for the three different sets of values of a and b. We can use the inverse method to solve these three systems by finding the inverse of the coefficient matrix. It is easy to show that the inverse matrix is:

$$\begin{bmatrix} \frac{2}{15} & -\frac{1}{15} \\ -\frac{1}{24} & \frac{1}{30} \end{bmatrix}$$

Then for week 1, $a = 1800$ and $b = 3300$, so

$$\begin{bmatrix} x \\ y \end{bmatrix} = \begin{bmatrix} \frac{2}{15} & -\frac{1}{15} \\ -\frac{1}{24} & \frac{1}{30} \end{bmatrix} \begin{bmatrix} 1800 \\ 3300 \end{bmatrix} = \begin{bmatrix} 20 \\ 35 \end{bmatrix}.$$

Thus, $x = 20$, $y = 35$, and during week 1, 20 standard model and 35 deluxe model tables should be produced.

Then for week 2, $a = 1400$ and $b = 2200$, so

$$\begin{bmatrix} x \\ y \end{bmatrix} = \begin{bmatrix} \frac{2}{15} & -\frac{1}{15} \\ -\frac{1}{24} & \frac{1}{30} \end{bmatrix} \begin{bmatrix} 1400 \\ 2200 \end{bmatrix} = \begin{bmatrix} 40 \\ 15 \end{bmatrix}.$$

Thus, $x = 40$, $y = 15$, and during week 2, 40 standard model and 15 deluxe model tables should be produced.

Then for week 3, $a = 2200$ and $b = 3950$, so

$$\begin{bmatrix} x \\ y \end{bmatrix} = \begin{bmatrix} \frac{2}{15} & -\frac{1}{15} \\ -\frac{1}{24} & \frac{1}{30} \end{bmatrix} \begin{bmatrix} 2200 \\ 3950 \end{bmatrix} = \begin{bmatrix} 30 \\ 40 \end{bmatrix}.$$

Thus, $x = 30$, $y = 40$, and during week 3, 30 standard model and 40 deluxe model tables should be produced.

38. Find A^{-1} if $A = \begin{bmatrix} 2 & -1 \\ 3 & 2 \end{bmatrix}$.

Form the augmented matrix, $[A|I]$.

$$\begin{bmatrix} 2 & -1 & | & 1 & 0 \\ 3 & 2 & | & 0 & 1 \end{bmatrix}$$

Multiply the second row by -1, and add the results to the first row.

$$\begin{bmatrix} -1 & -3 & | & 1 & -1 \\ 3 & 2 & | & 0 & 1 \end{bmatrix}$$

Multiply the first row by -1.

$$\begin{bmatrix} 1 & 3 & | & -1 & 1 \\ 3 & 2 & | & 0 & 1 \end{bmatrix}$$

Keep the first row, multiply it by -3, and add the results to the third row.

$$\begin{bmatrix} 1 & 3 & | & -1 & 1 \\ 0 & -7 & | & 3 & -2 \end{bmatrix}$$

Multiply the second row by $-\frac{1}{7}$.

$$\begin{bmatrix} 1 & 3 & | & -1 & 1 \\ 0 & 1 & | & -\frac{3}{7} & \frac{2}{7} \end{bmatrix}$$

Keep the second row, multiply it by -3, and add the results to the first row.

$$\begin{bmatrix} 1 & 0 & | & \frac{2}{7} & \frac{1}{7} \\ 0 & 1 & | & -\frac{3}{7} & \frac{2}{7} \end{bmatrix}$$

Thus,

$$A^{-1} = \begin{bmatrix} \frac{2}{7} & \frac{1}{7} \\ -\frac{3}{7} & \frac{2}{7} \end{bmatrix}.$$

39. Find A^{-1} if $A = \begin{bmatrix} 2 & -1 & 1 \\ 1 & 0 & 2 \\ 0 & 1 & -5 \end{bmatrix}$.

First form the augmented matrix $[A|I]$.

$$\begin{bmatrix} 2 & -1 & 1 & | & 1 & 0 & 0 \\ 1 & 0 & 2 & | & 0 & 1 & 0 \\ 0 & 1 & -5 & | & 0 & 0 & 1 \end{bmatrix}$$

Interchange the first two rows.

$$\begin{bmatrix} 1 & 0 & 2 & | & 0 & 1 & 0 \\ 2 & -1 & 1 & | & 1 & 0 & 0 \\ 0 & 1 & -5 & | & 0 & 0 & 1 \end{bmatrix}$$

Keep the first row (and the third row), multiply the first row by -2, and add the results to the second row.

$$\begin{bmatrix} 1 & 0 & 2 & | & 0 & 1 & 0 \\ 0 & -1 & -3 & | & 1 & -2 & 0 \\ 0 & 1 & -5 & | & 0 & 0 & 1 \end{bmatrix}$$

Add the second row to the third row to obtain the desired 0. Then multiply the second row by -1 to obtain the desired 1.

$$\begin{bmatrix} 1 & 0 & 2 & | & 0 & 1 & 0 \\ 0 & 1 & 3 & | & -1 & 2 & 0 \\ 0 & 0 & -8 & | & 1 & -2 & 1 \end{bmatrix}$$

Multiply the third row by $-\frac{1}{8}$.

$$\begin{bmatrix} 1 & 0 & 2 & | & 0 & 1 & 0 \\ 0 & 1 & 3 & | & -1 & 2 & 0 \\ 0 & 0 & 1 & | & -\frac{1}{8} & \frac{1}{4} & -\frac{1}{8} \end{bmatrix}$$

Keep the third row, multiply it by -2, and add the results to the first row. Then multiply the third row by -3, and add the results to the second row.

$$\begin{bmatrix} 0 & 0 & 1 & | & \frac{1}{4} & \frac{1}{2} & \frac{1}{4} \\ 0 & 1 & 0 & | & -\frac{5}{8} & \frac{5}{4} & \frac{3}{8} \\ 0 & 0 & 1 & | & -\frac{1}{8} & \frac{1}{4} & -\frac{1}{8} \end{bmatrix}$$

Thus,

$$A^{-1} = \begin{bmatrix} \frac{1}{4} & \frac{1}{2} & \frac{1}{4} \\ -\frac{5}{8} & \frac{5}{4} & \frac{3}{8} \\ -\frac{1}{8} & \frac{1}{4} & -\frac{1}{8} \end{bmatrix}$$

40. Solve using the inverse method.

$$\begin{aligned} 2x - y &= 11 \\ 3x + 2y &= -1 \end{aligned}$$

This system is equivalent to the matrix equation $AX = B$, where

$$A = \begin{bmatrix} 2 & -1 \\ 3 & 2 \end{bmatrix} \quad X = \begin{bmatrix} x \\ y \end{bmatrix} \quad B = \begin{bmatrix} 11 \\ -1 \end{bmatrix}.$$

Using the value of A^{-1} found in Exercise 38, we have:

$$\begin{bmatrix} x \\ y \end{bmatrix} = X = A^{-1}B = \begin{bmatrix} \frac{2}{7} & \frac{1}{7} \\ -\frac{3}{7} & \frac{2}{7} \end{bmatrix} \begin{bmatrix} 11 \\ -1 \end{bmatrix} = \begin{bmatrix} 3 \\ -5 \end{bmatrix}.$$

Thus, $x = 3$, $y = -5$, and the solution to the system is $(3, -5)$.

41. Solve using the inverse method.

$$\begin{aligned} 2x - y + z &= 3 \\ x + 2z &= 0 \\ y - 5z &= 5 \end{aligned}$$

This system is equivalent to the matrix equation $AX = B$, where

$$A = \begin{bmatrix} 2 & -1 & 1 \\ 1 & 0 & 2 \\ 0 & 1 & -5 \end{bmatrix} \quad X = \begin{bmatrix} x \\ y \\ z \end{bmatrix} \quad B = \begin{bmatrix} 3 \\ 0 \\ 5 \end{bmatrix}.$$

Using the value of A^{-1} found in Exercise 39, we have:

$$\begin{bmatrix} x \\ y \\ z \end{bmatrix} = X = A^{-1}B = \begin{bmatrix} \frac{1}{4} & \frac{1}{2} & \frac{1}{4} \\ -\frac{5}{8} & \frac{5}{4} & \frac{3}{8} \\ -\frac{1}{8} & \frac{1}{4} & -\frac{1}{8} \end{bmatrix} \begin{bmatrix} 3 \\ 0 \\ 5 \end{bmatrix} = \begin{bmatrix} 2 \\ 0 \\ -1 \end{bmatrix}.$$

Thus, $x = 2$, $y = 0$, $z = -1$, and the solution to the system is $(2, 0, -1)$.

42. Solve using the Gaussian method.

$$\begin{aligned} 2x - y &= 11 \\ 3x + 2y &= -1 \end{aligned}$$

First form the augmented matrix.

$$\begin{bmatrix} 2 & -1 & 11 \\ 3 & 2 & -1 \end{bmatrix}$$

Multiply the first row by -2, and add the results to the second row.

$$\begin{bmatrix} 2 & -1 & 11 \\ -1 & 4 & -23 \end{bmatrix}$$

Multiply the second row by -1, and interchange the two rows.

$$\begin{bmatrix} 1 & -4 & 23 \\ 2 & -1 & 11 \end{bmatrix}$$

Keep the first row, multiply it by -2, and add the results to the second row.

$$\begin{bmatrix} 1 & -4 & 23 \\ 0 & 7 & -35 \end{bmatrix}$$

Multiply the second row by $\frac{1}{7}$.

$$\begin{bmatrix} 1 & -4 & 23 \\ 0 & 1 & -5 \end{bmatrix}$$

Keep the second row, multiply it by 4, and add the results to the first row.

$$\begin{bmatrix} 1 & 0 & 3 \\ 0 & 1 & -5 \end{bmatrix}$$

Since this corresponds to $x = 3$ and $y = -5$, the solution to the system is $(3, -5)$.

43. Solve using the Gaussian method.

$$\begin{aligned} 2x - y + z &= 3 \\ x + 2z &= 0 \\ y - 5z &= 5 \end{aligned}$$

First form the augmented matrix.

$$\begin{bmatrix} 2 & -1 & 1 & 3 \\ 1 & 0 & 2 & 0 \\ 0 & 1 & -5 & 5 \end{bmatrix}$$

Interchange the first two rows.

$$\begin{bmatrix} 1 & 0 & 2 & 0 \\ 2 & -1 & 1 & 3 \\ 0 & 1 & -5 & 5 \end{bmatrix}$$

Keep the first row (and the third), multiply the first row by -2, and add the results to the second row.

$$\begin{bmatrix} 1 & 0 & 2 & 0 \\ 0 & -1 & -3 & 3 \\ 0 & 1 & -5 & 5 \end{bmatrix}$$

Add the second row to the third row. Then multiply the second row by -1.

$$\begin{bmatrix} 1 & 0 & 2 & 0 \\ 0 & 1 & 3 & -3 \\ 0 & 0 & -8 & 8 \end{bmatrix}$$

Multiply the third row by $-\frac{1}{8}$.

$$\begin{bmatrix} 1 & 0 & 2 & 0 \\ 0 & 1 & 3 & -3 \\ 0 & 0 & 1 & -1 \end{bmatrix}$$

Keep the third row, multiply it by -2, and add the results to the first row. Them multiply the third row by -3, and add the results to the second row.

$$\begin{bmatrix} 1 & 0 & 0 & 2 \\ 0 & 1 & 0 & 0 \\ 0 & 0 & 1 & -1 \end{bmatrix}$$

Since this corresponds to $x = 2$, $y = 0$, and $z = -1$, the solution to the system is $(2, 0, -1)$.

Use the following matrices in Exercises 44-64.

$$A = \begin{bmatrix} 2 & 3 \\ -1 & 4 \end{bmatrix} \quad B = \begin{bmatrix} 1 & 3 & 0 \\ -2 & 5 & 3 \end{bmatrix} \quad C = \begin{bmatrix} 2 \\ 1 \end{bmatrix} \quad D = \begin{bmatrix} 3 & 4 \end{bmatrix}$$

$$0 = \begin{bmatrix} 0 & 0 \\ 0 & 0 \end{bmatrix} \quad G = \begin{bmatrix} 3 & -2 & 4 \\ 1 & -4 & 0 \end{bmatrix} \quad H = \begin{bmatrix} x \\ y \end{bmatrix} \quad I = \begin{bmatrix} 1 & 0 \\ 0 & 1 \end{bmatrix}$$

44. $A + 0 = \begin{bmatrix} 2 & 3 \\ -1 & 4 \end{bmatrix} + \begin{bmatrix} 0 & 0 \\ 0 & 0 \end{bmatrix} = \begin{bmatrix} 2 & 3 \\ -1 & 4 \end{bmatrix} = A$

45. $B + G = \begin{bmatrix} 1 & 3 & 0 \\ -2 & 5 & 3 \end{bmatrix} + \begin{bmatrix} 3 & -2 & 4 \\ 1 & -4 & 0 \end{bmatrix}$

$= \begin{bmatrix} 1+3 & 3+(-2) & 0+4 \\ -2+1 & 5+(-4) & 3+0 \end{bmatrix}$

$= \begin{bmatrix} 4 & 1 & 4 \\ -1 & 1 & 3 \end{bmatrix}$

46. $2G - B = 2\begin{bmatrix} 3 & -2 & 4 \\ 1 & -4 & 0 \end{bmatrix} - \begin{bmatrix} 1 & 3 & 0 \\ -2 & 5 & 3 \end{bmatrix}$

$= \begin{bmatrix} 6 & -4 & 8 \\ 2 & -8 & 0 \end{bmatrix} - \begin{bmatrix} 1 & 3 & 0 \\ -2 & 5 & 3 \end{bmatrix}$

$= \begin{bmatrix} 5 & -7 & 8 \\ 4 & -13 & -3 \end{bmatrix}$

47. Since I is the identity matrix, $AI = A$.

48. Since I is the identity matrix, $IA = A$.

49. $AC = \begin{bmatrix} 2 & 3 \\ -1 & 4 \end{bmatrix}\begin{bmatrix} 2 \\ 1 \end{bmatrix} = \begin{bmatrix} (2)(2)+(3)(1) \\ (-1)(2)+(4)(1) \end{bmatrix} = \begin{bmatrix} 7 \\ 2 \end{bmatrix}$

50. The product CA is undefined since C has 1 column and A has 2 rows.

51. $AH = \begin{bmatrix} 2 & 3 \\ -1 & 4 \end{bmatrix}\begin{bmatrix} x \\ y \end{bmatrix} = \begin{bmatrix} 2x+3y \\ -x+4y \end{bmatrix}$

52. $AB = \begin{bmatrix} 2 & 3 \\ -1 & 4 \end{bmatrix}\begin{bmatrix} 1 & 3 & 0 \\ -2 & 5 & 3 \end{bmatrix}$

$= \begin{bmatrix} 2-6 & 6+15 & 0+9 \\ -1-8 & -3+20 & 0+12 \end{bmatrix}$

$= \begin{bmatrix} -4 & 21 & 9 \\ -9 & 17 & 12 \end{bmatrix}$

53. The product BA is undefined since B has 3 columns and A has 2 rows.

54. $DC = \begin{bmatrix} 3 & 4 \end{bmatrix}\begin{bmatrix} 2 \\ 1 \end{bmatrix} = [(3)(2)+(4)(1)] = [10]$

55. $CD = \begin{bmatrix} 2 \\ 1 \end{bmatrix}\begin{bmatrix} 3 & 4 \end{bmatrix} = \begin{bmatrix} (2)(3) & (2)(4) \\ (1)(3) & (1)(4) \end{bmatrix} = \begin{bmatrix} 6 & 8 \\ 3 & 4 \end{bmatrix}$

56. To find the transpose of A, interchange the rows and columns of A. Thus,

$A^T = \begin{bmatrix} 2 & -1 \\ 3 & 4 \end{bmatrix}.$

57. To find the transpose of B, interchange the rows and columns of B. Thus,

$B^T = \begin{bmatrix} 1 & -2 \\ 3 & 5 \\ 0 & 3 \end{bmatrix}.$

58. $|A| = \begin{vmatrix} 2 & 3 \\ -1 & 4 \end{vmatrix} = (2)(4) - (3)(-1) = 8 + 3 = 11$

59. $D \cdot C = \begin{bmatrix} 3 & 4 \end{bmatrix} \cdot \begin{bmatrix} 2 \\ 1 \end{bmatrix} = (3)(2)+(4)(1) = 10$

60. To find A^{-1}, begin with the augmented matrix $[A \mid I]$.

$\begin{bmatrix} 2 & 3 & \mid & 1 & 0 \\ -1 & 4 & \mid & 0 & 1 \end{bmatrix}$

Multiply the second row by -1, and interchange the result with the first row.

$\begin{bmatrix} 1 & -4 & \mid & 0 & -1 \\ 2 & 3 & \mid & 1 & 0 \end{bmatrix}$

Keep the first row, multiply it by -2, and add the results to the second row.

$\begin{bmatrix} 1 & -4 & \mid & 0 & -1 \\ 0 & 11 & \mid & 1 & 2 \end{bmatrix}$

Multiply the second row by $\frac{1}{11}$.

$\begin{bmatrix} 1 & -4 & \mid & 0 & -1 \\ 0 & 1 & \mid & \frac{1}{11} & \frac{2}{11} \end{bmatrix}$

Keep the second row, multiply it by 4, and add the results to the first row.

$\begin{bmatrix} 1 & 0 & \mid & \frac{4}{11} & -\frac{3}{11} \\ 0 & 1 & \mid & \frac{1}{11} & \frac{2}{11} \end{bmatrix}$

Thus,

$$A^{-1} = \begin{bmatrix} \frac{4}{11} & -\frac{3}{11} \\ \frac{1}{11} & \frac{2}{11} \end{bmatrix}.$$

61. Since A and A^{-1} are inverses, the product $AA^{-1} = I$.

62. Since B is not a square matrix, $|B|$ is not defined.

63. Since B is not a square matrix, B^{-1} is not defined.

64. If $H = C$, then

$$\begin{bmatrix} x \\ y \end{bmatrix} = \begin{bmatrix} 2 \\ 1 \end{bmatrix},$$

so that $x = 2$ and $y = 1$.

65. Solve for x **only** using Cramer's rule.

$$\begin{aligned} x - 2y + 4z &= 1 \\ x + y + z &= 4 \\ 2x - y + z &= 1 \end{aligned}$$

Since all we are to solve for is x, we must find $|A|$ and $|A_x|$. We will expand both determinants across the first row.

$$|A| = \begin{vmatrix} 1 & -2 & 4 \\ 1 & 1 & 1 \\ 2 & -1 & 1 \end{vmatrix}$$

$$= (1)(+1)\begin{vmatrix} 1 & 1 \\ -1 & 1 \end{vmatrix} + (-2)(-1)\begin{vmatrix} 1 & 1 \\ 2 & 1 \end{vmatrix}$$

$$+ (4)(+1)\begin{vmatrix} 1 & 1 \\ 2 & -1 \end{vmatrix}$$

$$= (1)[(1)(1)-(1)(-1)] + (2)[(1)(1)-(1)(2)]$$
$$\quad + (4)[(1)(-1)-(1)(2)]$$
$$= (1)[2] + (2)[-1] + (4)[-3]$$
$$= 2 - 2 - 12$$
$$= -12$$

Replace the coefficients of x in $|A|$ with the constants to find $|A_x|$.

$$|A_x| = \begin{vmatrix} 1 & -2 & 4 \\ 4 & 1 & 1 \\ 1 & -1 & 1 \end{vmatrix}$$

$$= (1)(+1)\begin{vmatrix} 1 & 1 \\ -1 & 1 \end{vmatrix} + (-2)(-1)\begin{vmatrix} 4 & 1 \\ 1 & 1 \end{vmatrix}$$

$$+ (4)(+1)\begin{vmatrix} 4 & 1 \\ 1 & -1 \end{vmatrix}$$

$$= (1)[(1)(1)-(1)(-1)] + (2)[(4)(1)-(1)(1)]$$
$$\quad + (4)[(4)(-1)-(1)(1)]$$
$$= (1)[2] + (2)[3] + (4)[-5]$$
$$= 2 + 6 - 20$$
$$= -12$$

Then,

$$x = \frac{|A_x|}{|A|} = \frac{-12}{-12} = 1.$$

CHAPTER 7 MATRICES AND DETERMINANTS

CHAPTER 7 Test

1. $2A - 3B = 2\begin{bmatrix} -2 & 0 \\ 4 & 1 \end{bmatrix} - 3\begin{bmatrix} 3 & -1 \\ 0 & 5 \end{bmatrix}$

    $= \begin{bmatrix} -4 & 0 \\ 8 & 2 \end{bmatrix} - \begin{bmatrix} 9 & -3 \\ 0 & 15 \end{bmatrix}$

    $= \begin{bmatrix} -13 & 3 \\ 8 & -13 \end{bmatrix}$

2. $\begin{bmatrix} -1 & 3 & 5 \end{bmatrix} \cdot \begin{bmatrix} 2 \\ 0 \\ -1 \end{bmatrix} = (-1)(2) + (3)(0) + (5)(-1)$

    $= -2 + 0 - 5$
    $= -7$

3. $AB = \begin{bmatrix} -1 & 3 \\ 4 & 2 \end{bmatrix}\begin{bmatrix} -2 & 0 \\ -1 & 5 \end{bmatrix}$

    $= \begin{bmatrix} (-1)(-2)+(3)(-1) & (-1)(0)+(3)(5) \\ (4)(-2)+(2)(-1) & (4)(0)+(2)(5) \end{bmatrix}$

    $= \begin{bmatrix} -1 & 15 \\ -10 & 10 \end{bmatrix}$

4. First find JC, which gives the total revenue produced in each store on the sale of the three items at the given prices.

    $JC = \begin{bmatrix} 27 & 48 & 74 \\ 35 & 43 & 61 \end{bmatrix}\begin{bmatrix} 100 \\ 20 \\ 15 \end{bmatrix}$

    $= \begin{bmatrix} (27)(100)+(48)(20)+(74)(15) \\ (35)(100)+(43)(20)+(61)(15) \end{bmatrix}$

    $= \begin{bmatrix} 4770 \\ 5275 \end{bmatrix}$

    Thus, Store 1 had total sales of $4770 and Store 2 had total sales of $5275 on these three items in January. The dot product of [1 1] with this matrix,

    $\begin{bmatrix} 1 & 1 \end{bmatrix} \cdot (JC) = \begin{bmatrix} 1 & 1 \end{bmatrix} \cdot \begin{bmatrix} 4770 \\ 5275 \end{bmatrix} = 4770 + 5275 = 10{,}045$

    represents the total sales of $10,045 on these items in the two stores in January.

5. Solve using the Gaussian method.

    $3x + y = 1$
    $-x + 2y = -5$

    First form the augmented matrix.

    $\begin{bmatrix} 3 & 1 & 1 \\ -1 & 2 & -5 \end{bmatrix}$

    Multiply the second row by -1, and interchange the result with the first row.

    $\begin{bmatrix} 1 & -2 & 5 \\ 3 & 1 & 1 \end{bmatrix}$

    Keep the first row, multiply it by -3, and add the results to the second row.

    $\begin{bmatrix} 1 & -2 & 5 \\ 0 & 7 & -14 \end{bmatrix}$

    Multiply the second row by $\frac{1}{7}$.

    $\begin{bmatrix} 1 & -2 & 5 \\ 0 & 1 & -2 \end{bmatrix}$

    Keep the second row, multiply it by 2, and add the results to the first row.

    $\begin{bmatrix} 1 & 0 & 1 \\ 0 & 1 & -2 \end{bmatrix}$

    Then this corresponds to $x = 1$ and $y = -2$, so the solution to the system is $(1, -2)$.

6. Solve using the Gaussian method.

    $2x - y + 3z = 10$
    $-x - y + 4z = 17$
    $5x + 3y - 2z = -13$

    Form the augmented matrix.

    $\begin{bmatrix} 2 & -1 & 3 & 10 \\ -1 & -1 & 4 & 17 \\ 5 & 3 & -2 & -13 \end{bmatrix}$

    Multiply the second row by -1, and interchange the result with the first row.

    $\begin{bmatrix} 1 & 1 & -4 & -17 \\ 2 & -1 & 3 & 10 \\ 5 & 3 & -2 & -13 \end{bmatrix}$

    Keep the first row, multiply it by -2, and add the results to the second row. Then multiply the first row by -5, and add the results to the third row.

$$\begin{bmatrix} 1 & 1 & -4 & -17 \\ 0 & -3 & 11 & 44 \\ 0 & -2 & 18 & 72 \end{bmatrix}$$

Multiply the third row by $-\frac{1}{2}$, and interchange the result with the second row.

$$\begin{bmatrix} 1 & 1 & -4 & -17 \\ 0 & 1 & -9 & -36 \\ 0 & -3 & 11 & 44 \end{bmatrix}$$

Keep the second row, multiply it by -1, and add the results to the first row. Then multiply the second row by 3, and add the results to the third row.

$$\begin{bmatrix} 1 & 0 & 5 & 19 \\ 0 & 1 & -9 & -36 \\ 0 & 0 & -16 & -64 \end{bmatrix}$$

Multiply the third row by $-\frac{1}{16}$.

$$\begin{bmatrix} 1 & 0 & 5 & 19 \\ 0 & 1 & -9 & -36 \\ 0 & 0 & 1 & 4 \end{bmatrix}$$

Keep the third row, multiply it by -5, and add the results to the first row. Then multiply the third row by 9, and add the results to the second row.

$$\begin{bmatrix} 1 & 0 & 0 & -1 \\ 0 & 1 & 0 & 0 \\ 0 & 0 & 1 & 4 \end{bmatrix}$$

Since this corresponds to $x = -1$, $y = 0$, and $z = 4$, the solution to the system is $(-1, 0, 4)$.

7. If $A = \begin{bmatrix} 1 & 3 \\ 2 & -1 \end{bmatrix}$, to find A^{-1}, begin with the augmented matrix $[A|I]$.

$$\begin{bmatrix} 1 & 3 & | & 1 & 0 \\ 2 & -1 & | & 0 & 1 \end{bmatrix}$$

Keep the first row, multiply it by -2, and add the results to the second row.

$$\begin{bmatrix} 1 & 3 & | & 1 & 0 \\ 0 & -7 & | & -2 & 1 \end{bmatrix}$$

Multiply the second row by $-\frac{1}{7}$.

$$\begin{bmatrix} 1 & 3 & | & 1 & 0 \\ 0 & 1 & | & \frac{2}{7} & -\frac{1}{7} \end{bmatrix}$$

Keep the second row, multiply it by -3, and add the results to the first row.

$$\begin{bmatrix} 1 & 0 & | & \frac{1}{7} & \frac{3}{7} \\ 0 & 1 & | & \frac{2}{7} & -\frac{1}{7} \end{bmatrix}$$

Thus,

$$A^{-1} = \begin{bmatrix} \frac{1}{7} & \frac{3}{7} \\ \frac{2}{7} & -\frac{1}{7} \end{bmatrix}.$$

8. Solve the system using the inverse method.

$$\begin{aligned} x + 3y &= 1 \\ 2x - y &= -5 \end{aligned}$$

This system is equivalent to the matrix equation $AX = B$, where

$$A = \begin{bmatrix} 1 & 3 \\ 2 & -1 \end{bmatrix} \quad X = \begin{bmatrix} x \\ y \end{bmatrix} \quad B = \begin{bmatrix} 1 \\ -5 \end{bmatrix}.$$

Using A^{-1}, found in Problem 7, we have

$$\begin{bmatrix} x \\ y \end{bmatrix} = X = A^{-1}B = \begin{bmatrix} \frac{1}{7} & \frac{3}{7} \\ \frac{2}{7} & -\frac{1}{7} \end{bmatrix} \begin{bmatrix} 1 \\ -5 \end{bmatrix} = \begin{bmatrix} -2 \\ 1 \end{bmatrix}$$

Since $x = -2$ and $y = 1$, the solution to the system is $(-2, 1)$.

9. If $A = \begin{bmatrix} 1 & 1 & 1 \\ 2 & 0 & 1 \\ 0 & 1 & -1 \end{bmatrix}$, to find A^{-1}, begin with the augmented matrix $[A|I]$.

$$\begin{bmatrix} 1 & 1 & 1 & | & 1 & 0 & 0 \\ 2 & 0 & 1 & | & 0 & 1 & 0 \\ 0 & 1 & -1 & | & 0 & 0 & 1 \end{bmatrix}$$

Keep the first row (and the third row), multiply the first row by -2, and add the results to the second row.

$$\begin{bmatrix} 1 & 1 & 1 & | & 1 & 0 & 0 \\ 0 & -2 & -1 & | & -2 & 1 & 0 \\ 0 & 1 & -1 & | & 0 & 0 & 1 \end{bmatrix}$$

Interchange the second and third rows.

$$\begin{bmatrix} 1 & 1 & 1 & | & 1 & 0 & 0 \\ 0 & 1 & -1 & | & 0 & 0 & 1 \\ 0 & -2 & -1 & | & -2 & 1 & 0 \end{bmatrix}$$

Keep the second row, multiply it by -1, and add the results to the first row. Then multiply the second row by 2, and add the results to the third row.

$$\begin{bmatrix} 1 & 0 & 2 & | & 1 & 0 & -1 \\ 0 & 1 & -1 & | & 0 & 0 & 1 \\ 0 & 0 & -3 & | & -2 & 1 & 2 \end{bmatrix}$$

Multiply the third row by $-\frac{1}{3}$.

$$\begin{bmatrix} 1 & 0 & 2 & | & 1 & 0 & -1 \\ 0 & 1 & -1 & | & 0 & 0 & 1 \\ 0 & 0 & 1 & | & \frac{2}{3} & -\frac{1}{3} & -\frac{2}{3} \end{bmatrix}$$

Keep the third row, and add it to the second row. Then multiply the third row by -2, and add the results to the first row.

$$\begin{bmatrix} 1 & 0 & 0 & | & -\frac{1}{3} & \frac{2}{3} & \frac{1}{3} \\ 0 & 1 & 0 & | & \frac{2}{3} & -\frac{1}{3} & \frac{1}{3} \\ 0 & 0 & 1 & | & \frac{2}{3} & -\frac{1}{3} & -\frac{2}{3} \end{bmatrix}$$

Thus,

$$A^{-1} = \begin{bmatrix} -\frac{1}{3} & \frac{2}{3} & \frac{1}{3} \\ \frac{2}{3} & -\frac{1}{3} & \frac{1}{3} \\ \frac{2}{3} & -\frac{1}{3} & -\frac{2}{3} \end{bmatrix}.$$

10. Solve using the inverse method.

$$\begin{aligned} x + y + z &= 1 \\ 2x + z &= 0 \\ y - z &= -2 \end{aligned}$$

Then this system is equivalent to the matrix equation $AX = B$, where

$$A = \begin{bmatrix} 1 & 1 & 1 \\ 2 & 0 & 1 \\ 0 & 1 & -1 \end{bmatrix} \quad X = \begin{bmatrix} x \\ y \\ z \end{bmatrix} \quad B = \begin{bmatrix} 1 \\ 0 \\ -2 \end{bmatrix}.$$

Using A^{-1}, found in Problem 9, we have

$$\begin{bmatrix} x \\ y \\ z \end{bmatrix} = X = A^{-1}B = \begin{bmatrix} -\frac{1}{3} & \frac{2}{3} & \frac{1}{3} \\ \frac{2}{3} & -\frac{1}{3} & \frac{1}{3} \\ \frac{2}{3} & -\frac{1}{3} & -\frac{2}{3} \end{bmatrix} \begin{bmatrix} 1 \\ 0 \\ -2 \end{bmatrix} = \begin{bmatrix} -1 \\ 0 \\ 2 \end{bmatrix}.$$

Then $x = -1$, $y = 0$, $z = 2$, and the solution to the system is $(-1,0,2)$.

11. Solve using Cramer's rule.

$$\begin{aligned} 2x - 3y &= 1 \\ 5x - 4y &= 20 \end{aligned}$$

The determinant of the coefficient matrix is:

$$|A| = \begin{vmatrix} 2 & -3 \\ 5 & -4 \end{vmatrix} = (2)(-4) - (-3)(5) = -8 + 15 = 7$$

Replace the coefficients of x with the constants to obtain:

$$|A_x| = \begin{vmatrix} 1 & -3 \\ 20 & -4 \end{vmatrix} = (1)(-4) - (-3)(20) = -4 + 60 = 56$$

Replace the coefficients of y with the constants to obtain:

$$|A_y| = \begin{vmatrix} 2 & 1 \\ 5 & 20 \end{vmatrix} = (2)(20) - (1)(5) = 40 - 5 = 35$$

Then

$$x = \frac{|A_x|}{|A|} = \frac{56}{7} = 8 \quad y = \frac{|A_y|}{|A|} = \frac{35}{7} = 5.$$

Thus, the solution to the system is $(8,5)$.

12. Evaluate the determinant

$$\begin{vmatrix} 3 & -4 & -3 \\ -1 & 5 & 6 \\ 2 & -2 & 4 \end{vmatrix}$$

by introducing two zeros in a row or column. Suppose we use the -1 in the first column to obtain 0's in the remaining positions in the first column. Keep the second row, multiply it by 3, and add the results to the first row. Then multiply the second row by 2, and add the results to the third row.

$$\begin{vmatrix} 0 & 11 & 15 \\ -1 & 5 & 6 \\ 0 & 8 & 16 \end{vmatrix}$$

Then we can expand down the first column to obtain:

$$\begin{vmatrix} 0 & 11 & 15 \\ -1 & 5 & 6 \\ 0 & 8 & 16 \end{vmatrix} = 0 + (-1)(-1)\begin{vmatrix} 11 & 15 \\ 8 & 16 \end{vmatrix} + 0$$

$$= (1)[(11)(16) - (15)(8)]$$
$$= (1)[176 - 120]$$
$$= (1)[56]$$
$$= 56$$

13. We know that

$$|A| = \begin{vmatrix} -3 & 2 & 1 \\ 0 & 0 & 0 \\ -5 & -7 & 8 \end{vmatrix} = 0$$

since all the elements in the second row are 0. Any determinant that has a row or column of zeros has value 0.

CHAPTER 8 TOPICS IN ANALYTIC GEOMETRY

SECTION 8.1 The Circle

1. Substitute $h = 7$, $k = 0$, and $r = \sqrt{13}$ in the standard form.

$$(x - h)^2 + (y - k)^2 = r^2$$
$$(x - 7)^2 + (y - 0)^2 = (\sqrt{13})^2$$
$$(x - 7)^2 + y^2 = 13 \quad \text{Standard form}$$

Expand to obtain the general form.

$$x^2 - 14x + 49 + y^2 = 13$$
$$x^2 + y^2 - 14x + 36 = 0 \quad \text{General form}$$

2. To find the standard form of the equation of the circle with general form

$$x^2 + y^2 + 4x - 6y + 4 = 0,$$

we complete the squares in both x and y. Group the terms and leave space as shown.

$$x^2 + 4x \quad + y^2 - 6y \quad = -4$$

To complete the square in x, add half the coefficient of x, squared, to both sides. That is, add $(2)^2 = 4$ to both sides. Similarly, to complete the square in y, add $(-3)^2 = 9$ to both sides.

$$x^2 + 4x + 4 + y^2 - 6y + 9 = -4 + 4 + 9$$
$$(x + 2)^2 + (y - 3)^2 = 9$$

Then the center of the circle is $(-2,3)$, the radius is 3, and the graph is given below.

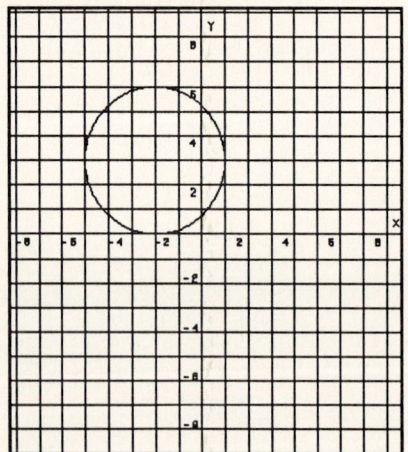

3. The tunnel is in the shape of a semicircle with equation

$$y = \sqrt{36 - x^2}.$$

Since the truck is 6 m wide, we must find the vertical clearance 3 m from the centerline, that is, find y when x is 3.

$$y = \sqrt{36 - x^2}$$
$$= \sqrt{36 - (3)^2}$$
$$= \sqrt{36 - 9}$$
$$= \sqrt{27}$$
$$\approx 5.196152423$$

Since the vertical clearance 3 m from the centerline is about 5.2 m, and since the height of the truck is 6 m, the truck cannot pass through the tunnel.

EXERCISES A

CHAPTER 8 TOPICS IN ANALYTIC GEOMETRY

SECTION 8.1 The Circle

1. Substitute $h = -3$, $k = 2$, and $r = 1$ in the standard form.

$$(x - h)^2 + (y - k)^2 = r^2$$
$$(x - (-3))^2 + (y - 2)^2 = (1)^2$$
$$(x + 3)^2 + (y - 2)^2 = 1$$

   Expand to obtain the general form.

$$x^2 + 6x + 9 + y^2 - 4y + 4 = 1$$
$$x^2 + y^2 + 6x - 4y + 13 = 1$$
$$x^2 + y^2 + 6x - 4y + 12 = 0$$

2. Substitute $h = \frac{1}{4}$, $k = -1$, and $r = 3$ in the standard form.

$$(x - h)^2 + (y - k)^2 = r^2$$
$$\left(x - \tfrac{1}{4}\right)^2 + (y - (-1))^2 = 3^2$$
$$\left(x - \tfrac{1}{4}\right)^2 + (y + 1)^2 = 9$$

   Expand to obtain the general form.

$$x^2 - \tfrac{1}{2}x + \tfrac{1}{16} + y^2 + 2y + 1 = 9$$
$$16x^2 - 8x + 1 + 16y^2 + 32y + 16 = 144$$
$$16x^2 + 16y^2 - 8x + 32y - 127 = 0$$

3. Substitute $h = 0$, $k = 0$, and $r = 1$ in the standard form.

$$(x - 0)^2 + (y - 0)^2 = 1^2$$
$$x^2 + y^2 = 1$$

   Then the general form is

$$x^2 + y^2 - 1 = 0.$$

4. Since the center of the circle is (0,0), and the circle passes through (0,5), the radius of the circle must be 5. Substitute $h = 0$, $k = 0$, and $r = 5$ in the standard form.

$$(x - 0)^2 + (y - 0)^2 = 5^2$$
$$x^2 + y^2 = 25$$

5. Since the center is (1,−5) and the circle passes through (7,3), the radius of the circle is the distance between these two points. Use the distance formula.

$$d = \sqrt{(x_1 - x_2)^2 + (y_1 - y_2)^2}$$
$$= \sqrt{(1 - 7)^2 + (-5 - 3)^2}$$
$$= \sqrt{(-6)^2 + (-8)^2}$$
$$= \sqrt{36 + 64}$$
$$= \sqrt{100}$$
$$= 10$$

   Substitute $h = 1$, $k = -5$, and $r = 10$ in the standard form.

$$(x - 1)^2 + (y - (-5))^2 = 10^2$$
$$(x - 1)^2 + (y + 5)^2 = 100$$

6. Substitute $h = 4$, $k = -1$, and $r = \sqrt{3}$ in the standard form.

$$(x - h)^2 + (y - k)^2 = r^2$$
$$(x - 4)^2 + (y - (-1))^2 = (\sqrt{3})^2$$
$$(x - 4)^2 + (y + 1)^2 = 3$$

7. Since the center is at (6,8), and the circle is tangent to the x-axis, the radius of the circle must be 8 (the point (6,8) is 8 units above the x-axis). Substitute $h = 6$, $k = 8$, and $r = 8$ in the standard form.

$$(x - h)^2 + (y - k)^2 = r^2$$
$$(x - 6)^2 + (y - 8)^2 = 8^2$$
$$(x - 6)^2 + (y - 8)^2 = 64$$

8. Since the circle has endpoints of a diameter at (1,−4) and (5,2), the center of the circle must be at the midpoint of the segment joining these two points. Use the midpoint formula.

$$(\bar{x},\bar{y}) = \left(\frac{x_1 + x_2}{2}, \frac{y_1 + y_2}{2}\right)$$
$$= \left(\frac{1 + 5}{2}, \frac{-4 + 2}{2}\right)$$
$$= (3,-1)$$

   Since the length of the radius is half the length of the diameter, we can use the distance formula to find the length of the diameter and divide the result by 2 to obtain the radius.

$$d = \sqrt{(x_1 - x_2)^2 + (y_1 - y_2)^2}$$
$$= \sqrt{(1-5)^2 + (-4-2)^2}$$
$$= \sqrt{(-4)^2 + (-6)^2}$$
$$= \sqrt{16 + 36}$$
$$= \sqrt{52}$$
$$= 2\sqrt{13}$$

Then the radius is $r = \dfrac{d}{2} = \dfrac{2\sqrt{13}}{2} = \sqrt{13}$. Substitute $h = 3$, $k = -1$, and $r = \sqrt{13}$ in the standard form.

$$(x - h)^2 + (y - k)^2 = r^2$$
$$(x - 3)^2 + (y - (-1))^2 = (\sqrt{13})^2$$
$$(x - 3)^2 + (y + 1)^2 = 13$$

9. Since the circle has endpoints of a diameter at $(-7, 8)$ and $(4, 7)$, the center of the circle must be at the midpoint of the segment joining these two points. Use the midpoint formula.

$$(\bar{x}, \bar{y}) = \left(\dfrac{x_1 + x_2}{2}, \dfrac{y_1 + y_2}{2} \right)$$
$$= \left(\dfrac{-7 + 4}{2}, \dfrac{8 + 7}{2} \right)$$
$$= \left(-\dfrac{3}{2}, \dfrac{15}{2} \right)$$

Since the length of the radius is half the length of the diameter, we can use the distance formula to find the length of the diameter and divide the result by 2 to obtain the radius.

$$d = \sqrt{(x_1 - x_2)^2 + (y_1 - y_2)^2}$$
$$= \sqrt{(-7 - 4)^2 + (8 - 7)^2}$$
$$= \sqrt{(-11)^2 + (1)^2}$$
$$= \sqrt{121 + 1}$$
$$= \sqrt{122}$$

Then the radius is $r = \dfrac{d}{2} = \dfrac{\sqrt{122}}{2}$. Substitute $h = -\dfrac{3}{2}$, $k = \dfrac{15}{2}$, and $r = \dfrac{\sqrt{122}}{2}$ in the standard form.

$$(x - h)^2 + (y - k)^2 = r^2$$
$$\left(x - \left(-\dfrac{3}{2}\right)\right)^2 + \left(y - \dfrac{15}{2}\right)^2 = \left(\dfrac{\sqrt{122}}{2}\right)^2$$
$$\left(x + \dfrac{3}{2}\right)^2 + \left(y - \dfrac{15}{2}\right)^2 = \dfrac{61}{2}$$

10. Graph. $x^2 + y^2 = 16$

The graph is a circle with center $(0,0)$ and radius 4 ($4^2 = 16$). The graph is given below.

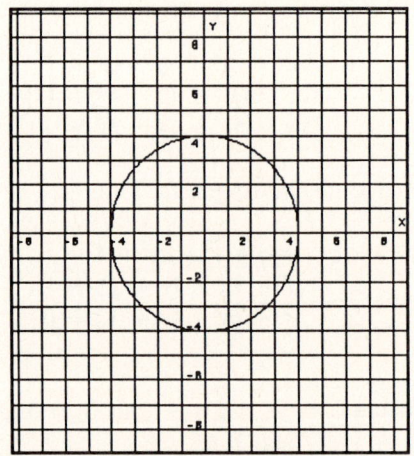

11. Graph. $(x - 1)^2 + (y + 4)^2 = 4$

The graph is a circle with center $(1, -4)$ and radius 2. The graph is given below.

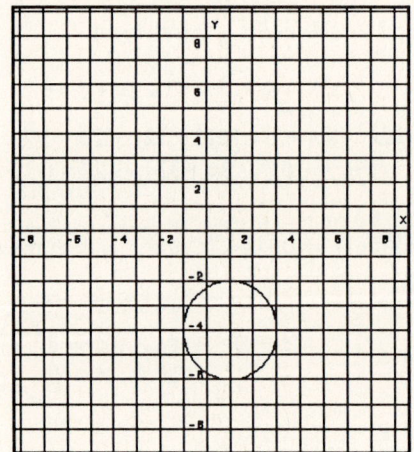

12. Graph. $x^2 + y^2 + 4y = 5$

First write the equation in standard form by completing the square on y.

$$x^2 + y^2 + 4y = 5$$
$$x^2 + y^2 + 4y + 4 = 5 + 4$$
$$x^2 + (y + 2)^2 = 9$$

The graph is a circle centered at $(0, -2)$ with radius 3. The graph is given below.

EXERCISES A

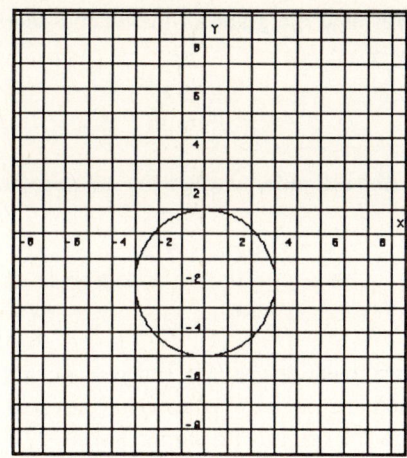

13. Graph. $x^2 + y^2 - 6x - 8y + 21 = 0$

    First write the equation in standard form by completing the squares on x and y.

    $$x^2 - 6x + y^2 - 8y = -21$$
    $$x^2 - 6x + 9 + y^2 - 8y + 16 = -21 + 9 + 16$$
    $$(x - 3)^2 + (y - 4)^2 = 4$$

    Then the graph is a circle centered at (3,4) with radius 2. The graph is given below.

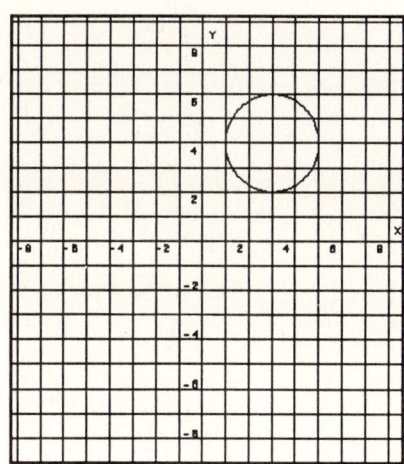

14. Write $x^2 + y^2 - 4x - 2y + 4 = 0$ in standard form.

    First rewrite the equation leaving space as shown.

    $$x^2 - 4x + y^2 - 2y = -4$$

    Complete the squares on x and y.

    $$x^2 - 4x + 4 + y^2 - 2y + 1 = -4 + 4 + 1$$

    Factor, and simplify the right side.

    $$(x - 2)^2 + (y - 1)^2 = 1 \quad \textit{Standard form}$$

15. Write $4x^2 + 4y^2 - 4x + 24y - 63 = 0$ in standard form.

    First rewrite the equation factoring out the coefficients of the squared terms.

    $$4(x^2 - x \phantom{+\tfrac{1}{4}}) + 4(y^2 + 6y) = 63$$

    Complete the square in both variables, and be careful to add the correct numbers to the left side.

    $$4\left(x^2 - x + \tfrac{1}{4}\right) + 4(y^2 + 6y + 9) = 63 + 4\left(\tfrac{1}{4}\right) + 4(9)$$
    $$4\left(x - \tfrac{1}{2}\right)^2 + 4(y + 3)^2 = 100$$
    $$\left(x - \tfrac{1}{2}\right)^2 + (y + 3)^2 = 25$$

16. Write $9x^2 + 9y^2 + 6x - 6y - 142 = 0$ in standard form.

    First rewrite the equation factoring out the coefficients of the squared variables.

    $$9\left(x^2 + \tfrac{2}{3}x \phantom{+\tfrac{1}{9}}\right) + 9\left(y^2 - \tfrac{2}{3}y \phantom{+\tfrac{1}{9}}\right) = 142$$

    Complete the square on x and y.

    $$9\left(x^2 + \tfrac{2}{3}x + \tfrac{1}{9}\right) + 9\left(y^2 - \tfrac{2}{3}y + \tfrac{1}{9}\right) = 142 + 9\left(\tfrac{1}{9}\right) + 9\left(\tfrac{1}{9}\right)$$
    $$9\left(x + \tfrac{1}{3}\right)^2 + 9\left(y - \tfrac{1}{3}\right)^2 = 144$$
    $$\left(x + \tfrac{1}{3}\right)^2 + \left(y - \tfrac{1}{3}\right)^2 = 16$$

17. A canal with cross section a semicircle is 10 ft deep at the center. Suppose we place the circle that contains the semicircle with center at the origin so that the desired semicircle is below the x-axis. Since the canal is 10 ft deep at the center, the radius of the circle is 10. Then using (0,0) as the center, $h = 0$, $k = 0$, and $r = 10$, so the equation of the circle is

    $$(x - 0)^2 + (y - 0)^2 = 10^2$$

    which simplifies to

    $$x^2 + y^2 = 100.$$

    To find the equation of the semicircle, solve for y.

    $$y^2 = 100 - x^2$$
    $$y = \pm\sqrt{100 - x^2}$$

    Since we only want that portion of the circle where y is negative, the lower semicircle, the equation of the semicircle is

$$y = -\sqrt{100 - x^2}.$$

To find the depth of the canal 4 ft from the edge, this will correspond to finding the value of y when $x = 6$ (note that 4 ft from the edge is 6 ft from the center since the radius is 10 ft).

$$\begin{aligned} y &= -\sqrt{100 - 6^2} \\ &= -\sqrt{100 - 36} \\ &= -\sqrt{64} \\ &= -8 \end{aligned}$$

Notice that when x is 6, y is -8, which corresponds to a depth of 8 ft at this point.

18. A pipe is circular with cross-sectional inside radius 20 cm. We want to determine whether it is possible to insert a board that is 25 cm thick and 31 cm wide as shown in the figure in the text. One way to answer this problem is to consider the equation of the circle. Assume that we place the center of the circle (the middle of the pipe) at the origin, then the radius of the circle is 20, and the equation of the circle is

$$x^2 + y^2 = 20^2 = 400.$$

Consider the upper right corner of the cross-section of the board as shown in the sketch. The coordinates of this point would be (12.5, 15.5), taking half the width and thickness. If the board will pass through the pipe, then $(12.5)^2 + (15.5)^2$ would have to be less than 400, that is, the point would have to be inside the circle. Since

$$(12.5)^2 + (15.5)^2 = 156.25 + 240.25 = 396.5,$$

we can see that the point is inside the circle, which means that the board will pass through the pipe.

19. There is no graph of the equation

$$x^2 + y^2 + 1 = 0$$

or equivalently,

$$x^2 + y^2 = -1,$$

since the equation has no solutions. Notice that for any values of x and y, the left side will always be nonnegative, and could never be -1. Since the equation is similar to the standard form for the equation of a circle, this degenerate conic is sometimes called an *imaginary circle*.

20. To determine the graph of the second-degree equation

$$x^2 - y^2 = 0,$$

factor the left side, and use the zero-product rule.

$$(x - y)(x + y) = 0$$
$$x - y = 0 \qquad x + y = 0$$
$$y = x \qquad\quad y = -x$$

The two equations, $y = x$ and $y = -x$, are the equations of intersecting straight lines passing through the origin with slopes 1 and -1, respectively.

CHAPTER 8 TOPICS IN ANALYTIC GEOMETRY

SECTION 8.2 The Ellipse

1. Find the equation of the ellipse with foci $(-4,0)$ and $(4,0)$ and x-intercepts $(-6,0)$ and $(6,0)$.

   Since the foci are on the x-axis, we know that $a = 6$. Thus,

   $$b^2 = a^2 - c^2 = (6)^2 - (4)^2 = 36 - 16 = 20,$$

   making

   $$b = \sqrt{20}.$$

   Substitute 6 for a and $\sqrt{20}$ for b in the standard form.

   $$\frac{x^2}{a^2} + \frac{y^2}{b^2} = 1$$
   $$\frac{x^2}{6^2} + \frac{y^2}{(\sqrt{20})^2} = 1$$
   $$\frac{x^2}{36} + \frac{y^2}{20} = 1$$

2. Graph the equation $x^2 + 9y^2 = 36$.

   Divide through by 36.

   $$\frac{x^2}{36} + \frac{9y^2}{36} = \frac{36}{36}$$
   $$\frac{x^2}{36} + \frac{y^2}{4} = 1 \quad \text{Standard form}$$

   Notice that since $36 > 4$, this is the standard form of an ellipse with horizontal major axis. Since $a^2 = 36$, $a = 6$, and since $b^2 = 4$, $b = 2$. Thus, the x-intercepts are $(6,0)$ and $(-6,0)$, and the y-intercepts are $(0,2)$ and $(0,-2)$. The graph is given below.

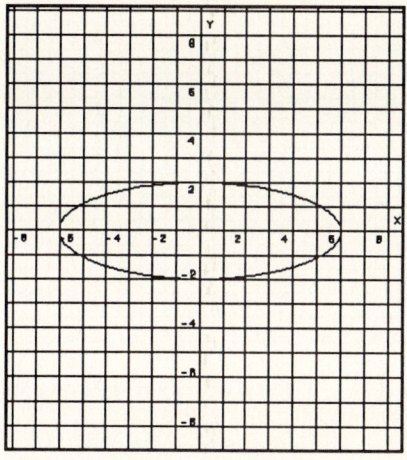

3. Since the space lab travels 26,100 mi in each revolution, and since one revolution is made every 24 hours, to find the rate in feet per second, we evaluate:

   $$\frac{26,100 \; mi}{24 \; hr} = \frac{(26,100)(5280) \; ft}{(24)(60)(60) \; sec}$$
   $$= \frac{137,808,000 \; ft}{86,400 \; sec} = 1595 \; ft/sec$$

4. Find the standard form and graph the ellipse that has $a = 5$ and foci at $(1,-1)$ and $(1,5)$.

   The center of the ellipse is the midpoint of the line segment joining the foci.

   $$h = \frac{1+1}{2} = 1 \quad \text{and} \quad k = \frac{-1+5}{2} = 2$$

   Thus, the center is $(h,k) = (1,2)$. Also,

   $$2c = 5 - (-1) = 6,$$

   so $c = 3$. Since $a = 5$, and $b^2 = a^2 - c^2$,

   $$b^2 = 5^2 - 3^2 = 25 - 9 = 16,$$

   so $b = 4$. The standard form of the ellipse is:

   $$\frac{(x-h)^2}{b^2} + \frac{(y-k)^2}{a^2} = 1$$
   $$\frac{(x-1)^2}{16} + \frac{(y-2)^2}{25} = 1$$

   The graph of the ellipse is given below.

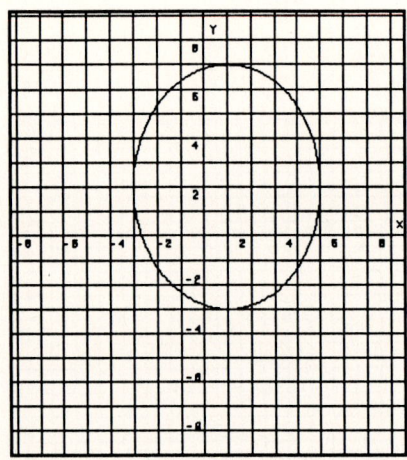

5. Write $4x^2 + y^2 + 40x - 12y + 100 = 0$ in standard form.

   Rewrite the equation leaving space as shown below.

   $$4x^2 + 40x \quad + y^2 - 12y \quad = -100$$

   Factor out the coefficients of the squared terms.

   $$4(x^2 + 10x \quad) + (y^2 - 12y \quad) = -100$$

   Complete the square in both variables and add the necessary values to both sides.

   $$4(x^2+10x+25) + (y^2-12y+36) = -100+4(25)+36$$

   Factor, and simplify the right side.

   $$4(x + 5)^2 + (y - 6)^2 = 36$$

   Divide through by 36 to obtain the standard form.

   $$\frac{4(x+5)^2}{36} + \frac{(y-6)^2}{36} = \frac{36}{36}$$

   $$\frac{(x+5)^2}{9} + \frac{(y-6)^2}{36} = 1$$

   Thus, the equation gives an ellipse with center $(-5,6)$, vertical major axis, $a = 6$, and $b = 3$.

EXERCISES A

CHAPTER 8 TOPICS IN ANALYTIC GEOMETRY

SECTION 8.2 The Ellipse

In Exercises 1-5, find the equation of the ellipse with center at the origin that satisfies the given conditions.

1. Intercepts $(\pm 7, 0)$ and $(0, \pm 6)$.

   Since $7 > 6$, this is an ellipse with horizontal major axis, $a = 7$, and $b = 6$. Substitute into the standard form.

   $$\frac{x^2}{a^2} + \frac{y^2}{b^2} = 1$$
   $$\frac{x^2}{7^2} + \frac{y^2}{6^2} = 1$$
   $$\frac{x^2}{49} + \frac{y^2}{36} = 1$$

2. Foci $(\pm 3, 0)$ and x-intercepts $(\pm 5, 0)$.

   Since the foci are on the x-axis, we know that $a = 5$. Thus,

   $$b^2 = a^2 - c^2 = (5)^2 - (3)^2 = 25 - 9 = 16,$$

   making $b = 4$. Substitute 5 for a and 4 for b in the standard form of an ellipse with horizontal major axis.

   $$\frac{x^2}{a^2} + \frac{y^2}{b^2} = 1$$
   $$\frac{x^2}{5^2} + \frac{y^2}{4^2} = 1$$
   $$\frac{x^2}{25} + \frac{y^2}{16} = 1$$

3. Foci $(0, \pm 3)$ and x-intercepts $(\pm 4, 0)$.

   Since the foci are on the y-axis, we know that $b = 4$. Thus,

   $$a^2 = b^2 + c^2 = (4)^2 + (3)^2 = 16 + 9 = 25,$$

   making $a = 5$. Substitute 5 for a and 4 for b in the standard form of an ellipse with vertical major axis.

   $$\frac{x^2}{b^2} + \frac{y^2}{a^2} = 1$$
   $$\frac{x^2}{4^2} + \frac{y^2}{5^2} = 1$$
   $$\frac{x^2}{16} + \frac{y^2}{25} = 1$$

4. Length of major axis 12 and y-intercepts $(0, \pm 5)$.

   Since the length of the major axis is 12, the x-intercepts are $(\pm 6, 0)$, and the major axis is horizontal. Substitute 6 for a and 5 for b in the standard form.

   $$\frac{x^2}{a^2} + \frac{y^2}{b^2} = 1$$
   $$\frac{x^2}{6^2} + \frac{y^2}{5^2} = 1$$
   $$\frac{x^2}{36} + \frac{y^2}{25} = 1$$

5. Major axis intercepts $(\pm 10, 0)$ and passing through the point $(5, \sqrt{3})$.

   Since the major axis intercepts are on the x-axis, the ellipse has a horizontal major axis, and $a = 10$. Substitute in the standard form.

   $$\frac{x^2}{a^2} + \frac{y^2}{b^2} = 1$$
   $$\frac{x^2}{10^2} + \frac{y^2}{b^2} = 1$$
   $$\frac{x^2}{100} + \frac{y^2}{b^2} = 1$$

   Since the ellipse passes through $(5, \sqrt{3})$, substitute these coordinates into the equation to find the value of b.

   $$\frac{(5)^2}{100} + \frac{(\sqrt{3})^2}{b^2} = 1$$
   $$\frac{25}{100} + \frac{3}{b^2} = 1$$
   $$\frac{1}{4} + \frac{3}{b^2} = 1$$
   $$b^2 + (3)(4) = 4b^2$$
   $$12 = 3b^2$$
   $$4 = b^2$$
   $$2 = b$$

   Then the desired equation is

   $$\frac{x^2}{100} + \frac{y^2}{4} = 1 \ .$$

6. Find the equation of the ellipse with center at (5,-2), $a = 4$, and foci (3,-2) and (7,-2).

Since the foci are on a horizontal line, the ellipse has a horizontal major axis and standard form:

$$\frac{(x-h)^2}{a^2} + \frac{(y-k)^2}{b^2} = 1$$

Since the center is (5,-2), $h = 5$, and $k = -2$. Since the distance from the center to each of the foci is 2, $c = 2$. Then

$$b^2 = a^2 - c^2 = 4^2 - 2^2 = 16 - 4 = 12.$$

Substituting 12 for b^2, 16 for a^2, 5 for h, and -2 for k, the equation is:

$$\frac{(x-5)^2}{16} + \frac{(y+2)^2}{12} = 1$$

7. Find the equation of the ellipse with foci (± 4,3) and y-intercepts (0,0) and (0,6).

Since the foci are on a horizontal line, the ellipse has a horizontal major axis and standard form:

$$\frac{(x-h)^2}{a^2} + \frac{(y-k)^2}{b^2} = 1$$

Since y-intercepts are (0,0) and (0,6) and the x-coordinate of these points is midway between the x-xoordinates of the foci, b must be 3. With $b = 3$ and $c = 4$, we have

$$a^2 = b^2 + c^2 = 3^2 + 4^2 = 9 + 16 = 25.$$

Also, the center is midway between the foci, so the center is (0,3). Substitute 0 for h, 3 for k, 25 for a^2, and 9 for b^2 to obtain the desired equation.

$$\frac{(x-0)^2}{25} + \frac{(y-3)^2}{9} = 1$$

$$\frac{x^2}{25} + \frac{(y-3)^2}{9} = 1$$

8. Find the equation of the ellipse with foci (7,-1) and (7,-7) and $a = 8$.

Since the foci are on a vertical line, the ellipse has a vertical major axis and standard form:

$$\frac{(x-h)^2}{b^2} + \frac{(y-k)^2}{a^2} = 1$$

Since the center is midway between the foci, the center is (7,-4). Since the distance from the center to each of the foci is 3, $c = 3$. Then

$$b^2 = a^2 - c^2 = 8^2 - 3^2 = 64 - 9 = 55.$$

Substituting 55 for b^2, 64 for a^2, 7 for h, and -4 for k, the equation is:

$$\frac{(x-7)^2}{55} + \frac{(y+4)^2}{64} = 1$$

9. Graph the ellipse.

$$\frac{x^2}{49} + \frac{y^2}{25} = 1$$

Since $49 > 25$, the ellipse has horizontal major axis. It is centered at the origin, and with $a = 7$ and $b = 5$, the intercepts are (± 7,0) and (0,± 5). The graph is given below.

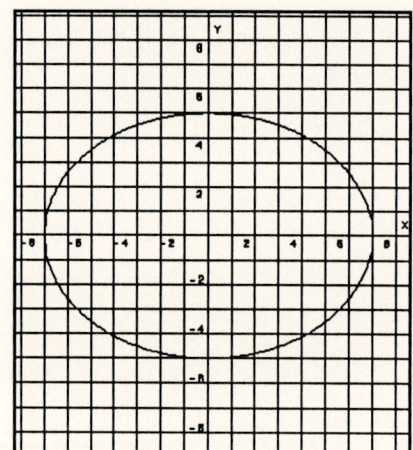

10. Graph the ellipse.

$$\frac{(x-2)^2}{16} + \frac{(y+2)^2}{4} = 1$$

Since $16 > 4$, the ellipse has horizontal major axis. The center is (2,-2), and $a = 4$ and $b = 2$. Place a new coordinate system at the point (2,-2), and move 4 units right and left to obtain the points (6,-2) and (-2,-2). Then move up and down 2 units to obtain the points (2,0) and (2,-4). These points act like the intercepts in this system. The graph is given below.

EXERCISES A

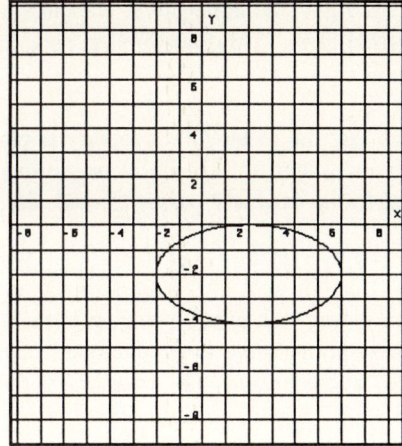

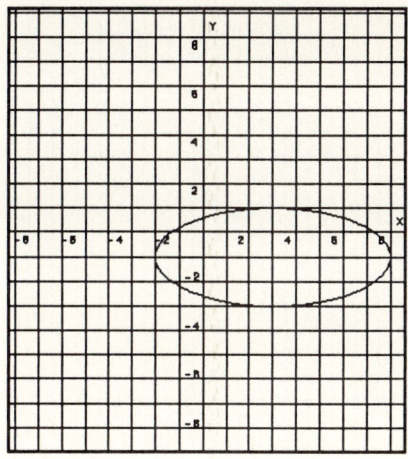

11. Graph the ellipse.

$$4x^2 + 25y^2 - 24x + 50y - 39 = 0$$

First write the equation in standard form by completing the squares in x and y.

$$4x^2 - 24x\ \ + 25y^2 + 50y\ \ = 39$$
$$4(x^2 - 6x\ \ ) + 25(y^2 + 2y\ \ ) = 39$$

Be sure to add correctly to the right side.

$$4(x^2 - 6x + 9) + 25(y^2 + 2y + 1) = 39 + 4(9) + 25(1)$$

Factoring and simplifying we obtain:

$$4(x - 3)^2 + 25(y + 1)^2 = 100$$

Dividing both sides by 100 gives the standard form.

$$\frac{(x-3)^2}{25} + \frac{(y+1)^2}{4} = 1$$

Since $25 > 4$, the ellipse has a horizontal major axis. The center is $(3,-1)$, and $a = 5$ and $b = 2$. Place a new coordinate system at the center and move right and left 5 units to obtain the points $(8,-1)$ and $(-2,-1)$. Then move up and down 2 units from the center to obtain the points $(3,1)$ and $(3,-3)$. These points act like the intercepts in this new system. The graph is given below.

12. Graph the ellipse.

$$16x^2 + 4y^2 + 32x - 20y + 5 = 0$$

Complete the square in x and y to obtain the standard form.

$$16x^2 + 32x\ \ + 4y^2 - 20y\ \ = -5$$
$$16(x^2 + 2x\ \ ) + 4(y^2 - 5y\ \ ) = -5$$

Be careful to add correctly on the right side.

$$16(x^2 + 2x + 1) + 4\left(y^2 - 5y + \frac{25}{4}\right) = -5 + 16 + 4\left(\frac{25}{4}\right)$$
$$16(x + 1)^2 + 4\left(y - \frac{5}{2}\right)^2 = 36$$
$$\frac{(x+1)^2}{\frac{9}{4}} + \frac{\left(y - \frac{5}{2}\right)^2}{9} = 1$$
$$\frac{(x+1)^2}{\left(\frac{3}{2}\right)^2} + \frac{\left(y - \frac{5}{2}\right)^2}{(3)^2} = 1$$

Then the center is $\left(-1, \frac{5}{2}\right)$, $a = 3$, and $b = \frac{3}{2}$.

Place a new coordinate system at the center and move left and right $\frac{3}{2}$ units to obtain $\left(-\frac{5}{2}, \frac{5}{2}\right)$ and $\left(\frac{1}{2}, \frac{5}{2}\right)$. Then move up and down 3 units to obtain $\left(-1, \frac{11}{2}\right)$ and $\left(-1, -\frac{1}{2}\right)$. These four points act like the vertices in the new coordinate system. The graph is given below.

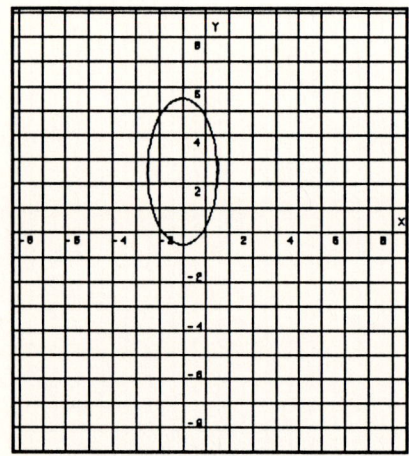

In Exercises 13-15, write each equation in the form

$$\frac{(x-h)^2}{m} + \frac{(y-k)^2}{n} = 1.$$

13. $8x^2 + 7y^2 - 56 = 0$

   Rewrite the equation with the constant on the right.

   $$8x^2 + 7y^2 = 56$$

   Divide both sides by 56.

   $$\frac{x^2}{7} + \frac{y^2}{8} = 1$$

14. $9x^2 + 50y^2 + 54x - 300y + 306 = 0$

   Complete the square in both x and y.

   $$9x^2 + 54x \quad +50y^2 - 300y \quad = -306$$
   $$9(x^2 + 6x \quad) + 50(y^2 - 6y \quad) = -306$$
   $$9(x^2 + 6x + 9) + 50(y^2 - 6y + 9) = -306 + 9(9) + 50(9)$$
   $$9(x+3)^2 + 50(y-3)^2 = 225$$
   $$\frac{9(x+3)^2}{225} + \frac{50(y-3)^2}{225} = \frac{225}{225}$$
   $$\frac{(x+3)^2}{25} + \frac{(y-3)^2}{\frac{9}{2}} = 1$$

15. $4x^2 + 9y^2 + 16x + 88 = 0$

   Complete the square on x.

   $$4x^2 + 16x \quad + 9y^2 = -88$$
   $$4(x^2 + 4x \quad) + 9y^2 = -88$$
   $$4(x^2 + 4x + 4) + 9y^2 = -88 + 4(4)$$
   $$4(x+2)^2 + 9y^2 = -72$$

   Since the right side is negative, we cannon obtain the desired form. This equation has no solution and no graph.

16. Assume that the rectangular property is placed in a coordinate system so that the rectangle has sides parallel to the axes and passes through the points $(3,0)$, $(-3,0)$, $\left(0,\frac{3}{2}\right)$, and $\left(0,-\frac{3}{2}\right)$. If the eliptical riding path is to be tangent to the property, then these points serve as the intercepts of an ellipse centered at the origin. The equation of the ellipse is:

   $$\frac{x^2}{3^2} + \frac{y^2}{\left(\frac{3}{2}\right)^2} = 1$$
   $$\frac{x^2}{9} + \frac{y^2}{\frac{9}{4}} = 1$$
   $$\frac{x^2}{9} + \frac{4y^2}{9} = 1$$

17. Assume that the center of the moon is the origin of a coordinate system. Since the moon has radius 900 km, and the satellite varies from 200 km to 300 km above the moon, we can use the four points $(1200,0)$, $(-1200,0)$, $(0,1100)$, and $(0,-1100)$ as the intercepts of the ellipse of the orbit which is centered at $(0,0)$. The equation of the ellipse is:

   $$\frac{x^2}{(1200)^2} + \frac{y^2}{(1100)^2} = 1$$

   Since this ellipse is nearly a circle, we could use the circumference of a circle to approximate the distance traveled by the satellite in one orbit. Since the farthest distance from the center is 1200, and the closest distance is 1100, suppose we take the average of these two numbers, 1150, and use this as the radius of the circle. Then the circumference of this circle is:

   $$C = 2\pi r = 2\pi(1150) \approx 7225.663103$$

   Thus, the satellite travels about 7200 km in one orbit.

18. The bridge is in the shape of a semiellipse, and the equation of the ellipse is

   $$\frac{x^2}{16} + \frac{y^2}{9} = 1.$$

EXERCISES A

Solve this equation for y and use the positive root only (we want the upper portion of the ellipse above the x-axis).

$$\frac{y^2}{9} = 1 - \frac{x^2}{16}$$

$$\frac{y^2}{9} = \frac{16 - x^2}{16}$$

$$y^2 = \frac{9}{16}(16 - x^2)$$

$$y = \frac{3}{4}\sqrt{16 - x^2}$$

To find the vertical clearance 3.0 m from the edge of the canal, we find the value of y when $x = 1$. (Note that 3 m from the edge is 1 m from the center of the canal.)

$$y = \frac{3}{4}\sqrt{16 - (1)^2}$$
$$= \frac{3}{4}\sqrt{15}$$
$$\approx 2.90473751$$

Thus, the clearance is about 2.9 m at this point.

19. If the circle has center $(2,-5)$ and passes through $(4,-1)$, then the distance between these two points is the radius of the circle. Use the distance formula.

$$d = \sqrt{(x_1 - x_2)^2 + (y_1 - y_2)^2}$$
$$= \sqrt{(2 - 4)^2 + (-5 - (-1))^2}$$
$$= \sqrt{(-2)^2 + (-4)^2}$$
$$= \sqrt{4 + 16}$$
$$= \sqrt{20}$$

Substitute 2 for h, -5 for k, and $\sqrt{20}$ for r in the standard form.

$$(x - 2)^2 + (y - (-5))^2 = \left(\sqrt{20}\right)^2$$
$$(x - 2)^2 + (y + 5)^2 = 20$$

20. Write $x^2 + y^2 + 4x - 16y + 59 = 0$ in standard form.

We must complete the square on x and y.

$$x^2 + 4x \quad + y^2 - 16y \quad = -59$$
$$x^2 + 4x + 4 + y^2 - 16y + 64 = -59 + 4 + 64$$
$$(x + 2)^2 + (y - 8)^2 = 9$$

21. Place a circle centered at the origin with diameter 30 ft. This circle contains the desired semicircle, the upper portion of the circle above the x-axis. The equation of this circle is

$$x^2 + y^2 = (15)^2$$

which is equivalent to

$$x^2 + y^2 = 225.$$

To find the equation of the desired semicircle, solve the above equation for y, and use the positive root.

$$y^2 = 225 - x^2$$
$$y = \sqrt{225 - x^2}$$

To find the vertical clearance 5 ft from the edge of the tunnel, substitute 10 for x and evaluate y. Note that 5 ft from the edge is the same as 10 ft from the center of the roadway.

$$y = \sqrt{225 - (10)^2}$$
$$= \sqrt{225 - 100}$$
$$= \sqrt{125}$$
$$\approx 11.18033989$$

Thus, the vertical clearance is about 11.2 ft at this point.

CHAPTER 8 TOPICS IN ANALYTIC GEOMETRY

SECTION 8.3 The Hyperbola

1. Find the equation of the hyperbola with foci $(\pm 3, 0)$ and $a = 2$.

   Since $c = 3$ and $a = 2$,
   $$b^2 = c^2 - a^2 = 3^2 - 2^2 = 9 - 4 = 5,$$
   so $b = \sqrt{5}$. The hyperbola has a horizontal transverse axis, vertices $(2,0)$ and $(-2,0)$, and equation
   $$\frac{x^2}{a^2} - \frac{y^2}{b^2} = 1$$
   $$\frac{x^2}{4} - \frac{y^2}{5} = 1.$$

2. Graph the hyperbola with equation
   $$\frac{x^2}{16} - \frac{y^2}{4} = 1.$$

   Then $a = 4$ and $b = 2$. Construct the rectangle with sides through the points $(4,0)$, $(0,2)$, $(-4,0)$, and $(0,-2)$. Draw the asymptotes through the corners of the rectangle. The vertices of the hyperbola, which opens left and right, are $(4,0)$ and $(-4,0)$. The graph is given below.

   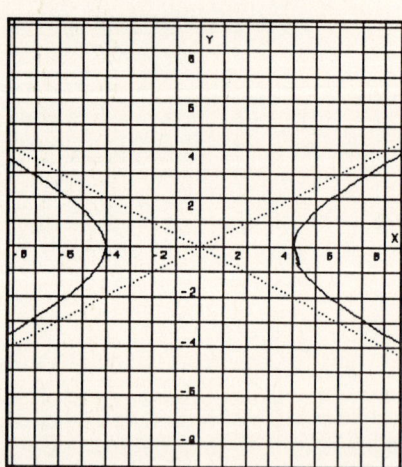

3. Graph the hyperbola with foci $(0, \pm\sqrt{29})$ and vertices $(0, \pm 2)$.

   Knowing that $c = \sqrt{29}$ and $a = 2$, find b.

   $$b^2 = c^2 - a^2 = 29 - 4 = 25$$

   Then $b = 5$. The equation of the hyperbola is
   $$\frac{y^2}{a^2} - \frac{x^2}{b^2} = 1$$
   $$\frac{y^2}{4} - \frac{x^2}{25} = 1,$$
   and the graph opens up and down. Construct the rectangle through the points $(5,0)$, $(0,2)$, $(-5,0)$, and $(0,-2)$, and draw the asymptotes of the graph, the diagonals of the rectangle. The graph is given below.

   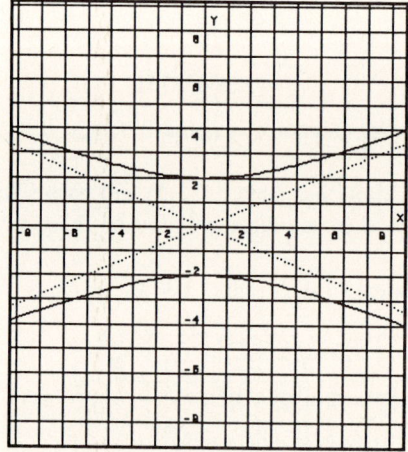

4. Graph the hyperbola
   $$\frac{(y-2)^2}{16} - \frac{(x+1)^2}{9} = 1.$$

   Then the center is $(h,k) = (-1,2)$, $a = 4$, and $b = 3$. Think of the lines $x = -1$ and $y = 2$ as new coordinate axes with the hyperbola centered at the origin of this system. Use $a = 4$ and $b = 3$ relative to these new axes to sketch the rectangle and asymptotes. Since the hyperbola has a vertical transverse axis, it passes through the points $(-1,6)$ and $(-1,-2)$. Using this information we obtain the graph shown below.

PRACTICE EXERCISES

SECTION 8.3   397

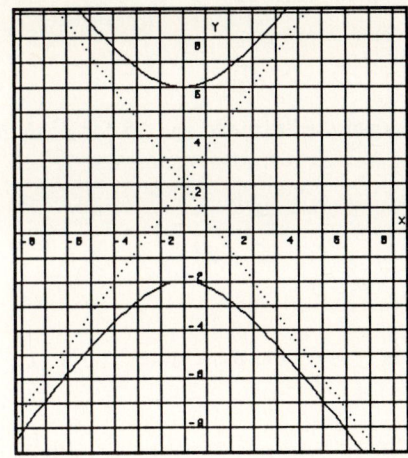

sketch of the graph as shown below.

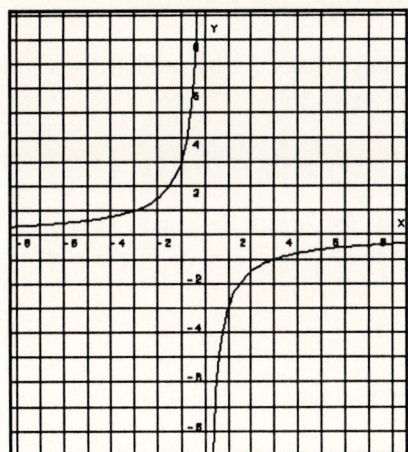

5. Write $-4x^2 + y^2 - 40x - 2y - 115 = 0$ in standard form.

First rewrite the equation leaving spaces as shown.

$$-4x^2 - 40x \quad + y^2 - 2y \quad = 115$$

Factor out the coefficients of the squared terms and group as shown.

$$-4(x^2 + 10x \quad) + (y^2 - 2y \quad) = 115$$

Complete the square in x and y, be sure to add the same expression to both sides.

$$-4(x^2+10x+25) + (y^2-2y+1) = 115+(-4)(25)+1$$

Factor, and simplify the right side.

$$-4(x + 5)^2 + (y - 1)^2 = 16$$

Divide through by 16 to obtain the standard form.

$$\frac{-4(x+5)^2}{16} + \frac{(y-1)^2}{16} = \frac{16}{16}$$

$$-\frac{(x+5)^2}{4} + \frac{(y-1)^2}{16} = 1$$

$$\frac{(y-1)^2}{16} - \frac{(x+5)^2}{4} = 1$$

Note that the hyperbola has a vertical transverse axis, opens up and down, center at $(-5,1)$, $a = 4$, and $b = 2$.

6. Graph $2xy + 6 = 0$.

Simplifying we have $xy = -3$. The asymptotes of the hyperbola are the coordinate axes, and the graph has two "branches" in the second and fourth quadrants. Plotting several points we can obtain a

CHAPTER 8 TOPICS IN ANALYTIC GEOMETRY

SECTION 8.3 The Hyperbola

Find the standard form of the equation of each hyperbola in Exercises 1-6.

1. Foci at $(\pm 5, 0)$ and vertices at $(\pm 3, 0)$.

    Since the foci are on the x-axis, the hyperbola has the equation

    $$\frac{x^2}{a^2} - \frac{y^2}{b^2} = 1.$$

    Since $c = 5$ and $a = 3$, we can find b.

    $$b^2 = c^2 - a^2 = 5^2 - 3^2 = 25 - 9 = 16$$

    Thus $b = 4$. Substitute 4 for b and 3 for a in the standard form.

    $$\frac{x^2}{9} - \frac{y^2}{16} = 1.$$

2. $a = b = 3$, transverse axis along the y-axis, and center at the origin.

    Since the transverse axis is along the y-axis, the transverse axis is vertical, and the equation has the form

    $$\frac{y^2}{a^2} - \frac{x^2}{b^2} = 1.$$

    Thus, we can substitute 3 for a and 3 for b to obtain the desired equation.

    $$\frac{y^2}{9} - \frac{x^2}{9} = 1.$$

3. Foci $(\pm 7, 0)$ and length of transverse axis is 10.

    Since the foci are on the x-axis, the transverse axis is horizontal, and the equation has the form:

    $$\frac{x^2}{a^2} - \frac{y^2}{b^2} = 1$$

    Since the length of the transverse axis is 10, $a = 5$. Then since $c = 7$,

    $$b^2 = c^2 - a^2 = 7^2 - 5^2 = 49 - 25 = 24.$$

    Then substitute 24 for b^2 and 25 for a^2 in the form above.

    $$\frac{x^2}{25} - \frac{y^2}{24} = 1$$

4. Vertices $(0, \pm 6)$ and asymptotes $y = \pm \frac{3}{4}x$.

    Since the vertices are on the y-axis, the hyperbola has a vertical transverse axis and equation of the form

    $$\frac{y^2}{a^2} - \frac{x^2}{b^2} = 1.$$

    Then $a = 6$, and since the equation for the assymptotes is

    $$y = \pm \frac{a}{b}x,$$

    and since we are given that the assymptotes are in fact

    $$y = \pm \frac{3}{4}x,$$

    setting $a = 6$ we can solve for b in the following equation.

    $$\frac{a}{b} = \frac{3}{4}$$
    $$\frac{6}{b} = \frac{3}{4}$$
    $$(6)(4) = (3)(b)$$
    $$24 = 3b$$
    $$8 = b$$

    Then substituting 6 for a and 8 for b in the form above, we obtain the desired equation.

    $$\frac{y^2}{36} - \frac{x^2}{64} = 1$$

5. Center at $(-3, 2)$, foci $(-3, 8)$ and $(-3, -4)$, and vertices $(-3, 6)$ and $(-3, -2)$.

    Since the foci are on a vertical line, the transverse axis is vertical, and the equation has the form:

    $$\frac{(y-k)^2}{a^2} - \frac{(x-h)^2}{b^2} = 1$$

EXERCISES A SECTION 8.3

Since the center is $(-3,2)$, $h = -3$ and $k = 2$.
Since the vertices are 4 units from the center,
$a = 4$. Since the foci are 6 units from the center,
$c = 6$. Then

$$b^2 = c^2 - a^2 = 6^2 - 4^2 = 36 - 16 = 20.$$

Substitute -3 for h, 2 for k, 16 for a^2, and 20 for b^2 in the form above to obtain the desired equation.

$$\frac{(y-2)^2}{16} - \frac{(x+3)^2}{20} = 1$$

6. Vertices $(\pm 4, 0)$ and passing through $(8, 6)$.

Since the vertices are evenly spaced about $(0,0)$ on the x-axis, the hyperbola has a horizontal transverse axis, is centered at the origin, and has equation of the form:

$$\frac{x^2}{a^2} - \frac{y^2}{b^2} = 1$$

Then $a = 4$, and substituting the equation becomes:

$$\frac{x^2}{16} - \frac{y^2}{b^2} = 1$$

Substitute 8 for x and 6 for y to find the value of b.

$$\frac{x^2}{16} - \frac{y^2}{b^2} = 1$$
$$\frac{64}{16} - \frac{36}{b^2} = 1$$
$$4 - \frac{36}{b^2} = 1$$
$$4b^2 - 36 = b^2$$
$$3b^2 = 36$$
$$b^2 = 12$$

Then substitute 12 for b^2 to obtain the desired equation.

$$\frac{x^2}{16} - \frac{y^2}{12} = 1$$

7. Graph the hyperbola.

$$\frac{x^2}{25} - \frac{y^2}{4} = 1$$

This is the form of a hyperbola with a horizontal transverse axis and center at the origin. The intercepts are $(5,0)$ and $(-5,0)$. The asymptotes are the diagonals of the rectangle with sides parallel to the axes and passing through the intercepts along

with $(0,2)$ and $(0,-2)$. The graph is given below.

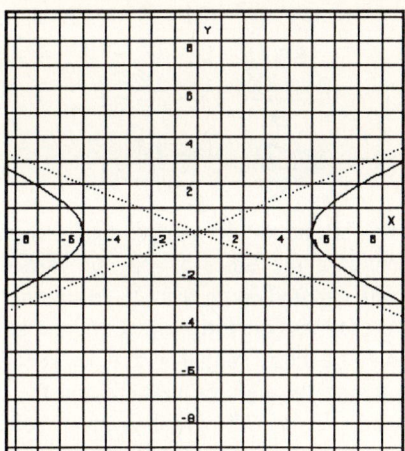

8. Graph the hyperbola.

$$\frac{(x+3)^2}{9} - \frac{(y+1)^2}{16} = 1$$

The center of the hyperbola is at the point $(-3,-1)$. Place a new coordinate system with origin at the center. This is the form of a hyperbola with horizontal transverse axis. Since $a = 3$, the vertices are 3 units to the right and left of the center at $(0,-1)$ and $(-6,-1)$. Also, $b = 4$, so the asymptotes are the lines through the diagonals of the rectangle with sides parallel to the axes and passing through the vertices and the points $(-3,3)$ and $(-3,-5)$. The graph is given below.

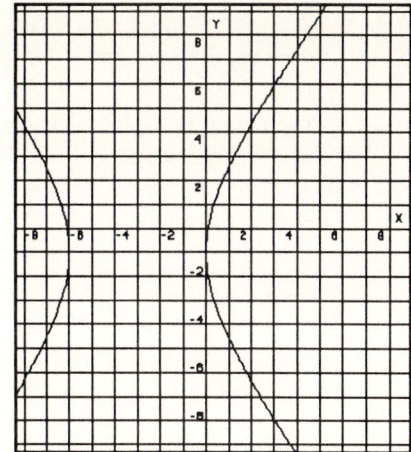

9. Graph the hyperbola.

$$x^2 - y^2 - 4x + 2y + 12 = 0$$

Complete the square on x and y.

$$(x^2 - 4x \quad) - (y^2 - 2y \quad) = -12$$
$$(x^2 - 4x + 4) - (y^2 - 2y + 1) = -12 + 4 - 1$$
$$(x - 2)^2 - (y - 1)^2 = -9$$
$$\frac{(x-2)^2}{-9} - \frac{(y-1)^2}{-9} = 1$$
$$\frac{(y-1)^2}{9} - \frac{(x-2)^2}{9} = 1$$

Then this is a hyperbola with vertical transverse axis and center (2,1). Since $a = 3$, the vertices are three units up and down from the center at (2,4) and (2,-2). The assymptotes are contained in the diagonals of the rectangle with sides parallel to the axes passing through the vertices and the points (5,1) and (-1,1). The graph is given below.

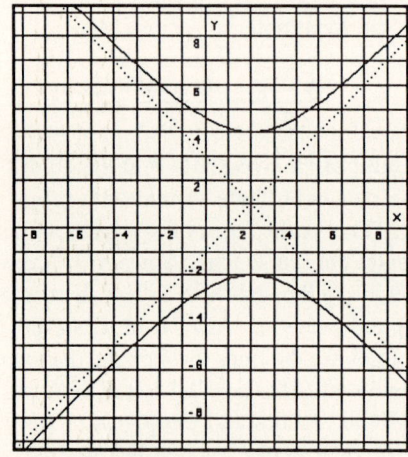

10. Graph the hyperbola.

$$xy = -2$$

This is a hyperbola with asymptotes the coordinate axes. Since the constant is negative, -2, the graph is found in quadrants II and IV. Plotting several points gives the graph as shown below.

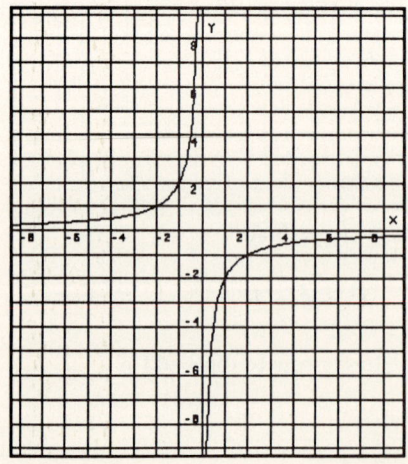

In Exercises 11-13, write each equation in the form

$$\frac{(x-h)^2}{m} - \frac{(y-k)^2}{n} = 1$$

or

$$\frac{(y-k)^2}{m} - \frac{(x-h)^2}{n} = 1.$$

11. $x^2 - y^2 + 25 = 0$

Isolate the constant on the right side.

$$x^2 - y^2 = -25$$

Divide both sides by -25.

$$\frac{x^2}{-25} - \frac{y^2}{-25} = \frac{-25}{-25}$$
$$\frac{y^2}{25} - \frac{x^2}{25} = 1$$

12. $16y^2 - 9x^2 - 192y - 54x + 351 = 0$

Complete the squares on x and y.

$$16(y^2 - 12y \quad) - 9(x^2 + 6x \quad) = -351$$
$$16(y^2 - 12y + 36) - 9(x^2 + 6x + 9) = -351 + 16(36) - 81$$
$$16(y - 6)^2 - 9(x + 3)^2 = 144$$
$$\frac{16(y-6)^2}{144} - \frac{9(x+3)^2}{144} = \frac{144}{144}$$
$$\frac{(y-6)^2}{9} - \frac{(x+3)^2}{16} = 1$$

13. $4x^2 - y^2 + 32x + 2y + 63 = 0$

Complete the squares on x and y.

$$4(x^2 + 8x \quad) - (y^2 - 2y \quad) = -63$$
$$4(x^2 + 8x + 16) - (y^2 - 2y + 1) = -63 + 4(16) - 1$$
$$4(x + 4)^2 - (y - 1)^2 = 0$$
$$\frac{(x+4)^2}{1} - \frac{(y-1)^2}{4} = 0$$

Since the right side is 0, we cannot write the equation in the form of a hyperbola. In fact, the equation represents two intersecting lines,

$$y = 2x + 9 \text{ and } y = -2x - 7,$$

and is sometimes called a degenerate hyperbola.

14. The roof of a building is in the shape of the hyperbola

EXERCISES A

$$y^2 - x^2 = 25,$$

where x and y are shown in meters in the figure in the text. The height of the outside walls, h, is the value of y when x is 6.0. Substitute to find y.

$$y^2 - (6.0)^2 = 25$$
$$y^2 - 36 = 25$$
$$y^2 = 61$$
$$y = \pm\sqrt{61}$$
$$y \approx \pm 7.810249676$$

Since y must be positive in this case, the height h is about 7.8 m.

15. The curve described is a hyperbola with foci at S_1 and S_2.

16. Start with the standard form of the equation,

$$\frac{(x-h)^2}{a^2} - \frac{(y-k)^2}{b^2} = 1$$

and clear the parentheses and collect like terms to obtain the form

$$Ax^2 + Bxy + Cy^2 + Dx + Ey + F = 0.$$

Begin by multiplying both sides by a^2b^2 to clear the fractions.

$$b^2(x-h)^2 - a^2(y-k)^2 = a^2b^2$$
$$b^2(x^2 - 2xh + h^2) - a^2(y^2 - 2yk + k^2) = a^2b^2$$
$$b^2x^2 - 2b^2hx + b^2h^2 - a^2y^2 + 2a^2ky - a^2k^2 = a^2b^2$$
$$[b^2]x^2 + [-a^2]y^2 + [-2b^2h]x +$$
$$[2a^2k]y + [b^2h^2 - a^2k^2 - a^2b^2] = 0$$

Thus, equating coefficients of like terms, we have $A = b^2$, $B = 0$, $C = -a^2$, $D = -2b^2h$, $E = 2a^2k$, and $F = b^2h^2 - a^2k^2 - a^2b^2$.

17. Find the standard form of the equation of the ellipse with foci $(0, \pm\sqrt{11})$ and length of minor axis 4.

Since the foci are on the y-axis, the ellipse has a vertical major axis and equation of the form

$$\frac{x^2}{b^2} + \frac{y^2}{a^2} = 1.$$

Since the length of the minor axis is 4, $b = 2$.

With $c = \sqrt{11}$, we can find a.

$$a^2 = b^2 + c^2 = 2^2 + \left(\sqrt{11}\right)^2 = 4 + 11 = 15$$

Substitute 15 for a^2 and 4 for b^2 in the form above.

$$\frac{x^2}{4} + \frac{y^2}{15} = 1$$

In Exercises 18-19, write each equation in standard form and identify the graph.

18. $8x^2 + 4y^2 - 24y + 4 = 0$

Complete the square on y.

$$8x^2 + 4(y^2 - 6y) = -4$$
$$8x^2 + 4(y^2 - 6y + 9) = -4 + 4(9)$$
$$8x^2 + 4(y-3)^2 = 32$$
$$\frac{8x^2}{32} + \frac{4(y-3)^2}{32} = \frac{32}{32}$$
$$\frac{x^2}{4} + \frac{(y-3)^2}{8} = 1$$

This is the standard form of the equation of an ellipse.

19. $x^2 + y^2 - 4x + 18y + 81 = 0$

Complete the square on x and y.

$$x^2 - 4x + y^2 + 18y = -81$$
$$x^2 - 4x + 4 + y^2 + 18y + 81 = -81 + 4 + 81$$
$$(x-2)^2 + (y+9)^2 = 4$$

This is the standard form of the equation of a circle.

CHAPTER 8 TOPICS IN ANALYTIC GEOMETRY

SECTION 8.4 The Parabola

1. Give the standard form of the equation of the parabola with vertex at the origin which opens left and has focus $(-3,0)$.

   By the theorem on standard forms of the equation of a parabola with vertex at the origin, we use the equation

   $$y^2 = 4px.$$

   Since the focus is $(-3,0)$, $p = -3$. Thus,

   $$y^2 = 4(-3)x$$
   $$y^2 = -12x.$$

2. Determine the vertex, focus, and directrix, and graph the parabola with equation $x^2 = -12y$.

   First write the equation in the form

   $$x^2 = 4py.$$

   Since $4p = -12$, $p = -3$. Thus, the equation is

   $$x^2 = 4(-3)y.$$

   From the theorem on standard forms, the vertex is at the origin, $(0,0)$, the focus is $(0,-3)$, and the directrix is $y = -p = -(-3) = 3$. The graph, which opens down, is shown below.

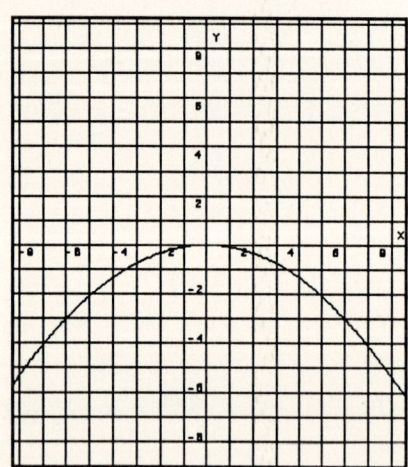

3. Determine the vertex, focus, and directrix, and graph the parabola with equation

   $$(y + 2)^2 = 8(x + 1).$$

   Rewrite the equation as

   $$(y - (-2))^2 = 8(x - (-1))$$

   to see that the vertex is $(h,k) = (-1,-2)$. Since

   $$4p = 8$$
   $$p = 2,$$

   so the focus is $(h+p,k) = (-1+2,-2) = (1,-2)$. That is, the focus is 2 units to the right of the vertex. The directrix is the vertical line $x = h - p = -1 - 2 = -3$. The graph opens to the right, and is shown below.

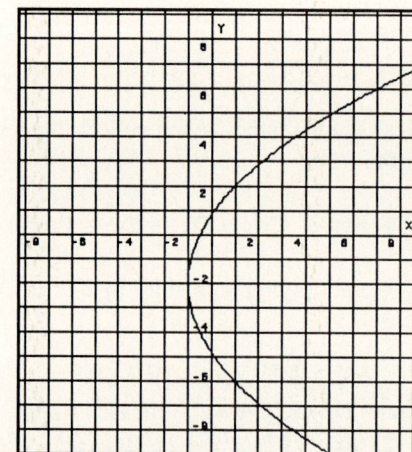

4. Determine the equation of the parabola with vertex $(-3,1)$, axis of symmetry $x = -3$, and passing through $\left(0, -\frac{5}{4}\right)$.

   Since the axis of symmetry $x = -3$ is parallel to the y-axis, the equation of the parabola has the form

   $$(x - h)^2 = 4p(y - k).$$

   Since the vertex is $(-3,1)$, $h = -3$, $k = 1$, and the equation becomes

   $$(x + 3)^2 = 4p(y - 1).$$

   Now the only unknown is p. Use the fact that the parabola passes through $\left(0, -\frac{5}{4}\right)$ and substitute to find p.

PRACTICE EXERCISES

$$(x + 3)^2 = 4p(y - 1)$$
$$(0 + 3)^2 = 4p\left(-\frac{5}{4} - 1\right)$$
$$9 = 4p\left(-\frac{9}{4}\right)$$
$$9 = p(-9)$$
$$-1 = p$$

Thus, the equation of the parabola is

$$(x + 3)^2 = 4(-1)(y - 1)$$

which simplifies to

$$(x + 3)^2 = -4(y - 1).$$

5. A domed parabolic ceiling is designed such that at a point 8.0 m down from the top of the dome, the ceiling is 20.0 m wide. The best location for the light source is at the focus of the equation of the parabola formed as the intersection of a plane with the ceiling. We can think of the parabola as having its vertex at (0,0) opening down. Then the equation of the parabola is

$$x^2 = 4py.$$

Since when $y = -8.0$ m, the width is 20.0 m, the points (10.0,−8.0) and (−10.0,−8.0) are on the curve. Use (10.0,−8.0) and substitute to find p.

$$x^2 = 4py$$
$$(10.0)^2 = 4p(8.0)$$
$$3.125 = p$$

Thus, the light source should be placed about 3.1 m down from the top of the ceiling.

6. Write $x^2 - 4x + 4y + 12 = 0$ in standard form and give the vertex, focus, and directrix. Graph the equation.

Complete the square on x. Rewrite the equation leaving space as shown.

$$x^2 - 4x \quad = -4y - 12$$

Add half the coefficient of x, squared, to both sides.

$$x^2 - 4x + 4 = -4y - 12 + 4$$

Factor, and simplify the right side.

$$(x - 2)^2 = -4(y + 2)$$

Then we recognize that the vertex is

$$(h,k) = (2,-2).$$

Since $4p = -4$, $p = -1$. The focus is

$$(h,k+p) = (2,-2-1) = (2,-3).$$

The directrix is

$$y = k - p = -2 - (-1) = -1.$$

The graph, which opens down, is given below.

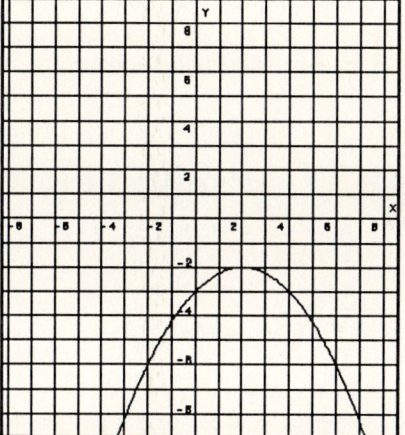

CHAPTER 8 TOPICS IN ANALYTIC GEOMETRY

SECTION 8.4 The Parabola

In Exercises 1-6 determine the equation of each parabola.

1. Vertex at the origin and focus $\left(0, \frac{3}{2}\right)$.

   Since the vertex is at the origin and the focus is on the y-axis, the equation has the form:

   $$x^2 = 4py$$

   Since $p = \frac{3}{2}$, substitute to obtain:

   $$x^2 = 4py$$
   $$x^2 = 4\left(\frac{3}{2}\right)y$$
   $$x^2 = 6y$$

2. Vertex at $(0,0)$ and focus $\left(\frac{7}{2}, 0\right)$.

   Since the vertex is at the origin and the focus is on the x-axis, the equation has the form:

   $$y^2 = 4px$$

   Since $p = \frac{7}{2}$, substitute to obtain:

   $$y^2 = 4\left(\frac{7}{2}\right)x$$
   $$y^2 = 14x$$

3. Directrix $x = -3$ and focus $(3,0)$.

   Since the focus is on the x-axis and $p = 3$, substitute into the form:

   $$y^2 = 4px$$
   $$y^2 = 4(3)x$$
   $$y^2 = 12x$$

4. Directrix $y = -1$ and focus $(4,5)$.

   Since the directrix is a horizontal line, the equation has the form:

   $$(x - h)^2 = 4p(y - k)$$

   The vertex is midway between the focus and the directrix, so the vertex has x-coordinate 4, and the y-coordinate must be

   $$\frac{5 + (-1)}{2} = \frac{4}{2} = 2$$

   Thus, the vertex is $(4,2)$. Then the distance from the focus to the vertex is 3 units, and since the parabola opens up (the focus is above the directrix), $p = 3$. Substitute 3 for p, 4 for h, and 2 for k in the form above.

   $$(x - 4)^2 = 4(3)(y - 2)$$
   $$(x - 4)^2 = 12(y - 2)$$

5. Vertex $(4,-6)$, axis of symmetry $x = 4$, and passing through $(0,-8)$.

   Since the axis of symmetry, $x = 4$, is a vertical line, the form of the parabola is:

   $$(x - h)^2 = 4p(y - k)$$

   Substitute 4 for h and -6 for k.

   $$(x - 4)^2 = 4p(y + 6)$$

   Since the parabola passes through $(0,-8)$, substitute 0 for x and -8 for y to find the value of p.

   $$(0 - 4)^2 = 4p(-8 + 6)$$
   $$16 = -8p$$
   $$-2 = p$$

   Substitute -2 for p in the above form:

   $$(x - 4)^2 = 4(-2)(y + 6)$$
   $$(x - 4)^2 = -8(y + 6)$$

6. Vertex $(8,-7)$, axis of symmetry parallel to the x-axis, and passing through $(6,-8)$.

   Since the axis of symmetry is parallel to the x-axis, the form is:

   $$(y - k)^2 = 4p(x - h)$$

   Substitute 8 for h and -7 for k.

   $$(y + 7)^2 = 4p(x - 8)$$

EXERCISES A

Since the parabola passes through (6,−8), substitute 6 for x and −8 for y to find the value of p.

$$(-8 + 7)^2 = 4p(6 - 8)$$
$$1 = -8p$$
$$-\frac{1}{8} = p$$

Substitute this value for p to obtain the desired equation:

$$(y + 7)^2 = -\frac{1}{2}(x - 8)$$

7. Graph the parabola.

$$y^2 = 2x$$

Since $4p = 2$, $p = \frac{1}{2}$. Since $p > 0$, the parabola opens to the right, and the vertex is at the origin. Two additional points on the parabola are (2,2) and (2,−2), and the graph is given below.

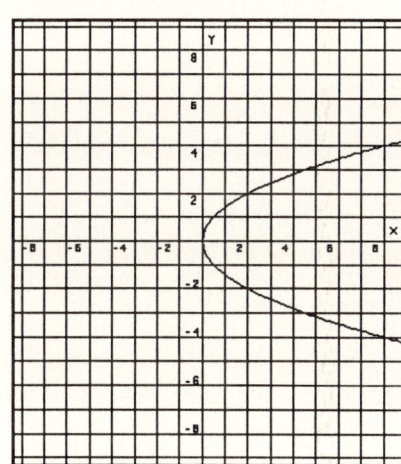

8. Graph the parabola.

$$(x - 2)^2 = -(y + 1)$$

Since $4p = -1$, $p = -\frac{1}{4}$. The vertex is (2,−1), and since $p < 0$, the graph opens down. Two additional points on the graph are (3,−2) and (1,−2). The graph is given below.

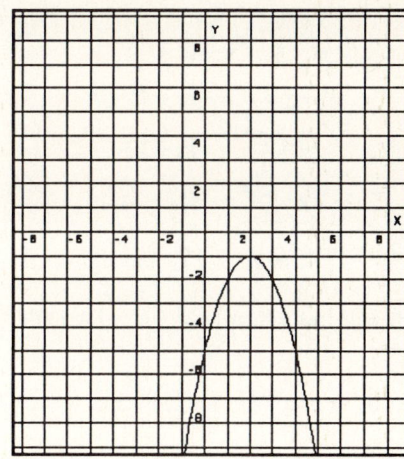

9. Graph the parabola.

$$x^2 - 4y + 8 = 0$$

First write the equation in standard form.

$$x^2 = 4y - 8$$
$$x^2 = 4(y - 2)$$

Then $4p = 4$, so $p = 1$, and the parabola opens up and has vertex (0,2). Two additional points on the parabola are (2,3) and (−2,3). The graph is given below.

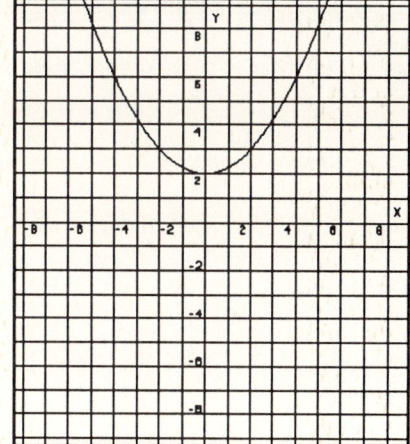

10. Graph the parabola.

$$y^2 + 4x + y - \frac{47}{4} = 0$$

First complete the square on y.

$$y^2 + y = -4x + \frac{47}{4}$$
$$y^2 + y + \frac{1}{4} = -4x + \frac{47}{4} + \frac{1}{4}$$
$$\left(y + \frac{1}{2}\right)^2 = -4x + 12$$
$$\left(y + \frac{1}{2}\right)^2 = -4(x - 3)$$

Then $4p = -4$, so $p = -1$, and the graph opens to the left and has vertex $\left(3, -\frac{1}{2}\right)$. Two additional points on the graph are $\left(-1, \frac{7}{2}\right)$ and $\left(-1, -\frac{9}{2}\right)$. The graph is given below.

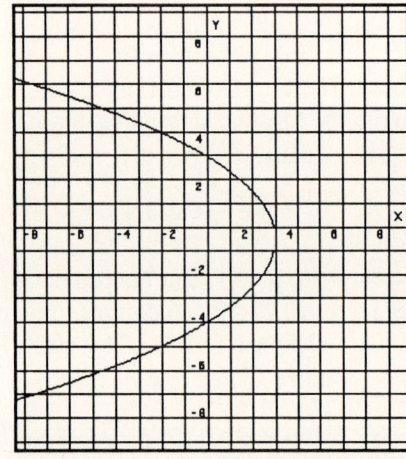

In Exercises 11-13 write each equation in the standard form

$$(y - k)^2 = 4p(x - h) \text{ or } (x - h)^2 = 4p(y - k),$$

and determine the vertex, focus, and directrix.

11. $y^2 - 12x + 24 = 0$

$$y^2 = 12x - 24$$
$$y^2 = 12(x - 2)$$

Then the vertex is $(2,0)$. Since $4p = 12$, $p = 3$, so the graph opens right, and the focus is $(2+3,0) = (5,0)$. The directrix is the vertical line 3 units left of the vertex, $x = -1$.

12. $x^2 + 8x + 9y + 25 = 0$

Complete the square on x.

$$x^2 + 8x = -9y - 25$$
$$x^2 + 8x + 16 = -9y - 25 + 16$$
$$(x + 4)^2 = -9y - 9$$
$$(x + 4)^2 = -9(y + 1)$$

Then the vertex is $(-4,-1)$. Since $4p = -9$, $p = -\frac{9}{4}$, and the parabola opens down. The focus is down from the vertex $\frac{9}{4}$ units at the point $\left(-4, -\frac{13}{4}\right)$. The directrix is the horizontal line up $\frac{9}{4}$ units from the vertex, the line $y = \frac{5}{4}$.

13. $4x^2 - 12x - 2y - 3 = 0$

Complete the square on x.

$$4x^2 - 12x = 2y + 3$$
$$4(x^2 - 3x) = 2y + 3$$
$$4\left(x^2 - 3x + \frac{9}{4}\right) = 2y + 3 + 4\left(\frac{9}{4}\right)$$
$$4\left(x - \frac{3}{2}\right)^2 = 2y + 12$$
$$\left(x - \frac{3}{2}\right)^2 = \frac{1}{2}y + 3$$
$$\left(x - \frac{3}{2}\right)^2 = \frac{1}{2}(y + 6)$$

Then the vertex is $\left(\frac{3}{2}, -6\right)$. Since $4p = \frac{1}{2}$, $p = \frac{1}{8}$, and the the parabola opens up. The focus is up from the vertex $\frac{1}{8}$ unit at the point $\left(\frac{3}{2}, -\frac{47}{8}\right)$. The directrix is the horizontal line down $\frac{1}{8}$ unit, the line $y = -\frac{49}{8}$.

In Exercises 14-17 complete the squares to determine if the equation represents a circle, ellipse, hyperbola, parabola, or degenerate conic.

EXERCISES A SECTION 8.4 407

14. $5x^2 - 2y^2 + x - 7 = 0$

$$5x^2 + x \quad -2y^2 = 7$$
$$5\left(x^2 + \tfrac{1}{5}x \quad\right) - 2y^2 = 7$$
$$5\left(x^2 + \tfrac{1}{5}x + \tfrac{1}{100}\right) - 2y^2 = 7 + 5\left(\tfrac{1}{100}\right)$$
$$5\left(x + \tfrac{1}{10}\right)^2 - 2y^2 = \tfrac{141}{20}$$
$$\frac{\left(x + \tfrac{1}{10}\right)^2}{\tfrac{141}{100}} - \frac{y^2}{\tfrac{141}{40}} = 1$$

This is the standard form of a hyperbola.

15. $-x^2 - y^2 + x + y + 12 = 0$

First multiply both sides by -1.

$$x^2 + y^2 - x - y - 12 = 0$$
$$x^2 - x \quad + y^2 - y \quad = 12$$
$$x^2 - x + \tfrac{1}{4} + y^2 - y + \tfrac{1}{4} = 12 + \tfrac{1}{4} + \tfrac{1}{4}$$
$$\left(x - \tfrac{1}{2}\right)^2 + \left(y - \tfrac{1}{2}\right)^2 = \tfrac{25}{2}$$

This is the standard form of the equation of a circle.

16. $4x^2 - 49y^2 + 8x + 4 = 0$

$$4x^2 + 8x \quad - 49y^2 = -4$$
$$4(x^2 + 2x \quad) - 49y^2 = -4$$
$$4(x^2 + 2x + 1) - 49y^2 = -4 + 4(1)$$
$$4(x+1)^2 - 49y^2 = 0$$
$$\frac{(x+1)^2}{49} - \frac{y^2}{4} = 0$$

This is a degenerate conic. It actually represents two intersecting straight lines.

17. $20x^2 + 10y^2 - 5x + 15y - 75 = 0$

$$20x^2 - 5x \quad + 10y^2 + 15y \quad = 75$$
$$20\left(x^2 - \tfrac{1}{4}x \quad\right) + 10\left(y^2 + \tfrac{3}{2}y \quad\right) = 75$$
$$20\left(x^2 - \tfrac{1}{4}x + \tfrac{1}{64}\right) + 10\left(y^2 + \tfrac{3}{2}y + \tfrac{9}{16}\right) = 75 + 20\left(\tfrac{1}{64}\right) + 10\left(\tfrac{9}{16}\right)$$
$$20\left(x - \tfrac{1}{8}\right)^2 + 10\left(y + \tfrac{3}{4}\right)^2 = \tfrac{1295}{16}$$
$$\frac{\left(x - \tfrac{1}{8}\right)^2}{\tfrac{259}{64}} + \frac{\left(y + \tfrac{3}{4}\right)^2}{\tfrac{259}{32}} = 1$$

This is the standard form of an ellipse.

18. From the information given and the figure in the text, we can see that the form of the parabola is

$$x^2 = 4py.$$

Since y is -32.5 when x is 12.1, substitute these values to find $4p$.

$$(12.1)^2 = 4p(-32.5)$$
$$-4.504923077 = 4p$$

Thus, to the nearest tenth, the equation is

$$x^2 = -4.5y.$$

19. Place a coordinate system over the tunnel so the origin is at the vertex of the parabola. Then the tunnel can be described by the equation

$$x^2 = 4py.$$

Since the maximum height of the tunnel is 12.8 m and the width at the base is 10.2 m, the point with coordinates $(5.1, -12.8)$ is on the parabola. Substitute to find the value of $4p$.

$$(5.1)^2 = 4p(-12.8)$$
$$-2.03203125 = 4p$$

Then the equation of the parabola is

$$x^2 = -2.03203125y.$$

To find the vertical clearance 1.5 m from the edge of the tunnel, substitute 3.6 for x (1.5 m from the edge is 3.6 m from the center of the base of the tunnel) and solve for y.

$$(3.6)^2 = -2.03203125y$$
$$-6.377854671 = y$$

Thus, the clearance is about 6.4 m.

20. Determine the equation of the hyperbola with center $(4, -2)$, foci $(4, -7)$ and $(4, 3)$, and length of the transverse axis 6.

Since the foci and center are on the same vertical line, the equation is of the form:

$$\frac{(y-k)^2}{a^2} - \frac{(x-h)^2}{b^2} = 1$$

Since the length of the transverse axis is 6, $a = 3$. Since the foci are each 5 units from the center, $c = 5$. Then

$$b^2 = c^2 - a^2 = 5^2 - 3^2 = 25 - 9 = 16.$$

Substitute 16 for b^2, 9 for a^2, 4 for h, and -2 for k in the above form.

$$\frac{(y+2)^2}{9} - \frac{(x-4)^2}{16} = 1$$

21. Write the equation in standard form and identify the graph.

$$6x^2 - 5y^2 - 24x - 30y - 51 = 0$$

$$6x^2 - 24x \quad - 5y^2 - 30y = 51$$
$$6(x^2 - 4x \quad) - 5(y^2 + 6y \quad) = 51$$
$$6(x^2 - 4x + 4) - 5(y^2 + 6y + 9) = 51 + 6(4) - 5(9)$$
$$6(x-2)^2 - 5(y+3)^2 = 30$$
$$\frac{(x-2)^2}{5} - \frac{(y+3)^2}{6} = 1$$

This is the standard form of a hyperbola.

In Exercises 22-23, sketch the graph of each equation.

22. $(x-2)^2 + y^2 = 4$

This is the equation of a circle with center (2,0) and radius 2. The graph is given below.

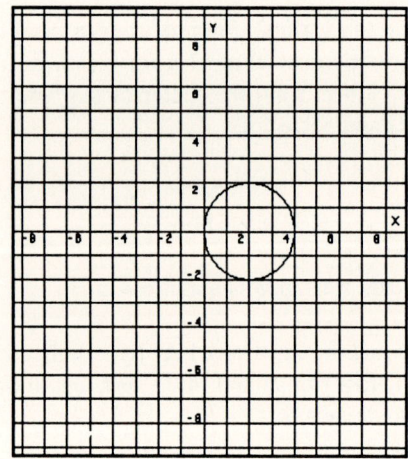

23. $y = \log_2 x$

Make a table of values. You might wish to use the equivalent exponential form: $2^y = x$. The graph is given below.

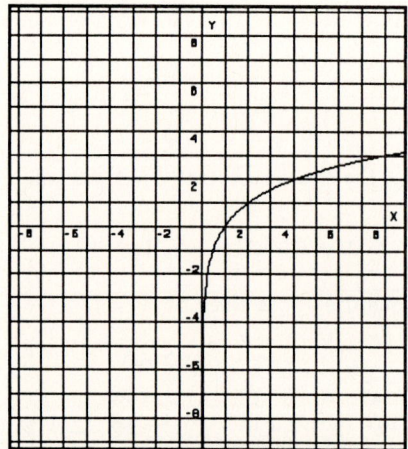

24. $y = |x - 1| + 1$

You may wish to start with the graph of the absolute value function, $y = |x|$, and shift this graph 1 unit to the right (using the -1) then 1 unit up (using the $+1$). The graph is given below.

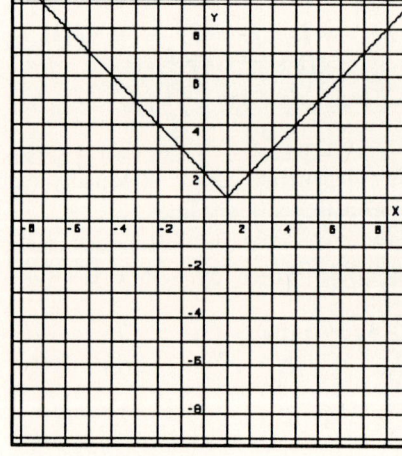

PRACTICE EXERCISES

SECTION 8.5   409

CHAPTER 8   TOPICS IN ANALYTIC GEOMETRY

SECTION 8.5   Nonlinear Systems

1. Solve the system.

$$x^2 + y^2 = 2$$
$$xy = 1$$

Note that the first equation is a circle and the second a hyperbola. Keep this in mind as you solve the system. Begin by solving the second equation for y,

$$y = \frac{1}{x},$$

and substitute this expression for y in the first equation.

$$x^2 + \left(\frac{1}{x}\right)^2 = 2$$
$$x^2 + \frac{1}{x^2} = 2$$

Multiply both sides by x^2 to clear the fraction.

$$x^4 + 1 = 2x^2$$
$$x^4 - 2x^2 + 1 = 0$$
$$(x^2 - 1)(x^2 - 1) = 0$$
$$x^2 - 1 = 0 \qquad x^2 - 1 = 0$$
$$x^2 = 1 \qquad x^2 = 1$$
$$x = \pm 1 \qquad x = \pm 1$$

Then there are two values for x, 1 and -1. The corresponding values for y are

$$y = \frac{1}{x} = \frac{1}{1} = 1 \quad \text{and} \quad y = \frac{1}{x} = \frac{1}{-1} = -1$$

Thus, the solutions to the system are $(1,1)$ and $(-1,-1)$.

2. Solve the system.

$$5x^2 - 2y^2 = 2$$
$$2x^2 + 3y^2 = 35$$

Multiply the first equation by 3,

$$15x^2 - 6y^2 = 6,$$

the second by 2,

$$4x^2 + 6y^2 = 70,$$

and add to eliminate y and obtain:

$$19x^2 = 76$$
$$x^2 = 4$$
$$x = \pm 2$$

Substitute 2 for x in the first equation to find the value of y.

$$5(2)^2 - 2y^2 = 2$$
$$20 - 2y^2 = 2$$
$$-2y^2 = -18$$
$$y^2 = 9$$
$$y = \pm 3$$

Thus two solutions are $(2,3)$ and $(2,-3)$. Then substituting -2 for x in the same equation also gives $y = \pm 3$. So two more solutions are $(-2,3)$ and $(-2,-3)$. Thus, the solution has four solutions, $(2,3)$, $(2,-3)$, $(-2,3)$, and $(-2,-3)$.

3. Solve the system.

$$x^2 + 2xy - 7y^2 = 4$$
$$x^2 - 4xy - y^2 = -2$$

Multiply the second equation by 2,

$$2x^2 - 8xy - 2y^2 = -4,$$

and add to the first equation to make the constant term 0.

$$3x^2 - 6xy - 9y^2 = 0$$

Divide both sides by the common factor 3.

$$x^2 - 2xy - 3y^2 = 0$$

Factor, and solve for x in terms of y.

$$(x - 3y)(x + y) = 0$$
$$x - 3y = 0 \qquad x + y = 0$$
$$x = 3y \qquad x = -y$$

Substitute $3y$ for x in the first original equation.

$$(3y)^2 + 2(3y)y - 7y^2 = 4$$
$$9y^2 + 6y^2 - 7y^2 = 4$$
$$8y^2 = 4$$
$$y^2 = \frac{1}{2}$$

$$y = \pm \frac{1}{\sqrt{2}} = \pm \frac{\sqrt{2}}{2}$$

Since $x = 3y$, $x = \pm \frac{3\sqrt{2}}{2}$. Thus, two solutions to the system are $\left(\frac{3\sqrt{2}}{2}, \frac{\sqrt{2}}{2}\right)$ and $\left(-\frac{3\sqrt{2}}{2}, -\frac{\sqrt{2}}{2}\right)$.

Then since we also have $x = -y$, substituting this expression for x in the first original equation will give us two more solutions.

$$(-y)^2 + 2(-y)y - 7y^2 = 4$$
$$y^2 - 2y^2 - 7y^2 = 4$$
$$-8y^2 = 4$$
$$y^2 = -\frac{1}{2}$$
$$y = \pm \frac{i\sqrt{2}}{2}$$

Then $x = \mp \frac{i\sqrt{2}}{2}$, and two more solutions, $\left(-\frac{i\sqrt{2}}{2}, \frac{i\sqrt{2}}{2}\right)$ and $\left(\frac{i\sqrt{2}}{2}, -\frac{i\sqrt{2}}{2}\right)$ are obtained.

4. Solve the system.

$$y + \ln x = 3$$
$$y - \ln x^2 = 0$$

Subtract the second equation from the first to eliminate y and obtain:

$$\ln x + \ln x^2 = 3$$

Use the power rule for logarithms.

$$\ln x + 2 \ln x = 3$$
$$3 \ln x = 3$$
$$\ln x = 1$$
$$x = e \qquad e^1 = x$$

Substitute e for x in the first equation and solve for y.

$$y + \ln e = 3$$
$$y + 1 = 3 \qquad \ln e = 1$$
$$y = 2$$

Thus, the solution to the system is $(e, 2)$.

5. Graph the system.

$$y \geq |x - 5|$$
$$y < \log_2 x$$

First graph the equation

$$y = |x - 5|$$

using a solid line (the inequality is $\geq$). One way to obtain the graph is to shift the graph of the absolute value function to the right 5 units. Then graph the equation

$$y = \log_2 x$$

using a dashed line (the inequality is $<$). This might be done using a table of values and the equivalent exponential equation $x = 2^y$. The test point $(0,0)$ relative to $y \geq |x - 5|$ gives

$$0 \geq |0 - 5| = 5 \quad \text{This is false}$$

which shows we shade the region **not** containing $(0,0)$, that is the region "inside the V" of the graph. The test point $(2,0)$ relative to $y < \log_2 x$ gives

$$0 < \log_2 2 = 1 \quad \text{This is true}$$

which shows that we shade the region containing $(2,0)$, that is the region "below" the curve.

The overlap of the two regions is the final region shaded, corresponding to the graph of the system, given in Figure 8.36 in the text.

EXERCISES A   SECTION 8.5   411

CHAPTER 8 TOPICS IN ANALYTIC GEOMETRY

SECTION 8.5 Nonlinear Systems

1. Solve the system.

$$x^2 + y^2 = 10$$
$$x - y = 2$$

Solve the second equation for x, $x = y + 2$, and substitute this expression for x in the first equation.

$$(y + 2)^2 + y^2 = 10$$
$$y^2 + 4y + 4 + y^2 = 10$$
$$2y^2 + 4y - 6 = 0$$
$$y^2 + 2y - 3 = 0$$
$$(y + 3)(y - 1) = 0$$
$$y + 3 = 0 \quad\quad y - 1 = 0$$
$$y = -3 \quad\quad\quad y = 1$$

When $y = -3$, $x = y + 2 = -3 + 2 = -1$.
When $y = 1$, $x = y + 2 = 1 + 2 = 3$.

Thus, the solutions to the system are $(-1,-3)$ and $(3,1)$.

2. Solve the system.

$$x + y = 3$$
$$x^2 - y^2 = 3$$

Solve the first equation for x, $x = 3 - y$, and substitute this expression for x in the second equation.

$$(3 - y)^2 - y^2 = 3$$
$$9 - 6y + y^2 - y^2 = 3$$
$$-6y = -6$$
$$y = 1$$

When $y = 1$, $x = 3 - y = 3 - 1 = 2$.

Thus, the solution to the system is $(2,1)$.

3. Solve the system.

$$5x^2 + xy - y^2 = -1$$
$$y - 2x = 1$$

Solve the second equation for y, $y = 2x + 1$, and substitute into the first equation.

$$5x^2 + x(2x + 1) - (2x + 1)^2 = -1$$
$$5x^2 + 2x^2 + x - 4x^2 - 4x - 1 = -1$$
$$3x^2 - 3x = 0$$
$$x^2 - x = 0$$
$$x(x - 1) = 0$$

$$x = 0 \quad\quad x - 1 = 0$$
$$\quad\quad\quad\quad x = 1$$

When $x = 0$, $y = 2x + 1 = 2(0) + 1 = 1$.
When $x = 1$, $y = 2x + 1 = 2(1) + 1 = 3$.

Thus, the solutions are $(0,1)$ and $(1,3)$.

4. Solve the system.

$$x^2 + 3y^2 = 37$$
$$2x^2 - y^2 = 46$$

Multiply the first equation by -2,

$$-2x^2 - 6y^2 = -74,$$

and add the result to the second equation eliminating x and obtaining:

$$-7y^2 = -28$$
$$y^2 = 4$$
$$y = \pm 2$$

When $y = 2$,
$$x^2 + 3(2)^2 = 37$$
$$x^2 + 12 = 37$$
$$x^2 = 25$$
$$x = \pm 5.$$

Thus, two solutions are $(5,2)$ and $(-5,2)$. Similarly, when $y = -2$, we obtain $x = \pm 5$ giving solutions $(5,-2)$ and $(-5,-2)$.

Thus, the four solutions are $(5,2)$, $(5,-2)$, $(-5,2)$, and $(-5,-2)$.

5. Solve the system.

$$x^2 + y^2 = 5$$
$$xy = 2$$

Solve the second equation for x, $x = \frac{2}{y}$, and substitute into the first equation.

$$\left(\frac{2}{y}\right)^2 + y^2 = 5$$
$$\frac{4}{y^2} + y^2 = 5$$
$$4 + y^4 = 5y^2$$
$$y^4 - 5y^2 + 4 = 0$$
$$(y^2 - 1)(y^2 - 4) = 0$$

Setting each factor equal to 0 using the zero-product rule, we obtain:

$$y^2 - 1 = 0 \qquad y^2 - 4 = 0$$
$$y^2 = 1 \qquad y^2 = 4$$
$$y = \pm 1 \qquad y = \pm 2$$

When $y = 1$, $x = \dfrac{2}{y} = \dfrac{2}{1} = 2$.

When $y = -1$, $x = \dfrac{2}{y} = \dfrac{2}{-1} = -2$.

When $y = 2$, $x = \dfrac{2}{y} = \dfrac{2}{2} = 1$.

When $y = -2$, $x = \dfrac{2}{y} = \dfrac{2}{-2} = -1$.

Thus, the solutions are $(2,1)$, $(-2,-1)$, $(1,2)$, and $(-1,-2)$.

6. Solve the system.

$$x^2 + 2xy + y^2 = 9$$
$$x^2 - 2xy - y^2 = 9$$

Adding the two equations we obtain

$$2x^2 = 18$$
$$x^2 = 9$$
$$x = \pm 3$$

Substitute 3 for x in the first equation and solve for y.

$$(3)^2 + 2(3)y + y^2 = 9$$
$$9 + 6y + y^2 = 9$$
$$y^2 + 6y = 0$$
$$y(y + 6) = 0$$
$$y = 0 \qquad y + 6 = 0$$
$$y = -6$$

Thus two solutions are $(3,0)$ and $(3,-6)$. Now substitute -3 for x in the first equation and solve for y.

$$(-3)^2 + 2(-3)y + y^2 = 9$$
$$9 - 6y + y^2 = 9$$
$$y^2 - 6y = 0$$
$$y(y - 6) = 0$$
$$y = 0 \qquad y - 6 = 0$$
$$y = 6$$

Then two more solutions are $(-3,0)$ and $(-3,6)$.

Thus the four solutions to the system are $(3,0)$, $(3,-6)$, $(-3,0)$, and $(-3,6)$.

7. Solve the system.

$$x^2 + 2xy + 2y^2 = 10$$
$$2x^2 + xy + 22y^2 = 50$$

Multiply the first equation by -5 and add the result to the second equation to eliminate the constant term and obtain

$$-3x^2 - 9xy + 12y^2 = 0.$$

Divide through by -3.

$$x^2 + 3xy - 4y^2 = 0$$

Factor and use the zero-product rule.

$$(x + 4y)(x - y) = 0$$
$$x + 4y = 0 \qquad x - y = 0$$
$$x = -4y \qquad x = y$$

Substitute $-4y$ for x in the first equation.

$$(-4y)^2 + 2(-4y)y + 2y^2 = 10$$
$$16y^2 - 8y^2 + 2y^2 = 10$$
$$10y^2 = 10$$
$$y^2 = 1$$
$$y = \pm 1$$

When $y = 1$, $x = -4y = -4(1) = -4$.
When $y = -1$, $x = -4(-1) = 4$.
Then two solutions are $(-4,1)$ and $(4,-1)$.

Then substitute y for x in the first equation.

$$y^2 + 2(y)y + 2y^2 = 10$$
$$y^2 + 2y^2 + 2y^2 = 10$$
$$5y^2 = 10$$
$$y^2 = 2$$
$$y = \pm\sqrt{2}$$

When $y = \sqrt{2}$, $x = y = \sqrt{2}$.

When $y = -\sqrt{2}$, $x = y = -\sqrt{2}$.

The two more solutions are $(\sqrt{2},\sqrt{2})$ and $(-\sqrt{2},-\sqrt{2})$.

EXERCISES A

Thus the four solutions are $(4,-1)$, $(-4,1)$, $(\sqrt{2},\sqrt{2})$, and $(-\sqrt{2},-\sqrt{2})$.

8. Solve the system.

$$10^x + y = 11$$
$$10^x + 2y = 12$$

Subtract the first equation from the second to obtain

$$y = 1.$$

Substitute 1 for y in the first equation to find the value of x.

$$10^x + 1 = 11$$
$$10^x = 10$$
$$x = 1$$

Thus, the solution to the system is $(1,1)$.

9. Solve the system.

$$y + \log_4 (x+3) = 6$$
$$y - \log_4 x = 5$$

Solve the second equation for y, $y = \log_4 x + 5$, and substitute into the first equation.

$$\log_4 x + 5 + \log_4 (x + 3) = 6$$
$$\log_4 x + \log_4 (x + 3) = 1$$
$$\log_4 x(x + 3) = 1$$
$$x(x + 3) = 4^1$$
$$x(x + 3) = 4$$
$$x^2 + 3x = 4$$
$$x^2 + 3x - 4 = 0$$
$$(x + 4)(x - 1) = 0$$
$$x + 4 = 0 \qquad x - 1 = 0$$
$$x = -4 \qquad x = 1$$

When $x = -4$, $y = \log_4(-4) + 5$ is undefined. Thus, -4 must be discarded.
When $x = 1$, $y = \log_4(1) + 5 = 0 + 5 = 5$.

Thus, the only solution is $(1,5)$.

10. Solve the system.

$$|y| + x = 7$$
$$2|y| - x = 5$$

Add the two equations to eliminate x and obtain:

$$3|y| = 12$$
$$|y| = 4$$
$$y = 4 \quad \text{or} \quad y = -4$$

When $y = 4$,

$$|y| + x = 7$$
$$|4| + x = 7$$
$$4 + x = 7$$
$$x = 3.$$

When $y = -4$,

$$|-4| + x = 7$$
$$4 + x = 7$$
$$x = 3.$$

Thus, the two solutions are $(3,4)$ and $(3,-4)$.

11. Graph the system.

$$y \geq x^2$$
$$y \leq x + 2$$

First graph the parabola $y = x^2$ using a solid line. The vertex is at $(0,0)$, and graph opens up, and passes through the two points $(1,1)$ and $(-1,1)$. The text point $(0,1)$ shows that we shade the region above the graph that contains the test point. Then graph the line $y = x + 2$ using a solid line. The intercepts are $(0,2)$ and $(-2,0)$. The test point $(0,0)$ shows that we shade the region below the line containing the test point. The combined region, the region showing the graph of the system, is shown below.

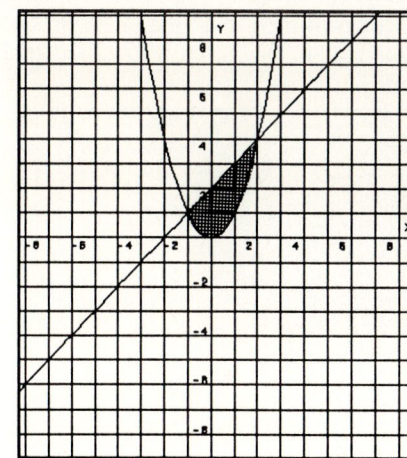

12. Graph the system.

$$y - |x| \geq 0$$
$$y + |x| < 4$$

First graph the absolute function $y = |x|$ using a solid line. The test point $(1,0)$ shows that we shade the region above the graph, the region "inside the V." Then graph $y = 4 - |x|$ using a dashed line.

This graph can be obtained from the graph of the absolute value function by first reflecting in the x-axis then sliding the result up 4 units. The test point (0,0) shows that we shade the region below the graph, "the region inside the V." The combination of these, the graph of the system, is shown below.

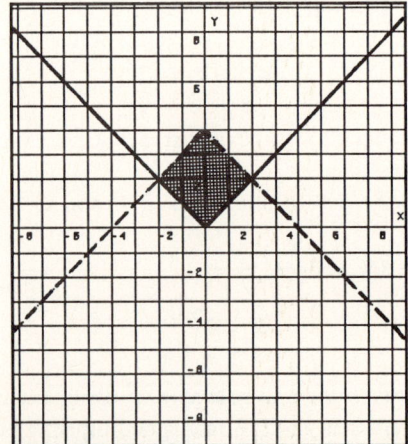

13. Graph the system.

$$y - 2^x \geq 0$$
$$y - x \geq 2$$

First graph the exponential function $y = 2^x$ using a solid line. The test point (0,0) shows that we shade the region above the curve. Then graph the line $y = x + 2$ using a solid line. The test point (0,0) shows that we shade the region above the line. The combination of these two, the graph of the system, is shown below.

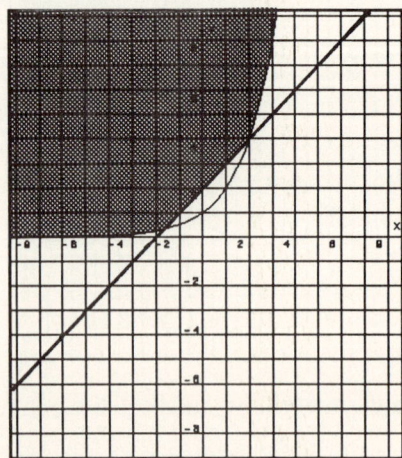

14. Graph the system.

$$y + |x - 2| < 0$$
$$x - |y + 1| > 0$$

First graph the function $y = -|x - 2|$ by reflecting the graph of the absolute value function in the x-axis then shifting the result 2 units to the right. The test point (0,0) shows that we shade the region below the graph, "the region inside the V." The graph is also a dashed line since the inequality is $<$. Then graph the relation $x = |y + 1|$. This is a "V-shaped" curve "opening to the right." The test point (0,0) shows that we shade the region "inside the V." The graph is also a dashed line. The combined region, the graph of the system, is shown below.

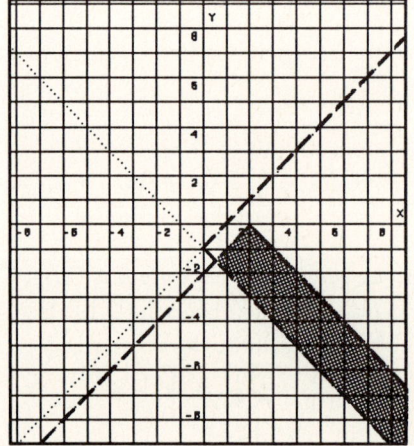

15. Graph the system.

$$y \geq 0$$
$$y \leq \sqrt{x - 1}$$
$$y > x - 3$$

The first inequality describes the points in the upper half plane and the x-axis. Graph the function

$y = \sqrt{x - 1}$ using a solid line. The graph is the upper branch of a parabola that opens to the right with vertex (1,0). The test point (2,0) shows that we shade the region below the curve. Then graph the line $y = x - 3$ using a dashed line and intercepts (0,-3) and (3,0). The test point (0,0) shows that we shade the region above the line. Putting these three regions together, we obtain the graph of the system shown below.

EXERCISES A   SECTION 8.5   415

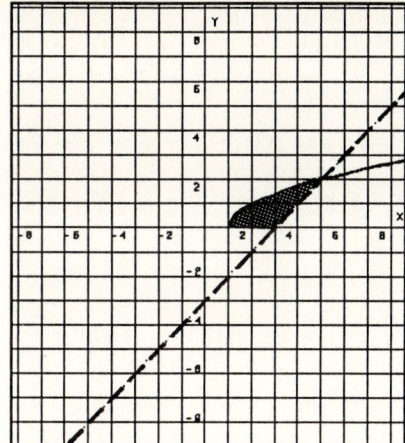

16. Graph the system.

$$x^2 + y^2 > 4$$
$$x^2 + y^2 < 16$$

First graph the circle $x^2 + y^2 = 4$ using a dashed line. The center is at the origin, and the radius is 2. The test point (0,0) shows that we shade the region "outside the circle." Then graph the circle $x^2 + y^2 = 16$ using a dashed line. The center is at the origin, and the radius is 4. The test point (0,0) shows that we shade the region "inside the circle." The combination of these two, the graph of the system, is shown below.

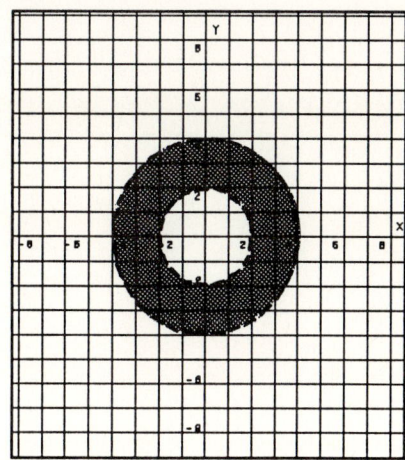

17. Let x = the length of the garden,
    y = the width of the garden.

Since the perimeter is 20 yd, we have

$$2x + 2y = 20,$$

which is equivalent to

$$x + y = 10.$$

Since the area is 24 yd², we have

$$xy = 24.$$

Thus, we obtain the following system.

$$x + y = 10$$
$$xy = 24$$

Solve the second equation for y, $y = \frac{24}{x}$, and substitute this expression into the first equation.

$$x + \frac{24}{x} = 10$$
$$x^2 + 24 = 10x$$
$$x^2 - 10x + 24 = 0$$
$$(x - 4)(x - 6) = 0$$

Use the zero-product rule.

$$x - 4 = 0 \qquad x - 6 = 0$$
$$x = 4 \qquad x = 6$$

Then

$$y = \frac{24}{x} = \frac{24}{4} = 6 \quad \text{and} \quad y = \frac{24}{x} = \frac{24}{6} = 4.$$

Thus, we obtain two solutions to the system, (4,6) and (6,4). But in the context of this problem, we actually only obtain one solution, the rectangle must have dimensions 6 yd by 4 yd.

18. The demand equation is

$$xy = 100,$$

and the supply equation is

$$x^2 - xy = 44.$$

The *market equilibrium point* occurs at a point where
the supply curve intersects the demand curve. Thus, we must solve the system of equations. Since $xy = 100$, substitute 100 for xy in the second equation.

$$x^2 - 100 = 44$$
$$x^2 = 144$$
$$x = \pm 12$$

Since x represents the number of people willing to pay y dollars for a product, x cannot be negative. Thus, we discard -12. When $x = 12$,

$$y = \frac{100}{x} = \frac{100}{12} = 8.\overline{3}.$$

Thus, the price people are willing to pay is about $8.33.

19. The total weekly cost of making and selling x grills is given by

$$C = 50x + 1000,$$

and the total weekly revenue produced is given by

$$R = 100x - 0.2x^2.$$

The *break-even points* are the values of x for which $C = R$. Thus, we must solve:

$$50x + 1000 = 100x - 0.2x^2$$
$$0.2x^2 - 50x + 1000 = 0$$
$$2x^2 - 500x + 10{,}000 = 0$$
$$x^2 - 250x + 5000 = 0$$

Use the quadratic formula to solve for x.

$$\begin{aligned}x &= \frac{-b \pm \sqrt{b^2 - 4ac}}{2a}\\ &= \frac{-(-250) \pm \sqrt{(-250)^2 - 4(1)(5000)}}{2(1)}\\ &= \frac{250 \pm \sqrt{62{,}500 - 20{,}000}}{2}\\ &= \frac{250 \pm \sqrt{42{,}500}}{2}\\ &\approx 228.0776406,\ 21.92235936\end{aligned}$$

Thus, the break-even points are approximately 228 grills and 22 grills.

20. Give the vertex, focus, directrix, and line of symmetry of the parabola with equation

$$(x - 1)^2 = -16(y + 3).$$

This is the standard form of a parabola opening down, since $4p = -16$ making $p = -4$. The vertex is $(1,-3)$. The focus is 4 units below the vertex at $(1,-7)$. The directrix is the horizontal line 4 units above the vertex, the line $y = 1$. The line of symmetry is the vertical line through the vertex, the line $x = 1$.

21. Given $f(n) = 2n + 1$.

$$f(1) = 2(1) + 1 = 2 + 1 = 3$$
$$f(3) = 2(3) + 1 = 6 + 1 = 7$$
$$f(10) = 2(10) + 1 = 20 + 1 = 21$$

22. Given $f(n) = (-1)^n n^2$.

$$f(1) = (-1)^1(1)^2 = (-1)(1) = -1$$
$$f(3) = (-1)^3(3)^2 = (-1)(9) = -9$$
$$f(10) = (-1)^{10}(10)^2 = (+1)(100) = 100$$

CHAPTER 8 TOPICS IN ANALYTIC GEOMETRY

CHAPTER 8 Review Exercises

1. Find the equation of the circle with center $\left(-3, \frac{2}{3}\right)$ and radius 4.

   Substitute -3 for h, $\frac{2}{3}$ for k, and 4 for r in the standard form.

   $$(x - h)^2 + (y - k)^2 = r^2$$
   $$(x - (-3))^2 + \left(y - \frac{2}{3}\right)^2 = (4)^2$$
   $$(x + 3)^2 + \left(y - \frac{2}{3}\right)^2 = 16$$

2. Sketch the graph.

   $$x^2 + y^2 = 9$$

   This is the standard form of the equation of a circle with center (0,0) and radius 3. The graph is given below.

   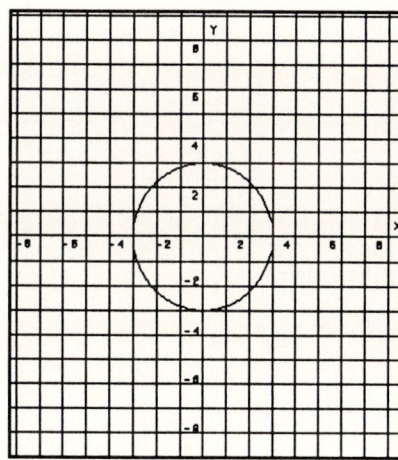

3. Complete the squares to determine if the equation represents a circle, ellipse, hyperbola, parabola, or degenerate conic.

   $$3x^2 + 3y^2 - 6x + 12y - 40 = 0$$

   $$3x^2 - 6x + 3y^2 + 12y = 40$$
   $$3(x^2 - 2x) + 3(y^2 + 4y) = 40$$
   $$3(x^2 - 2x + 1) + 3(y^2 + 4y + 4) = 40+3(1)+3(4)$$
   $$3(x - 1)^2 + 3(y + 2)^2 = 55$$

   $$(x - 1)^2 + (y + 2)^2 = \frac{55}{3}$$

   This is the standard form of the equation of a circle.

4. Since the depth of the canal at the center is 15.0 ft, we can use 15.0 for the radius of the circle centered at the origin that contains the desired semicircle as the lower portion of the circle below the x-axis. The equation is:

   $$x^2 + y^2 = (15.0)^2 = 225$$

   Solve this equation for y and use the negative root to obtain the equation of the semicircle.

   $$y^2 = 225 - x^2$$
   $$y = -\sqrt{225 - x^2}$$

   To find the depth of the water in the canal 3.5 ft from the edge of the canal, substitute 11.5 for x (3.5 ft from the edge is 11.5 ft from the center of the canal) and evaluate y.

   $$y = -\sqrt{225 - (11.5)^2} \approx -9.630680142$$

   Thus, the depth 3.5 ft from the edge is about 9.6 ft.

5. Find the equation of the ellipse centered at the origin with intercepts $(\pm 3, 0)$ and $(0, \pm 7)$.

   Since $7 > 3$, the ellipse has a vertical major axis, $a = 7$, and $b = 3$. Substitute these values into the standard form.

   $$\frac{x^2}{b^2} + \frac{y^2}{a^2} = 1$$
   $$\frac{x^2}{3^2} + \frac{y^2}{7^2} = 1$$
   $$\frac{x^2}{9} + \frac{y^2}{49} = 1$$

6. Sketch the graph.

   $$\frac{x^2}{25} + \frac{y^2}{4} = 1$$

   Since $25 > 4$, the ellipse has a horizontal major axis, $a = 5$, and $b = 2$. The intercepts are $((0, \pm 2)$ and $(\pm 5, 0)$. The graph is given below.

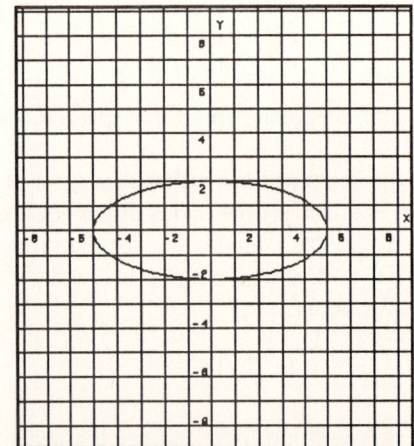

7. Complete the squares to determine if the equation represents a circle, ellipse, hyperbola, parabola, or degenerate conic.

$$3x^2 + 2y^2 - 6x + 12y - 60 = 0$$

$$3x^2 - 6x + 2y^2 + 12y = 60$$
$$3(x^2 - 2x\ ) + 2(y^2 + 6y\ ) = 60$$
$$3(x^2 - 2x + 1) + 2(y^2 + 6y + 9) = 60+3(1)+2(9)$$
$$3(x - 1)^2 + 2(y + 3)^2 = 81$$

Dividing both sides by 81, we obtain:

$$\frac{(x-1)^2}{27} + \frac{(y+3)^2}{\frac{81}{2}} = 1$$

This is the standard form of the equation of an ellipse.

8. Consider the ellipse centered at the origin so that the semiellipse corresponding to the tunnel is that portion of the ellipes above the x-axis. Since the maximum height of the tunnel is 12.8 and the width is 16.2, we have one intercept as (0,12.8) and the two intercepts on the x-axis are (8.1,0) and (−8.1,0). Thus the ellipse has a vertical major axis, $a = 12.8$, and $b = 8.1$. The equation of the ellipse is:

$$\frac{x^2}{b^2} + \frac{y^2}{a^2} = 1$$
$$\frac{x^2}{(8.1)^2} + \frac{y^2}{(12.8)^2} = 1$$
$$\frac{x^2}{65.61} + \frac{y^2}{163.84} = 1$$

The equation of the semiellipse is found by solving for y and using the positive root.

$$\frac{y^2}{163.84} = 1 - \frac{x^2}{65.61}$$
$$y^2 = \frac{163.84}{65.61}(65.61 - x^2)$$
$$y \approx 1.58\sqrt{65.61 - x^2}$$

To find the vertical clearance 4.4 m from the edge of the tunnel, substitute 3.7 for x (4.4 m from the edge is 3.7 m from the center of the roadway) and evaluate y.

$$y = 1.58\sqrt{65.61 - (3.7)^2} = 11.38477439$$

Thus, the vertical clearance is about 11.4 m at this point.

9. Find the equation of the hyperbola with foci (±4,0) and length of the transverse axis 6.

Since the foci are on the x-axis, evenly spaced about the origin, the hyperbola is centered at (0,0) and has horizontal transverse axis. Since the length of the transverse axis is 6, $a = 3$. With $c = 4$, we have

$$b^2 = c^2 - a^2 = 4^2 - 3^2 = 16 - 9 = 7.$$

Substitute 9 for a^2 and 7 for b^2 in the standard form.

$$\frac{x^2}{a^2} - \frac{y^2}{b^2} = 1$$
$$\frac{x^2}{9} - \frac{y^2}{7} = 1$$

10. Sketch the graph.

$$y^2 - x^2 = 4$$

This is equivalent to

$$\frac{y^2}{4} - \frac{x^2}{4} = 1.$$

This is the equation of a hyperbola with vertical transverse axis, center at (0,0), and vertices (0,±2). The asymptotes of the hyperbola contain the diagonals of the rectangle with sides parallel to the axes and passing through the vertices and the points (±2,0). The graph is given below.

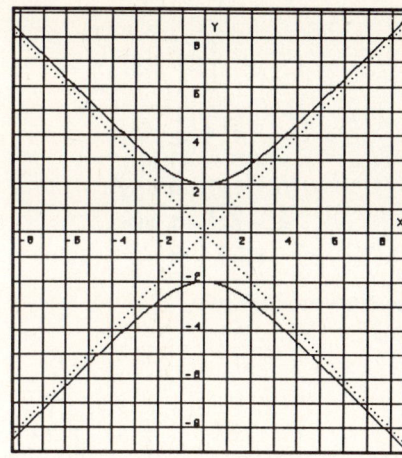

11. Complete the squares to determine if the equation represents a circle, ellipse, hyperbola, parabola, or degenerate conic.

$$3x^2 - 3y^2 - 6x + 12y - 80 = 0$$

$$3x^2 - 6x - 3y^2 + 12y = 80$$
$$3(x^2 - 2x) - 3(y^2 - 4y) = 80$$
$$3(x^2 - 2x + 1) - 3(y^2 - 4y + 4) = 80 + 3(1) + 3(4)$$
$$3(x - 1)^2 - 3(y - 2)^2 = 95$$

Divide both sides by 95.

$$\frac{(x - 1)^2}{\frac{95}{3}} - \frac{(y - 2)^2}{\frac{95}{3}} = 1$$

This is the standard form of the equation of a hyperbola.

12. The roof of the arena is in the shape of a hyperbola with equation

$$y^2 - x^2 = 1600$$

where x and y are in feet, as shown in the figure in the text. To find the height h of the outside walls, substitute 120 for x and solve for y.

$$y^2 - (120)^2 = 1600$$
$$y^2 = 1600 + (120)^2$$
$$y^2 = 16{,}000$$
$$y = \pm\sqrt{16{,}000}$$
$$y = \pm 126.4911064$$

Since we are only using the upper branch of the hyperbola, we can discard the negative solution for y. Thus, the height h of the outside walls is about 126.5 ft.

13. Find the equation of the parabola with directrix $x = -3$ and focus $(5, 2)$.

Since the directrix is a vertical line, the form of the parabola is:

$$(y - k)^2 = 4p(x - h).$$

Since the vertex is on the horizontal line through the focus and located midway between the focus and the directrix, the vertex is $(1, 2)$. Since the focus is to the right of the directrix, the parabola opens to the right and $p > 0$. The distance between the focus and the vertex is p, and this value is 4. Substitute 1 for h, 2 for k, and 4 for p in the above form,

$$(y - 2)^2 = 4(4)(x - 1),$$

which simplifies to

$$(y - 2)^2 = 16(x - 1).$$

14. Sketch the graph.
$$x^2 + 2y = 0$$
This can be written in the form $x^2 = -2y$. We recognize the graph as a parabola, centered at the origin, and with $p < 0$, the parabola opens down.

The focus is $\left(0, -\frac{1}{2}\right)$, and two additional points on the graph are $\left(1, -\frac{1}{2}\right)$ and $\left(-1, -\frac{1}{2}\right)$. the graph is given below.

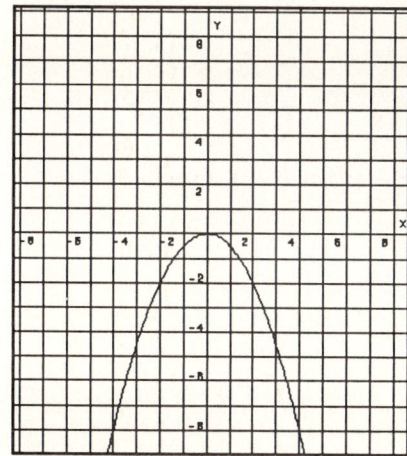

15. Complete the square to determine if the equation represents a circle, ellipse, hyperbola, parabola, or degenerate conic.

$$3x^2 - 6x + 12y - 48 = 0$$

Divide through by 3.

$$x^2 - 2x + 4y - 16 = 0$$

Complete the square on x.

$$\begin{aligned} x^2 - 2x &= -4y + 16 \\ x^2 - 2x + 1 &= -4y + 16 + 1 \\ (x - 1)^2 &= -4y + 17 \end{aligned}$$

At this point we can recognize that the equation is that of a parabola.

16. Place a coordinate system over the cross-section of the ceiling so the vertex of the dome is at the point (0,20). With this arrangement, the parabola opens down, has axis of symmetry the y-axis, and standard form equation

$$(x - h)^2 = 4p(y - k).$$

Substitute 0 for h and 20 for k,

$$(x - 0)^2 = 4p(y - 20),$$

which simplifies to

$$x^2 = 4p(y - 20).$$

Since the plans call for a 16 ft beam across the dome at a point 6 ft below the top, when $x = 8$, $y = 14$. Substitute these values to find the value of $4p$.

$$\begin{aligned} 8^2 &= 4p(14 - 20) \\ 64 &= 4p(-6) \\ \tfrac{64}{-6} &= 4p \\ -\tfrac{32}{3} &= 4p \end{aligned}$$

Then the equation of the parabola is

$$x^2 = -\tfrac{32}{3}(y - 20).$$

17. Solve the system.

$$\begin{aligned} x^2 + y^2 &= 5 \\ xy &= 2 \end{aligned}$$

Solve the second equation for y, $y = \tfrac{2}{x}$, and substitute this expression for y in the first equation.

$$x^2 + \left(\tfrac{2}{x}\right)^2 = 5$$
$$x^2 + \tfrac{4}{x^2} = 5$$
$$x^4 + 4 = 5x^2$$
$$x^4 - 5x^2 + 4 = 0$$
$$(x^2 - 1)(x^2 - 4) = 0$$

Use the zero-product rule.

$$\begin{array}{ll} x^2 - 1 = 0 & x^2 - 4 = 0 \\ x^2 = 1 & x^2 = 4 \\ x = \pm 1 & x = \pm 2 \end{array}$$

When $x = 1$, $y = \tfrac{2}{x} = \tfrac{2}{1} = 2$.

When $x = -1$, $y = \tfrac{2}{x} = \tfrac{2}{-1} = -2$.

When $x = 2$, $y = \tfrac{2}{x} = \tfrac{2}{2} = 1$.

When $x = -2$, $y = \tfrac{2}{x} = \tfrac{2}{-2} = -1$.

Thus, the four solutions are $(1,2)$, $(-1,-2)$, $(2,1)$, and $(-2,-1)$.

18. Solve the system.

$$\begin{aligned} 2x^2 - 6xy + 12y^2 &= 8 \\ x^2 - xy + 6y^2 &= 8 \end{aligned}$$

Subtract the second equation from the first to obtain an equation with the constant equal to 0.

$$x^2 - 5xy + 6y^2 = 0$$

Factor and use the zero-product rule.

$$\begin{array}{ll} (x - 2y)(x - 3y) = 0 & \\ x - 2y = 0 & x - 3y = 0 \\ x = 2y & x = 3y \end{array}$$

Substitute $2y$ for x in the second original equation and solve for y.

$$\begin{aligned} (2y)^2 - (2y)y + 6y^2 &= 8 \\ 4y^2 - 2y^2 + 6y^2 &= 8 \\ 8y^2 &= 8 \\ y^2 &= 1 \\ y &= \pm 1 \end{aligned}$$

When $y = 1$, $x = 2y = 2(1) = 2$.
When $y = -1$, $x = 2y = 2(-1) = -2$.

Thus, two solutions are $(2,1)$ and $(-2,-1)$.

Now substitute $3y$ for x in the second equation and solve for y.

$$(3y)^2 - (3y)y + 6y^2 = 8$$
$$9y^2 - 3y^2 + 6y^2 = 8$$
$$12y^2 = 8$$
$$y^2 = \frac{2}{3}$$
$$y = \pm\sqrt{\frac{2}{3}} = \pm\frac{\sqrt{6}}{3}$$

When $y = \frac{\sqrt{6}}{3}$, $x = 3y = \sqrt{6}$.

When $y = -\frac{\sqrt{6}}{3}$, $x = 3y = -\sqrt{6}$.

Thus, two more solutions are $\left(\sqrt{6}, \frac{\sqrt{6}}{3}\right)$ and $\left(-\sqrt{6}, -\frac{\sqrt{6}}{3}\right)$.

19. Graph the system.

$$x \leq 4 - y^2$$
$$x > |y| - 2$$

First graph the parabola $x = 4 - y^2$, or $y^2 = -x + 4$, with vertex $(4,0)$ opening left. Use a solid curve. The text point $(0,0)$ shows that we shade the region inside the parabola to the left. Then graph $x = |y| - 2$ using a dashed line. This is a V-shaped curve opening to the right. The text point $(0,0)$ shows that we shade the region inside the V. The combination of these two, the graph of the system is given below.

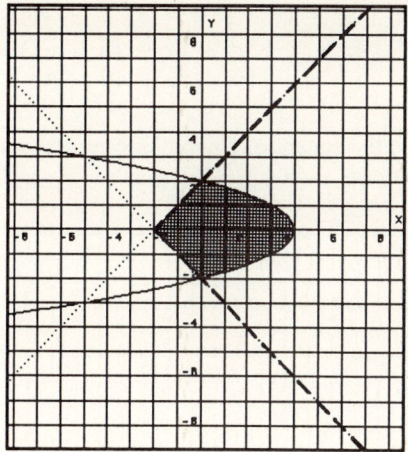

20. Graph the system.

$$x^2 + y^2 \geq 4$$
$$\frac{x^2}{25} + \frac{y^2}{4} \leq 1$$

First graph $x^2 + y^2 = 4$, a circle centered at the origin with radius 2, using a solid curve. The test point $(0,0)$ shows that we shade the region outside the circle. Then graph $\frac{x^2}{25} + \frac{y^2}{4} = 1$, an ellipse centered at the origin with intercepts $(\pm 5, 0)$ and $(0, \pm 2)$, using a solid curve. The test point $(0,0)$ shows that we shade the region inside the ellipse. The combination of these two, the graph of the system is given below.

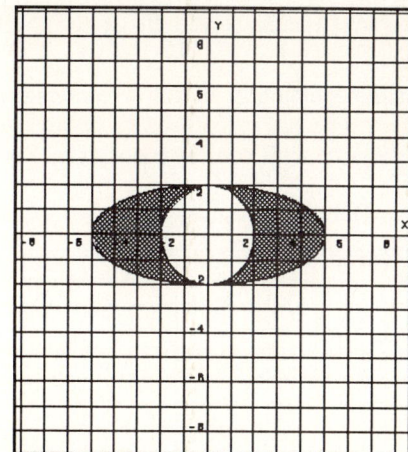

21. Graph the conic.

$$x^2 + y^2 - 4x + 8y + 11 = 0$$

Complete the square on x and y.

$$x^2 - 4x + y^2 + 8y = -11$$
$$x^2 - 4x + 4 + y^2 + 8y + 16 = -11 + 4 + 16$$
$$(x-2)^2 + (y+4)^2 = 9$$

This is the equation of a circle, centered at the point $(2,-4)$ with radius 3. The graph is given below.

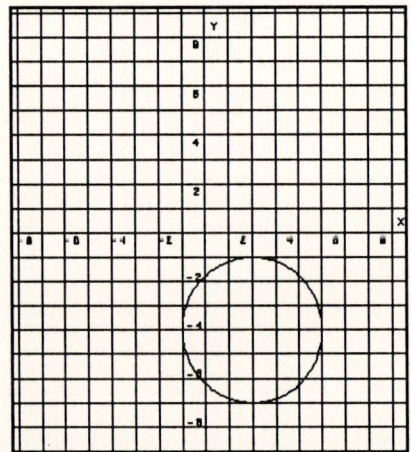

22. Graph the conic.

$$4x^2 + y^2 + 4x - 6y - 6 = 0$$

Complete the square on x and y.

$$4(x^2 + x \quad) + (y^2 - 6y \quad) = 6$$
$$4\left(x^2 + x + \frac{1}{4}\right) + (y^2 - 6y + 9) = 6 + 4\left(\frac{1}{4}\right) + 9$$
$$4\left(x + \frac{1}{2}\right)^2 + (y-3)^2 = 16$$
$$\frac{\left(x + \frac{1}{2}\right)^2}{4} + \frac{(y-3)^2}{16} = 1$$

This is the equation of an ellipse centered at the point $\left(-\frac{1}{2}, 3\right)$ with vertical major axis, $a = 4$, and $b = 2$. The graph is given below.

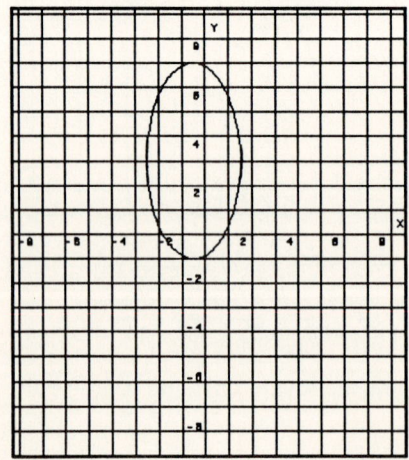

23. Graph the conic.

$$4x^2 - y^2 + 16x - 2y - 21 = 0$$

Complete the square on x and y.

$$4(x^2 + 4x \quad) - (y^2 + 2y \quad) = 21$$
$$4(x^2 + 4x + 4) - (y^2 + 2y + 1) = 21 + 4(4) - 1$$
$$4(x+2)^2 - (y+1) = 36$$
$$\frac{(x+2)^2}{9} - \frac{(y+1)^2}{36} = 1$$

This is the equation of a hyperbola centered at $(-2,-1)$ opening left and right with vertices $(1,-1)$ and $(-5,-1)$. The asymptotes contain the diagonals of the rectangle with sides parallel to the axes passing through the vertices and the points $(-2,5)$ and $(-2,-7)$. The graph is given below.

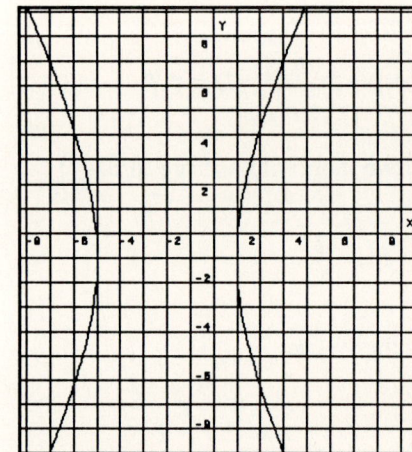

24. Graph the conic.

$$y^2 + 6x - 4y + 22 = 0$$

Complete the square on y.

$$y^2 - 4y = -6x - 22$$
$$y^2 - 4y + 4 = -6x - 22 + 4$$
$$(y-2)^2 = -6x - 18$$
$$(y-2)^2 = -6(x+3)$$
$$(y-2)^2 = 4\left(-\frac{3}{2}\right)(x+3)$$

This is the equation of a parabola, with vertex $(-3,2)$, opening to the left. The focus is $\left(-\frac{9}{2}, 2\right)$ and the directrix is $x = -\frac{3}{2}$. The graph is given below.

CHAPTER 8 REVIEW

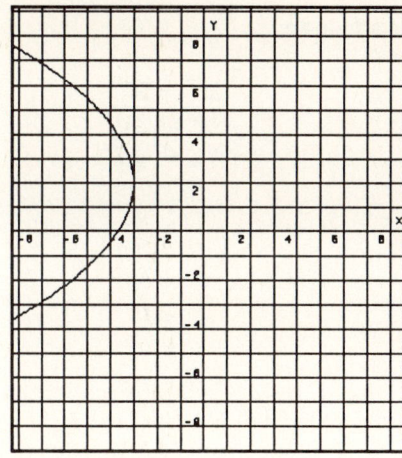

25. Solve the system.

$$2x^2 + y^2 = 3$$
$$x - y = 2$$

Solve the second equation for x, $x = y + 2$, and substitute into the first equation.

$$2(y+2)^2 + y^2 = 3$$
$$2y^2 + 8y + 8 + y^2 = 3$$
$$3y^2 + 8y + 5 = 0$$
$$(3y+5)(y+1) = 0$$
$$3y + 5 = 0 \qquad y + 1 = 0$$
$$3y = -5 \qquad y = -1$$
$$y = -\tfrac{5}{3}$$

When $y = -\tfrac{5}{3}$, $x = y + 2 = -\tfrac{5}{3} + 2 = \tfrac{1}{3}$.

When $y = -1$, $x = y + 2 = (-1) + 2 = 1$.

Thus, the solutions are $\left(\tfrac{1}{3}, -\tfrac{5}{3}\right)$ and $(1,-1)$.

26. Solve the system.

$$x^2 + y^2 = 20$$
$$xy = 8$$

Solve the second equation for y, $y = \tfrac{8}{x}$, and substitute into the first equation.

$$x^2 + \left(\tfrac{8}{x}\right)^2 = 20$$
$$x^2 + \tfrac{64}{x^2} = 20$$
$$x^4 + 64 = 20x^2$$
$$x^4 - 20x^2 + 64 = 0$$
$$(x^2 - 4)(x^2 - 16) = 0$$

Use the zero-product rule.

$$x^2 - 4 = 0 \qquad x^2 - 16 = 0$$
$$x^2 = 4 \qquad x^2 = 16$$
$$x = \pm 2 \qquad x = \pm 4$$

When $x = 2$, $y = \tfrac{8}{x} = \tfrac{8}{2} = 4$.

When $x = -2$, $y = \tfrac{8}{x} = \tfrac{8}{-2} = -4$.

When $x = 4$, $y = \tfrac{8}{x} = \tfrac{8}{4} = 2$.

When $x = -4$, $y = \tfrac{8}{x} = \tfrac{8}{-4} = -2$.

Thus, the four solutions are $(2,4)$, $(-2,-4)$, $(4,2)$, and $(-4,-2)$.

27. Solve the system.

$$3x^2 + xy - 6y^2 = 8$$
$$x^2 + 2xy - 4y^2 = 4$$

Multiply the second equation by -2,

$$-2x^2 - 4xy + 8y^2 = -8,$$

and add to the first equation to obtain 0 for the constant.

$$x^2 - 3xy + 2y^2 = 0$$
$$(x - y)(x - 2y) = 0$$
$$x - y = 0 \qquad x - 2y = 0$$
$$x = y \qquad x = 2y$$

Substitute x for y in the second original equation.

$$x^2 + 2x(x) - 4(x)^2 = 4$$
$$x^2 + 2x^2 - 4x^2 = 4$$
$$-x^2 = 4$$
$$x^2 = -4$$
$$x = \pm 2i$$

When $x = 2i$, $y = x = 2i$.
When $x = -2i$, $y = x = -2i$.

Thus, two solutions are $(2i, 2i)$ and $(-2i, -2i)$.

Substitute $2y$ for x in the second original equation.

$$(2y)^2 + 2(2y)y - 4y^2 = 4$$
$$4y^2 + 4y^2 - 4y^2 = 4$$
$$4y^2 = 4$$
$$y^2 = 1$$
$$y = \pm 1$$

When $y = 1$, $x = 2y = 2(1) = 2$.
When $y = -1$, $x = 2y = 2(-1) = -2$.

Thus, two more solutions are $(2,1)$ and $(-2,-1)$.

28. Solve the system.

$$3^x + y = 10$$
$$3^x - y = 8$$

Add the two equations to eliminate y.

$$2(3^x) = 18$$
$$3^x = 9$$
$$3^x = 3^2$$
$$x = 2$$

Substitute 2 for x in the first original equation.

$$3^2 + y = 10$$
$$9 + y = 10$$
$$y = 1$$

Thus, the solution is $(2,1)$.

29. Solve the system.

$$x^2 + y^2 = 169$$
$$x + y = 17$$

Solve the second equation for y, $y = 17 - x$, and substitute into the first equation.

$$x^2 + (17 - x)^2 = 169$$
$$x^2 + 289 - 34x + x^2 = 169$$
$$2x^2 - 34x + 120 = 0$$
$$x^2 - 17x + 60 = 0$$
$$(x - 5)(x - 12) = 0$$
$$x - 5 = 0 \qquad x - 12 = 0$$
$$x = 5 \qquad x = 12$$

When $x = 5$, $y = 17 - x = 17 - 5 = 12$.
When $x = 12$, $y = 17 - x = 17 - 12 = 5$.

Thus, the solutions are $(5,12)$ and $(12,5)$.

30. Solve the system.

$$x^2 + y^2 = 25$$
$$x^2 - y^2 = 25$$

Add to eliminate y.

$$2x^2 = 50$$
$$x^2 = 25$$
$$x = \pm 5$$

When $x = 5$,

$$5^2 + y^2 = 25$$
$$y^2 = 0$$
$$y = 0$$

Thus, one solution is $(5,0)$.

When $x = -5$,

$$(-5)^2 + y^2 = 25$$
$$y^2 = 0$$
$$y = 0$$

Thus, another solution is $(-5,0)$.

Find the equation of each conic in Exercises 31-34.

31. Circle with endpoints of a diameter at $(-2,-3)$ and $(6,7)$.

The center of the circle must be at the midpoint of the segment joining the two endpoints of the diameter. Use the midpoint formula.

$$(\bar{x}, \bar{y}) = \left(\frac{x_1+x_2}{2}, \frac{y_1+y_2}{2}\right)$$
$$= \left(\frac{-2+6}{2}, \frac{-3+7}{2}\right)$$
$$= \left(\frac{4}{2}, \frac{4}{2}\right)$$
$$= (2,2)$$

The radius of the circle is half the diameter. Use the distance formula.

$$d = \sqrt{(x_2-x_1)^2 + (y_2-y_1)^2}$$
$$= \sqrt{(-2-6)^2 + (-3-7)^2}$$
$$= \sqrt{64+100}$$
$$= \sqrt{164}$$
$$= 2\sqrt{41}$$

Then $r = \dfrac{d}{2} = \dfrac{2\sqrt{41}}{2} = \sqrt{41}$.

Substitute in the standard form for a circle.

$$(x-2)^2 + (y-2)^2 = 41$$

CHAPTER 8 REVIEW

32. Ellipse with foci $(2,-5)$ and $(2,-1)$, and $b = 1$.

Since the foci are on the same vertical line, the form of the ellipse is

$$\frac{(x-h)^2}{b^2} + \frac{(y-k)^2}{a^2} = 1.$$

The center of the ellipse is at the midpoint of the segment joining the foci, at $(2,-3)$, and $c = 2$. Since $b = 1$, we have

$$a^2 = b^2 + c^2 = 1^2 + 2^2 = 1 + 4 = 5.$$

Substitute 5 for a^2, 1 for b^2, 2 for h, and -3 for k in the above form.

$$\frac{(x-2)^2}{1} + \frac{(y+3)^2}{5} = 1.$$

33. Hyperbola with vertices $(0,\pm 5)$ and passing through $(3,10)$.

Since the vertices are on the y-axis, equally spaced about the origin, the hyperbola has a vertical transverse axis and its equation is of the form:

$$\frac{y^2}{a^2} - \frac{x^2}{b^2} = 1$$

Also, we have $a = 5$, so the equation becomes:

$$\frac{y^2}{25} - \frac{x^2}{b^2} = 1$$

Since the hyperbola passes through $(3,10)$, substitute 3 for x and 10 for y to find the value of b^2.

$$\frac{100}{25} - \frac{9}{b^2} = 1$$

$$4 - \frac{9}{b^2} = 1$$

$$4b^2 - 9 = b^2$$

$$3b^2 = 9$$

$$b^2 = 3$$

Thus, the equation of the hyperbola is

$$\frac{y^2}{25} - \frac{x^2}{3} = 1.$$

34. Parabola with vertex $(1,-2)$, axis of symmetry $x = 1$, and passing through $(5,2)$.

Since the axis of symmetry is a vertical line, and since the vertex is $(1,-2)$, the equation is:

$$(x - 1)^2 = 4p(y + 2)$$

Since the parabola passes through the point $(5,2)$, substitute 5 for x and 2 for y to find $4p$.

$$(5 - 1)^2 = 4p(2 + 2)$$
$$16 = 4p(4)$$
$$4 = 4p$$

Thus, the equation of the parabola is

$$(x - 1)^2 = 4(y + 2).$$

35. Graph the system.

$$x^2 + y^2 \leq 16$$
$$y > |x|$$

First graph the circle $x^2 + y^2 = 16$, centered at $(0,0)$ with radius 4, using a solid curve. The test point $(0,0)$ shows that we shade the region inside the circle. Then graph the absolute value function $y = |x|$ using a dashed curve. The test point $(0,1)$ shows that we shade the region above the graph, inside the V. Combining these two, we obtain the graph of the system shown below.

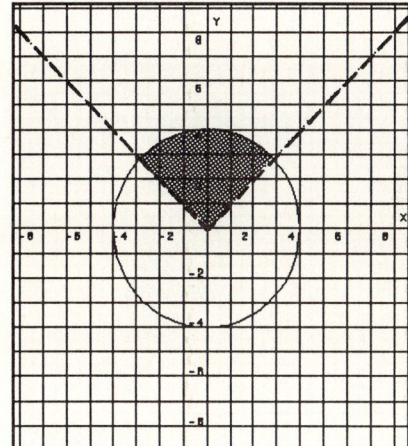

36. Graph the system.

$$y - 3^x \geq 0$$
$$3x - 4y \geq -9$$

First graph the exponential function $y = 3^x$ using a solid curve. The test point $(0,0)$ shows that we shade the region above the graph. Then graph the line $3x - 4y = -9$ using a solid line. The test point $(0,0)$ shows that we shade the region below the line. Combining these two together, we obtain the graph of the system shown below.

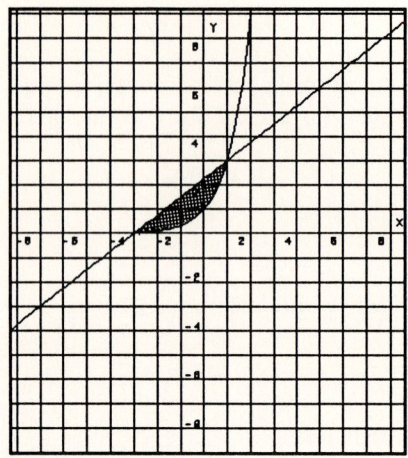

37. Graph the system.

$$|y| - x \leq 0$$
$$x^2 + y^2 < 16$$

First graph the relation $|y| = x$, a V-shaped curve opening to the right, using a solid line. The test point $(1,0)$ shows that we shade the region inside the V. Then graph the circle $x^2 + y^2 = 16$, centered at the origin with radius 4, using a dashed curve. The test point $(0,0)$ shows that we shade the region inside the circle. Combining these two together, we obtain the graph of the system as shown below.

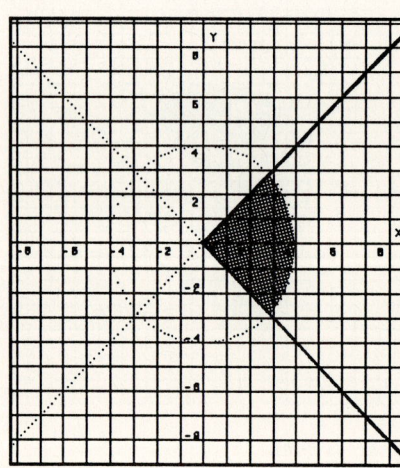

38. Placing the vertex of the parabola corresponding to the shape of the bridge at the origin of a coordinate system, the form of the equation of the parabola is

$$x^2 = 4py.$$

In this configuration, the point $(16,-4)$ is on the curve. Substitute to find the value of $4p$.

$$(16)^2 = 4p(-4)$$
$$-64 = 4p$$

Thus the equation of the parabola is

$$x^2 = -64y$$

which could also be written in the form

$$y = -\frac{1}{64}x^2.$$

In Exercises 39-40, complete the squares to determine if the equation represents a circle, ellipse, hyperbola, parabola, or degenerate conic.

39. $3x^2 + 3y^2 - 6x - 6y + 6 = 0$

$$3(x^2 - 2x) + 3(y^2 - 2y) = -6$$
$$3(x^2 - 2x + 1) + 3(y^2 - 2y + 1) = -6+3(1)+3(1)$$
$$3(x - 1)^2 + 3(y - 1)^2 = 0$$
$$(x - 1)^2 + (y - 1)^2 = 0$$

This is a degenerate conic. The only solution is the point $(1,1)$. This is sometimes called a *point circle*.

40. $9x^2 - 4y^2 - 18x - 16y - 43 = 0$

$$9(x^2 - 2x) - 4(y^2 + 4y) = 43$$
$$9(x^2 - 2x + 1) - 4(y^2 + 4y + 4) = 43+9(1)-4(4)$$
$$9(x - 1)^2 - 4(y + 2)^2 = 36$$

Divide both sides by 36 to obtain

$$\frac{(x - 1)^2}{4} - \frac{(y + 2)^2}{9} = 1,$$

which is the standard form of the equation of a hyperbola.

CHAPTER 8 TOPICS IN ANALYTIC GEOMETRY

CHAPTER 8 Test

1. Find the standard form of the equation of a circle with center $(-2,4)$ and radius 5.

   Substitute -2 for h, 4 for k, and 5 for r in the standard form.

   $$(x - h)^2 + (y - k)^2 = r^2$$
   $$(x - (-2))^2 + (y - 4)^2 = 5^2$$
   $$(x + 2)^2 + (y - 4)^2 = 25$$

2. Graph $x^2 + y^2 - 2x + 4y - 4 = 0$.

   Complete the squares on x and y.

   $$x^2 - 2x \quad + y^2 + 4y \quad = 4$$
   $$x^2 - 2x + 1 + y^2 + 4y + 4 = 4+1+4$$
   $$(x - 1)^2 + (y + 2)^2 = 9$$

   This is the standard form of the equation of a circle centered at $(1,-2)$ with radius 3. The graph is given below.

   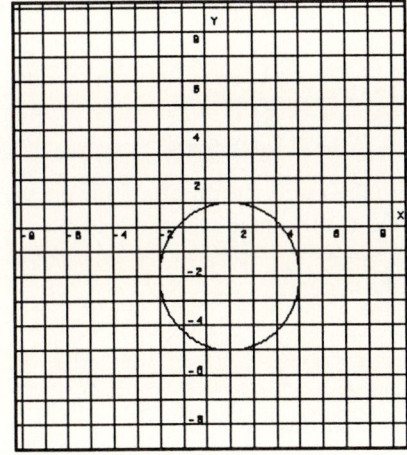

3. Find the equation of the ellipse with $a = 3$ and foci $(2,-1)$ and $(6,-1)$.

   Since the foci are on the same horizontal line, the ellipse has a horizontal major axis and equation of the form:

   $$\frac{(x - h)^2}{a^2} + \frac{(y - k)^2}{b^2} = 1$$

   Since the center of the ellipse is midway between the foci, the center is $(4,-1)$, and $c = 2$. Since $a = 3$, we have

   $$b^2 = a^2 - c^2 = 3^2 - 2^2 = 9 - 4 = 5.$$

   Substitute 4 for h, -1 for k, 9 for a^2, and 5 for b^2 in the form above.

   $$\frac{(x - 4)^2}{9} + \frac{(y + 1)^2}{5} = 1$$

4. Graph $\frac{(x - 1)^2}{25} + \frac{(y + 2)^2}{4} = 1$.

   This is the standard form of the equation of an ellipse centered at $(1,-2)$. Place a new coordinate system with origin at this center, and move 5 units right and left to obtain the points $(6,-2)$ and $(-4,-2)$. Then move up and down 2 units to obtain the points $(1,0)$ and $(1,-4)$. These four points act like the intercepts in the new system. The graph of the ellipse is given below.

   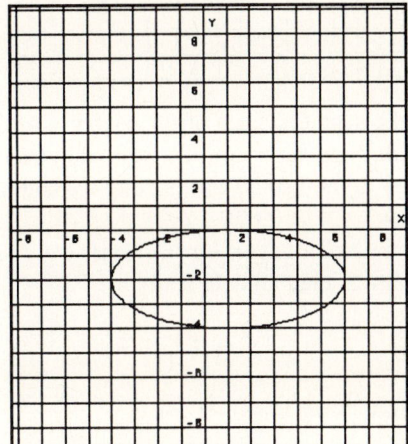

5. Find the standard form of the equation of the hyperbola with general form

   $$8y^2 - x^2 - 16y - 4x - 4 = 0.$$

   Complete the squares on x and y.

   $$8(y^2 - 2y \quad) - (x^2 + 4x \quad) = 4$$
   $$8(y^2 - 2y + 1) - (x^2 + 4x + 4) = 4+8(1)-4$$
   $$8(y - 1)^2 - (x + 2)^2 = 8$$

   Divide both sides by 8 to obtain the standard form.

   $$\frac{(y - 1)^2}{1} - \frac{(x + 2)^2}{8} = 1$$

6. Graph $\frac{(x-1)^2}{25} - \frac{(y+2)^2}{4} = 1$.

This is the standard form of a hyperbola centered at $(1,-2)$ opening left and right. The vertices are $(6,-2)$ and $(-4,-2)$. The asymptotes contain the diagonals of the rectangle with sides parallel to the axes passing through the vertices and the points $(1,0)$ and $(1,-4)$. The graph is given below.

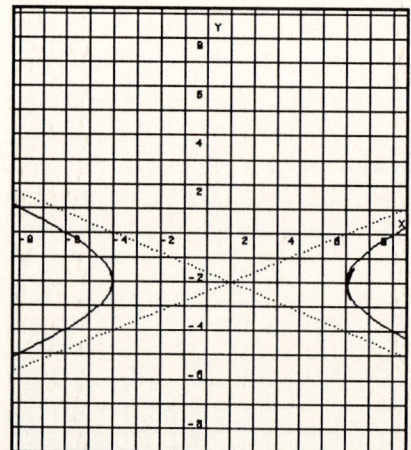

7. Find the standard form of the equation of the parabola with vertex $(4,-5)$, axis of symmetry parallel to the x-axis, and passing through $(0,-7)$.

Since the axis of symmetry is parallel to the x-axis, the form of the parabola is:

$$(y - k)^2 = 4p(x - h)$$

Substitute 4 for h and -5 for k.

$$(y + 5)^2 = 4p(x - 4)$$

Since the parabola passes through the point $(0,-7)$, substitute 0 for x and -7 for y to find the value of $4p$.

$$(-7 + 5)^2 = 4p(0 - 4)$$
$$4 = 4p(-4)$$
$$-1 = 4p$$

Substitute -1 for $4p$ in the form above.

$$(y + 5)^2 = -(x - 4)$$

8. Place the arch in a coordinate system in such a way that the vertex is at the point $(0,40)$. Then the equation of the parabola is

$$(x - 0)^2 = 4p(y - 40)$$

or

$$x^2 = 4p(y - 40).$$

Since the arch is 18 ft wide at the base, when $x = 9$, $y = 0$. Substitute these values for x and y to find $4p$.

$$9^2 = 4p(0 - 40)$$
$$81 = 4p(-40)$$
$$-\frac{81}{40} = 4p$$

Thus, the equation of the parabola is

$$x^2 = -\frac{81}{40}(y - 40).$$

To find the vertical clearance 2.5 ft from the center of the base of the arch, substitute 2.5 for x and find the value of y.

$$(2.5)^2 = -\frac{81}{40}(y - 40)$$
$$(6.25)(40) = -81(y - 40)$$
$$-\frac{(6.25)(40)}{81} = y - 40$$
$$40 - \frac{(6.25)(40)}{81} = y$$
$$36.91358025 = y$$

Thus, to the nearest tenth of a foot, the vertical clearance at this point is about 36.9 ft.

9. Solve the system.

$$x^2 + y^2 = 81$$
$$x^2 - y^2 = 81$$

Add to eliminate y.

$$2x^2 = 162$$
$$x^2 = 81$$
$$x = \pm 9$$

When $x = 9$,

$$(9)^2 + y^2 = 81$$
$$81 + y^2 = 81$$
$$y^2 = 0$$
$$y = 0$$

Thus, one solution is $(9,0)$.

When $x = -9$,

$$(-9)^2 + y^2 = 81$$
$$81 + y^2 = 81$$
$$y^2 = 0$$
$$y = 0$$

Thus, another solution is $(-9, 0)$.

10. Solve the system.

$$y + e^x = 2$$
$$y - e^x = 0$$

Add to eliminate the term involving x.

$$2y = 2$$
$$y = 1$$

Substitute 1 for y in the first equation.

$$1 + e^x = 2$$
$$e^x = 1$$
$$x = 0$$

Thus, the solution to the system is $(0, 1)$.

11. Graph the system.

$$x^2 + y^2 \leq 4$$
$$y \geq x^2$$

First graph the circle $x^2 + y^2 = 4$, centered at $(0,0)$ with radius 2, using a solid curve. The test point $(0,0)$ shows that we shade the region inside the circle. Then graph the parabola $y = x^2$ with vertex $(0,0)$ opening up. Use a solid curve. The test point $(0,1)$ shows that we shade the region above the curve, inside the U. Combining these two, we obtain the graph of the system as shown below.

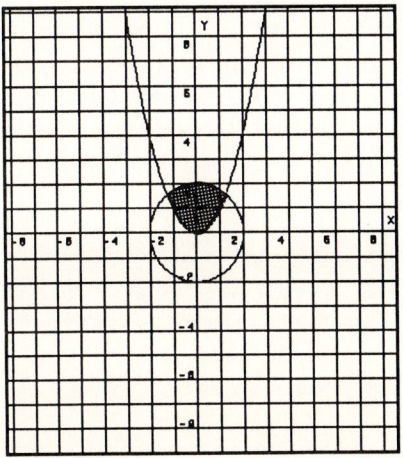

12. Graph the system.

$$y \geq 2^x$$
$$y \leq x^2 + 3$$

First graph the exponential function $y = 2^x$ using a solid curve. The test point $(0,0)$ shows that we shade the region above the curve. Then graph the parabola $y = x^2 + 3$ using a solid line. The vertex is at $(0,3)$ and the parabola opens up. The test point $(0,0)$ shows that we shade the region below the curve, outside the U. Combining these we obtain the graph of the system as shown below.

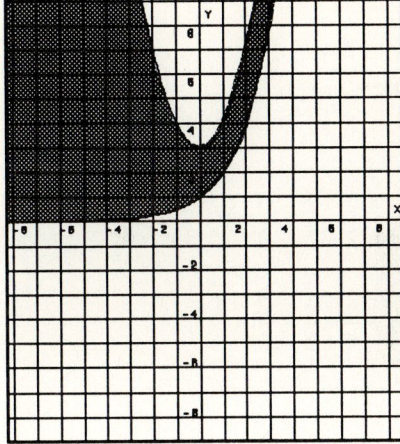

CHAPTER 9 SEQUENCES, SERIES, AND PROBABILITY

SECTION 9.1 Sequences and Series

1. Find the second, fourth, fifth, and seventh terms of each sequence.

   (a) $x_n = 2n^2 + 3$

   $x_2 = 2(2)^2 + 3 = 2(4) + 3 = 8 + 3 = 11$
   $x_4 = 2(4)^2 + 3 = 2(16) + 3 = 32 + 3 = 35$
   $x_5 = 2(5)^2 + 3 = 2(25) + 3 = 50 + 3 = 53$
   $x_7 = 2(7)^2 + 3 = 2(49) + 3 = 98 + 3 = 101$

   (b) $a_n = \dfrac{n-2}{n+2}$

   $a_2 = \dfrac{2-2}{2+2} = \dfrac{0}{4} = 0$

   $a_4 = \dfrac{4-2}{4+2} = \dfrac{2}{6} = \dfrac{1}{3}$

   $a_5 = \dfrac{5-2}{5+2} = \dfrac{3}{7}$

   $a_7 = \dfrac{7-2}{7+2} = \dfrac{5}{9}$

   (c) $b_n = (-1)^{n+1}(3n - 5)$

   $b_2 = (-1)^{2+1}(3(2) - 5)$
   $= (-1)(6 - 5)$
   $= (-1)(1) = -1$
   $b_4 = (-1)^{4+1}(3(4) - 5)$
   $= (-1)^5(12 - 5)$
   $= (-1)(7) = -7$
   $b_5 = (-1)^{5+1}(3(5) - 5)$
   $= (-1)^6(15 - 5)$
   $= (1)(10) = 10$
   $b_7 = (-1)^{7+1}(3(7) - 5)$
   $= (-1)^8(21 - 5)$
   $= (1)(16) = 16$

2. Determine a formula for a_n.

   (a) 7, 12, 17, 22, ...

   Notice that the terms seem to be increasing by 5. When $n = 1$, $a_1 = 7$, so we might try $5(n) + 2$. Note when $n = 1$, $5(1) + 2$ does indeed equal 7. When $n = 2$, $5(2) + 2$ does indeed equal 12. Similarly this expression works for 17 and 22. Thus, we conclude that

   $a_n = 5n + 2.$

   (b) 3, -6, 9, -12, ...

   Since the terms alternate in sign, we will need a factor of (-1) to some power. Since the first term is positive, we try $(-1)^{n+1}$. Also the terms appear to be increasing by a multiple of 3 (ignoring the signs), so we might try another factor of $3n$. Thus, we try

   $a_n = (-1)^{n+1}3n.$

   Notice that $a_1 = (-1)^{1+1}3(1) = (1)(3) = 3$. Also, $a_2 = (-1)^{2+1}3(2) = (-1)(6) = -6$. Similarly, we can obtain the third and fourth terms. Thus, we conclude that the expression for a_n given above is a formula that will work.

3. Let $a_{n+1} = 5 - a_n$ and $a_1 = 15$.

   $a_2 = 5 - a_1 = 5 - 15 = -10$
   $a_3 = 5 - a_2 = 5 - (-10) = 5 + 10 = 15$
   $a_4 = 5 - a_3 = 5 - 15 = -10$
   $a_5 = 5 - a_4 = 5 - (-10) = 5 + 10 = 15$
   $a_6 = 5 - a_5 = 5 - 15 = -10$
   $a_7 = 5 - a_6 = 5 - (-10) = 5 + 10 = 15$

4. Determine the value of each series.

   (a) $\displaystyle\sum_{k=1}^{7}(5k - 6)$

   $= (5(1) - 6) + (5(2) - 6) + (5(3) - 6) + (5(4) - 6)$
   $\quad + (5(5) - 6) + (5(6) - 6) + (5(7) - 6)$
   $= (-1) + (4) + (9) + (14) + (19) + (24) + (29)$
   $= 98$

   (b) $\displaystyle\sum_{k=3}^{5}(-1)^k(k^2 + 2)$

   $= (-1)^3(3^2 + 2) + (-1)^4(4^2 + 2) + (-1)^5(5^2 + 2)$
   $= (-1)(11) + (1)(18) + (-1)(27)$
   $= -11 + 18 - 27$
   $= -20$

EXERCISES A

CHAPTER 9 SEQUENCES, SERIES, AND PROBABILITY

SECTION 9.1 Sequences and Series

In Exercises 1-4 give the first five terms, the eighth term, and the twelfth term of each sequence.

1. $a_n = 4n$

   $a_1 = 4(1) = 4$
   $a_2 = 4(2) = 8$
   $a_3 = 4(3) = 12$
   $a_4 = 4(4) = 16$
   $a_5 = 4(5) = 20$
   $a_8 = 4(8) = 32$
   $a_{12} = 4(12) = 48$

2. $x_n = \left(-\frac{1}{2}\right)^n$

   $x_1 = \left(-\frac{1}{2}\right)^1 = -\frac{1}{2}$
   $x_2 = \left(-\frac{1}{2}\right)^2 = \frac{1}{4}$
   $x_3 = \left(-\frac{1}{2}\right)^3 = -\frac{1}{8}$
   $x_4 = \left(-\frac{1}{2}\right)^4 = \frac{1}{16}$
   $x_5 = \left(-\frac{1}{2}\right)^5 = -\frac{1}{32}$
   $x_8 = \left(-\frac{1}{2}\right)^8 = \frac{1}{256}$
   $x_{12} = \left(-\frac{1}{2}\right)^{12} = \frac{1}{4096}$

3. $b_n = (-1)^n(n^2 + 5)$

   $b_1 = (-1)^1(1^2 + 5) = (-1)(6) = -6$
   $b_2 = (-1)^2(2^2 + 5) = (1)(9) = 9$
   $b_3 = (-1)^3(3^2 + 5) = (-1)(14) = -14$
   $b_4 = (-1)^4(4^2 + 5) = (1)(21) = 21$
   $b_5 = (-1)^5(5^2 + 5) = (-1)(30) = -30$
   $b_8 = (-1)^8(8^2 + 5) = (1)(69) = 69$
   $b_{12} = (-1)^{12}(12^2 + 5) = (1)(149) = 149$

4. $x_n = (-1)^n + (-1)^{n+1}$

   $x_1 = (-1)^1 + (-1)^{1+1} = (-1) + 1 = 0$
   $x_2 = (-1)^2 + (-1)^{2+1} = 1 + (-1) = 0$
   $x_3 = (-1)^3 + (-1)^{3+1} = (-1) + 1 = 0$
   $x_4 = (-1)^4 + (-1)^{4+1} = 1 + (-1) = 0$
   $x_5 = (-1)^5 + (-1)^{5+1} = (-1) + 1 = 0$
   $x_8 = (-1)^8 + (-1)^{8+1} = 1 + (-1) = 0$
   $x_{12} = (-1)^{12} + (-1)^{12+1} = 1 + (-1) = 0$

Find a formula for a_n given the first few terms of the sequence in Exercises 5-8.

5. 2, 6, 10, 14, ...

   The terms all appear to be increasing by a multiple of 4. The multiples of 4, $4n$, however give us 4, 8, 12, ... So if we subtract 2 from each of these terms, we would obtain the given sequence. Thus,

   $$a_n = 4n - 2.$$

6. 1, -4, 9, -16, ...

   Since the terms alternate in sign, we must have a factor of (-1) raised to some power. Since the first term is positive, we try $(-1)^{n+1}$. Also, ignoring the signs of the terms, we recognize that each term is a perfect square, n^2. Thus,

   $$a_n = (-1)^{n+1}n^2.$$

7. $\frac{1}{2}, \frac{2}{3}, \frac{3}{4}, \frac{4}{5}, ...$

   Since the numerators are 1, 2, 3, 4, ..., the numerator is n. Since each denominator is 1 more than the numerator, the denominator is $n + 1$. Thus,

   $$a_n = \frac{n}{n+1}.$$

8. log 3, log 6, log 9, log 12, ...

   Each term is the logarithm of some number. Since the numbers are multiples of 3, $3n$, we have

   $$a_n = \log 3n.$$

In Exercises 9-11 determine the second, third, fourth, and fifth terms of each sequence.

9. $a_1 = 3;\ a_{n+1} = 4a_n$

   $a_2 = 4a_1 = 4(3) = 12$
   $a_3 = 4a_2 = 4(12) = 48$
   $a_4 = 4a_3 = 4(48) = 192$
   $a_5 = 4a_4 = 4(192) = 768$

10. $b_1 = 8$; $b_n = -3b_{n-1}$

$$b_2 = -3b_1 = -3(8) = -24$$
$$b_3 = -3b_2 = -3(-24) = 72$$
$$b_4 = -3b_3 = -3(72) = -216$$
$$b_5 = -3b_4 = -3(-216) = 648$$

11. $x_1 = \frac{2}{3}$; $x_{n+1} = 9x_n + 5$

$$x_2 = 9x_1 + 5 = 9\left(\frac{2}{3}\right) + 5 = 6 + 5 = 11$$
$$x_3 = 9x_2 + 5 = 9(11) + 5 = 99 + 5 = 104$$
$$x_4 = 9x_3 + 5 = 9(104) + 5 = 936 + 5 = 941$$
$$x_5 = 9x_4 + 5 = 9(941) + 5 = 8469 + 5 = 8474$$

Write out each series in Exercises 12-14

12. $\sum_{k=1}^{5} x_k = x_1 + x_2 + x_3 + x_4 + x_5$

13. $\sum_{m=1}^{4} \frac{1}{2m+1}$

$$= \frac{1}{2(1)+1} + \frac{1}{2(2)+1} + \frac{1}{2(3)+1} + \frac{1}{2(4)+1}$$
$$= \frac{1}{3} + \frac{1}{5} + \frac{1}{7} + \frac{1}{9}$$

14. $\sum_{k=0}^{6} \frac{(-1)^{k+1}}{k+1}$

$$= \frac{(-1)^1}{0+1} + \frac{(-1)^2}{1+1} + \frac{(-1)^3}{2+1} + \frac{(-1)^4}{3+1}$$
$$+ \frac{(-1)^5}{4+1} + \frac{(-1)^6}{5+1} + \frac{(-1)^7}{6+1}$$
$$= \frac{-1}{1} + \frac{1}{2} + \frac{-1}{3} + \frac{1}{4} + \frac{-1}{5} + \frac{1}{6} + \frac{-1}{7}$$
$$= -1 + \frac{1}{2} - \frac{1}{3} + \frac{1}{4} - \frac{1}{5} + \frac{1}{6} - \frac{1}{7}$$

In Exercises 15-18 write each series using sigma summation notation.

15. $1 + 2 + 3 + 4 + \ldots + n = \sum_{k=1}^{n} k$

16. $1 + \frac{1}{\sqrt{2}} + \frac{1}{\sqrt{3}} + \ldots + \frac{1}{\sqrt{n}} = \sum_{k=1}^{n} \frac{1}{\sqrt{k}}$

17. $2 + 4 + 6 + 8 = \sum_{k=1}^{4} 2k$

Evaluate each series in Exercises 18-23.

18. $\sum_{k=1}^{6} (2k-1) = (2(1)-1) + (2(2)-1) + (2(3)-1)$
$$+ (2(4)-1) + (2(5)-1) + (2(6)-1)$$
$$= 1 + 3 + 5 + 7 + 9 + 11$$
$$= 36$$

19. $\sum_{i=0}^{3} (i^2 + 1) = (0^2+1) + (1^2+1) + (2^2+1) + (3^2+1)$
$$= 1 + 2 + 5 + 10$$
$$= 18$$

20. $\sum_{m=1}^{5} [1+(-1)^m]m^2$
$$= [1+(-1)^1]1^2 + [1+(-1)^2]2^2 + [1+(-1)^3]3^2$$
$$+ [1+(-1)^4]4^2 + [1+(-1)^5]5^2$$
$$= [0](1) + [2](4) + [0](9) + [2](16) + [0](25)$$
$$= 0 + 8 + 0 + 32 + 0$$
$$= 40$$

21. Using the sum of constants theorem, we have

$$\sum_{k=1}^{50} 3 = (50)(3) = 150.$$

22. $\sum_{k=1}^{4} (k+1)(k-2) = (1+1)(1-2) + (2+1)(2-2)$
$$+ (3+1)(3-2) + (4+1)(4-2)$$
$$= (2)(-1) + (3)(0) + (4)(1) + (5)(2)$$
$$= -2 + 0 + 4 + 10$$
$$= 12$$

23. $\sum_{m=1}^{5} \left(\frac{1}{m} - \frac{1}{m+1}\right) = \left(\frac{1}{1} - \frac{1}{2}\right) + \left(\frac{1}{2} - \frac{1}{3}\right) + \left(\frac{1}{3} - \frac{1}{4}\right)$
$$+ \left(\frac{1}{4} - \frac{1}{5}\right) + \left(\frac{1}{5} - \frac{1}{6}\right)$$
$$= 1 - \frac{1}{6}$$
$$= \frac{5}{6}$$

24. Compare the two series:

$\sum_{k=2}^{5} (k+1)(k+2) = (3)(4) + (4)(5) + (5)(6) + (6)(7)$
$$= 12 + 20 + 30 + 42$$
$$= 104$$

EXERCISES A

$$\sum_{k=1}^{4}(k+2)(k+3) = (3)(4)+(4)(5)+(5)(6)+(6)(7)$$
$$= 12+20+30+42$$
$$= 104$$

Both series have value 104. Notice that the terms of the series are identical even though they are formed using different expressions.

25. Determine the approximate value of a_{30} by repeated use of the square root key on a calculator if $a_1 = 6$ and $a_{n+1} = \sqrt{a_n}$.

Since this is a recursive definition of the sequence, enter 6 on your calculator, and press the square root button to obtain the second term (2.449489743), then press the square root button again to obtain the third term (1.56508458), then again to obtain the third term (1.251033405), and so forth. By the time you get to the 30th term, the number in the display will be very close to, if not exactly, 1.

26. Use your calculator to find a_{1000} if

$$a_n = \left(1 + \frac{1}{n}\right)^n.$$

First find $1 + \dfrac{1}{1000} = 1.001$ and use the y^x button to raise this number to the 1000th power. The result is approximately

2.7169.

27. Use your calculator to find the value of

$$1 + 1 + \frac{1}{2} + \frac{1}{6} + \frac{1}{24} + \frac{1}{120}$$

and compare the result with Exercise 26.

Converting to decimals, the sum above is

$1 + 1 + 0.5 + 0.166666666 + 0.041666666 + 0.008333$

which is approximately equal to

2.7167.

Notice that the two results, here and in Exercise 26, are very close.

Let S_n denote the sum of the first n terms of a sequence. In Exercises 28-29 use a calculator to determine S_5 to four decimal places.

28. $a_n = n + \sqrt{n}$

Find the first five terms.

$$a_1 = 1 + \sqrt{1} = 2$$
$$a_2 = 2 + \sqrt{2} = 3.414213562$$
$$a_3 = 3 + \sqrt{3} = 4.732050808$$
$$a_4 = 4 + \sqrt{4} = 6$$
$$a_5 = 5 + \sqrt{5} = 7.236067978$$

Then $S_5 = a_1 + a_2 + a_3 + a_4 + a_5$, which is approximately 23.3823, correct to four decimal places.

29. $b_1 = 0.5;\ b_{n+1} = 2b_n + b_n^2$

Find the first five terms.

$$b_1 = 0.5$$
$$b_2 = 2(0.5) + (0.5)^2 = 1.25$$
$$b_3 = 2(1.25) + (1.25)^2 = 4.0625$$
$$b_4 = 2(4.065) + (4.065)^2 = 24.654225$$
$$b_5 = 2(24.654225) + (24.654225)^2$$
$$= 657.1392604$$

Then $S_5 = b_1 + b_2 + b_3 + b_4 + b_5$ which is approximately 687.6060, correct to four decimal places.

In Exercises 30-31 find a simplified formula for S_n, the sum of the first n terms.

30. $a_k = \dfrac{1}{k} - \dfrac{1}{k+1}$

$$\sum_{k=1}^{n}\left(\frac{1}{k}-\frac{1}{k+1}\right) = \frac{1}{1}-\frac{1}{2}+\frac{1}{2}-\frac{1}{3}+\frac{1}{3}+$$
$$\frac{1}{4}\ldots+\frac{1}{n}-\frac{1}{n+1}$$
$$= \frac{1}{1}-\frac{1}{n+1}$$
$$= 1-\frac{1}{n+1}$$

Notice that all the terms except the first and last subtract out in the above expression giving us the desired result.

31. $b_k = \sqrt{k+1} - \sqrt{k}$

$$\sum_{k=1}^{n} (\sqrt{k+1} - \sqrt{k}) = \sqrt{2} - \sqrt{1} + \sqrt{3} - \sqrt{2} + \sqrt{4} - \sqrt{3} + \ldots$$
$$+ \sqrt{n+1} - \sqrt{n}$$
$$= -\sqrt{1} + \sqrt{n+1}$$
$$= \sqrt{n+1} - 1$$

Notice that most of the terms subtract out leaving us with just the two terms in the desired expression.

If $\sum_{k=1}^{n} k = \dfrac{n(n+1)}{2}$ and $\sum_{k=1}^{n} k^2 = \dfrac{n(n+1)(2n+1)}{6}$,

determine each sum in Exercises 32-34.

32. $\sum_{k=1}^{6} k = \dfrac{6(6+1)}{2} = \dfrac{6(7)}{2} = 21$

33.
$$\sum_{k=1}^{n} (k^2 + k) = \sum_{k=1}^{n} k^2 + \sum_{k=1}^{n} k$$
$$= \dfrac{n(n+1)(2n+1)}{6} + \dfrac{n(n+1)}{2}$$
$$= \dfrac{n(n+1)(2n+1)}{6} + \dfrac{3n(n+1)}{6}$$
$$= n(n+1)\left(\dfrac{2n+1}{6} + \dfrac{3}{6}\right)$$
$$= n(n+1)\left(\dfrac{2n+1+3}{6}\right)$$
$$= n(n+1)\left(\dfrac{2n+4}{6}\right)$$
$$= n(n+1)\left(\dfrac{n+2}{3}\right)$$
$$= \dfrac{n(n+1)(n+2)}{3}$$

34.
$$\sum_{k=1}^{n} (2k^2 + 3k) = 2\sum_{k=1}^{n} k^2 + 3\sum_{k=1}^{n} k$$
$$= 2\left(\dfrac{n(n+1)(2n+1)}{6}\right) + 3\left(\dfrac{n(n+1)}{2}\right)$$
$$= n(n+1)\left[\dfrac{2n+1}{3} + \dfrac{3}{2}\right]$$
$$= n(n+1)\left[\dfrac{4n+2+9}{6}\right]$$
$$= n(n+1)\left[\dfrac{4n+11}{6}\right]$$
$$= \dfrac{n(n+1)(4n+11)}{6}$$

35.
$$\sum_{k=1}^{n} (a_k - b_k) = \sum_{k=1}^{n} (a_k + (-1)b_k)$$
$$= \sum_{k=1}^{n} a_k + \sum_{k=1}^{n} (-1)b_k$$
$$= \sum_{k=1}^{n} a_k + (-1)\sum_{k=1}^{n} b_k$$
$$= \sum_{k=1}^{n} a_k - \sum_{k=1}^{n} b_k$$

This proof makes use of the other two summation formulas given in the theorem in the text. We could also prove the result directly using an approach similar to that given in the text for the sum of a series of terms that are themselves sums of two terms.

36. Given that $\bar{x} = \dfrac{1}{n}\sum_{k=1}^{n} x_k$, determine $\bar{x}$ when the data is 1, 2, 3, ..., n.

Since $x_k = k$, for $k = 1, 2, 3, \ldots, n$, we have:
$$\bar{x} = \dfrac{1}{n}\sum_{k=1}^{n} x_k = \dfrac{1}{n}\sum_{k=1}^{n} k$$
$$= \dfrac{1}{n}\left[\dfrac{n(n+1)}{2}\right]$$
$$= \dfrac{n+1}{2}$$

In Exercises 37-38 use the sequence defined by

$$a_1 = a_2 = 1 \text{ and } a_{k+1} = a_k + a_{k-1}.$$

This is a well-known sequence in mathematics called the Fibonacci sequence.

37. Determine the first eight terms of the sequence.

$a_1 = 1$
$a_2 = 1$
$a_3 = a_2 + a_1 = 1 + 1 = 2$
$a_4 = a_3 + a_2 = 2 + 1 = 3$
$a_5 = a_4 + a_3 = 3 + 2 = 5$
$a_6 = a_5 + a_4 = 5 + 3 = 8$
$a_7 = a_6 + a_5 = 8 + 5 = 13$
$a_8 = a_7 + a_6 = 13 + 8 = 21$

EXERCISES A

38. If $r_k = \dfrac{a_{k+1}}{a_k}$, determine r_8 to three decimal places.

First find a_9. Use the results of Exercise 37.

$$a_9 = a_8 + a_7 = 21 + 13 = 34$$

Then since $a_8 = 21$ and $a_9 = 34$,

$$r_8 = \dfrac{a_9}{a_8} = \dfrac{34}{21} \approx 1.619.$$

As k increases the sequence r_k gives a better and better approximation for the *golden ratio*, the ratio of the sides of a rectangle thought to be most pleasing to the eye.

In Exercises 39-40 determine the nature of the conic section by completing the square.

39. $5x^2 + 10y^2 - 10x + 20y = 0$

$$5(x^2 - 2x) + 10(y^2 + 2y) = 0$$
$$5(x^2 - 2x + 1) + 10(y^2 + 2y + 1) = 0+1+1$$
$$5(x - 1)^2 + 10(y + 1)^2 = 2$$

Divide both sides by 2 to obtain

$$\dfrac{(x-1)^2}{\frac{2}{5}} + \dfrac{(y+1)^2}{\frac{1}{5}} = 1,$$

which is the standard form for the equation of an ellipse.

40. $5x^2 - 10y^2 - 10x + 20y = 0$

$$5(x^2 - 2x) - 10(y^2 - 2y) = 0$$
$$5(x^2 - 2x + 1) - 10(y^2 - 2y + 1) = 0+1+1$$
$$5(x - 1)^2 - 10(y - 1)^2 = 2$$

Divide both sides by 2 to obtain

$$\dfrac{(x-1)^2}{\frac{2}{5}} - \dfrac{(y-1)^2}{\frac{1}{5}} = 1,$$

which is the standard form of the equation of a hyperbola.

CHAPTER 9 SEQUENCES, SERIES, AND PROBABILITY

SECTION 9.2 Arithmetic Sequences and Series

1. Find the seventh and tenth terms of the sequence

   $-2, 4, 10, 16, 22, \ldots$.

   We have $a_1 = -2$ and $d = 6$.

   $$a_7 = a_1 + (7 - 1)d$$
   $$= -2 + (6)(6)$$
   $$= -2 + 36$$
   $$= 34$$

   $$a_{10} = a_1 + (10 - 1)d$$
   $$= -2 + (9)(6)$$
   $$= -2 + 54$$
   $$= 52$$

2. Find x so that $2x$, $x + 3$, and $3x - 3$ form a three-term arithmetic sequence in the given order.

   Use the fact that the difference between successive terms is equal to the common difference d.

   $(x + 3) - 2x = d$ and $(3x - 3) - (x + 3) = d$

   Set both expressions for d equal to each other.

   $$(x + 3) - 2x = (3x - 3) - (x + 3)$$
   $$x + 3 - 2x = 3x - 3 - x - 3$$
   $$3 - x = 2x - 6$$
   $$-3x = -9$$
   $$x = 3$$

   Then $2x = 2(3) = 6$, $x + 3 = 3 + 3 = 6$, and $3x - 3 = 3(3) - 3 = 6$. Thus the sequence is

   $6, 6, 6$.

3. Find a_{12} and S_{12} for the sequence

   $15, 11, 7, 3, -1, \ldots$.

   We have $a_1 = 15$ and $d = -4$.

   $$a_{12} = a_1 + (12 - 1)d$$
   $$= 15 + (11)(-4)$$
   $$= 15 - 44$$
   $$= -29$$

   $$S_{12} = \tfrac{n}{2}[2a_1 + (n - 1)d]$$
   $$= \tfrac{12}{2}[2(15) + (12 - 1)(-4)]$$
   $$= 6[30 + (11)(-4)]$$
   $$= 6[30 - 44]$$
   $$= 6[-14]$$
   $$= -84$$

4. If $a_{11} = 20$ and $S_9 = 72$, find d, the first five terms, and S_{20}.

   Since $a_{11} = 20$, we have

   $$20 = a_1 + (11 - 1)d$$
   $$20 = a_1 + 10d$$

   Since $S_9 = 72$, we have

   $$72 = \tfrac{9}{2}[2a_1 + (9 - 1)d]$$
   $$= \tfrac{9}{2}[2a_1 + 8d]$$
   $$72 = 9a_1 + 36d$$
   $$8 = a_1 + 4d$$

   Then we must solve the following system:

   $$a_1 + 4d = 8$$
   $$a_1 + 10d = 20$$

   Subtract the top equation from the bottom equation.

   $$6d = 12$$
   $$d = 2$$

   Substitute 2 for d in the first equation to find the value of a_1.

   $$a_1 + 4(2) = 8$$
   $$a_1 + 8 = 8$$
   $$a_1 = 0$$

   $$a_2 = a_1 + (2 - 1)d$$
   $$= 0 + (1)(2)$$
   $$= 0 + 2$$
   $$= 2$$

   $$a_3 = a_1 + (3 - 1)d$$
   $$= 0 + (2)(2)$$
   $$= 0 + 4$$
   $$= 4$$

PRACTICE EXERCISES **SECTION 9.2 437**

$$a_4 = a_1 + (4 - 1)d$$
$$= 0 + (3)(2)$$
$$= 0 + 6$$
$$= 6$$

$$a_5 = a_1 + (5 - 1)d$$
$$= 0 + (4)(2)$$
$$= 0 + 8$$
$$= 8$$

$$S_{20} = \frac{20}{2}[2a_1 + (20 - 1)d]$$
$$= 10[2(0) + (19)(2)]$$
$$= 10[38]$$
$$= 380$$

5. We calculate the sum of the increases over the 8 years with $a_1 = 0.03$, $a_2 = 0.05$, $a_3 = 0.07,\ldots$ and $d = 0.02$.

$$S_8 = \frac{8}{2}[2a_1 + (8 - 1)d]$$
$$= 4[2(0.03) + (7)(0.02)]$$
$$= 4[0.06 + 0.14]$$
$$= 4[0.20]$$
$$= 0.8$$

Thus, the total of the increases is $0.8 = 80\%$. The new enrollment is the former enrollment plus the increase. Thus, the new enrollment is

$$10{,}300 + (0.8)(10{,}300) = 18{,}540.$$

6. Insert five arithmetic means between 0.5 and 9.5.

Since five arithmetic means are to be inserted between 0.5 and 9.5, there will be seven terms with $a_1 = 0.5$ and $a_7 = 9.5$. Use the formula for a_7.

$$a_7 = a_1 + (7 - 1)d$$
$$9.5 = 0.5 + (6)d$$
$$9 = 6d$$
$$1.5 = d$$

Then the five arithmetic means are:

$$a_2 = a_1 + d = 0.5 + 1.5 = 2.0$$

$$a_3 = a_2 + d = 2.0 + 1.5 = 3.5$$

$$a_4 = a_3 + d = 3.5 + 1.5 = 5.0$$

$$a_5 = a_4 + d = 5.0 + 1.5 = 6.5$$

$$a_6 = a_5 + d = 6.5 + 1.5 = 8.0$$

CHAPTER 9 SEQUENCES, SERIES, AND PROBABILITY

SECTION 9.2 Arithmetic Sequences and Series

In Exercises 1-3 determine if each sequence is arithmetic. If it is, give the common difference.

1. 4, 8, 12, 16,...

   Since $8 - 4 = 4$, $12 - 8 = 4$, $16 - 12 = 4$, this sequence is arithmetic with common difference 4.

2. 9, -1, -11, -21,...

   Since $-1 - 9 = -10$, $-11 - (-1) = -10$, $-21 - (-11) = -10$, this sequence is arithmetic with common difference -10.

3. 1, -1, 1, -1,...

   Since $-1 - 1 = -2$, $1 - (-1) = 2$, the difference between successive terms is not the same so the sequence is not arithmetic.

Find the first six terms of each arithmetic sequence in Exercises 4-6.

4. $a_1 = 2$ and $d = 7$

   $a_1 = 2$
   $a_2 = a_1 + (2 - 1)d = 2 + (1)(7) = 9$
   $a_3 = a_1 + (3 - 1)d = 2 + (2)(7) = 16$
   $a_4 = a_1 + (4 - 1)d = 2 + (3)(7) = 23$
   $a_5 = a_1 + (5 - 1)d = 2 + (4)(7) = 30$
   $a_6 = a_1 + (6 - 1)d = 2 + (5)(7) = 37$

5. $a_1 = -2$ and $a_2 = 5$

   Then $d = a_2 - a_1 = 5 - (-2) = 5 + 2 = 7$.

   $a_1 = -2$
   $a_2 = a_1 + (2 - 1)d = -2 + (1)(7) = 5$
   $a_3 = a_1 + (3 - 1)d = -2 + (2)(7) = 12$
   $a_4 = a_1 + (4 - 1)d = -2 + (3)(7) = 19$
   $a_5 = a_1 + (5 - 1)d = -2 + (4)(7) = 26$
   $a_6 = a_1 + (6 - 1)d = -2 + (5)(7) = 33$

6. $a_1 = \sqrt{3}$ and $d = 4\sqrt{3}$

   $a_1 = \sqrt{3}$
   $a_2 = a_1 + (2-1)d = \sqrt{3} + (1)(4\sqrt{3}) = 5\sqrt{3}$
   $a_3 = a_1 + (3-1)d = \sqrt{3} + (2)(4\sqrt{3}) = 9\sqrt{3}$
   $a_4 = a_1 + (4-1)d = \sqrt{3} + (3)(4\sqrt{3}) = 13\sqrt{3}$
   $a_5 = a_1 + (5-1)d = \sqrt{3} + (4)(4\sqrt{3}) = 17\sqrt{3}$
   $a_6 = a_1 + (6-1)d = \sqrt{3} + (5)(4\sqrt{3}) = 21\sqrt{3}$

7. Find x so that x, $x + 4$, and $2x$ form a three-term arithmetic sequence in the given order. Determine the sequence.

   Use the fact that successive terms have the same common difference. That is,

   $$x + 4 - x = d \text{ and } 2x - (x + 4) = d.$$

   Then since both expressions are equal to d, they are equal.

   $$x + 4 - x = 2x - (x + 4)$$
   $$4 = 2x - x - 4$$
   $$8 = x$$

   Then $x = 8$, $x + 4 = 8 + 4 = 12$, and $2x = 2(8) = 16$. Thus, the sequence is

   $$8, 12, 16.$$

In Exercises 8-10 find the indicated sum of the arithmetic sequence.

8. $-8, -1, 6, 13,...$; S_{10}

   We have $a_1 = -8$. Since the difference between successive terms is 7, $d = 7$. Then

   $$S_n = \frac{n}{2}[2a_1 + (n-1)d]$$
   $$S_{10} = \frac{10}{2}[2(-8) + (10-1)(7)]$$
   $$= 5[-16 + (9)(7)]$$
   $$= 5[-16 + 63]$$
   $$= 5[47]$$
   $$= 235.$$

9. $a_4 = 6$ and $a_8 = 26$; S_{12}

   Since

   $$6 = a_4 = a_1 + (4-1)d = a_1 + 3d$$

   and

   $$26 = a_8 = a_1 + (8-1)d = a_1 + 7d,$$

   we must solve the following system.

   $$a_1 + 3d = 6$$
   $$a_1 + 7d = 26$$

EXERCISES A

Subtract the top equation from the bottom equation to eliminate a_1 and obtain

$$4d = 20$$
$$d = 5$$

Substitute 5 for d in the first equation.

$$a_1 + 3(5) = 6$$
$$a_1 + 15 = 6$$
$$a_1 = -9$$

Then

$$S_n = \frac{n}{2}[2a_1 + (n-1)d]$$
$$S_{12} = \frac{12}{2}[2(-9) + (12-1)(5)]$$
$$= 6[-18 + (11)(5)]$$
$$= 6[-18 + 55]$$
$$= 6[37]$$
$$= 222.$$

10. $\sum_{k=1}^{6}(k+2) = 3+4+5+6+7+8 = 33$

In Exercises 11-16, some of the numbers n, a_1, a_n, d, and S_n are given. Find the missing ones.

11. $a_1 = 2$, $n = 17$, $d = 3$

We must find a_{17} and S_{17}. Then

$$a_{17} = a_1 + (17-1)d = 2 + (16)(3) = 50,$$

and

$$S_{17} = \frac{17}{2}[a_1 + a_{17}]$$
$$= \frac{17}{2}[2 + 50]$$
$$= \frac{17}{2}[52]$$
$$= 17(26)$$
$$= 442.$$

12. $a_n = 27$, $S_n = 63$, $a_1 = -9$

We must find n and d.

$$S_n = \frac{n}{2}[a_1 + a_n]$$
$$63 = \frac{n}{2}[-9 + 27]$$
$$63 = \frac{n}{2}[18]$$
$$63 = 9n$$
$$7 = n$$

$$a_n = a_1 + (n-1)d$$
$$27 = -9 + (7-1)d$$
$$27 = -9 + 6d$$
$$36 = 6d$$
$$6 = d$$

13. $a_1 = \frac{5}{3}$, $d = \frac{1}{6}$, $n = 12$

We must find $a_n = a_{12}$ and $S_n = S_{12}$.

Then

$$a_n = a_1 + (n-1)d$$
$$a_{12} = \frac{5}{3} + (12-1)\frac{1}{6}$$
$$= \frac{5}{3} + (11)\frac{1}{6}$$
$$= \frac{5}{3} + \frac{11}{6}$$
$$= \frac{10}{6} + \frac{11}{6}$$
$$= \frac{21}{6}$$
$$= \frac{7}{2}$$

and

$$S_n = \frac{n}{2}[a_1 + a_n]$$
$$S_{12} = \frac{12}{2}\left[\frac{5}{3} + \frac{7}{2}\right]$$
$$= 6\left[\frac{10+21}{6}\right]$$
$$= 6\left[\frac{31}{6}\right]$$
$$= 31.$$

14. $a_{15} = 4$, $S_{15} = 30$

Note that we are also given that $n = 15$, so we must find a_1 and d. Then

$$S_n = \frac{n}{2}[a_1 + a_n]$$
$$S_{15} = \frac{15}{2}[a_1 + a_{15}]$$
$$30 = \frac{15}{2}[a_1 + 4]$$
$$60 = 15[a_1 + 4]$$
$$60 = 15a_1 + 60$$
$$0 = 15a_1$$
$$0 = a_1$$

and

$$a_n = a_1 + (n-1)d$$
$$a_{15} = a_1 + (15-1)d$$
$$4 = 0 + (14)d$$
$$4 = 14d$$
$$\frac{4}{14} = d$$
$$\frac{2}{7} = d.$$

15. $a_1 = -9$, $a_7 = 21$

We are also given that $n = 7$, so we must find d and $S_n = S_7$. Then

$$a_7 = a_1 + (7-1)d$$
$$21 = -9 + (6)d$$
$$30 = 6d$$
$$5 = d$$

and

$$S_7 = \frac{7}{2}[a_1 + a_7]$$
$$= \frac{7}{2}[-9 + 21]$$
$$= \frac{7}{2}[12]$$
$$= 7(6)$$
$$= 42.$$

16. $a_1 = \log 7$, $d = \log 49$, $n = 5$

We must find $a_n = a_5$ and $S_n = S_5$. Notice that

$$d = \log 49 = \log 7^2 = 2\log 7.$$

Then

$$a_5 = a_1 + (5-1)d$$
$$= \log 7 + (4)(2\log 7)$$
$$= \log 7 + 8\log 7$$
$$= 9\log 7$$

and

$$S_5 = \frac{5}{2}[a_1 + a_5]$$
$$= \frac{5}{2}[\log 7 + 9\log 7]$$
$$= \frac{5}{2}[10\log 7]$$
$$= 25\log 7.$$

17. Insert six arithmetic means between 11 and 32.

To insert six arithmetic means between 11 and 32, we are really considering an eight term arithmetic sequence with $a_1 = 11$ and $a_8 = 32$. We must find the remaining six terms. First find d.

$$a_8 = a_1 + (8-1)d$$
$$32 = 11 + 7d$$
$$21 = 7d$$
$$3 = d$$

Then the six arithmetic means are:

$$a_2 = a_1 + d = 11 + 3 = 14$$
$$a_3 = a_2 + d = 14 + 3 = 17$$
$$a_4 = a_3 + d = 17 + 3 = 20$$
$$a_5 = a_4 + d = 20 + 3 = 23$$
$$a_6 = a_5 + d = 23 + 3 = 26$$
$$a_7 = a_6 + d = 26 + 3 = 29$$

18. How many integers between 39 and 146 are divisible by 5?

Since 40 is the first integer between 39 and 146 that is divisible by 5 and 145 is the last integer divisible by 5, we could solve this problem by considering the number of terms in the arithmetic sequence with $a_1 = 40$, $a_n = 145$, and $d = 5$.

$$a_n = a_1 + (n-1)d$$
$$145 = 40 + (n-1)5$$
$$105 = 5n - 5$$
$$110 = 5n$$
$$22 = n$$

Thus, there are 22 terms between 39 and 146 that are divisible by 5.

19. Find the sum of all the even integers between 1 and 201.

We can solve this problem by first finding the number of even integers between 1 and 201. The first such integer is 2 and the last is 200. Let $a_1 = 2$, $a_n = 200$, $d = 2$, and find n.

$$a_n = a_1 + (n-1)d$$
$$200 = 2 + (n-1)2$$
$$198 = 2n - 2$$
$$200 = 2n$$
$$100 = n$$

Then find $S_n = S_{100}$.

EXERCISES A

$$S_{100} = \frac{100}{2}[a_1 + a_n]$$
$$= \frac{100}{2}[2 + 200]$$
$$= \frac{100}{2}[202]$$
$$= 100(101)$$
$$= 10{,}100$$

Thus, the sum of the even integers between 1 and 201 is 10,100.

20. Prove that the sum of the sequence

$$2, 4, 6, \ldots, 2n$$

is $n^2 + n$.

This is an arithmetic sequence with $a_1 = 2$, $d = 2$, $n = n$, and $a_n = 2n$. Then

$$S_n = \frac{n}{2}[a_1 + a_n]$$
$$= \frac{n}{2}[2 + 2n]$$
$$= \frac{n}{2}[2(1 + n)]$$
$$= n(1 + n)$$
$$= n + n^2$$
$$= n^2 + n.$$

21. The depreciations form an arithmetic sequence

$$2.1\%, 1.8\%, 1.5\%, \ldots$$

or

$$0.021, 0.018, 0.015, \ldots$$

With first term 0.021 and common difference $d = -0.003$. First find the sum of the first twelve terms of the sequence.

$$S_{12} = \frac{12}{2}[2a_1 + (12-1)d]$$
$$= 6[2(0.021) + 11(-0.003)]$$
$$= 6[0.042 - 0.033]$$
$$= 6[0.009]$$
$$= 0.054$$

Then the total of the depreciations is $0.054 = 5.4\%$. Since the car cost \$8400, the amount of depreciation is 5.4% of \$8400 which is

$$0.054(8400) = 453.60.$$

Then the value of the car is

$$8400 - 453.6 = 7946.40,$$

which to the nearest dollar is \$7946.

22. The number of seats can be found by finding the sum of the first 40 terms ($n = 40$) of the arithmetic sequence with $a_1 = 20$ and $d = 3$.

$$S_{40} = \frac{40}{2}[2a_1 + (40-1)d]$$
$$= 20[2(20) + (39)3]$$
$$= 20[40 + 117]$$
$$= 20[157]$$
$$= 3140$$

Thus, the theater has 3140 seats.

23. We must find the sum of the first 30 terms (September has 30 days) of the arithmetic sequence with $a_1 = 1$, $a_{30} = 30$, and $d = 1$.

$$S_{30} = \frac{30}{2}[a_1 + a_{30}]$$
$$= 15[1 + 30]$$
$$= 15[31]$$
$$= 465$$

Thus, she will have \$465 in the account at the end of September.

24. During the first second, the rock falls $a_1 = 16$ ft. During the second second, the rock falls $a_2 = 48$ ft. During the third second, the rock falls $a_3 = 80$ ft. This is an arithmetic sequence with $d = 32$. We must find a_8, the distance the rock falls during the eighth second.

$$a_8 = a_1 + (8-1)d$$
$$= 16 + (7)32$$
$$= 16 + 224$$
$$= 240$$

Thus, the rock falls 240 ft during the eighth second.

In Exercises 25-26 determine the fourth and fifth terms of each general sequence.

25. $x_n = (-1)^{n+1} 3^{-n}$

$$x_4 = (-1)^{4+1} 3^{-4} = (-1)\frac{1}{3^4} = -\frac{1}{81}$$
$$x_5 = (-1)^{5+1} 3^{-5} = (1)\frac{1}{3^5} = \frac{1}{243}$$

26. $x_1 = 6$ and $x_{k+1} = 1 - x_k$

Since this is a recursive definition of the sequence, to find the fourth and fifth terms, we must find the preceding terms.

$$x_1 = 6$$
$$x_2 = 1 - x_1 = 1 - 6 = -5$$
$$x_3 = 1 - x_2 = 1 - (-5) = 6$$
$$x_4 = 1 - x_3 = 1 - 6 = -5$$
$$x_5 = 1 - x_4 = 1 - (-5) = 6$$

27. Write

$$\frac{1}{2} + \frac{2}{5} + \frac{3}{10} + \ldots + \frac{n}{n^2+1}$$

using sigma summation notation.

$$\frac{1}{2} + \frac{2}{5} + \frac{3}{10} + \ldots + \frac{n}{n^2+1} = \sum_{k=1}^{n} \frac{k}{k^2+1}$$

28. Evaluate. $\sum_{k=2}^{4} \frac{1}{3k-2}$

$$\sum_{k=2}^{4} \frac{1}{3k-2} = \frac{1}{3(2)-2} + \frac{1}{3(3)-2} + \frac{1}{3(4)-2}$$
$$= \frac{1}{4} + \frac{1}{7} + \frac{1}{10}$$
$$= \frac{35}{140} + \frac{20}{140} + \frac{14}{140}$$
$$= \frac{69}{140}$$

PRACTICE EXERCISES   SECTION 9.3   443

CHAPTER 9 SEQUENCES, SERIES, AND PROBABILITY

SECTION 9.3 Geometric Sequences

1. Find the fifth and ninth terms of the sequence

   $$1, -3, 9, -27, \ldots .$$

   We have $a_1 = 1$ and $r = -3$.

   $$\begin{aligned} a_5 &= a_1 r^{5-1} \\ &= (1)(-3)^4 \\ &= (1)(81) \\ &= 81 \end{aligned}$$

   $$\begin{aligned} a_9 &= a_1 r^{9-1} \\ &= (1)(-3)^8 \\ &= (1)(6561) \\ &= 6561 \end{aligned}$$

2. Find y and the sequence if $3y + 2$, $y + 2$, and y form a three-term geometric sequence in the given order.

   Use the fact that the ratio of successive terms is equal to the common ratio r, hence are equal.

   $$\frac{y+2}{3y+2} = \frac{y}{y+2}$$
   $$(y+2)^2 = y(3y+2)$$
   $$y^2 + 4y + 4 = 3y^2 + 2y$$
   $$0 = 2y^2 - 2y - 4$$
   $$0 = y^2 - y - 2$$
   $$0 = (y-2)(y+1)$$

   Using the zero-product rule we obtain:

   $$\begin{array}{ll} y - 2 = 0 & y + 1 = 0 \\ y = 2 & y = -1 \end{array}$$

   When $y = 2$:

   $$\begin{aligned} 3y + 2 &= 3(2) + 2 = 6 + 2 = 8 \\ y + 2 &= 2 + 2 = 4 \\ y &= 2 \end{aligned}$$

   Thus, one sequence is 8, 4, 2.

   When $y = -1$:

   $$\begin{aligned} 3y + 2 &= 3(-1) + 2 = -3 + 2 = -1 \\ y + 2 &= -1 + 2 = 1 \\ y &= -1 \end{aligned}$$

   Thus, a second sequence is $-1, 1, -1$.

3. Find a_{11} and S_{11} for the sequence

   $$3, -6, 12, -24, \ldots .$$

   We have $a_1 = 3$ and $r = -2$.

   $$\begin{aligned} a_{11} &= a_1 r^{11-1} \\ &= (3)(-2)^{10} \\ &= (3)(1024) \\ &= 3072 \end{aligned}$$

   $$\begin{aligned} S_{11} &= \frac{a_1 - a_1 r^{11}}{1 - r} \\ &= \frac{3 - (3)(-2)^{11}}{1 - (-2)} \\ &= \frac{3 - (3)(-2048)}{1 + 2} \\ &= \frac{3 + 6144}{3} \\ &= \frac{6147}{3} \\ &= 2049 \end{aligned}$$

4. If in a geometric sequence $r = 4$ and $S_4 = 425$, find a_1, a_6 and S_6.

   First find a_1 using the sum formula and the value of S_4.

   $$S_4 = \frac{a_1 - a_1 r^4}{1 - r}$$
   $$425 = \frac{a_1 - a_1(4)^4}{1 - 4}$$
   $$425 = \frac{a_1 - a_1(256)}{-3}$$
   $$-1275 = a_1 - 256a_1$$
   $$-1275 = -255a_1$$
   $$5 = a_1$$

   Now find a_6.

   $$\begin{aligned} a_6 &= a_1 r^{6-1} \\ &= (5)(4)^5 \\ &= (5)(1024) \\ &= 5120 \end{aligned}$$

   Use the value of a_6 in the second formula for S_6.

$$S_6 = \frac{a_1 - ra_6}{1-r}$$
$$= \frac{5 - (4)(5120)}{1-4}$$
$$= \frac{5 - 20{,}480}{-3}$$
$$= \frac{-20{,}475}{-3}$$
$$= 6825$$

5. One way to solve this problem is to notice that over the period of three years, the original $1000 will be worth

$$1000(1.08)^3.$$

The $1000 that was put in the second year, for the period of two years, will be worth

$$1000(1.08)^2.$$

Finally, the $1000 that was put in the third year, for the period of one year, will be worth

$$1000(1.08)^1.$$

Adding these together, we obtain the total value of the account.

$$1000(1.08)^1 + 1000(1.08)^2 + 1000(1.08)^3$$

Factor out the common factor 1000.

$$1000[(1.08)^1 + (1.08)^2 + (1.08)^3]$$

Inside the brackets we have a geometric sequence with $a_1 = 1.08$, $r = 1.08$, and $n = 3$. The sum of this sequence is:

$$S_3 = \frac{1.08 - (1.08)(1.08)^3}{1 - 1.08}$$
$$= 3.506112 \quad \textit{Using } y^x \textit{ button}$$

Multiply this number by 1000 to obtain the value of the account, $3506.11, correct to the nearest cent.

EXERCISES A   SECTION 9.3   445

CHAPTER 9 SEQUENCES, SERIES, AND PROBABILITY

SECTION 9.3 Geometric Sequences

Find the first six terms of each geometric sequence in Exercises 1-3.

1. $a_1 = 4$ and $r = 2$

    $a_1 = 4$
    $a_2 = a_1 r^{2-1} = (4)(2)^1 = (4)(2) = 8$
    $a_3 = a_1 r^{3-1} = (4)(2)^2 = (4)(4) = 16$
    $a_4 = a_1 r^{4-1} = (4)(2)^3 = (4)(8) = 32$
    $a_5 = a_1 r^{5-1} = (4)(2)^4 = (4)(16) = 64$
    $a_6 = a_1 r^{6-1} = (4)(2)^5 = (4)(32) = 128$

2. $a_1 = -16$ and $r = -\frac{1}{2}$

    $a_1 = -16$
    $a_2 = a_1 r^{2-1} = (-16)\left(-\frac{1}{2}\right)^1 = (-16)\left(-\frac{1}{2}\right) = 8$
    $a_3 = a_1 r^{3-1} = (-16)\left(-\frac{1}{2}\right)^2 = (-16)\left(\frac{1}{4}\right) = -4$
    $a_4 = a_1 r^{4-1} = (-16)\left(-\frac{1}{2}\right)^3 = (-16)\left(-\frac{1}{8}\right) = 2$
    $a_5 = a_1 r^{5-1} = (-16)\left(-\frac{1}{2}\right)^4 = (-16)\left(\frac{1}{16}\right) = -1$
    $a_6 = a_1 r^{6-1} = (-16)\left(-\frac{1}{2}\right)^5 = (-16)\left(-\frac{1}{32}\right) = \frac{1}{2}$

3. $a_1 = \sqrt{2}$ and $r = -\sqrt{2}$

    $a_1 = \sqrt{2}$
    $a_2 = a_1 r^{2-1} = (\sqrt{2})(-\sqrt{2})^1 = -(\sqrt{2})(\sqrt{2}) = -2$
    $a_3 = a_1 r^{3-1} = (\sqrt{2})(-\sqrt{2})^2 = (\sqrt{2})(2) = 2\sqrt{2}$
    $a_4 = a_1 r^{4-1} = (\sqrt{2})(-\sqrt{2})^3 = (\sqrt{2})(-2\sqrt{2}) = -(2)(2) = -4$
    $a_5 = a_1 r^{5-1} = (\sqrt{2})(-\sqrt{2})^4 = (\sqrt{2})(4) = 4\sqrt{2}$
    $a_6 = a_1 r^{6-1} = (\sqrt{2})(-\sqrt{2})^5 = (\sqrt{2})(-4\sqrt{2}) = -(4)(2) = -8$

4. Find x so that $x + 7$, $x - 3$, and $x - 8$ form a three-term geometric sequence in the given order. Determine the sequence.

    Use the fact that in a geometric sequence, the ratio of successive terms is constant. That is

    $$\frac{x-3}{x+7} = r \quad \text{and} \quad \frac{x-8}{x-3} = r.$$

    Set these two ratios equal and solve for x.

    $$\frac{x-3}{x+7} = \frac{x-8}{x-3}$$
    $(x-3)(x-3) = (x-8)(x+7)$
    $x^2 - 6x + 9 = x^2 - x - 56$
    $-6x + 9 = -x - 56$
    $-5x = -65$
    $x = 13$

    Then $x + 7 = 13 + 7 = 20$, $x - 3 = 13 - 3 = 10$, and $x - 8 = 13 - 8 = 5$, so the desired sequence is

    $$20, 10, 5.$$

In Exercises 5-10 find the indicated sum of each geometric sequence.

5. $\frac{1}{6}, \frac{1}{12}, \frac{1}{24}, \ldots ; S_9$

    We have that $a_1 = \frac{1}{6}$ and $r = \frac{1}{2}$. Then

    $$S_n = \frac{a_1 - a_1 r^n}{1 - r}$$

    $$S_9 = \frac{\frac{1}{6} - \left(\frac{1}{6}\right)\left(\frac{1}{2}\right)^9}{1 - \frac{1}{2}}$$

    $$= \frac{\frac{1}{6}\left[1 - \frac{1}{512}\right]}{\frac{1}{2}}$$

    $$= \left(\frac{1}{6}\right)\left(\frac{511}{512}\right)\left(\frac{2}{1}\right)$$

    $$= \frac{511}{1536}$$

6. $4, 24, 144, \ldots ; S_6$

    We have that $a_1 = 4$ and $r = 6$. Then

    $$S_6 = \frac{a_1 - a_1 r^6}{1 - r}$$

    $$= \frac{4 - 4(6)^6}{1 - 6}$$

    $$= \frac{4(1 - 46{,}656)}{-5}$$

    $$= \frac{4(-46{,}655)}{-5}$$

    $$= 37{,}324$$

7. $a_3 = \frac{1}{6}$ and $a_5 = \frac{1}{24}$; S_5

First we must find r.

$$\frac{1}{6} = a_3 = a_1 r^2$$
$$\frac{1}{24} = a_5 = a_1 r^4$$

Then we can divide the bottom equation by the top equation to obtain:

$$\frac{\frac{1}{24}}{\frac{1}{6}} = \frac{a_1 r^4}{a_1 r^2}$$
$$\frac{6}{24} = r^2$$
$$\frac{1}{4} = r^2$$
$$\pm \frac{1}{2} = r$$

When $r = \frac{1}{2}$:

$$a_3 = \frac{1}{6} = a_1 r^2 = a_1 \left(\frac{1}{2}\right)^2$$
$$\frac{1}{6} = a_1 \left(\frac{1}{4}\right)$$
$$\frac{4}{6} = a_1$$
$$\frac{2}{3} = a_1$$

$$S_5 = \frac{a_1 - ra_5}{1 - r}$$
$$= \frac{\frac{2}{3} - \left(\frac{1}{2}\right)\left(\frac{1}{24}\right)}{1 - \frac{1}{2}}$$
$$= \frac{\frac{2}{3} - \frac{1}{48}}{\frac{1}{2}}$$
$$= \frac{\frac{32}{48} - \frac{1}{48}}{\frac{1}{2}}$$
$$= \left(\frac{31}{48}\right)\left(\frac{2}{1}\right)$$
$$= \frac{31}{24}$$

When $r = -\frac{1}{2}$:

$$a_3 = \frac{1}{6} = a_1 r^2 = a_1 \left(-\frac{1}{2}\right)^2$$
$$\frac{1}{6} = a_1 \left(\frac{1}{4}\right)$$
$$\frac{4}{6} = a_1$$
$$\frac{2}{3} = a_1$$

$$S_5 = \frac{\frac{2}{3} - \left(-\frac{1}{2}\right)\left(\frac{1}{24}\right)}{1 - \left(-\frac{1}{2}\right)}$$
$$= \frac{\frac{2}{3} + \frac{1}{48}}{\frac{3}{2}}$$
$$= \frac{\frac{32}{48} + \frac{1}{48}}{\frac{3}{2}}$$
$$= \left(\frac{33}{48}\right)\left(\frac{2}{3}\right)$$
$$= \frac{11}{24}$$

8. $a_2 = 10$ and $a_4 = 40$; S_8

First find r.

$$40 = a_4 = a_1 r^3$$
$$10 = a_2 = a_1 r$$

Dividing we have

$$4 = r^2$$
$$\pm 2 = r$$

When $r = 2$:

$$10 = a_2 = a_1 r = a_1(2)$$
$$10 = 2a_1$$
$$5 = a_1$$

$$S_8 = \frac{a_1 - a_1 r^8}{1 - r}$$
$$= \frac{5 - 5(2)^8}{1 - 2}$$
$$= \frac{5 - 1280}{-1}$$
$$= \frac{-1275}{-1}$$
$$= 1275$$

When $r = -2$:

$$10 = a_2 = a_1 r = a_1(-2)$$
$$10 = -2a_1$$
$$-5 = a_1$$

EXERCISES A

$$S_8 = \frac{a_1 - a_1 r^8}{1 - r}$$
$$= \frac{-5 - (-5)(-2)^8}{1 - (-2)}$$
$$= \frac{-5 + 1280}{3}$$
$$= \frac{1275}{3}$$
$$= 425$$

9. $\sum_{k=1}^{7} \left(\frac{5}{6}\right)^k = \frac{5}{6} + \left(\frac{5}{6}\right)^2 + \left(\frac{5}{6}\right)^3 + \ldots + \left(\frac{5}{6}\right)^7$

This is a geometric sequence with $a_1 = \frac{5}{6}$, $n = 7$, and $r = \frac{5}{6}$. We must find S_7.

$$S_7 = \frac{a_1 - a_1 r^7}{1 - r}$$
$$= \frac{\frac{5}{6} - \left(\frac{5}{6}\right)\left(\frac{5}{6}\right)^7}{1 - \frac{5}{6}}$$
$$= \frac{\frac{5}{6} - \left(\frac{5}{6}\right)^8}{\frac{1}{6}}$$
$$\approx 3.6046 \quad \text{Using a calculator}$$

10. $\sum_{n=2}^{5} 4^n = 4^2 + 4^3 + 4^4 + 4^5$

We have $a_1 = 4^2 = 16$, $r = 4$, and $n = 4$. We must find S_4.

$$S_4 = \frac{a_1 - a_1 r^4}{1 - r}$$
$$= \frac{16 - 16(4)^4}{1 - 4}$$
$$= \frac{-4080}{-3}$$
$$= 1360$$

In Exercises 11–13 some of the numbers n, a_1, a_n, r, and S_n are given. Find the missing ones.

11. $a_1 = 2$, $n = 6$, $r = 2$

We must find $a_n = a_6$ and $S_n = S_6$.

$a_6 = a_1 r^5 = (2)(2)^5 = 2^6 = 64.$

$$S_6 = \frac{a_1 - r a_6}{1 - r}$$
$$= \frac{2 - (2)(64)}{1 - 2}$$
$$= \frac{2 - 128}{-1}$$
$$= \frac{-126}{-1}$$
$$= 126$$

12. $r = \frac{1}{2}$, $a_9 = 1$

Then we are also given $n = 9$, so we must find a_1 and $S_n = S_9$.

$$a_9 = 1 = a_1 r^8 = a_1 \left(\frac{1}{2}\right)^8$$
$$1 = a_1 \left(\frac{1}{256}\right)$$
$$256 = a_1$$

$$S_9 = \frac{a_1 - r a_n}{1 - r}$$
$$= \frac{256 - \left(\frac{1}{2}\right)(1)}{1 - \frac{1}{2}}$$
$$= \frac{\frac{511}{2}}{\frac{1}{2}}$$
$$= 511$$

13. $a_7 = \frac{1}{5}$, $r = \frac{1}{5}$

Then we are also given $n = 7$, so we must find a_1 and $S_n = S_7$.

$$a_7 = \frac{1}{5} = a_1 r^6 = a_1 \left(\frac{1}{5}\right)^6$$
$$\frac{1}{5} = a_1 \left(\frac{1}{5^6}\right)$$
$$\frac{5^6}{5} = a_1$$
$$5^5 = a_1$$
$$3125 = a_1$$

$$S_7 = \frac{3125 - \left(\frac{1}{5}\right)\left(\frac{1}{5}\right)}{1 - \frac{1}{5}}$$
$$= \frac{\frac{78{,}124}{25}}{\frac{4}{5}} = \frac{78{,}124}{25} \cdot \frac{5}{4} = \frac{19{,}531}{5}$$

14. Insert four geometric means between 5 and -160.

We can solve this problem by thinking of a six-term geometric sequence with $a_1 = 5$ and $a_6 = -160$. The middle four terms are the desired means. First find r.

$$a_6 = a_1 r^5$$
$$-160 = 5r^5$$
$$-32 = r^5$$
$$-2 = r$$

Then the desired means are:

$$a_2 = a_1 r = (5)(-2) = -10$$
$$a_3 = a_2 r = (-10)(-2) = 20$$
$$a_4 = a_3 r = (20)(-2) = -40$$
$$a_5 = a_4 r = (-40)(-2) = 80$$

15. Find all possible values of r if $a_1 = 3$ and $S_3 = 9$.

$$S_3 = \frac{a_1 - a_1 r^3}{1 - r}$$
$$9 = \frac{3 - 3r^3}{1 - r}$$
$$9 - 9r = 3 - 3r^3$$
$$3r^3 - 9r + 6 = 0$$
$$r^3 - 3r + 2 = 0$$

This polynomial equation is of degree 3 so there are 3 roots. Using the rational root theorem, and synthetic division, we can show that 1 is one solution.

```
1 |  1 + 0 - 3 + 2
        + 1 + 1 - 2
     ─────────────────
     1 + 1 - 2 + 0
```

The remaining roots are solutions to the quadratic equation

$$r^2 + r - 2 = 0.$$
$$(r + 2)(r - 1) = 0$$
$$r + 2 = 0 \qquad r - 1 = 0$$
$$r = -2 \qquad r = 1$$

Then 1 is a double root, and another root is -2. Thus, the possible values for r are 1 and -2.

16. Let $a_1 = 6400$, the initial value of the car. Then since the car depreciates 20% of its value each year, at the end of 1 year, the car is worth

$$a_2 = 6400 - (0.20)(6400) = (0.80)(6400).$$

At the end of the second year, the car is worth

$$a_3 = (0.80)(6400) - (0.20)[(0.80)(6400)]$$
$$= (0.80)(0.80)(6400)$$
$$= (0.80)^2(6400)$$

Continuing in this manner, the value of the car at the end of 6 years is given by

$$a_7 = (0.80)^6(6400).$$

Notice that the term is the *seventh* term, not the sixth term of the sequence. Evaluating we have that the car is worth $1678, correct to the nearest dollar.

17. Let $a_1 = 2000$. Then at the end of 1 year, she would have to pay

$$a_2 = 2000 + (0.11)(2000) = (1.11)(2000).$$

At the end of 2 years she would have to pay

$$a_3 = (1.11)(2000) + (0.11)(1.11)(2000)$$
$$= (1.11)(1.11)(2000)$$
$$= (1.11)^2(2000).$$

Continuing in this way, we can see that at the end of four years she would have to pay

$$a_5 = (1.11)^4(2000).$$

Notice that the term is the *fifth term* not the fourth term as you might expect initially. The value of this is

$$\$3036.14,$$

found using the y^x button on a calculator.

18. In effect, we must find the sum of the first 30 terms of the series

$$1, 2, 4, 8, 16, 32, \ldots$$

for which $a_1 = 1$, $r = 2$, and $n = 30$.

$$S_{30} = \frac{a_1 - a_1 r^{30}}{1 - r}$$
$$= \frac{1 - (1)(2)^{30}}{1 - 2}$$
$$= \frac{1 - 2^{30}}{-1} = 1{,}073{,}741{,}823$$

Converting from cents back to dollars, the salary for the month is about $10,737,418!

EXERCISES A

SECTION 9.3 449

19. The tip of a pendulum sweeps out an arc of 20 cm on the first pass. Let $a_1 = 20$. On the second pass, it passes through an arc of $\left(\frac{4}{5}\right)20$ cm. On the third pass, it travels $\left(\frac{4}{5}\right)^2 20$ cm. And on the fourth pass, it travels $\left(\frac{4}{5}\right)^3 20$ cm. We must add these together.

Notice that $r = \frac{4}{5}$.

$$S_4 = \frac{a_1 - a_1 r^4}{1 - r}$$

$$= \frac{20 - 20\left(\frac{4}{5}\right)^4}{1 - \frac{4}{5}}$$

$$= \frac{20\left(1 - \frac{256}{625}\right)}{\frac{1}{5}}$$

$$= 100\left(\frac{369}{625}\right)$$

$$= \frac{1476}{25} \approx 59.0 \text{ cm}$$

20. On the first drop, the ball falls 12.0 ft. It then rebounds and falls $\frac{3}{4}(12.0)$ ft and hits the ground the second time. Then it rebounds and falls $\left(\frac{3}{4}\right)^2(12.0)$ ft and hits the ground the third time, and so forth. We must add up the following numbers:

$$12.0 + \qquad \text{Hits ground 1st time}$$
$$\left(\tfrac{3}{4}\right)(12.0) + \left(\tfrac{3}{4}\right)(12.0) + \qquad \text{Hits ground 2nd time}$$
$$\left(\tfrac{3}{4}\right)^2(12.0) + \left(\tfrac{3}{4}\right)^2(12.0) + \qquad \text{Hits ground 3rd time}$$
$$\vdots$$
$$\left(\tfrac{3}{4}\right)^7(12.0) + \left(\tfrac{3}{4}\right)^7(12.0) \qquad \text{Hits ground 8th time}$$

If we find the sum

$$\left(\tfrac{3}{4}\right)(12.0) + \left(\tfrac{3}{4}\right)^2(12.0) + \left(\tfrac{3}{4}\right)^3(12.0) + \ldots + \left(\tfrac{3}{4}\right)^7(12.0)$$

and double the result and add 12.0, we will obtain the total distance traveled. Considering the series above, $a_1 = \left(\frac{3}{4}\right)(12.0) = 9.0$, $n = 7$, and $r = \frac{3}{4}$.

Thus, the sum is:

$$S_7 = \frac{a_1 - a_1 r^7}{1 - r}$$

$$= \frac{9 - 9\left(\frac{3}{4}\right)^7}{1 - \frac{3}{4}}$$

$$\approx 31.19458008 \quad \textit{Using } y^x \textit{ button}$$

Multiply this number by 2 and add 12.0 to obtain

74.38916016.

Thus, the total distance traveled by the ball is about 74.4 ft.

21. One way to solve this problem is to notice that over the period of 10 years, the original $1000 will be worth

$$1000(1.08)^{10}.$$

The $1000 that was put in the second year, for a period of 9 years, will be worth

$$1000(1.08)^9.$$

The $1000 that was put in the third year, for a period of 8 years, will be worth

$$1000(1.08)^8,$$

and so forth, until we have the last year when the $1000 put in will be worth

$$1000(1.08)^1.$$

Adding these together we obtain the total value of the account, which can be expressed in the following form after factoring out the common factor 1000.

$$1000[(1.08)^1 + (1.08)^2 + (1.08)^3 + \ldots + (1.08)^{10}]$$

Inside the brackets we have a geometric sequence with $a_1 = 1.08$, $r = 1.08$, and $n = 10$. The sum of this sequence is:

$$S_{10} = \frac{1.08 - (1.08)(1.08)^{10}}{1 - 1.08}$$

$$\approx 15.64548746 \quad \textit{Using } y^x \textit{ button}$$

Multiply this number by 1000 to obtain the value of the account, $15,645.49, correct to the nearest cent.

22. Find the first five terms of the arithmetic sequence with $a_1 = 7$ and $d = -4$.

$a_1 = 7$
$a_2 = a_1 + d = 7 + (-4) = 3$
$a_3 = a_2 + d = 3 + (-4) = -1$
$a_4 = a_3 + d = -1 + (-4) = -5$
$a_5 = a_4 + d = -5 + (-4) = -9$

23. Find the sum of the first nine terms of the arithmetic sequence

$$-\tfrac{1}{2},\ 0,\ \tfrac{1}{2},\ 1,\ \tfrac{3}{2},\ \ldots.$$

We have $a_1 = -\tfrac{1}{2}$, $n = 9$, and $d = \tfrac{1}{2}$. Then

$$S_9 = \tfrac{9}{2}[2a_1 + (9-1)d]$$
$$= \tfrac{9}{2}\left[2\left(-\tfrac{1}{2}\right) + (8)\left(\tfrac{1}{2}\right)\right]$$
$$= \tfrac{9}{2}[-1 + 4]$$
$$= \tfrac{9}{2}[3]$$
$$= \tfrac{27}{2}$$

24. Insert four arithmetic means between 12 and -13.

Think of a six-term arithmetic sequence with $a_1 = 12$ and $a_6 = -13$. First find the value of d.

$a_6 = a_1 + (6-1)d$
$-13 = 12 + 5d$
$-25 = 5d$
$-5 = d$

Then the four arithmetic means are:

$a_2 = a_1 + d = 12 + (-5) = 7$
$a_3 = a_2 + d = 7 + (-5) = 2$
$a_4 = a_3 + d = 2 + (-5) = -3$
$a_5 = a_4 + d = -3 + (-5) = -8$

25. A theater has 50 rows with 15 seats in the first row, 17 in the second, 19 in the third, and so forth. To find the number of seats in the theater we must add the first 50 numbers in the sequence:

$$15 + 17 + 19 + 21 + \ldots$$

This is an arithmetic sequence with $a_1 = 15$, $d = 2$, and the 50 rows tells us that $n = 50$.

$$S_{50} = \tfrac{50}{2}[2a_1 + (50-1)d]$$
$$= 25[2(15) + (49)(2)]$$
$$= 25[30 + 98]$$
$$= 25[128]$$
$$= 3200$$

Thus, the theater has 3200 seats.

PRACTICE EXERCISES SECTION 9.4 451

CHAPTER 9 SEQUENCES, SERIES, AND PROBABILITY

SECTION 9.4 Infinite Geometric Sequences and Series

1. Find the first five terms of the infinite geometric sequence with $a_1 = 8$ and $S = \frac{2}{3}$.

   Substitute into the formula.
   $$S = \frac{a_1}{1-r}$$
   $$\frac{2}{3} = \frac{8}{1-r}$$
   $$2(1-r) = 24$$
   $$2 - 2r = 24$$
   $$-2r = 22$$
   $$r = -11$$

   Since $r = -11$, $|r| = |-11| = 11 > 1$, so there is no series with these properties.

2. Convert $5.\overline{63}$ to a fraction.

   First write $5.\overline{63}$ as follows:

   $$5.\overline{63} = 5 + 0.63 + 0.0063 + 0.000063 + \ldots$$

   Then consider the series

   $$0.63 + 0.0063 + 0.000063 + \ldots .$$

   This is an infinite geopetric series with $a_1 = 0.63$, and $r = 0.01$. Then the sum is:

   $$S = \frac{a_1}{1-r}$$
   $$= \frac{0.63}{1 - 0.01}$$
   $$= \frac{0.63}{0.99}$$
   $$= \frac{63}{99}$$
   $$= \frac{7}{11}$$

   Then add on the 5 to obtain

   $$5.\overline{63} = 5 + \frac{7}{11} = \frac{62}{11}.$$

3. The tip of a pendulum sweeps out an arc of 84 cm on the first pass. On each succeeding pass the distance traveled is $\frac{7}{9}$ of the preceding pass. Then the distance traveled by the tip is

   $$84 + \left(\frac{7}{9}\right)84 + \left(\frac{7}{9}\right)^2 84 + \left(\frac{7}{9}\right)^3 84 + \ldots$$

   This is an infinite geometric series with $a_1 = 84$ and $r = \frac{7}{9}$. Find the sum.

   $$S = \frac{a_1}{1-r}$$
   $$= \frac{84}{1 - \frac{7}{9}}$$
   $$= \frac{84}{\frac{2}{9}}$$
   $$= \frac{(9)(84)}{2}$$
   $$= 378$$

   Thus, the distance traveled by the tip of the pendulum is 378 cm.

CHAPTER 9 SEQUENCES, SERIES, AND PROBABILITY

SECTION 9.4 Infinite Geometric Sequences and Series

In Exercises 1-3 determine if each sequence has a sum.

1. 100, 10, 1, 0.1, ...

   Since $r = 0.1$, and $|r| < 1$, the sequence does have a sum.

2. $\frac{1}{64}, \frac{1}{16}, \frac{1}{4}, 1, \ldots$

   Since $r = 4$, and $|r| \geq 1$, the sequence does not have a sum.

3. 125, −75, 45, −27, ...

   Since $r = -\frac{3}{5}$, and $|r| < 1$, the sequence does have a sum.

In Exercises 4-9 find the sum of each infinite geometric sequence.

4. $\frac{1}{2}, \frac{1}{4}, \frac{1}{8}, \frac{1}{16}, \ldots$

   We have $a_1 = \frac{1}{2}$ and $r = \frac{1}{2}$. Thus the sum is:
   $$S = \frac{a_1}{1-r} = \frac{\frac{1}{2}}{1-\frac{1}{2}} = \frac{\frac{1}{2}}{\frac{1}{2}} = 1$$

5. $\frac{1}{16}, \frac{1}{8}, \frac{1}{4}, \frac{1}{2}, \ldots$

   Since $r = 2$ and $|r| \geq 1$, the sequence has no sum.

6. $-14, 8, -\frac{32}{7}, \frac{128}{49}, \ldots$

   We have $a_1 = -14$ and $r = -\frac{4}{7}$. Thus the sum is:
   $$S = \frac{a_1}{1-r} = \frac{-14}{1-\left(-\frac{4}{7}\right)} = \frac{-14}{\frac{11}{7}} = -\frac{98}{11}$$

7. 15, 1.5, 0.15, 0.015, ...

   We have $a_1 = 15$ and $r = 0.1$. Since $|r| < 1$, the sum is:
   $$S = \frac{a_1}{1-r} = \frac{15}{1-0.1} = \frac{15}{0.9} = \frac{150}{9} = \frac{50}{3}$$

8. −64, 48, −36, 27, ...

   We have $a_1 = -64$ and $r = -\frac{3}{4}$. Since $|r| < 1$, the sum is:
   $$S = \frac{a_1}{1-r} = \frac{-64}{1-\left(-\frac{3}{4}\right)} = \frac{-64}{\frac{7}{4}} = -\frac{256}{7}$$

9. $8, 4\sqrt{2}, 4, 2\sqrt{2}, \ldots$

   We have $a_1 = 8$ and $r = \frac{\sqrt{2}}{2}$. Since $|r| < 1$, the sum is:
   $$S = \frac{a_1}{1-r} = \frac{8}{1-\frac{\sqrt{2}}{2}}$$
   $$= \frac{8}{\frac{2-\sqrt{2}}{2}}$$
   $$= \frac{16}{2-\sqrt{2}}$$
   $$= \frac{16(2+\sqrt{2})}{(2-\sqrt{2})(2+\sqrt{2})}$$
   $$= \frac{16(2+\sqrt{2})}{4-2}$$
   $$= \frac{16(2+\sqrt{2})}{2}$$
   $$= 8(2+\sqrt{2})$$
   $$= 16 + 8\sqrt{2}$$

In Exercises 10-12 find the indicated sum.

10. $\sum_{k=1}^{\infty} \left(\frac{3}{4}\right)^k$

EXERCISES A

When $k = 1$, we obtain $a_1 = \frac{3}{4}$. The common ratio is $r = \frac{3}{4}$. Since $|r| < 1$, the sum is:

$$S = \frac{a_1}{1-r} = \frac{\frac{3}{4}}{1-\frac{3}{4}} = \frac{\frac{3}{4}}{\frac{1}{4}} = \frac{3}{4} \cdot \frac{4}{1} = 3$$

11. $\sum_{k=1}^{\infty} 4^{-k} = \sum_{k=1}^{\infty} \frac{1}{4^k} = \sum_{k=1}^{\infty} \left(\frac{1}{4}\right)^k$

When $k = 1$, we obtain $a_1 = \frac{1}{4}$. The common ratio is $r = \frac{1}{4}$. Since $|r| < 1$, the sum is:

$$S = \frac{a_1}{1-r} = \frac{\frac{1}{4}}{1-\frac{1}{4}} = \frac{\frac{1}{4}}{\frac{3}{4}} = \frac{1}{4} \cdot \frac{4}{3} = \frac{1}{3}$$

12. $\sum_{n=1}^{\infty} (0.2)^n$

When $n = 1$, we obtain $a_1 = 0.2$. The common ratio is $r = 0.2$. Since $|r| < 1$, the sum is:

$$S = \frac{a_1}{1-r} = \frac{0.2}{1-0.2} = \frac{0.2}{0.8} = \frac{2}{8} = \frac{1}{4}$$

In Exercises 13-15 find the first six terms of the geometric sequence.

13. $a_1 = 25$ and $S = 125$

We must find the common ratio r.

$$S = \frac{a_1}{1-r}$$
$$125 = \frac{25}{1-r}$$
$$125(1-r) = 25$$
$$125 - 125r = 25$$
$$-125r = -100$$
$$r = \frac{100}{125} = \frac{4}{5}$$

Then the first six terms of the sequence are:

$a_1 = 25$
$a_2 = a_1 r = 25\left(\frac{4}{5}\right) = 20$
$a_3 = a_2 r = 20\left(\frac{4}{5}\right) = 16$
$a_4 = a_3 r = 16\left(\frac{4}{5}\right) = \frac{64}{5}$
$a_5 = a_4 r = \left(\frac{64}{5}\right)\left(\frac{4}{5}\right) = \frac{256}{25}$
$a_6 = a_5 r = \left(\frac{256}{25}\right)\left(\frac{4}{5}\right) = \frac{1024}{125}$

14. $a_1 = 4$ and $S = -7$

First find the common ratio r.

$$S = \frac{a_1}{1-r}$$
$$-7 = \frac{4}{1-r}$$
$$-7(1-r) = 4$$
$$-7 + 7r = 4$$
$$7r = 11$$
$$r = \frac{11}{7}$$

Since $|r| \geq 1$, there is no sequence with the given properties.

15. $a_1 = -4$ and $S = -5$

First find the common ratio r.

$$S = \frac{a_1}{1-r}$$
$$-5 = \frac{-4}{1-r}$$
$$-5(1-r) = -4$$
$$-5 + 5r = -4$$
$$5r = 1$$
$$r = \frac{1}{5}$$

Then the first six terms of the sequence are:

$a_1 = -4$
$a_2 = a_1 r = (-4)\left(\frac{1}{5}\right) = -\frac{4}{5}$
$a_3 = a_2 r = \left(-\frac{4}{5}\right)\left(\frac{1}{5}\right) = -\frac{4}{25}$
$a_4 = a_3 r = \left(-\frac{4}{25}\right)\left(\frac{1}{5}\right) = -\frac{4}{125}$
$a_5 = a_4 r = \left(-\frac{4}{125}\right)\left(\frac{1}{5}\right) = -\frac{4}{625}$
$a_6 = a_5 r = \left(-\frac{4}{625}\right)\left(\frac{1}{5}\right) = -\frac{4}{3125}$

In Exercises 16-19 convert each decimal to a fraction.

16. $0.\overline{3} = 0.3 + 0.03 + 0.003 + 0.0003 + \ldots$

    Then $a_1 = 0.3$ and $r = 0.1$.

    $$S = \frac{a_1}{1-r} = \frac{0.3}{1-0.1} = \frac{0.3}{0.9} = \frac{3}{9} = \frac{1}{3}$$

    Thus, $0.\overline{3} = \frac{1}{3}$.

17. $0.\overline{21} = 0.21 + 0.0021 + 0.000021 + \ldots$

    Then $a_1 = 0.21$ and $r = 0.01$.

    $$S = \frac{a_1}{1-r} = \frac{0.21}{1-0.01} = \frac{0.21}{0.99} = \frac{21}{99} = \frac{7}{33}$$

    Thus, $0.\overline{21} = \frac{7}{33}$.

18. $0.\overline{123} = 0.123 + 0.000123 + 0.000000123 + \ldots$

    Then $a_1 = 0.123$ and $r = 0.001$.

    $$S = \frac{a_1}{1-r} = \frac{0.123}{1-0.001} = \frac{0.123}{0.999} = \frac{123}{999} = \frac{41}{333}$$

    Thus, $0.\overline{123} = \frac{41}{333}$.

19. $2.1\overline{5} = 2.1 + 0.05 + 0.005 + 0.0005 + \ldots$

    We ignore the first term, 2.1, and concentrate on the geometric sequence that follows. When we have found the value of the sequence, we will add on the 2.1 to obtain the desired fraction. Then $a_1 = 0.05$ and $r = 0.1$.

    $$S = \frac{a_1}{1-r} = \frac{0.05}{1-0.1} = \frac{0.05}{0.9} = \frac{5}{90} = \frac{1}{18}$$

    Then

    $$\begin{aligned} 2.1\overline{5} &= 2.1 + \frac{1}{18} \\ &= \frac{21}{10} + \frac{1}{18} \\ &= \frac{189}{90} + \frac{5}{90} \\ &= \frac{194}{90} \\ &= \frac{97}{45} \end{aligned}$$

20. Prove that $0.999\ldots = 1$.

    Since

    $$0.999\ldots = 0.9 + 0.09 + 0.009 + 0.0009 + \ldots,$$

    This is a geometric sequence with $a_1 = 0.9$ and $r = 0.1$. The sum of this sequence is:

    $$S = \frac{a_1}{1-r} = \frac{0.9}{1-0.1} = \frac{0.9}{0.9} = 1$$

    Thus, $0.999\ldots = 1$.

21. We use an infinite geometric sequence to solve this problem even though in reality, friction will cause the swing to stop eventually. Let $a_1 = 22$, corresponding to the first pass. Then on the second pass, the child travels a distance of $a_2 = \left(\frac{5}{7}\right)(22)$. On the third pass, the distance traveled is $a_3 = \left(\frac{5}{7}\right)^2(22)$. Continuing in this way, we obtain a geometric sequence with $r = \frac{5}{7}$, the sum of which is:

    $$S = \frac{a_1}{1-r} = \frac{22}{1-\frac{5}{7}} = \frac{22}{\frac{2}{7}} = 22 \cdot \frac{7}{2} = 77$$

    Thus, the child will travel about 77 ft before coming to rest.

22. Since the ball initially falls 40 ft, and then rebounds and falls half this distance, then rebounds and falls half of the previous distance, and so forth, to find the total distance traveled up and down, we must add the following numbers:

    $$40 + 20 + 20 + 10 + 10 + 5 + 5 + \ldots$$

    This sum can also be written as:

    $$40 + 2[20 + 10 + 5 + \ldots]$$

    Then inside the brackets we have a geometric sequence with $a_1 = 20$ and $r = \frac{1}{2}$, the sum of which is:

    $$S = \frac{a_1}{1-r} = \frac{20}{1-\frac{1}{2}} = \frac{20}{\frac{1}{2}} = 40$$

EXERCISES A   SECTION 9.4   455

Then the total distance traveled by the ball is:

$$40 + 2[40] \quad \text{Substitute 40 inside brackets}$$

Thus, the ball traveled $40 + 2[40] = 120$ ft.

If we were to find the total distance traveled when the ball hits the ground for the hundredth time, the calculation using the formula for the sum of the first 100 terms of a geometric sequence would be more difficult to use than the infinite sum formula, and the two results would not differ by much. As a result we often use the easier formula for infinite sums to approximate finite sums with a large number of terms.

23. We can approximate the total amount of money received using the infinite geometric sequence:

$$5000 + \left(\frac{3}{5}\right)5000 + \left(\frac{3}{5}\right)^2 5000 + \left(\frac{3}{5}\right)^3 5000 + \ldots$$

Then $a_1 = 5000$ and $r = \frac{3}{5}$. The sum is:

$$S = \frac{a_1}{1-r} = \frac{5000}{1-\frac{3}{5}} = \frac{5000}{\frac{2}{5}} = 12{,}500$$

Thus, he will receive about $12,500 in his lifetime.

24. Refer to the figure in the text. The largest outside square has area 64 in^2, so the sides of the square must be 8 inches. To find the sides of the next square, note that they are the hypotenuse of a right triangle with legs 4 inches. Call the side a and use the Pythagorean theorem.

$$a^2 = 4^2 + 4^2$$
$$a^2 = 2 \cdot 4^2$$
$$a = 4\sqrt{2}$$

Note that we only use the positive root. Then the area of this square is 32 in^2.

The next smaller square has sides that are the hypotenuse of a triangle with legs of length

$$b = \frac{a}{2} = \frac{4\sqrt{2}}{2} = 2\sqrt{2}.$$

Using the Pythagorean theorem again,

$$b^2 = (2\sqrt{2})^2 + (2\sqrt{2})^2$$
$$b^2 = 16$$
$$b = 4$$

Then the area of this square is 16 in^2. Continuing in this manner, we can see that the sum of the areas of the squares is given by

$$64 + 32 + 16 + 8 + 4 + \ldots.$$

This is a geometric series with $a_1 = 64$ and $r = \frac{1}{2}$. The sum is:

$$S = \frac{a_1}{1-r} = \frac{64}{1-\frac{1}{2}} = \frac{64}{\frac{1}{2}} = 128$$

Thus, the sum of the areas of the squares is 128 in^2.

25. We have that

$$a_1 = 1{,}000{,}000$$

represents the salaries paid, and thus the amount spent in the country initially. Then 80% of this amount,

$$a_2 = (0.80)(1{,}000{,}000)$$

will be spent in the country. Then 80% of this amount will again be spent in the country giving

$$a_3 = (0.80)^2(1{,}000{,}000).$$

Then since 80% of this amount will again be spent in the country, we have

$$a_4 = (0.80)^3(1{,}000{,}000).$$

This process continues, and the sum of these terms gives the total amount of spending generated by any one year of company salaries. This is a geometric sequence with $a_1 = 1{,}000{,}000$ and $r = 0.80$. The sum is:

$$S = \frac{a_1}{1-r} = \frac{1{,}000{,}000}{1-0.80} = \frac{1{,}000{,}000}{0.20} = 5{,}000{,}000$$

Thus, this economic multiplier effect results in a total of $5,000,000 being spent in the country by virtue of the original $1,000,000 payroll.

26. Find the first four terms of the geometric sequence with $a_1 = -5$ and $r = 0.08$.

$$a_1 = -5$$
$$a_2 = a_1 r = (-5)(0.08) = -0.4$$
$$a_3 = a_2 r = (-0.4)(0.08) = -0.032$$
$$a_4 = a_3 r = (-0.032)(0.08) = -0.00256$$

27. Find a_6 for the sequence

$$-54, 36, -24, 16, \ldots$$

Since the ratio of successive terms is

$$r = -\tfrac{2}{3},$$

the next term (the fifth) would be

$$\left(-\tfrac{2}{3}\right)(16) = -\tfrac{32}{3},$$

and the next term (the sixth) would be

$$\left(-\tfrac{2}{3}\right)\left(-\tfrac{32}{3}\right) = \tfrac{64}{9}.$$

28. Find S_6 for the sequence

$$-54, 36, -24, 16, \ldots$$

In Exercise 27, we found the common ratio r and the sixth term a_6. Use these in the sum formula to find S_6.

$$S_6 = \frac{a_1 - ra_6}{1 - r}$$

$$= \frac{-54 - \left(-\tfrac{2}{3}\right)\left(\tfrac{64}{9}\right)}{1 - \left(-\tfrac{2}{3}\right)}$$

$$= \frac{-54 + \tfrac{128}{27}}{\tfrac{5}{3}}$$

$$= \frac{-\tfrac{1330}{27}}{\tfrac{5}{3}}$$

$$= -\frac{1330}{27} \cdot \frac{3}{5} = -\frac{266}{9}$$

29. If $a_2 = -5$ and $a_5 = 0.04$, find r and S_6.

$$-5 = a_2 = a_1 r$$
$$0.04 = a_5 = a_1 r^4$$

Dividing the bottom equation by the top equation we obtain:

$$-0.008 = r^3$$
$$-0.2 = r$$

Then substitute this value for r in the expression for a_2 to find a_1, $a_1 = 25$. Then find S_6.

$$S_6 = \frac{a_1 - a_1 r^6}{1 - r}$$

$$= \frac{25 - 25(-0.2)^6}{1 - (-0.2)}$$

$$= 20.832 \quad \text{Using a calculator}$$

30. Insert three geometric means between 250 and 0.025.

Then $a_1 = 250$ and $a_5 = 0.025$, and we must find $a_2, a_3,$ and a_4. First find r.

$$a_5 = a_1 r^4$$
$$0.025 = 250 r^4$$
$$0.0001 = r^4$$
$$\pm 0.1 = r$$

When $r = 0.1$:

$$a_2 = a_1 r = (250)(0.1) = 25$$
$$a_3 = a_2 r = (25)(0.1) = 2.5$$
$$a_4 = a_3 r = (2.5)(0.1) = 0.25$$

When $r = -0.1$:

$$a_2 = a_1 r = (250)(-0.1) = -25$$
$$a_3 = a_2 r = (-25)(-0.1) = 2.5$$
$$a_4 = a_3 r = (2.5)(-0.1) = -0.25$$

PRACTICE EXERCISES — SECTION 9.5

CHAPTER 9 SEQUENCES, SERIES, AND PROBABILITY

SECTION 9.5 Mathematical Induction

1. Prove that
$$1+4+7+\ldots+(3n-2) = \frac{n(3n-1)}{2}.$$

   Verification: Clearly this is true when $n = 1$, since in this case we have
$$1 = \frac{(1)(3(1)-1)}{2} = \frac{3-1}{2} = \frac{2}{2} = 1$$

   Induction: Assume that the statement is true when $n = k$.
$$1+4+7+\ldots+(3k-2) = \frac{k(3k-1)}{2}$$

   We must show that the statement is true when $n = k + 1$. That is, we must show that
$$1+4+7+\ldots+(3k-2)+[3(k+1)-2] = \frac{(k+1)(3(k+1)-1)}{2}$$

   is true. The left side of this equation can be written as
$$[1 + 4 + 7 +\ldots+ (3k-1)] + [3(k+1)-2].$$

   Substitute $\frac{k(3k-1)}{2}$ for
$$[1 + 4 + 7 +\ldots+(3k-1)].$$

$$1+4+7+\ldots+(3k-2)+[3(k+1)-2]$$
$$= \frac{k(3k-1)}{2} + 3(k+1)-2$$
$$= \frac{3k^2-k}{2} + 3k + 3 - 2$$
$$= \frac{3k^2-k}{2} + 3k + 1$$
$$= \frac{3k^2-k}{2} + \frac{2(3k+1)}{2}$$
$$= \frac{3k^2-k+6k+2}{2}$$
$$= \frac{3k^2+5k+2}{2}$$
$$= \frac{(k+1)(3k+2)}{2}$$
$$= \frac{(k+1)(3(k+1)-1)}{2}$$

   Then using the induction assumption, we have shown that the statement is true for $n = k + 1$ under the assumption it is true for $n = k$.

   Conclusion: Since the statement is true for $n = 1$ (by verification), and it is true for $n = k + 1$ when it is true for $n = k$, we know by the principle of mathematical induction that the statement is true for every positive integer n.

2. Prove that
$$3+3^2+3^3+\ldots+3^n = \frac{3}{2}(3^n - 1).$$

   Verification: When $n = 1$, the left side is 3 (there is only one term on the left) and the right side is
$$\frac{3}{2}(3^1 - 1) = \frac{3}{2}(2) = 3.$$

   Thus, the statement is true when $n = 1$.

   Induction: Assume the statement is true for $n = k$. That is, assume that
$$3+3^2+3^3+\ldots+3^k = \frac{3}{2}(3^k - 1)$$

   is true. Then we must show using this that the statement is true when $n = k + 1$. That is, we must show that
$$3+3^2+3^3+\ldots+3^k+3^{k+1} = \frac{3}{2}(3^{k+1} - 1)$$

   is true. Substitute $\frac{3}{2}(3^k - 1)$ for all the terms on the left side up to the last one. Then the left side becomes:

$$\frac{3}{2}(3^k - 1) + 3^{k+1} = \frac{3 \cdot 3^k}{2} - \frac{3}{2} + 3^{k+1}$$
$$= \frac{3^{k+1}}{2} - \frac{3}{2} + \frac{2 \cdot 3^{k+1}}{2}$$
$$= \frac{1}{2}(3^{k+1} - 3 + 2 \cdot 3^{k+1})$$
$$= \frac{1}{2}(3 \cdot 3^{k+1} - 3)$$
$$= \frac{3}{2}(3^{k+1} - 1)$$

   Thus, we have shown that the statement is true for $n = k + 1$ under the assumption that it is true for $n = k$.

Conclusion: By the principle of mathematical induction, the statement is true for every positive integer n.

3. Prove that $(ab)^n = a^n b^n$.

   Verification: When $n = 1$, the left side is $(ab)^1 = ab$, and the right side is $a^1 b^1 = ab$. Thus the statement is true for $n = 1$.

   Induction: Assume the statement is true for $n = k$, that is that

   $$(ab)^k = a^k b^k$$

   is true. Under this assumption we must show that

   $$(ab)^{k+1} = a^{k+1} b^{k+1}$$

   is true. Start with the left side.

   $$\begin{aligned}(ab)^{k+1} &= (ab)^k (ab) \\ &= a^k b^k (ab) \quad \text{Induction assumption} \\ &= a^k a b^k b \quad \text{Commutative law} \\ &= a^{k+1} b^{k+1}\end{aligned}$$

   Thus, the statement is true for $n = k + 1$ under the assumption it is true for $n = k$.

   Conclusion: By the principle of mathematical induction, the statement is true for every positive integer n.

4. Prove that $3^n < 3^{n+1}$.

   Verification: When $n = 1$,

   $$3^1 = 3 < 3^{1+1} = 3^2 = 9,$$

   so the statement is true for $n = 1$.

   Induction: Assume the statement is true when $n = k$, that is assume that

   $$3^k < 3^{k+1}$$

   is true. Then we must show that

   $$3^{k+1} < 3^{(k+1)+1} = 3^{k+2}$$

   is true. Start with

   $$3^k < 3^{k+1}$$

   and multiply both sides by 3.

   $$3 \cdot 3^k < 3 \cdot 3^{k+1}$$

   Then we have

   $$3^{k+1} < 3^{(k+1)+1},$$

   which is the desired statement. Thus, the statement is true when $n = k + 1$ under the assumption it is true for $n = k$.

   Conclusion: By the principle of mathematical induction, the statement is true for all positive integers n.

EXERCISES A SECTION 9.5 459

CHAPTER 9 SEQUENCES, SERIES, AND PROBABILITY

SECTION 9.5 Mathematical Induction

In Exercises 1-3 the statements are not true for all n. Show that S(1), S(2), and S(3) are true, then find the first positive integer n for which S(n) is false.

1. $n < 5$

   $S(1)$: $1 < 5$ is clearly true.
   $S(2)$: $2 < 5$ is clearly true.
   $S(3)$: $3 < 5$ is clearly true.

   However, when $n = 5$, $5 < 5$ is false.

2. n divides 420

   $S(1)$: 1 divides 420 is true since $420 = 1 \cdot 420$.
   $S(2)$: 2 divides 420 is true since $420 = 2 \cdot 210$.
   $S(3)$: 3 divides 420 is true since $420 = 3 \cdot 140$.

   Also, 4, 5, 6, and 7 also divide 420, but 8 does not.

3. $\dfrac{|n-10|}{n-10} = -1$

   $S(1)$: $\dfrac{|1-10|}{1-10} = \dfrac{9}{-9} = -1$ *is true*.

   $S(2)$: $\dfrac{|2-10|}{2-10} = \dfrac{8}{-8} = -1$ *is true*.

   $S(3)$: $\dfrac{|3-10|}{3-10} = \dfrac{7}{-7} = -1$ *is true*.

   Also, the statement is true when n is 4, 5, 6, 7, 8, and 9. However, when $n = 10$, we have 0 in the denominator making the expression undefined.

In Exercises 4-6 write out the statements for S(1), S(k), and S(k+1).

4. $S(n)$: $1+3+5+\ldots+(2n-1) = n^2$

   $S(1)$: $1 = 1^2$

   $S(k)$: $1+3+5+\ldots+(2k-1) = k^2$

   $S(k+1)$: $1+2+3+\ldots+(2k-1)+[2(k+1)-1] = (k+1)^2$

5. $S(n)$: $n < 2^n$

   $S(1)$: $1 < 2^1$

   $S(k)$: $k < 2^k$

   $S(k+1)$: $k+1 < 2^{k+1}$

6. $S(n)$: 3 divides $n^3 - n + 3$

   $S(1)$: 3 divides $1^3 - 1 + 3 = 3$

   $S(k)$: 3 divides $k^3 - k + 3$

   $S(k+1)$: 3 divides $(k+1)^3 - (k+1) + 3$

In Exercises 7-9 prove that each statement in Exercises 4-6 is true for all positive integers n.

7. $S(n)$: $1+3+5+\ldots+(2n-1) = n^2$

   Verification: Since $S(1)$: $1 = 1^2$ is true, the statement is true when $n = 1$.

   Induction: Assume that

   $$S(k): 1+3+5+\ldots+(2k-1) = k^2$$

   is true. We must show that

   $S(k+1)$: $1+2+3+\ldots+(2k-1)+[2(k+1)-1] = (k+1)^2$

   is true. Replace the first k terms in the statement $S(k+1)$ with k^2.

   $$\begin{aligned} k^2 + [2(k+1)-1] &= k^2 + [2k+2-1] \\ &= k^2 + 2k + 1 \\ &= (k+1)^2 \end{aligned}$$

   Since we obtain the left side of the equation in $S(k+1)$, we have shown that $S(k+1)$ is true under the assumption that $S(k)$ is true.

   Conclusion: Then $S(n)$ is true for every positive integer n by the principle of mathematical induction.

8. $S(n)$: $n < 2^n$

   Verification: Since $S(1)$: $1 < 2^1$ is true, the statement is true when $n = 1$.

Induction: Assume that $S(k)$: $k < 2^k$ is true. We must show that $S(k+1)$: $k+1 < 2^{k+1}$ is true. Since $k < 2^k$, multiply both sides by 2 to obtain

$$2k < 2 \cdot 2^k$$

which is the same as

$$2k < 2^{k+1}.$$

Since $1 \leq k$, we have that $k + 1 \leq k + k$, so that

$$k + 1 \leq 2k.$$

Thus,

$$k + 1 \leq 2k < 2^{k+1},$$

which shows that the statement $S(k+1)$ is true under the assumption that $S(k)$ is true.

Conclusion: By the principle of mathematical induction, the statement is true for every positive integer n.

9. $S(n)$: 3 divides $n^3 - n + 3$

Verification: Since $S(1)$: 3 divides $1^3 - 1 + 3 = 3$ is true, the statement is true when $n = 1$.

Induction: Assume that

$$S(k)\text{: 3 divides } k^3 - k + 3$$

is true, we must show that

$$S(k+1)\text{: 3 divides } (k+1)^3 - (k+1) + 3$$

is true. Expanding, we have:

$$\begin{aligned}(k+1)^3 &- (k+1) + 3 \\ &= k^3 + 3k^2 + 3k + 1 - k - 1 + 3 \\ &= k^3 - k + 3 + 3k^2 + 3k \\ &= (k^3 - k + 3) + 3(k^2 + k)\end{aligned}$$

Since 3 divides the first term by the induction hypothesis, and clearly 3 divides the second term, we can conclude that

$$3 \text{ divides } (k+1)^3 - (k+1) + 3$$

under the assumption that 3 divides $k^3 - k + 3$.

Conclusion: By the principle of mathematical induction, 3 divides $n^3 - n + 3$ for every positive integer n.

In Exercises 10-16 prove that each statement is true for all positive integers n.

10. $5 + 10 + 15 + \ldots + 5n = \frac{5}{2}n(n+1)$

Verification: When $n = 1$, we have

$$5 = \frac{5}{2}(1)(1+1) = \frac{5}{2}(2) = 5$$

which is true.

Induction: Assume that

$$5 + 10 + 15 + \ldots + 5k = \frac{5}{2}k(k+1)$$

is true. We must show that

$$5 + 10 + 15 + 5k + 5(k+1) = \frac{5}{2}(k+1)((k+1)+1)$$

is true. Using the induction hypothesis we have:

$$\begin{aligned}5 + 10 &+ 15 + \ldots + 5k + 5(k+1) \\ &= \frac{5}{2}k(k+1) + 5(k+1) \\ &= (k+1)\left(\frac{5}{2}k + 5\right) \\ &= \frac{5}{2}(k+1)(k+2) \\ &= \frac{5}{2}(k+1)((k+1)+1)\end{aligned}$$

Thus the statement is true for $n = k + 1$ when it is true for $n = k$.

Conclusion: By the principle of mathematical induction, the statement is true for every positive integer n.

11. $1^2 + 2^2 + 3^2 + \ldots + n^2 = \frac{n}{6}(n+1)(2n+1)$

Verification: When $n = 1$,

$$1^2 = \frac{1}{6}(1+1)(2(1)+1) = \frac{1}{6}(2)(3) = 1$$

so the statement is true.

Induction: Assume that

$$1^2 + 2^2 + 3^2 + \ldots + k^2 = \frac{k}{6}(k+1)(2k+1)$$

is true. We must show that

$$1^2 + 2^2 + \ldots + k^2 + (k+1)^2 = \frac{k+1}{6}((k+1)+1)(2(k+1)+1)$$

Replace the first k terms on the left side using the

EXERCISES A

induction hypothesis.

$$\frac{k}{6}(k+1)(2k+1) + (k+1)^2$$
$$= (k+1)\left[\frac{k}{6}(2k+1) + (k+1)\right]$$
$$= \frac{1}{6}(k+1)[k(2k+1) + 6(k+1)]$$
$$= \frac{(k+1)}{6}[2k^2 + k + 6k + 6]$$
$$= \frac{(k+1)}{6}[2k^2 + 7k + 6]$$
$$= \frac{(k+1)}{6}(k+2)(2k+3)$$
$$= \frac{(k+1)}{6}((k+1)+1)(2(k+1)+1)$$

Thus the statement is true for $n = k + 1$ under the assumption that it is true for $n = k$.

Conclusion: By the principle of mathematical induction, the statement is true for every positive integer n.

12. $2 \leq 2^n$

Verification: When $n = 1$, $2 \leq 2^1 = 2$ is true.

Induction: Assume that $2 \leq 2^k$ is true. We must show that $2 \leq 2^{k+1}$. Multiply both sides of $2 \leq 2^k$ by 2.

$$2 \cdot 2 \leq 2 \cdot 2^k = 2^{k+1}$$

Since $2 \leq 2 \cdot 2 = 4$, we have that

$$2 \leq 2^{k+1}.$$

Thus the statement is true for $n = k + 1$ when it is true for $n = k$.

Conclusion: By the principle of mathematical induction, the statement is true for all positive integers n.

13. $1 + \frac{1}{2} + \frac{1}{2^2} + \dots + \frac{1}{2^{n-1}} = \frac{2^n - 1}{2^{n-1}}$

Verification: When $n = 1$,

$$1 = \frac{2^1 - 1}{2^{1-1}} = \frac{2-1}{2^0} = \frac{1}{1} = 1.$$

So the statement is true in this case.

Induction: Assume that

$$1 + \frac{1}{2} + \frac{1}{2^2} + \dots + \frac{1}{2^{k-1}} = \frac{2^k - 1}{2^{k-1}}$$

is true. We must show that

$$1 + \frac{1}{2} + \frac{1}{2^2} + \dots + \frac{1}{2^{k-1}} + \frac{1}{2^k} = \frac{2^{k+1} - 1}{2^{(k+1)-1}}$$

is true. Replace the first k terms of the expression on the right using the induction hypothesis to obtain:

$$\frac{2^k - 1}{2^{k-1}} + \frac{1}{2^k} = \frac{2(2^k - 1)}{2 \cdot 2^{k-1}} + \frac{1}{2^k}$$
$$= \frac{2^{k+1} - 2}{2^k} + \frac{1}{2^k}$$
$$= \frac{2^{k+1} - 2 + 1}{2^k}$$
$$= \frac{2^{k+1} - 1}{2^{(k+1)-1}}$$

Thus, the statement is true for $n = k + 1$ when it is true for $n = k$.

Conclusion: By the principle of mathematical induction, the statement is true for every positive integer n.

14. $\left(1 + \frac{1}{1}\right)\left(1 + \frac{1}{2}\right)\dots\left(1 + \frac{1}{n}\right) = n + 1$

Verification: When $n = 1$,

$$\left(1 + \frac{1}{1}\right) = 2 = 1 + 1.$$

Thus, the statement is true for $n = 1$.

Induction: Assume that

$$\left(1 + \frac{1}{1}\right)\left(1 + \frac{1}{2}\right)\dots\left(1 + \frac{1}{k}\right) = k + 1$$

is true. We must show that

$$\left(1 + \frac{1}{1}\right)\left(1 + \frac{1}{2}\right)\dots\left(1 + \frac{1}{k}\right)\left(1 + \frac{1}{k+1}\right) = (k+1) + 1$$

is true. Replace the first k factors on the left with $(k+1)$ using the induction hypothesis.

$$(k+1)\left(1 + \frac{1}{k+1}\right) = (k+1) + (k+1)\frac{1}{k+1}$$
$$= (k+1) + 1$$

Thus, the statement is true for $n = k + 1$ under the assumption that it is true for $n = k$.

Conclusion: By the principle of mathematical induction, the statement is true for every positive integer n.

15. $1 + 2n < 3^n$, $n \geq 2$

Verification: This time we begin with $n = 2$ instead of $n = 1$. When $n = 2$,
$$1 + 2(2) = 5 < 3^2 = 9,$$
which is true.

Induction: Assume that
$$1 + 2k < 3^k \quad (k \geq 2)$$
is true. We must showe that
$$1 + 2(k+1) < 3^{k+1}$$
is true. Start with the left side.

$$\begin{aligned} 1 + 2(k+1) = (1 + 2k) + 2 &< 3^k + 2 \\ &< 3^k + 3^k \\ &= 2 \cdot 3^k \\ &< 3 \cdot 3^k \\ &= 3^{k+1} \end{aligned}$$

Thus, the statement is true for $n = k + 1$ under the assumption that it is true for $n = k$.

Conclusion: By the principle of mathematical induction, the statement is true for every positive integer n.

16. 2 divides $n^2 + n$

Verification: When $n = 1$, $n^2 + n = 1^2 + 1 = 2$, so 2 divides $1^2 + 1$.

Induction: Assume that
$$2 \text{ divides } k^2 + k,$$
we must show that
$$2 \text{ divides } (k+1)^2 + (k+1).$$

Since
$$\begin{aligned} (k+1)^2 + (k+1) &= k^2 + 2k + 1 + k + 1 \\ &= k^2 + k + 2k + 2 \\ &= (k^2 + k) + 2(k + 1), \end{aligned}$$

by the induction hypothesis, 2 divides the first term, and clearly 2 divides the second term. Thus 2 divides the expression when $n = k+1$, assuming that 2 divides the expression when $n = k$.

Conclusion: By the principle of mathematical induction, the statement is true for every positive integer n.

17. Convert $3.6\overline{25}$ to a fraction.

Note that
$$3.6\overline{25} = 3.6 + 0.025 + 0.00025 + 0.0000025 + ...$$

The terms after the first, 3.6, form an infinite geometric series with $a_1 = 0.025$ and $r = 0.01$. Find the sum of this series.

$$S = \frac{a_1}{1-r} = \frac{0.025}{1-0.01} = \frac{0.025}{0.99} = \frac{25}{990}$$

Then we have
$$\begin{aligned} 3.6\overline{25} &= 3.6 + \frac{25}{990} \\ &= \frac{36}{10} + \frac{25}{990} \\ &= \frac{3564}{990} + \frac{25}{990} \\ &= \frac{3589}{990} \end{aligned}$$

18. The total distance traveled by the tip of the pendulum is given by the infinite geometric series:

$$18 + \left(\frac{7}{9}\right)18 + \left(\frac{7}{9}\right)^2 18 + \left(\frac{7}{9}\right)^3 18 + ...$$

Then $a_1 = 18$ and $r = \frac{7}{9}$. Since $|r| < 1$, the series has a sum given by:

$$S = \frac{a_1}{1-r} = \frac{18}{1-\frac{7}{9}} = \frac{18}{\frac{2}{9}} = (18)\left(\frac{9}{2}\right) = 81$$

Thus, the tip of the pendulum travels a distance of 81 inches before coming to rest.

PRACTICE EXERCISES

CHAPTER 9 SEQUENCES, SERIES, AND PROBABILITY

SECTION 9.6 Permutations and Combinations

1. Consider making four-digit numbers from

   1, 2, 3, 4, 5, 6, and 7.

   (a) How many ways can this be done if repetition is allowed?

   Since there are seven choices for each digit, each of the four blanks can be filled in seven ways with repetition allowed.

   $$\underline{7}\ \underline{7}\ \underline{7}\ \underline{7}$$

   Thus, there are $7 \cdot 7 \cdot 7 \cdot 7 = 2401$ ways to make a four-digit number.

   (b) How many four-digit numbers are possible if repetition is not allowed?

   This time, there are seven choices for the first digit, then only six choices for the second digit (the digit used in the first position is no longer available), then only five choices for the third digit, and only 4 choices for the fourth digit. The four blanks can be filled in the following way.

   $$\underline{7}\ \underline{6}\ \underline{5}\ \underline{4}$$

   Thus, there are $7 \cdot 6 \cdot 5 \cdot 4 = 840$ ways to form the four-digit number.

2. How many license plates are there with one letter followed by five digits?

   Since there are 26 letters, the first position can be filled in 26 ways. Since there are 10 digits, each of the next five positions can be filled in 10 ways.

   $$\underline{26}\ \underline{10}\ \underline{10}\ \underline{10}\ \underline{10}\ \underline{10}$$

   Thus, there are

   $$26 \cdot 10 \cdot 10 \cdot 10 \cdot 10 \cdot 10 = 2{,}600{,}000$$

   ways to make the license plates.

3. **(a)** $_7P_7 = \dfrac{7!}{(7-7)!} = \dfrac{7!}{0!} = \dfrac{7!}{1} = 7! = 5040$

   (b) $_9P_6 = \dfrac{9!}{(9-6)!} = \dfrac{9!}{3!} = 60{,}480$

   (c) $_7P_0 = \dfrac{7!}{(7-0)!} = \dfrac{7!}{7!} = 1$

4. The number of ways that seven different prizes can be given to three different people is

   $$_7P_3 = \dfrac{7!}{(7-3)!} = \dfrac{7!}{4!} = \dfrac{7\cdot 6\cdot 5\cdot 4!}{4!} = 7\cdot 6\cdot 5 = 210.$$

5. Find the number of distinguishable permutations of the letters in the word ELEMENT.

   There are seven letters: three E's, one L, one M, one N, and one T. The number of distinguishable permutations is:

   $$\dfrac{7!}{3!\,1!\,1!\,1!\,1!} = \dfrac{7\cdot 6\cdot 5\cdot 4\cdot 3!}{3!\cdot 1\cdot 1\cdot 1\cdot 1}$$
   $$= 7\cdot 6\cdot 5\cdot 4$$
   $$= 840$$

6. In how many ways could nine different plates be placed around a circular table?

   In this case, $n = 9$, so that

   $$(n-1)! = (9-1)! = 8! = 40{,}320$$

   is the number of possible circular permutations of the nine plates.

7. **(a)** $_7C_4 = \dfrac{7!}{(7-4)!\,4!} = \dfrac{7!}{3!\,4!} = \dfrac{7\cdot 6\cdot 5\cdot 4!}{4!\cdot 3\cdot 2\cdot 1} = 35$

   (b) $_7C_7 = \dfrac{7!}{(7-7)!\,7!} = \dfrac{7!}{0!\,7!} = \dfrac{7!}{1\cdot 7!} = 1$

   (c) $_7C_0 = \dfrac{7!}{(7-0)!\,0!} = \dfrac{7!}{7!\cdot 1} = 1$

8. How many five-card poker hands are possible with 2 aces, 2 kings, and a card that is not an ace or a king?

There are

$$_4C_2 = \frac{4!}{(4-2)!2!} = \frac{4!}{2!2!} = \frac{4 \cdot 3 \cdot 2!}{2 \cdot 2!} = 2 \cdot 3 = 6$$

ways to get 2 of the 4 aces, and

$$_4C_2 = \frac{4!}{(4-2)!2!} = \frac{4!}{2!2!} = \frac{4 \cdot 3 \cdot 2!}{2 \cdot 2!} = 2 \cdot 3 = 6$$

ways to get 2 of the 4 kings, and

$$_{44}C_1 = \frac{44!}{(44-1)!1!} = \frac{44 \cdot 43!}{43! \cdot 1} = 44$$

ways to get 1 card from the 44 that are not aces nor kings. Thus the total number of ways to get this poker hand is:

$$_4C_2 \cdot {}_4C_2 \cdot {}_{44}C_1 = 6 \cdot 6 \cdot 44 = 1584$$

EXERCISES A SECTION 9.6

CHAPTER 9 SEQUENCES, SERIES, AND PROBABILITY

SECTION 9.6 Permutations and Combinations

1. $_7P_3 = \dfrac{7!}{(7-3)!} = \dfrac{7 \cdot 6 \cdot 5 \cdot 4!}{4!} = 7 \cdot 6 \cdot 5 = 210$

2. $_{10}P_2 = \dfrac{10!}{(10-2)!} = \dfrac{10 \cdot 9 \cdot 8!}{8!} = 10 \cdot 9 = 90$

3. $_9C_0 = \dfrac{9!}{(9-0)!0!} = \dfrac{9!}{9!(1)} = \dfrac{9!}{9!} = 1$

4. $_{10}C_7 = \dfrac{10!}{(10-7)!7!} = \dfrac{10 \cdot 9 \cdot 8 \cdot 7!}{3!7!} = 10 \cdot 3 \cdot 4 = 120$

5. Since there are 5 choices to go from *A* to *B*, and 3 choices to go from *B* to *C*, the total number of choices to go from *A* to *C* via *B* is

    $\underline{5} \cdot \underline{3}$

    which is $5 \cdot 3 = 15$.

6. License plates consisting of two letters followed by four digits are to be made. If repetition of both letters and digits is allowed, the number of choices is

    $\underline{26} \cdot \underline{26} \cdot \underline{10} \cdot \underline{10} \cdot \underline{10} \cdot \underline{10}$

    which is $26^2 10^4 = 6{,}760{,}000$.

    If repetition of letters is possible, but repetition of digits is not, the number of possible plates is

    $\underline{26} \cdot \underline{26} \cdot \underline{10} \cdot \underline{9} \cdot \underline{8} \cdot \underline{7}$

    which is $3{,}407{,}040$.

    If neither letters nor digits can be repeated, the number of possible plates is

    $\underline{26} \cdot \underline{25} \cdot \underline{10} \cdot \underline{9} \cdot \underline{8} \cdot \underline{7}$

    which is $3{,}276{,}000$.

7. There are three slots to fill, corresponding to the three flags on the flagpole. Since repetition is not possible, the number of ways to fly the possible signals is

    $\underline{8} \cdot \underline{7} \cdot \underline{6}$

    which is equal to 336.

8. Since all the letters in the word SHELF are different, the number of ways of arranging the five letters is $_5P_5 = 5! = 120$.

9. The number of ways of arranging eight suspects in a lineup is $_8P_8 = 8! = 40{,}320$.

10. Determine the number of ways that the manager of a baseball team (9 players) can arrange the batting order under the following conditions.

    (a) Any player can bat in any position. Fill the nine slots in the following way.

    $\underline{9} \cdot \underline{8} \cdot \underline{7} \cdot \underline{6} \cdot \underline{5} \cdot \underline{4} \cdot \underline{3} \cdot \underline{2} \cdot \underline{1}$

    Thus there are $9! = 362{,}880$ ways to set the lineup in this case.

    (b) The pitcher must bat last. Since there is only one choice for the last position, there are only eight to choose from for the first position, 7 for the second, and so forth. The slots are filled in the following way.

    $\underline{8} \cdot \underline{7} \cdot \underline{6} \cdot \underline{5} \cdot \underline{4} \cdot \underline{3} \cdot \underline{2} \cdot \underline{1} \cdot \underline{1}$

    Thus, there are $8! \cdot 1 = 8! = 40{,}320$ ways to set the lineup in this case.

    (c) The pitcher must bat last, and the centerfielder must bat in the clean-up position (fourth). Then the slots must be filled in the following way:

    $\underline{7} \cdot \underline{6} \cdot \underline{5} \cdot \underline{1} \cdot \underline{4} \cdot \underline{3} \cdot \underline{2} \cdot \underline{1} \cdot \underline{1}$

    Thus, there are $7! \cdot 1 \cdot 1 = 7! = 5040$ ways to set the lineup in this case.

    (d) The pitcher must bat last, the three outfielders must bat in the first three positions. Then the slots must be filled in the following way:

    $\underline{3} \cdot \underline{2} \cdot \underline{1} \cdot \underline{5} \cdot \underline{4} \cdot \underline{3} \cdot \underline{2} \cdot \underline{1} \cdot \underline{1}$

Thus, there are $3!5! \cdot 1 = 3!5! = 720$ ways to set the lineup in this case.

11. How many distinguishable permutations of the letters in the given word are possible.

    (a) SHEEP

    Since there are a total of 5 letters, with 1 S, 1 H, 2 E's, and 1 P, we have:

    $$\frac{5!}{1!1!2!1!} = \frac{5!}{2} = \frac{5 \cdot 4 \cdot 3 \cdot 2 \cdot 1}{2} = 5 \cdot 4 \cdot 3 = 60$$

    (b) LETTER

    Since there are a total of 6 letters, with 1 L, 2 E's, 2 T's, and 1 R, we have:

    $$\frac{6!}{1!2!2!1!} = \frac{6!}{4} = \frac{6 \cdot 5 \cdot 4 \cdot 3 \cdot 2 \cdot 1}{4} = 6 \cdot 5 \cdot 3 \cdot 2 = 180$$

    (c) TENNESSEE

    Since there are a total of 9 letters, with 1 T, 4 E's, 2 N's, and 2 S's, we have:

    $$\frac{9!}{1!4!2!2!} = \frac{9 \cdot 8 \cdot 7 \cdot 6 \cdot 5 \cdot 4!}{4! \cdot 4} = 9 \cdot 2 \cdot 7 \cdot 6 \cdot 5 = 3780$$

    (d) AARDVARK

    Since there are a total of 8 letters, with 3 A's, 2 R's, 1 D, 1 V, and 1 K, we have:

    $$\frac{8!}{3!2!1!1!1!} = \frac{8 \cdot 7 \cdot 6 \cdot 5 \cdot 4 \cdot 3!}{3! \cdot 2} = 4 \cdot 7 \cdot 6 \cdot 5 \cdot 4 = 3360$$

12. The number of distinct signals made with fourteen flags when five are white, four are red, three are green, and two are black is the number of distinguishable permutations of 14 objects when 5 are the same, 4 are the same, 3 are the same, and 2 are the same. This number is given by:

    $$\frac{14!}{5!4!3!2!} = \frac{14 \cdot 13 \cdot 12 \cdot 11 \cdot 10 \cdot 9 \cdot 8 \cdot 7 \cdot 6 \cdot 5!}{5! \cdot 4 \cdot 3 \cdot 2 \cdot 1 \cdot 3 \cdot 2 \cdot 1 \cdot 2 \cdot 1}$$
    $$= 14 \cdot 13 \cdot 11 \cdot 10 \cdot 9 \cdot 2 \cdot 7$$
    $$= 2{,}522{,}520$$

13. The number of ways to seat eight people around a circular table is a curcular permutation problem with $n = 8$. Thus the number of ways is:
    $$(n-1)! = 7! = 5040.$$

14. The number of ways to select 5 soldiers from a group of 14 is:

    $$_{14}C_5 = \frac{14!}{(14-5)!5!} = \frac{14 \cdot 13 \cdot 12 \cdot 11 \cdot 10 \cdot 9!}{9! \cdot 5 \cdot 4 \cdot 3 \cdot 2 \cdot 1}$$
    $$= 14 \cdot 13 \cdot 11$$
    $$= 2002$$

    Notice that in this case, order of choice is **not** important. That is, it is a combination problem.

15. The number of ways to select 5 soldiers from a group of 14 and present them with 5 distinct awards is:

    $$_{14}P_5 = \frac{14!}{(14-5)!} = \frac{14 \cdot 13 \cdot 12 \cdot 11 \cdot 10 \cdot 9!}{9!}$$
    $$= 14 \cdot 13 \cdot 12 \cdot 11 \cdot 10$$
    $$= 240{,}240$$

    Notice that in this case, order of choice is important. That is, it is a permutation problem.

16. In how many ways can a committee of 5 be chosen from 8 seniors and 10 juniors under the following conditions?

    (a) The committee must consist of exactly 4 seniors.

    This means that there must be 4 seniors and 1 junior on the committee. We must select 4 seniors from the 8 seniors and 1 junior from the 10 juniors. This is a combination problem. The number of ways of doing this is:

    $$_8C_4 \cdot {_{10}C_1} = \frac{8!}{(8-4)!4!} \cdot \frac{10!}{(10-1)!1!}$$
    $$= \frac{8!}{4!4!} \cdot \frac{10!}{9!1!}$$
    $$= 70 \cdot 10$$
    $$= 700$$

    (b) The committee must consist of at least 4 seniors. This means that the committee must consist of 4 seniors and 1 junior, or it must consist of 5 seniors. The number of ways of choosing 5 seniors is:

    $$_8C_5 = \frac{8!}{(8-5)!5!} = \frac{8!}{3!5!} = 56$$

    In part (c) we found the number of ways of choosing 4 seniors and 1 junior, 700. Thus, the number of ways of having one or the other of these two is the sum of the number of ways for each, $700 + 56 = 756$.

EXERCISES A SECTION 9.6 467

(c) The committee must consist of exactly 3 juniors.

This means that the committee must consist of 3 juniors and 2 seniors. The number of ways of making this selection is:

$$_{10}C_3 \cdot {_8C_2} = \frac{10!}{(10-3)!3!} \cdot \frac{8!}{(8-2)!2!}$$
$$= 120 \cdot 28$$
$$= 3360$$

(d) The committee must consist of at least 3 juniors.

This means that the committee consists of:

3 juniors and 2 seniors

or

4 juniors and 1 senior

or

5 juniors and 0 seniors.

The number of ways of making this selection is:

$$_{10}C_3 \cdot {_8C_2} + {_{10}C_4} \cdot {_8C_1} + {_{10}C_5} \cdot {_8C_0}$$
$$= \frac{10!}{(10-3)!3!} \cdot \frac{8!}{(8-2)!2!} + \frac{10!}{(10-4)!4!} \cdot \frac{8!}{(8-1)!1!}$$
$$+ \frac{10!}{(10-5)!5!} \cdot \frac{8!}{(8-0)!0!}$$
$$= (120)(28) + (210)(8) + (252)(1)$$
$$= 3360 + 1680 + 252$$
$$= 5292$$

17. Since 2 points determine a line, we must choose 2 points from the 7 noncollinear points. This is a combination problem, and the total number of lines possible is:

$$_7C_2 = \frac{7!}{(7-2)!2!} = \frac{7 \cdot 6 \cdot 5!}{5! \cdot 2} = 7 \cdot 3 = 21$$

18. The number of ways choosing 7 of the first 10 questions is $_{10}C_7$, and the number of ways of choosing 4 of the next 5 questions is $_5C_4$. Notice that order of selection is not important, so this is a combination problem, not a permutation problem. The total number of ways to select the questions on the test is $_{10}C_7 \cdot {_5C_4} = (120)(5) = 600$.

19. How many 5-card poker hands dealt from a 52-card deck are possible if they must consist of the following cards?

(a) 5 hearts

The number of ways of choosing 5 hearts from the 13 hearts is:

$$_{13}C_5 = \frac{13!}{(13-5)!5!} = 1287$$

(b) 4 aces and a seven

The number of ways of choosing 4 aces from the four aces and 1 seven from the four sevens is:

$$_4C_4 \cdot {_4C_1} = \frac{4!}{(4-4)!4!} \cdot \frac{4!}{(4-1)!1!} = (1)(4) = 4$$

(c) 3 aces and two cards that are not aces

The number of ways of choosing 3 aces from the 4 aces and 2 cards from the 48 cards that are not aces is:

$$_4C_3 \cdot {_{48}C_2} = \frac{4!}{(4-3)!3!} \cdot \frac{48!}{(48-2)!2!}$$
$$= (4)(1128) = 4512$$

(d) 3 hearts and 2 clubs

The number of ways of choosing 3 hearts from the 13 hearts and 2 clubs from the 13 clubs is:

$$_{13}C_3 \cdot {_{13}C_2} = \frac{13!}{(13-3)!3!} \cdot \frac{13!}{(13-2)!2!}$$
$$= (286)(78) = 22,308$$

(e) 3 face cards and 2 sevens

The number of ways of choosing 3 face cards from the 12 face cards (jacks, queens, and kings) and 2 of the 4 sevens is:

$$_{12}C_3 \cdot {_4C_2} = \frac{12!}{(12-3)!3!} \cdot \frac{4!}{(4-2)!2!} = (220)(6) = 1320$$

20. The number of ways of choosing 1 of the 3 models and 1 of the 4 colors and 1 of the 5 upholstery colors is:

$$_3C_1 \cdot {_4C_1} \cdot {_5C_1} = (3)(4)(5) = 60$$

21. Give an example to show that

$$(n!)(m!) \neq (n \cdot m)!.$$

One example that will do this is $n = 2$ and $m = 3$. Then

$$(n!)(m!) = (2!)(3!) = (2)(6) = 12$$

but

$$(n \cdot m)! = (2 \cdot 3)! = 6! = 720.$$

22. Since

$$_nC_r = \frac{n!}{(n-r)!r!}$$

and

$$_nC_{n-r} = \frac{n!}{(n-(n-r))!(n-r)!} = \frac{n!}{r!(n-r)!}$$

we can see that

$$_nC_r = {_nC_{n-r}}.$$

In Exercises 23-24 use mathematical induction to prove the statement.

23. $3 + 3^2 + 3^3 + \ldots + 3^n = \frac{3}{2}(3^n - 1)$

    Verification: When $n = 1$, we have

    $$3 = \frac{3}{2}(3^1 - 1) = \frac{3}{2}(2) = 3.$$

    Thus, the statement is true when $n = 1$.

    Induction: Assume that

    $$3 + 3^2 + 3^3 + \ldots + 3^k = \frac{3}{2}(3^k - 1)$$

    is true. We must show that

    $$3 + 3^2 + \ldots + 3^k + 3^{k+1} = \frac{3}{2}(3^{k+1} - 1)$$

    is true. Replace the first k terms of the expression on the left using the induction hypothesis.

    $$\frac{3}{2}(3^k - 1) + 3^{k+1} = \frac{1}{2}(3^{k+1} - 3) + \frac{2}{2}3^{k+1}$$
    $$= \frac{3}{2}3^{k+1} - \frac{3}{2}$$
    $$= \frac{3}{2}(3^{k+1} - 1)$$

    Then the statement is true for $n = k + 1$ under the assumption that it is true for $n = k$.

    Conclusion: Thus, the statement is true for every positive integer n by the principle of mathematical induction.

24. 6 divides $7^n - 1$

    Verification: When $n = 1$, since $7^1 - 1 = 6$, the statement 6 divides $7^1 - 1$ is true.

    Induction: Assume that

    $$6 \text{ divides } 7^k - 1$$

    is true. We must show that

    $$6 \text{ divides } 7^{k+1} - 1$$

    is true. Since $7^{k+1} - 1 = 7 \cdot 7^k - 7 + 7 - 1$, we have that

    $$7^{k+1} - 1 = 7(7^k - 1) + 6.$$

    Since 6 divides the first term by the induction hypothesis, and clearly 6 divides the second term, we have that 6 divides their sum. That is,

    $$6 \text{ divides } 7^{k+1} - 1.$$

    Then the statement is true for $n = k + 1$ under the assumption that it is true for $n = k$.

    Conclusion: By the principle of mathematical induction, 6 divides $7^n - 1$ for every positive integer n.

PRACTICE EXERCISES

CHAPTER 9 SEQUENCES, SERIES, AND PROBABILITY

SECTION 9.7 The Binomial Theorem

1. Expand $(2x - y)^5$.

   In this problem, $n = 5$, $a = 2x$, and $b = -y$. Row 5 of Pascal's triangle has the following numbers for the coefficients.

   $$1 \quad 5 \quad 10 \quad 10 \quad 5 \quad 1$$

   We thus substitute $a = 2x$ and $b = -y$ into

   $$a^5 + 5a^4b + 10a^3b^2 + 10a^2b^3 + 5ab^4 + b^5$$

   to obtain:

   $$(2x)^5 + 5(2x)^4(-y) + 10(2x)^3(-y)^2 + 10(2x)^2(-y)^3 + 5(2x)(-y)^4 + (-y)^5$$

   $$= 32x^5 - 80x^4y + 80x^3y^2 - 40x^2y^3 + 10xy^4 - y^5$$

2. Expand $(3x + 2)^4$.

   In this problem, $n = 4$, $a = 3x$, and $b = 2$. Row 4 of Pascal's triangle is:

   $$1 \quad 4 \quad 6 \quad 4 \quad 1$$

   Substitute $3x$ for a and 2 for b in

   $$a^4 + 4a^3b + 6a^2b^2 + 4ab^3 + b^4.$$

   $$(3x)^4 + 4(3x)^3(2) + 6(3x)^2(2)^2 + 4(3x)(2)^3 + (2)^4$$

   $$= 81x^4 + 216x^3 + 216x^2 + 96x + 16$$

3. (a) Find the third term in the expansion of

   $$(x + 2y)^7.$$

   In this case, $n = 7$, $a = x$, $b = 2y$, and since we want the third term, r is 2.

   $$\binom{n}{r}a^{n-r}b^r = \binom{7}{2}(x)^{7-2}(2y)^2$$

   $$= \frac{7!}{(7-2)!2!}x^5(4y^2)$$

   $$= \frac{7 \cdot 6 \cdot 5!}{5! \cdot 2}4x^5y^2$$

   $$= 7 \cdot 3 \cdot 4x^5y^2$$

   $$= 84x^5y^2$$

   (b) Find the 4th term in the expansion of

   $$(a^2 - 3y)^{10}.$$

   In this case $n = 10$, $a = a^2$, $b = -3y$, and since we want the 4th term, r is 3.

   $$\binom{n}{r}a^{n-r}b^r = \binom{10}{3}(a^2)^{10-3}(-3y)^3$$

   $$= \frac{10!}{(10-3)!3!}(a^2)^7(-27y^3)$$

   $$= \frac{10 \cdot 9 \cdot 8 \cdot 7!}{7!3 \cdot 2 \cdot 1}a^{14}(-27y^3)$$

   $$= 10 \cdot 3 \cdot 4(-27)a^{14}y^3$$

   $$= -3240a^{14}y^3$$

CHAPTER 9 SEQUENCES, SERIES, AND PROBABILITY

SECTION 9.7 The Binomial Theorem

In Exercises 1-3 evaluate each binomial coefficient.

1. $\binom{4}{2} = \dfrac{4!}{(4-2)!2!} = \dfrac{4!}{2!2!} = \dfrac{4 \cdot 3 \cdot 2!}{2 \cdot 1 \cdot 2!} = 6$

2. $\binom{5}{0} = \dfrac{5!}{(5-0)!0!} = \dfrac{5!}{5! \cdot 1} = \dfrac{5!}{5!} = 1$

3. $\binom{10}{8} = \dfrac{10!}{(10-8)!8!} = \dfrac{10!}{2!8!} = \dfrac{10 \cdot 9 \cdot 8!}{2 \cdot 1 \cdot 8!} = 45$

Expand each binomial in Exercises 4-9.

4. $(x + y)^5$

$(x+y)^5$
$= \sum_{r=0}^{5} \binom{5}{r} x^{5-r} y^r$
$= \binom{5}{0} x^{5-0} y^0 + \binom{5}{1} x^{5-1} y^1 + \binom{5}{2} x^{5-2} y^2 +$
$\quad \binom{5}{3} x^{5-3} y^3 + \binom{5}{4} x^{5-4} y^4 + \binom{5}{5} x^{5-5} y^5$
$= (1)x^5(1) + (5)x^4 y^1 + (10)x^3 y^2 +$
$\quad (10)x^2 y^3 + (5)xy^4 + (1)x^0 y^5$
$= x^5 + 5x^4 y + 10x^3 y^2 + 10x^2 y^3 + 5xy^4 + y^5$

5. $(3a - 1)^5$

$(3a-1)^5$
$= \sum_{r=0}^{5} \binom{5}{r} (3a)^{5-r} (-1)^r$
$= \binom{5}{0} (3a)^{5-0}(-1)^0 + \binom{5}{1}(3a)^{5-1}(-1)^1 + \binom{5}{2}(3a)^{5-2}(-1)^2 +$
$\quad \binom{5}{3}(3a)^{5-3}(-1)^3 + \binom{5}{4}(3a)^{5-4}(-1)^4 + \binom{5}{5}(3a)^{5-5}(-1)^5$
$= (1)(3a)^5(1) + (5)(3a)^4(-1) + (10)(3a)^3(1) +$
$\quad (10)(3a)^2(-1) + (5)(3a)(1) + (1)(3a)^0(-1)$
$= 243a^5 - 405a^4 + 270a^3 - 90a^2 + 15a - 1$

6. $(3x - y)^4$

$(3x - y)^4$
$= \sum_{r=0}^{4} \binom{4}{r}(3x)^{4-r}(-y)^r$
$= \binom{4}{0}(3x)^{4-0}(-y)^0 + \binom{4}{1}(3x)^{4-1}(-y)^1 + \binom{4}{2}(3x)^{4-2}(-y)^2 +$
$\quad \binom{4}{3}(3x)^{4-3}(-y)^3 + \binom{4}{4}(3x)^{4-4}(-y)^4$
$= (1)(3x)^4(1) + (4)(3x)^3(-y) + (6)(3x)^2(-y)^2 +$
$\quad (4)(3x)(-y)^3 + (1)(1)(-y)^4$
$= 81x^4 - 108x^3 y + 54x^2 y^2 - 12xy^3 + y^4$

7. $(u^2 + v^2)^6$

$(u^2 + v^2)^6$
$= \sum_{r=0}^{6} \binom{6}{r}(u^2)^{6-r}(v^2)^r$
$= \binom{6}{0}(u^2)^{6-0}(v^2)^0 + \binom{6}{1}(u^2)^{6-1}(v^2)^1 + \binom{6}{2}(u^2)^{6-2}(v^2)^2 +$
$\quad \binom{6}{3}(u^2)^{6-3}(v^2)^3 + \binom{6}{4}(u^2)^{6-4}(v^2)^4 +$
$\quad \binom{6}{5}(u^2)^{6-5}(v^2)^5 + \binom{6}{6}(u^2)^{6-6}(v^2)^6$
$= (1)(u^2)^6(1) + (6)(u^2)^5(v^2)^1 + (15)(u^2)^4(v^2)^2 +$
$\quad (20)(u^2)^3(v^2)^3 + (15)(u^2)^2(v^2)^4 +$
$\quad (6)(u^2)^1(v^2)^5 + (1)(1)(v^2)^6$
$= u^{12} + 6u^{10}v^2 + 15u^8 v^4 + 20u^6 v^6 + 15u^4 v^8 + 6u^2 v^{10} + v^{12}$

8. $(a + a^{-1})^7$

$(a + a^{-1})^7$
$= \sum_{r=0}^{7} \binom{7}{r} a^{n-r}(a^{-1})^r$
$= \binom{7}{0} a^{7-0}(a^{-1})^0 + \binom{7}{1} a^{7-1}(a^{-1})^1 + \binom{7}{2} a^{7-2}(a^{-1})^2 +$
$\quad \binom{7}{3} a^{7-3}(a^{-1})^3 + \binom{7}{4} a^{7-4}(a^{-1})^4 + \binom{7}{5} a^{7-5}(a^{-1})^5 +$
$\quad \binom{7}{6} a^{7-6}(a^{-1})^6 + \binom{7}{7} a^{7-7}(a^{-1})^7$
$= (1)a^7(1) + (7)a^6 a^{-1} + (21)a^5 a^{-2} + (35)a^4 a^{-3} +$
$\quad (35)a^3 a^{-4} + (21)a^2 a^{-5} + (7)aa^{-6} + (1)a^{-7}$
$= a^7 + 7a^5 + 21a^3 + 35a + 35a^{-1} + 21a^{-3} + 7a^{-5} + a^{-7}$

EXERCISES A   SECTION 9.7   471

9. $(x^{1/2} - y^{1/2})^4$

$(x^{1/2} - y^{1/2})^4$

$= \sum_{r=0}^{4} \binom{4}{r}(x^{1/2})^{4-r}(-y^{1/2})^r$

$= \binom{4}{0}(x^{1/2})^{4-0}(-y^{1/2})^0 + \binom{4}{1}(x^{1/2})^{4-1}(-y^{1/2})^1 +$
$\binom{4}{2}(x^{1/2})^{4-2}(-y^{1/2})^2 + \binom{4}{3}(x^{1/2})^{4-3}(-y^{1/2})^3 +$
$\binom{4}{4}(x^{1/2})^{4-4}(-y^{1/2})^4$

$= (1)(x^{1/2})^4 + (4)(x^{1/2})^3(-y^{1/2}) + (6)(x^{1/2})^2(-y^{1/2})^2 +$
$(4)(x^{1/2})(-y^{1/2})^3 + (1)(-y^{1/2})^4$

$= x^2 - 4x^{3/2}y^{1/2} + 6xy - 4x^{1/2}y^{3/2} + y^2$

In Exercises 10-15 find the indicated term in each binomial expansion.

10. 3rd; $(x + 2)^5$

In this case, $a = x$, $b = 2$, $n = 5$, and $r = 2$ (r is 1 less than the designated term 3).

$\binom{n}{r}a^{n-r}b^r = \binom{5}{2}x^{5-2}(2)^2$

$= \frac{5!}{(5-2)!2!}x^3(4)$

$= (10)x^3(4)$

$= 40x^3$

11. 4th; $(x + 2y)^5$

In this case, $a = x$, $b = 2y$, $n = 5$, and $r = 3$.

$\binom{n}{r}a^{n-r}b^r = \binom{5}{3}x^{5-3}(2y)^3$

$= \frac{5!}{(5-3)!3!}x^2(8y^3)$

$= (10)x^2(8y^3)$

$= 80x^2y^3$

12. 4th; $(x + y^2)^5$

In this case, $a = x$, $b = y^2$, $n = 5$, and $r = 3$.

$\binom{n}{r}a^{n-r}b^r = \binom{5}{3}x^{5-3}(y^2)^3$

$= \frac{5!}{(5-3)!3!}x^2(y^6)$

$= (10)x^2(y^6)$

$= 10x^2y^6$

13. 6th; $(3a - b)^7$

In this case, $a = 3a$, $b = -b$, $n = 7$, and $r = 5$.

$\binom{n}{r}a^{n-r}b^r = \binom{7}{5}(3a)^{7-5}(-b)^5$

$= \frac{7!}{(7-5)!5!}(3a)^2(-b)^5$

$= (21)(9a^2)(-b^5)$

$= -189a^2b^5$

14. middle; $(4a - 2)^6$

In this case, $a = 4a$, $b = -2$, $n = 6$, and the middle term corresponds to $r = 3$ ($r = 0, 1, 2, 3, 4, 5, 6$).

$\binom{n}{r}a^{n-r}b^r = \binom{6}{3}(4a)^{6-3}(-2)^3$

$= \frac{6!}{(6-3)!3!}(4a)^3(-2)^3$

$= (20)(64a^3)(-8)$

$= -10,240a^3$

15. 6th; $(a - a^{-1})^8$

In this case $a = a$, $b = -a^{-1}$, $n = 8$, and $r = 5$.

$\binom{n}{r}a^{n-r}b^r = \binom{8}{5}(a)^{8-5}(-a^{-1})^5$

$= \frac{8!}{(8-5)!5!}(a)^3(-a^{-1})^5$

$= (56)(a^3)(-a^{-5})$

$= -56a^3a^{-5}$

$= -56a^{-2}$

In Exercises 16-17 find the value of the real or complex number raised to a power.

16. $(1 + 1)^7$

Clearly this is $2^7 = 128$. However, suppose we expand using the binomial theorem to verify this result. Then $a = 1$, $b = 1$, and $n = 7$.

$(1 + 1)^7$

$= \sum_{r=0}^{7} \binom{7}{r}(1)^{7-r}(1)^r$

$= \binom{7}{0}(1)^7(1)^0 + \binom{7}{1}(1)^6(1)^1 + \binom{7}{2}(1)^5(1)^2 +$
$\binom{7}{3}(1)^4(1)^3 + \binom{7}{4}(1)^3(1)^4 + \binom{7}{5}(1)^2(1)^5 +$
$\binom{7}{6}(1)^1(1)^6 + \binom{7}{7}(1)^0(1)^7$

$= \binom{7}{0} + \binom{7}{1} + \binom{7}{2} + \binom{7}{3} + \binom{7}{4} + \binom{7}{5} + \binom{7}{6} + \binom{7}{7}$

$= 1 + 7 + 21 + 35 + 35 + 21 + 7 + 1$

$= 128$

17. $(1 + i)^5$

In this case, $a = 1$, $b = i$, and $n = 5$.

$$(1+i)^5$$
$$= \sum_{r=0}^{5} \binom{5}{r}(1)^{5-r}(i)^r$$
$$= \binom{5}{0}(1)^5(i)^0 + \binom{5}{1}(1)^4(i)^1 + \binom{5}{2}(1)^3(i)^2 +$$
$$\binom{5}{3}(1)^2(i)^3 + \binom{5}{4}(1)^1(i)^4 + \binom{5}{5}(1)^0(i)^5$$
$$= (1)(1)(1) + (5)(1)(i) + (10)(1)(-1) +$$
$$(10)(1)(-i) + (5)(1)(1) + (1)(1)(i)$$
$$= 1 + 5i - 10 - 10i + 5 + i$$
$$= -4 - 4i$$

18. Find the term containing x^5 in $(3x - 5y)^6$.

The term containing x^5 corresponds to

$$n - r = 6 - r = 5,$$

so $r = 1$. Also, $a = 3x$ and $b = -5y$.

$$\binom{6}{1}(3x)^{6-1}(-5y)^1 = (6)(3x)^5(-5y)^1 = -7290x^5 y$$

19. Use the first four terms of $(1 - 0.1)^6$ to approximate $(0.9)^6$. Compare the result with a calculator.

In this case, $a = 1$, $b = -0.1$, $n = 6$, and we want $r = 0, 1, 2,$ and 3.

$$\binom{6}{0}(1)^6(-0.1)^0 + \binom{6}{1}(1)^5(-0.1)^1 +$$
$$\binom{6}{2}(1)^4(-0.1)^2 + \binom{6}{3}(1)^3(-0.1)^3$$
$$= (1)(1)(1) + (6)(1)(-0.1) + (15)(1)(-0.1)^2 +$$
$$(20)(1)(-0.1)^3$$
$$= 1 - 0.6 + 0.15 - 0.02$$
$$= 0.53$$

Using a calculator, we obtain

$$(0.9)^6 \approx 0.531441.$$

20. $\binom{n}{0} = \dfrac{n!}{(n-0)!0!} = \dfrac{n!}{n!0!} = \dfrac{n!}{n!} = 1$

$\binom{n+1}{0} = \dfrac{(n+1)!}{((n+1)-0)!0!} = \dfrac{(n+1)!}{(n+1)!0!} = \dfrac{(n+1)!}{(n+1)!} = 1$

Thus,

$$\binom{n}{0} = \binom{n+1}{0}.$$

$\binom{n}{n} = \dfrac{n!}{(n-n)!n!} = \dfrac{n!}{0!n!} = \dfrac{n!}{n!} = 1$

$\binom{n+1}{n+1} = \dfrac{(n+1)!}{((n+1)-(n+1))!(n+1)!} = \dfrac{(n+1)!}{0!(n+1)!}$
$= \dfrac{(n+1)!}{(n+1)!}$
$= 1$

Thus,

$$\binom{n}{n} = \binom{n+1}{n+1}.$$

21. The number of ways that five members of a basketball team can be introduced to the crowd if the center must be introduced first is:

$$\underline{1} \cdot \underline{4} \cdot \underline{3} \cdot \underline{2} \cdot \underline{1} = 4! = 24$$

22. The number of ways of selecting 4 light bulbs from 24 light bulbs is a combination problem (order of selection is not important). This number is:

$$_{24}C_4 = \dfrac{24!}{(24-4)!4!} = \dfrac{24 \cdot 23 \cdot 22 \cdot 21 \cdot 20!}{20! \cdot 4 \cdot 3 \cdot 2 \cdot 1} = 10{,}626$$

PRACTICE EXERCISES

CHAPTER 9 SEQUENCES, SERIES, AND PROBABILITY

SECTION 9.8 Probability

1. The sample space for the experiment of flipping two fair coins is

   $$S = \{hh, ht, th, tt\}.$$

   (a) If E is the event of obtaining two tails, then

   $$E = \{tt\}.$$

   Then

   $$P(E) = \frac{n(E)}{n(S)} = \frac{1}{4}.$$

   (b) If F is the event of obtaining no tails, then

   $$F = \{hh\}.$$

   Then

   $$P(F) = \frac{n(F)}{n(S)} = \frac{1}{4}.$$

2. Use the sample space for rolling two dice given in the text.

   (a) If E is the event of rolling a total of 10, then

   $$E = \{(6,4),(5,5),(4,6)\}.$$

   Then

   $$P(E) = \frac{n(E)}{n(S)} = \frac{3}{36} = \frac{1}{12}.$$

   (b) If F is the event of rolling a total of 5, then

   $$F = \{(4,1),(3,2),(2,3),(1,4)\}.$$

   Then

   $$P(F) = \frac{n(F)}{n(S)} = \frac{4}{36} = \frac{1}{9}.$$

3. What is the probability of drawing a five-card hand of 3 aces and 2 kings.

   Let E represent this event. There are $_4C_3$ ways of drawing three aces, $_4C_2$ ways of drawing two kings, and a total of $_{52}C_5$ ways of drawing five cards. Thus,

   $$P(E) = \frac{_4C_3 \cdot _4C_2}{_{52}C_5} = \frac{\frac{4!}{1!3!} \cdot \frac{4!}{2!2!}}{\frac{52!}{47!5!}} = \frac{1}{108,290}.$$

4. Use the sample space of drawing one card from a well-shuffled deck of 52 cards. Find the probability of drawing:

   (a) a jack or a queen

   $$P(\text{jack or queen}) = P(\text{jack}) + P(\text{queen}) - P(\text{jack and})$$
   $$= \frac{4}{52} + \frac{4}{52} - \frac{0}{52}$$
   $$= \frac{8}{52}$$
   $$= \frac{2}{13}$$

   (b) a ten or a black card

   $$P(\text{ten or black}) = P(\text{ten}) + P(\text{black}) - P(\text{ten and black})$$
   $$= \frac{4}{52} + \frac{26}{52} - \frac{2}{52} \quad \text{Two black tens}$$
   $$= \frac{28}{52}$$
   $$= \frac{7}{13}$$

   (c) a five and a seven

   $$P(\text{five and seven}) = 0$$

   since the two events are mutually exclusive.

5. If a couple plans to have three children, the sample space is

   $$S = \{ggg, ggb, gbg, gbb, bgg, bgb, bbg, bbb\}.$$

   The event of having one girl and two boys is

   $$E = \{gbb, bgb, bbg\}.$$

   Then

   $$P(E) = \frac{n(E)}{n(S)} = \frac{3}{8}.$$

The event of having no more than one boy is

$$F = \{ggg, bgg, gbg, ggb\}.$$

Then

$$P(F) = \frac{n(F)}{n(S)} = \frac{4}{8} = \frac{1}{2}.$$

6. Given that the probability of contracting measles is 0.012, the probability of contracting mumps is 0.031, and the probability of contracting both is 0.005.

   (a) The probability of the event E of not contracting both measles and mumps is

   $$P(E) = 1 - P(\text{both}) = 1 - 0.005 = 0.995.$$

   (b) The event of *not* contracting measles and *not* contracting mumps is the same as the event of *not* contracting either of the diseases. The probability of this event, found in Example 6 (b), is 0.962.

7. Use the data in the histogram given in Figure 9.5 in the text.

   (a) What is the probability that a person is shorter than 72 inches?

   We add the class frequencies for the class intervals 66-68, 68-70, and 70-72.

   $$4 + 12 + 24 = 40$$

   Since there are 60 people in the sample space, the probability that a person is shorter than 72 inches is

   $$\frac{40}{60} = \frac{2}{3}.$$

   (b) What is the probability that a person is at least 68 inches but less than 74 inches?

   We add the class frequencies for the class intervals 68-70, 70-72, and 72-74.

   $$12 + 24 + 18 = 54$$

   Then the probability that a person is at least 68 inches but less than 74 inches is

   $$\frac{54}{60} = \frac{9}{10}.$$

EXERCISES A SECTION 9.8 475

CHAPTER 9 SEQUENCES, SERIES, AND PROBABILITY

SECTION 9.8 Probability

In Exercises 1-3 answer yes if the number could be a probability and no if it could not.

1. Yes; 0 can be a probability.

2. No; $-\frac{1}{2}$ cannot be a probability since a probability is a number between 0 and 1, inclusive.

3. Yes; 0.00001 can be a probability.

Let $S = \{1, 2, 3, 4, 5, 6\}$ be the sample space of rolling one unloaded die. Determine the probability of each event in Exercises 4-6.

4. Since the event of rolling a 5 is $E = \{5\}$,
$$P(E) = \frac{n(E)}{n(S)} = \frac{1}{6}.$$

5. Since the event of rolling a number greater than 0 is $E = \{1, 2, 3, 4, 5, 6\}$,
$$P(E) = \frac{n(E)}{n(S)} = \frac{6}{6} = 1. \quad E \text{ is certain}$$

6. Since the event of rolling a number $n \geq 4$ is $E = \{4, 5, 6\}$,
$$P(E) = \frac{n(E)}{n(S)} = \frac{3}{6} = \frac{1}{2}.$$

Let S be the sample space for the experiment of rolling two unloaded dice. Find the probability of each event in Exercises 7-9.

7. Since the event of rolling a total of seven is
$$E = \{(6,1),(5,2),(4,3),(3,4),(2,5),(1,6)\}$$
we have that
$$P(E) = \frac{n(E)}{n(S)} = \frac{6}{36} = \frac{1}{6}.$$

8. Since the event of rolling a total of 7 or 11 is
$$E = \{(6,1),(5,2),(4,3),(3,4),(2,5),(1,6),(6,5),(5,6)\}$$
we have that
$$P(E) = \frac{n(E)}{n(S)} = \frac{8}{36} = \frac{2}{9}.$$

9. Since the event of rolling a total less than or equal to 3 consists of the three elements $(1,1),(2,1)$, and $(1,2)$, the event E of rolling a total greater than 3 must consist of 33 elements ($36 - 3 = 33$). Thus,
$$P(E) = \frac{n(E)}{n(S)} = \frac{33}{36} = \frac{11}{12}.$$

In the game of craps, on the first roll a player wins with a total of 7 or 11 and loses with a total of 2, 3, or 12. Any other total becomes his point, and he must continue to roll until he rolls his point and wins, or rolls a 7 and loses. Find the probability of each event in Exercises 10-11.

10. The probability of winning on the first roll is the probability of rolling a 7 or 11 on the first roll. We found this probability to be $\frac{2}{9}$ in Exercise 8.

11. The event of a total other than a winning or losing total, that is, a total of 4, 5, 6, 8, 9, or 10, has 24 elements. Thus, the probability of this event is
$$\frac{24}{36} = \frac{2}{3}.$$

A roulette wheel contains 38 slots numbered 00, 0, 1, 2, 3, ..., 35, 36. Eighteen of the slots numbered 1 through 36 are colored black and the rest are colored red. The 00 and 0 slots are colored green and are called house numbers. The wheel is spun and a ball is rolled around the rim in the opposite direction. Eventually the ball falls into a slot. Find the probability of each event in Exercises 12-15.

12. The sample space contains 38 elements, and the number 3 slot is one of these 38. Thus, the probability that the ball falls into the 3-slot is:
$$\frac{1}{38}.$$

13. Since there are 18 slots that are red out of the total of 38 slots, the probability that the ball falls into a red slot is
$$\frac{18}{38} = \frac{9}{19}.$$

14. Since there are no slots that are blue, the probability that the ball falls into a blue slot is

$$\frac{0}{38} = 0,$$

that is, this event is impossible.

15. Since there are 18 of the 36 numbered slots that are odd (excluding the 2 that are house numbers, 00 and 0), the probability that the ball falls into an odd numbered slot is

$$\frac{18}{38} = \frac{9}{19}.$$

A card is drawn from a well-shuffled deck of 52 cards. In Exercises 16-22 find the probability of each event.

16. Since there are 4 aces out of the 52 cards in the deck, the probability of drawing an ace is

$$\frac{4}{52} = \frac{1}{13}.$$

17. Since there are 26 black cards out of the 52 cards in the deck (the spades and clubs), the probability of drawing a black card is

$$\frac{26}{52} = \frac{1}{2}.$$

18. Since there is only 1 queen of hearts out of the 52 cards in the deck, the probability of drawing the queen of hearts is

$$\frac{1}{52}.$$

19. Since there are 4 sevens and 4 queens, for a total of 8 cards that are favorable to this event, out of the 52 cards in the deck, the probability of drawing a seven or a queen is

$$\frac{8}{52} = \frac{2}{13}.$$

20. There are 13 diamonds and 4 sevens in a deck. However, the seven of diamonds is counted twice if we simply add 13 and 4. Consider the 13 diamonds (one of which is the seven of diamonds) and add to this total the 3 sevens that are not diamonds, and we obtain a total of 16 cards that are favorable to this event. Thus, the probability of drawing a seven or a diamond is

$$\frac{16}{52} = \frac{4}{13}.$$

21. There are 12 face cards (jacks, queens, and kings) and 13 hearts in a deck. However if we simply add 12 and 13, the 3 heart face cards will have been counted twice. Start with the 13 hearts, add to this total the 9 face cards that are not hearts, and we obtain a total of 22 cards that are favorable to this event. Thus, the probability of drawing a face card or a heart is

$$\frac{22}{52} = \frac{11}{26}.$$

22. The event of drawing a king *and* a heart consists of only one card, the king of hearts. Thus the probability of drawing a king and a heart is

$$\frac{1}{52}.$$

If 5 cards are dealt form a well-shuffled deck of 52 cards, find the probability of the hands in Exercises 23-25.

23. The number of ways of obtaining 5 hearts can be thought of as a combination problem, the number of ways of selecting 5 hearts from the 13 hearts. This number of ways is:

$$_{13}C_5 = 1287$$

The number of ways of selecting 5 cards from 52 cards, the number of objects in the sample space of being dealt five cards, is:

$$_{52}C_5 = 2,598,960$$

Thus, the probability of being dealt 5 hearts is:

$$\frac{_{13}C_5}{_{52}C_5} = \frac{1287}{2,598,960} = \frac{33}{66,640}$$

24. The number of ways of being dealt 4 aces and 1 card that is not an ace is

$$_4C_4 \cdot {}_{48}C_1 = (1)(48) = 48$$

Then the probability of being dealt such a hand is:

$$\frac{_4C_4 \cdot {}_{48}C_1}{_{52}C_5} = \frac{48}{2,598,960} = \frac{1}{54,145}$$

EXERCISES A

25. The number of ways of being dealt 3 queens and 2 jacks is:

$$_4C_3 \cdot {_4C_2} = (4)(6) = 24$$

Thus, the probability of being dealt such a hand is:

$$\frac{_4C_3 \cdot {_4C_2}}{_{52}C_5} = \frac{24}{2{,}598{,}960} = \frac{1}{108{,}290}$$

26. If the probability of winning a game is $\frac{4}{7}$, assuming that there is no possibility of a tie, the probability of losing the game is

$$1 - \frac{4}{7} = \frac{3}{7}.$$

27. The number of ways of dialing seven digits can be thought of as the number of ways of filling seven slots with any of 10 choices (the 10 digits). The total number of ways this can be done is:

$$10 \cdot 10 \cdot 10 \cdot 10 \cdot 10 \cdot 10 \cdot 10 = 10^7$$

Since only one of these choices can be the child's own telephone number, the probability of dialing his own number is:

$$\frac{1}{10^7} = 10^{-7}$$

28. Since there are a total of 2^{48} different possible genetic combinations that a child of one couple can have, the probability that a second child has the same genetic make-up as a first is:

$$\frac{1}{2^{48}} = 2^{-48}$$

Use the bar graph given in the text to find the probabilities requested in Exercises 29-32.

29. The total number of students in the sample is 1400, and this number is the number of elements in the sample space. Since there are 300 math majors and 400 science majors, for a total of 700 students out of the population of 1400, the probability that a particular student is a math major or science major is:

$$\frac{700}{1400} = \frac{1}{2} = 0.5$$

30. Since the number of students that are *not* history majors, the total of the other four categories, is 1100, the probability that a student is not a history major is:

$$\frac{1100}{1400} = \frac{11}{14} \approx 0.79$$

31. The total number of students in math, English, or history is 750, so the probability that a student has one of these majors is:

$$\frac{750}{1400} = \frac{15}{28} \approx 0.54$$

32. If a student is not in math, not in science, and not in history, then the student is among the 400 in English or some other major, and the probability of this is:

$$\frac{400}{1400} = \frac{2}{7} \approx 0.29$$

Use the histogram given in the text to find the requested probabilities in Exercises 33-36.

33. The total number of professors at the university is 130, which is the number elements in the sample space. The number of professors with less than 4 or more than 24 years of experience is $5 + 5 = 10$. Thus, the probability that a particular professor is in this category is:

$$\frac{10}{130} = \frac{1}{13} \approx 0.08$$

34. The number of professors with 8 or more years of experience is $25 + 40 + 30 + 15 + 5 = 115$. Thus, the probability that a particular professor is in this category is:

$$\frac{115}{130} = \frac{23}{26} \approx 0.88$$

35. The number of professors that do not have at least 8 years of experience, that is that have less than 8 years of experience, is $5 + 10 = 15$. Thus, the probability that a particular professor is in this category is:

$$\frac{15}{130} = \frac{3}{26} \approx 0.12$$

36. The number of professors that have 4-7 or 20-24 years of experience is $10 + 15 = 25$. Thus, the

probability that a professor is in this category is:

$$\frac{25}{130} = \frac{5}{26} \approx 0.19$$

37. Use the binomial theorem to expand $(2x - 5y)^4$.

    In this case, $a = 2x$, $b = -5y$, and $n = 4$.

    $$(2x - 5y)^4$$
    $$= \sum_{r=0}^{4} \binom{4}{r}(2x)^{4-r}(-5y)^r$$
    $$= \binom{4}{0}(2x)^{4-0}(-5y)^0 + \binom{4}{1}(2x)^{4-1}(-5y)^1 +$$
    $$\binom{4}{2}(2x)^{4-2}(-5y)^2 + \binom{4}{3}(2x)^{4-3}(-5y)^3 +$$
    $$\binom{4}{4}(2x)^{4-4}(-5y)^4$$
    $$= (1)(2x)^4(1) + (4)(2x)^3(-5y) + (6)(2x)^2(-5y)^2 +$$
    $$(4)(2x)(-5y)^3 + (1)(1)(-5y)^4$$
    $$= 16x^4 - 160x^3y + 600x^2y^2 - 1000xy^3 + 625y^4$$

38. Find the 5th term of $(x^2 - 2y)^7$.

    In this case, $a = x^2$, $b = -2y$, $n = 7$ and $r = 4$.

    $$\binom{n}{r}a^{n-r}b^r = \binom{7}{4}(x^2)^{7-4}(-2y)^4$$
    $$= \frac{7!}{(7-4)!4!}(x^2)^3(-2y)^4$$
    $$= (35)x^6(16y^4)$$
    $$= 560x^6y^4$$

CHAPTER 9 SEQUENCES, SERIES, AND PROBABILITY

CHAPTER 9 Review Exercises

In Exercises 1-2 the nth term of a sequence is given. Find the first five terms and the eighth term.

1. $a_n = \frac{n^2-1}{n}$

$$a_1 = \frac{(1)^2-1}{1} = \frac{0}{1} = 0$$
$$a_2 = \frac{(2)^2-1}{2} = \frac{3}{2}$$
$$a_3 = \frac{(3)^2-1}{3} = \frac{8}{3}$$
$$a_4 = \frac{(4)^2-1}{4} = \frac{15}{4}$$
$$a_5 = \frac{(5)^2-1}{5} = \frac{24}{5}$$
$$a_8 = \frac{(8)^2-1}{8} = \frac{63}{8}$$

2. $x_n = \frac{(-1)^n}{3n+1}$

$$x_1 = \frac{(-1)^1}{3(1)+1} = \frac{-1}{4} = -\frac{1}{4}$$
$$x_2 = \frac{(-1)^2}{3(2)+1} = \frac{1}{7}$$
$$x_3 = \frac{(-1)^3}{3(3)+1} = \frac{-1}{10} = -\frac{1}{10}$$
$$x_4 = \frac{(-1)^4}{3(4)+1} = \frac{1}{13}$$
$$x_5 = \frac{(-1)^5}{3(5)+1} = \frac{-1}{16} = -\frac{1}{16}$$
$$x_8 = \frac{(-1)^8}{3(8)+1} = \frac{1}{25}$$

3. $\sum_{k=0}^{3} \sqrt{k^2+1}$

$$= \sqrt{0^2+1} + \sqrt{1^2+1} + \sqrt{2^2+1} + \sqrt{3^2+1}$$
$$= \sqrt{1} + \sqrt{2} + \sqrt{5} + \sqrt{10}$$
$$= 1 + \sqrt{2} + \sqrt{5} + \sqrt{10}$$

4. $\frac{1}{2} + \frac{4}{3} + \frac{9}{4} + \frac{16}{5} + \ldots + \frac{n^2}{n+1} = \sum_{k=1}^{n} \frac{k^2}{k+1}$

5. If $a_1 = -3$ and $a_{n+1} = 1 - 2a_n$, then the second and third terms of the sequence are:

$$a_2 = 1 - 2a_1 = 1 - 2(-3) = 1 + 6 = 7$$
$$a_3 = 1 - 2a_2 = 1 - 2(7) = 1 - 14 = -13$$

6. Find the formula for the nth term of

3, 6, 11, 18, 27, ...

Note that each term is 2 more than the square of a counting number. Thus,

$$a_n = n^2 + 2.$$

7. Evaluate the series. $\sum_{k=1}^{5} (-1)^{k+1} 2^k$

$$\sum_{k=1}^{5} (-1)^{k+1} 2^k$$
$$= (-1)^{1+1}2^1 + (-1)^{2+1}2^2 + (-1)^{3+1}2^3 + (-1)^{4+1}2^4 + (-1)^{5+1}2^5$$
$$= (1)(2) + (-1)(4) + (1)(8) + (-1)(16) + (1)(32)$$
$$= 2 - 4 + 8 - 16 + 32$$
$$= 22$$

8. Evaluate the series. $\sum_{m=3}^{6} \frac{m+5}{m-1}$

$$\sum_{m=3}^{6} \frac{m+5}{m-1} = \frac{3+5}{3-1} + \frac{4+5}{4-1} + \frac{5+5}{5-1} + \frac{6+5}{6-5}$$
$$= \frac{8}{2} + \frac{9}{3} + \frac{10}{4} + \frac{11}{1}$$
$$= 4 + 3 + \frac{5}{2} + 11$$
$$= \frac{41}{2}$$

9. Write the first six terms of an arithmetic sequence with $a_1 = -9$ and $d = 4$, and find S_{12}.

$$a_1 = -9$$
$$a_2 = a_1 + d = -9 + 4 = -5$$
$$a_3 = a_2 + d = -5 + 4 = -1$$
$$a_4 = a_3 + d = -1 + 4 = 3$$
$$a_5 = a_4 + d = 3 + 4 = 7$$
$$a_6 = a_5 + d = 7 + 4 = 11$$

$$S_{12} = \frac{12}{2}[2a_1 + (12-1)d]$$
$$= 6[2(-9) + (11)(4)]$$
$$= 6[-18 + 44]$$
$$= 6[26]$$
$$= 156$$

10. For an arithmetic sequence, $a_1 = 5$, $a_n = 19$, and $S_n = 96$. Find n and d.

$$S_n = \frac{n}{2}[a_1 + a_n]$$
$$96 = \frac{n}{2}[5 + 19]$$
$$192 = n[24]$$
$$8 = n$$

$$a_n = a_1 + (n-1)d$$
$$19 = 5 + (8-1)d$$
$$14 = 7d$$
$$2 = d$$

11. To insert three arithmetic means between 17 and 5, consider finding a five-term sequence with $a_1 = 17$ and $a_5 = 5$. First find n.

$$a_5 = a_1 + (5-1)d$$
$$5 = 17 + 4d$$
$$-12 = 4d$$
$$-3 = d$$

Then the three arithmetic means are:

$$a_2 = a_1 + d = 17 + (-3) = 14$$
$$a_3 = a_2 + d = 14 + (-3) = 11$$
$$a_4 = a_3 + d = 11 + (-3) = 8$$

12. For $x+1$, $2x+3$, $4x+1$ to be a three-term arithmetic sequence, the difference of successive terms must be equal (to the common difference d). Thus, solve for x:

$$(2x + 3) - (x + 1) = (4x + 1) - (2x + 3)$$
$$2x + 3 - x - 1 = 4x + 1 - 2x - 3$$
$$x + 2 = 2x - 2$$
$$4 = x$$

When $x = 4$:

$$x+1 = 5 \quad 2x+3 = 11 \quad 4x+1 = 17$$

Thus, the desired sequence is

$$5, 11, 17.$$

13. The depreciations, 24%, 20%, 16%, ..., form an arithmetic sequence with $a_1 = 24\% = 0.24$ and $d = -0.04$. The sum of the first five terms gives the total depreciation.

$$S_5 = \frac{5}{2}[2(0.24) + (5-1)(-0.04)]$$
$$= \frac{5}{2}[0.48 - 0.16]$$
$$= \frac{5}{2}[0.32]$$
$$= 0.80$$

Then the total depreciation is

$$0.08(9000) = \$7200,$$

and the value of the car is

$$\$9000 - \$7200 = \$1800.$$

14. To find the total number of dimes in the collection, we must find the sum of the twenty terms of the arithmetic sequence

$$20 + 19 + 18 + \ldots + 1,$$

for which $a_1 = 20$, $d = -1$, $n = 20$, and $a_{20} = 1$.

$$S_{20} = \frac{20}{2}[a_1 + a_{20}]$$
$$= \frac{20}{2}[20 + 1]$$
$$= 10[21]$$
$$= 210$$

Then the value of the collection of dimes is

$$210(\$0.10) = \$21.00.$$

15. Write the first six terms of a geometric sequence with $a_1 = \frac{1}{6}$ and $r = 2$, and find S_6.

$$a_1 = \frac{1}{6}$$
$$a_2 = a_1 r = \left(\frac{1}{6}\right)(2) = \frac{2}{6} = \frac{1}{3}$$
$$a_3 = a_2 r = \left(\frac{1}{3}\right)(2) = \frac{2}{3}$$
$$a_4 = a_3 r = \left(\frac{2}{3}\right)(2) = \frac{4}{3}$$
$$a_5 = a_4 r = \left(\frac{4}{3}\right)(2) = \frac{8}{3}$$
$$a_6 = a_5 r = \left(\frac{8}{3}\right)(2) = \frac{16}{3}$$

$$S_6 = \frac{a_1 - ra_6}{1-r}$$

$$= \frac{\frac{1}{6} - (2)\left(\frac{16}{3}\right)}{1-2}$$

$$= \frac{\frac{1}{6} - \frac{32}{3}}{-1}$$

$$= \frac{\frac{1}{6} - \frac{64}{6}}{-1}$$

$$= \frac{-\frac{63}{6}}{-1} = \frac{63}{6} = \frac{21}{2}$$

16. To find the sum of the first eight terms of the geometric sequence with $a_1 = \frac{1}{27}$ and $a_8 = 81$, first find r.

$$a_8 = a_1 r^{8-1}$$
$$81 = \frac{1}{27} r^7$$
$$(81)(27) = r^7$$
$$3^7 = r^7$$
$$3 = r$$

Then the sum of the first eight terms is:

$$S_8 = \frac{a_1 - 2a_8}{1-r}$$

$$= \frac{\frac{1}{27} - (3)(81)}{1-3}$$

$$= \frac{\frac{1}{27} - 243}{-2}$$

$$= \frac{\frac{1}{27} - \frac{6561}{27}}{-2}$$

$$= \frac{-\frac{6560}{27}}{-2}$$

$$= \frac{3280}{27}$$

17. To insert four geometric means between -12 and $\frac{3}{8}$, think of a six term geometric sequence with $a_1 = -12$ and $a_6 = \frac{3}{8}$. First find r.

$$a_6 = a_1 r^5$$
$$\frac{3}{8} = -12 r^5$$
$$-\frac{1}{32} = r^5$$
$$\left(-\frac{1}{2}\right)^5 = r^5$$
$$-\frac{1}{2} = r$$

Then the four geometric means are:

$$a_2 = a_1 r = (-12)\left(-\frac{1}{2}\right) = 6$$
$$a_3 = a_2 r = (6)\left(-\frac{1}{2}\right) = -3$$
$$a_4 = a_3 r = (-3)\left(-\frac{1}{2}\right) = \frac{3}{2}$$
$$a_5 = a_4 r = \left(\frac{3}{2}\right)\left(-\frac{1}{2}\right) = -\frac{3}{4}$$

18. To find a positive number x so that $3x-1, x+3, x-2$ forms a three-term geometric sequence, use the fact that the ratio of successive terms is constant (the common ratio), and solve the equation for x.

$$\frac{x+3}{3x-1} = \frac{x-2}{x+3}$$
$$(x+3)(x+3) = (x-2)(3x-1)$$
$$x^2 + 6x + 9 = 3x^2 - 7x + 2$$
$$0 = 2x^2 - 13x - 7$$
$$0 = (2x+1)(x-7)$$

Using the zero-product rule we obtain:

$$2x + 1 = 0 \qquad x - 7 = 0$$
$$2x = -1 \qquad x = 7$$
$$x = -\frac{1}{2}$$

Since x was to be positive, we discard the negative solution and have that $x = 7$. Then

$$3x-1 = 20 \quad x+3 = 10 \quad x-2 = 5.$$

Thus the sequence is

$$20, 10, 5.$$

19. Let $a_1 = 1500$. At the end of the first year, the amount that would have to be paid is

$$a_2 = 1500 + (0.09)(1500) = (1.09)(1500).$$

At the end of the second year, the amount to be paid would be

$$a_3 = (1.09)(1500) + (0.09)(1.09)(1500)$$
$$= (1.09)^2(1500)$$

Continuing in this manner, at the end of the fifth year, the amount to be paid is

$$a_6 = (1.09)^5(1500).$$

Note that we have the *sixth* term of the sequence, not the *fifth* term. Then the amount to be paid back is

$$a_6 = \$2307.94. \quad \text{Use } y^x \text{ button}$$

20. On the first drop, the ball falls 30.0 ft. It then rebounds and falls $\frac{4}{5}(30.0)$ ft and hits the ground the second time. Then it rebounds and falls $\left(\frac{4}{5}\right)^2(30.0)$ ft and hits the ground the third time, and so forth. We must consider the following sum:

$$30.0 + \quad \text{Hits ground 1st time}$$
$$\left(\tfrac{4}{5}\right)(30.0) + \left(\tfrac{4}{5}\right)(30.0) + \quad \text{Hits ground 2nd time}$$
$$\left(\tfrac{4}{5}\right)^2(30.0) + \left(\tfrac{4}{5}\right)^2(30.0) + \quad \text{Hits ground 3rd time}$$
$$\vdots$$
$$\left(\tfrac{4}{5}\right)^5(30.0) + \left(\tfrac{4}{5}\right)^5(30.0) \quad \text{Hits ground 6th time}$$

Then the fifth rebound, the ball travels a distance of:

$$\left(\tfrac{4}{5}\right)^5(30.0) \approx 9.8 \text{ ft} \quad \textit{Use calculator}$$

If we find the sum

$$\left(\tfrac{4}{5}\right)(30.0) + \left(\tfrac{4}{5}\right)^2(30.0) + \left(\tfrac{4}{5}\right)^3(30.0) + \ldots + \left(\tfrac{4}{5}\right)^5(30.0)$$

and double the result and add 30.0, we will obtain the total distance traveled. Considering the series above, $a_1 = \left(\tfrac{4}{5}\right)(30.0) = 24.0$, $n = 5$, and

$$r = \tfrac{4}{5}.$$

Thus, the sum is:

$$S_5 = \frac{a_1 - a_1 r^5}{1-r}$$
$$= \frac{24 - 24\left(\tfrac{4}{5}\right)^5}{1 - \tfrac{4}{5}}$$
$$\approx 80.6784 \quad \textit{Using } y^x \textit{ button}$$

Multiply this number by 2 and add 30.0 to obtain

$$191.3568.$$

Thus, the total distance traveled by the ball is about 191.4 ft.

21. Find the sum of the infinite geometric sequence

$$8, -2, \tfrac{1}{2}, -\tfrac{1}{8}, \ldots.$$

We have that $a_1 = 8$ and $r = -\tfrac{1}{4}$. Since $|r| < 1$, the sum exists and is given by:

$$S = \frac{a_1}{1-r} = \frac{8}{1-\left(-\tfrac{1}{4}\right)} = \frac{8}{\tfrac{5}{4}} = \frac{32}{5}$$

22. To find the first five terms of an infinite geometric sequence with $a_1 = 30$ and $S = 36$, first find the value of r.

$$S = \frac{a_1}{1-r}$$
$$36 = \frac{30}{1-r}$$
$$36(1-r) = 30$$
$$36 - 36r = 30$$
$$-36r = -6$$
$$r = \tfrac{1}{6}$$

$$a_1 = 30$$
$$a_2 = a_1 r = (30)\left(\tfrac{1}{6}\right) = 5$$
$$a_3 = a_2 r = (5)\left(\tfrac{1}{6}\right) = \tfrac{5}{6}$$
$$a_4 = a_3 r = \left(\tfrac{5}{6}\right)\left(\tfrac{1}{6}\right) = \tfrac{5}{36}$$
$$a_5 = a_4 r = \left(\tfrac{5}{36}\right)\left(\tfrac{1}{6}\right) = \tfrac{5}{216}$$

23. $1.\overline{2} = 1 + 0.2 + 0.02 + 0.002 + 0.0002 + \ldots$

First, ignore the first term, 2, and find the sum of the remaining terms which form an infinite geometric series with $a_1 = 0.2$ and $r = 0.1$.

$$S = \frac{a_1}{1-r} = \frac{0.2}{1-0.1} = \frac{0.2}{0.9} = \frac{2}{9}$$

Then we have that

$$1.\overline{2} = 1 + \frac{2}{9} = \frac{9}{9} + \frac{2}{9} = \frac{11}{9}.$$

24. $5.\overline{15} = 5 + 0.15 + 0.0015 + 0.000015 + ...$

First, ignore the first term, 5, and find the sum of the remaining terms which form an infinite geometric series with $a_1 = 0.15$ and $r = 0.01$.

$$S = \frac{a_1}{1-r} = \frac{0.15}{1-0.01} = \frac{0.15}{0.99} = \frac{15}{99} = \frac{5}{33}$$

Then we have that

$$5.\overline{15} = 5 + \frac{5}{33} = \frac{165}{33} + \frac{5}{33} = \frac{170}{33}.$$

25. The distances traveled by the child on the swing form the following geometric series.

$$8 + \left(\frac{8}{9}\right)(8) + \left(\frac{8}{9}\right)^2(8) + \left(\frac{8}{9}\right)^3(8) + ...$$

Then $a_1 = 8$, $r = \frac{8}{9}$, and

$$S = \frac{a_1}{1-r} = \frac{8}{1-\frac{8}{9}} = \frac{8}{\frac{1}{9}} = 72.$$

Thus, the child will travel about 72 m before coming to rest.

26. To find the total distance traveled by the ball, up and down, consider the following sum:

$$27 + \left(\frac{2}{3}\right)27 + \left(\frac{2}{3}\right)^2 27 + \left(\frac{2}{3}\right)^2 27 + \left(\frac{2}{3}\right)^2 27 + ...$$
$$= 27 + 2\left[\left(\frac{2}{3}\right)27 + \left(\frac{2}{3}\right)^2 27 + \left(\frac{2}{3}\right)^3 27 + ...\right]$$

The terms inside the brackets form an infinite geometric series with $a_1 = \frac{2}{3}(27) = 18$, and $r = \frac{2}{3}$. The sum of this series is:

$$S = \frac{a_1}{1-r} = \frac{18}{1-\frac{2}{3}} = \frac{18}{\frac{1}{3}} = 54$$

Double this number, and add 27, to find the total distance traveled by the ball, 135 ft.

In Exercises 27-28 use the principle of mathematical induction to prove that each statement is true for every positive integer n.

27. $S(n)$: $4 + 8 + 12 + ... + 4n = 2n(n + 1)$

Verification: When $n = 1$, we have

$$4 = 2(1)(1 + 1) = 2(2) = 4,$$

so the statement is true for $n = 1$.

Induction: Assume that

$$4 + 8 + 12 + ... + 4k = 2k(k + 1)$$

is true. We must show that

$$4 + 8 + ... + 4k + 4(k + 1) = 2(k+1)((k+1) + 1)$$

is true. Use the induction hypothesis and replace the first k terms of the left side of the above expression with $2k(k + 1)$

$$2k(k + 1) + 4(k + 1) = 2(k + 1)[k + 2]$$
$$= 2(k + 1)((k + 1) + 1)$$

Thus the statement is true for $n = k+1$ under the assumption that it is true for $n = k$.

Conclusion: By the principle of mathematical induction, the statement $S(n)$ is true for every positive integer n.

28. $1 \cdot 2 + 2 \cdot 3 + 3 \cdot 4 + ... + n(n+1) = \frac{1}{3}n(n+1)(n+2)$

Verification: When $n = 1$,

$$1 \cdot 2 = \frac{1}{3}(1)(1+1)(1+2) = \frac{1}{3}(2)(3) = 2$$

so the statement is true in this case.

Induction: Assume that

$$1 \cdot 2 + 2 \cdot 3 + 3 \cdot 4 + ... + k(k+1) = \frac{1}{3}k(k+1)(k+2)$$

is true, we must show that

$$1 \cdot 2 + 2 \cdot 3 + ... + k(k+1) + (k+1)(k+2)$$
$$= \frac{1}{3}(k+1)(k+2)(k+3)$$

Use the induction hypothesis and replace the first k terms in the expression on the left side.

$$\frac{1}{3}k(k+1)(k+2) + (k+1)(k+2)$$
$$= (k+1)(k+2)[\tfrac{1}{3}k + 1]$$
$$= (k+1)(k+2)\tfrac{1}{3}(k+3)$$
$$= \tfrac{1}{3}(k+1)(k+2)(k+3)$$
$$= \tfrac{1}{3}(k+1)((k+1)+1)((k+1)+2)$$

Thus, the statement is true for $n = k+1$ under the assumption that it is true for $n = k$.

Conclusion: By the principle of mathematical induction, the statement is true for every positive integer n.

29. $\;_3P_2 = \dfrac{3!}{(3-2)!} = \dfrac{3!}{1!} = 3! = 6$

30. $\;_3C_2 = \dfrac{3!}{(3-2)!2!} = \dfrac{3!}{1!2!} = \dfrac{3\cdot 2\cdot 1}{1\cdot 2\cdot 1} = 3$

31. $\;_5P_5 = \dfrac{5!}{(5-5)!} = \dfrac{5!}{0!} = \dfrac{5!}{1} = 5! = 120$

32. $\;_5C_5 = \dfrac{5!}{(5-5)!5!} = \dfrac{5!}{0!5!} = \dfrac{5!}{1\cdot 5!} = 1$

33. $\;_8P_0 = \dfrac{8!}{(8-0)!} = \dfrac{8!}{8!} = 1$

34. $\;_8C_0 = \dfrac{8!}{(8-0)!0!} = \dfrac{8!}{8!\cdot 1} = 1$

35. How many serial numbers formed by a letter followed by a three-digit numeral are possible under the following conditions?

   (a) When no restrictions are imposed, the number is:

   $\underline{26\cdot 10\cdot 10\cdot 10} = 26{,}000$

   (b) If the letter cannot be an A (leaving 25 choices) and the digits cannot be repeated, the number is:

   $\underline{25\cdot 10\cdot 9\cdot 8} = 18{,}000$

   (c) If the letter must be an A, B, or C (three choices) and the first digit cannot be zero (nine choices) the number is:

   $\underline{3\cdot 9\cdot 10\cdot 10} = 2700$

36. The number of ways to choose four officers from an organization of thirteen members is a permutation problem since the set of four chosen can be the officers in various ways. Thus the number of ways of doing this is:

   $_{13}P_4 = \dfrac{13!}{(13-4)!} = \dfrac{13!}{9!} = 13\cdot 12\cdot 11\cdot 10 = 17{,}160$

37. The number of ways to place nine people in a line, a permutation problem, is:

   $9! = 362{,}880$

38. The number of ways to seat nine people around a table, a circular permutation problem, is:

   $(n-1)! = (9-1)! = 8! = 40{,}320$

39. Since FLAGSTAFF has 9 letters, with 3 F's, 1 L, 2 A's, 1 G, 1 S, and 1 T, the number of distinguishable permutations of the nine letters is:

   $\dfrac{9!}{3!1!2!1!1!1!} = \dfrac{9!}{3!2!} = \dfrac{9!}{12} = 30{,}240$

40. Since two points determine a line, we must select two of the six (noncollinear) points for each line. The number of ways of doing this is:

   $_6C_2 = \dfrac{6!}{(6-2)!2!} = \dfrac{6!}{4!2!} = 15$

41. The number of ways to choose a committee of three men and two women from eight men and eleven women is:

   $_8C_3\cdot\,_{11}C_2 = \dfrac{8!}{(8-3)!3!} + \dfrac{11!}{(11-2)!2!} = (56)(55) = 3080$

42. The number of ways of choosing a committee of four from an organization of thirteen members is:

   $_{13}C_4 = \dfrac{13!}{(13-4)!4!} = \dfrac{13\cdot 12\cdot 11\cdot 10\cdot 9!}{9!\cdot 4\cdot 3\cdot 2\cdot 1} = 715$

CHAPTER 9 REVIEW — REVIEW EXERCISES

43. Expand $(3a - z)^5$.

$$(3a-z)^5$$
$$= \sum_{r=0}^{5} \binom{5}{r}(3a)^{5-r}(-z)^r$$
$$= \binom{5}{0}(3a)^{5-0}(-z)^0 + \binom{5}{1}(3a)^{5-1}(-z)^1 +$$
$$\binom{5}{2}(3a)^{5-2}(-z)^2 + \binom{5}{3}(3a)^{5-3}(-z)^3 +$$
$$\binom{5}{4}(3a)^{5-4}(-z)^4 + \binom{5}{5}(3a)^{5-5}(-z)^5$$
$$= (1)(3a)^5(-z)^0 + (5)(3a)^4(-z) + (10)(3a)^3(-z)^2 +$$
$$(10)(3a)^2(-z)^3 + (5)(3a)(-z)^4 + (1)(3a)^0(-z)^5$$
$$= 243a^5 - 405a^4z + 270a^3z^2 - 90a^2z^3 + 15az^4 - z^5$$

44. Expand $(y^{-1} + y)^4$

$$(y^{-1}+y)^4$$
$$= \sum_{r=0}^{4} \binom{4}{r}(y^{-1})^{4-r}(y)^r$$
$$= \binom{4}{0}(y^{-1})^{4-0}(y)^0 + \binom{4}{1}(y^{-1})^{4-1}(y)^1 +$$
$$\binom{4}{2}(y^{-1})^{4-2}(y)^2 + \binom{4}{3}(y^{-1})^{4-3}(y)^3 +$$
$$\binom{4}{4}(y^{-1})^{4-4}(y)^4$$
$$= (1)(y^{-4})(1) + (4)(y^{-3})(y) + (6)(y^{-2})(y^2) +$$
$$(4)(y^{-1})(y^3) + (1)(1)(y^4)$$
$$= y^{-4} + 4y^{-2} + 6 + 4y^2 + y^4$$

45. Find the 4th term in the expansion of $(2x - y^2)^6$.

We have $a = 2x$, $b = -y^2$, $n = 6$, and $r = 3$.

$$\binom{n}{r}a^{n-r}b^r = \binom{6}{3}(2x)^{6-3}(-y^2)^3$$
$$= \frac{6!}{(6-3)!3!}(2x)^3(-y^2)^3$$
$$= (20)(8x^3)(-y^6)$$
$$= -160x^3y^6$$

46. Find the 5th term in the expansion of $(a - 3b)^7$.

We have $a = a$, $b = -3b$, $n = 7$, and $r = 4$.

$$\binom{n}{r}a^{n-r}b^r = \binom{7}{4}(a)^{7-4}(-3b)^4$$
$$= \frac{7!}{(7-4)!4!}(a)^3(-3b)^4$$
$$= (35)(a^3)(81b^4)$$
$$= 2835a^3b^4$$

One marble is selected, sight unseen, from a bag containing 8 red and 7 black marbles. In Exercises 47-50 determine the probability of the given event.

47. Since there are 15 marbles in the bag, the sample space for this experiment contains 15 elements. Since 8 marbles are red, the probability of selecting a red marble is:

$$\frac{8}{15}$$

48. The probability of selecting a black marble is:

$$\frac{7}{15}$$

49. Since there are no blue marbles in the bag, the probability of selecting a blue marble is:

$$\frac{0}{15} = 0$$

That is, this event is impossible.

50. Since all marbles in the bag are red or black, any marble selected will *not* be blue, so the probability of selecting a marble that is not blue is:

$$\frac{15}{15} = 1$$

That is, this event is certain.

A single card is drawn from a well-shuffled deck of 52 cards. In Exercises 51-56 determine the probability of the event.

51. Drawing a spade

Since there are 13 spades in the deck of 52 cards, the probability of drawing a spade is:

$$\frac{13}{52} = \frac{1}{4}$$

52. Drawing a face card

Since there are 12 face cards (4 jacks, 4 queens, and 4 kings) in the deck of 52 cards, the probability of drawing a face card is:

$$\frac{12}{52} = \frac{3}{13}$$

53. Drawing the jack of hearts

Since there is only 1 jack of hearts in the deck, the probability of drawing the jack of hearts is:

$$\frac{1}{52}$$

54. Drawing a red ace

Since there are 2 red aces (the ace of hearts and the ace of diamonds) in the deck, the probability of drawing a red ace is:

$$\frac{2}{52} = \frac{1}{26}$$

55. Drawing a ten or a queen

Since there are 4 tens and 4 queens, there are 8 cards favorable to this event. Thus, the probability of drawing a ten or a queen is:

$$\frac{8}{52} = \frac{2}{13}$$

56. Drawing a ten and a queen

Since 1 card cannot be both a ten and a queen, this is impossible. Thus, the probability of drawing a ten and a queen is 0.

57. If the probability of winning first prize in a contest is 10^{-4}, the probability of not winning is:

$$1 - 10^{-4} = 1 - \frac{1}{10^4} = \frac{10^4 - 1}{10^4} = \frac{9999}{10,000}$$

58. Given that:

$$P(\text{passing English}) = 0.92$$

$$P(\text{passing math}) = 0.85$$

$$P(\text{passing both}) = 0.78$$

$P(\text{passing English } or \text{ passing math})$
$= P(\text{passing English}) + P(\text{passing math}) - P(\text{passing both})$
$= 0.92 + 0.85 - 0.78$
$= 0.99$

The probability of failing both is the complement of passing one or the other. Thus, the probability of failing both is:

$$1 - 0.99 = 0.01.$$

Use the bar graph given in the text to determine the probabilities in Exercises 59-60.

59. There are a total of 150 people in the survey, so the number of elements in the sample space is 150. The number of people who preferred food other than American, Mexican, or Chinese is 20. Thus the probability that a person selected at random falls into this category is:

$$\frac{20}{150} = \frac{2}{15}$$

60. The number of people who prefer Mexican or Chinese is $40 + 30 = 70$. Thus the probability that a person selected at random falls into this category is:

$$\frac{70}{150} = \frac{7}{15}$$

61. Find the sum of the infinite geometric sequence

$$16, -8, 4, \ldots .$$

We have that $a_1 = 16$ and $r = -\frac{1}{2}$. Since $|r| < 1$, the sum exists and is:

$$S = \frac{a_1}{1-r} = \frac{16}{1-\left(-\frac{1}{2}\right)} = \frac{16}{\frac{3}{2}} = \frac{32}{3}$$

62. Find a formula for the general term of the general sequence

$$4, 9, 14, 19, \ldots .$$

Since each term is 1 less than a multiple of 5, the general term is

$$a_n = 5n - 1.$$

63. The number of ways of choosing 3 of the 4 aces *and* 2 of the 4 eights to obtain the desired hand is

$$_4C_3 \cdot {}_4C_2 = \frac{4!}{(4-3)!3!} \cdot \frac{4!}{(4-2)!2!} = (4)(6) = 24$$

Since the number of ways of selecting a 5-card hand is $_{52}C_5$, the probability of being dealt 3 aces and 2 eights is:

$$\frac{_4C_3 \cdot {_4C_2}}{_{52}C_5} = \frac{24}{2{,}598{,}960} = \frac{1}{108{,}290}$$

64. Find the sum of the integers divisible by 4 between −25 and 125.

The first integer is −24 and the last integer is 124. Think of the arithmetic sequence

$$-24, -20, -16, -12, \ldots, 124$$

with $a_1 = -24$, $a_n = 124$, and $d = 4$. First find n, the number of terms in this sequence.

$$\begin{aligned} a_n &= a_1 + (n-1)d \\ 124 &= -24 + (n-1)(4) \\ 148 &= 4n - 4 \\ 152 &= 4n \\ 38 &= n \end{aligned}$$

Then the sum of the integers divisible by 4 between −25 and 125 is:

$$\begin{aligned} S_{38} &= \tfrac{38}{2}[a_1 + a_{38}] \\ &= 19[-24 + 124] \\ &= 19[100] \\ &= 1900 \end{aligned}$$

65. If $x_1 = \tfrac{1}{3}$ and $x_n = 9x_{n-1} - 5$, then the second and third terms of the sequence are:

$$x_2 = 9x_1 - 5 = 9\left(\tfrac{1}{3}\right) - 5 = 3 - 5 = -2$$

$$x_3 = 9x_2 - 5 = 9(-2) - 5 = -18 - 5 = -23$$

66. The number of ways of choosing 10 problems out of 15 problems on a test is a combination problem (order of selection is not important). The number of ways of doing this is:

$$_{15}C_{10} = \frac{15!}{(15-10)!10!} = \frac{15!}{5!10!} = 3003$$

A single card is drawn from a well-shuffled deck of 52 cards. In Exercises 67-72 determine the probability of the event.

67. Drawing a black card

Since there are 26 black cards (13 spades and 13 clubs), the probability of drawing a black card is:

$$\frac{26}{52} = \frac{1}{2}$$

68. Drawing a card that is not a face card

Since there are 12 face cards (4 jacks, 4 queens, and 4 kings), the number of cards that are *not* face cards is 52 - 12 = 40. Thus, the probability of drawing a card that is not a face card is:

$$\frac{40}{52} = \frac{10}{13}$$

69. Drawing a joker

Since there are 52 cards in the deck, the jokers have been removed. Since there are 0 jokers in the deck, the probability of drawing a joker is:

$$\frac{0}{52} = 0$$

That is, this event is impossible.

70. Drawing a heart or a club

Since there are 13 hearts and 13 clubs, the total number of cards favorable to this event is 26. Thus, the probability of drawing a heart or a club is:

$$\frac{26}{52} = \frac{1}{2}$$

71. Drawing a red card or an ace

Since there are 26 red cards (13 hearts and 13 diamonds) and 4 aces, we might be tempted to think there are 30 (26 + 4) favorable cards. However, the two red aces have been counted twice. Think of taking the 26 red cards and adding the 2 black aces for a total of 28 favorable cards. Thus, the probability of drawing a red card or an ace is:

$$\frac{28}{52} = \frac{7}{13}$$

72. Drawing a face card or a club

Since there are 12 face cards and 13 clubs, we might think there are 25 favorable cards. However, the 3 club face cards have been counted twice. Think of taking the 13 clubs and adding the 9 face cards that are not clubs. This total of 22 cards makes up the favorable cards. Thus the probability of drawing a face card or a club is:

$$\frac{22}{52} = \frac{11}{26}$$

73. The present population is $a_1 = 8500$. In 1 year, the population will be

$$a_2 = 8500 - 400 = 8500 + (1)(-400).$$

In 2 years, the population will be

$$a_3 = 8500 + (2)(-400).$$

Continuing in this manner, we see that we are obtaining an arithmetic sequence with $d = -400$, and the population in 9 years will be given by

$$a_{10} = 8500 + (10-1)(-400).$$

Notice that the term is the 10th term, not the ninth. Thus, the population will be:

$$8500 + 9(-400) = 4900$$

74. $_6C_4 = \dfrac{6!}{(6-4)!4!} = \dfrac{6 \cdot 5 \cdot 4!}{2!4!} = \dfrac{6 \cdot 5}{2} = 15$

75. $_6P_4 = \dfrac{6!}{(6-4)!} = \dfrac{6 \cdot 5 \cdot 4 \cdot 3 \cdot 2!}{2!} = 6 \cdot 5 \cdot 4 \cdot 3 = 360$

76. $_{12}C_{10} = \dfrac{12!}{(12-10)!10!} = \dfrac{12 \cdot 11 \cdot 10!}{2!10!} = 6 \cdot 11 = 66$

77. Find the 4th term of $(x^2 - y^2)^6$.

We have that $a = x^2$, $b = -y^2$, $n = 6$, and $r = 3$.

$$\binom{n}{r}a^{n-r}b^r = \binom{6}{3}(x^2)^{6-3}(-y^2)^3$$
$$= \frac{6!}{(6-3)!3!}(x^2)^3(-y^2)^3$$
$$= (20)(x^6)(-y^6)$$
$$= -20x^6y^6$$

78. In a geometric sequence $a_1 = 2$ and $a_4 = 16$. Find S_5.

First find r.

$$a_4 = a_1 r^{4-1}$$
$$16 = (2)r^3$$
$$8 = r^3$$
$$2 = r$$

$$S_5 = \frac{a_1 - a_1 r^5}{1 - r}$$
$$= \frac{2 - (2)(2)^5}{1 - 2}$$
$$= \frac{2 - 64}{-1}$$
$$= \frac{-62}{-1}$$
$$= 62$$

79. Convert $0.\overline{78}$ to a fraction.

Since

$$0.\overline{78} = 0.78 + 0.0078 + 0.000078 + \ldots$$

we have an infinite geometric series with $a_1 = 0.78$ and $r = 0.01$. The sum of this series is:

$$S = \frac{a_1}{1-r} = \frac{0.78}{1 - 0.01} = \frac{0.78}{0.99} = \frac{78}{99} = \frac{26}{33}$$

Thus, $0.\overline{78} = \frac{26}{33}$.

80. Prove the statement is true for every positive integer n using mathematical induction.

$$8 \text{ divides } (9^n - 1)$$

Verification: When $n = 1$, $9^1 - 1 = 9 - 1 = 8$, so 8 divides $9^1 - 1$ is true.

Induction: Assume that 8 divides $(9^k - 1)$ is true. We must show that 8 divides $(9^{k+1} - 1)$ is true. Since $9^{k+1} - 1 = 9 \cdot 9^k - 9 + 8 = 9(9^k - 1) + 8$, and 8 divides the first term by hypothesis and clearly divides the second term (8), we have that the statement is true for $n = k+1$ under the assumption it is true for $n = k$.

Conclusion: Thus, the statement is true for every positive integer by the principle of mathematical induction.

CHAPTER 9 Test

CHAPTER 9 SEQUENCES, SERIES, AND PROBABILITY

1. Give the sixth term of the sequence

   $$a_n = (-1)^{n+1}(2n - 1).$$

   Substitute 6 for n throughout.

   $$\begin{aligned}a_6 &= (-1)^{6+1}(2(6) - 1)\\ &= (-1)^7(12 - 1)\\ &= (-1)(11)\\ &= -11\end{aligned}$$

2. Find the second and third terms of the sequence.

   $$a_1 = 5;\ a_{n+1} = 2a_n - 4$$

   $$a_2 = 2a_1 - 4 = 2(5) - 4 = 10 - 4 = 6$$

   $$a_3 = 2a_2 - 4 = 2(6) - 4 = 12 - 4 = 8$$

3. Use

   $$\sum_{k=1}^{n} k = \frac{n(n+1)}{2}$$

   and

   $$\sum_{k=1}^{n} k^2 = \frac{n(n+1)(2n+1)}{6}$$

   to determine the sum.

   $$\sum_{k=1}^{4}(k+k^2)$$

   $$\begin{aligned}\sum_{k=1}^{4}(k+k^2) &= \sum_{k=1}^{4}k + \sum_{k=1}^{4}k^2\\ &= \frac{4(4+1)}{2} + \frac{4(4+1)(2(4)+1)}{6}\\ &= \frac{20}{2} + \frac{180}{6}\\ &= 10 + 30\\ &= 40\end{aligned}$$

4. Given that $a_{10} = 24$ and $S_{10} = 55$ for an arithmetic sequence. Note that $n = 10$ is also given. Find d and a_1.

   First find a_1 using one of the sum formulas.

   $$\begin{aligned}S_{10} &= \tfrac{10}{2}[a_1 + a_{10}]\\ 55 &= 5[a_1 + 24]\\ 11 &= a_1 + 24\\ -13 &= a_1\end{aligned}$$

   Then find d.

   $$\begin{aligned}a_n &= a_1 + (n - 1)d\\ 24 &= -13 + (10 - 1)d\\ 37 &= 9d\\ \tfrac{37}{9} &= d\end{aligned}$$

5. Insert four arithmetic means between -5 and 10.

   We can think of a six-term arithmetic sequence with $a_1 = -5$ and $a_6 = 10$. First find d.

   $$\begin{aligned}a_6 &= a_1 + (6 - 1)d\\ 10 &= -5 + 5d\\ 15 &= 5d\\ 3 &= d\end{aligned}$$

   Then the four arithmetic means are:

   $$\begin{aligned}a_2 &= a_1 + d = -5 + 3 = -2\\ a_3 &= a_2 + d = -2 + 3 = 1\\ a_4 &= a_3 + d = 1 + 3 = 4\\ a_5 &= a_4 + d = 4 + 3 = 7\end{aligned}$$

6. Stacking the logs can be thought of in terms of the arithmetic sequence with first term $a_1 = 1$, $a_2 = 2$, $a_3 = 3$, and so forth. The common difference is $d = 1$, and the sum of the terms is $S_n = 120$. We must determine n, the number of terms, and ultimately, a_n, the number of terms to be placed on the bottom row to begin the stack. Use one of the sum formulas to find n.

   $$\begin{aligned}S_n &= \tfrac{n}{2}[2a_1 + (n-1)d]\\ 120 &= \tfrac{n}{2}[2(1) + (n-1)(1)]\\ 240 &= n[2 + (n-1)]\\ 240 &= n[n+1]\\ 240 &= n^2 + n\\ 0 &= n^2 + n - 240\\ 0 &= (n - 15)(n + 16)\end{aligned}$$

   Use the zero-product rule.

$$n - 15 = 0 \quad n + 16 = 0$$
$$n = 15 \quad n = -16$$

We can discard the negative solution since the number of rows cannot be negative. Thus $n = 15$. Then the number of logs to place in the base row is:

$$a_{15} = a_1 + (15-1)d$$
$$= 1 + (14)(1)$$
$$= 1 + 14$$
$$= 15$$

7. Given $a_1 = 3$, $a_n = -\frac{1}{729}$, and $S_n = \frac{1640}{729}$ for a geometric sequence. Find n and r. First find r using one of the sum formulas.

$$S_n = \frac{a_1 - ra_n}{1-r}$$
$$\frac{1640}{729} = \frac{3 - r\left(-\frac{1}{729}\right)}{1-r}$$
$$1640(1-r) = 729\left[3 + \frac{r}{729}\right]$$
$$1640 - 1640r = 2187 + r$$
$$-547 = 1641r$$
$$\frac{-547}{1641} = r$$
$$-\frac{1}{3} = r$$

Now find n.

$$a_n = a_1 r^{n-1}$$
$$-\frac{1}{729} = 3\left(-\frac{1}{3}\right)^{n-1}$$
$$-\frac{1}{2187} = \left(-\frac{1}{3}\right)^{n-1}$$
$$\left(-\frac{1}{3}\right)^7 = \left(-\frac{1}{3}\right)^{n-1}$$
$$7 = n - 1$$
$$8 = n$$

8. Find the sum of the infinite geometric sequence.

$$27, 18, 12, 8, \ldots$$

We have that $a_1 = 27$ and $r = \frac{2}{3}$. Since $|r| < 1$, the sum exists, and is given by:

$$S = \frac{a_1}{1-r} = \frac{27}{1-\frac{2}{3}} = \frac{27}{\frac{1}{3}} = 81$$

9. Let $a_1 = 10{,}000$. If the loan were to be paid at the end of 1 year, the amount to be paid would be:

$$a_2 = 10{,}000 + (0.12)(10{,}000) = (1.12)(10{,}000)$$

At the end of two years, the amount to be paid would be

$$a_3 = (1.12)^2(10{,}000).$$

Continuing in this manner, we can see that we obtain a geometric sequence with $r = 1.12$, and the amount to be paid at the end of 5 years is:

$$a_6 = (1.12)^5(10{,}000)$$
$$= \$17{,}623.42 \quad \textit{Use } y^x \textit{ button}$$

10. Use mathematical induction to prove that the statement is true for all positive integers n.

$$4 \text{ divides } 5^n - 1$$

Verification: When $n = 1$, $5^1 - 1 = 5 - 1 = 4$, so clearly 4 divides $5^1 - 1$ is true.

Induction: Assume that 4 divides $5^k - 1$ is true. We must show that 4 divides $5^{k+1} - 1$ is true. Since

$$5^{k+1} - 1 = 5^{k+1} - 5 + 5 - 1$$
$$= 5(5^k - 1) + 4$$

and 4 divides the first term by assumption and 4 clearly divides the second term 4, we have that 4 divides $5^{k+1} - 1$ under the assumption that 4 divides $5^k - 1$.

Conclusion: By the principle of mathematical induction, 4 divides $5^n - 1$ for every positive integer n.

11. $_5P_2 = \dfrac{5!}{(5-2)!} = \dfrac{5 \cdot 4 \cdot 3!}{3!} = 5 \cdot 4 = 20$

12. To find the number of three-digit numbers possible using the digits 1, 3, 5, 7, and 9, fill three slots with any of five possibilities. Thus the number of possibilities is:

$$\underline{5} \cdot \underline{5} \cdot \underline{5} = 5^3 = 125$$

CHAPTER 9 REVIEW

13. The number of ways to form a committee of three from a club with twelve members is a combination problem (order of selection is not important). The number of possible committees is:

$$_{12}C_3 = \frac{12!}{(12-3)!3!} = \frac{12 \cdot 11 \cdot 10 \cdot 9!}{9! \cdot 3 \cdot 2 \cdot 1} = 220$$

14. Since the word CRITICS has 7 letters, 2 C's, 1 R, 2 I's, 1 T, and 1 S, the number of distinguishable permutations of the letters in this word is:

$$\frac{7!}{2!1!2!1!1!} = \frac{7!}{2!2!} = \frac{7!}{4} = 1260$$

15. Find the sixth term in the binomial expansion of $(2x + y)^7$.

    We have that $a = 2x$, $b = y$, $n = 7$, and $r = 5$. (Remember that r is always 1 less than the specified term.)

$$\binom{n}{r}a^{n-r}b^r = \binom{7}{5}(2x)^{7-5}(y)^5$$
$$= \frac{7!}{(7-2)!2!}(2x)^2(y)^5$$
$$= (21)(4x^2)(y^5)$$
$$= 84x^2y^5$$

16. A single card is drawn from a deck of 52 cards. What is the probability that it is a spade or a face card?

    Since there are 52 cards in the deck, the number of elements in the sample space is 52. Since there are 13 spades and 12 face cards, we might be tempted to say that there are $13 + 12 = 25$ favorable cards. However, this is wrong since then the 3 spade face cards (jack of spades, queen of spades, king of spades) have been counted twice. Start with the 13 spades and add to them the 9 face cards that are not spades to obtain a total of 22 cards that are favorable. Then the probability of drawing a card that is a spade or a face card is:

$$\frac{22}{52} = \frac{11}{26}$$

17. If the probability of rain is $\frac{3}{5}$, the probability that it *will not* rain is

$$1 - \frac{3}{5} = \frac{2}{5}.$$

FINAL REVIEW EXERCISES

1. The set of integers is:

   $\{\ldots, -3, -2, -1, 0, 1, 2, 3, \ldots\}$

2. The numbers formed by taking quotients of integers with division by zero excluded are the rational numbers.

3. The property illustrated by

   $6(3 + x) = 18 + 6x$

   is the distributive law.

4. The product or quotient of two numbers with opposite signs is always a negative number.

5. A trinomial has three terms.

6. $a^3 - b^3 = (a - b)(a^2 + ab + b^2)$

7. If k is even, then $\sqrt[k]{a^k} = |a|$.

8. If k is odd, then $\sqrt[k]{a^k} = a$.

9. $-3 + (-2) = -(3 + 2) = -5$

10. $(-8) - (-5) = -8 + 5 = -(8 - 5) = -3$

11. $\left(\dfrac{3}{4}\right)\left(-\dfrac{16}{3}\right) = -\dfrac{3 \cdot 4 \cdot 4}{4 \cdot 3} = -4$

12. $\left(-\dfrac{7}{6}\right) \div \left(\dfrac{14}{3}\right) = \left(-\dfrac{7}{6}\right) \cdot \left(\dfrac{3}{14}\right) = -\dfrac{7 \cdot 3}{2 \cdot 3 \cdot 7 \cdot 2} = -\dfrac{1}{4}$

13. $|(-2)(-5) - (-6)| = |10 + 6| = 16$

14. $2x - (-3x + 5) = 2x + 3x - 5 = 5x - 5$

15. $(8x^2y^3 - 6xy^2 + 2xy) + (-4xy^2 - 3xy + 5) - (6x^2y^3 + 2xy^2 - 7)$
    $= 8x^2y^3 - 6xy^2 + 2xy - 4xy^2 - 3xy + 5 - 6x^2y^3 - 2xy^2 + 7$
    $= 2x^2y^3 - 12xy^2 - xy + 12$

16. $\left(\dfrac{3^0 x^{-6}}{2y^3}\right)^{-2} = \dfrac{1 \cdot x^{12}}{2^{-2} y^{-6}} = 2^2 x^{12} y^6 = 4x^{12} y^6$

17. $\left(\dfrac{4x^2 y^{-3}}{x^{-4} y^3}\right)^{-1} = \left(\dfrac{4x^6}{y^6}\right)^{-1} = \dfrac{4^{-1} x^{-6}}{y^{-6}} = \dfrac{y^6}{4x^6}$

18. $0.0000159 = 1.59 \times 10^{-5}$

19. $(8x - 3y)(5x + 4y)$
    $= (8x)(5x) + (8x)(4y) - (3y)(5x) - (3y)(4y)$
    $= 40x^2 + 32xy - 15xy - 12y^2$
    $= 40x^2 + 17xy - 12y^2$

20. $(3x - 7y)^2$
    $= (3x)^2 - 2(3x)(7y) + (7y)^2$
    $= 9x^2 - 42xy + 49y^2$

21. $(5a - 4b)(5a + 4b)$
    $= (5a)^2 - (4b)^2$
    $= 25a^2 - 16b^2$

22. $(2a + 9b)^2$
    $= (2a)^2 + 2(2a)(9b) + (9b)^2$
    $= 4a^2 + 36ab + 81b^2$

23. $x^2 - 6xy + 8y^2$

    The factors of 8 that add to give -6 are -2 and -4. Thus,

    $x^2 - 6xy + 8y^2 = (x - 2y)(x - 4y)$.

24. $8x^2 - 2y^4$

    First remove the common factor 2.

    $2(4x^2 - y^4)$

    The remaining factor is the difference of squares.

    $8x^2 - 2y^4 = 2(2x - y^2)(2x + y^2)$

25. $5a^2 + 13ab - 6b^2$

    The factors of 5 and -6 that combine to give 13 are 5,1 and -2,3. Thus,

    $5a^2 + 13ab - 6b^2 = (5a - 2b)(a + 3b)$.

26. $-6a^2 + 36ab - 54b^2$

    First remove the common factor -6.

    $-6(a^2 - 6ab + 9b^2)$

FINAL REVIEW

The remaining factor is a perfect square.

$$-6a^2 + 36ab - 54b^2 = -6(a - 3b)^2$$

27. $125u^3 + 27v^3$

This is the sum of cubes. Use the sum of cubes formula with $a = 5u$ and $b = 3v$.

$$125u^3 + 27v^3 = (5u + 3v)(25u^2 - 15uv + 9v^2)$$

28. $24u^3 - 81v^3$

First remove the common factor 3.

$$24u^3 - 81v^3 = 3(8u^3 - 27v^3)$$

The remaining factor is the difference of cubes. Use the difference of cubes formula with $a = 2u$ and $b = 3v$.

$$24u^3 - 81v^3 = 3(2u - 3v)(4u^2 + 6uv + 9v^2)$$

29. $\sqrt{125} = \sqrt{25 \cdot 5} = \sqrt{25}\sqrt{5} = 5\sqrt{5}$

30. $\sqrt[4]{32x^6y^4} = \sqrt[4]{16 \cdot x^4 \cdot y^4 \cdot 2x^2} = 2xy\sqrt[4]{2x^2}$

31. $5\sqrt{75} - 3\sqrt{48} = 5\sqrt{25 \cdot 3} - 3\sqrt{16 \cdot 3}$
$= 5 \cdot 5\sqrt{3} - 3 \cdot 4\sqrt{3}$
$= 25\sqrt{3} - 12\sqrt{3}$
$= (25 - 12)\sqrt{3}$
$= 13\sqrt{3}$

32. $\sqrt[3]{\dfrac{48u^5v}{3uv^2}} = \sqrt[3]{\dfrac{16u^4}{v}}$
$= \sqrt[3]{\dfrac{8 \cdot u^3 \cdot 2uv^2}{v^3}}$
$= \dfrac{2u\sqrt[3]{2uv^2}}{v}$

33. $\left(\dfrac{8^{2/3}a^{-3/2}b^3}{3a^{-5/2}b^{1/2}}\right)^{-2} = \dfrac{8^{-4/3}a^3b^{-6}}{3^{-2}a^5b^{-1}}$
$= \dfrac{3^2}{8^{4/3}a^2b^5}$
$= \dfrac{9}{\left(\sqrt[3]{8}\right)^4 a^2b^5}$
$= \dfrac{9}{2^4 a^2b^5}$
$= \dfrac{9}{16a^2b^5}$

34. $\dfrac{\sqrt{27}-\sqrt{5}}{\sqrt{3}+\sqrt{5}} = \dfrac{(\sqrt{27}-\sqrt{5})(\sqrt{3}-\sqrt{5})}{(\sqrt{3}+\sqrt{5})(\sqrt{3}-\sqrt{5})}$
$= \dfrac{\sqrt{27}\sqrt{3}-\sqrt{5}\sqrt{3}-\sqrt{27}\sqrt{5}+\sqrt{5}\sqrt{5}}{\sqrt{3}\sqrt{3}-\sqrt{5}\sqrt{5}}$
$= \dfrac{9-\sqrt{15}-3\sqrt{15}+5}{3-5}$
$= \dfrac{14-4\sqrt{15}}{-2}$
$= 2\sqrt{15}-7$

35. $\dfrac{x^2-3x+2}{x^2-2x+1} \cdot \dfrac{x^2-x}{x^2-4}$
$= \dfrac{(x-1)(x-2)}{(x-1)(x-1)} \cdot \dfrac{x(x-1)}{(x-2)(x+2)}$
$= \dfrac{x(x-1)(x-2)(x-1)}{(x-1)(x-1)(x-2)(x+2)}$
$= \dfrac{x}{x+2}$

36. $\dfrac{4x^2+4x+1}{2x^2-9x-5} \div \dfrac{x^2+5x+25}{x^3-125}$
$= \dfrac{(2x+1)(2x+1)}{(2x+1)(x-5)} \cdot \dfrac{(x-5)(x^2+5x+25)}{x^2+5x+25}$
$= 2x+1$

37. $\dfrac{x+5}{x^2+7x+10} - \dfrac{x-2}{x^2+5x+6}$
$= \dfrac{x+5}{(x+5)(x+2)} - \dfrac{x-2}{(x+2)(x+3)}$
$= \dfrac{1}{x+2} - \dfrac{x-2}{(x+2)(x+3)}$
$= \dfrac{1 \cdot (x+3)}{(x+2)(x+3)} - \dfrac{x-2}{(x+2)(x+3)}$
$= \dfrac{x+3-x+2}{(x+2)(x+3)}$
$= \dfrac{5}{(x+2)(x+3)}$

38. $\dfrac{x + \dfrac{x}{y}}{x - \dfrac{x}{y}} = \dfrac{\left[x + \dfrac{x}{y}\right]y}{\left[x - \dfrac{x}{y}\right]y}$

$= \dfrac{xy + x}{xy - x}$

$= \dfrac{x(y+1)}{x(y-1)}$

$= \dfrac{y+1}{y-1}$

39. When solving an inequality, if an inequality such as $5 > 7$ is obtained, the original inequality has no solution.

40. When solving $|x| = a$, $a > 0$, we solve the two equations

$$x = a \text{ and } x = -a.$$

41. An equation stating that two ratios are equal is called a proportion.

42. When multiplying both sides of an inequality by a negative number, the inequality must be reversed.

43. Solve. $x - (3 - 2x) = -2x + 7$

$x - (3 - 2x) = -2x + 7$
$x - 3 + 2x = -2x + 7$
$3x - 3 = -2x + 7$
$5x = 10$
$x = 2$

44. Solve. $\sqrt[3]{x - 2} + 3 = 2$

$\sqrt[3]{x-2} + 3 = 2$
$\sqrt[3]{x-2} = -1$
$\left(\sqrt[3]{x-2}\right)^3 = (-1)^3$
$x - 2 = -1$
$x = 1$

45. Solve. $\sqrt{x-2} - \sqrt{x+5} = -1$

$\sqrt{x-2} - \sqrt{x+5} = -1$
$\sqrt{x-2} = \sqrt{x+5} - 1$
$\left(\sqrt{x-2}\right)^2 = \left(\sqrt{x+5} - 1\right)^2$

$x - 2 = x + 5 - 2\sqrt{x+5} + 1$
$-8 = -2\sqrt{x+5}$
$4 = \sqrt{x+5}$
$(4)^2 = \left(\sqrt{x+5}\right)^2$
$16 = x + 5$
$11 = x$

Since 11 does check, the solution is 11.

46. Solve. $\dfrac{2}{x-3} - \dfrac{3}{x-2} = \dfrac{5}{x^2 - 5 + 6}$

$\dfrac{2}{x-3} - \dfrac{3}{x-2} = \dfrac{5}{(x-2)(x-3)}$

$(x-3)(x-2)\left[\dfrac{2}{x-3} - \dfrac{3}{x-2}\right] = (x-2)(x-3)\left[\dfrac{5}{(x-2)(x-3)}\right]$

$2(x-2) - 3(x-3) = 5$
$2x - 4 - 3x + 9 = 5$
$-x + 5 = 5$
$-x = 0$
$x = 0$

Since 0 does check, the solution is 0.

47. Let $x =$ the former salary.

Then since the raise is a 7% raise, the amount of the raise is $0.07x$. Then the present salary is the former salary plus the raise, which is $x + 0.07x$. Since the present salary is $40,660, we must solve:

$$x + 0.07x = 40{,}660$$
$$1.07x = 40{,}660$$
$$x = \dfrac{40{,}660}{1.07} = 38{,}000$$

Thus, the former salary was $38,000.

48. Let $x =$ number of days to do the job together,
    $6 =$ number of days for Henry to do the job,
    $10 =$ number of days for Joe to do the job.

Then $\dfrac{1}{x} =$ amount done together in 1 day,

$\dfrac{1}{6} =$ amount done by Henry in 1 day,

$\dfrac{1}{10} =$ amount done by Joe in 1 day.

Since the amount done together in 1 day is equal to the sum of the amounts done individually in 1 day we must solve:

$$\frac{1}{x} = \frac{1}{6} + \frac{1}{10}$$
$$30x\left[\frac{1}{x}\right] = 30x\left[\frac{1}{6} + \frac{1}{10}\right]$$
$$30 = 5x + 3x$$
$$30 = 8x$$
$$\frac{30}{8} = x$$
$$\frac{15}{4} = x$$

Thus, it takes $\frac{15}{4}$ days, or 3.75 days, to do the job together.

49. Let x = the speed of the plane in still air,
    100 = the speed of the wind,
    $x+100$ = speed of plane with the wind,
    $x-100$ = speed of plane against the wind,
    300 = distance traveled with the wind,
    200 = distance traveled against the wind.

Use the distance formula, $d = rt$, solved for t to obtain two expressions for the time with and the time against the wind.

$\frac{300}{x+100}$ = time traveled with the wind,

$\frac{200}{x-100}$ = time traveled against the wind.

Since the time with the wind is equal to the time against the wind, we must solve:

$$\frac{300}{x+100} = \frac{200}{x-100}$$
$$300(x-100) = 200(x+100)$$
$$300x - 30{,}000 = 200x + 20{,}000$$
$$100x = 50{,}000$$
$$x = 500$$

Thus, the plane can fly 500 km/hr in still air.

50. Solve and graph $|2x - 3| < 5$.

This inequality translates to the compound inequality

$$-5 < 2x - 3 < 5.$$

Add 3 throughout.

$$-2 < 2x < 8$$

Divide by 2 throughout.

$$-1 < x < 4$$

The graph is given below.

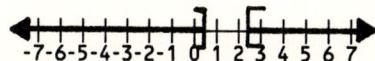

51. Solve and graph $|3 - 2x| \geq 2$.

This inequality translates to the compound inequality:

$$3 - 2x \leq -2 \quad \text{or} \quad 3 - 2x \geq 2$$

Subtract 3 on both sides.

$$-2x \leq -5 \quad \text{or} \quad -2x \geq -1$$

Divide both sides by -2 and remember to reverse the inequality symbols.

$$x \geq \frac{5}{2} \quad \text{or} \quad x \leq \frac{1}{2}$$

The graph is given below.

52. Solve $\frac{1}{a} + \frac{1}{b} = \frac{1}{c}$ for b.

$$\frac{1}{a} + \frac{1}{b} = \frac{1}{c}$$
$$abc\left[\frac{1}{a} + \frac{1}{b}\right] = abc\left[\frac{1}{c}\right]$$
$$bc + ac = ab$$
$$ac = ab - bc$$
$$ac = b(a - c)$$
$$\frac{ac}{a-c} = b$$

53. The mean proportional, x, between 6 and 24 satisfies the following equation.

$$\frac{6}{x} = \frac{x}{24}$$
$$x^2 = 144$$
$$x = \pm\sqrt{144} = \pm 12$$

54. If the methods of factoring or taking roots fail when solving a quadratic equation, the next best method to use is the quadratic formula.

55. For the imaginary number i, $i^2 = -1$.

56. The conjugate of $9+5i$ is $9-5i$.

57. $(5-7i) - (-3-6i) = 5-7i+3+6i = 8-i$.

58. $(3+4i)(8-7i) = 24 + 32i - 21i - 28i^2$
    $= 24 + 11i - 28(-1)$
    $= 24 + 11i + 28$
    $= 52+11i$

59. $\dfrac{5-2i}{-3+4i} = \dfrac{(5-2i)(-3-4i)}{(-3+4i)(-3-4i)}$
    $= \dfrac{-15+6i-20i+8i^2}{9-16i^2}$
    $= \dfrac{-15-14i-8}{9+16}$
    $= \dfrac{-23-14i}{25}$

60. $\dfrac{1}{2+3i} = \dfrac{1(2-3i)}{(2+3i)(2-3i)}$
    $= \dfrac{2-3i}{4-9i^2}$
    $= \dfrac{2-3i}{4+9}$
    $= \dfrac{2-3i}{13}$

61. Solve. $3y^2 + y - 10 = 0$

    $3y^2 + y - 10 = 0$
    $(3y - 5)(y + 2) = 0$
    $3y - 5 = 0 \qquad y + 2 = 0$
    $3y = 5 \qquad y = -2$
    $y = \dfrac{5}{3}$

62. Solve. $2y^2 + 4y = 3$.

    This is equivalent to $2y^2 + 4y - 3 = 0$. Since the equation will not factor, we use the quadratic formula.

    $y = \dfrac{-b \pm \sqrt{b^2 - 4ac}}{2a}$
    $= \dfrac{-4 \pm \sqrt{4^2 - 4(2)(-3)}}{2(2)}$
    $= \dfrac{-4 \pm \sqrt{16 + 24}}{4}$
    $= \dfrac{-4 \pm \sqrt{40}}{4} = \dfrac{-4 \pm 2\sqrt{10}}{4} = \dfrac{-2 \pm \sqrt{10}}{2}$

63. Solve. $(u^2+5)^2 - 7(u^2+5) + 12 = 0$

    Substitute x for u^2+5.

    $x^2 - 7x + 12 = 0$
    $(x - 4)(x - 3) = 0$
    $x - 4 = 0 \qquad x - 3 = 0$
    $x = 4 \qquad x = 3$

    Backsubstitute u^2+5 to find the values of u.

    $u^2 + 5 = 4 \qquad u^2 + 5 = 3$
    $u^2 = -1 \qquad u^2 = -2$

    $u = \pm i \qquad u = \pm i\sqrt{2}$

64. Solve. $\dfrac{2u}{u-2} = -\dfrac{8}{u^2-4} + \dfrac{u}{u+2}$

    $\dfrac{2u}{u-2} = -\dfrac{8}{u^2-4} + \dfrac{u}{u+2}$

    $\dfrac{2u}{u-2} = -\dfrac{8}{(u-2)(u+2)} + \dfrac{u}{u+2}$

    $(u-2)(u+2)\left[\dfrac{2u}{u-2}\right] =$
    $(u-2)(u+2)\left[-\dfrac{8}{(u-2)(u+2)} + \dfrac{u}{u+2}\right]$
    $2u(u+2) = -8 + u(u-2)$
    $2u^2 + 4u = -8 + u^2 - 2u$
    $u^2 + 6u + 8 = 0$
    $(u+2)(u+4) = 0$

    Use the zero-product rule.

    $u + 2 = 0 \qquad u + 4 = 0$
    $u = -2 \qquad u = -4$

    But -2 does not check (since -2 makes two of the original denominators 0). The only solution is -4.

65. Solve. $\sqrt{4x+1} - \sqrt{x-2} = 3$

    $\sqrt{4x+1} = \sqrt{x-2} + 3$
    $(\sqrt{4x+1})^2 = (\sqrt{x-2} + 3)^2$
    $4x + 1 = x - 2 + 6\sqrt{x-2} + 9$
    $3x - 6 = 6\sqrt{x-2}$
    $x - 2 = 2\sqrt{x-2}$
    $(x-2)^2 = (2\sqrt{x-2})^2$
    $x^2 - 4x + 4 = 4(x-2)$
    $x^2 - 4x + 4 = 4x - 8$
    $x^2 - 8x + 12 = 0$
    $(x-2)(x-6) = 0$

Use the zero product rule.

$$x - 2 = 0 \qquad x - 6 = 0$$
$$x = 2 \qquad x = 6$$

Since both 2 and 6 do check in the original equation, the solutions are 2 and 6.

66. Solve. $\dfrac{1}{x+4} + \dfrac{1}{x} = \dfrac{1}{5}$

$$\dfrac{1}{x+4} + \dfrac{1}{x} = \dfrac{1}{5}$$
$$5x(x+4)\left[\dfrac{1}{x+4} + \dfrac{1}{x}\right] = 5x(x+4)\left[\dfrac{1}{5}\right]$$
$$5x + 5(x+4) = x(x+4)$$
$$5x + 5x + 20 = x^2 + 4x$$
$$0 = x^2 - 6x - 20$$

$$x = \dfrac{-b \pm \sqrt{b^2 - 4ac}}{2a}$$
$$= \dfrac{6 \pm \sqrt{6^2 - 4(1)(-20)}}{2(1)}$$
$$= \dfrac{6 \pm \sqrt{36 + 80}}{2}$$
$$= \dfrac{6 \pm \sqrt{116}}{2}$$
$$= \dfrac{6 \pm 2\sqrt{29}}{2}$$
$$= 3 \pm \sqrt{29}$$

67. Let $n =$ the number of men planning to go initially,
$c =$ the original cost per man.

Since the total cost of the trip is $240, we have that

$$nc = 240.$$

When two men decide not to go, this leaves $n-2$ men, and the cost per share rises $4, so the new cost per man is $c+4$. Since the total cost remains at $240, we have that

$$(n-2)(c+4) = 240.$$

Since we are asked to find the number of men, n, solve each of these equations for c,

$$c = \dfrac{240}{n} \quad \text{and} \quad c = \dfrac{240}{n-2} - 4$$

and set the results equal to obtain an equation in the desired variable n.

$$\dfrac{240}{n} = \dfrac{240}{n-2} - 4$$
$$n(n-2)\left[\dfrac{240}{n}\right] = n(n-2)\left[\dfrac{240}{n-2} - 4\right]$$
$$240(n-2) = 240n - 4n(n-2)$$
$$240n - 480 = 240n - 4n^2 + 8n$$
$$4n^2 - 8n - 480 = 0$$
$$n^2 - 2n - 120 = 0$$
$$(n+10)(n-12) = 0$$

Use the zero-product rule.

$$n + 10 = 0 \qquad n - 12 = 0$$
$$n = -10 \qquad n = 12$$

Since the number of men cannot be negative, we discard -10. Thus the number of men in the group is 12.

68. Let $x =$ speed of one boat,
$x + 5 =$ speed of second boat.

After 2 hours, the distance that each travels is $2x$ and $2(x + 5)$. Since the boats are traveling at right angles to each other, the distances traveled form the legs of a right triangle. Since the boats are 50 mi apart after 2 hours, the hypotenuse of the right triangle is 50. Use the pythagorean theorem to obtain the followint equation.

$$(2x)^2 + [2(x + 5)]^2 = 50^2$$
$$4x^2 + 4(x + 5)^2 = 2500$$
$$4x^2 + 4(x^2 + 10x + 25) = 2500$$
$$4x^2 + 4x^2 + 40x + 100 = 2500$$
$$8x^2 + 40x - 2400 = 0$$
$$x^2 + 5x - 300 = 0$$
$$(x - 15)(x + 20) = 0$$
$$x - 15 = 0 \qquad x + 20 = 0$$
$$x = 15 \qquad x = -20$$

We can discard the negative solution since a rate of speed cannot be negative. Thus the speed of one boat is 15 mph, and the speed of the other is 20 mph ($x + 5 = 15 + 5 = 20$).

69. Find the intercepts and graph $2x + 3y = 12$.

When $x = 0$, $3y = 12$ making $y = 4$. When $y = 0$, $2x = 12$ making $x = 6$. Thus, the intercepts are (0,4) and (6,0), and the graph is given below.

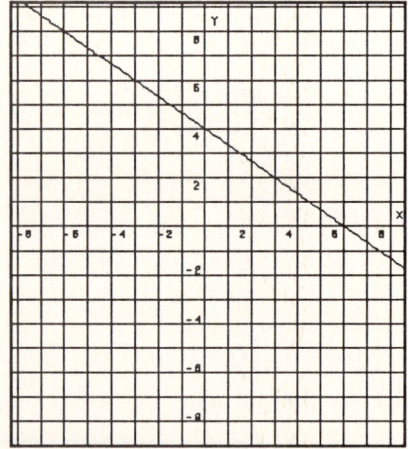

70. Find the vertex, x-intercepts, and graph of $f(x) = -x^2 + 4x + 5$.

Since this is a quadratic function, the graph is a parabola that opens down ($a = -1 < 0$). The x-coordinate of the vertex is:

$$-\frac{b}{2a} = -\frac{4}{2(-1)} = 2$$

The y-coordinate of the vertex is:

$$f(2) = -2^2 + 4(2) + 5 = 9$$

Thus the vertex is $(2,9)$. To find the x-intercepts, solve the following equation.

$$-x^2 + 4x + 5 = 0$$
$$x^2 - 4x - 5 = 0$$
$$(x - 5)(x + 1) = 0$$
$$x - 5 = 0 \quad x + 1 = 0$$
$$x = 5 \quad x = -1$$

Thus, the x-intercepts are $(5,0)$ and $(-1,0)$. The graph is given below.

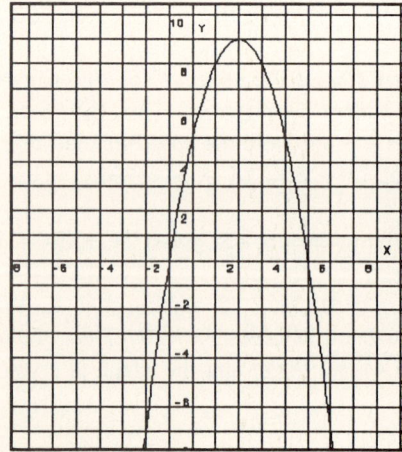

71. Find the general form of the equation of the line passing through $(4,-1)$ and $(6,3)$.

First find the slope of this line.

$$m = \frac{y_2 - y_1}{x_2 - x_1} = \frac{3-(-1)}{6-4} = \frac{4}{2} = 2$$

Use the point-slope form with $m = 2$ and $(x_1, y_1) = (6,3)$.

$$y - y_1 = m(x - x_2)$$
$$y - 3 = 2(x - 6)$$
$$y - 3 = 2x - 12$$
$$0 = 2x - y - 9$$

Thus, the general form of the line is

$$2x - y - 9 = 0.$$

72. To find the equation of the line passing through $(5,-2)$ and perpendicular to the line $2x + 5y = 7$, first find the slope of this line by writing the equation in slope-intercept form.

$$5y = -2x + 7$$
$$y = -\frac{2}{5}x + \frac{7}{5}$$

Since the slope of this line is $-\frac{2}{5}$, the slope of a line perpendicular to it is $\frac{5}{2}$. Use the point-slope form.

$$y - (-2) = \frac{5}{2}(x - 5)$$
$$2(y + 2) = 5(x - 5)$$
$$2y + 4 = 5x - 25$$
$$0 = 5x - 2y - 29$$

Thus, the general form of the desired line is

$$5x - 2y - 29 = 0.$$

73. To find the slope and y-intercept of the line with equation $3x + 7y - 2 = 0$, write the equation in slope-intercept form, that is, solve for y.

$$3x + 7y - 2 = 0$$
$$7y = -3x + 2$$
$$y = -\frac{3}{7}x + \frac{2}{7}$$

Thus, the slope is $-\frac{3}{7}$ and the y-intercept is $\left(0, \frac{2}{7}\right)$.

FINAL REVIEW

74. Find the distance between the points (−2,5) and (6,1).

    Use the distance formula.

    $$d = \sqrt{(x_2-x_1)^2 + (y_2-y_1)^2}$$
    $$= \sqrt{(6-(-2))^2 + (1-5)^2}$$
    $$= \sqrt{(8)^2 + (-4)^2}$$
    $$= \sqrt{64 + 16}$$
    $$= \sqrt{80} = \sqrt{16 \cdot 5} = 4\sqrt{5}$$

75. Find the midpoint of the line segment joining (7,4) and (−2,3).

    Use the midpoint formula.

    $$(\overline{x},\overline{y}) = \left(\frac{x_1+x_2}{2}, \frac{y_1+y_2}{2}\right)$$
    $$= \left(\frac{7+(-2)}{2}, \frac{4+3}{2}\right)$$
    $$= \left(\frac{5}{2}, \frac{7}{2}\right)$$

76. Refer to the figures in the text.

    (a) This does represent a function since no element on the left is paired with two elements on the right.

    (b) This does not represent a function since the element 4 on the left is paired with two elements (1 and 2) on the right.

77. Refer to the figures in the text.

    (a) This is the graph of a function since the graph passes the vertical line test, that is, every vertical line passes through only one point on the graph.

    (b) This is not the graph of a function since it is possible to draw a vertical line that passes through two points on the graph.

78. Consider the graph of $f(x) = -x^2$ given below. From the graph, it is clear that when $x \leq 0$ the graph is increasing, and when $x \geq 0$ the graph is decreasing.

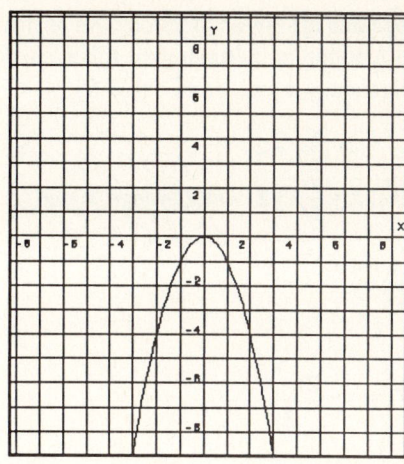

79. Determine the symmetry of the following relations.

    (a) $x^2 + y^2 = 4$

    If x is replaced with $-x$, the equation is unchanged. Thus, the graph is symmetric with respect to the y-axis.

    Similarly, if y is replaced with $-y$, the equation is unchanged making the graph symmetric with respect to the x-axis.

    And when x is replaced with $-x$ and y is replaced with $-y$ at the same time, the equation is unchanged making the graph symmetric with respect to the origin.

    (b) $xy = 6$

    If x is replaced with $-x$, the equation is changed to $xy = -6$. Thus, the graph is not symmetric with respect to the y-axis.

    Similarly, if y is replaced with $-y$, the equation is changed to $xy = -6$. Thus, the graph is not symmetric with respect to the x-axis.

    But when x is replaced with $-x$ and y is replaced with $-y$ at the same time, the equation is unchanged making the graph symmetric with respect to the origin.

80. Determine the inverse of each of the following functions.

    (a) $f(x) = \frac{3}{2}x + 2$

Since this is a linear function with positive slope, the function is increasing so it has an inverse. Interchange x and y and solve for y.

$$y = \frac{3}{2}x + 2$$
$$x = \frac{3}{2}y + 2 \quad \text{Interchange } x \text{ and } y$$
$$2x = 3y + 4 \quad \text{Solve for } y$$
$$2x - 4 = 3y$$
$$\frac{2x-4}{3} = y$$

$$f^{-1}(x) = \frac{2x-4}{3}$$

(b) $g(x) = x^2 - 5, \; x \geq 0$

Since the graph of g is the "right branch" of the parabola opening up with vertex at $(0,-5)$, the function is increasing and has an inverse. Interchange x and y and solve for y.

$$y = x^2 - 5 \quad x \geq 0$$
$$x = y^2 - 5 \quad y \geq 0 \quad \text{Interchange } x \text{ and } y$$
$$x + 5 = y^2$$
$$\sqrt{x+5} = y \quad \text{Positive root only } (y \geq 0)$$

$$g^{-1}(x) = \sqrt{x+5}$$

81. First translate "y varies directly as the square of x and inversely as the cube root of z."

$$y = \frac{cx^2}{\sqrt[3]{z}}$$

Substitute 5 for y, 4 for x, and 27 for z and solve for c.

$$5 = \frac{c(4)^2}{\sqrt[3]{27}}$$
$$5 = \frac{16c}{3}$$
$$15 = 16c$$
$$\frac{15}{16} = c$$

Then the equation of variation is:

$$y = \frac{\frac{15}{16}x^2}{\sqrt[3]{z}} \quad \text{or} \quad y = \frac{15x^2}{16\sqrt[3]{z}}$$

82. The table gives the three points $(0,20)$, $(1,21)$, and $(3,29)$. If we plot these three points, we can see that as x increases, so does y. Suppose we place the vertex of the parabola through these points at $(0,20)$, then equation of the function would be $f(x) = x^2 + 20$. Note that the two pairs $(1,21)$ and $(3,29)$ satisfy this equation. Using this model, we can predict the cost of operation on a day when 10 units are produced by finding $f(10)$.

$$f(10) = (10)^2 + 20 = 100 + 20 = 120$$

Thus, the cost on such a day would be $120.

[*NOTE:* The function above could be found more directly by recognizing that it is a quadratic function of the form $f(x) = ax^2 + bx + c$, substituting the three sets of ordered pairs into the form, and solving the resulting system of three equations in the three unknowns a, b, and c. This technique, although more precise than the method given above, involves solving a system, a topic not discussed until Chapter 6.]

83. If $P(x)$ is a polynomial and $P(b) = 0$, b is called a zero of the polynomial $P(x)$.

84. The remainder when $P(x)$ is divided by $x + 5$ is $P(-5)$.

85. The theorem used to answer Exercise 84 is the remainder theorem.

86. If $P(-8) = 0$, then one factor of $P(x)$ is $x + 8$.

87. The theorem used to answer Exercise 86 is the factor theorem.

88. If $\frac{p}{q}$ is a solution to the polynomial equation

$$5x^3 - 4x^2 + 3x - 6 = 0,$$

then p is a factor of the constant term -6.

89. Counting multiplicities, a polynomial of degree n has exactly n zeros.

90. If r is a zero of $Q(x)$ in rational function

$$f(x) = \frac{P(x)}{Q(x)},$$

then $x = r$ is a vertical asymptote of the graph of the rational function.

91. If $4-3i$ is a zero of polynomial $P(x)$, with real coefficients, then another zero of $P(x)$ is $4+3i$.

FINAL REVIEW

92. If 3 and $1-3\sqrt{2}$ are zeros of a polynomial $P(x)$, with rational coefficients, then $1+3\sqrt{2}$ is also a zero of $P(x)$. Thus, the least degree that $P(x)$ can have is three.

93. Divide $4x^4 - 2x^3 + x - 9$ by $x - 2$ using synthetic division.

Do not forget to use a 0 for the missing x^2 term.

```
2 |  4 - 2 +  0 +  1 -  9
  |    + 8 + 12 + 24 + 50
  _____
     4 + 6 + 12 + 25 + 41
```

Since the remainder is 41, $P(2) = 41$.

94. Use Descarte's rule of signs to find the possible number of negative, positive, and nonreal solutions to

$$-x^4 + 2x^3 - 3x^2 + x - 7 = 0.$$

Since the degree of the polynomial is 4, there are 4 solutions. The number of sign changes in the polynomial is 4 so there are either 4, 2, or 0 real solutions. Consider

$$P(-x) = -x^4 - 2x^3 - 3x^2 - x - 7.$$

Since there are 0 sign changes in $P(x)$, there are 0 negative solutions. Putting this information together, we have either 0,0,4, 0,2,2, or 0,4,0 solutions; where the first number is the negative solutions, the second the positive solutions, and the third the nonreal solutions.

95. Use the upper and lower bound test to find the smallest positive integer upper bound and the largest negative integer lower bound for the real solutions to

$$4x^3 - 16x^2 + 11x + 10 = 0.$$

It is easy to see that the signs in the bottom row are not all positive when the polynomial is divided by 1 and 2. Suppose we try 3.

```
3 |  4 - 16 + 11 + 10
  |    + 12 - 12 -  3
  _____
     4 -  4 -  1 +  7
```

Thus, 3 is not an upper bound. Try 4.

REVIEW EXERCISES

```
4 |  4 - 16 + 11 + 10
  |    + 16 +  0 + 44
  _____
     4 +  0 + 11 + 54
```

Since the signs are now all positive, we know that 4 is an upper bound for the positive solutions, the smallest integer upper bound. Now try the negative integers. Start with –1.

```
-1 | 4 - 16 + 11 + 10
   |   -  4 + 20 - 31
  _____
     4 - 20 + 31 - 21
```

Since the signs alternate in the bottom row, –1 is a lower bound for the negative solutions, the largest integer lower bound.

96. Find all solutions of $2x^3 + 7x^2 + 2x - 6 = 0$.

By the rational root theorem, if $\frac{p}{q}$ is a solution, then p divides –6 and q divides 2. Thus, the possibilities for p are $\pm 1, \pm 2, \pm 3, \pm 6$, and the possibilities for q are $\pm 1, \pm 2$. Then the possible rational solutions are:

$$\frac{p}{q}: \pm 1, \pm 2, \pm 3, \pm 6, \pm \frac{1}{2}, \pm \frac{3}{2}$$

By trial and error along with the upper and lower bound test, we can eliminate all of the integer possibilities. Suppose we try $-\frac{3}{2}$.

```
-3/2 | 2 + 7 + 2 - 6
     |   - 3 - 6 + 6
  _____
       2 + 4 - 4 + 0
```

Thus, $-\frac{3}{2}$ is one solution, and the remaining solutions must solve the quadratic equation

$$2x^2 + 4x - 4 = 0$$

which is equivalent to

$$x^2 + 2x - 2 = 0.$$

Use the quadratic formula.

$$x = \frac{-b \pm \sqrt{b^2 - 4ac}}{2a}$$

$$= \frac{-2 \pm \sqrt{2^2 - 4(1)(-2)}}{2(1)}$$

$$= \frac{-2 \pm \sqrt{4 + 8}}{2}$$

$$= \frac{-2 \pm \sqrt{12}}{2}$$

$$= \frac{-2 \pm 2\sqrt{3}}{2}$$

$$= -1 \pm \sqrt{3}$$

Thus, the three solutions are

$$-\tfrac{3}{2},\ -1 + \sqrt{3},\ \text{and}\ -1 - \sqrt{3}.$$

97. Graph $P(x) = x^3 + x^2 - 2x$, and give the following information.

(a) The maximum number of turning points is 2 since the degree of $P(x)$ is $n = 3$, and $n - 1 = 2$.

(b) To find the x-intercepts, solve:

$$x^3 + x^2 - 2x = 0$$
$$x(x^2 + x - 2) = 0$$
$$x(x + 2)(x - 1) = 0$$

Using the zero-product rule, we obtain $x = 0$, -2, and 1. Thus, the x-intercepts are $(0,0)$, $(-2,0)$, and $(1,0)$.

(c) Since the degree of the polynomial is 3 (odd) and $a_3 = 1 > 0$, for large positive values of x to the right, the graph will eventually go up.

(d) Since the degree of the polynomial is 3 (odd) and $a_3 = 1 > 0$, for small negative values of x to the left, the graph will eventually go down.

Using this information along with a few additional points, we obtain the graph given below.

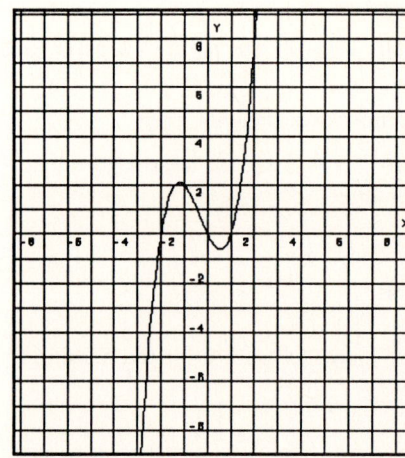

98. Graph $f(x) = \dfrac{1}{x^2 + 2x + 1}$ and give the information requested.

(a) To find the vertical asymptotes, set the denominator equal to 0 and solve for x.

$$x^2 + 2x + 1 = 0$$
$$(x + 1)(x + 1) = 0$$
$$x + 1 = 0 \qquad x + 1 = 0$$
$$x = -1 \qquad x = -1$$

Thus, there is only one vertical asymptote, the line $x = -1$.

(b) Since the degree of the numerator is less than the degree of the denominator, the x-axis, the line $y = 0$, is a horizontal asymptote of the graph.

(c) There is no oblique asymptote since the degree of the numerator is not 1 more than the degree of the denominator.

(d) Since $f(x)$ is never 0 (the numerator is never 0), there are no x-intercepts.

(e) Since $f(0) = 1$, the y-intercept is $(0,1)$.

(f) There are no symmetries of the graph, the tests for symmetry all fail.

The graph of the function is given below.

FINAL REVIEW

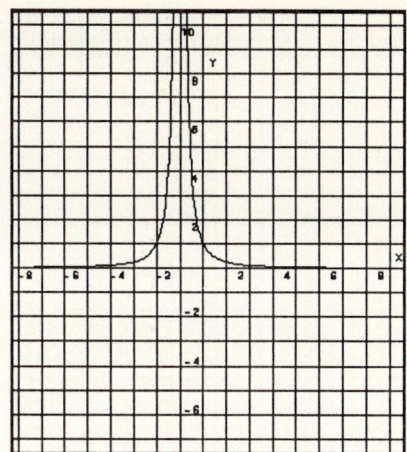

Solve and graph each inequality in Exercises 99-100.

99. $2x^2 + 9x - 5 < 0$

First solve the quadratic equation

$$2x^2 + 9x - 5 = 0$$

to find the critical points.

$$2x^2 + 9x - 5 = 0$$
$$(2x - 1)(x + 5) = 0$$
$$2x - 1 = 0 \quad\quad x + 5 = 0$$
$$2x = 1 \quad\quad\quad x = -5$$
$$x = \tfrac{1}{2}$$

These two critical points divide the number line into the three intervals:

$$(-\infty, -5), \quad \left(-5, \tfrac{1}{2}\right), \quad \left(\tfrac{1}{2}, \infty\right).$$

Using a test point from each interval, it is easy to see that the solution to the inequality is

$$\left(-5, \tfrac{1}{2}\right) \quad \text{or} \quad -5 < x < \tfrac{1}{2}.$$

The graph of the solution is given below.

100. $\dfrac{3x - 1}{2x - 3} \geq 0$

Setting the numerator equal to zero and the denominator equal to zero we obtain the two critical points $\tfrac{1}{3}$ and $\tfrac{3}{2}$. These critical points divide the number line into the three intervals

$$\left(-\infty, \tfrac{1}{3}\right), \quad \left(\tfrac{1}{3}, \tfrac{3}{2}\right), \quad \left(\tfrac{3}{2}, \infty\right).$$

Test points from each of these intervals show that the solution to the inequality includes the intervals

$$\left(-\infty, \tfrac{1}{3}\right) \quad \text{or} \quad \left(\tfrac{3}{2}, \infty\right).$$

Since the inequality is $\geq$, we include the endpoint that makes the rational expression $= 0$, but do not include the endpoint that makes the denominator 0, where the rational expression is undefined. Thus, the solution to the inequality is

$$\left(-\infty, \tfrac{1}{3}\right] \quad \text{or} \quad \left(\tfrac{3}{2}, \infty\right)$$

or, using inequalities,

$$x \leq \tfrac{1}{3} \quad \text{or} \quad x > \tfrac{3}{2}.$$

The graph of the solution is given below.

101. The exponential equation $x = a^y$ is equivalent to the logarithmic equation $y = \log_a x$.

102. For any base a, $\log_a 1 = 0$ ($a^0 = 1$).

103. For any base a, $\log_a a = 1$ ($a^1 = a$).

104. The graph of $y = \log_a x$ can be obtained from the graph of $y = a^x$ by reflecting it across the line $y = x$ since these two functions are inverses.

105. Relative to the equation $\log n = x$, n is called the antilogarithm of x.

106. If a and b are bases and $x > 0$, by the base conversion formula,

$$\frac{\log_a x}{\log_a b} = \log_b x.$$

Graph each function in Exercises 107-109.

107. $f(x) = 3^x$

Make a table of values, plot the points, and join them with a smooth curve to obtain the graph of the function given below.

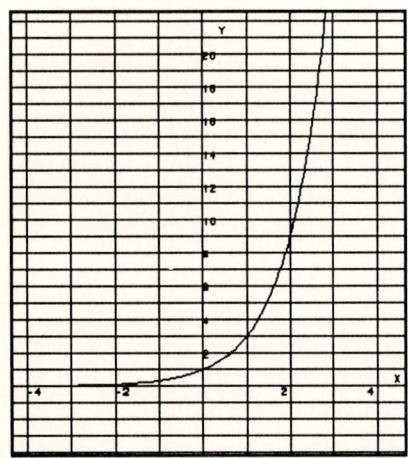

108. $f(x) = \log_3 x$

Make a table of values using the equivalent exponential form of the equation, $3^y = x$. Alternatively, the graph can be obtained from the graph of the function in Exercise 107 by reflecting the graph in the line $y = x$ since the two functions are inverses. The graph is given below.

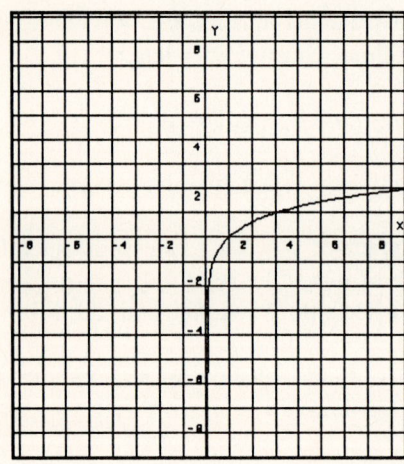

109. $f(x) = e^{x+1}$

Make a table of values using a calculator. For example, when $x = 1$, $f(1) \approx 7.4$ and when $x = 0$, $f(0) \approx 2.7$. Plot the points and connect them with a smooth curve to obtain the graph given below.

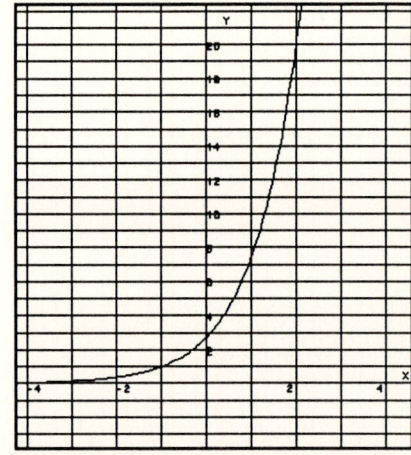

Solve for x in Exercises 110-111.

110. $8^{2/3} = x$

$$x = 8^{2/3} = \left(\sqrt[3]{8}\right)^2 = 2^2 = 4$$

111. $\log_4 \frac{1}{64} = x$

Convert to exponential form.

$$4^x = \frac{1}{64}$$
$$4^x = \frac{1}{4^3}$$
$$4^x = 4^{-3}$$
$$x = -3$$

112. Expand $\log_a \dfrac{x^2}{(y^2z)^3}$ using the properties of logarithms.

$$\log_a \frac{x^2}{(y^2z)^3} = \log_a x^2 - \log_a (y^2z)^3$$
$$= 2\log_a x - 3\log_a y^2 z$$
$$= 2\log_a x - 3[\log_a y^2 + \log_a z]$$
$$= 2\log_a x - 3[2\log_a y + \log_a z]$$
$$= 2\log_a x - 6\log_a y - 3\log_a z$$

113. Express $2\log_a xyz - \frac{1}{3}\log_a xy^2 + 4\log_a z$ as a single logarithm and simplify.

FINAL REVIEW REVIEW EXERCISES 505

$$2\log_a xyz - \tfrac{1}{3}\log_a xy^2 + 4\log_a z$$
$$= \log_a (xyz)^2 - \log_a (xy^2)^{1/3} + \log_a z^4$$
$$= \log_a \frac{(xyz)^2 z^4}{(xy^2)^{1/3}}$$
$$= \log_a \frac{x^2 y^2 z^6}{x^{1/3} y^{2/3}}$$
$$= \log_a x^{5/3} y^{4/3} z^6$$

Use a calculator to find the value of n in each expression in Exercises 114-119.

114. $n = \log 3.421$

Enter 3.421 and press the LOG button.

$n = 0.5342$ *Correct to four decimal places*

115. $\log n = 0.3271$

Enter 0.3271 and press INV and LOG buttons.

$n = 2.12$ *Correct to three significant digits*

116. $n = \ln 0.00351$

Enter 0.00351 and press $\ln x$ button.

$n = -5.6521$ *Correct to four decimal places*

117. $\ln n = 2.5378$

Enter 2.5378 and press INV and $\ln x$ buttons.

$n = 12.7$ *Correct to three significant digits*

118. $n = e^{3.2564}$

Enter 3.2564 and press INV and $\ln x$ buttons.

$n = 26.0$ *Correct to three dignificant digits*

119. $\log_5 482$

Use the base conversion formula.

$$\log_5 482 = \frac{\log 482}{\log 5} = 3.8386$$

Be sure to divide the two logarithms, do not subtract. The answer is given correct to four decimal places.

Solve each equation in Exercises 120-123.

120. $8^{3x+2} = 4^{x-1}$

Write both sides as powers of base 2 and equate the exponents.

$$8^{3x+2} = 4^{x-1}$$
$$(2^3)^{3x+2} = (2^2)^{x-1}$$
$$2^{9x+6} = 2^{2x-2}$$
$$9x + 6 = 2x - 2$$
$$7x = -8$$
$$x = -\tfrac{8}{7}$$

121. $\log_2 (2x + 2) - \log_2 (x - 2) = 2$

Use the quotient rule to write the left side as a single logarithm, then convert to exponential form.

$$\log_2 \frac{2x+2}{x-2} = 2$$
$$\frac{2x+2}{x-2} = 2^2$$
$$2x + 2 = 4(x-2)$$
$$2x + 2 = 4x - 8$$
$$10 = 2x$$
$$5 = x$$

Since 5 does check in the original equation, the solution is 5.

122. $2^x = 5^{x-1}$

Take the common logarithm of both sides and solve for x.

$$\log 2^x = \log 5^{x-1}$$
$$x \log 2 = (x-1) \log 5$$
$$x \log 2 = x \log 5 - \log 5$$
$$x \log 2 - x \log 5 = -\log 5$$
$$x(\log 2 - \log 5) = -\log 5$$
$$x = \frac{-\log 5}{\log 2 - \log 5}$$
$$\approx 1.756470797$$

Thus, correct to three significant digits, the solution is 1.76.

123. $\log (\log x) = 0$

First convert to exponential form.

$$\log x = 10^0 = 1$$

Then convert to exponential form again.

$$x = 10^1 = 10$$

124. Let 1970 correspond to year $t = 0$. then

$$120 = ce^{k(0)} = ce^0 = c(1) = c$$

Then the equation becomes $y = 120e^{kt}$. In 1990, when $t = 20$, we have $y = 180$. Substitute to find the value of k.

$$180 = 120e^{k(20)}$$
$$\frac{180}{120} = e^{20k}$$
$$\ln\left(\frac{180}{120}\right) = \ln e^{20k} = 20k$$
$$\frac{\ln\left(\frac{180}{120}\right)}{20} = k$$
$$0.020273255 = k$$

Use this value of k with $t = 30$ (corresponding to the year 2000) and find the value of y.

$$y = 120e^{0.020273255(30)} \approx 220.4540769$$

Thus, the population will be about 220 in the year 2000.

125. Use the compound interest formula,

$$A = P\left(1 + \frac{r}{k}\right)^{kt}$$

with $A = 1000$, $P = 300$, $r = 0.12$, and $k = 4$ to find the value of $4t$, the number of compounding periods.

$$1000 = 300\left(1 + \frac{0.12}{4}\right)^{4t}$$
$$\frac{1000}{300} = (1.03)^{4t}$$
$$\log\left(\frac{10}{3}\right) = \log(1.03)^{4t}$$
$$\log\left(\frac{10}{3}\right) = 4t\log(1.03)$$
$$\frac{\log\left(\frac{10}{3}\right)}{\log 1.03} = 4t$$
$$40.73144759 = 4t$$

Thus, the number of compounding periods is about 41.

126. To find the decibel level of a noise with intensity 0.45 watt/m², substitute 0.45 for S, 10^{-12} for S_0 in the formula $D = 10 \log \frac{S}{S_0}$ and evaluate D.

$$D = 10\log\frac{0.45}{10^{-12}} = 10(11.65321251) \approx 116.5$$

Thus, the level of the noise is about 116.5 decibels.

127. Substitute $1.8 \times 10^7 A_0$ for A in the Richter scale formula $M = \log\frac{A}{A_0}$ and evaluate M.

$$M = \log\frac{1.8 \times 10^7 A_0}{A_0} = \log(1.8 \times 10^7) \approx 7.255272505$$

Thus, the earthquake measured about 7.3 on the Richter scale.

128. When the slopes of the lines in a system of two linear equations are equal and the y-intercepts are also equal, the lines coincide and the system has infinitely many solutions.

129. When the slopes of the lines in a system of two linear equations are unequal, the lines intersect and the system has exactly one solution.

130. When solving a system of equations, if we obtain an equation such as $5 = 0$, that is a contradiction, we know that the system has no solution (the lines are parallel).

Solve each system in Exercises 131-134.

131. $\begin{array}{l} 3x - 2y = -12 \\ x + 4y = 10 \end{array}$

Solve the second equation for x, $x = 10 - 4y$, and substitute this expression into the first equation.

$$3(10 - 4y) - 2y = -12$$
$$30 - 12y - 2y = -12$$
$$-14y = -42$$
$$y = 3$$

Substitute 3 for y in $x = 10 - 4y$ to find the value of x.

$$x = 10 - 4(3)$$
$$x = 10 - 12$$
$$x = -2$$

Thus, the solution to the system is $(-2, 3)$.

132. $\begin{array}{l} 5x + 3y = 5 \\ 2x - 4y = 28 \end{array}$

Divide through the second equation by 2 to obtain

$$x - 2y = 14,$$

then solve this equation for x, $x = 14 + 2y$, and substitute into the first equation.

$$5(14 + 2y) + 3y = 5$$
$$70 + 10y + 3y = 5$$
$$13y = -65$$
$$y = -5$$

Substitute -5 for y in $x = 14 + 2y$ to find the value of x.

$$x = 14 + 2(-5)$$
$$x = 14 - 10$$
$$x = 4$$

Thus, the solution to the system is $(4,-5)$.

133.
$$x + 3y + 2z = -3$$
$$3x + 2y - 3z = 13$$
$$-4x - 3y + 3z = -14$$

Suppose we eliminate x. Multiply the first equation by -3 and add the result to the second to obtain:

$$-7y - 9z = 22$$

Multiply the first equation by 4 and add the result to the third to obtain:

$$9y + 11z = -26$$

Now solve this system of two equations in the two variables y and z. Multiply the first equation by 9 to obtain:

$$-63y - 81z = 198$$

Then multiply the second equation by 7 to obtain:

$$63y + 77z = -182$$

Adding these two equations will eliminate y and give:

$$-4z = 16$$
$$z = -4$$

Substitute -4 for z in $-7y - 9z = 22$ to find the value of y.

$$-7y - 9(-4) = 22$$
$$-7y + 36 = 22$$
$$-7y = -14$$
$$y = 2$$

Substitute 2 for y and -4 for z in the first original equation to obtain the value of x.

$$x + 3(2) + 2(-4) = -3$$
$$x + 6 - 8 = -3$$
$$x = -1$$

Thus, the solution to the system is $(-1,2,-4)$.

134.
$$2x + y - z = 5$$
$$x - 4y + z = 4$$

Since we have two equations in three variables, there will either be no solution or infinitely many solutions. If we add the two equations, we will eliminate z and obtain:

$$3x - 3y = 9$$
$$x - y = 3$$

Since we do not obtain a contradiction, there are infinitely many solutions. Solve this equation for y in terms of x, $y = x - 3$. Substitute this expression for y in the second original equation to write z in terms of x.

$$x - 4(x - 3) + z = 4$$
$$x - 4x + 12 + z = 4$$
$$z = 3x - 8$$

Thus, the solutions to the system can be written in the form $(x,x-3,3x-8)$, for x any real number.

135. We must find the measures of three angles in a triangle.

Let x = measure of first angle,
y = measure of second angle,
z = measure of third angle.

Since the sum of the measures of the angles of a triangle is 180°, one equation is:

$$x + y + z = 180$$

Since the second angle measures three times the first, we have a second equation:

$$y = 3x$$

or equivalently,

$$3x - y = 0$$

Since the third angle has measure one-fourth the difference of the measures of the first two, a third equation is:

$$z = \tfrac{1}{4}(y - x)$$

or equivalently,

$$x - y + 4z = 0$$

Thus, we obtain the following system of equations.

$$\begin{aligned} x + y + z &= 180 \\ 3x - y &= 0 \\ x - y + 4z &= 0 \end{aligned}$$

Multiply the first equation by -4,

$$-4x - 4y - 4z = -720$$

and add the result to the third equation to eliminate z and obtain:

$$-3x - 5y = -720$$

Pair this equation with the second original equation to obtain the following system of two equations in x and y.

$$\begin{aligned} 3x - y &= 0 \\ -3x - 5y &= -720 \end{aligned}$$

Add these two equations to eliminate x and obtain:

$$\begin{aligned} -6y &= -720 \\ y &= 120 \end{aligned}$$

Substitute 120 for y in $3x - y = 0$ to find the value of x.

$$\begin{aligned} 3x - 120 &= 0 \\ 3x &= 120 \\ x &= 40 \end{aligned}$$

Substitute 40 for x and 120 for y in the first original equation to find the value of z.

$$\begin{aligned} 40 + 120 + z &= 180 \\ 160 + z &= 180 \\ z &= 20 \end{aligned}$$

Thus, the angles of the triangle are $40°$, $120°$, and $20°$.

Graph each system in Exercises 136-137.

136. $\begin{aligned} x + y &> 5 \\ 2x - y &\le 4 \end{aligned}$

First graph the equation $x + y = 5$ using intercepts $(0,5)$ and $(5,0)$ and a dashed line (the inequality is $>$). The test point $(0,0)$ shows that we shade the region that does not contain $(0,0)$, that is, the region above the line. Then graph $2x - y = 4$ using intercepts $(0,-4)$ and $(2,0)$ and a solid line (the inequality is $\le$). The test point $(0,0)$ shows that we shade the region cntaining the test point $(0,0)$, that is the region above the line. Putting this information together, we obtain the graph of the system shown below.

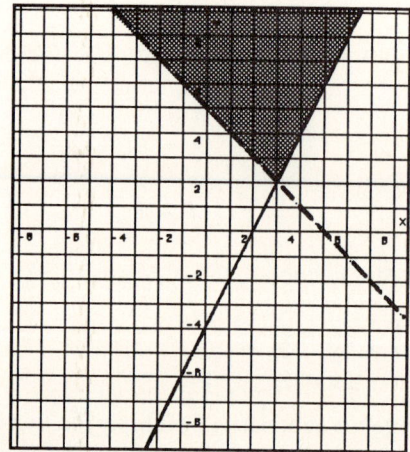

137. $\begin{aligned} x &\ge 0 \\ y &\ge 0 \\ 3y &< -2x + 6 \end{aligned}$

The first two inequalities describe the points in quadrant I together with the points on the positive x-axis and the positive y-axis. Thus, we can concentrate on graphing the third inequality. First graph the line $3y = -2x + 6$ using intercepts $(0,2)$ and $(3,0)$ and a dashed line (the inequality is $<$). The test point $(0,0)$ shows that we shade the region containing $(0,0)$, that is, the region below the line. Putting this information together, we obtain the graph of the system shown below.

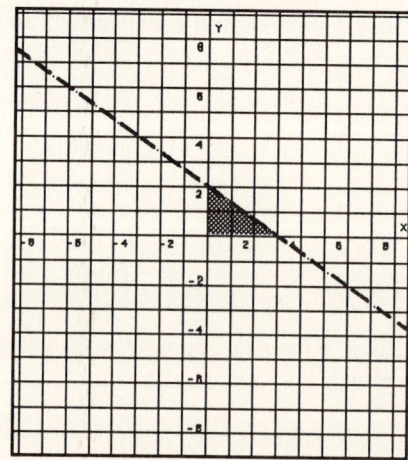

138. The maximum and minimum values of the objective function $P = 15x + 4y$ subject to the constraints

graphed in the text must occur at the vertices of the region. However, since the region is unbounded, we can make the objective function as large as we please so there will be no maximum value. The minimum value, however, must occur at a vertex. We summarize the possibilities in the following table.

Vertex	$P = 15x + 4y$	
(1,3)	27	Minimum Value
(2,2)	38	
(6,1)	94	

Thus, the minimum value is 27, and there is no maximum value.

139. Let $x =$ the number of Model A units to make,
$y =$ the number of Model B units to make.

Then two obvious inequalities forming the constraints are:

$$x \geq 0 \quad \text{and} \quad y \geq 0$$

Since the company can produce up to a total of 50 detectors each day, a third constraint is:

$$x + y \leq 50$$

Since it takes 5 hours to make one Model A and 3 hours to make one Model B, and up to a total of 200 man-hours are available each day, a fourth constraint is:

$$5x + 3y \leq 200$$

Since one Model A returns a profit of $60 and one Model B returns a profit of $40, the objective function to maximize subject to the constraints is:

$$P = 60x + 40y$$

The feasible region, the graph of the system of constraints is given below. The vertices of the region are (0,0), (0,50), (25,25), and (40,0), where the point (25,25) is obtained by solving the system of equations:

$$x + y = 50$$
$$5x + 3y = 200$$

Calculating P at each vertex shows that (25,25) gives a maximum profit of $2500.

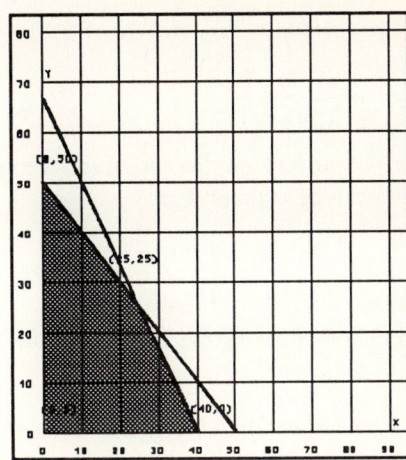

In Exercises 140-141, find the partial fraction decomposition of the given rational function.

140. $f(x) = \dfrac{7x-1}{x^2+x-2}$

First factor the denominator of the function.

$$f(x) = \frac{7x-1}{(x-1)(x+2)}$$

Then by the partial fraction decomposition theorem,

$$f(x) = \frac{7x-1}{(x-1)(x+2)} = \frac{A}{x-1} + \frac{B}{x+2}$$

Multiply both sides of this equation by the LCD, $(x - 1)(x + 2)$.

$$7x - 1 = A(x+2) + B(x-1)$$
$$7x - 1 = (A+B)x + (2A-B)$$

Then equating coefficients of like terms we obtain the following system:

$$A + B = 7$$
$$2A - B = -1$$

Adding these two equations will eliminate B and give:

$$3A = 6$$
$$A = 2$$

Substitute 2 for A in the first equation to find the value of B.

$$2 + B = 7$$
$$B = 5$$

Thus, the partial fraction decomposition of the given function is:

$$f(x) = \frac{2}{x-1} + \frac{5}{x+2}$$

141. $f(x) = \dfrac{7x^2 - 6x + 5}{(x-3)(x^2+1)}$

By the partial fraction decomposition theorem, the function can be expressed as:

$$f(x) = \dfrac{7x^2 - 6x + 5}{(x-3)(x^2+1)} = \dfrac{Ax+B}{x^2+1} + \dfrac{C}{x-3}$$

Multiply both sides by the LCD and collect like terms to obtain:

$$7x^2 - 6x + 5 = (A+C)x^2 + (-3A+B)x + (-3B+C)$$

Equating coefficients of like terms gives the following system:

$$\begin{aligned} A + C &= 7 \\ -3A + B &= -6 \\ -3B + C &= 5 \end{aligned}$$

Subtract the third equation from the first to eliminate C and obtain:

$$A + 3B = 2$$

Pair this equation with the second original equation to obtain the following system.

$$\begin{aligned} A + 3B &= 2 \\ -3A + B &= -6 \end{aligned}$$

Multiply the first equation by 3,

$$3A + 9B = 6,$$

And add to the second equation to obtain:

$$\begin{aligned} 10B &= 0 \\ B &= 0 \end{aligned}$$

Substitute 0 for B in $A + 3B = 2$ to find the value of A.

$$\begin{aligned} A + 3(0) &= 2 \\ A &= 2 \end{aligned}$$

Substitute 2 for A in $A + C = 7$ to find the value of C.

$$\begin{aligned} 2 + C &= 7 \\ C &= 5 \end{aligned}$$

Then the partial fraction decomposition of the given function is:

$$f(x) = \dfrac{2x}{x^2+1} + \dfrac{5}{x-3}$$

142. The order of the matrix

$$\begin{bmatrix} 2 & 3 & 5 \\ -1 & 5 & 2 \end{bmatrix}$$

is 2×3 since there are 2 rows and 3 columns.

143. An $n \times n$ matrix is called a square matrix.

144. No; to add two matrices, they must have the same dimension.

145. The identity matrix of order 3 is

$$\begin{bmatrix} 1 & 0 & 0 \\ 0 & 1 & 0 \\ 0 & 0 & 1 \end{bmatrix}.$$

146. If the rows and columns of an $m \times n$ matrix are interchanged, the resulting $n \times m$ matrix is called the transpose of the original matrix.

147. The matrix A^{-1} with the property that

$$A^{-1}A = AA^{-1} = I$$

is called the inverse of matrix A.

148. The coefficient matrix of the system

$$\begin{aligned} x + y &= 3 \\ 3x - 2y &= 1 \end{aligned}$$

is

$$\begin{bmatrix} 1 & 1 \\ 3 & -2 \end{bmatrix}.$$

149. The augmented matrix of the system

$$\begin{aligned} x + y &= 3 \\ 3x - 2y &= 1 \end{aligned}$$

is

$$\begin{bmatrix} 1 & 1 & 3 \\ 3 & -2 & 1 \end{bmatrix}.$$

Perform the indicated operations in Exercises 150-155.

FINAL REVIEW

150.

$$\begin{bmatrix} 2 & 3 \\ -1 & 5 \end{bmatrix} \begin{bmatrix} 4 & -1 & 3 \\ 0 & 2 & -1 \end{bmatrix}$$

$$= \begin{bmatrix} (2)(4)+(3)(0) & (2)(-1)+(3)(2) & (2)(3)+(3)(-1) \\ (-1)(4)+(5)(0) & (-1)(-1)+(5)(2) & (-1)(3)+(5)(-1) \end{bmatrix}$$

$$= \begin{bmatrix} 8+0 & -2+6 & 6-3 \\ -4+0 & 1+10 & -3-5 \end{bmatrix}$$

$$= \begin{bmatrix} 8 & 4 & 3 \\ -4 & 11 & -8 \end{bmatrix}$$

151. The product

$$\begin{bmatrix} 4 & -1 & 3 \\ 0 & 2 & -1 \end{bmatrix} \begin{bmatrix} 2 & 3 \\ -1 & 5 \end{bmatrix}$$

is not defined since the matrix on the left has 3 columns and the matrix on the right has only 2 rows.

152. $\begin{bmatrix} 4 & 0 & -2 \\ 0 & -3 & 7 \end{bmatrix} + \begin{bmatrix} -2 & 0 & 1 \\ 4 & -1 & 5 \end{bmatrix} = \begin{bmatrix} 2 & 0 & -1 \\ 4 & -4 & 12 \end{bmatrix}$

153. If $A = \begin{bmatrix} 3 & 2 & 1 \\ -1 & 7 & 8 \end{bmatrix}$ then A^T is formed by interchanging the rows and columns of A.

$$A^T = \begin{bmatrix} 3 & -1 \\ 2 & 7 \\ 1 & 8 \end{bmatrix}$$

154. Find A^{-1} if $A = \begin{bmatrix} 3 & -1 \\ 2 & 1 \end{bmatrix}$.

First form the augmented matrix $[A|I]$.

$$\begin{bmatrix} 3 & -1 & | & 1 & 0 \\ 2 & 1 & | & 0 & 1 \end{bmatrix}$$

Keep the second row, multiply it by -1, and add the result to the first row.

$$\begin{bmatrix} 1 & -2 & | & 1 & -1 \\ 2 & 1 & | & 0 & 1 \end{bmatrix}$$

Keep the first row, multiply it by -2, and add the result to the second row.

$$\begin{bmatrix} 1 & -2 & | & 1 & -1 \\ 0 & 5 & | & -2 & 3 \end{bmatrix}$$

Multiply the second row by $\frac{1}{5}$.

$$\begin{bmatrix} 1 & -2 & | & 1 & -1 \\ 0 & 1 & | & -\frac{2}{5} & \frac{3}{5} \end{bmatrix}$$

Keep the second row, multiply it by 2, and add the result to the first row.

$$\begin{bmatrix} 1 & 0 & | & \frac{1}{5} & \frac{1}{5} \\ 0 & 1 & | & -\frac{2}{5} & \frac{3}{5} \end{bmatrix}$$

Thus,

$$A^{-1} = \begin{bmatrix} \frac{1}{5} & \frac{1}{5} \\ -\frac{2}{5} & \frac{3}{5} \end{bmatrix}.$$

155. Find A^{-1} if $A = \begin{bmatrix} 1 & 0 & 2 \\ 2 & -1 & 1 \\ 0 & 1 & -5 \end{bmatrix}$.

First form the augmented matrix $[A|I]$.

$$\begin{bmatrix} 1 & 0 & 2 & | & 1 & 0 & 0 \\ 2 & -1 & 1 & | & 0 & 1 & 0 \\ 0 & 1 & -5 & | & 0 & 0 & 1 \end{bmatrix}$$

Keep the first and third rows, multiply the first row by -2, and add the result to the second row.

$$\begin{bmatrix} 1 & 0 & 2 & | & 1 & 0 & 0 \\ 0 & -1 & -3 & | & -2 & 1 & 0 \\ 0 & 1 & -5 & | & 0 & 0 & 1 \end{bmatrix}$$

Interchange the second and third rows.

$$\begin{bmatrix} 1 & 0 & 2 & | & 1 & 0 & 0 \\ 0 & 1 & -5 & | & 0 & 0 & 1 \\ 0 & -1 & -3 & | & -2 & 1 & 0 \end{bmatrix}$$

Keep the first and second rows, add the second row to the third.

$$\begin{bmatrix} 1 & 0 & 2 & | & 1 & 0 & 0 \\ 0 & 1 & -5 & | & 0 & 0 & 1 \\ 0 & 0 & -8 & | & -2 & 1 & 1 \end{bmatrix}$$

Multiply the third row by $-\frac{1}{8}$.

$$\begin{bmatrix} 1 & 0 & 2 & | & 1 & 0 & 0 \\ 0 & 1 & -5 & | & 0 & 0 & 1 \\ 0 & 0 & 1 & | & \frac{1}{4} & -\frac{1}{8} & -\frac{1}{8} \end{bmatrix}$$

Keep the third row. Multiply it by -2 and add to the first row. Then multiply the third row by 5 and add to the second row.

$$\begin{bmatrix} 1 & 0 & 0 & | & \frac{1}{2} & \frac{1}{4} & \frac{1}{4} \\ 0 & 1 & 0 & | & \frac{5}{4} & -\frac{5}{8} & \frac{3}{8} \\ 0 & 0 & 1 & | & \frac{1}{4} & -\frac{1}{8} & -\frac{1}{8} \end{bmatrix}$$

Thus,

$$A^{-1} = \begin{bmatrix} \frac{1}{2} & \frac{1}{4} & \frac{1}{4} \\ \frac{5}{4} & -\frac{5}{8} & \frac{3}{8} \\ \frac{1}{4} & -\frac{1}{8} & -\frac{1}{8} \end{bmatrix}.$$

156. Write the system as a matrix equation and solve using the inverse method.

$$3x - y = -2$$
$$2x + y = -3$$

This system of equations is equivalent to the matrix equation $AX = B$, where

$$A = \begin{bmatrix} 3 & -1 \\ 2 & 1 \end{bmatrix}, \quad X = \begin{bmatrix} x \\ y \end{bmatrix}, \quad \text{and } B = \begin{bmatrix} -2 \\ -3 \end{bmatrix}.$$

Notice that the matrix A^{-1} was found in Exercise 154. Using this result, we have

$$X = \begin{bmatrix} x \\ y \end{bmatrix} = A^{-1}B = \begin{bmatrix} \frac{1}{5} & \frac{1}{5} \\ -\frac{2}{5} & \frac{3}{5} \end{bmatrix} \begin{bmatrix} -2 \\ -3 \end{bmatrix} = \begin{bmatrix} -1 \\ -1 \end{bmatrix}.$$

Thus, $x = -1$ and $y = -1$, so the solution to the system is $(-1, -1)$.

157. Write the system as a matrix equation and solve using the inverse method.

$$x + 2z = 4$$
$$2x - y + z = 1$$
$$ y - 5z = -9$$

This system of equations is equivalent to the matrix equation $AX = B$, where

$$A = \begin{bmatrix} 1 & 0 & 2 \\ 2 & -1 & 1 \\ 0 & 1 & -5 \end{bmatrix}, \quad X = \begin{bmatrix} x \\ y \\ z \end{bmatrix}, \quad \text{and } B = \begin{bmatrix} 4 \\ 1 \\ -9 \end{bmatrix}.$$

Notice that the matrix A^{-1} was found in Exercise 155. Using this result, we have

$$X = \begin{bmatrix} x \\ y \\ z \end{bmatrix} = A^{-1}B = \begin{bmatrix} \frac{1}{2} & \frac{1}{4} & \frac{1}{4} \\ \frac{5}{4} & -\frac{5}{8} & \frac{3}{8} \\ \frac{1}{4} & -\frac{1}{8} & -\frac{1}{8} \end{bmatrix} \begin{bmatrix} 4 \\ 1 \\ -9 \end{bmatrix} = \begin{bmatrix} 0 \\ 1 \\ 2 \end{bmatrix}.$$

Thus, $x = 0$, $y = 1$, and $z = 2$, so the solution to the system is $(0, 1, 2)$.

Solve using the Gaussian method in Exercises 158-159.

158. $3x - y = -2$
$2x + y = -3$

First form the augmented matrix of the system.

$$\begin{bmatrix} 3 & -1 & -2 \\ 2 & 1 & -3 \end{bmatrix}$$

Keep the second row, multiply it by -1, and add the results to the first row.

$$\begin{bmatrix} 1 & -2 & 1 \\ 2 & 1 & -3 \end{bmatrix}$$

Keep the first row, multiply it by -2, and add the result to the second row.

$$\begin{bmatrix} 1 & -3 & 4 \\ 0 & 5 & -5 \end{bmatrix}$$

Multiply the second row by $\frac{1}{5}$.

$$\begin{bmatrix} 1 & -2 & 1 \\ 0 & 1 & -1 \end{bmatrix}$$

Keep the second row, multiply it by 2, and add the results to the first row.

$$\begin{bmatrix} 1 & 0 & -1 \\ 0 & 1 & -1 \end{bmatrix}$$

Since this matrix corresponds to $x = -1$ and $y = -1$, the solution to the system is $(-1, -1)$.

159.
$$x + 2z = 4$$
$$2x - y + z = 1$$
$$y - 5z = -9$$

First form the augmented matrix of the system.

$$\begin{bmatrix} 1 & 0 & 2 & 4 \\ 2 & -1 & 1 & 1 \\ 0 & 1 & -5 & -9 \end{bmatrix}$$

Keep the first and third rows. Multiply the first row by -2, and add the results ot the third row.

$$\begin{bmatrix} 1 & 0 & 2 & 4 \\ 0 & -1 & -3 & -7 \\ 0 & 1 & -5 & -9 \end{bmatrix}$$

Interchange the second and third rows.

$$\begin{bmatrix} 1 & 0 & 2 & 4 \\ 0 & 1 & -5 & -9 \\ 0 & -1 & -3 & -7 \end{bmatrix}$$

Keep the first and second rows. Add the second row to the third row.

$$\begin{bmatrix} 1 & 0 & 2 & 4 \\ 0 & 1 & -5 & -9 \\ 0 & 0 & -8 & -16 \end{bmatrix}$$

Multiply the third row by $-\frac{1}{8}$.

$$\begin{bmatrix} 1 & 0 & 2 & 4 \\ 0 & 1 & -5 & -9 \\ 0 & 0 & 1 & 2 \end{bmatrix}$$

Keep the third row. Multiply the third row by -2, and add the results to the first row. Then multiply the third row by 5, and add the results to the second row.

$$\begin{bmatrix} 1 & 0 & 0 & 0 \\ 0 & 1 & 0 & 1 \\ 0 & 0 & 1 & 2 \end{bmatrix}$$

This corresponds to $x = 0$, $y = 1$, and $z = 2$. Thus, the solution to the system is $(0,1,2)$.

160. The matrix $J + A$ represents the total number of each of the items sold during the two-month period of July and August. Then $C \cdot (J + A)$ represents the total revenue produced on the sale of these three items during the two-month period.

$$C \cdot (J+A) = [100 \quad 350 \quad 325] \cdot \left(\begin{bmatrix} 3 \\ 2 \\ 4 \end{bmatrix} + \begin{bmatrix} 5 \\ 1 \\ 3 \end{bmatrix} \right)$$

$$= [100 \quad 350 \quad 325] \cdot \begin{bmatrix} 8 \\ 3 \\ 7 \end{bmatrix}$$

$$= (100)(8) + (350)(3) + (325)(7)$$
$$= 800 + 1050 + 2275$$
$$= 4125$$

Thus, the total revenue produced on the sale of the three items during the two-month period is $4125.

Evaluate the determinants in Exercises 161-163.

161. $\begin{vmatrix} a & b \\ c & d \end{vmatrix} = ad - bc$

162. $\begin{vmatrix} 2 & -5 \\ -3 & 4 \end{vmatrix} = (2)(4) - (-5)(-3) = 8 - 15 = -7$

163. Evaluate the determinant along the first row.

$$\begin{vmatrix} 1 & -2 & 3 \\ 2 & 7 & 4 \\ 4 & 5 & 3 \end{vmatrix}$$

$$= (1)(+1)\begin{vmatrix} 7 & 4 \\ 5 & 3 \end{vmatrix} + (-2)(-1)\begin{vmatrix} 2 & 4 \\ 4 & 3 \end{vmatrix}$$
$$ (3)(+1)\begin{vmatrix} 2 & 7 \\ 4 & 5 \end{vmatrix}$$
$$= (1)[21 - 20] + (2)[6 - 16] + (3)[10 - 28]$$
$$= (1)[1] + (2)[-10] + (3)[-18]$$
$$= 1 - 20 - 54$$
$$= -73$$

Solve using Cramer's rule in Exercises 164-165.

164. $2x - 5y = -2$
$-3x + 4y = 10$

$$|A| = \begin{vmatrix} 2 & -5 \\ -3 & 4 \end{vmatrix} = (2)(4) - (-5)(-3) = 8 - 15 = -7$$

Replace the coefficients of x with the constants.

$$|A_x| = \begin{vmatrix} -2 & -5 \\ 10 & 4 \end{vmatrix} = (-2)(4) - (-5)(10) = -8 + 50 = 42$$

Replace the coefficients of y with the constants.

$$|A_y| = \begin{vmatrix} 2 & -2 \\ -3 & 10 \end{vmatrix} = (2)(10) - (-2)(-3) = 20 - 6 = 14$$

Then we have:

$$x = \frac{A_x}{A} = \frac{42}{-7} = -6 \qquad y = \frac{A_y}{A} = \frac{14}{-7} = -2$$

Thus, the solution to the system is $(-6, -2)$.

165. $\begin{aligned} x - 2y + 3z &= 9 \\ 2x + 7y + 4z &= -6 \\ 4x + 5y + 3z &= 1 \end{aligned}$

$$|A| = \begin{vmatrix} 1 & -2 & 3 \\ 2 & 7 & 4 \\ 4 & 5 & 3 \end{vmatrix}$$

$$= (1)(+1)\begin{vmatrix} 7 & 4 \\ 5 & 3 \end{vmatrix} + (-2)(-1)\begin{vmatrix} 2 & 4 \\ 4 & 3 \end{vmatrix}$$

$$+ (3)(+1)\begin{vmatrix} 2 & 7 \\ 4 & 5 \end{vmatrix}$$

$$= (1)[21 - 20] + (2)[6 - 16] + (3)[10 - 28]$$
$$= (1)[1] + (2)[-10] + (3)[-18]$$
$$= 1 - 20 - 54$$
$$= -73$$

Replace the coefficients of x with the constants.

$$|A_x| = \begin{vmatrix} 9 & -2 & 3 \\ -6 & 7 & 4 \\ 1 & 5 & 3 \end{vmatrix}$$

$$= (9)(+1)\begin{vmatrix} 7 & 4 \\ 5 & 3 \end{vmatrix} + (-2)(-1)\begin{vmatrix} -6 & 4 \\ 1 & 3 \end{vmatrix}$$

$$+ (3)(+1)\begin{vmatrix} -6 & 7 \\ 1 & 5 \end{vmatrix}$$

$$= (9)[21 - 20] + (2)[-18 - 4] + (3)[-30 - 7]$$
$$= (9)[1] + (2)[-22] + (3)[-37]$$
$$= 9 - 44 - 111$$
$$= -146$$

Replace the coefficients of y with the constants.

$$|A_y| = \begin{vmatrix} 1 & 9 & 3 \\ 2 & -6 & 4 \\ 4 & 1 & 3 \end{vmatrix}$$

$$= (1)(+1)\begin{vmatrix} -6 & 4 \\ 1 & 3 \end{vmatrix} + (9)(-1)\begin{vmatrix} 2 & 4 \\ 4 & 3 \end{vmatrix}$$

$$+ (3)(+1)\begin{vmatrix} 2 & -6 \\ 4 & 1 \end{vmatrix}$$

$$= (1)[-18 - 4] + (-9)[6 - 16] + (3)[2 + 24]$$
$$= (1)[-22] + (-9)[-10] + (3)[26]$$
$$= -22 + 90 + 78$$
$$= 146$$

Replace the coefficients of z with the constants.

$$|A_z| = \begin{vmatrix} 1 & -2 & 9 \\ 2 & 7 & -6 \\ 4 & 5 & 1 \end{vmatrix}$$

$$= (1)(+1)\begin{vmatrix} 7 & -6 \\ 5 & 1 \end{vmatrix} + (-2)(-1)\begin{vmatrix} 2 & -6 \\ 4 & 1 \end{vmatrix}$$

$$+ (9)(+1)\begin{vmatrix} 2 & 7 \\ 4 & 5 \end{vmatrix}$$

$$= (1)[7 + 30] + (2)[2 + 24] + (9)[10 - 28]$$
$$= (1)[37] + (2)[26] + (9)[-18]$$
$$= 37 + 52 - 162$$
$$= -73$$

Then

$$x = \frac{A_x}{A} = \frac{-146}{-73} = 2 \qquad y = \frac{A_y}{A} = \frac{146}{-73} = -2$$

$$z = \frac{A_z}{A} = \frac{-73}{-73} = 1$$

Thus, the solution to the system is $(2, -2, 1)$.

Without evaluating the determinants in Exercises 166-167, tell why each statement is true.

166. $\begin{vmatrix} 0 & 3 \\ 0 & 8 \end{vmatrix} = 0$

The determinant is 0 because a column is all zeros.

167. $\begin{vmatrix} 1 & 2 & -3 \\ 2 & 0 & 5 \\ 1 & 2 & -3 \end{vmatrix} = 0$

The determinant is 0 because two rows (the first and third) are the same.

168. Solve for x. $\begin{vmatrix} x & 0 & 0 \\ 0 & x & 1 \\ 0 & 1 & 1 \end{vmatrix} = 2$

Evaluate the determinant along the first row.

$$\begin{vmatrix} x & 0 & 0 \\ 0 & x & 1 \\ 0 & 1 & 1 \end{vmatrix} = x\begin{vmatrix} x & 1 \\ 1 & 1 \end{vmatrix} = x(x - 1)$$

Thus, we are to solve the equation $x(x - 1) = 2$.

$$x(x - 1) = 2$$
$$x^2 - x = 2$$

$$x^2 - x - 2 = 0$$
$$(x - 2)(x + 1) = 0$$
$$x - 2 = 0 \quad x + 1 = 0$$
$$x = 2 \quad x = -1$$

Thus, the solutions to the equation are 2 and -1.

169. Since circles, ellipses, parabolas, and hyperbolas can be obtained by intersecting a plane with a cone, they are often called conic sections.

170. The collection of all points in a plane, the difference of whose distances form two different points is constant, is called a hyperbola.

171. The collection of all points in a plane which are located a constant distance form a fixed point in that plane is called a circle.

172. The collection of all points in a plane, the sum of whose distances from two different fixed points is constant, is called an ellipse.

173. The graph of
$$(x - 2)^2 + (y + 1)^2 = 1$$
is a circle (centered at (2,-1) with radius 1).

174. The graph of
$$(x - 2)^2 - (y + 1)^2 = 1$$
is a hyperbola.

175. The vertex of the parabola with standard-form equation
$$(y + 2)^2 = 8(x - 3)$$
is (3,-2).

176. The equation
$$2x^2 + y^2 - 4x + 2y - 30 = 0$$
has as its graph an ellipse. This can be seen by writing the equation in the standard form.

177. The equation $xy = -7$ has as its graph a hyperbola (with asymptotes the coordinate axes).

178. The equation
$$2x^2 - 3y^2 - 4x + 2y - 35 = 0$$
has as its graph a hyperbola. This can be seen by writing the equation in the standard form.

179. The standard form of the equation of a circle centered at $\left(3, -\frac{1}{2}\right)$ with radius $r = 2$ is:
$$(x - 3)^2 + \left(y + \frac{1}{2}\right)^2 = 2^2 = 4$$

180. The standard form of the equation of the ellipse centered at the origin with intercepts $(\pm 3, 0)$ and $(0, \pm 4)$ is:
$$\frac{x^2}{9} + \frac{y^2}{16} = 1$$

Graph the conics in Exercises 181-183.

181. $4x^2 + 9y^2 - 8x + 36y + 4 = 0$

Complete the square on x and y to obtain the stand form.
$$4(x^2 - 2x \quad) + 9(y^2 + 4y \quad) = -4$$
$$4(x^2 - 2x + 1) + 9(y^2 + 4y + 4) = -4 + 4 + 36$$
$$4(x - 1)^2 + 9(y + 2)^2 = 36$$

Divide both sides by 36 to obtain the standard form.
$$\frac{(x-1)^2}{9} + \frac{(y+2)^2}{4} = 1$$

This is an ellipse centered at the point (1,-2). Placing a new coordinate system centered at this point, we can see that the intercepts of the ellipse (relative to this system) are at (4,-2), (1,0), (-2,-2), and (1,-4). The graph is given below.

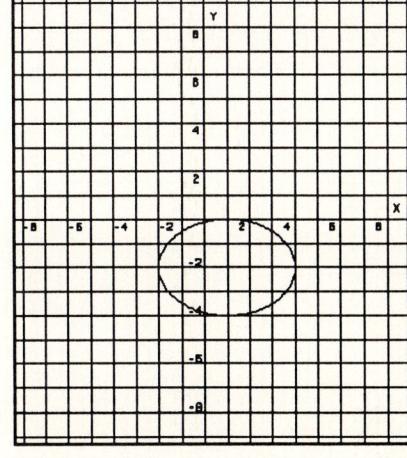

182. $9x^2 - 4y^2 - 18x - 16y - 43 = 0$

Complete the square on x and y to find the standard form.

$$9(x^2 - 2x) - 4(y^2 + 4y) = 43$$
$$9(x^2 - 2x + 1) - 4(y^2 + 4y + 4) = 43 + 9 - 16$$
$$9(x-1)^2 - 4(y+2)^2 = 36$$

Divide both sides by 36.

$$\frac{(x-1)^2}{4} - \frac{(y+2)^2}{9} = 1$$

This is the standard form of a hyperbola, centered at $(1,-2)$, opening left and right. The asymptotes of the hyperbola contain the diagonals of the rectangle with sides parallel to the axes and passing through the vertices $(3,-2)$ and $(-1,-2)$ and the two points $(1,1)$ and $(1,-5)$. The graph is given below.

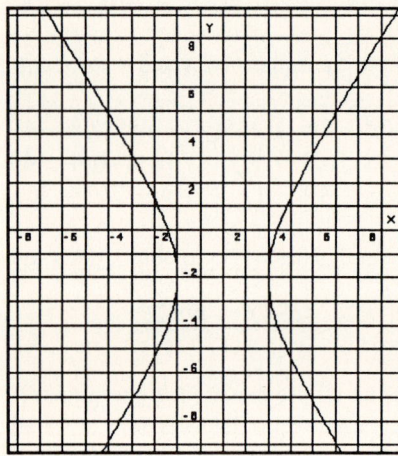

183. $xy = -6$

The graph is a hyperbola with "branches" located in quadrants II and IV. Make a table of values and plot several points to obtain the graph given below.

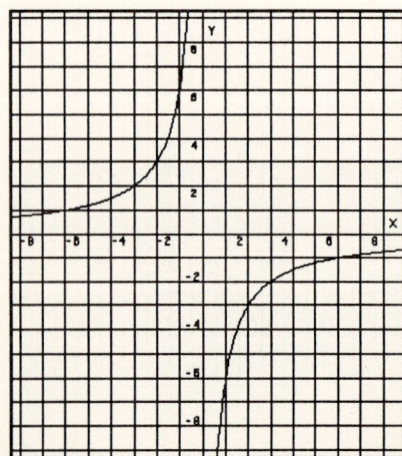

Solve the systems in Exercises 184-186.

184. $\begin{aligned} x^2 + y^2 &= 13 \\ x + y &= 5 \end{aligned}$

Solve the second equation for x, $x = 5 - y$, and substutute this expression in the first equation.

$$(5-y)^2 + y^2 = 13$$
$$25 - 10y + y^2 + y^2 = 13$$
$$2y^2 - 10y + 12 = 0$$
$$y^2 - 5y + 6 = 0$$
$$(y-2)(y-3) = 0$$
$$y - 2 = 0 \quad\quad y - 3 = 0$$
$$y = 2 \quad\quad\quad y = 3$$

When $y = 2$: $x = 5 - y = 5 - 2 = 3$.
When $y = 3$: $x = 5 - y = 5 - 3 = 2$.

Thus, the solutions to the system are $(3,2)$ and $(2,3)$.

185. $\begin{aligned} 4x^2 + y^2 &= 17 \\ x^2 - 3y^2 &= 1 \end{aligned}$

Multiply the first equation by 3,

$$12x^2 + 3y^2 = 51,$$

and add the result to the second equation to eliminate y and obtain:

$$13x^2 = 52$$
$$x^2 = 4$$
$$x = \pm 2$$

When $x = 2$:
$$(2)^2 - 3y^2 = 1$$
$$4 - 3y^2 = 1$$
$$-3y^2 = -3$$
$$y^2 = 1$$
$$y = \pm 1$$

Similarly, when $x = -2$, $y = \pm 1$.

Thus, the solutions to the system are $(2,1)$, $(2,-1)$, $(-2,1)$, and $(-2,-1)$.

186. $\begin{aligned} 4^x + y &= 1 \\ 4^x - y &= 1 \end{aligned}$

Add the two equations to eliminate y.

$$2 \cdot 4^x = 2$$
$$4^x = 1$$
$$4^x = 4^0$$
$$x = 0$$

When $x = 0$:

$$4^0 + y = 1$$
$$1 + y = 1$$
$$y = 0$$

Thus, the solution to the system is (0,0).

187. Consider the circle centered at the origin with the portion corresponding to the ditch the semicircle below the x-axis. Since the maximum depth of the ditch is 7.5 ft, this is the radius of the circle, and the circle has equation

$$x^2 + y^2 = (7.5)^2.$$

Then the semicircle corresponding to the ditch has equation

$$y = -\sqrt{(7.5)^2 - x^2}.$$

To find the depth of the water 2.0 ft from the edge of the ditch, substitute 5.5 for x (2.0 ft from the edge is 5.5 ft from the center of the water level in the ditch).

$$y = -\sqrt{(7.5) - (5.5)^2} \approx -5.099019514$$

Thus, the depth of the water at this point is about 5.1 ft.

Graph each nonlinear system in Exercises 188-189.

188. $\begin{array}{l} x^2 + y^2 \leq 25 \\ x^2 - y^2 \geq 4 \end{array}$

First graph the circle, $x^2 + y^2 = 25$, with center at (0,0) and radius 5, using a solid curve. The test point (0,0) shows that we shade the region inside the circle. Then graph the hyperbola, $x^2 - y^2 = 4$, centered at (0,0) and opening left and right, using a solid curve. The test point (0,0) shows that we shade the region left and right of the curve. Putting this information together, we obtain the graph of the system shown below.

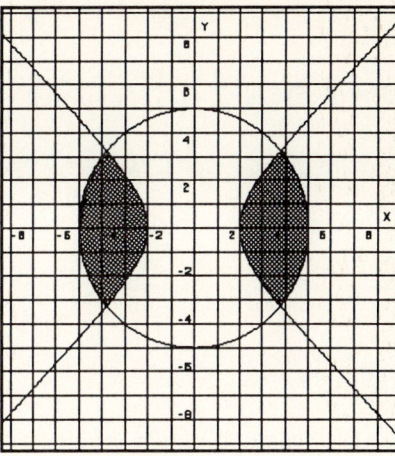

189. $\begin{array}{l} y \leq \ln x \\ (x-1)^2 + y^2 \leq 4 \end{array}$

First graph the logarithm function $y = \ln x$ using a solid curve. The test point (2,0) shows that we shade the region below the curve. Then graph the circle $(x - 1)^2 + y^2 = 4$, centered at (1,0) with radius 2, using a solid curve. The test point (1,0) shows that we shade the region inside the circle. Putting this information together, we obtain the graph of the system shown below.

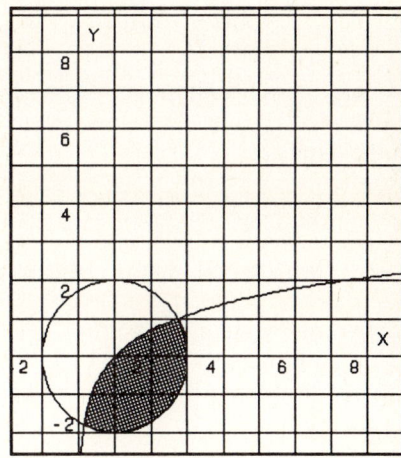

190. A sequence in which each term after the first is obtained by adding a fixed number to the preceding term is called an arithmetic sequence.

191. A sequence in which each term after the first is obtained by multiplying the preceding term by a fixed number is called a geometric sequence.

192. The formula for calculating the sum of an infinite geometric sequence only applies when the common ratio r satisfies $|r| < 1$.

193.

$$\sum_{k=1}^{3}(k^3-5) = (1-5)+(8-5)+(27-5) = -4+3+22 = 21$$

194.

$$\frac{1}{2}+\frac{8}{3}+\frac{27}{4}+\ldots+\frac{n^3}{n+1}+\ldots = \sum_{n=1}^{\infty}\frac{n^3}{n+1}$$

195. Given the sequence

$$7, 4, 1, -2, \ldots\,.$$

Since each term after the first is obtained from the preceding one by adding -3, this is an arithmetic sequence. Thus, we have $a_1 = 7$ and $d = -3$.

$$a_8 = a_1 + (8-1)d$$
$$a_8 = 7 + (7)(-3)$$
$$= 7 - 21$$
$$= -14$$

$$S_8 = \frac{8}{2}[a_1+a_8]$$
$$= (4)[7+(-14)]$$
$$= (4)[-7]$$
$$= -28$$

196. Given the sequence

$$8, -4, 2, -1, \frac{1}{2}, \ldots.$$

Since each term after the first is obtained from the preceding term by multiplying it by $-\frac{1}{2}$, this is a geometric sequence. We have $a_1 = 8$ and $r = -\frac{1}{2}$.

$$a_7 = ar^{7-1}$$
$$= (8)\left(-\frac{1}{2}\right)^6$$
$$= \frac{8}{64}$$
$$= \frac{1}{8}$$

$$S_7 = \frac{a_1 - ra_7}{1-r}$$
$$= \frac{8 - \left(-\frac{1}{2}\right)\left(\frac{1}{8}\right)}{1-\left(-\frac{1}{2}\right)}$$
$$= \frac{8 + \frac{1}{16}}{\frac{3}{2}} = \frac{\frac{129}{16}}{\frac{3}{2}} = \frac{129}{16}\cdot\frac{2}{3} = \frac{43}{8}$$

197. To find the number of coins in the array, we must find the following sum:

$$30 + 29 + 28 + 27 + \ldots + 1$$

This is an arithmetic sequence sith $d = -1$, $a_1 = 30$, $n = 30$, and $a_n = 1$.

$$S_{30} = \frac{30}{2}[a_1 + a_{30}]$$
$$= (15)[30+1]$$
$$= (15)[31]$$
$$= 465$$

Thus, the number of coins is 465, and the value of the coins is

$$(465)(\$0.10) = \$46.50.$$

198. Let $a_1 = 1200$. The amount that would have to be paid back at the end of 1 year is:

$$a_2 = 1200 + (0.10)(1200) = (1.10)(1200)$$

The amount to be paid back at the end of 2 years is:

$$a_3 = (1.10)^2(1200)$$

Continuing in this manner, we can see that at the end of 5 years, the amount to be paid back is:

$$a_6 = (1.10)^5(1200) = \$1932.61$$

199. To find the total distance traveled on the swing, we must add the following:

$$20 + \left(\frac{9}{10}\right)(20) + \left(\frac{9}{10}\right)^2(20) + \ldots$$

This is an infinite geometric sequence with $a_1 = 20$ and $r = \frac{9}{10}$. Since $|r| < 1$, the sum exists and is given by:

$$S = \frac{a_1}{1-r} = \frac{20}{1-\left(\frac{9}{10}\right)} = \frac{20}{\frac{1}{10}} = 200$$

Thus, the total distance traveled is 200 ft.

200. $3.\overline{72} = 3 + 0.72 + 0.0072 + 0.000072 + \ldots$

Ignore the first term 3, then the remaining terms form an infinite geometric sequence with $a_1 = 0.72$ and $r = 0.01$. Find the sum of this series.

$$S = \frac{0.72}{1-0.01} = \frac{0.72}{0.99} = \frac{72}{99} = \frac{8}{11}$$

Then we have

$$3.\overline{72} = 3 + \frac{8}{11} = \frac{41}{11}.$$

201. Use the principle of mathematical induction to prove that

$$7 + 14 + 21 + \ldots + 7n = \frac{7}{2}n(n+1)$$

is true for every positive integer n.

Verification: When $n = 1$,

$$7 = \frac{7}{2}(1)(1+1) = \frac{7}{2}(2) = 7$$

so the statement is true in this case.

Induction: Assume that

$$7 + 14 + 21 + \ldots + 7k = \frac{7}{2}k(k+1)$$

is true. We must show that

$$7 + 14 + \ldots + 7k + 7(k+1) = \frac{7}{2}(k+1)((k+1)+1)$$

is true. Replace the first k terms in the above expression using the induction assumption.

$$\frac{7}{2}k(k+1) + 7(k+1) = \frac{7}{2}[k(k+1) + 2(k+1)]$$
$$= \frac{7}{2}[(k+1)(k+2)]$$
$$= \frac{7}{2}(k+1)((k+1)+1)$$

Thus, the statement is true for $n = k+1$ under the assumption that it is true for $n = k$.

Conclusion: By the principle of mathematical induction, the statement is true for every positive integer n.

202. When an experiment is performed, the set of all possible outcomes is called the sample space of the experiment.

203. A subset of the sample space in an experiment is called an event.

204. When working with permutations, order *is* important.

205. Evaluate.

(a) $_3P_2 = \frac{3!}{(3-2)!} = \frac{3!}{1!} = 3! = 6$

(b) $_3C_2 = \frac{3!}{(3-2)!2!} = \frac{3!}{1!2!} = \frac{3 \cdot 2 \cdot 1}{1 \cdot 2 \cdot 1} = 3$

(c) $_5P_5 = \frac{5!}{(5-5)!} = \frac{5!}{0!} = 5! = 120$

(d) $_5C_5 = \frac{5!}{(5-5)!5!} = \frac{5!}{0!5!} = \frac{5!}{5!} = 1$

206. If a single card is drawn from a well-shuffled deck of 52 cards, what is the probability of drawing

(a) a red card?

Since there are 26 red cards (13 hearts and 13 diamonds) out of the 52 cards in the deck, the probability of drawing a red card is:

$$\frac{26}{52} = \frac{1}{2}$$

(b) a spade?

Since there are 13 spades out of the 52 cards in the deck, the probability of drawing a spade is:

$$\frac{13}{52} = \frac{1}{4}$$

(c) a three?

Since there are 4 threes in the deck, the probability of drawing a three is:

$$\frac{4}{52} = \frac{1}{13}$$

(d) a heart or a face card?

Since there are 13 hearts and 9 face cards that are not hearts, the total number of favorable cards is 22. Thus, the probability of drawing a heart or a face card is:

$$\frac{22}{52} = \frac{11}{26}$$

(e) a heart and a face card?

Since there are three hearts that are face cards (the jack of hearts, the queen of hearts, and the king of hearts), there are 3 favorable cards. Thus, the probability of drawing a heart and a face card is:

$$\frac{3}{52}$$

(f) a black card or an ace?

Since there are 26 black cards and 2 aces that are red cards, the total number of favorable cards is 28. Thus, the probability of drawing a black card or an ace is:

$$\frac{28}{52} = \frac{7}{13}$$

207. If E and F are complementary events and $P(E) = \frac{5}{6}$, then

$$P(F) = 1 - P(E) = 1 - \frac{5}{6} = \frac{1}{6}.$$

208. To find the number of five-digit numbers that can be formed from the digits 0, 1, 2, 3, 4, 5, 6, 7 if zero cannot come first and no digit may be repeated, we fill slots in the following way.

$$\underline{7} \cdot \underline{7} \cdot \underline{6} \cdot \underline{5} \cdot \underline{4} = 5880$$

Thus, 5880 such numbers can be formed.

209. The number of ways of choosing a president, vice president, and secretary from the 10 members of a club is a permutation problem. Notice that order is specified, that is, order is important. The number of ways of doing this is:

$$_{10}P_3 = \frac{10!}{(10-3)!} = \frac{10!}{7!} = 10 \cdot 9 \cdot 8 = 720$$

210. The number of ways of obtaining a 5-card poker hand consisting of 3 kings and 2 jacks is given by:

$$_4C_3 \cdot {_4C_2} = \frac{4!}{(4-3)!3!} \cdot \frac{4!}{(4-2)!2!} = (4)(6) = 24$$

The number of possible 5-card poker hands is:

$$_{52}C_5 = \frac{52!}{47!5!} = 2,598,960$$

Thus, the probability of being dealt a hand consisting of 3 kings and 2 jacks is:

$$\frac{_4C_3 \cdot {_4C_2}}{_{52}C_5} = \frac{24}{2,598,960} = \frac{1}{108,290}$$

211. The number of ways to place 6 people in a line is $6! = 720$. The number of ways to place 6 people around a circle is $(6-1)! = 5! = 120$.

212. Note that there are 8 letters in the word CALCULUS, 2 C's, 1 A, 2 L's, 2 U's, and 1 S. Thus, the number of distinguishable permutations of these letters is:

$$\frac{8!}{2!1!2!2!1!} = \frac{8!}{8} = 7! = 5040$$

213. If A and B are mutually exclusive, then $P(A \text{ and } B) = 0$. Thus,

$$P(A \text{ or } B) = P(A) + P(B) - P(A \text{ and } B)$$
$$= \frac{2}{5} + \frac{1}{3} - 0 = \frac{6}{15} + \frac{5}{15} = \frac{11}{15}$$

214.
$$P(\text{ma or hi}) = P(\text{ma}) + P(\text{hi}) - P(\text{ma and hi})$$
$$= \frac{2}{3} + \frac{4}{5} - \frac{8}{15} = \frac{14}{15}$$

215. To find the fourth term in the expansion of $(2x - y^2)^6$, notice that $n = 6$, $a = 2x$, $b = -y^2$, and $r = 3$. Thus, the fourth term is:

$$\binom{n}{r}a^{n-r}b^r = \binom{6}{3}(2x)^{6-3}(-y^2)^3$$
$$= \frac{6!}{(6-3)!3!}(2x)^3(-y^2)^3$$
$$= (20)(8x^3)(-y^6)$$
$$= -160x^3y^6$$